# Galactic Astronomy

James Binney and Michael Merrifield

# GALACTIC ASTRONOMY

PRINCETON UNIVERSITY PRESS
Princeton, New Jersey

**Library of Congress Cataloging-in-Publication Data**

Binney, James, 1950–
Galactic astronomy / James Binney and Michael Merrifield.
p.  cm.—(Princeton series in astrophysics)
Includes bibliographical references and index.
ISBN 0-691-00402-1 (cloth : alk. paper) —ISBN 0-691-02565-7 (pbk. : alk. paper)
1. Galaxies.  2. Milky Way.  3. Stars.  I. Merrifield, Michael, 1964–  .  II. Title.
III. Series.
QB857.B522   1998
523.1—dc21   98-24385

The publisher would like to acknowledge the authors of this volume for providing the
camera-ready copy from which this book was printed

Princeton University Press books are printed on acid-free paper and meet the guidelines
for permanence and durability of the Committee on Production Guidelines for Book
Longevity of the Council on Library Resources

http://pup.princeton.edu

Printed in the United States of America

10  9  8  7  6  5  4  3  2  1

10  9  8  7
(Pbk.)

ISBN-13:  978-0-691-02565-0  (pbk)

ISBN-10:  0-691-02565-7  (pbk)

# Contents

# Preface

This book is intended as a replacement for *Galactic Astronomy* by Mihalas & Binney, and a companion to *Galactic Dynamics* by Binney & Tremaine, which we henceforth refer to as BT.

Nearly 19 years of rapid scientific progress have elapsed since Mihalas & Binney went to press. In these years, there have been fundamental changes in the way that we approach the study of galaxies. Spectacular advances in telescope and detector technology, and in the scale of space observatories, have enormously enriched the quality of the data that are available, particularly for external galaxies. Although some details of galactic structure are still best studied using our uniquely close-up view of the Milky Way, the over-all picture can be seen most clearly from the perspective afforded by observations of other galaxies. The emphasis in studies of galactic structure has therefore shifted toward exploiting the wealth of extragalactic observations that is now available, and the layout of this book has been altered from that of Mihalas & Binney to reflect this change.

It remains important to understand the historical development of the subject, and essential to have a good grasp of fundamental astronomical concepts, so the first three chapters still cover the history of galactic astronomy, astronomical measurements and the phenomenology of stars. We then turn directly to the morphology of external galaxies – Chapter 4 gives a broad perspective on the different types of galaxies that populate the Universe. Chapter 5 reviews the theory of stellar structure and nucleosynthesis insofar as it bears on the ages and evolutionary histories of galaxies. The simplest applications of this theory are to star clusters, which form the subject of Chapter 6. Chapter 7 covers the sometimes tortuous lines of reasoning that are used to deduce distances in the Universe, and hence estimate such basic parameters as the sizes of galaxies and the distances between them. Chapters 8 and 9 are concerned with the structure and dynamics of the interstellar medium, in external galaxies and in the Milky Way, respectively. The interstellar medium has provided a wealth of information on the dynamics and the chemical evolution of galaxies. It also determines the rate at which stars form, which, in turn, profoundly influences the optical appearance of galaxies. Our study of the Milky Way gets under way in earnest with Chapter 9 and continues into Chapter 10, which complements the interstellar studies of Chapter 9 with a coherent picture of the distribution of the Galaxy's stars. The study of stars near the Sun is particularly rewarding because we can analyze their motions and chemical compositions in great detail. A wealth of information about the dynamics and history of the Galaxy can be gleaned from such analyses. Chapter 11 completes the book by covering our knowledge of the motions of stars within external galaxies.

In a subject area as broad as galactic astronomy, we have inevitably had to compromise on the material that we have included in the book. Moreover, it is impossible to be familiar with every relevant publication, and we have surely based some discussions on older or less satisfactory papers than we might have done. We apologize to the colleagues we have unfairly neglected, and trust that they will forgive us for not chasing the unachievable goal of a truly comprehensive work.

Throughout, we have tried to present basic observational data, and let readers decide for themselves how firmly they believe the conclusions that have been drawn from them. Since one cannot avoid drawing on a good deal of theory for the interpretation of observational data, we review any theoretical details that are likely to be unfamiliar to a student who has a solid background in undergraduate physics. We have, however, sought to ensure that theory and observation are kept clearly separated, to enable students to distinguish raw facts from their model-dependent interpretation.

In a few places, we have borrowed text directly from BT; the intention is to drop borrowed material from any revised edition of that work. At many points, we refer the reader to BT for the discussion of more theoretical topics such as the derivation of a dynamical result, or the explanation for a dynamical process. As far as possible, the notation in this book is the same as that in BT.

For more than thirty years now, most undergraduate physics courses have employed SI units. Feeling that it is time that the astronomical literature reflected this development, we decided – with some trepidation – to use SI units. Our hope is that the present generation of graduate students will finally start to publish in SI units, and, by doing so, make life less perplexing for generations to come.

JJB wishes to thank the Director and staff of the Mount Stromlo and Siding Spring Observatories for their warm hospitality during an important phase in the writing of this book. We thank D. Hartmann, H. Morrison and M. Strauss for carefully commenting on some draft chapters and R. van der Marel for a superb list of corrigenda. Finally, we thank the numerous colleagues who have shared with us their thoughts, ideas, data, and figures. Many of our 330 figure captions mention a specific case of such debt. Although we shoulder full responsibility for any shortcomings and inaccuracies in the book, any credit must be widely shared across the entire astronomical community. We hope that this text represents a fitting tribute to the progress that the community has made in the study of galaxies, and that it will help in taking stock of the many important tasks that still lie ahead.

1999 September 29

James Binney
Michael Merrifield

# Galactic Astronomy

# 1

# Galaxies: an overview

## 1.1 Introduction

The Sun is located towards the outskirts of the **Milky Way**, a gravitation-ally-bound collection of stars, or **galaxy**, similar to countless other such systems.[1] These systems, in turn, are arranged into bound clusters and still larger structures, but it is the galaxies themselves that are usually consid-ered to be the fundamental building blocks of the Universe. Part of the reason that galaxies occupy this pride of place is historic: until well into the twentieth century, it was by no means clear that any objects existed beyond the confines of the Milky Way, and so no larger structures could be stud-ied. The huge contrast in brightness between galaxies and their surroundings also picks them out as basic constituents of the Universe, but it should be borne in mind that this distinction also represents a human bias: if our eyes were tuned to the X-ray part of the spectrum rather than optical light, then clusters of galaxies would stand out as the most impressive individual struc-tures. Perhaps the best explanation for the enduring appeal of galaxies to astronomers is even more strongly anthropocentric: with their rich variety of

---

[1] The Milky Way is often referred to simply as "the Galaxy." We therefore use the adjective "Galactic" to mean "belonging to the Milky Way," whereas "galactic" refers to galaxies in general.

shapes and intricate spiral patterns, they provide the most visually stunning phenomena in the night sky.

Our goal in this book is to present the fullest description possible of these beautiful objects. To this end, we will discuss the current understanding of the individual elements, such as stars and gas, that make up galaxies, and the way in which these elements are arranged to form complete systems. Generally speaking, small-scale phenomena are best observed within our own galaxy, the Milky Way, where they are sufficiently nearby to be seen clearly. Large-scale galactic structures, on the other hand, are often best observed in external galaxies where we have a clear perspective on the whole system; when we try to study the large-scale properties of the Milky Way, we have considerable difficulty in seeing the wood for the trees. Analyses of the properties of the Milky Way and external systems are thus complementary in our quest for a complete description of the properties of galaxies, and so in this text we draw together these disparate strands and attempt to weave a coherent picture from them.

As will become clear through the course of the book, our picture of galaxies and their structure is still far from complete. By its very nature, astronomy is an observational science: we cannot tune the physical parameters of a galaxy to see how they alter its appearance; nor can we change our vantage point to get a better idea of its three-dimensional shape. Instead, we must synthesize all the fragmentary data that are observationally accessible in order to make the best sense that we can of our limited information. One key tool in this synthesis is the **cosmological principle**, which expresses our belief that the laws of physics are the same throughout the Universe. By applying the laws of physics as we know them locally to objects on the scales of galaxies, we seek to make sense of the observed properties of these systems.

The remainder of this chapter is devoted to a brief description of the historical development of galactic astronomy. It is not intended to provide a comprehensive history of the subject, but rather it seeks to show how the various ideas that make up our understanding of galactic astronomy developed, and to place the material in subsequent chapters into some sort of context.

## 1.2 A brief history of galactic astronomy

On a moon-less summer night away from city lights, a swathe of light can be seen stretching across the sky from horizon to horizon. This dramatic sight has intrigued people since ancient times, and has been the subject of many myths and legends. Its appearance as a stream of diffuse white light led the ancient Greeks to describe it as a river of milk flowing from the breast of Hera, wife of Zeus – the very word "galaxy" comes from the Greek word for milk. The Romans, too, saw this path across the sky as a *Via Lactea*, or

Milky Way. It was only in 1610 when Galileo first turned his telescope on the Milky Way that it was discovered that this band is not made up from a luminous "celestial fluid," but rather consists of huge numbers of faint stars which could not be resolved with the naked eye. Thus, the Milky Way was discovered to be primarily a stellar system.

The next major development in the scientific study of the Milky Way came in the mid-eighteenth century, when Immanuel Kant published his treatise, *General Natural History and Theory of the Heavens*. In it, he demonstrated how the planar structure of the Solar System arose naturally from the attractive force of gravity from the Sun which bound the system, and the ordered rotation of the planets which prevented its collapse. Drawing on the earlier work of Thomas Wright, Kant went on to point out that the apparent structure of the Milky Way could arise if this stellar system were similar in arrangement to the Solar System, but on a huge scale. He reasoned that the force of gravity acts between stars just as it does between Sun and planets, and so the stellar system should take on a disk-like structure if it possesses a systematic rotational motion to balance the inward gravitational pull. The central plane of this distribution is usually referred to as the **Galactic plane** or the **plane** for short. From our location within it, a disk-shaped stellar distribution would be seen as a band of stars stretching across the sky in a great circle, just as the Milky Way appears to us. On account of the huge scale of the structure, the period of its rotation would be so long that the motions of stars on the sky would be immeasurably small. Drawing an analogy with comets in the solar system, Kant also pointed out that the small number of stars found a long way from the band of the Milky Way cannot share the ordered motion of the major component of the system, but must lie on more randomly distributed orbits. Finally, Kant suggested that the Milky Way might not be the only such stellar system, and that some of the **nebulae** – faint, fuzzy, approximately elliptical patches of light seen in the sky – might be complete **island universes**, similar in structure to the Milky Way but viewed from large distances and at a variety of angles to the line of sight. Given the small amount of observational evidence which lay behind it, the remarkable prescience of this whole line of argument is a tribute to Kant's powers of reasoning.

Towards the end of the eighteenth century, increasingly powerful telescopes led to more systematic studies of nebulae. The comet hunter Charles Messier compiled a catalog of 109 of the brightest nebulae in the northern sky, primarily so that he could distinguish between these permanent diffuse patches of light and the transient comets that he was seeking. This list contains the most dramatic of the nebulae that are visible from the northern hemisphere, and prominent celestial objects are still frequently described by their number in Messier's catalog. The Great Nebula in Andromeda, for example, is referred to as Messier 31, usually abbreviated to M31.

In a far more extensive survey, William Herschel, his sister Caroline and his son John made a study of the entire sky as visible from both the northern

**Figure 1.1** Lord Rosse's sketch of the Whirlpool Nebula, M51,
c. 1850. [Reproduced from Berry (1898)]

and southern hemispheres. In the course of these observations the Herschels
compiled a catalog of nearly 5000 nebulae. Their telescopes were also able
to resolve the light from some of the closest nebulae – now known to be star
clusters associated with the Milky Way – into individual stars. This discovery convinced William that many of the still-unresolved nebulae were Kant's
island universes which, given sufficiently good observations, would be shown
to be made up from individual stars. However, he was also struck by the
appearance of **planetary nebulae**, some of which consist of a continuous
ring of glowing material surrounding what appears to be a single normal
star. These systems, Herschel argued, were fundamentally different from the
island universes, and thus the distinction between the truly gaseous nebulae
and unresolved stellar systems was recognized. However, it was only in the
late nineteenth century that pioneering studies of the spectra of nebulae by
William Huggins allowed the distinction between stellar and gaseous systems
to be quantified unambiguously. The Herschels' original list of both gaseous
and stellar nebulae was steadily added to over the course of the nineteenth
century, until Dreyer (1888) produced a compilation of 7840 objects in his
*New General Catalogue*. Subsequently, he supplemented this list with a further 5086 objects forming the *Index Catalogue* (Dreyer 1895, Dreyer 1908).
To this day, most reasonably bright non-stellar objects are identified by their
numbers in these catalogs, abbreviated as the **NGC** or the **IC**, respectively.

The nineteenth century saw continued improvements in observations resulting from advances in telescope technology. In 1845, William Parsons,
Third Earl of Rosse, finished construction of a telescope with the then-
enormous diameter of 72 inches (a size not surpassed until the completion

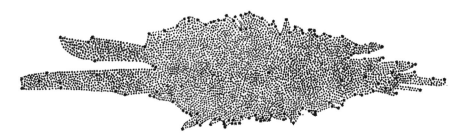

**Figure 1.2** Map of the Milky Way derived from William Herschel's "star gauging." The bright star near the middle marks the location of the Sun. [Reproduced from Herschel (1785)]

of the 100-inch Mount Wilson telescope in 1917). With this unprecedented light collecting area, Lord Rosse was able to observe the faint nebulae in much greater detail than had previously been possible. His examination of the nebulae cataloged by the Herschels revealed that many of the objects fell into two distinct categories: some appeared as completely featureless, very regular elliptical distributions of light; while others were less symmetric, displaying a distinctive spiral structure (see Figure 1.1). The very appearance of the whirlpool-like shapes in these latter **spiral nebulae** added weight to Kant's suggestion that these systems rotate about an axis perpendicular to their planes. Furthermore, Lord Rosse was able to use his powerful telescope to resolve individual point sources within the spiral nebulae. Although these objects were probably giant gaseous emitting regions rather than individual stars, the fact that at least some of the emission from nebulae could be resolved into individual objects supported Kant's conjecture that the nebulae were actually island universes made up of many distinct sources.

At the end of the nineteenth century, the application of photography to astronomy revolutionized the subject. By exposing photographic plates for long periods, it became possible to observe much fainter objects than were accessible to visual observations. Moreover, photographs could record images of hundreds of thousands of objects on a single plate, and the brightnesses of the individual images could be measured much more accurately from a photograph than was possible through the eyepiece of a telescope. Thus, a new age of quantitative astronomy was born.

### 1.2.1 Photometric models of the Milky Way

In the early years of the twentieth century, the detailed structure of the Milky Way provided an obvious target for study using photographic techniques. Previously, only crude studies of the shape of the Milky Way had been possible: Herschel (1785) had attempted to determine the shape of the system using a technique he termed **star gauging** in which he laboriously counted the number of stars that he could observe to successive limits of apparent brightness in 683 different regions of the sky. He then assumed that all the stars have approximately the same intrinsic brightness; that they are arranged approximately uniformly through the body of the Milky Way; and that he could see the stars all the way to the edge of the system. On the basis of these assumptions, he was able to map out the distribution of stars in the Milky Way: he concluded that the Sun lies close to the middle of stellar distribution, and that the distribution is flattened such that it extends approximately five times further in the plane of the Milky Way than in the direction perpendicular to the plane (see Figure 1.2). Since Herschel had no measure of the intrinsic luminosities of the stars that he observed, he was unable to put an absolute scale on the size of this system.

In order to refine such studies by making use of the large amounts of information that could be obtained from photographic plates, Jacobus Kapteyn decided to study in great detail 200 **selected areas** distributed across the sky. He coordinated a large collaboration of astronomers from all over the world to obtain the necessary photographs, and to analyze them by measuring the number of stars of different brightnesses and their small shifts in apparent position (**proper motions**) from year to year. He also used photographic plates to record the spectra of the stars in order to determine their types and their line-of-sight velocities (from the Doppler shifts in characteristic lines in the spectra). From an analysis of the proper-motion data, Kapteyn was able to estimate average distances for stars at various apparent brightness levels, and, from an analysis of the star-count data, he inferred the complete three-dimensional distribution of stars in space.

The final picture which emerged from this immense undertaking (Kapteyn & van Rhijn 1920, Kapteyn 1922) is usually referred to as the **Kapteyn Universe**. In agreement with Herschel's work, Kapteyn found that we are located close to the center of an approximately oblate spheroidal distribution of stars which extends about five times as far in the plane as perpendicular to it. He also demonstrated that the density of stars drops uniformly with distance from the center of the Milky Way. Moreover, Kapteyn was able to use the proper motion data to provide the first estimate of the absolute scale for the size of the Milky Way: he concluded that the density of stars dropped to half its maximum value at a radius of 800 parsecs[2] in the plane of the galaxy (and thus, from the measured flattening, the density dropped to half its maximum value at a distance $800/5 \sim 150\,\mathrm{pc}$ from the center perpendicular to the plane). In the plane, the density dropped to 10% of its central value at a radius of 2800 pc and 1% of its central value at a radius of 8500 pc.

Kapteyn's analysis also indicated that the Sun was located slightly out of the plane of the Milky Way at a distance of just 650 pc from the center. This proximity to the center provides an uncomfortably heliocentric feature

---

[2] A **parsec**, usually abbreviated to **pc**, is 3.26 light years or $3.1 \times 10^{16}$ m. It is roughly the distance from the Sun to its nearest neighbor star. Astronomers usually use kiloparsecs (kpc) and megaparsecs (Mpc) to measure larger distances.

in the Kapteyn Universe. Less than 10 percent of all the stars in this model lie within 700 pc of the center of the Milky Way; since presumably we could have evolved on a planet orbiting any of our galaxy's stars, it is statistically rather unlikely that we should find ourselves so close to the center. Kapteyn himself was well aware that there was an alternative explanation for the data: if there was an absorbing **interstellar medium** between the stars, then the light from distant stars would be dimmed by the absorbing medium. If this dimming were incorrectly interpreted as a distance effect, then the stars would be erroneously placed at excessive distances, leading to a spurious systematic falloff in stellar density in all directions. If this effect were strong, then we would appear to lie close to the center of the distribution whatever the true arrangement of stars.

There was plenty of evidence that some regions are, indeed, obscured: a dark rift along the central plane of the Milky Way is plainly visible to the naked eye, and numerous other dark patches where there are no stars show up clearly on photographs. If these regions were simply local voids in the distribution of stars, we should at least be able to see fainter, more distant stars beyond them. Stellar voids could account for the total absence of stars in some directions, but only if we were located at the end of a system of long, straight tunnels through the Milky Way, which are entirely empty of stars. It seemed far more likely that the absence of stars results from a nearby cloud of obscuring material which blocks the light from all the more distant objects. Kapteyn therefore expended considerable effort in trying to determine whether these dark clouds were isolated phenomena, or whether a more general absorbing medium pervades the Milky Way.

In order to assess the significance of absorption, Kapteyn sought to understand the physical processes that might be responsible for the obscuration. At the time, it was already well known that gas atoms can deflect light rays by the Rayleigh scattering process. If interstellar space were filled with gas, then light traveling to us from a distant star would have a high probability of being scattered in a random direction out of our line of sight, greatly diminishing the star's brightness. Thus, Rayleigh scattering provided a sensible mechanism which might produce apparent absorption in the Milky Way. To test this hypothesis, Kapteyn noted that the Rayleigh scattering process is much more efficient for blue light than for red light. We would therefore expect the light from stars to be more efficiently dimmed in the blue part of the spectrum than in the red, and so more distant stars should appear systematically redder – an effect known as **reddening**. Kapteyn (1909) looked for this effect by comparing the apparent brightnesses of stars as recorded by photographic plates to those estimated visually. Since photographic plates are more sensitive to blue light than the human eye, reddening of distant stars should have produced systematically greater brightnesses for the visual estimates than for the photographic data. Kapteyn found only small amounts of reddening in his data, and concluded that obscuration was unimportant.

In fact, we now know that the dominant source of obscuration is absorption by interstellar dust rather than Rayleigh scattering. The wavelength dependence of dust absorption is much smaller than that of Rayleigh scattering, and so absorption by dust reddens stars much less strongly than does Rayleigh scattering. With this more uniform absorption across the spectrum, the small amount of reddening detected by Kapteyn corresponds to a much larger total amount of obscuration.

The true degree to which dust obscures our view of the Milky Way was only recognized when Trumpler (1930) studied a sample of **open clusters** – loose concentrations of typically a few hundred stars found close to the plane of the Milky Way. He estimated the distances to these clusters by measuring their angular extents and assuming that they all are intrinsically the same size. He was then able to show that stars in remote clusters are systematically far fainter than would be predicted from their estimated distances. He thus demonstrated the existence of a strongly-absorbing interstellar medium which is responsible for this excess dimming. In fact, the amount of absorption implied by Trumpler's analysis is sufficient to invalidate Kapteyn's analysis of star counts entirely, and so we now know that his heliocentric picture of the Milky Way is erroneous.

Even at the time of Kapteyn's analysis, evidence was mounting which suggested that his model of the Universe was incorrect. In a classic series of papers (Shapley 1918b, Shapley 1918a, Shapley 1919c, Shapley 1919b, Shapley 1919a), Harlow Shapley presented a radically different picture of the Milky Way. Shapley had undertaken a detailed study of **globular clusters**. These approximately spherical systems, originally classified as nebulae, were readily resolved with telescopes into aggregates of between $10^4$ and $10^6$ stars. Unlike the stars of the Milky Way, globular clusters are not restricted to a narrow band in the sky, but are distributed throughout the sky. However, Shapley demonstrated that this distribution is not uniform: although there are roughly equal numbers of clusters on either side of the plane, they are not distributed uniformly in longitude along the plane; instead, they show a marked concentration toward the great star clouds in Sagittarius, which also define the brightest section of the Milky Way. Shapley argued that the massive globular clusters must be a major structural element of the Milky Way, and one would expect such a major element to be distributed symmetrically around the center of the system. The large asymmetry in the distribution of globular clusters implied that we are not near the center of the Milky Way, in contradiction to Kapteyn's analysis. Shapley went on to estimate the distances to the globular clusters using the apparent brightnesses of variable stars (with known intrinsic luminosities) and the apparent size and brightness of each cluster as a whole (assuming they are all intrinsically of comparable sizes and luminosities). On the basis of these measurements, he concluded that the Sun was some 15 kpc from the center of the globular cluster distribution, and hence, presumably, from the center of the Milky Way. Thus Shapley's picture of the Milky Way differed radically from the

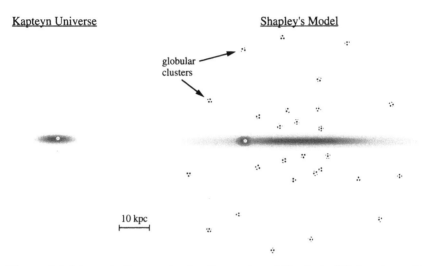

**Figure 1.3** Schematic diagram showing Kapteyn's and Shapley's differing views of the size and structure of the Milky Way. Kapteyn placed the Sun (⊙) close to the center of the stellar system (represented as a grayscale), whereas Shapley used the distribution of globular clusters to conclude that we lay far from the true center of the Milky Way, but possibly in a local stellar density enhancement.

Kapteyn Universe. Shapley further estimated that the whole globular cluster system was close to 100 kpc across, almost ten times larger than the Kapteyn model.

As Figure 1.3 illustrates, there was clearly a major inconsistency between Shapley's view of the Milky Way and Kapteyn's Universe. In retrospect, we know that these models can only be reconciled if we allow for the effects of interstellar absorption. The absorbing dust in the Milky Way is strongly concentrated to the plane, and so the apparent stellar brightnesses are strongly dimmed leading to the illusion that we lie close to the center of a relatively small stellar distribution. With the approximately spherical distribution of globular clusters, on the other hand, only the small fraction of the systems that lie close to the plane of the Milky Way will appear dimmed by dust absorption, so the overall derived distribution will not be significantly distorted. In this context, it is interesting to note that Shapley himself pointed out that no globular clusters are visible within about a kiloparsec of the plane of the Milky Way. Since he was unaware of the large amount of dimming which afflicts objects close to the plane, Shapley argued that the absence of detected globular clusters in this part of the sky arose because strong gravitational forces close to the plane of the Milky Way would have disrupted any clusters originally in this region.

In order to try to reconcile his picture of the Milky Way with the more

heliocentric Kapteyn Universe, Shapley also suggested that the stellar analysis had picked out a local concentration of stars which was, indeed, centered
close to the Sun, but that the global center of the distribution was at the
same distance as the center of the globular cluster distribution, some 15 kpc
away. There is certainly a small element of truth in this argument, since the
Sun does lie close to the center of a local loose cluster of stars referred to as
**Gould's Belt**. However, the ultimate reconciliation between the disparate
views of the Milky Way had to await the recognition that the apparent stellar
distribution is dominated by the effects of absorption. In fact, it was only
much more recently, with the development of computer programs capable
of numerically evaluating the effects of extinction on stellar number counts
[such as that written by Bahcall & Soneira (1980)], that quantitative sense
could be made of such data.

### 1.2.2 The nature of the spiral nebulae

Shapley's radical model of a large Milky Way must have played a key role in
the development of his ideas regarding the nature of spiral nebulae. To him,
the suggestion that the nebulae were independent island universes similar
to the Milky Way was quite implausible. The extent of his new Milky Way
model expanded the scale of the known Universe so far that it was hard
to believe that yet more remote objects could exist. Further, the spiral
nebulae have quite small angular sizes; if they were comparable in size to
Shapley's enlarged Milky Way, then they would have to lie at what were
then inconceivably large distances.

Shapley's ideas were by no means universally accepted, and many still
adhered to Kapteyn's smaller model of the Milky Way and concluded that
spiral nebulae were other similar systems. This split in the astronomical
community led George Ellery Hale (director of the Mount Wilson Observatory) to suggest that the issues might be debated at the National Academy
of Science in Washington as part of a lecture series in memory of his father,
William Ellery Hale. Shapley himself was the obvious choice of advocate for
the new ideas, and the role of his opponent fell to Heber Curtis, who had
studied spiral nebulae in great detail and was convinced of their extragalactic
nature. Their public confrontation took place at the National Academy in
April 1920, and has since become known as the **Great Debate**.

In fact, this title is something of a misnomer since the meeting did not
take the form of a debate, but rather consisted of two short lectures in which
each party presented his case [Hoskins (1976) gives a detailed account of the
meeting]. Shapley's contribution concentrated primarily on his model of the
Milky Way, and seems to have made only passing mention of it implications
for the nature of spiral nebulae. Curtis, on the other hand, used much of his
allotted time to discuss the evidence for and against the island universe hypothesis, unsurprisingly concluding that the hypothesis appears valid. Since

each talk lasted only about half an hour and concentrated on different aspects of the issue, there was no real winner in this "debate," although the consensus seems to have been that Curtis was the better speaker and gave a clearer exposition of his ideas.

More important than the debate itself were the expanded accounts of their addresses, which Shapley and Curtis published the following year (Shapley 1921; Curtis 1921). These papers went into greater technical detail than had been possible in their face-to-face encounter, and provided the opportunity for each to rebut the other's arguments. Curtis' paper cast doubt on the very difficult measurements that Shapley had used to calculate the distances to nearby stars. In particular, he was unconvinced by the large values of the distances that Shapley attributed to the Cepheid variable stars which were used to calibrate the distances to similar stars in globular clusters. Curtis also presented an analysis in which he incorrectly assumed that the brightest stars seen in globular clusters were comparable to the most luminous nearby stars; in fact, the brightest globular cluster members are giant stars with much higher intrinsic luminosities, and so Curtis radically underestimated their distances.

On the issue of the nature of spiral nebulae, Curtis advanced a wide range of arguments in support of his thesis that they were external to the Milky Way. In particular, he pointed out that spiral nebulae have angular sizes ranging from more than two degrees for the Great Andromeda Nebula, M31, down to a few arcseconds for the smallest nebulae which had been photographed at that time. If these objects are of comparable intrinsic size, then their distances must differ by more than a factor of a thousand. Even if the Andromeda Nebula were within the Milky Way, other apparently smaller nebulae would have to lie at huge distances well beyond the bounds of even Shapley's extended Milky Way.

He also noted that quite a number of novae – bright stellar sources which appear briefly before fading back to obscurity – had been detected in the direction of M31. The large number of novae in this small region of sky implied that they must be physically associated with M31. Moreover, the novae in the direction of M31 were very much fainter than those seen from elsewhere in the Milky Way, implying that they were located much further away. Invoking the inverse-square law for the variation in apparent brightness with distance, Curtis calculated that M31 was at a distance of around 100 kpc. At this distance, the angular size of M31 implied that its linear size was around 3 kpc, comparable in size to Kapteyn's model of the Milky Way. With hindsight, we know that Curtis made two mistakes in his analysis which conveniently canceled out: by confusing Galactic novae with much brighter supernovae in M31, he underestimated the distance to the nebula by about a factor of five; however, the Kapteyn model of the Milky Way is also too small by a similar factor. He thus correctly (if fortuitously) concluded that M31 and the Milky Way are comparable systems of similar linear extent.

**Figure 1.4** Optical image of the edge-on galaxy, NGC 891, originally discovered by Caroline Herschel. Note the prominent dust lane in the plane of the galaxy, and the presence of both extensive disk and central spheroidal stellar components. [DSS image from the Palomar/National Geographic Society Sky Survey, reproduced by permission]

The dispersion of the light from spiral nebulae into spectra provided two further pieces of evidence in favor of the idea that they were distinct stellar systems. First, the characteristic absorption lines in the spectra of many nebulae are the same as those seen in normal stars, and the total spectra resemble closely the spectrum that would be obtained from the integrated starlight of the Milky Way, which suggests that they are similar systems. Second, the lines in the spectra of both stars and galaxies are shifted relative to their rest wavelengths. These shifts can be understood if they arise from the Doppler shift due to the line-of-sight motions of the emitting body. However, the inferred velocities for the nebulae are many times larger than those of the stars in the Milky Way, suggesting that the nebulae and stars do not form a single dynamical entity. Moreover, although the nebulae have large line-of-sight velocities, their positions on the sky have not changed detectably with time. Unless the nebulae all happen to be moving exactly away from us, which seems most unlikely, the lack of any apparent motions transverse to the line of sight implies that they must be at very large distances. However, no explanation was offered as to why the vast majority of spiral nebulae have Doppler shifts which imply that they are receding from the Milky Way. This systematic effect could easily be interpreted as suggesting that the Milky Way is repelling the nebulae, implying that they must be physically associated with each other. The true significance of this phenomenon, the expansion of the Universe initiated by the Big Bang, would only be uncovered almost a decade later.

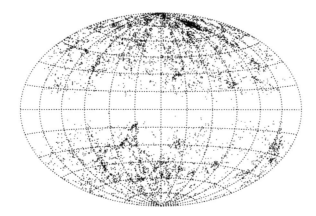

**Figure 1.5** Map showing the distribution of New General Catalogue (NGC) and Index Catalogue (IC) objects which have been identified as spiral or elliptical nebulae. In this Aitoff projection, the plane of the Milky Way runs horizontally through the center of the map. Note the dearth of objects in the "zone of avoidance" within fifteen degrees of the plane.

One last particularly elegant piece of reasoning presented by Curtis came from the observation that edge-on spiral nebulae often contain dark bands through their centers (see Figure 1.4). Curtis interpreted this band as a ring of absorbing material surrounding the systems. He further pointed out that the **zone of avoidance** – a region close to the plane of the Milky Way where no spiral nebulae are observed (see Figure 1.5) – could be explained naturally if the Milky Way were encircled by a similar absorbing ring, and the spiral nebulae all lay beyond this ring. He thus neatly brought together the ideas that the Milky Way would look similar to a spiral nebula if viewed from the outside, and that the nebulae must lie beyond the fringes of the Milky Way. The only slight flaw in this reasoning was the assumption that the absorbing material lies in a single ring at large radii; as we now know, absorption arises from dust which is distributed throughout the plane of the Milky Way. Curtis was forced to assume that the absorbing material lay only at large distances since he adhered to Kapteyn's analysis of the stellar distribution in the Milky Way which relied on their being no absorption of stellar light within the system.

For his part, Shapley produced several strong points in favor of the hypothesis that spiral nebulae are intrinsically different from the Milky Way. Foremost amongst these arguments was his belief that the Milky Way was a very large system, which would imply that the relatively small spiral nebulae would have to be at tremendous distances to be comparable systems. However, he also pointed to a number of analyses which imply that the Milky Way has other properties that distinguish it from the spiral nebulae. The Milky Way has a lower surface brightness (i.e. luminosity per unit area) than any of the spiral nebulae, and many spiral nebulae were observed to be significantly bluer than stars in the Milky Way. These distinctions implied that the nebulae and the Milky Way could not be intrinsically similar systems. With hindsight, we can put the differences down to the effects of the absorption of which both Shapley and Curtis were unaware: since we see the Milky Way edge-on, stars on the far side of the Galaxy are viewed through much absorbing material, and so the surface brightness of the system is greatly reduced when compared to observations of more face-on systems; similarly, light from the bright blue stars which populate the spiral arms in the Milky Way is strongly attenuated, and so we preferentially see the more-plentiful nearby faint red stars, which makes the system appear redder than external galaxies.

Perhaps the most telling argument reported by Shapley was the result of a set of measurements by Adriaan van Maanen, which compared images of several spiral nebulae taken over a number of years. The comparisons implied that the spiral structure in these systems rotates at a perceptible rate, with a rotational period of only around $10^5$ years. If the nebulae had radii in excess of around 5 kpc, then their outer parts would have to be rotating at velocities in excess of the speed of light in order to maintain this apparent rotation rate! The absurdity of this conclusion implied that the spiral nebulae must be very much smaller than 5 kpc, and so could not be comparable in size to the Milky Way. The reasoning here was impeccable, and Curtis conceded that if the observations were confirmed then the whole island universe hypothesis would have to be abandoned. Van Maanen was a well-respected observer, and so his results carried a great deal of weight at the time. It was only several years later, when Lundmark re-measured van Maanen's photographic plates, that it was conclusively demonstrated that the detection of rotational motion was spurious. We will probably never know where van Maanen went wrong in this set of measurements.

It is clear that both Shapley and Curtis had grasped important aspects of the truth as we now perceive it. In her review of their encounter, Trimble (1995) makes the insightful point that each party made best use of data that they had collected themselves: Shapley's analysis of his globular cluster observations produced the fundamental change in ideas about the location of the Sun and the size of the Milky Way, and Curtis' understanding of the true nature of the spiral nebulae came from his long study of these systems. Most of the confusion arose when other peoples' analyses, such as Kapteyn's model of the Milky Way and van Maanen's rotation measurements, were added to the debate.

Within five years, Edwin Hubble had resolved the controversy as to the nature of spiral nebulae once and for all. Using the superior optics of the recently-completed 100-inch telescope at Mount Wilson, he was able to resolve the outer parts of two nearby spiral nebulae into swarms of very faint objects indistinguishable from stars. If these objects were assumed to be comparable to the brightest stars in the Milky Way, then the nebulae must be at large distances. The clinching observation came in late 1923 when Hubble established that the brightnesses of a few of these myriad stellar images in the Andromeda Nebula, M31, varied in the characteristic periodic manner of **Cepheid variable stars**. It had already been established that the intrinsic luminosities of these particular stars can be determined directly from their periods of variability, and so by measuring the periods of the Cepheids in M31, Hubble was able to estimate their luminosities. Measuring their apparent brightnesses then yielded a direct estimate for their distances. Using this method, Hubble (1922) obtained a value for the distance to M31 of some 300 kpc. Although more recent calibrations of the luminosities of Cepheids imply that this estimate is more than a factor of two too small, it nonetheless firmly demonstrated that the Andromeda Nebula is not an element within the Milky Way, but is a comparable stellar system in its own right. Thus, it was finally established that the Milky Way is but one galaxy amongst its peers. The spiral nebulae are thus actually **spiral galaxies**, while the more featureless elliptical nebulae are mostly **elliptical galaxies**.

### 1.2.3 Kinematic models of the Milky Way

At about the same time that Hubble was demonstrating that spiral nebulae are indeed separate galaxies, Bertil Lindblad was producing new insights into the properties of the Milky Way using an entirely different approach. He calculated the total mass of Kapteyn's model by adding up the contributions from all the stars. He then used the Doppler shifts in the lines seen in spectra of globular clusters to show that these objects move with velocities as high as $250 \, \text{km s}^{-1}$. Such speeds are significantly higher than the escape velocity from Kapteyn's Universe; the relatively small total mass of this system results in a gravitational field that is too weak to retain the globular clusters as gravitationally bound members of the system. The fact that a large number of globular clusters are associated with the Milky Way implies either that the true gravitational forces are stronger than those predicted by the Kapteyn model, thus permanently binding the clusters to the Milky Way, or that globular clusters continuously form at a sufficient rate to replace their escaping kin. Since globular clusters each contain up to a million stars, there is not enough mass in the Kapteyn Universe to provide the raw materials required to replenish the continually-escaping globular clusters. These dynamical arguments provided perhaps the last nail in the coffin of the Kapteyn Universe, and suggested that Shapley's model for the Milky Way with its larger mass (and correspondingly greater escape velocity) was closer to the truth.

Lindblad (1927) went on to develop a more detailed kinematic model of our galaxy, which sought to explain its apparent structure in terms of the motions of its constituents under their mutual gravitational attraction. He proposed that the Milky Way might be divided into a number of subsystems, each of which was symmetric about the central axis of the whole system (which, in agreement with Shapley, he placed a considerable distance from the Sun). Each component was further assumed to rotate about this symmetry axis with some characteristic speed. Echoing the ideas of Kant two centuries earlier, Lindblad noted that the degree to which each subsystem is flattened would depend on whether its motions were dominated by rotational or random motions: the most slowly rotating subsystems would be made up of objects on largely random orbits and hence would display little in the way of flattening.

Since Shapley's analysis had shown that the distribution of globular clusters in the Milky Way is approximately spherical, Lindblad proposed that these objects constitute a subsystem with almost no rotational motion. Furthermore, since almost all the stars in the solar neighborhood possess very little velocity relative to the Sun, the random component of these stars' motions must be small, and so Lindblad concluded that they must follow well-ordered circular orbits and thus produce the highly-flattened disk of the Milky Way. By measuring the velocity of the Sun and its neighboring stars relative to the mean velocity of the non-rotating globular cluster population, Lindblad was able to demonstrate that the nearby disk stars rotate around the Milky Way with a velocity of between 200 and 300 km s$^{-1}$.

These ideas were developed into a complete theory of Galactic stellar kinematics by Jan Oort (Oort 1927, Oort 1928). Amongst the phenomena explained by this theory were the so-called **high-velocity stars**. These stars make up an asymmetric tail in the distribution of stellar velocities, traveling at velocities such that they lag behind the rotational motion of the Sun. Thus, although these stars have high velocities relative to the Sun, they are somewhat misnamed since they are actually moving at very low speeds, with their high relative velocities arising almost entirely from the motion of the Sun around the Milky Way. Oort pointed out that such a population would arise if there were a stellar component of our galaxy with little net rotation. The members of this population would, while in the solar neighborhood, have insufficient tangential velocities to maintain them on circular orbits against the gravitational pull of the Milky Way, and thus their orbits should carry them radially inward toward the center of the system. Since this population of stars has little net rotation, we should expect them to form an approximately spherical concentration toward the center of the Milky Way. Oort's analysis of the remaining stars in the solar neighborhood showed that their kinematics were exactly what would be expected if they were following approximately circular orbits in a differentially rotating disk (i.e., one in which stars nearer the center have a faster angular rotation rate than those at larger radii) with the Sun located far from the disk's center. Images of edge-on spiral galaxies (see Figure 1.4) reveal the presence of both relatively unflattened spheroidal distributions of stars at their centers and highly-flattened disks that extend to larger radii. Thus, Oort's stellar kinematic analysis confirmed the idea that the Milky Way is indistinguishable in structure from these external galaxies.

A great advance in dynamical studies of the Milky Way and other galaxies came from the discovery that the gas in these systems emits at radio wavelengths. Karl Jansky established in 1932 that the Milky Way emits a broad spectrum of radio waves, but the major breakthrough came from the discovery that it also emits strongly in a spectral line at 21 cm. The existence of this emission line, arising from a hyperfine transition in atomic hydrogen, was predicted by H.C. van de Hulst in 1944, but it was not until 1951 that this prediction was observationally confirmed by Ewan and Purcell at Harvard, Christiansen in Australia, and Muller and Oort in the Netherlands. This radio emission provided the ideal tool for studying the large-scale kinematics of the Milky Way: the distinctive line enabled astronomers to measure line-of-sight motions of atomic hydrogen via Doppler shifts in the line's detected wavelength. Further, radiation at such long wavelengths is entirely unaffected by dust, and so the absorption which limits our ability to study the stellar structure of the Milky Way at optical wavelengths ceases to be a problem. Observations of 21 cm emission revealed that atomic hydrogen is a major component of the Milky Way, permeating its disk out to distances beyond twice the solar radius; it therefore provides us with a tracer of the properties of the disk of the Milky Way over a wide range of radii.

By combining 21 cm observations obtained from both the northern and southern hemispheres, Oort, Kerr & Westerhout (1958) were able to produce the first complete map of the distribution of atomic hydrogen throughout much of the Galaxy (see Figure 1.6). These and subsequent observations showed that the gas is strongly concentrated toward the plane, that it is distributed fairly uniformly in azimuth, and that it travels on approximately circular orbits about the Galactic center (as would be expected for such a highly flattened component).

More detailed examination of the gas distribution revealed that such a simple axisymmetric model is an oversimplification. Even on the largest scales, the atomic hydrogen in the Milky Way is not arranged uniformly in azimuth: more gas lies on one side of the system than on the other. The gas disk is not flat, either, but warps away from the plane of the Milky Way at large radii. As Figure 1.6 shows, there are also long, narrow regions in which the density of gas is significantly enhanced, and Oort, Kerr & Westerhout (1958) were quick to identify these features with the spiral structure found in other galaxies (see Figure 1.1). Since the arms are associated with density enhancements in the disk, it is not surprising that these asymmetric perturbations in the gravitational potential produce non-circular motions in the orbiting gas. The map presented in Figure 1.6 was constructed by assuming that the gas follows circular orbits, and the presence of non-circular motions distorts its appearance. It is this phenomenon which is responsible for the spurious manner in which the arm features appear to converge on the Sun. Thus, although mapping out the Galaxy through its 21 cm emission represented a major advance in our study of the structure of the Milky Way,

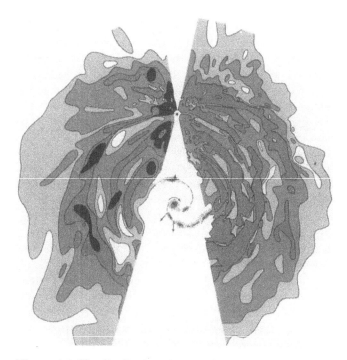

**Figure 1.6** The distribution of atomic hydrogen in the plane of
the Milky Way as inferred from early 21 cm observations. The Sun
is located at the point marked "⊙." The empty cone toward the
Galactic Center and anti-center cannot be mapped out using the
kinematic technique employed here. [Reproduced from Oort, Kerr
& Westerhout (1958), by permission]

the complexity of interpreting non-circular motions meant that the picture
was still ambiguous.

The strongest signs of non-circular gas orbits in the Milky Way come
from close to the Galactic center, where 21 cm emission has been observed
to be Doppler shifted from its expected circular velocity by hundreds of kilometers per second. These anomalous velocities were originally interpreted
as implying that the gas is being flung from the center of our galaxy, but it
was subsequently realized that they could equally be interpreted as resulting
from non-circular orbits in a strongly non-axisymmetric gravitational potential (Peters 1975). Since at least one third of external spiral galaxies are
observed to contain a bar-like asymmetric structure at their centers (see Figure 1.7), we might reasonably conclude that the non-circular motions in the
Milky Way are induced by a similar **galactic bar**.

The early radio observations used single dish antennae, which gave a
spatial resolution of only a degree or so, and this limitation made it difficult
to study even the large-scale distribution of hydrogen in external galaxies.
Far greater resolution was achieved with the construction of **interferom-**

**Figure 1.7** Optical image of the galaxy, NGC 3992. Note the prominent bar in this system in addition to the spheroidal and disk stellar components. [DSS image from the Palomar/National Geographic Society Sky Survey, reproduced by permission]

**eters**, which consist of a set of moderate-sized radio dishes arrayed over several kilometers. The angular resolution of such an instrument is equivalent to that of a single dish the size of the whole array. Data obtained from these interferometers, such as the Westerbork Synthesis Radio Telescope in the Netherlands and the Very Large Array in New Mexico, revealed that other spiral galaxies also have hydrogen disks in which the gas follows approximately circular orbits, but with the same complications due to warping and non-circular streaming as the Milky Way. These disks could be traced to large radii, well beyond the observed optical edges of the galaxies.

Since it is the gravitational pull of a galaxy that provides the force which retains the gas on its approximately circular orbits, it is straightforward to translate the observed orbital velocities in a gas disk into an estimate of the distribution of mass in the galaxy. We can compare the mass derived in this way to the sum of the masses of the individual components such as the stars in the disk and spheroid, and the hydrogen gas itself. The surprising result of such comparisons in the 1970s was that the total mass inferred dynamically exceeds the sum of all the known components, and that the discrepancy becomes greater as we look to larger radii in a galaxy. From these observations it was inferred that galaxies are embedded in huge **dark halos**, which contain considerably more mass than the visible galaxies themselves. This unexpected discovery was originally dubbed the "missing mass problem", but this title is somewhat confusing, since we know from the kinematics that there is an excess of mass rather than a deficit. The real problem is that we do not know what form the excess mass takes, since it evidently is not made up of directly-visible material. Our inability to identify the nature of the material which dominates the mass of many systems is therefore now usually referred to as the "dark matter problem." Its resolution remains one of the major goals of galactic astronomy.

### 1.2.4 Stellar populations

The recognition by Lindblad and Oort that spiral galaxies could be broken down kinematically into separate spheroidal and disk components ties in closely with the key notion of distinctive **stellar populations**. Taking advantage of the wartime blackout in Los Angeles, Baade (1944) used the 100-inch Mount Wilson telescope to resolve individual stars in the inner regions of several nearby spiral galaxies, where the spheroidal component dominates. He also obtained resolved stellar images in a couple of nearby elliptical galaxies. By analyzing the colors and brightnesses of the stellar images, Baade realized that the brightest stars in both elliptical galaxies and the spheroidal components of spiral galaxies are **red giants**; these stars are quite different in appearance from the **blue supergiants** which dominate the spiral arms in the disks of galaxies. These observations suggested to Baade the existence of two characteristic stellar populations: **population I**, which contains luminous blue stars, accompanied by dust and gas; and **population II**, which is dominated by luminous red stars in an essentially gas- and dust-free environment. Open clusters and stellar disks comprise population I material, while globular clusters, galactic spheroids and elliptical galaxies are made up of population II stars.

Detailed studies of the colors and brightnesses of stars in the different types of star cluster confirmed the clear distinction between the two stellar populations. In the early part of the twentieth century, Ejnar Hertzsprung and Henry Norris Russell independently discovered that, in a plot of color versus luminosity, stars are not randomly scattered, but rather are concentrated within tightly defined bands. Such a plot was termed an **HR diagram** after its inventors, although in its modern form it is more usually described as a **color–magnitude**, or **CM diagram**. When CM diagrams were constructed for stellar clusters, the bands which are populated by the stars of open clusters were found to differ from those which are populated by globular-cluster stars. From this discovery it followed that the stellar populations of these two types of cluster contain very different types of stars.

The location of a star in a CM diagram depends on the rate at which it is generating energy by nuclear fusion in its core, and on the structure of the star itself. The two decades up to about 1965 saw advances in understanding of the energy generation processes and the development of computers capable of calculating detailed models of stellar structure. It thus became possible to predict where a star should appear on the CM diagram, and how its location changes with time as the star's structure evolves. From these calculations, it has become clear that the difference between open and globular clusters arises from their ages: younger stars occupy the regions of CM diagrams populated by open-cluster stars, whereas globular-cluster stars fall only in regions of the CM diagram that are occupied by old stars. Thus, the distinction between populations I and II is also one of age, with population I stars having formed more recently than those of population II.

Confirmation of this distinction has come from detailed spectral analysis of stars from the two populations. These studies allow one to determine the chemical compositions of stars from the strengths of characteristic spectral absorption features. Population II stars are found to be very deficient in all elements heavier than helium (or **metals** as they are somewhat confusingly termed by astronomers), whereas population I stars have rather higher metal abundances, comparable to the solar fraction of heavy elements (which is not very surprising since the Sun is itself a population I star). Some of the heavy elements found in stars are only produced in the supernova explosions which end many stars' lives. Hence, the presence of these elements in a star indicates that the star must contain material synthesized by supernovae in a previous stellar generation. The difference between the metal abundances in the two stellar populations thus once again points to a difference in ages. A star that formed late in the history of a galaxy will be made from material that has had time to be processed through several earlier generations of stars, and will thus generally have high metal abundances; the star will therefore be classified as a population I object. A star that formed much earlier will have been produced from material whose chemical abundances lie much closer to the primordial composition of the Universe with only very small amounts of metals, and so the star will be a population II object.

It therefore seems highly probable that the globular clusters and spheroid of the Milky Way, being made up of population II stars, are the parts of the Milky Way that formed first, with the population I stars in the disk forming later. This evolutionary sequence is confirmed by the fact that stars are observed to form in the disk even today, whereas no star-forming regions can be seen away from the Galactic plane. It is notable that even the most metal-poor stars that have been observed in the Galaxy contain trace amounts of heavy elements, which they could not have synthesized themselves. The idea has therefore been mooted that there might have been a still-earlier generation of **population III** stars which produced these elements, but the existence of such objects lacks any direct observational confirmation.

## 1.2.5 More recent developments

As should be apparent from the above discussion, much of the history of astronomy has been dictated by the technology available to astronomers. We have, for example, been able to form images of progressively fainter objects as larger and larger telescopes have been built. Similarly, the quantitative analysis of astronomical images only became practicable with the invention of photographic detectors which could be used to obtain a permanent record of the data.

Technological advances continue to play a key role in the development of the subject. For example, the rapid developments of computer technology in the latter part of the twentieth century have allowed astronomers to handle large quantities of data and to automate much of the analysis of these data. With these techniques, it has been possible to undertake enormous surveys which automatically monitor the light from literally millions of stars on a nightly basis. The purpose of these experiments is to look for the magnification of light from a distant star that will occur if a massive object forms a temporary gravitational lens as it passes through our line of sight to the star. If the dark halo of the Milky Way is made up of compact massive objects, then their presence could be detected by this phenomenon. The probability of a single star being magnified in this way is very small, and so it is only by using computers to monitor automatically very large numbers of stars that we stand a good chance of detecting the effects and thus determining the content of the Milky Way's dark halo.

Observational techniques have also continued to benefit from advances in technology. The introduction of very efficient **charge coupled devices** (**CCDs**) to replace photographic plates as light detectors has revolutionized optical astronomy. These detectors are similar to those found in camcorders, except that they are cooled with liquid nitrogen to reduce the level of thermal noise. Since CCDs are some ten times more efficient than photographs at detecting light, they have enabled astronomers to study very faint phenomena such as the outermost parts of galaxies for the first time. Developments in detector technology have also expanded the region of the spectrum over which observations can be made. Of particular importance has been the development in the 1980s of infrared detector arrays analogous to the optical CCDs. These efficient digital devices have made it possible to study galaxies in the near infrared, where the absorbing effects of dust are small, providing a clearer picture of the underlying stellar light distribution.

In addition to the improvements in light-gathering potential provided by the new detector technologies, advances have also been made through the construction of ever bigger telescopes. A number of telescopes with diameters of around eight meters are currently at various stages of planning or construction. However, an even larger telescope is already in operation: the Keck Telescope on Mauna Kea in Hawaii consists of 36 hexagonal mirror segments each 1.8 meters across, mounted together in a honeycomb pattern. The exact alignment of all the mirror segments is maintained by computer, allowing the system to act as a single telescope with a light-collecting power that is equivalent to that of a single mirror 10 meters in diameter. A duplicate of this amazing piece of technology is being built adjacent to the original, so that these two huge telescopes will be able to operate in concert.

Significant advances have also been made in the quality of the images recorded. Motions in the earth's atmosphere distort the light from astronomical objects, causing the familiar twinkling of stars. These distortions blur out any structure in objects which have angular scales of less than about an arcsecond. This degradation can be ameliorated by furnishing telescopes with moving mirrors that compensate for the motions of the atmosphere in order to reduce its distorting influence. Such **active optics** can decrease the amount of blurring by more than a factor of two, but still greater gains can be made by getting above the earth's atmosphere entirely. The Hubble Space Telescope (HST) is a satellite telescope with a 2.4 m mirror, which was designed to produce images that can resolve structure down to a scale of 0.1 arcseconds. Initially, a flaw in the shape of the mirror prevented this goal from being achieved, but the introduction of new optics to correct its myopia has enabled the HST to exploit fully its location above the earth's atmosphere. In addition to studying small-scale structures such as the very cores of galaxies in unprecedented detail, the improved image quality and lower background light level in space have allowed the HST to image very faint objects; such studies have, for example, demonstrated that the dark matter halo of the Milky Way cannot be made up from intrinsically faint low-mass stars.

Observations from space have proved important in several other areas of the study of galactic structure. The Hipparcos Satellite, for example, contained a small (29 cm diameter) optical telescope dedicated to measuring accurate stellar positions (a field of study known as **astrometry**). Over its four year lifetime, this satellite measured the positions of more than 100 000 stars to an accuracy of $\sim$ 0.002 arcseconds. In addition to the absolute positions of the target stars on the sky, the satellite also measured the small periodic shift in apparent location of the stars that arises from the changing position of the earth as it orbits around the Sun. The magnitude of this effect, known as **parallax**, depends on the distance to the star, and so the Hipparcos observations provide a measure of the distances to all the target stars out to distances of several hundred parsecs (at which point the parallax becomes too small to measure). Thus, Hipparcos measured the full three-dimensional spatial coordinates for a large sample of relatively nearby stars. The satellite also recorded the shifts in the location of the stars due to their own orbital motions; combining these observations with the line-of-sight velocities inferred from Doppler shifts in spectra obtained from the ground, we can reconstruct the full three-dimensional velocities of the target stars. Thus, Hipparcos has provided us with a wealth of information on the spatial and kinematic properties of the Milky Way which has yet to be fully exploited.

The development of astronomical satellites has also opened up parts of the electromagnetic spectrum that are unobservable from the ground due to their absorption by the earth's atmosphere. The whole field of X-ray astronomy, for example, only became possible with the advent of space flight. Solar observations were made from rockets as early as the 1940s, and these experiments were followed by rocket observations of the moon and other bright X-ray sources in the 1960s. However, X-ray astronomy only really came of age with the development of satellite technology, and particularly the launch

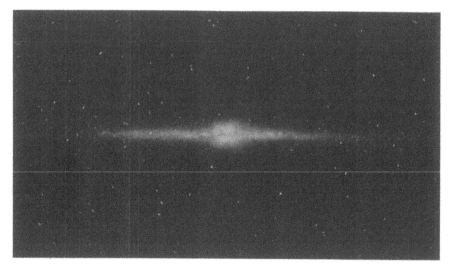

**Figure 1.8** Infrared image of the Milky Way obtained by the COBE satellite. [Reproduced courtesy of NASA Goddard Space Flight Center and the COBE Science Working Group]

of the Einstein Observatory in 1978. This mission and its successors have produced images of some of the most energetic phenomena in the Universe, including the ultra-luminous **active galactic nuclei**, believed to be powered by super-massive black holes, which lurk at the centers of some galaxies. X-ray observations have also revealed an entire new component of galaxies: a halo of gas which is so hot that it emits at these short wavelengths. The discovery of this component in elliptical galaxies was particularly surprising since, as we have discussed above, it had long been supposed that these systems were essentially gas-free.

One last example of space-based observations of importance to the study of galaxies comes from the infrared part of the spectrum. At the longer infrared wavelengths, the earth's atmosphere is opaque and so little could be learned from the ground. The Infrared Astronomical Satellite (IRAS), launched in 1983, provided the first comprehensive survey at these wavelengths. Cool dust in galaxies re-radiates the light that it absorbs at these long infrared wavelengths, and so this survey gave us the first comprehensive map of the distribution of dust in the Milky Way. These observations revealed that the far infrared emission from dust is strongly concentrated toward the plane of the Galaxy, which ties in with the large amount of optical extinction in this region. However, it also turned up the unexpected discovery of diffuse patches of emission – termed **cirrus** by analogy with the terrestrial thin clouds – far from the plane of the Milky Way. Dust is clearly more pervasive in the Galaxy than had been expected.

A more recent study of the Milky Way at infrared wavelengths has been made by the Cosmic Background Explorer (COBE) satellite. Although primarily intended to test cosmological theories, the telescopes on this satellite also recorded the strong flux from both stars (at near infrared wavelengths) and dust (at far infrared wavelengths) in the Milky Way. At the near infrared wavelengths, little light is absorbed by dust, and so the all-sky image produced by COBE at these wavelengths provides us with a uniquely clear picture of the large-scale stellar distribution in the Milky Way. The COBE image of the plane of the Milky Way is presented in Figure 1.8. Comparing this image to the picture of NGC 891 in Figure 1.4, it is strikingly apparent that – once the confusing effects of dust obscuration have been removed – the Milky Way is indistinguishable from the multitude of other galaxies which make up the cosmos.

# 2

# Astronomical Measurements

Our basic observations of astronomical objects yield information about their positions, their motions on the sky, and the distribution of the energy emitted in their spectra as a function of frequency and time. Essentially everything else we wish to know about astronomical bodies must follow from deductions made from analyses of these data, using the laws of physics established in terrestrial laboratories. It is actually rather remarkable that we know as much about the stars and galaxies as we do, given the severe limitation of having at our disposal only the photons that they casually emit.

Our knowledge of astrophysical objects has been painstakingly acquired by observing them with delicate and complex equipment, usually followed by an elaborate set of reductions to eliminate extraneous effects – see, for example, Budding (1993), Sterken & Manfroid (1992), or Appenzeller, Habing & Léna (1989). To communicate the results of such measurements and reductions, astronomers have developed a framework of concepts that are peculiar to astronomy. In this chapter we develop this framework and introduce its associated vocabulary.

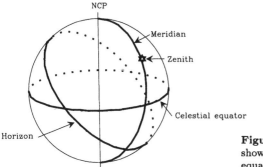

**Figure 2.1** The celestial sphere, showing the celestial poles, celestial equator, meridian, zenith point, and horizon.

## 2.1 Positions, motions and coordinate systems

Positional astronomy, or **astrometry**, is the oldest branch of astronomy. Catalogs of the locations of astronomical objects on the sky have come down to us from antiquity. It is probably fair to say that astrometry was the central problem of observational astronomy through the early twentieth century. Later in the twentieth century, astrophysical observations attracted more interest and generated more excitement than astrometry, to the extent that students sometimes have the (mistaken) impression that positional astronomy is old fashioned and even unimportant. Such an opinion is grievously wrong, not only because the data obtained by astrometric measurements lie at the foundation of all astrophysics, but also because in recent years the subject has undergone a renaissance. Technological advances that got under way around 1980, have spectacularly improved the accuracy, homogeneity, and extent of the available astrometric data. In 1997 the field was revolutionized by the publication of the Hipparcos Catalogue, which is based on data taken by the first astrometric satellite; the catalog contains data of unprecedented homogeneity and accuracy for more than 100 000 nearby stars. These new data will over time have a profound influence on almost every branch of astrophysics, from stellar astronomy to cosmology.

The issues that positional astronomy deals with are anything but trivial. Among them is the task of establishing in practical terms, an inertial reference frame. Extensive discussions of this problem can be found in Green (1985), Perryman *et al.* (1997), Seidelmann (1992) and Murray (1983). In our treatment of this subject we start by studying how astronomers report the positions and motions of objects on the sky.

### 2.1.1 The equatorial system

Any mention of the position of an object implies that a coordinate system has been chosen and established observationally. The fundamental coordinate

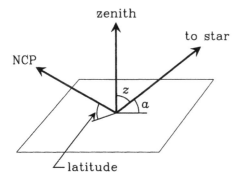

**Figure 2.2** The zenith distance of a star is the angle $z$, and its altitude is the angle $\alpha$.

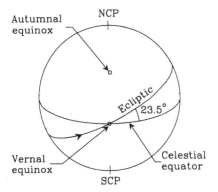

**Figure 2.3** The relationship between the celestial equator, the ecliptic and the equinoxes. The angle between the ecliptic and the equator is called the obliquity of the ecliptic.

system for observations made from the surface of the Earth is the **equatorial system**. The Earth rotates eastward on its axis once a day, and, as a consequence, the sky appears to rotate westward about the Earth. The extension of the Earth's axis to the **celestial sphere** (an imaginary sphere of infinite radius centered on the earth) defines the north and south **celestial poles** (**NCP** and **SCP**), and the extension of the Earth's equatorial plane determines the **celestial equator** (see Figure 2.1). The equator is a **great circle**; that is, a circle on the celestial sphere defined by the intersection of a plane passing through the sphere's center with the surface of the sphere.

For each observer on the Earth's surface, the direction of gravity fixes the direction of the **local vertical**; that point at which the extended vertical line intersects the celestial sphere is the **zenith**. The great circle passing through the celestial poles and the zenith is the **meridian**. The **horizon** is the great circle whose pole is the zenith (and thus lies 90° from the zenith).

The **zenith distance** $z$ of a star is its angular distance from the zenith (Figure 2.2). The complement of $z$ is the star's **altitude** $a \equiv 90° - z$. The altitude of the NCP equals the observer's latitude on the surface of the Earth.

As the Earth revolves annually in its orbit around the Sun, the Sun

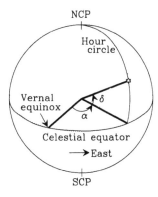

**Figure 2.4** The position of the star in the equatorial system is specified by its right ascension $\alpha$ and its declination $\delta$.

appears to move from west to east around the celestial sphere on a path called the **ecliptic** – see Figure 2.3. The ecliptic is the great circle in which the Earth's orbital plane intersects the celestial sphere. The Earth's rotation axis is inclined away from the normal to its orbit by an angle of about $23°27'$; hence the ecliptic is also inclined to the celestial equator by this angle, which is called the **obliquity of the ecliptic**. The ecliptic and the celestial equator intersect at two points (separated by 180°) called the **vernal** and **autumnal equinoxes**. The Sun passes through the vernal equinox (VE) on approximately March 21, moving from south to north of the celestial equator. About six months later, it passes through the autumnal equinox traveling from north to south. The time interval between successive returns of the Sun to the VE is called the **tropical year** – see Figure 2.3.

The great circle through the celestial poles and a star's position is that star's **hour circle**. The star's **hour angle** is the angle around the celestial equator between the meridian and its hour circle. The star's **right ascension** $\alpha$ is the arc of the celestial equator from the VE to the star's hour circle. Right ascensions increase from west to east so that stars with large right ascensions rise later than those with small ones (see Figure 2.4). The **declination** $\delta$, is the angular distance measured from the equator along a star's hour circle to the star, and it is positive northward and negative southward. For example, the NCP is at $\delta = +90°$, the SCP is at $\delta = -90°$, and the declination of the celestial equator is $0°$. $\alpha$ and $\delta$ play the roles for the celestial sphere that longitude and latitude fulfill for the Earth.

Angles on the celestial sphere can always be expressed in radians or in degrees, minutes, and seconds of arc. Because a star's right ascension is very nearly equal to the time between the meridian transit of the star and that of the VE,[1] right ascensions are usually expressed in terms of hours, minutes, and seconds of time. The conversions from time to angular measure are $24^\mathrm{h} = 360°$, $1^\mathrm{h} = 15°$, $1^\mathrm{m} = 15'$, and $1^\mathrm{s} = 15''$. The **local sidereal time**

---

[1] The right ascension differs from this time difference only because the Earth does not rotate quite uniformly – see §2.1.5.

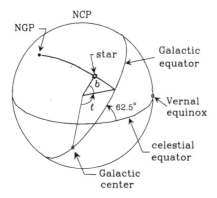

**Figure 2.5** The Galactic plane intersects the sky in a great circle that is inclined by about 62.5° to the celestial equator. Galactic longitude $l$ is the angle along this great circle measured from the direction to the Galactic center.

is the right ascension of the meridian expressed in units of time. Hence a star's hour angle, also expressed in units of time, is the difference between the local sidereal time and the star's right ascension, and is equal to the time since the star crossed the meridian.

Directions on the sky are specified by giving the angle between the given direction and the direction to the NCP. This **position angle** is conventionally measured positive from north through east. Astronomical images and maps are usually shown with north upwards; this orientation places east on the left, as one may convince oneself by facing to the south and looking up at the sky. Since right ascension increases from west to east, it increases from right to left on astronomical maps.

### 2.1.2 Galactic coordinates

Although the equatorial system constitutes the fundamental observable system, it is clearly geocentric, and, as such, provides an inappropriate viewpoint for problems of Galactic structure and dynamics. It is useful, therefore, to set up a Galactic system of coordinates that has a direct physical connection with the structure of our galaxy. The **Galactic equator** is chosen to be the great circle that most closely approximates the plane of the Milky Way. This plane is inclined at an angle of 62.87° to the celestial equator. The north pole of the Galactic system [the **north Galactic pole (NGP)**] is located at $(\alpha_{\rm GP}, \delta_{\rm GP}) = (192.859\,48°, 27.128\,25°) \simeq (12^{\rm h}51^{\rm m}, +27°7.7')$.[2] The **Galactic latitude** $b$ of a star is the angle from the Galactic equator to the star along the great circle through the star and the Galactic poles; for example, the NGP is at $b = +90°$, and the south Galactic pole (**SGP**) is at $b = -90°$ (see Figure 2.5).

**Galactic longitude** $l$ is measured with respect to the direction to the Galactic center. (The compact radio source, Sgr A*, now believed to mark

---

[2] The cited values $(\alpha_{\rm GP}, \delta_{\rm GP})$ are for epoch 2000 – see §2.1.5.

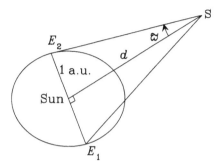

**Figure 2.6** The parallax $\varpi$ of a star is the angle subtended at 1 AU. (Not to scale)

the position of the Galactic nucleus, actually lies about $5'$ away from this point.) Thus, the direction to the Galactic center is $l = 0°, b = 0°$, which corresponds to $(\alpha, \delta) = (266.405°, -28.936°) \simeq (17^h45.6^m, -28°56.2')$.[3] In texts on spherical astrometry (for example, Green 1985) it is shown that the Galactic coordinates $(l, b)$ of a star with equatorial coordinates $(\alpha, \delta)$ may be obtained by solving these equations:

$$\sin b = \sin \delta_{\mathrm{GP}} \sin \delta + \cos \delta_{\mathrm{GP}} \cos \delta \cos(\alpha - \alpha_{\mathrm{GP}})$$
$$\cos b \sin(l_{\mathrm{CP}} - l) = \cos \delta \sin(\alpha - \alpha_{\mathrm{GP}}) \tag{2.1}$$
$$\cos b \cos(l_{\mathrm{CP}} - l) = \cos \delta_{\mathrm{GP}} \sin \delta - \sin \delta_{\mathrm{GP}} \cos \delta \cos(\alpha - \alpha_{\mathrm{GP}}).$$

Here $l_{\mathrm{CP}} = 123.932°$ is the longitude of the NCP. The inverse transformation is given by

$$\sin \delta = \sin \delta_{\mathrm{GP}} \sin b + \cos \delta_{\mathrm{GP}} \cos b \cos(l_{\mathrm{CP}} - l)$$
$$\cos \delta \sin(\alpha - \alpha_{\mathrm{GP}}) = \cos b \sin(l_{\mathrm{CP}} - l) \tag{2.2}$$
$$\cos \delta \cos(\alpha - \alpha_{\mathrm{GP}}) = \cos \delta_{\mathrm{GP}} \sin b - \sin \delta_{\mathrm{GP}} \cos b \cos(l_{\mathrm{CP}} - l).$$

### 2.1.3 Parallax

As the Earth revolves around the Sun, the vantage point from which we view the stars continually changes. Thus, their apparent directions also change slightly. Consider Figure 2.6, and suppose the Earth is initially at $E_1$; the direction to a star $S$ is then along the line $E_1S$. Six months later, when the Earth is at $E_2$, diametrically opposite the Sun, the star will appear in

---

[3] Actually, there have been two systems of Galactic coordinates. The old system $(l^{\mathrm{I}}, b^{\mathrm{I}})$ measured Galactic longitude $l^{\mathrm{I}}$ from one of the points where the Galactic equator intersects the celestial equator. This choice, which lacked physical motivation, was superseded in 1958 (Blaauw *et al.* 1959). In some older work, the $l$ and $b$ defined here are referred to as $l^{\mathrm{II}}$ and $b^{\mathrm{II}}$. The cited values of $(\alpha, \delta)$ are for epoch 2000.

the direction $E_2S$. During the course of the year, the apparent position of a star traces out an elliptical path called the **parallactic ellipse**. The lines $E_1S$ and $E_2S$ contain an angle at $S$ that is defined to be $2\varpi$, where $\varpi$ is the **parallax** of the star. If $r$ is the radius of the Earth's orbit, and $d$ is the distance from the Sun to the star, then, because $d \gg r$, $\varpi$ is a small angle, and

$$\frac{r}{d} = \tan \varpi \approx \varpi \text{ rad}. \tag{2.3}$$

If we convert to seconds of arc,

$$\varpi'' = 206\,265\varpi \text{ rad}, \tag{2.4}$$

and define the **astronomical unit** (AU) such that $r = 1\,\text{AU}$, then we have

$$d = \frac{206\,265}{\varpi''} \text{ AU}. \tag{2.5}$$

Our fundamental unit of distance, the **parsec** (pc), is defined as the distance at which a star would have a parallax of $1''$. We have

$$1\,\text{pc} = 206\,265\,\text{AU} = 3.086 \times 10^{13}\,\text{km} = 3.26\,\text{light-years}, \tag{2.6}$$

and the distance to a star with observed parallax $\varpi''$ is

$$d = \frac{1}{\varpi''} \text{ pc}. \tag{2.7}$$

For Galactic-structure work, it is convenient to use the **kiloparsec** (kpc) $= 10^3$ pc as a distance unit, and, for the discussion of distances between galaxies and in the Universe at large, the **megaparsec** (Mpc) $= 10^6$ pc is employed. As is obvious from equation (2.7), the greater the distance to an object, the smaller is its parallax.

Equation (2.7) shows that a star's parallax is the reciprocal of its distance. On account of this close relation, the word 'parallax' has for astronomers come to be nearly synonymous with 'distance'. In fact, the distance to a star that has been obtained by measuring the star's brightness (§3.5.2) is often called a **photometric parallax**, while a distance that has been directly obtained by observing a parallactic ellipse is called a **trigonometric parallax**. In general one is liable to cause needless confusion by using the term 'parallax' for anything other than an inverse distance.

Because stars are far away, their parallaxes are small. For several hundred years, attempts to measure them to "prove" the heliocentric theory of the solar system were frustrated by the difficulty of measuring shifts in position of less than $1''$. It was not until 1838 that F. W. Bessel finally determined the parallax of the star 61 Cygni to be $0''.29$. It is now possible to measure parallaxes out to distances of the order of 200 pc. The corresponding parallaxes are conveniently measured in milli-arcseconds (**mas**).

The classical procedure for determining a parallax is in principle simple. An image of star $S$ is taken when the Earth is at $E_1$, and another is taken when at $E_2$. The position of star $S$ is measured on each image with respect to faint background stars (which are presumably very distant), and the difference in these positions is then twice the **relative parallax**. Naturally, the procedure is more complicated in practice. Since the atmosphere blurs the images of stars into splodges one arcsec or so in radius,[4] many images are required and great care has to be taken in making the measurements. Moreover, the reference stars do not actually lie at infinite distances, and they must therefore have finite (if small) parallaxes themselves. They will thus move on parallactic ellipses similar to that of star $S$ (but, ideally, of much smaller size), and a correction has to be applied to reduce the measured relative parallax to an **absolute parallax**, which refers to a frame truly fixed in space. The reduction to absolute parallax is usually made by applying a statistical correction, which depends on the brightness of the reference stars and Galactic coordinates of the field – see Green (1985) or McNally (1974) for details.

The scope for error in all these reductions and corrections is very great, and classical trigonometric parallaxes are notoriously beset by large systematic errors which one could neither eliminate nor even reliably estimate. The entire field has recently been revolutionized by a special-purpose astrometric satellite, called Hipparcos.

Hipparcos, which collected data from November 1989 to March 1993, was able to measure directly the angles between stars with unprecedented accuracy because in space: (i) there is no atmospheric refraction; (ii) a telescope is weightless and therefore does not flex as it slews over the sky;[5] and (iii) it is possible to obtain diffraction-limited images. As a result of these advantages of working in space, Hipparcos measured the angles between stars that are widely separated on the sky with errors that are often only $\sim 0.5$ mas. Hipparcos was able to determine accurate positions and trigonometric parallaxes for over 120 000 stars. However, since it was working with a relatively small telescope, these stars were all comparatively bright.[6]

The largest known parallax is $0''.75$ for $\alpha$ Centauri, which places it at a distance of 1.3 pc. The second largest is $0''.55$ for Barnard's star, which is at a distance of 1.8 pc. The typical errors in Hipparcos parallaxes are 1 mas, so the distance of a star at 200 pc ($\varpi = 5$ mas) is uncertain by about

---

[4] This blurring is called 'seeing' – see §3.2.3.

[5] Hipparcos measured the angles between stars that are widely separated on the sky by using rigid optics to superimpose two different fields in a common focal plane.

[6] The Hipparcos catalogue (Perryman *et al.* 1997) is complete only for apparent visual magnitude (§2.3) $V \lesssim 8$, although it does contain some parallaxes for stars as faint as $V = 12$.

20%, which implies a 40% uncertainty in the determination of its intrinsic brightness (see §2.3.3).

The limitation that accurate trigonometric parallaxes are available only for nearby stars is a severe one, for there are types of stars of great astrophysical importance that are too rare to be found within the surveyable volume. To calibrate the intrinsic brightness of such stars, we must develop alternative methods for estimating their distances. Some of these methods are geometric in nature and hence are in some sense as fundamental as trigonometric measures (though in most cases they invoke additional hypotheses). However, for many stellar types, the calibrations can be made only by comparing the apparent brightnesses of these stars with those of stars whose brightnesses have already been calibrated geometrically and with which they are known to be physically associated (for example, in binaries and clusters). In fact, trigonometric parallaxes for nearby stars play a key role in fixing the intrinsic brightness of practically every other kind of astrophysical object and thus ultimately in setting the distance scale for the entire Universe (Chapter 7).

### 2.1.4 Proper motions

Because of their intrinsic motions in space, the stars slowly change position with respect to one another on the sky. The component of a star's motion across the line of sight, after the motion of the Earth around the Sun has been removed, produces an angular rate of change in position termed **proper motion**. This angular rate of change is directly proportional to the star's linear velocity perpendicular to the line of sight and inversely proportional to star's distance. Therefore, if we can measure a star's proper motion and estimate its distance, we can determine its **transverse velocity** – two components of its three-dimensional **space velocity** relative to the Sun (the other component is the radial velocity; see §2.2.1). The space motions of stars can be analyzed (see §10.3) to give information about the distribution of stellar velocities about the center of the Galaxy. Hence proper motions are an essential component in the development of our knowledge of the kinematics – and thus, ultimately, the dynamics – of the Galaxy.

Proper motion is a vector quantity and has both a magnitude $\mu$, typically measured in mas per year, and a direction, which may be specified by its position angle $\theta$. In the equatorial system, the motion is usually resolved into two components: $\mu \cos \theta = \mu_\delta$, which is perpendicular to the celestial equator, and $\mu \sin \theta = \mu_\alpha \cos \delta$, which is parallel to the celestial equator. Here, $\mu_\alpha$ is the annual rate of change in right ascension (often expressed in seconds of *time* instead of seconds of *arc*) and the factor $\cos \delta$ comes from the convergence of hour circles to the celestial poles.

Proper motions are usually tiny; the largest known is that of Barnard's star, which has a motion of $10''.3$ per year. Typical measured motions are only a few hundredths of a second of arc per year. Since the phenomenon of 'seeing' (see §3.2.3) blurs the images of stars to finite-sized splodges, it is clear that many years must elapse before these small motions accumulate to an accurately-measurable effect. Nonetheless, it is now possible to measure the proper motions of stars as far away as the Galactic center (Spaenhauer, Jones & Whitford 1992).

The secret of getting a good proper motion is obviously to compare the positions of stars at widely different epochs. Since Hipparcos only observed the sky for about three years, proper motions obtained from Hipparcos data alone are not remarkably accurate despite the unprecedented accuracy of Hipparcos positions. Often a more accurate proper motion can be obtained by combining a Hipparcos position with a less accurate position from old plate material. This fact and the ability of computers to make short work of extensive observational reductions have revived interest in the **Astrographic Catalogue**, an ambitious sky-mapping program that was carried on around the World in the years 1892–1950.[7] The U.S. Naval Observatory is re-reducing all the data, starting from the $(x, y)$ coordinates of stars on individual plates. The final result will be an all-sky catalogue that contains the positions at epoch $\sim 1900$ and accurate to $\sim 0''.5$, of seven million of stars to magnitude $V \simeq 12.5$. These data will play a vital rôle in complementing the Hipparcos positions with accurate proper motions, which are essential if we are to carry the positions of Hipparcos stars forward in time.

### 2.1.5 Precession and nutation

Both Newtonian physics and (for an effectively isolated system) Einstein's theory of relativity recognize the existence of a privileged class of 'inertial' frames of reference. For a variety of reasons, it is highly desirable to record the positions of astronomical objects relative to such an inertial frame. Unfortunately, the system of equatorial coordinates described in §2.1.1 is not inertial because its two defining directions, the NCP and the VE, both move on complicated paths over the celestial sphere, with the result that the equatorial coordinate system is constantly rotating with respect to inertial space.

In view of this motion, astronomers do not actually record positions relative to the NCP and VE, but relative to the directions that on a given date nearly coincided with the directions of the NCP and VE. To explain how these reference directions are defined, we have to digress briefly into the dynamics of the solar system.

Since we are interested in changes that occur over many years, or hundreds and thousands of months, we can treat the Moon's mass as if it were uniformly distributed over the ring around the Earth that the Moon sweeps out in the course of a month. This ring is inclined by about 5° to the plane of the ecliptic, and the Sun and planets apply a torque to it. It responds to these torques by precessing like a top. That is, its axis slowly moves in

---

[7] The AC was probably the first great international scientific collaboration.

a small circle around the ecliptic pole with a period of 18.6 years. Hence, averaged over times far in excess of 18.6 years, the Moon's mass acts like a band around the Earth whose axis points to the ecliptic pole.

The Earth is not spherical but flattened. The Earth's flattening allows the Sun and the Moon to exert torques on the Earth, and, because the Earth spins like a top, its axis of rotation precesses about the ecliptic pole. The period of this **luni-solar precession** is long, about 25 800 years.

Finally, the shape and orientation of the Earth's orbit about the Sun is changing on a time scale of tens of millions of years in response to the forces applied by the planets. Since the VE is defined as the intersection on the celestial sphere of the great circles of the ecliptic and the equator, any change in the direction of the ecliptic is liable to change the VE.

By the late 19th century the theory of all these motions had been worked out in great detail and put into definitive form by Simon Newcomb. Newcomb obtained Fourier series for the positions of the NCP and the VE as functions of time. These series contain terms with periods of order 18 years and shorter, and terms with periods of order 26 000 years and longer. The motion described by the short-period terms is called **nutation**. The positions predicted for the NCP and VE when these short-period terms are dropped are called the positions of the **mean NCP** and **mean VE**. They are the positions of the NCP and the VE when averaged over a time long compared to 18.6 years and short compared to several millennia, and are the positions actually used by astronomers when measuring $(\alpha, \delta)$.

Of course these mean positions are still moving with respect to inertial space, so it is essential to specify the **epoch** at which $(\alpha, \delta)$ were measured. In practice only a few epochs are employed: 1900, 1950 and the current standard, 2000. That is, quoted values of $(\alpha, \delta)$ are positions relative to the expected locations of the mean NCP and mean VE at the very beginning of 1900, 1950 or 2000. One speaks of **precessing** equatorial coordinates from one epoch to another because this coordinate transformation largely consists in allowing for the precession of the NCP about the pole of the ecliptic. At present the VE moves by $50''\!.25$ per year.

Unfortunately, our knowledge of the mass distribution within the Earth does not allow us to calculate the luni-solar precession with sufficient accuracy. Thus, it has to be determined by observing the apparent motion of celestial objects to which it gives rise. The natural objects to study for this purpose are distant galaxies and quasars, because these objects are so distant that, for any plausible transverse velocities, their motions relative to an inertial frame will be undetectably small. Unfortunately, such objects appear very faint, and it has only recently become feasible to use them to define an inertial system – see below. Observations of them currently yield for the value of the luni-solar precession constant $50''\!.3851 \pm 0''\!.0004\,\mathrm{yr}^{-1}$.

### 2.1.6 Astrometric systems

Ground-based optical observations can determine the relative positions of objects that are close to one another on the sky very much more easily than the angles between widely separated objects: the angles between all objects that lie within a single field of view of a telescope can be determined simply by measuring the distances between their images on a photographic plate or CCD. However, atmospheric refraction, telescope flexure and other effects make it very difficult to determine the exact location of the field. Consequently, astrometrists proceed by first determining the **absolute positions** of certain **fundamental stars** that are distributed all over the sky, and then relate the positions of other objects to the nearest fundamental object; such positions are called **relative positions**. For example, the first step in the re-reduction of the Astrographic Catalogue plates at the U.S. Naval Observatory was the determination of absolute positions of several stars on each plate – the so-called **ACRS stars**.[8]

An **astrometric system** comprises a catalog of objects with absolute positions and proper motions. These data are then used to convert positions and proper motions measured relative to the fundamental objects of the astrometric system to absolute values *in that system*.

The classical procedure for determining the absolute position of a star is to note its time of passage through the meridian, while simultaneously measuring its zenith distance. These measurements are taken with a special-purpose telescope, called a **meridian circle**, which is mounted so that it can point only along the meridian. Each year the Carlsberg Automated Meridian Circle on La Palma measures the positions of about 12 000 stars – see Appendix B.

It is unfortunate that the reference stars observed by meridian-circles are bright, because bright stars tend to be nearby, and thus to have large proper motions. Hence, our astrometric reference objects are moving all over the place, and it is essential to determine what these motions are so that one can say where any given reference point will be at any given epoch. But, how is one to determine these motions, given that the reference directions of the coordinate system one uses are themselves moving in an unknown way? Theory tells us the direction of the luni-solar precession, even if it cannot tell us its magnitude. The standard escape from this dilemma is thus to *assume* that any apparent average motion of the bright stars in the direction of the luni-solar precession is caused by precession, and hence to determine the magnitude of the precession.

The most accurate astrometric system based on meridian-circle observations is the FK5 system[9] (see Appendix B), in which positions and proper motions of the reference stars have typical accuracies 0.03 arcsec and $\sim 0.8\,\mathrm{mas\,yr^{-1}}$, respectively. Position and proper-motion catalogs that were

---

[8] ACRS is the acronym of Astrographic Catalog Reference Stars.

[9] FK5 stands for the fifth Fundamental Katalog.

published before 1997 were usually tied to either the FK5 system or one of its predecessors.

A superior astrometric system, called the **International Celestial Reference System (ICRS)**, has now been established by observing extragalactic objects (Arias *et al.* 1995). These objects are so distant that their proper motions are almost certainly unobservably small. They are too faint for their absolute positions to be determined with a meridian circle, but many of them are radio sources, and radio interferometers can measure accurately the angles between sources that are widely separated on the sky. Hence by radio interferometry one can establish, without reference to the Earth, accurate relative positions for a network of extragalactic objects that are distributed throughout the sky. Since the true proper motions of the ICRS objects must be negligible, they form a reference system that is inherently superior to any that is based on Galactic stars.

As was mentioned in §2.1.3, Hipparcos, like a radio telescope, could measure accurately the angles between stars that are widely separated on the sky. This capability made it possible to put every star in the Hipparcos catalogue on the ICRS by adding to the Hipparcos observing list just a few objects whose positions could be determined on the ICRS by radio interferometry. Significant errors were uncovered in the FK5 system, especially at southern latitudes, where errors as large as 0.16 arcsec have been detected.

Another important catalog with positions on the ICRS is the Northern Proper Motion catalog (NPM), the first part of which was recently completed after decades of work at the Lick Observatory (Klemola, Jones & Hanson 1987, Hanson, Klemola & Jones 1994). It contains the positions and proper motions of 149 000 rather faint stars located throughout the portion of the sky that lies north of the Milky Way (Appendix B). It nicely complements the Hipparcos catalogue, which is complete for fairly bright stars.

## 2.2 Distances determined from velocities

We have seen that the proper motions can now be measured for stars that are as far away as the Galactic centre. If one knows both the distance and the proper motion of a star, then its transverse velocity can be calculated. To complete one's knowledge of the star's velocity vector $\mathbf{v}$ it remains only to determine the component $v_r$ of $\mathbf{v}$ along the line of sight. Traditionally $\mathbf{v}$ is called the star's **space velocity** and $v_r$ is called its **radial velocity**. The sign of $v_r$, is taken to be positive if the source moves away from the observer and negative if it approaches.

### 2.2.1 Radial velocities

When a source of radiation moves toward or away from an observer, the observed frequencies of photons will be different from their emitted frequencies. Quantitatively, if $v_r$ is the radial velocity of a source that emits photons that have frequency $\nu_0$ in the rest frame of the source, then these photons will be detected at frequency

$$\nu = (1 - \beta)\gamma\nu_0, \quad \text{where} \quad \begin{cases} \beta \equiv v_r/c, \\ \gamma \equiv (1 - \beta^2)^{-1/2}. \end{cases} \tag{2.8}$$

Here $c$ is the velocity of light. To the lowest order in $v/c$, $\gamma = 1$ and the frequency shift is

$$\Delta\nu \equiv \nu - \nu_0 = -\beta\nu_0. \tag{2.9}$$

In terms of wavelengths $\lambda = c/\nu$ we have to this order in $v/c$ that $\Delta\lambda/\lambda_0 = -\Delta\nu/\nu_0$ so

$$\Delta\lambda = \frac{v_r}{c}\lambda_0, \tag{2.10}$$

Thus $\Delta\lambda$ for a receding source is positive, and the observed spectrum is **redshifted** relative to its rest wavelength; for an approaching source, $\Delta\lambda$ is negative and the observed spectrum is **blueshifted**. In general, the **redshift** of an object is defined to be

$$z \equiv \Delta\lambda/\lambda_0. \tag{2.11}$$

When we examine the spectrum of a star, we can often identify the observed spectral lines with those produced by various chemical elements in definite states of excitation and ionization; from terrestrial experiments we know the rest wavelengths of these lines, and by measuring the positions of the lines in the star's spectrum, we can determine their wavelength shift and hence the radial velocity of the object with respect to the spectrograph. After correcting for any components of the Earth's orbital velocity $(30 \, \text{km s}^{-1})$ and rotation velocity $(0.5 \, \text{km s}^{-1})$ along the line of sight we obtain the **heliocentric radial velocity** of the object.

One often requires values of $v_r$ that are accurate to $\pm 1 \, \text{km s}^{-1} \simeq 3 \times 10^{-6}c$. In this case the wavelength must be determined to a few parts in a million. Fortunately, algorithms have been developed that enable surprisingly accurate radial velocities to be extracted from mediocre spectra – see §11.1.1. Since these algorithms are computationally intensive, historically many of the best radial velocities in the literature have been determined with instruments that implement the algorithms in hardware (Griffin 1967, Baranne, Mayor & Poncet 1979). With such instruments, velocities can be rather easily measured to an accuracy of better than $\pm 1 \, \text{km s}^{-1}$. In fact, the velocities of reasonably bright stars can be measured to the limit in accuracy,

$\sim \pm 0.1\,\mathrm{km\,s^{-1}}$, that is set by turbulence in stellar photospheres (e.g., Griffin *et al.* 1988).

Some stars are observed to have variable radial velocities. These may be caused by the orbital motion of a star around a companion or by pulsation of the star – see §§3.1.2 and 5.1.10. In either case, the variations in radial velocity yield astrophysically important information.

### 2.2.2 Distances from the moving-cluster method

Imagine watching a time-lapse movie of a cluster of stars as it recedes from the Sun. At first it occupies a appreciable patch of the sky, but gradually it shrinks to no more than dot of light in the distance. Let us assume that the cluster has a constant physical diameter, $d$. Then its angular diameter is

$$\theta = \frac{d}{D}, \tag{2.12}$$

where $D$ is the cluster's distance. Taking natural logarithms of both sides of this equation and differentiating with respect to time, we easily find that

$$D = -\frac{\theta}{\dot{\theta}}v_r, \tag{2.13}$$

where $v_r = \dot{D}$ is the cluster's radial velocity. Hence, by measuring both $v_r$ and the rate $-\dot{\theta}/\theta$ at which the cluster is apparently shrinking, we can determine the distance $D$. Clearly, the same logic applies to the case of a cluster that is approaching us. If the cluster is not travelling directly away from us, it will move across the sky as it shrinks or expands.

Clearly, the proper motions of individual stars within a receding cluster will be directed towards the point on the sky that the whole cluster will occupy as it shrinks to a point. Hence, this point is known as the **convergent point** of the cluster. In the case of an approaching cluster, the convergent point is the point from which the cluster has expanded. In fact, many clusters have been discovered by searching proper-motion catalogs for stars whose proper-motion vectors all intersect in a single point, the convergent point of a **moving cluster** (Figure 2.7).

The convergent point is defined by the proper motions of cluster stars but it contains key information about the *space velocity* **v** of the stars: it defines the *direction* **v**/|**v**| of this velocity. To see this, imagine the situation when the stars in a receding cluster have finally reached the convergent point on the sky. At this time they cease to move across the sky because they are moving directly away from us.

We now show that once a cluster's convergent point has been identified, we can use it to determine distances to individual cluster stars. Figure 2.8 shows the plane that contains the Sun, a cluster star and the velocity vector

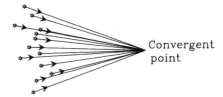

**Figure 2.7** Schematic diagram of a moving cluster. When the proper motions of the cluster stars are extended, they appear to intersect at a point of convergence on the sky.

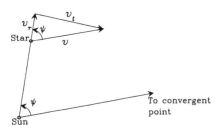

**Figure 2.8** The space motion of a star in a moving cluster is assumed to be parallel to the line of sight from the Sun to the apparent convergent point of the cluster. Then the angle $\psi$ is both the angle between the positions of the star and of the convergent point on the sky, and the angle between the star's space-velocity vector and the Sun-star line of sight.

$\mathbf{v}$ of the cluster. As we have seen, $\mathbf{v}$ is in the direction of the convergent point. In the notation of the figure, the star's radial velocity is $v_r = v \cos \psi$, and its velocity transverse to the line of sight is $v_t = v \sin \psi$. Thus

$$v_t = v_r \tan \psi, \tag{2.14}$$

which gives $v_t$ in terms of directly observable quantities. But $v_t = \mu d$, where $\mu$ is the magnitude of the star's proper motion and $d$ is its distance. Hence

$$\mu = \frac{v_r \tan \psi}{d} \quad \Leftrightarrow \quad d = \frac{v_r \tan \psi}{\mu}. \tag{2.15}$$

Equation (2.15) now yields

$$\frac{\varpi}{\text{mas}} = \frac{4.74}{\tan \psi} \left( \frac{v_r}{\text{km s}^{-1}} \right)^{-1} \frac{\mu}{\text{mas yr}^{-1}}. \tag{2.16}$$

From these equations we can obtain the distance (or parallax) of each cluster member from its observed proper motion, radial velocity, and angular separation from the convergent point.

In practice, the procedure is somewhat more complicated than that just outlined. Details are described in Hanson (1975) and the references cited therein. The method works best when the angular size of the cluster is sufficiently large that the proper motions of the individual cluster stars are oriented in substantially different directions and thus define a convergent point accurately. Furthermore, the cluster must be close enough that the proper motions are large enough to be measured accurately, the angle $\psi$

must be such that the space motion makes a significant contribution to both $v_r$ and $v_t$. Finally, cluster members must be reliably identifiable so that foreground and background field stars can be excluded.

The moving-cluster method has been applied to a number of clusters, including: (i) the Hyades, which contains about 200 stars at an average distance $d \approx 45\,$pc (van Altena 1974, Hanson 1975, Hanson 1980, Schwan 1991, Perryman *et al.* 1995); (ii) the Ursa Major group, which contains about 60 stars at an average distance of about 24 pc (Eggen 1958, 1960); (iii) the Pleiades, which contains about 600 stars at an average distance of about 115 pc (van Leeuwen & Hansen Ruiz 1997); and (iv) the Scorpio-Centaurus group, which contains about 100 stars at an average distance of about 170 pc (Bertiau 1958). Historically, the distance to the Hyades cluster was of enormous importance because the cluster includes types of stars whose distances, and hence whose absolute magnitudes, could not be measured by the trigonometric parallax method. In fact, until very recently, the distance to the Hyades set the scale for essentially all Galactic and extragalactic distance measurements, and revisions of it have had the most far-reaching repercussions. Schwan (1991) determined $45.3 \pm 2.1\,$pc from the classical moving cluster method, while Gunn *et al.* (1988) obtained $47.9 \pm 0.9\,$pc from a modified version of the moving-cluster method that employs radial velocities. The Hyades cluster is less important now that trigonometric parallaxes can be obtained at its distance and beyond. Indeed, parallax and proper-motion data are now so accurate that it is possible to determine which stars lie on the near side and which on the far side of the Hyades, so the whole concept of a single distance to the Hyades is outdated. Of course, from individual distances one can determine the distance to the cluster's barycenter: Brown *et al.* (1997) find $\bar{d} = 46.3 \pm 0.3\,$pc.

### 2.2.3 Secular parallaxes

In a frame in which the mean of the velocities of the nearby stars vanishes, the Sun, like other stars, has a non-zero space velocity, $\mathbf{v}_\odot$. Let us assume that we know the Sun's space velocity $\mathbf{v}_\odot$, and that it has magnitude $|\mathbf{v}_\odot| \simeq 20\,\text{km s}^{-1}$.

The trigonometric parallax of a star is half the angular displacement on the sky produced when the star is viewed from opposite sides of the Earth's orbit. As we have seen, this effect cannot be reliably measured for stars more distant than $d \simeq 200\,$pc. Obviously, we could extend the range of our distance determinations if we could use a longer baseline than the 2 AU diameter of the Earth's orbit. The Sun's non-zero space velocity makes this extension possible: the Sun's motion at around $20\,\text{km s}^{-1}$ relative to the average of nearby stars implies that the solar system moves about 4 AU per year, so observations separated by, say, twenty years give us an effective baseline forty times larger than that provided by the Earth's motion around the Sun. The method of secular parallaxes exploits this longer baseline.

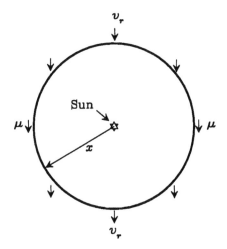

**Figure 2.9** The motion relative to the Sun of stars on a sphere of radius $x$ gives rise to both radial velocities and proper motions. In some directions the proper motion vanishes, while in others the radial velocity vanishes.

We focus on stars that lie within a spherical shell, radius $x$, centered on the Sun – we can pick out such stars by selecting stars that all have roughly the same luminosity (see §3.6.4) and similar apparent brightnesses. For the moment we assume that these stars have a common space velocity $\mathbf{v}$ relative to the Sun. At most points on the sky $\mathbf{v}$, contributes to both radial velocities and proper motions, but in the direction of $\mathbf{v}$ proper motions will vanish and radial velocities will be equal to the magnitude of $\mathbf{v}$ (Figure 2.9). Hence, by measuring radial velocities over the sky, we can determine $\mathbf{v}$. The radius of the sphere can now be determined from the magnitude $\mu_\perp$ of the proper motions in directions at which the radial velocities vanish (Figure 2.9). Specifically, we have

$$x = v/\mu_\perp. \tag{2.17}$$

Equation (2.17) is in practice not very useful for two reasons. First, the only proper motions it uses are those of stars that lie in directions exactly perpendicular to $\mathbf{v}$. Second, it assumes that all stars have a common space velocity, whereas in reality the velocities of individual stars will scatter around some mean motion. By formulating our idea more mathematically, we can resolve both these problems.

Let $\mathbf{v}_i$ be the velocity of the $i^{\text{th}}$ star in the frame in which the mean velocity of the stars of our group is zero: $\sum_i \mathbf{v}_i = 0$. The heliocentric velocity of the $i^{\text{th}}$ star is

$$\mathbf{u}_i \equiv \mathbf{v}_i - \mathbf{v}_\odot, \tag{2.18}$$

where $\mathbf{v}_\odot$ is the **solar motion** with respect to the stars under investigation. In principle, $\mathbf{v}_\odot$ should be determined by measuring the radial velocities of these stars. Since measuring large numbers of radial velocities is very telescope-time consuming, a value for $\mathbf{v}_\odot$ is frequently adopted from some other sample of stars. This procedure is acceptable provided that the mean velocities of the two samples are similar.

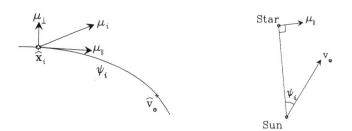

**Figure 2.10** Left: on the sky, the $i^{\text{th}}$ star is at $\hat{\mathbf{x}}_i$, while the Sun is moving in the direction $\hat{\mathbf{v}}_\odot$. The angle between these directions is $\psi_i = \arccos(\hat{\mathbf{x}}_i \cdot \hat{\mathbf{v}}_\odot)$. The proper motion vector $\boldsymbol{\mu}_i$ is decomposed into components parallel and perpendicular to the great circle joining $\hat{\mathbf{x}}_i$ to $\hat{\mathbf{v}}_\odot$. Right: the plane containing the Sun the star and $\hat{\mathbf{v}}_\odot$.

The proper motion $\boldsymbol{\mu}_i$ of the $i^{\text{th}}$ star is proportional to the component of $\mathbf{u}_i$ that is perpendicular to the position-vector $\mathbf{x}_i$ of the star. With the convention that for any vector $\mathbf{b}$, $\hat{\mathbf{b}} \equiv \mathbf{b}/b$ is the unit vector in the direction of $\mathbf{b}$, we have

$$\boldsymbol{\mu}_i = \frac{(\mathbf{u}_i \times \hat{\mathbf{x}}_i) \times \hat{\mathbf{x}}_i}{x_i} = \frac{((\mathbf{v}_i - \mathbf{v}_\odot) \times \hat{\mathbf{x}}_i) \times \hat{\mathbf{x}}_i}{x_i}. \tag{2.19}$$

$\boldsymbol{\mu}_i$ is a vector on the sky based on the location $\hat{\mathbf{x}}_i$ of the $i^{\text{th}}$ star – see Figure 2.10. We now resolve it into its component $\boldsymbol{\mu}_{\|i}$ parallel to the line (great circle) that joins $\hat{\mathbf{x}}_i$ to $\hat{\mathbf{v}}_\odot$, and its perpendicular component, $\boldsymbol{\mu}_{\perp i}$.[10] Consideration of the geometry of the plane containing the Sun, the star and $\hat{\mathbf{v}}_\odot$ (see Figure 2.10) shows that

$$\begin{aligned}
\mu_{\|i} &= \frac{\boldsymbol{\mu}_i \cdot \hat{\mathbf{v}}_\odot}{\sin \psi_i} \\
&= \frac{(((\mathbf{v}_i - \mathbf{v}_\odot) \times \hat{\mathbf{x}}_i) \times \hat{\mathbf{x}}_i) \cdot \hat{\mathbf{v}}_\odot}{x_i \sin \psi_i} \\
&= \frac{(\hat{\mathbf{x}}_i \times \hat{\mathbf{v}}_\odot) \cdot ((\mathbf{v}_i - \mathbf{v}_\odot) \times \hat{\mathbf{x}}_i)}{x_i \sin \psi_i}.
\end{aligned} \tag{2.20}$$

Here we have used the vector identity $(\mathbf{a} \times \mathbf{b}) \cdot \mathbf{c} = (\mathbf{b} \times \mathbf{c}) \cdot \mathbf{a}$ to get from the second to the third line. Solving for $x_i$ and bearing in mind that $\psi_i$ is the angle between $\hat{\mathbf{x}}_i$ and $\hat{\mathbf{v}}_\odot$, we find

$$\begin{aligned}
x_i &= \frac{v_\odot |\hat{\mathbf{x}}_i \times \hat{\mathbf{v}}_\odot|^2}{\mu_{\|i} \sin \psi_i} - \frac{(\hat{\mathbf{x}}_i \times \hat{\mathbf{v}}_\odot) \cdot (\hat{\mathbf{x}}_i \times \mathbf{v}_i)}{\mu_{\|i} \sin \psi_i} \\
&= v_\odot \frac{\sin^2 \psi_i}{\mu_{\|i} \sin \psi_i} - \frac{(\hat{\mathbf{x}}_i \times \hat{\mathbf{v}}_\odot) \cdot (\hat{\mathbf{x}}_i \times \mathbf{v}_i)}{\mu_{\|i} \sin \psi_i}.
\end{aligned} \tag{2.21}$$

---

[10] In the literature, the symbols $\|$ and $\perp$ are denoted $v$ (upsilon) and $\tau$ (tau). These names are not mnemonic, and only people with good eyesight can distinguish $v$ from $v$.

The idea now is to average over all the $N$ stars in our group by summing over $i$ and dividing by $N$. The second term in the second line of equation (2.21) then vanishes, since at any point on the sky, $\mathbf{v}_i$ averages to zero by our choice of reference frame. However, we have only a finite number of stars, and it pays to take care how we do this averaging. First we multiply through by $\sin\psi_i$ so that small denominators do not bother us. Then, since we are seeking a parallax, rather than a distance, we multiply through by $\varpi_i \equiv 1/x_i$. Finally, we average each term and discard the term proportional to $\mathbf{v}_i$, to find

$$\langle\varpi\rangle = \frac{\langle\mu_{\|i}\sin\psi_i\rangle}{v_\odot\langle\sin^2\psi_i\rangle}. \tag{2.22}$$

A parallax obtained from equation (2.22) is called a **secular parallax**.

### 2.2.4 Statistical parallaxes

We now introduce an additional assumption: we assume that the velocities $\mathbf{v}_i$ are isotropically distributed. Specifically, we assume that the average value of $|\hat{\mathbf{e}}\cdot\mathbf{v}_i|$ is independent of the unit vector $\hat{\mathbf{e}}$. If this assumption holds, we can derive from the perpendicular components of proper motion $\boldsymbol{\mu}_{\perp i}$ defined in the last subsection, an independent estimate of the distance to the sample of stars.

The observed radial velocity of the $i^{\text{th}}$ star is

$$\begin{aligned} u_{ri} &= \hat{\mathbf{x}}_i\cdot(\mathbf{v}_i - \mathbf{v}_\odot) \\ &= \hat{\mathbf{x}}_i\cdot\mathbf{v}_i - v_\odot\cos\psi_i, \end{aligned} \tag{2.23}$$

where the angle $\psi_i$ is defined in Figure 2.10. The component of $\mathbf{v}_i$ perpendicular to the plane containing the Sun, the star and $\mathbf{v}_\odot$ is $x_i\mu_{\perp i}$, and, by our hypothesis that the mean magnitude of any component of $\mathbf{v}_i$ is the same, we have that

$$\langle|\hat{\mathbf{x}}_i\cdot\mathbf{v}_i|\rangle = \langle|x_i\boldsymbol{\mu}_{\perp i}|\rangle. \tag{2.24}$$

Substituting this equality into equation (2.23), and assuming that all stars are at the same distance $\bar{x} = 1/\bar{\varpi}$, we find

$$\bar{\varpi} = \frac{\langle|\mu_{\perp i}|\rangle}{\langle u_{ri} + v_\odot\cos\psi_i\rangle}. \tag{2.25}$$

A parallax obtained from equation (2.25) is called a **statistical parallax**.

A more sophisticated analysis, which takes into account the ellipsoidal shape of the random velocity distribution (see §10.3.2) is described in §3.73 of Trumpler & Weaver (1953). Also, when we apply the secular and statistical parallax methods to distant stars, the equations given here should, strictly speaking, be generalized to allow for the effects of differential Galactic rotation. This generalization is in principle straightforward, but lies outside the scope of our discussion.

If $\langle |\hat{\mathbf{x}} \cdot \mathbf{v}| \rangle$ is greater than $v_\odot$, the statistical parallax is probably more reliable than the secular one, because the latter exploits the solar motion, while the former depends essentially on the stars' random velocities. Conversely, if $v_\odot$ is greater than $\langle |\hat{\mathbf{x}} \cdot \mathbf{v}| \rangle$, the secular parallax is probably the more reliable.

The secular and statistical parallax methods have been used to estimate the distances, and hence the absolute magnitudes, of rare but luminous stars. Until Hipparcos completed its work, these stars, which play an important rôle in setting the scale of the Universe, were found only beyond the range of the trigonometric parallax method ($d \simeq 20\,\mathrm{pc}$), whereas the secular and statistical parallax methods could determine their distances out to $d \simeq 500\,\mathrm{pc}$.

At the present, anomalous, moment in astronomy, when Hipparcos has provided positions and parallaxes at one epoch for which no comparable data exist from earlier epochs, the secular and statistical parallax methods are less important than they were, although the volume around the Sun within which they work is still an order of magnitude larger than that accessible to trigonometric parallaxes. They will probably suffer a renaissance if a second astrometric satellite flies early in the next century, and extremely high-precision proper motions become available.

## 2.3 Magnitudes and colors

Every astronomical body emits radiation over much of the electromagnetic spectrum. Ideally, we would like to measure the complete spectral distribution of this radiation. Then we could determine the energy received in terms of the flux per unit frequency interval $f_\nu$ in $\mathrm{W\,m^{-2}\,s^{-1}\,Hz^{-1}}$, or the flux per unit wavelength interval $f_\lambda$ in $\mathrm{W\,m^{-2}\,s^{-1}\,nm^{-1}}$, over the entire spectrum. In practice, the measurement of such **absolute energy distributions** is difficult, for two reasons. First, the absolute response of the observing equipment must be determined. Second, different frequencies of electromagnetic energy penetrate the Earth's atmosphere to different depths. Consequently, at many frequencies observations can be made only from special sites such as the South Pole (Lane & Stark 1996) or even from space – see Figure 2.11. Hence, to obtain the absolute energy distribution of an object one has to combine data from many different instruments, and usually data for some frequency ranges will be lacking.

For many purposes, we do not need to find the complete energy distribution of an object but merely the total energy received by a detector in some definite range of frequencies, that is, the integrated radiation flux $f$ measured in units of $\mathrm{W\,m^{-2}}$, contained in a particular frequency range $\Delta\nu$.

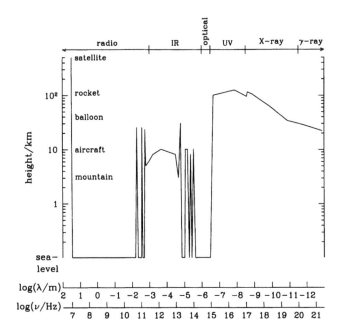

**Figure 2.11** The depth in the Earth's atmosphere to which electromagnetic radiation penetrates as a function of frequency. The line indicates the altitude at which the flux of an astronomical source is attenuated by approximately a factor of e = 2.718.... The altitudes for various observing locations and the standard division of the spectrum into wavebands are marked.

In particular, we can use detectors such a CCDs to measure the apparent brightnesses of objects in various bands in the optical and infrared regions of the spectrum; this procedure is called **astronomical photometry** – see Budding (1993), Sterken & Manfroid (1992) or Hayes, Pasinetti & Davis Philip (1985).

## 2.3.1 Apparent magnitudes

The apparent brightnesses of stars are expressed in terms of their apparent magnitudes. The ancient Greek astronomers divided stars into six magnitude groups judged to be separated by equal steps in brightness. The brighter stars were of the first magnitude, and the faintest that could be seen by the naked eye were of the sixth magnitude. From subsequent physiological studies, it was learned that equal steps of brightness sensed by the eye correspond fairly well to equal ratios of radiant energy; that is, the response of the eye to stimulus by light is essentially logarithmic in intensity. Thus, if $m_1$ and $m_2$ denote the magnitudes assigned to stars with energy fluxes $f_1$ and $f_2$, then

$$m_1 - m_2 = -k \log_{10} \left( \frac{f_1}{f_2} \right), \tag{2.26}$$

where the minus sign is chosen so as to assign smaller numerical values to brighter stars.

Photometric studies in the nineteenth century showed that sixth magnitude stars are about a hundred times fainter than first-magnitude stars. Hence, following the suggestion of N. Pogson, the magnitude system was defined such that a difference of 5 mag corresponds exactly to a factor of one hundred in the ratio of radiation fluxes. Thus, for $(f_1/f_2) = 100$, $m_2 - m_1 = 5$. Hence, in equation (2.26), $k = 2.5$, and, in general,

$$m_1 - m_2 = -2.5 \log_{10} \left( \frac{f_1}{f_2} \right), \tag{2.27}$$

or

$$\frac{f_1}{f_2} = 10^{-0.4(m_1 - m_2)}. \tag{2.28}$$

Interestingly, $m_1 - m_2 = 0.921 \log_e(f_2/f_1)$, so that the scale defined by the ancient Greeks was essentially one based on natural logarithms. In particular, if $(f_1/f_2) = 1 + \Delta f$ is close to unity ($\Delta f \ll 1$), then

$$\Delta m \equiv m_2 - m_1 \approx 1.086 \Delta f; \tag{2.29}$$

that is, the magnitude difference (when small) between two objects is about equal to the fractional difference in their relative brightnesses. It follows from equation (2.28) that a magnitude difference of 1 mag corresponds to a flux ratio of $10^{0.4} \approx 2.512$, and 2.5 mag corresponds to a flux ratio of 10.

Since the atmosphere absorbs some photons, and others are lost in the telescope systems and are not detected, we do not directly measure the flux $f_\nu$ that reaches the solar system from the star. Instead, we measure

$$f \equiv \int_0^\infty f_\nu T_\nu F_\nu R_\nu \, d\nu, \tag{2.30}$$

where $T_\nu$ is the transmission of the atmosphere, $F_\nu$ is the transmission of any filter, which has been put into the system to isolate a particular range of frequencies, and $R_\nu$ is the efficiency of the telescope system; that is, $R_\nu = $ (energy detected/incident energy)$_\nu$.

The transmission of the atmosphere satisfies $T_\nu \propto e^{-a}$, where $a$ is proportional to the column density of air along the line of sight. This is obviously least at the zenith and increases towards the horizon. The ratio of the actual value of $a$ to its value $a_0$ at the zenith is called the **air mass** $\equiv a/a_0$. For a plane-parallel atmosphere, air mass $= \sec z$, where $z$ is the zenith distance of the object being observed. Thus **atmospheric extinction** produces a dimming of starlight according to the relation

$$m(z) = k \sec z + \text{constant}, \tag{2.31}$$

where $k$ is a constant and $m(z)$ is the observed stellar magnitude at zenith distance $z$. We can correct for atmospheric extinction (at wavelengths where the atmosphere is not totally opaque) by observing the magnitude of a star at several values of $z$. A fit to these data then yields the constant $k$ in equation (2.31) and allows us to correct not only to unit air mass at $z = 0$ but also to extrapolate to zero air mass (that is, "$\sec z = 0$") and thus obtain the magnitude outside the Earth's atmosphere – see Sterken & Manfroid (1992) for details.

The filter transmission $F_\nu$ in equation (2.30) is readily determined and can, in fact, be chosen at will to measure the energy contained in definite wavelength intervals. Most photometric systems employ several different filters, and the filter band used must always be stated when giving an apparent magnitude. A plot of $F_\nu$ versus wavelength will typically show a hump that may be crudely characterized by a central or **effective wavelength** $\lambda_{\text{eff}}$ and a width. The latter is conventionally given as the wavelength difference between the points at which $F_\nu$ falls to half its peak value – this quantity is called the **full width at half max** or **FWHM** for short. Tables 2.1 and 2.3 list these parameters for some widely used filter systems.

The instrumental efficiency $R_\nu$ in equation (2.30) is a composite of the efficiency of the telescope's optical system and the sensitivity of the photon receiver itself. While mirror reflectivities can be measured fairly easily, the other two factors just mentioned are extremely difficult to determine accurately. In practice, therefore, the system must be calibrated by measuring its response to a source of known brightness. At most observatories this is a standard star, whose brightness has been determined (with great difficulty) at some other observatory by using a telescope to observe both the star and a terrestrial source of known characteristics – see below. The determination of the brightnesses of standard stars is difficult, and the results correspondingly uncertain, so we know the *relative* brightnesses of stars much more accurately than their *absolute* brightnesses.

Until the 1950s, most astronomical photometry was done photographically with two systems: the blue-violet-sensitive **international photographic system** giving magnitudes $m_{\text{pg}}$, and the **photovisual system**, $m_{\text{pv}}$ whose wavelength sensitivity simulates that of the eye. Unfortunately, the photographic plate has a nonlinear response to different levels of light intensity, and the relation between incident intensity and photographic darkening must be determined empirically by calibrating each plate. Furthermore, the dynamic response range of a plate is only about a factor of twenty in intensity; to cover larger ranges, a set of several plates must be used. Given such properties of the detector, it becomes very difficult to do absolute photometry and to extend a magnitude scale over a wide range of intensity without making large systematic errors. Many of the results in the older photographic photometric catalogs (see Appendix B) are, unfortunately, seriously affected by such errors. On the other hand, if some stars on the plate have accurately known brightnesses, then it is usually possible to make differential measurements to $\pm 0.1$ mag, which is adequate for many purposes. In the best cases, such differential measurements may be accurate to $\pm 0.03$ mag.

Astronomical photometry was revolutionized with the advent of photoelectric photometers. These instruments have strictly linear responses, they have an enormous dynamic range, and they are capable of factor-of-ten better precision than photographic plates. Consequently, they yield magnitudes accurate to $\pm 0.01$ mag and magnitude differences often accurate to $\pm 0.002$ mag. At this level of accuracy, color-magnitude diagrams of star clusters become sensitive diagnostic tools (see §3.5). The advent in the early 1980s of CCD detectors was a further important advance since, like photoelectric detectors, CCDs are linear, but, unlike photoelectric detectors, they are imaging devices. Thus, with a CCD many stars can be measured simultaneously with high accuracy. Moreover, CCD photometry can be done for stars in a very crowded field, such as the centre of a globular cluster, which are inaccessible to conventional photoelectric photometry.

The standard photometric system today is based on the **ultraviolet-blue-visual (*UBV*) system** of Johnson & Morgan (1953). As measurements in the infrared became more important, this system was extended by adding seven bands reaching out to beyond 1000 nm – see Table 2.1. Magnitudes in the extended system are denoted by the capital letters designating the filter; thus $V$ magnitudes are now the standard visual magnitudes, $K$ magnitudes the most important near-infrared ones, and $M$ magnitudes the longest-wavelength ones in common use.

A photometric system is ideally defined by a particular filter-detector combination. Using this equipment an observer anywhere can reproduce accurately a given spectral response. Of course different observers have telescopes of varying quality, so no two observers would measure exactly the same brightness for a given star. However, so long as they use the same filter-detector combination, they should always agree on the ratio of the brightnesses of any two given stars. Consequently, a photometric system such as the *UBV* system is in practice established by one particular observer assigning magnitudes to stars that are well distributed over the sky. Any other observer can then measure the brightness of any star by using the standard filter-detector combination to compare the star of interest to a nearby standard star.

Unfortunately, no photometric system satisfies the ideal just described because the steady advance of technology soon renders any given filter-detector combination obsolete. Consequently, astronomers are forever trying to measure, say, Johnson–Morgan $B$-band magnitudes with equipment that does not have the $B$-band spectral response implicitly specified by Johnson

& Morgan (1953). Strictly speaking, this measurement is possible only if the star being observed has *exactly* the same spectrum as one of the Johnson & Morgan standard stars; the contemporary astronomer's equipment and that of Johnson & Morgan would measure different portions of the spectra of the program and standard stars, and so intensity ratios measured with each instrument will agree only if the spectra of the two stars are proportional to one another.

In practice *UBV*-values accurate to $\sim 0.01$ magnitudes can be obtained by choosing a filter that matches a contemporary detector to the Johnson–Morgan filter-detector combination.[11] Moreover, brightnesses measured with several different filters usually provide sufficient knowledge of any deviations between the spectra of the program and standard stars to enable accurate *UBV*-values to be determined even when none of the adopted filters provides a match to the Johnson–Morgan filter-detector combination. Suppose an object has been observed through filters that do not precisely coincide with ones that match the standard filter-detector combinations, but provide reasonable approximations to those combinations. Let magnitudes in the observed bands be denoted $U', B', V'$ etc. and those in the *UBV* system are called $U, B, V$, etc.[12] These magnitudes are 'transformed' to standard *UBV*-magnitudes by relations of the type

$$V = \alpha_V V' + \beta_V (B' - V') + \gamma_V. \qquad (2.32)$$

The constants $\alpha_V$, $\beta_V$ and $\gamma_V$ are determined such that the transformed *UBV*-magnitudes of standard stars agree as closely as possible with the published *UBV*-magnitudes of these stars. Leggett (1992) gives transformation equations of this type between different photometric systems.

The extension of the *UBV*-system into the infrared has not proceeded smoothly, with the result that the bands longward of $V$ are not as well defined as is desirable. Johnson & Mitchell (1962), Johnson *et al.* (1966), Johnson (1965) added bands *RIJKLMN* that cover the range $700 - 10\,040$ nm. Cousins (1976) later redefined the *RI* bands, and when $R$ and $I$ magnitudes are now cited, they are usually (but not invariably) on the Cousins rather than the Johnson system. Glass (1974) defined an $H$ band that lies between the $J$ and $K$ bands, rather than coming before $J$. Other small adjustments to the Johnson infrared bands have been made by different observers with the result that there is now no truly standard system. A useful definition of *JHKL* magnitudes will be found in Elias *et al.* (1982). In the infrared, the Earth's atmosphere is only transparent within certain wavelength 'windows,' and the choices of central wavelengths and bandpasses for the infrared extensions of

---

[11] Mis-matched filters are a significant cause of confusion, however. See Bessell (1993) for a discussion of this problem.

[12] We use the standard convention that $V \equiv m_V$, etc., where $m_V$ is an object's apparent magnitude in the $V$ band.

the *UBV* system have, to a large extent, been dictated by the locations of
these windows.

Table 2.1 gives the effective wavelengths and the full widths at half maximum sensitivity for both the extended *UBV*-system and several other widely
used filter systems. The Thuan-Gunn system was designed to avoid wavelengths at which the spectrum of the night-sky shows prominent emission
lines. The Hipparcos band was defined to be exceptionally broad so as to
maximize the flux received from stars of every color.

Table 2.2 lists the parameters of some of the filters used for observations
with HST. These filters, which correspond approximately to Johnson bandpasses, are the most widely used of the large set that is available on HST
instruments. For the optical imager, WFPC2, any filter may be identified by
a name such as F555W, which indicates a filter centered on 555 nm that has
a wide bandpass (other possible suffixes being 'M' for medium bandpass, 'N'
for narrow bandpass, and 'LP' for a filter which passes all the light longward
of a particular wavelength). The infrared imager, NICMOS, has a similar
system for designating its filters except that the wavelength scale is shifted
by a factor of ten: thus, for example, F175W in this instrument is a wide
bandpass filter centered on 1.75 $\mu$m. Since HST is not restricted to parts
of the infrared where the Earth's atmosphere is transparent, the NICMOS
filters give more complete spectral coverage than the extended *UBV* system
in the near infrared.

The zero-points of each scale are usually chosen such that the star Vega
has $U = B = V = Hp = V_T \ldots = 0$. Examples of $V$ magnitudes are: Sun,
$V = -26.74$; Sirius (apparently brightest star), $V = -1.45$; faintest stars
measured, $V \approx 27$.[13] The total range from the Sun to the faintest measurable
stars is about 54 mag or a ratio of $3 \times 10^{21}$ in apparent brightness!

### 2.3.2 Colors

Suppose we have a photometric system with several filter bands at different
wavelengths. Then, by taking the difference in magnitudes measured in two
different bands, we can form a **color**, or **color index**. That is, if $A$ and $B$
denote two different filters, we write

$$(C.I.)_{AB} \equiv m_A - m_B = \text{const.} - 2.5 \log \frac{\int_0^\infty d\lambda \, S_\lambda\,(A)\,f_\lambda}{\int_0^\infty d\lambda \, S_\lambda\,(B)\,f_\lambda}, \qquad (2.33)$$

where $S_\lambda$ denotes the combined telescope-receiver-filter sensitivity.[14] A color
index is usually written using the letters that denote the different filters

---

[13] Remarkably, for historical reasons, the apparent magnitude of Vega, $V = +0.04$,
differs significantly from zero.

[14] $S$, which is usually normalized so that its peak value is 1, is tabulated for several
systems in Lamla (1982).

**Table 2.1** Filter characteristics of broad-band photometric systems

| System | Band | $\lambda_{\text{eff}}$ nm | FWHM nm | $f_X(A0V)$ Jy | $\frac{L_\odot}{10^{25}\,\text{W}}$ | $M_\odot$ |
|---|---|---|---|---|---|---|
| *UBVRI* | *U* | 365 | 66 | 1780 | 1.86 | 5.61 |
| | *B* | 445 | 94 | 4000 | 4.67 | 5.48 |
| | *V* | 551 | 88 | 3600 | 4.64 | 4.83 |
| | *R* | 658 | 138 | 3060 | 6.94 | 4.42 |
| | *I* | 806 | 149 | 2420 | 4.71 | 4.08 |
| | *J* | 1220 | 213 | 1570 | 2.49 | 3.64 |
| | *H* | 1630 | 307 | 1020 | 1.81 | 3.32 |
| | *K* | 2190 | 390 | 636 | 0.82 | 3.28 |
| | *L* | 3450 | 472 | 281 | 0.17 | 3.25 |
| | *M* | 4750 | 460 | 154 | – | – |
| Hipparcos | *Hp* | 550 | 225 | – | – | – |
| Tycho | $B_T$ | 420 | 75 | – | – | – |
| | $V_T$ | 510 | 100 | – | – | – |
| Thuan–Gunn | *g* | 512 | 120 | – | – | – |
| | *r* | 668 | 100 | – | – | – |
| | *i* | 792 | 150 | – | – | – |
| | *z* | 912 | 140 | – | – | – |
| SDSS | *u′* | 352 | 63 | – | – | – |
| | *g′* | 480 | 141 | – | – | – |
| | *r′* | 625 | 139 | – | – | – |
| | *i′* | 769 | 154 | – | – | – |
| | *z′* | 911 | 141 | – | – | – |

SOURCE: From data published in Bessell (1993), Perryman *et al.* (1997), Thuan & Gunn (1976), Schneider, Gunn & Hoessel (1983) and Fukugita *et al.* (1996). The *UBVRI* bands are for the combined Johnson–Cousins–Glass system. $f_X(A0V)$ is the *X*-band flux density [equation (2.40)] of a $V = 0$ A0V star. The solar luminosities and solar magnitudes to the red of *V* were derived from Buser & Kurucz (1992) solar models by W. Vacca. Note that $1\,\text{Jy} = 10^{-26}\,\text{W}\,\text{m}^{-2}\,\text{Hz}^{-1}$

involved; that is, $(A - B)$ for the hypothetical example just given or $(B - V)$ and $(U - B)$ for the standard *UBVRI* system. As is clear from equation (2.33), a color index essentially measures the ratio of stellar fluxes near the effective wavelengths of the two bands. Since this quantity depends only on a ratio of fluxes, it measures a property of an object's spectrum that is independent of its distance (which is highly desirable, given the uncertainties in the distances to many astronomical objects).

The detailed shape of the energy distribution emitted by a star is determined by a few basic physical parameters such as the stellar temperature,

**Table 2.2**   Broad-band HST filters

| Instrument | Name | $\lambda_{\text{eff}}$/ nm | FWHM/ nm | $UBV$ analog |
|------------|------|---------------------------|----------|--------------|
| WFPC2 | F336W | 332.7 | 37.1 | $U$ |
|  | F439W | 429.2 | 46.4 | $B$ |
|  | F555W | 525.2 | 122.3 | $V$ |
|  | F675W | 673.5 | 88.9 | $R$ |
|  | F814W | 826.9 | 175.8 | $I$ |
|  | F300W | 292.4 | 72.8 | Wide $U$ |
|  | F450W | 444.5 | 92.5 | Wide $B$ |
|  | F606W | 584.3 | 157.9 | Wide $V$ |
| NICMOS | F110W | 1100 | 300 | $J$ |
|  | F140W | 1400 | 500 |  |
|  | F160W | 1600 | 200 | Narrow $H$ |
|  | F175W | 1750 | 550 | Wide $H$ |
|  | F187W | 1875 | 125 |  |
|  | F205W | 2050 | 300 |  |

SOURCE: From Biretta *et al.* (1996) and MacKenty *et al.* (1997); see also
http://www.stsci.edu/instrument-news/handbooks/

**Table 2.3**   Filter characteristics of intermediate-band
photometric systems

| System | Band | $\lambda_{\text{eff}}$/ nm | FWHM/ nm |
|--------|------|---------------------------|----------|
| Strömgren | $u$ | 349 | 30 |
|  | $v$ | 411 | 19 |
|  | $b$ | 467 | 18 |
|  | $y$ | 547 | 23 |
|  | $\beta_w$ | 489 | 15 |
|  | $\beta_n$ | 486 | 3 |
| DDO | 45 | 451.7 | 7.6 |
|  | 42 | 425.7 | 7.3 |
|  | 41 | 416.6 | 8.3 |
|  | 38 | 380.0 | 17.2 |
|  | 35 | 349.0 | 37.0 |

SOURCE: From data published in Crawford (1966) and
McClure & van den Bergh (1968)

surface gravity, and chemical composition. By measuring broad-band colors,
we can, to some extent, probe these physical properties. For example, since
cool stars are red and hot stars are blue, a color index such as $(B - V)$ that
measures the flux ratio between a short and a long wavelength provides a
measure of the stellar temperature. As another example, the ultraviolet spec-

tra of some stars contain large numbers of absorption lines whose strengths reflect the abundances of the heavier elements in their atmospheres, while the $B$ band is relatively line free. Thus, $(U - B)$ is sensitive to stellar chemical composition. Unfortunately, there is another important physical parameter that also affects the observed value of $(U - B)$. There is a sharp drop in continuum intensity in some hot stars at wavelengths shorter than about 370 nm; this drop arises from the sudden onset of continuum absorption from the $n = 2$ level of hydrogen, and it is called the **Balmer jump**. Since 370 nm lies right in the middle of the $U$ band, $(U - B)$ is sensitive to the strength of the Balmer jump as well as to chemical composition.

As this example illustrates, broad-band filters provide too crude a measure of a star's spectral energy distribution to distinguish between the effects of quite disparate physical phenomena. In order to lift such degeneracies, we need to use more finely-tuned filters, which pick out individual features in a star's spectrum. Although it is tempting to design a filter set with very narrow bandpasses in order to isolate spectral features, such filters will transmit so little light that their use is not practicable for any but the brightest stars. One is therefore forced to compromise, and adopt intermediate-bandpass systems with filter bandwidths in the range 10 nm $\lesssim$ FWHM $\lesssim$ 30 nm. Once such a filter system has been constructed, we can derive correlations between the chosen color indices and the physical parameters of stars, so that we can ultimately use the observed colors as diagnostic tools to infer the physical properties of stellar atmospheres. Table 2.3 describes two important intermediate-band filter sets, the Strömgren and DDO systems, which have been developed for this purpose.

The Strömgren system [Strömgren (1966), Crawford (1966)], has four intermediate-band filters, $uvby$, and a narrow-band pair, $(\beta_n, \beta_w)$, measuring the H$\beta$-line and adjacent continuum. The usual color indices employed in this system are $(b-y)$, $(u-b)$, $c_1 \equiv (u-v)-(v-b)$, and $m_1 \equiv (u-b)-(b-y)$. Both $(b-y)$ and $(u-b)$ serve as temperature indicators; $c_1$ is a temperature indicator for hot stars and a luminosity indicator for cooler stars; $(\beta_n - \beta_w)$ is a luminosity indicator for hot stars and a temperature indicator for cooler stars. For cooler stars $m_1$ measures the amount of **line-blanketing**, the dimming of the blue part of the spectrum caused by millions of heavy-element absorption lines. The parameter $m_1$ also acts as an indicator of spectral peculiarity in some stars. For recent work on the calibration of this system, see Strömgren (1987), Olsen (1987), Edvardsson *et al.* (1993).

The DDO system employs five filters. The $C(42 - 45)$ color index is primarily a measure of effective temperature (§3.4), while the $C(45 - 48)$ color is sensitive to surface gravity. The $C(41 - 42)$ index is sensitive to the strength of spectral lines due to CN molecules and Fe atoms, and is thus a measure of the abundances of heavier elements. Details of the calibration of the DDO system can be found in Claria *et al.* (1994).

Finally, it must be remembered that the effects of interstellar extinction on colors have been ignored in this discussion. Because the interstellar medium scatters light more efficiently at short wavelengths than at long wavelengths, transmitted starlight is not only dimmed by interstellar extinction but is also **reddened**. It is therefore necessary to correct observed colors for interstellar reddening in order to derive intrinsic colors. As we shall see in §3.7.1, it is fortunate that interstellar extinction is accompanied by reddening, for it is possible to devise combinations of color indices that allow us to determine the amount of reddening present, and hence to correct for both reddening and extinction.

### 2.3.3 Absolute magnitudes

The energy flux we receive at the Earth from an object depends on both its intrinsic brightness and its distance. If $F$ is the flux received when the object is at distance $D$, the flux $f$ that would be received if it were at some other distance $d$ is given by the inverse square law,

$$f = \left(\frac{D}{d}\right)^2 F. \qquad (2.34)$$

Obviously, the farther away an object is, the fainter it will appear, and to obtain information about the relative intrinsic brightnesses of objects, we must account for differences in their distances from us. We therefore define the **absolute magnitude** $M$ to be the apparent magnitude an object would have if it were located at some standard distance $D$. From equations (2.27) and (2.34), we see that

$$m - M = -2.5 \log\left(\frac{f}{F}\right) = 5 \log\left(\frac{d}{D}\right). \qquad (2.35)$$

The standard distance $D$ is always taken to be $10 \, \mathrm{pc}$, so if $d$ is measured in parsecs, then

$$m - M = 5 \log d - 5. \qquad (2.36)$$

The quantity $(m - M)$ is called the **distance modulus** of the object. If we know $m$ and $d$, we can immediately correct for the nonstandard distance of the object and reduce the apparent magnitude $m$ to the absolute magnitude $M$ via equation (2.36). Conversely, if we know $m$ and $M$, we can infer $d$.

Absolute magnitudes are normally quoted for the visual band and denoted $M_V$. It should be noted that the absolute magnitude is *not* a direct measure of the total energy output (**luminosity**) of an object, but only of the energy in the $V$ band. To measure total energy output, we use so-called 'bolometric magnitudes,' which will be discussed in §2.3.4.

The distance from the Earth to the Sun is $1 \, \mathrm{AU} = (1/206,265) \, \mathrm{pc}$. Thus, we know immediately that the distance modulus of the Sun is $-31.57$ mag and therefore that the absolute magnitude of the Sun $M_V(\odot) = +4.83$. We shall see in §3.5 that the Sun is a star of rather average intrinsic brightness –

**Table 2.4** Redshifts at which one *UBV* band is shifted to another

|   | U | B | V | R | I | J | H | K | L | M |
|---|---|---|---|---|---|---|---|---|---|---|
| U | 0 | 0.22 | 0.51 | 0.80 | 1.21 | 2.34 | 3.47 | 5.00 | 8.45 | 12.01 |
| B |   | 0 | 0.24 | 0.48 | 0.81 | 1.74 | 2.66 | 3.92 | 6.75 | 9.67 |
| V |   |   | 0 | 0.19 | 0.46 | 1.21 | 1.96 | 2.97 | 5.26 | 7.62 |
| R |   |   |   | 0 | 0.22 | 0.85 | 1.48 | 2.33 | 4.24 | 6.22 |

the most luminous stars are about $10^6$ times brighter, and the least luminous are more than $10^4$ times fainter.

Finally, we must caution that absorption and scattering of photons in interstellar space cause stars to appear dimmer than they would be from distance effects alone, and hence increase their apparent magnitudes – see §3.7.1. This dimming is usually expressed in magnitudes. If there are $A$ magnitudes of interstellar extinction, equation (2.36) must be rewritten as

$$m - M = 5 \log d - 5 + A. \tag{2.37}$$

In §3.7.1 we shall see that the value of $A$ depends strongly on waveband in the sense that it decreases towards longer wavelengths.

If the object is moving rapidly away from us, perhaps because it lies at a cosmologically-significant distance, the photons received in the $V$ band, for example, will have been emitted at shorter wavelengths. Consequently, the observed value of $V$ is not directly connected to the value $M_V$ that we would measure if the object were stationary at 10 pc. If we know the shape of the object's spectrum, we can infer a relationship between $V$ and $M_V$, however. This relationship is written

$$m - M = 5 \log d - 5 + A + K, \tag{2.38}$$

where $K$ is called the **K-correction**. For several different photometric systems and galaxy types, Frei & Gunn (1994) tabulate a quantity $k(z)$, which is related to $K$ by $K = k + 2.5 \log(1 + z)$, where $z$ is the redshift of the galaxy [equation (2.11)].

Since we have good spectral information only for nearby galaxies and the spectra of galaxies evolve in time, K-corrections are reliable only for nearby galaxies, for which $k \simeq 0$. In reality we should compare the brightnesses of a distant galaxy with the brightness of a nearby galaxy in the waveband at which the photons collected from the distant galaxy were emitted. Table 2.4 gives the redshift $z$ for which light emitted at the center of either the $U$, $B$, $V$ or $R$ band will be observed to lie at the center of another band in the $UBV$ system. For example, that we make $J$-band measurements of a galaxy at $z = 1.21$. Then the collected photons were emitted in the $V$-band and we can accurately infer the $V$ magnitude that would be measured by an observer located at the Sun for whom the galaxy was at rest.

### 2.3.4 Absolute energy distributions and bolometric magnitudes

Color indices provide a good deal of useful information about stellar energy distributions, but, as was mentioned earlier, we would ideally like to know the detailed variation of the energy flux per unit wavelength (or frequency) interval throughout the entire spectrum. If we have such observations, we can apply the theory of stellar atmospheres, and analyze the data to infer physical characteristics of stars [see Mihalas (1978), §7-4]. Furthermore, we can use absolute energy distributions to obtain total luminosities and stellar temperatures – see §3.4.

The basic problem in this work is the measurement of the absolute efficiency of the telescope-receiver system; that is, the determination of the amount of energy that must be put into the system in order to produce unit instrumental response. The only practical way to derive the efficiency is to use the instrument to observe a source whose absolute energy distribution is known in advance, for then we know the amount of incident energy that produces the measured response. There are only two types of sources for which we can specify the rate of energy emission from unchallengeable theory: (i) a blackbody source at a known temperature, the energy distribution of which is given by the Planck function $B_\nu(T)$, and (ii) a synchrotron-radiation source, from which the emission by relativistic electrons can be calculated as a function of their energy. The procedure is to observe such an absolute reference source with a given telescope-spectrometer system and then, with the same system, to observe a standard star. Observations of the absolute reference source calibrate the instrumental efficiency at each wavelength, and thus provide the factor required to convert the observed responses to the stellar radiation into an absolute energy emission from the star (after correction, of course, for atmospheric extinction).

The literature now contains measurements of the absolute energy distribution of $\alpha$ Vir,[15] $\eta$ UMa, and $\alpha$ Leo in the range 137 nm $\leq \lambda \leq$ 292 nm (Bless et al. 1976), of $\alpha$ Lyr in the range 130 nm $\leq \lambda \leq$ 200 nm (Tanaka et al. 1984), of Vega in the range 330 nm $\leq \lambda \leq$ 1080 nm (Hayes & Latham 1975), and of $\alpha$ Tau in the range 1200 $\leq \lambda \leq$ 35 000 nm (Cohen, Walker & Witteborn 1992). The ultraviolet and infrared data are usually given directly in Janskys.[16] Optical data are often given in terms of a relative absolute energy distribution, written in monochromatic magnitudes,

$$m_\nu \equiv -2.5 \log\left(f_\nu / f_{\nu_0}\right), \tag{2.39}$$

plus the absolute flux $f_{\nu_0}$ at the reference frequency $\nu_0$. With $\nu_0$ chosen to correspond to $\lambda = 555.6$ nm the present best estimate of $f_{\nu_0}$ for Vega is

---

[15] Bright stars are designated the name of the constellation within which they lie, preceded by a Greek letter; an $\alpha$ indicates that the star is the brightest in its constellation, $\beta$ the second-brightest, etc.

[16] See the Notes to Table 2.1 for the definition of a Jansky.

$f_{\nu_0} = 3560 \pm 25 \, \mathrm{Jy}$ (Mégessier 1997). This format is chosen because relative monochromatic magnitudes are easier to determine, and probably more accurately known, than the absolute flux normalization. They are therefore likely to remain unchanged even if the absolute flux requires future adjustment.

Once the absolute energy distribution of even a single star is known precisely, it can henceforth be used as a reference standard directly on the sky. Observations of the absolute energy distributions of other stars are then relatively simple – one merely measures program stars and the reference star with the same equipment, corrects for differences in atmospheric extinction, and then uses the known absolute distribution of the standard to obtain absolute distributions for the program stars. Catalogs of spectrophotometric data for hundreds of stars of various spectral types are available (see Appendix B). These data are indispensable for performing astrophysical analyses both of individual stars and of the integrated energy distributions from stellar systems (clusters and galaxies), which are composites of contributions from many different spectral types.

One often wants to know the flux from a star averaged over one of the standard photometric bands. In some band $X$ ($X = U, B, V, \ldots$), the $X$-band flux

$$f_X \equiv \frac{\int_0^\infty d\nu \, S_X(\nu) f_\nu}{\int_0^\infty d\nu \, S_X(\nu)}, \tag{2.40}$$

where $S_X(\nu)$ is the response of the standard $X$-band filter–photometer combination. This quantity is determined by observing sources of known spectral energy distribution as described above in connection with the determination of $f_\nu$. Once $f_X$ has been determined from (2.40) for a single standard star, we can immediately infer $f_X$ for any other star through

$$\log\left[f_X(*)\right] = \log\left[f_X(\text{standard})\right] - 0.4\left[X(*) - X(\text{standard})\right]. \tag{2.41}$$

$f_X$ is tabulated for an A0V star in Table 2.1.

The integral of the absolute energy distribution over all frequencies is an important quantity and is usually reported as a magnitude by defining the **apparent bolometric magnitude** $m_{\mathrm{bol}}$ as

$$m_{\mathrm{bol}} \equiv -2.5 \log\left(\int_0^\infty d\nu f_\nu\right) + C_{\mathrm{bol}}. \tag{2.42}$$

Here $C_{\mathrm{bol}}$ is a constant that is determined by convention. Unfortunately, two conventions are in widespread use, as will be explained below. Independent of the choice of $C_{\mathrm{bol}}$, the difference between the absolute bolometric magnitudes of two stars gives the ratios of their luminosities.

The difference between $m_{\mathrm{bol}}$ and the star's magnitude in a band $X$, $V$, is called the **bolometric correction** $\mathrm{BC}_X$:

$$\mathrm{BC}_X \equiv m_{\mathrm{bol}} - X. \tag{2.43}$$

If two stars have identical spectra, they will have identical values of $BC_X$, even if their luminosities are different. Consequently, we may consider $BC_X$ to be a function of spectral type only, and $BC_X$ only needs to be determined for one standard of each spectral type and the results tabulated, as in Table 3.7.

The definition (2.43) of BC involves the constant $C_{bol}$ of equation (2.42). One scale for bolometric magnitudes and corrections is determined by choosing $C_{bol}$ such that the Sun's $V$-band bolometric correction is zero: $BC_V(\odot) = 0$. In MKS units, the corresponding value of $C_{bol}$ turns out to be $C_{bol} = -18.90$ (Problem 2.3). In the other widespread convention, $C_{bol}$ is chosen such that all bolometric corrections are negative and $BC_V(\odot) = -0.19$.

### 2.3.5 Mass-to-light ratios

The brightness of an astronomical object is often quantified by stating how many Suns, placed at the object's distance, would be required to produce the observed flux. For the waveband $X$, this number is

$$N = L_X/L_{\odot X} = 10^{-0.4(M_X - M_{\odot X})}, \tag{2.44}$$

where $L_X$ and $M_X$ are, respectively, luminosity and absolute magnitude in the $X$ band. If the object's mass, $\mathcal{M}$, can be somehow measured, it is interesting to compare $\mathcal{M}$ with the mass $N\mathcal{M}_\odot$ by calculating the **mass-to-light ratio**

$$\Upsilon_X \equiv \frac{\mathcal{M}}{N\mathcal{M}_\odot} = \frac{\mathcal{M}/\mathcal{M}_\odot}{L_X/L_{\odot X}} = 10^{0.4(M_X - M_{\odot X})}\frac{\mathcal{M}}{\mathcal{M}_\odot}. \tag{2.45}$$

If the object's spectral energy distribution is the same as that of the Sun, $\Upsilon_X$ is the same for every waveband $X$. Equation (2.45) shows that in the general case $\Upsilon_X$ varies with $X$. Consider, for example, the case of a giant elliptical galaxy, which might have $B - K = 3.3$, whereas the Sun has $B - K = 2.2$. Thus for this galaxy $\Upsilon_B/\Upsilon_K = 10^{0.4(3.3-2.2)} = 2.75$.

The wave-band dependence of $\Upsilon_X$ has been the source of considerable confusion. Consequently astronomers sometimes try to estimate the bolometric mass-to-light ratio

$$\Upsilon_{bol} \equiv \frac{\mathcal{M}/\mathcal{M}_\odot}{L_{bol}/L_{\odot bol}}. \tag{2.46}$$

Unfortunately, $L_{bol}$ can only be directly measured by observing over the entire electromagnetic spectrum, which is rarely possible. Hence estimates of $\Upsilon_{bol}$ tend to rely on assumptions about the object's spectral energy distribution that may be seriously in error.

### 2.3.6 Surface brightness and isophotal radii

To this point we have been concerned with total brightnesses and luminosities. Galaxies, unlike most stars, are resolved objects, so that in addition to measuring the total radiative flux from a galaxy, we can, in principle, measure for each point in its image the **surface brightness**, which is the radiative flux per unit solid angle of the image. To a first approximation, the surface brightness of an extended object is independent of its distance from us. As one removes the object to a distance $r$ from the observer, the amount of light received falls off as $1/r^2$, but the solid angle subtended by the object, over which its light is distributed, falls off in the same proportion, so that the amount of light per square second of arc remains constant. Thus the surface brightness at a given point on the surface of the galaxy is a well-defined, distance-independent quantity.[17]

Optical astronomers measure surface brightnesses in magnitudes per square arcsecond. The following example will demonstrate that these are strange units. Suppose that the surface brightness $I_1$ at a given point in the image of a nearby elliptical galaxy is $21 \, \mathrm{mag \, arcsec^{-2}}$. Then a square one arcsecond on a side around that point emits as much light as a star of magnitude 21. Now suppose that the this galaxy lies in front of a more distant spiral galaxy and that the given point coincides with the center of that galaxy, where the surface brightness $I_2$ is also $21 \, \mathrm{mag \, arcsec^{-2}}$. Then what is the combined surface brightness $I_1 + I_2$ of the two galaxies? The answer is not $42 \, \mathrm{mag \, arcsec^{-2}}$ but $20.25 \, \mathrm{mag \, arcsec^{-2}}$ because the amount of light coming from the two galaxies is $2 \times 10^{-21/2.5} = 10^{-20.25/2.5}$ arbitrary units.

In this book we employ the notation $I = 21 \, \mu_B$ to indicate a surface brightness in the $B$ band equal to $21 \, \mathrm{mag \, arcsec^{-2}}$.

Radio astronomers frequently measure surface brightnesses in degrees Kelvin, and say that they have measured the object's 'brightness temperature' $T_B$. If $T_B = 20 \, \mathrm{K}$, then the object's surface brightness at the wavelength of the measurement is the same as that of a black body whose temperature is $20 \, \mathrm{K}$ – see §8.1.4 for more details.[18]

Since galaxies do not have sharp edges, it is usual to characterize the physical size of a galaxy by quoting an **isophotal radius** or **isophotal diameter**, that is, the radius or diameter at which some particular surface-brightness level is reached. For example, the *Reference Catalogues* (de Vaucouleurs, de Vaucouleurs & Corwin 1976, de Vaucouleurs *et al.* 1991) cite $D_{25}$, the diameter that one estimates the $I = 25 \, \mu_B$ isophote would have if the galaxy were seen face on and unobscured by dust. We shall frequently cite values of $R_{25} \equiv \frac{1}{2} D_{25}$, which is frequently called the **de Vaucouleurs**

---

[17] In fact, the greater the distance to a galaxy the greater its redshift, and the redshift tends to cause the surface brightness at a given wavelength to decrease with distance.

[18] §8.1.4 also explains the other surface-brightness units commonly employed by Radio astronomers: $\mathrm{Jy \, km \, s^{-1}}$.

**radius**. An older, but still sometimes cited isophotal radius is the **Holm-berg radius**, which is the semimajor axis of the $I = 26.5\,\mu_{pg}$ isophote.

## 2.4 Gravitational lensing

Two of the fundamental problems of astronomy are (i) determining the phys-ical scale of the Universe and of the objects such as galaxies that lie within it, and (ii) determining the masses of these objects. Any physical effect that enables us to make progress with either of these tasks is precious. The phe-nomenon of gravitational lensing, which was first observed in 1979 (Walsh *et al.* 1979), enables us to make progress with *both* of these tasks. In view of this fact, a great deal of effort has been directed towards discovering instances of gravitational lensing over the last twenty years, and towards understanding the physical implications of such instances. In the last few years, these efforts have begun to bear a rich harvest so that gravitational lensing must now be reckoned to be a powerful probe of the structure of the Milky Way (§10.1) and of the Universe at large (§7.2.3). In the coming years, the astronomical importance of this phenomenon will surely grow. In this section we describe the theoretical framework that underlies these advances.

A gravitational field can deflect a beam of light. In practical applica-tions this deflection is best understood by considering that gravity causes the refractive index of the vacuum to differ from unity. Quantitatively, the refractive index of the vacuum is given by (see Appendix A)

$$n = 1 + \frac{2|\Phi|}{c^2}. \tag{2.47}$$

When a ray passes through a region within which the refractive index $n$ is a function of position, it is deflected through some angle $\alpha$ – see Figure 2.12, which illustrates the simple case in which source and observer both lie within the plane of the ray. Quantitatively, $\alpha$ is given by (Appendix A)

$$\boldsymbol{\alpha}(\mathbf{x}_\perp) = -\frac{4}{c^2}\boldsymbol{\nabla}_\perp\Phi_2(\mathbf{x}_\perp), \tag{2.48}$$

where $\mathbf{x}_\perp$ is the ray's impact parameter and the function $\Phi_2$ is related to the deflecting mass density $\rho(\mathbf{x})$ by

$$\Phi_2(\mathbf{x}_\perp) \equiv G \int d^2\mathbf{x}'_\perp \Sigma(\mathbf{x}'_\perp) \ln|(\mathbf{x} - \mathbf{x}')_\perp|,$$
$$\Sigma(\mathbf{x}'_\perp) \equiv \int dz'\,\rho(\mathbf{x}'). \tag{2.49}$$

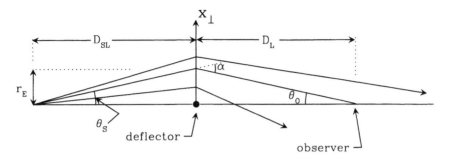

**Figure 2.12** The definition of the Einstein radius $r_{\rm E}$.

Equation (2.49) has a simple physical interpretation: the deflection is given by the gradient of the gravitational potential that is generated (in two dimensions) by the surface density that one obtains by integrating the deflecting mass density along the line of sight.

An important special case is that in which the matter distribution is effectively that of a point mass $\mathcal{M}$ – that is, the deflecting matter distribution is confined well inside the impact parameter $\mathbf{x}_\perp$ of every ray of interest. Then $\Phi_2(\mathbf{x}_\perp) = G\mathcal{M} \ln |\mathbf{x}_\perp|$ and

$$\alpha = \frac{4G\mathcal{M}}{c^2 x_\perp}. \tag{2.50}$$

The situation when the source, mass and observer all lie on a straight line is described by Figure 2.12: rays that encounter the mass at small impact parameter cross the source–observer line in front of the observer, while rays that encounter the mass at large impact parameters cross behind her. The ray that encounters the mass with critical impact parameter $r_{\rm E}$ reaches the observer. Since, in the notation of Figure 2.12, $\theta_{\rm S} \simeq r_{\rm E}/D_{\rm SL}$, $\theta_{\rm O} \simeq r_{\rm E}/D_{\rm L}$ and $\alpha = \theta_{\rm S} + \theta_{\rm O}$, it follows with a little algebra from (2.50) that

$$r_{\rm E} = \sqrt{\frac{4G\mathcal{M}}{c^2}} \sqrt{\frac{D_{\rm SL} D_{\rm L}}{D_{\rm SL} + D_{\rm L}}}.$$

The critical radius $r_{\rm E}$ is called the **Einstein radius** of the deflector. Note that it depends on the relative positions of source, observer and deflector in addition to the deflector's mass. If the source, deflector and observer are collinear as in Figure 2.12, the observer sees a bright ring of radius $r_{\rm E}$ around the deflector. The angular radius of this ring is the **Einstein angle**

$$\theta_{\rm E} \equiv \sqrt{\frac{4G\mathcal{M}}{c^2}} \sqrt{\frac{D_{\rm SL}}{D_{\rm L}(D_{\rm SL} + D_{\rm L})}}. \tag{2.51}$$

The Einstein radius is a dimensionally important quantity because lensing significantly modifies the appearance of a source that lies within about $r_{\rm E}$ of

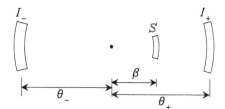

**Figure 2.13** Images of a patch $S$ on the surface of a source are formed at $I_+$ and $I_-$.

the deflector–observer line, while a source that lies further than $r_{\rm E}$ from this line will be seen very much as it would be if the deflector were not present. Objects that lie within $r_{\rm E}$ of a deflector are said to be **strongly lensed**, while any distortion of the images of objects that lie further out is ascribed to **weak lensing**.

Imagine observing stars that lie at a common distance $D_{\rm SL}+D_{\rm L}$ through a screen of point-mass deflectors of mass $\mathcal{M}$ that all lie at distance $D_{\rm L}$. The probability that a given source is lensed is the probability that it lies on the sky within one of the disks of radius $\theta_{\rm E}$ that we may draw around each deflecting mass. For astronomically interesting surface-densities $N$ of deflectors, these disks have a negligible probability of overlapping, so the fraction of the sky that they cover is $N\pi\theta_{\rm E}^2$, and this is the probability that any given source will be lensed. When we use equation (2.51) to eliminate $\theta_{\rm E}$ from our expression for this probability, we find the latter is

$$\tau = \frac{4\pi G N \mathcal{M}}{c^2 D_{\rm L}} \frac{D_{\rm SL}}{D_{\rm SL} + D_{\rm L}}. \tag{2.52}$$

The probability $\tau$ is called the **optical depth to gravitational lensing** by the screen of deflectors. Notice that it is proportional to the surface density of mass, $\Sigma \equiv N\mathcal{M}$, contributed by the deflectors; for given mass surface density, $\tau$ is the same for a large number of low-mass deflectors as for a smaller number of more massive deflectors.

What *does* depend on the mass of the deflectors is the angle through which light is typically deflected when lensing takes place. For example, for $D_{\rm SL} = D_{\rm L} = 10\,{\rm kpc}$, $\theta_{\rm E} = 0.9(\mathcal{M}/\mathcal{M}_\odot)^{1/2}\,{\rm mas}$. Thus when the deflector is a star and the source and deflector lie within the Milky Way, the deflection angle is too small to be measured even with HST. We can none the less detect such a lensing event by monitoring the luminosity of the star as it is lensed: we shall show that this temporarily increases when a deflector passes close to the line of sight to the source in very much the same way that a burning glass locally increases the brightness of the Sun.

We now calculate the amount by which a point-mass deflector brightens an unresolved background star. The key to the calculation is the fact that gravitational deflection leaves unchanged the surface brightness of a lensed object. Consequently, the luminosity of an image is simply proportional to the area of the sky covered by it. Figure 2.13 shows the relevant geometry when a source is imaged by a spherically-symmetric deflector such as a point

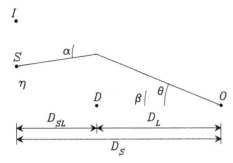

Figure 2.14 A ray from the source
$S$ is deflected through angle $\alpha$ by the
gravitational field of the object $D$.

mass. The true position and angular extent of a portion the source is shown
at $S$, while $I_+$ and $I_-$ mark the locations and extents of the two images of $S$.
Each of these images is formed by taking each point of $S$ and moving it along
the line through that point and the center of the deflector to a new location.
This location is determined by consideration of Figure 2.14, which shows the
plane defined by the source point $S$, the deflector D and the observer O.
Since the marked angles are in reality very small, we have[19]

$$\eta = D_S\beta = D_S\theta - D_{SL}\alpha \quad \Rightarrow \quad \beta = \theta - \frac{D_{SL}}{D_S}\alpha(\theta) \qquad (2.53)$$

Here $\alpha(\theta)$ is the function obtained by substituting $x_\perp = D_L\theta$ into equation
(2.50). Making this substitution and then using (2.51) we find

$$\begin{aligned} \beta &= \theta - \frac{D_{SL}}{D_S D_L}\frac{4G\mathcal{M}}{c^2\theta} \\ &= \theta - \frac{\theta_E}{\theta}. \end{aligned} \qquad (2.54)$$

For given source location $\beta$ this is a quadratic equation in $\theta$ with solutions

$$\theta_\pm = \tfrac{1}{2}\left(\beta \pm \sqrt{\beta^2 + 4\theta_E^2}\right). \qquad (2.55)$$

Returning to Figure 2.13, light from the patch $S$ of the source is received
in the two patches $I_+$ and $I_-$, which correspond to the two roots $\theta_\pm$ of
equation (2.55). Their areas $A_\pm$ are seen to be related to the area $A_S$ of $S$
by

$$\begin{aligned} \frac{A_\pm}{A_S} &= \frac{\theta_\pm \, d\theta_\pm}{\beta \, d\beta} \\ &= \frac{\theta_\pm}{2\beta}\left(1 \pm \frac{\beta}{\sqrt{\beta^2 + 4\theta_E^2}}\right). \end{aligned} \qquad (2.56)$$

---

[19] Equation (2.53) is often called the **lensing equation**.

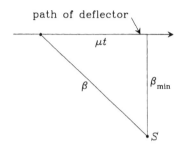

Figure 2.15 The geometry on the sky of a lensing event.

When we use (2.55) to express everything as a function of $u \equiv \beta/\theta_E$, we find

$$\frac{A_\pm}{A_S} = \frac{u \pm \sqrt{u^2+4})}{4u}\left(\frac{\sqrt{u^2+4} \pm u}{\sqrt{u^2+4}}\right)$$
$$= \tfrac{1}{2} \pm \frac{u^2+2}{2u\sqrt{u^2+4}}. \tag{2.57}$$

The second term on the right of equation (2.57) is numerically larger than $\tfrac{1}{2}$, so $A_-/A_S$ is negative. This result arises because the image at $\theta_-$ is inverted: the points that lie nearest the deflector are imaged to points that lie furthest from the deflector. Consequently, $d\theta_-/d\beta < 0$ and equation (2.56) yields a negative value for $A_-/A_S$.

Since gravitational lensing preserves surface brightness, in each image lensing amplifies the intrinsic brightness of the source $S$ by a factor $a_\pm = |A_\pm/A_S|$. When the individual images are unresolved, the observed magnification is the sum

$$a = a_+ + a_- = \frac{u^2+2}{u\sqrt{u^2+4}}. \tag{2.58}$$

As a deflector passes close to our line of sight to a background source, $\beta$ decreases from a large value and then increases again, with the result that the brightness of the source is amplified by a time-dependent factor $a(t)$. From Figure 2.15, which shows the path on the sky of the deflector relative to the source, it is clear that

$$\beta(t) = [(\mu t)^2 + \beta_{min}^2]^{1/2}, \tag{2.59}$$

where $\mu$ is the magnitude of the relative proper motion, $\beta_{min}$ is the closest approach of the deflector to the source, and $t$ is the time to the point of closest approach. Figure 2.16 shows plots of the amplification factor $a(t)$ that one obtains by inserting (2.59) into equation (2.58) for several values of $u_{min} = \beta_{min}/\theta_E$.

Each year of order a hundred stars are observed to suffer episodes of brightening that can be fitted by the function $a(t)$ of Figure 2.16 for suitable

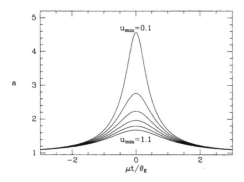

**Figure 2.16** As a deflector drifts past the line of sight to a background star, the star brightens by the factor $a(t)$ that is shown here for several values of $u_{min} = \beta_{min}/\theta_E$.

values of $\mu$ and $u_{min}$. We can be pretty sure that these stars are being lensed gravitationally because the light curves of a given star in different wavebands can be fitted by the *same* values of $\mu$ and $u_{min}$ just as lensing theory predicts. By contrast, intrinsic variability in stars is almost always associated with color, with the consequence that each waveband yields a different light curve.

## 2.5 Archival data and catalogs

The raw materials of any astronomical study are observational data. From these data, the first interpretive step is the production of **data products**. For example, from raw images of star fields, one could derive a catalog of stellar positions, magnitudes, colors, and so on. Such data products form the backbone of most astronomical research, and the importance of the painstaking work required for their assembly is impossible to overestimate. Since astronomical catalogs may have applications to a wide range of research problems, it is important that they are widely disseminated. Such catalogs also frequently have a very long "shelf-life" – the NGC and IC catalogs of galaxies and star clusters are more than a hundred years old, but still form the basis for many studies. The long-term archiving of these products is therefore also an issue. The original observations from which the products were derived can also sometimes be put to uses other than those for which they were obtained. It is therefore important that the raw data, as well as the processed results, be archived for future use.

In this section, we discuss some of the archival resources available to astronomers. A list of astronomical catalogs is given in Appendix B; this list is by no means complete, but includes the more important catalogs that the reader is likely to encounter in the literature. Some of these data products are included primarily for historical reasons, but it is important to realize that historic catalogs can remain of some importance even when more current data are available. For example, even though a more recent spectral classification may well have been made, a star is still commonly identified by the HD number that tags it in the Henry Draper Catalog. One area where historical precedent is particularly important is in specifying an object's coordinates. As we saw in §2.1, there have been a number of revisions in the way in which the celestial coordinate system is defined. An object's position is usually calculated by measuring its offset from an object in one of the catalogs in Appendix B. The coordinates adopted in these surveys therefore define the celestial coordinates of subsequent observations. Indeed, some of the revisions to the astrometric catalogs listed in Appendix B, such as FK4 to FK5, were motivated by the need to improve the consistency and accuracy of the celestial coordinate system. The older catalogs will remain important even after the world-wide adoption of the ICRS (§2.1.6) as the standard astrometric system, since we will still need to be able to cross-reference between the various coordinate systems to convert older observations to the new system.

Simple data products such as positions and magnitudes of stars can be straightforwardly presented in tabular form. However, when it comes to more complex data such as spectra, tables do not provide a very efficient medium for presenting the results, so such data are usually published as graphical atlases. Similarly, the mass of information contained in an image of a galaxy cannot readily be reduced to a few numbers, so a catalog of the morphology of galaxies is often supplemented by an atlas of images. Table 2.5 gives a short list of atlases that contain galaxy images; browsing through such atlases is probably the best way to obtain an overall impression of the range of structures that one finds in galaxies, their similarities and differences. Our location within the Milky Way means that studies of the large-scale structure of our own galaxy need data from all around the sky. Table 2.6 lists some of the more important optical sky surveys for the study of Galactic structure, while Table 2.7 lists some of the all-sky data sets available at other wavelengths (further details on HI and CO surveys of the Milky Way can be found in Tables 9.1 and 9.2, respectively).

It is clear from these tables that there is a vast amount of data available for studies of galactic structure, and the amount of information is growing extremely rapidly. The main reason for this "data explosion" has been the advent of digital techniques for data acquisition and storage. Digital data are easy to reproduce and disseminate, and the ease with which such data can be manipulated by computer makes it possible to analyze and interpret even the largest catalogs.

The proliferation of large data products led to the setting up in the 1970s of a number of specialized institutes to collect and store data. First amongst these **astronomical data centers** (**ADCs**) was the *Centre de Données Astronomiques de Strasbourg* (CDS), and this institution still serves as the primary clearing house for data, culling results from the literature and distributing them to users.

Another consequence of the widespread sharing of data has been the need to develop standard formats for data files, so that data written by

**Table 2.5** Galaxy atlases

| Atlas | Principal content | Number of objects | Reference |
|---|---|---|---|
| Hubble Atlas of Galaxies | Description, classification, photographs of galaxies | 180 | Sandage (1961) |
| Atlas of peculiar galaxies | Photographs of peculiar interacting galaxies | 338 | Arp (1966) |
| Atlas of southern peculiar galaxies | Photographs of peculiar interacting galaxies | 6445 | Arp, Madore & Roberton (1987) |
| NASA Atlas of Galaxies | Photographs of spiral galaxies useful for establishing the cosmic distance scale | 322 | Sandage & Bedke (1988) |
| Carnegie Atlas of Galaxies | Description, classification, photographs of galaxies | 1225 | Sandage & Bedke (1994) |
| Digital Catalog of Galaxies | CCD images in several colors of nearby galaxies | 113 | Frei *et al.* (1996) |

**Table 2.6** Optical sky surveys

| Survey | Telescope | Sky Coverage | Limiting Magnitude (and band) |
|---|---|---|---|
| Franklin-Adams Survey, 1911 | 10-inch Astrograph | All-sky | 15 |
| Palomar–National Geographic Society Sky Survey (POSS-I), 1960 | Palomar 48-inch Schmidt | $-33° \leq \delta \leq 90°$ | $21(B)$, $20(R)$ |
| ESO Atlas of Southern Milky Way, 1969 | Boyden 10-inch Astrograph | Southern Milky Way | 16 |
| Canterbury Sky Atlas, 1972 | Ross 5-inch Astrograph | $-90° \leq \delta \leq 22°$ | 16 |
| ESO/SERC Southern Sky Survey, 1980 | UK 1.2 m Schmidt & ESO 1.0 m Schmidt | $-90° \leq \delta \leq -17°$ | $23(B)$, $22(R)$ |
| SERC Equatorial *J* Extension, 1991 (*R*-band on-going) | UK 1.2 m Schmidt | $-17° \leq \delta \leq 3°$ | $23(B)$ $22(R)$ |
| Second Palomar Sky Survey (POSS-II), on-going | Palomar 48-inch Schmidt | $-3° \leq \delta \leq 90°$ | $22.5(B)$ $21(R)$ $19.5(I)$ |
| Second Southern Sky Survey, on-going | UK 1.2 m Schmidt | $-90° \leq \delta \leq -17°$ | $21.5(R)$ |
| Sloan Digital Sky Survey (SDSS), on-going | SDSS 2.5 m (5 band imaging + spectra for redshifts | $b > 30°$, all $l$ plus a $2° \times 50°$ strip through SGP | $23–25(R)$ |

NOTES: Since many of these surveys were recorded photographically through non-Johnson filters, the bandpasses listed should only be taken as indicative.

**Table 2.7**  Non-optical sky surveys

| Waveband | Instrument | Sky Coverage | Reference |
|---|---|---|---|
| HI (21 cm) | Hat Creek 85-foot dish | $-10° < b < 10°$ $10° < l < 250°$ | Weaver & Williams (1973) |
| | CSIRO 18 m dish | $-10° < b < 10°$ $240° < l < 350°$ | Kerr *et al.* (1986) |
| | AT&T 20-foot horn | $\delta > -40°$ | Stark *et al.* (1992) |
| | Dwingeloo 25 m dish | $\delta > -30°$ | Hartmann & Burton (1997) |
| 1.4 Ghz Radio continuum | Very Large Array (VLA) | $\delta > -40°$ | Condon *et al.* (1997) |
| | Very Large Array (VLA) | Same as SDSS | Becker *et al.* (1995) |
| CO (2.6 mm) | 1.2 m dishes (USA & Chile) | $-15° < b < 15°$ all $l$ | Dame *et al.* (1987) |
| Infrared | Infrared Astronomical Satellite (IRAS) | All-sky | Wheelock *et al.* (1991) |
| | Cosmic Background Explorer (COBE) | All-sky | Hauser *et al.* (1997) |
| Ultraviolet | Roentgen Satellite (ROSAT) | All-sky | Pounds *et al.* (1993) |
| | Extreme Ultraviolet Explorer (EUVE) | All-sky | Bowyer *et al.* (1994) |
| X-ray | High Energy Astrophysics Observatory (HEAO-I) | All-sky | Wood *et al.* (1984) |
| | Roentgen Satellite (ROSAT) | All-sky | Snowden *et al.* (1995) Voges *et al.* (1996) |
| $\gamma$-ray | COS-B | Galactic plane | Bloemen (1989) |
| | Compton Gamma-Ray Observatory (CGRO) | All-sky | Fichtel *et al.* (1994) Schönfelder *et al.* (1996) |

one computer can be read by another. The most commonly-used format for astronomical data is the **Flexible Image Transport System**, or **FITS** protocol. As its name suggests, this file format was originally created to allow images to be moved between machines, but the format is now widely used to store and transmit a wide variety of types of data, including catalogs and other tables (Schlesinger 1994).

Increasing computer power and storage capacity have given a number of older resources a new lease of life. The Space Telescope Science Institute (STScI) took on the mammoth task of digitizing plates from the POSS-I and SERC optical surveys (see Table 2.6) to produce a Digitized Sky Survey (DSS) of the entire sky. The amount of data involved is huge – some 600 Gbytes; even after passing the images through sophisticated data compression algorithms, the DSS still occupies more than a hundred CD-ROMs. A second-generation DSS is now being produced by scanning the plates of the POSS-II, the Second Southern Sky Survey, and the red equatorial extension (see Table 2.6). The second-generation DSS will have higher resolution than its predecessor (1 arcsec per pixel as opposed to 1.7 arcsec per pixel), so it will occupy over 3 Terabytes.

Full exploitation of the second-generation DSS requires photometric calibration of the original survey plates (Lasker *et al.* 1990). This calibration is tricky because the blackness of a photographic plate is a highly non-linear function of the intensity of the light to which it has been exposed. This difficult calibration will be largely avoided by the next generation of surveys, such as the Sloan Digital Sky Survey (SDSS), which will use large-format CCDs rather than photographic plates to image the sky.

Archival data are also of critical importance to satellite-based observations. Due to the great expense and limited lifetimes of such missions, data obtained by satellites are even more valuable than data from ground-based instruments, and careful archiving of data is vital. Fortunately, their archiving is facilitated by the constraints that are inherent in satellite observatories: data must be digital so that they can be relayed to the ground, and the details of each observation are inevitably carefully documented when the observation is scheduled. Further, since the space-based instruments have to work reliably in a "hands-off" environment, the modes in which they operate are relatively simple and well-documented. Consequently, data taken with satellites can readily be interpreted by an astronomer who was not involved in the original observations.

By contrast, ground-based instruments have lagged in producing useful data archives. Particularly at optical wavelengths, there has been a tradition of the data "belonging" to the observer. Thus, even though most facilities now release all data after a proprietary period (typically a year), relatively little use has been made of such public-domain data. The more interactive nature of ground-based observing can make it difficult to interpret data taken by someone else, since it is sometimes hard to reconstruct exactly how the data were taken. Nonetheless, the importance of archival data as a resource is also being recognized at ground-based facilities, and many are now producing usable data archives.

**On-line resources**    Happily, the huge expansion in the quantity of archival data available has been accompanied by a similar explosion in the use of the Internet. It is no longer necessary for researchers to maintain their own copies of all the databases that they may use in the course of their research, since they can access the information across the network from wherever in the world it may be physically stored.

Although the Internet has been used by astronomers for e-mail since the early 1980s, it was initially not much used to access data remotely; files could be transferred with the **file transfer protocol** (**ftp**), but information could not be extracted from a file without downloading the whole thing. Further, unless one knew the exact name and location of the data file in question, it was almost impossible to track down the information one required.

The situation improved somewhat with the setting up of **telnet** services, which enable a user to log on to a remote machine and query databases stored there. One of the most widely used of these services, called SIM-BAD[20], is provided by the CDS to give access to the basic information and bibliographic references that they hold on more than a million astronomical objects. A similar service for purely extragalactic objects is provided by the NASA/IPAC Extragalactic Database (NED). These services allow quite sophisticated searches of their databases, but the user interface is a fairly rudimentary text-based one.

The use of on-line resources really took off in the early 1990s with the development of the **World Wide Web (WWW)**. This paradigm for accessing data on the Internet emerged from the high-energy physics community, but is also well suited to the needs of astronomers for several reasons. First, the organization of information into structured **hypertext** documents rather than simple data files makes it straightforward to navigate one's way to the information required, even when the location and name of the file containing the relevant data is initially unknown. Second, the WWW allows sophisticated interactions between the client (the person/machine that is accessing a database) and the server (the machine that holds the data), so one can easily extract just the data that one requires and in a form that is suited to one's project. Third, the ability of the WWW to transfer and display graphical information is ideally suited to image-based astronomical data.

New on-line resources and tools are currently being developed at a dramatic rate. For example, the Java programming language, designed for running complete programs across the Internet, offers a whole new mechanism for the remote manipulation of data. Any attempt to provide a definitive description of the electronic resources available is therefore doomed to be out of date long before it gets into print. In fact, the only way to keep abreast of developments is by using the existing resources regularly and trying out new ones as they emerge. A good starting place for finding the available astronomical resources is provided by the AstroWeb Database,[21] which contains information on thousands of astronomical resources that are available over the WWW. The various search engines[22] that automatically index the entire WWW are also valuable tools for locating particular pieces of information.

Although we make no attempt to provide a definitive list of on-line resources, Table 2.8 gives the URLs for a number of services which have proved particularly useful, and which are sufficiently well-established that they are unlikely to disappear overnight. This table also gives some indication of the

---

[20] The name is the rather contrived acronym formed from 'Set of Identifications, Measurements and Bibliography for Astronomical Data.'

[21] Either http://ecf.hq.eso.org/astroweb/yp_astro_resources.html or http://www.cv.nrao.edu/fits/www/astronomy.html – these **Unified Resource Locators (URLs)** give the location of the information being accessed, and the protocol by which it is to be accessed. In this text, URLs are set in Courier font.

[22] For example, http://home.netscape.com/home/internet-search.html.

**Table 2.8**  Selection of on-line astronomical research resources

| Resource | Description | Unified Resource Locator (URL) |
|---|---|---|
| | Data resources | |
| SIMBAD | Basic data and bibliographic references for $\sim 10^6$ objects | telnet://simbad.u-strasbg.fr/<br>http://cdsweb.u-strasbg.fr/Simbad.html |
| NASA/IPAC Extragalactic Database (NED) | Basic data and bibliographic references for $\sim 750,000$ extragalactic objects | telnet://ned@ned.ipac.caltech.edu/<br>http://nedwww.ipac.caltech.edu/ |
| ADC catalog databases | On-line access to $\sim 2000$ catalogs of data and tables from journals | http://vizier.u-strasbg.fr/<br>http://adc.gsfc.nasa.gov/<br>http://adac.mtk.nao.ac.jp/ |
| HEASARC archive browser | Archive of data and catalogs, mostly obtained by high energy satellites | http://heasarc.gsfc.nasa.gov/w3browse/ |
| SkyView | On-line access to sky survey images at a wide range of wavelengths | http://skyview.gsfc.nasa.gov/ |
| STScI Digitized Sky Survey (DSS) | On-line access to the optical sky surveys digitized by STScI | http://stdatu.stsci.edu/dss/<br>http://archive.eso.org/cgi-bin/dss/<br>http://cadcwww.dao.nrc.ca/cadcbin/getdss |
| | Literature resources | |
| Los Alamos National Laboratory (LANL) e-print archive | Electronic astronomy preprint archive. | http://xxx.lanl.gov/archive/astro-ph<br>(see http://xxx.lanl.gov/servers.html for worldwide mirror services) |
| Astrophysics Data System (ADS) Abstract Service | Abstracts and complete texts of many astronomical journals | http://adsabs.harvard.edu/abstract_service.html<br>http://cdsads.u-strasbg.fr/abstract_service.html<br>http://ads.nao.ac.jp/abstract_service.html |
| | Miscellaneous resources | |
| CADC meetings database | Listing of up-coming conferences. | http://cadcwww.dao.nrc.ca/meetings/meetings.html |
| AAS job register | Astronomical situations vacant. | http://www.aas.org/JobRegister/aasjobs.html |

NOTES: An on-line version of this table can be found at http://www.astro.soton.ac.uk/astrores.html

range of services now available. As well·as the basic information provided by services such as SIMBAD and NED, it is now possible to access a great deal of raw astronomical data from various instruments. A good example of an on-line data archive is the High Energy Astrophysics Science Archive Research Center (HEASARC), which provides easy access to the data from a large number of X-ray and $\gamma$-ray satellites. Access to the astronomical literature has also been revolutionized by the WWW. The ADS Abstract Service allows one to search through all the common astronomical journals for particular text in their abstracts or for particular objects, and even to retrieve the complete texts of many articles. The LANL E-print Archive maintains an on-line database of preprint copies of astronomical papers, and by browsing through the recent additions one can keep up with current developments in the field worldwide.

This burgeoning of on-line resources has caused radical changes in the way in which astronomical research is undertaken. For any project, it is now straightforward to see what applicable data already exist in the public domain, and what work has already been done on the subject. It is also practicable to sift through vast amounts of data reasonably quickly, so large survey-type projects can be undertaken with relative ease. Finally, because these on-line resources can be accessed from anywhere in the World, it is now possible to do leading-edge research away from the large institutions which have traditionally dominated the astronomical community. Quite where these changes will end remains to be seen, but it is already clear that a major revolution is underway.

## Problems

**2.1** In projection, a point in the bulge of M31 lies at a small distance from the center of a spherical globular cluster. The measured surface brightness at the given point is $19 \, \mathrm{mag\, arcsec}^{-2}$. From the structure of the globular cluster, it is estimated that the intrinsic surface brightness of the cluster at that point is $20 \, \mathrm{mag \, arcsec}^{-2}$. What is the surface brightness of the bulge at that point?

**2.2** $I_B$ is the $B$-band surface brightness of a galactic disk measured in $L_\odot \, \mathrm{pc}^{-2}$. Show that, when viewed face-on, the $B$-band surface brightness of the disk will be measured to be $\mu_B$ magnitudes per square arcsecond, where

$$\mu_B = 27.05 - 2.5 \log(I_B). \qquad (2.60)$$

[You will need to use the fact that the $B$-band absolute luminosity of the Sun is $M_B(\odot) = 5.48$.]

**2.3** From equation (2.42), show that

$$C_{\mathrm{bol}} = M_{\mathrm{bol}} + 2.5 \log(L/\mathrm{W}) - 90.195, \qquad (2.61)$$

where $L/\mathrm{W}$ is the star's luminosity in watts. Show further that, if the Sun's $V$-band bolometric correction is taken to be zero, $C_{\mathrm{bol}} = -18.90$.

**2.4** When bolometric corrections are defined such that $BC_V(\odot) = -0.19$, show that

$$\begin{aligned} M_{\text{bol}} &= -2.5 \log L/L_\odot + 4.64 \\ &= -5 \log R/R_\odot - 10 \log T_{\text{eff}}/T_{\text{eff}\odot} + 4.64. \end{aligned} \tag{2.62}$$

# 3

# The Properties of Stars

The study of galaxies is founded upon our knowledge of stars. Stars provide nearly all the light by which we see galaxies, and through their births and deaths they profoundly influence the structures of the galaxies that they inhabit. In this chapter we discuss the phenomenology of stars: their masses, radii, the characteristics of their spectra. We describe how these observable quantities are connected to a star's chemical composition and the stage in its life-cycle that a star has reached. Then we take census of the stars near the Sun: how common are stars of each type? Finally, we explain how the knowledge of the intrinsic properties of stars that we have gathered in this chapter enable us to correct for the insidious effects of the interstellar dust through which we are obliged to peer at many remote objects.

## 3.1 The masses of stars

Fundamental determinations of stellar masses are almost invariably based on an application of **Kepler's third law** to the orbits of binary stars. This law states that

$$G(\mathcal{M}_1 + \mathcal{M}_2)P^2 = 4\pi^2 a^3, \tag{3.1}$$

where $\mathcal{M}_1$ and $\mathcal{M}_2$ are the masses of two bodies in mutual revolution on a relative orbit with semimajor axis $a$ and period $P$, and $G$ is the gravitational constant.

Kepler's third law applies both to planets revolving around the Sun and to stars revolving around one another. Indeed, we can use the Earth's motion around the Sun to make a convenient choice of units for binary stars. Because $\mathcal{M}_\odot$, the mass of the Sun, is much larger than $\mathcal{M}_\oplus$, the mass of the Earth, we can write $\mathcal{M}_\odot + \mathcal{M}_\oplus \simeq \mathcal{M}_\odot$ to a high degree of approximation. Therefore,

$$G\,\mathcal{M}_\odot P_\oplus^2 = 4\pi^2 a_\oplus^3, \tag{3.2}$$

where $P_\oplus$ is the period of revolution of the Earth around the Sun (a year), and $a_\oplus$ is the mean distance from the Earth to the Sun (an astronomical unit). Taking the ratio of equation (3.1) to (3.2), the numerical constants cancel, and we can write

$$\frac{(\mathcal{M}_1 + \mathcal{M}_2)P^2}{\mathcal{M}_\odot P_\oplus^2} = \frac{a^3}{a_\oplus^3}, \tag{3.3}$$

so that, if we express masses in solar masses, periods in years, and distances in astronomical units, we have simply

$$(\mathcal{M}_1 + \mathcal{M}_2)P^2 = a^3. \tag{3.4}$$

By the definition of star's parallax $\varpi$ (§2.1.3), we have

$$\frac{a}{\mathrm{AU}} = \frac{a''}{\varpi/\mathrm{arcsec}}, \tag{3.5}$$

where $a''$ is the observed semimajor axis in arcseconds, and $\varpi$ is the parallax of the star.

### 3.1.1 The Mass of the Sun

To convert stellar masses expressed in solar units to physical units, we obviously need to know the mass of the Sun. We first determine the ratio of the mass of the Sun to the mass of the Earth by applying equation (3.3) to the orbit of the Moon around the Earth. Thus

$$\frac{(\mathcal{M}_\oplus + \mathcal{M}_-)P_-^2}{\mathcal{M}_\odot P_\oplus^2} = \frac{a_-^3}{a_\oplus^3}, \tag{3.6}$$

or

$$\frac{\mathcal{M}_\odot}{\mathcal{M}_\oplus} = \left(\frac{P_-}{P_\oplus}\right)^2 \left(\frac{a_\oplus}{a_-}\right)^3 \left(1 + \frac{\mathcal{M}_-}{\mathcal{M}_\oplus}\right), \tag{3.7}$$

where terms subscripted with a dash refer to the moon. All the terms on the right-hand side are known from observation, including the ratio $(\mathcal{M}_-/\mathcal{M}_\oplus)$, which is obtained by measuring the parallax in the position of a nearby planet that is produced by the Earth's monthly motion around the center of mass of the Earth-Moon system [see Abell (1964), Chapter 8]. Using the result $(\mathcal{M}_-/\mathcal{M}_\oplus) = 1/81.3$ and inserting the measured values for $(P_-/P_\oplus)$ and $(a_\oplus/a_-)$ into equation (3.7), we find $(\mathcal{M}_\odot/\mathcal{M}_\oplus) = 332\,945$. The mass of the Earth can be measured directly [see Abell (1964), Chapter 5] and is found to be $5.98 \times 10^{24}\,\mathrm{kg}$; hence we derive $\mathcal{M}_\odot = 1.99 \times 10^{30}\,\mathrm{kg}$.

### 3.1.2 Masses of binary stars

Binary and multiple stars are quite common: at least 50% of the "stars" within 5 pc of the Sun are actually double or multiple, so that a large majority of stars are members of such systems. For our purposes it is important to distinguish (i) **visual binaries**, in which the individual components can be seen directly, (ii) **spectroscopic binaries**, which reveal their orbital motion by shifts in the wavelengths of their spectral lines, and (iii) **eclipsing binaries**, which show evidence of orbital motion by periodic variations in their apparent brightnesses due to one star passing in front of the other. We can obtain different kinds of astrophysical information from each of these groups (which are not mutually exclusive). An excellent general discussion of binary and multiple stars can be found in Batten (1973), while an important review is that of Popper (1980).

**Visual binaries**   When stars are examined telescopically, relatively close pairs are frequently found. Some of these stars are mere **optical doubles**; that is, two stars that happen to lie, by coincidence, along nearly the same line of sight. Herschel searched for optical doubles in the hope of measuring the parallax of the nearer star relative to the further one but in 1781 he concluded that most of the systems he had found were actually **physical binaries**: stars that orbit around one another. Careful measurements of the separation and position angle of the two stars make it possible to determine the projection onto the plane of the sky of the relative orbit of one star around the other. The true orbit is an ellipse, and one can show that the projected orbit is also an ellipse.

Because the two components of a visual binary must be resolved, we tend to select systems that are widely separated in linear distance. This separation implies that they will have long periods, and in fact many known visual binaries have completed only a fraction of an orbital revolution in the entire time they have been observed (sometimes a century or more). The phenomenon of 'seeing' (§3.2.3) hinders measurement of binary-star separations smaller than about 0.15 arcsec, although interferometric techniques (see §3.2) now allow smaller separations to be studied. For example, Pan *et al.* (1992) have resolved the orbit of the spectroscopic binary $\alpha$ And, which has a major-axis separation of 24 mas.

From a geometric analysis of the apparent orbit [for details see Green (1985), Chapter 19], one can obtain the 'elements' that describe the shape and orientation of the true relative orbit of a visual binary. In particular, one can measure the semimajor axis $a''$ in arcseconds. Now, if we know the distance to the system and hence $\varpi$, we can convert the observed value of $a''$ to linear measure via equation (3.5), and, knowing $P$, we can find the sum of the masses $(\mathcal{M}_1 + \mathcal{M}_2)$ in solar units from Kepler's third law, equation (3.4). In practice, the lack of reliable parallaxes, which enter as the third power in the expression for $(\mathcal{M}_1 + \mathcal{M}_2)$, poses a very severe difficulty for the determination of accurate stellar masses.

To find the individual masses ($\mathcal{M}_1$ and $\mathcal{M}_2$), we need to know not only the relative orbit of the two stars but also the orbit of each of the components around their common center of mass. The center of mass of a binary system moves nearly on a straight line. The ratio of the two masses is easily determined as the inverse of the ratio of the amplitudes of the individual motions relative to the straight-line path of the center of mass.

Reliable masses are available for only about fifty visual binaries because of the difficulties of obtaining accurate distances and accurate apparent orbits – an easily resolved system usually has a long period, and hence only a fraction of its orbit has been observed, whereas systems with short periods are often at the limit of resolution. Lists of well-determined visual-binary masses and a discussion of the practical difficulties are given in Popper (1980) and Henry & McCarthy (1993).

We have just noted that masses found from visual binaries depend sensitively on the parallax of the system. This fact suggests that it can be advantageous to turn the procedure around and use it to estimate the parallax instead of the stellar masses. Thus, if a reasonably good guess can be made for $\mathcal{M}_1$ and $\mathcal{M}_2$, for example by assuming that the star's masses are normal for their spectral types, then we can determine the distance $d$ to the system. Clearly, if the system's semi-major axis $a$ subtends an angle $\theta$ at the Sun, then $d = a/\theta$. If $\mathcal{M}_1$ and $\mathcal{M}_2$ are given in solar units and the period $P$ in years, equation (3.4) gives $a$ in AU. Hence

$$\frac{d}{\mathrm{pc}} = \frac{(\mathcal{M}_1 + \mathcal{M}_2)^{1/3} P^{2/3}}{(\theta/\mathrm{arcsec})} \tag{3.8}$$

Such a distance estimate is referred to as a **dynamical distance**. Because only the cube root of the total mass enters in this formula, even a crude estimate of $\mathcal{M}_1$ and $\mathcal{M}_2$ will yield $d$ with good accuracy. Once we have an estimate of $d$ we can estimate the absolute magnitude of each star. We can then obtain a more refined estimate of $\mathcal{M}_1$ and $\mathcal{M}_2$ from the appropriate mass–luminosity relation (see §3.5.2). With improved values for the masses, a better value for $d$ can be derived from equation (3.8). If necessary, the procedure may be iterated; it converges rapidly.

Distances obtained from the dynamical parallax method are not fundamental as they are based on the assumption that the stars in the binary are normal. Nevertheless, they are often very useful estimates and occasionally provide an important check on other methods. For example, in the 1960s, the dynamical parallaxes of binaries in the Hyades were discordant with the distance determined by the moving-cluster method (see §2.2.2), and ultimately led to a revision of the accepted distance to the Hyades.

**Spectroscopic binaries**   If the orbital velocities of the components of a binary are large enough, we can measure them by the Doppler effect with reasonable accuracy (§2.2.1). From these measurements we can make deductions regarding the masses of the components. Generally, such binaries have short periods and small separations because these imply large mutual velocities. The periods may be as short as 0.1 day, in which case the stars may touch one another, or they can be as long as 7500 days, in which case the stars may be so widely separated as to be resolved as visual binaries. However, most of the stars for which we can determine velocities spectroscopically are too close to be resolved from the ground. Such systems are called **spectroscopic binaries**.

There are two types of spectroscopic binary. In the spectrum of a **double-line spectroscopic binary** the lines split into two components. These components show a periodic back-and-forth shift in wavelength but are in antiphase with one another. In these systems we can measure the velocities of both components. In the spectrum of a **single-line spectroscopic binary** all the measurable lines move in phase with one another. In these systems one component is so much brighter than the other (typically, $\Delta m \gtrsim 2.5$ mag) that it swamps the spectrum of the secondary and renders it invisible. For a given $\Delta m$, it is generally easier to detect both spectra if the two components are of very different colors.

From an analysis of the radial velocity of one or both of the components as a function of time, one can determine the elements of the binary orbit [see Green (1985), Chapter 19 for detailed methods]. The orbital plane is inclined to the plane of the sky by some angle $i$, which in general is unknown and cannot be determined from the spectroscopic data alone, because the observed radial velocity $v_r$ yields only the projection of the orbital velocity **v** along the line of sight; that is, $v_r = v \sin i$. As we shall see, this fact limits the information we can obtain about the system unless $i$ can be determined in some other way.

Individual masses can be determined only for the favorable case of a double-line binary, so we restrict our attention to this case. Suppose for simplicity that the orbits of both stars are circular. Then, from the observed radial velocities, we can immediately determine the projected radii (in linear units, that is, kilometers or astronomical units) of the two orbits: equating distance travelled to speed times time, we have $2\pi a_i = v_i P$, or

$$a_1 = \frac{v_{r1} P}{2\pi \sin i} \quad , \quad a_2 = \frac{v_{r2} P}{2\pi \sin i}, \tag{3.9}$$

where $v_{r1}$ is the amplitude of the observed oscillations in the radial velocity of the first star, etc. The mass ratio is thus

$$\frac{\mathcal{M}_1}{\mathcal{M}_2} = \frac{a_2}{a_1} = \frac{v_{r2}}{v_{r1}}. \tag{3.10}$$

Because we do not know the semi-major axis, $a \equiv a_1 + a_2$ but only its projection $a \sin i = (v_{r1} + v_{r2})P/(2\pi)$, we cannot determine $(\mathcal{M}_1 + \mathcal{M}_2)$ from Kepler's third law [equation (3.1)] but only

$$(\mathcal{M}_1 + \mathcal{M}_2) \sin^3 i = \frac{(v_{r1} + v_{r2})^2 P}{2\pi G}. \tag{3.11}$$

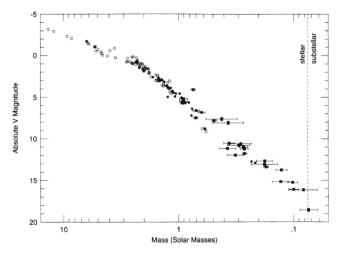

**Figure 3.1** Absolute visual magnitude versus mass for components of nearby binary stars. Visual and eclipsing binaries are marked by squares and circles, respectively. [After Latham (1998) courtesy of D. Latham]

Thus from equations (3.10) and (3.11) we can find $\mathcal{M}_1 \sin^3 i$ and $\mathcal{M}_2 \sin^3 i$ separately, but we obtain $\mathcal{M}_1$ and $\mathcal{M}_2$ only if we can determine the inclination by some other means.

Because we know that $\sin^3 i$ must always be less than or equal to unity, from spectroscopic data alone we can determine a lower bound on the masses of the binary components. A statistical correction for inclination effects can be made by using the result that $\langle \sin^3 i \rangle = 0.59$ for orbital planes oriented randomly with respect to the plane of the sky (Problem 3.1). This procedure is open to question, however, as selection effects obviously favor the discovery of systems with large values of $\sin i$: these have larger radial velocities for a given orbital velocity.

In this section we have studied the constraints one can impose on the masses of components of binary stars. The problem with visual binaries is that the masses depend sensitively on the system's distance. If the radial velocity $v_{r1}$ of one component is measurable, we can determine the physical length $a_1$ from equation (3.9), and thus obtain the distance $d$ from our astrometric measurement of the angle $a_1/d$.

The problem with spectroscopic binaries is ignorance of $\sin i$. In the case of an eclipsing binary, $\sin i$ can be measured: if the stars are reasonably well separated, the occurrence of eclipses implies that the orbital plane lies nearly in the line of sight, and hence we know that $i \approx 90°$ and $\sin i \simeq 1$; in close binaries, eclipses can still occur for inclinations significantly smaller than $90°$, but even for these systems an analysis of high-quality photometric data enables one to determine $i$.

From visual binaries with accurate distances and eclipsing binaries, accurate masses have been determined for over 100 stars. Figure 3.1 shows these masses plotted against $M_V$. It is clear that these two variables are tightly correlated.

## 3.2 The radii of stars

The radius of a star is a parameter of great importance because (i) for a given mass, it sets the stellar surface gravity and mean density (which have important implications in terms of stellar structure and evolution – see Chapter 5), and (ii) for a given luminosity, it determines the effective temperature of stellar atmosphere (see §3.4) and hence the appearance of the star's spectrum. To measure radii, we can either (i) measure a star's angular diameter, which, when combined with the star's distance, gives its physical size, or (ii) use observations of the light curves and orbital velocities in an eclipsing spectroscopic binary to obtain the component stars' radii in physical units, independent of knowledge of the system's distance. McAlister (1985) reviews this topic.

If we can resolve the disk of a star, we can measure its angular diameter directly. At present, this can be done for precisely one star: the Sun. The Sun has an angular diameter of $0.5331°$, which implies that $R_\odot = 0.004652\,\mathrm{AU} = 6.96 \times 10^5$ km. Viewed from a distance of 2 pc, the Sun would have an angular diameter of only 5 mas, which is much smaller than the lower limit, $\sim 0''.3$, on the diameter of telescope images that is set by turbulence in the Earth's atmosphere. Hence, the diameters of stars that are as small as the Sun cannot be measured by simple imaging. Some stars have radii of the order of a few hundred solar radii (see §3.5 below) but typically they lie at distances of 10 to 20 pc. Hence, even they have angular diameters $\lesssim 50$ mas which cannot be measured by ground-based imaging (although they are in principle within the range of the HST). However, we now discuss a number of cunning ploys that have enabled astronomers to measure the diameters of some dozens of nearby stars.

### 3.2.1 Phase interferometry

Around 1920, A.A. Michelson and F.G. Pease measured the angular diameters of several nearby stars by observing interference patterns produced in an interferometer formed by mounting a 25-foot ($\sim 8$ m) track bearing two small mirrors (equivalent to the slits of a Young's double-slit experiment) on the 100-inch Mount Wilson telescope. The visibility of the fringes set up when this system is directed at a star of angular diameter $\theta$ declines as the distance $D$ between the entrance mirrors is increased. The visibility effectively vanishes when the fringes produced by points on opposite sides of the star get $180°$ out of phase; this happens when $D \simeq \frac{1}{2}\lambda/\theta \simeq 5$ m for

$\theta = 20$ mas. In this way, Michelson and Pease obtained the angular diameters of about ten red giants. These angular diameters, when combined with distances, give stellar radii in physical units. A purpose-built optical interferometer operates on Mount Wilson and has measured angular diameters to an accuracy of $\lesssim 0.1$ mas (Mozurkewich *et al.* 1991). Another purpose-built optical interferometer is operated by the University of Sydney (Davis 1994). Since 'seeing' (§3.2.3) is a smaller effect at infrared wavelengths, it is advantageous to measure angular diameters at the longest feasible wavelengths. Di Benedetto & Rabbia (1987) have used an infrared interferometer to measure angular diameters.

### 3.2.2 Intensity interferometry

Brown & Twiss (1956) developed a new interferometric technique that exploits the fact that the fluctuations in the intensities of the signals received from a star by two instruments separated by distance $D$ are correlated in a way that depends on $D$ and the angular diameter of the star. In intensity interferometry, phase coherence is not required, and it becomes possible to use separate telescopes and large baselines (Brown 1968). An interferometer consisting of two 6.5 m mosaic "light-buckets" separated by baselines of length 10 to 190 m operated for several years at Narrabri, Australia, and produced angular-diameter measurements for about thirty stars (Brown *et al.* 1974b, Brown *et al.* 1974a).

Angular diameters as small as 0.5 mas have been measured with this technique. However, the method is restricted to hot stars, which (with the Narrabri interferometer) must be brighter than $V \approx 2.5$. The list of such stars observable from the southern hemisphere has already been exhausted, and new results will emerge only from northern-hemisphere observations, or from the use of a larger instrument that will reach fainter stars.

### 3.2.3 Speckle interferometry

In 1970 A. Labeyrie pointed out that, at any instant, the pattern of light produced by a star at the focus of a large telescope is speckled. Each **speckle** is an independent, diffraction-limited stellar image. Turbulence in the air above the telescope causes these images to dance around by introducing fluctuations in the phase of the wave front entering the telescope. In an exposure lasting longer than a fraction of a second, all these speckles are superimposed to produce a single splodge of brightness. One says that the image has been degraded by the phenomenon of 'seeing', and refers to the splodge's full width at half maximum as the **seeing**.

The technique of **speckle interferometry** involves making a very large number of short exposures of the field. An individual exposure is so short that the speckle pattern does not move appreciably as it is taken, and thus contains information on the angular size of the star. On the other hand, even when the imaged star is fairly bright, the number of photons in an individual exposure is not great, and therefore the quality of the image it provides is poor. However, by combining information from a large number of exposures, it is possible to compensate for their poor individual quality. With a 5 m telescope imaging at optical wavelengths, this method yields reliable angular diameters down to about 20 mas. A review of this method and its accomplishments is given in Labeyrie (1978).

### 3.2.4 Lunar occultations

As the Moon (or a planet) moves on the sky, it occasionally passes in front of a star and **occults** it. Typically, the light from a star is extinguished within a few milliseconds. To the extent that we can idealize a star as a point source and the Moon's disk as a semi-infinite plane, we expect a Fresnel diffraction pattern to sweep over the Earth as the moon passes in front of the star. This pattern can be measured by recording the brightness of the star as a function of time during the occultation. Because the star is not actually a point source but has a finite (if small) angular diameter, the diffraction pattern differs slightly from a Fresnel pattern, and, by a careful analysis of the data, one can infer the star's angular diameter (e.g., Brown 1968, Nather & McCants 1970). White & Feierman (1987) list diameters obtained in this way for 124 stars, including several fainter than $V = 7$ and with angular diameters $\lesssim 1$ mas. The method is obviously restricted to stars near the plane of the ecliptic, and it is limited in part by imperfect knowledge of the irregularities of the Moon's limb.

### 3.2.5 Eclipsing binaries

It sometimes happens that the orbital plane of a binary system is observed nearly edge-on, so that, as the stars revolve around their common center of mass, one star passes in front of the other (see Figure 3.2) and hence produces an eclipse. There will then be periodic variations in the light received from the system; a plot of the system's apparent magnitude as a function of time is called its **light curve**. The first eclipsing binary to be discovered was Algol ($\beta$ Per), which was noticed to be variable in 1670 and explained physically by J. Goodricke in 1782. Most known eclipsing systems have short periods (90% less than 10 days), but a few have very long periods (for example, $\varepsilon$ Aur has period $P \approx 27$ years). The probability of discovering a long-period eclipsing system is small, both because the orbital plane of a widely separated system must be inclined almost exactly 90° to the plane of the sky and because the eclipses occupy only a tiny fraction of the orbital period.

The **primary eclipse** occurs when the star having the higher surface brightness (hence higher effective temperature – see §3.4) is eclipsed, and the **secondary eclipse** occurs when the component having the lower surface

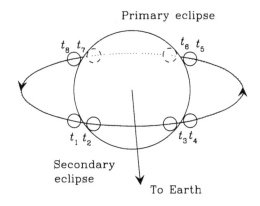

Primary eclipse

Secondary
eclipse

To Earth

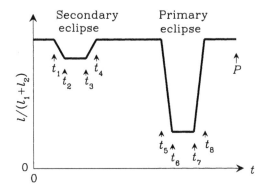

Secondary
eclipse

Primary
eclipse

**Figure 3.2** If the orbital plane of a
binary happens to be nearly edge-on
as seen from the Earth, then eclipses
will occur as one of the components
passes in front of the other. The
eclipse geometry is shown above, and
a schematic light curve for the case
in which the smaller star has the
greater surface brightness is shown
below. The binary period is $P$.

brightness is eclipsed. These phenomena are illustrated in Figure 3.2 for the
case in which the smaller star is a hot dwarf and the larger star a cool giant.

The simplest case to analyze occurs if the orbits are circular, the orbital
plane is inclined at $90°$ to the plane of the sky. Then the eclipses are total
(primary eclipse in Figure 3.2) and annular (secondary eclipse in Figure 3.2).
Let $d_1$ be the diameter of the larger star, and $d_2$ the diameter of the smaller.
These diameters in units of the orbital radius $a$ can be determined from
the duration of the eclipses. For example, if $t_1, \ldots, t_4$ denote times of first
contact,...,fourth contact during the secondary eclipse (Figure 3.2), then

$$\frac{d_1}{2\pi a} = \frac{(t_3 - t_1) + (t_4 - t_2)}{2P} \quad \text{and} \quad \frac{d_2}{2\pi a} = \frac{(t_2 - t_1) + (t_4 - t_3)}{2P}, \quad (3.12)$$

where for convenience we have assumed that $d_1 \ll a$ and $d_2 \ll a$. Similar
formulae can be written for the primary eclipse, and they provide a check on
the results.

Furthermore, it is clear that, if $\ell_1$ and $\ell_2$ are the apparent brightness of
the larger and smaller stars, respectively, and $h_p$ is the residual brightness

of the system at primary eclipse in units of the total brightness outside of eclipse, then for the case illustrated in Figure 3.2,

$$\frac{\ell_1}{\ell_1 + \ell_2} = h_p \quad \text{and} \quad \frac{\ell_2}{\ell_1 + \ell_2} = 1 - h_p. \tag{3.13}$$

Again, a check on the results can be obtained from the secondary eclipse because then we know that

$$\frac{\ell_1(1 - k^2) + \ell_2}{\ell_1 + \ell_2} = h_s \quad \text{and} \quad \frac{k^2 \ell_1}{\ell_1 + \ell_2} = 1 - h_s, \tag{3.14}$$

where $h_s$ is the residual brightness at secondary eclipse and $k^2 \equiv (d_2/d_1)^2$; $k^2$ should be found to be consistent with the results of equations (3.12).

    If, in addition to measuring the light curve, we can also measure the radial velocities of the two stars (that is, the star is an eclipsing spectroscopic binary), then we can determine the orbital radius in linear units as $a = (v_{\max}P/2\pi)$, where $v_{\max}$ is the observed maximum relative radial velocity of the two components. We can then re-express the stellar radii in linear units. In this favorable case, we determine both masses (§3.1.2) and radii (from the eclipses) for the stars.

    More generally, a detailed analysis of the light curve yields complete orbital elements for the system, including parameters that specify the shape and orientation of the orbit (in particular the inclination $i$) and that describe the radii and surface-brightness distributions of the two component stars [see Green (1985), Chapter 19 for methods]. In practice, the analysis is difficult because the relationships between stellar and orbital parameters and the properties of the observed light curve are complex. The quality of the results depends sensitively on the accuracy of the data. Moreover, numerous physical complications can occur. For example, the eclipses may be only partial, the orbits may be elliptical, the stars may be deformed (oblate or even ellipsoidal), the light curves may contain reflection effects (light from one star illuminating the other), or mass exchange may occur in the system. Most of these complications occur in close binaries, which are therefore often extremely troublesome to analyze [see Kopal (1959) for details]. In view of the complexity of the problem, it is perhaps not surprising that accurate results are available for only a few hundred systems (Karetnikov 1991).

### 3.2.6 Astrophysical estimates

For many astrophysically interesting stellar types, none of the direct methods just described can be applied, and yet even an approximate estimate of the radii of these stars can be of enormous importance for an understanding of stellar structure and evolution. In such cases, we can estimate radii by means of Stefan's law for black-body radiation if we know a star's luminosity (from its apparent magnitude and distance) and its effective temperature (from spectroscopic evidence) – see Blackwell & Shallis (1977). Such estimates, although not fundamental measurements, can nevertheless be quite accurate and useful.

## 3.3 Classification of stars

When a star is mentioned in a conversation, we obviously want to know at the outset what *type* of star it is. Is it luminous or faint? Is it blue or red? Is it old or young? Is it a variable star? Does it have unusual emission at, say, X-ray wavelengths or in the infrared? The answers to many of these questions can be most concisely communicated by relating the star under discussion to other, better known stars. A powerful way of establishing such relations is to group stars into classes according to readily observable properties. Then, when we tell a colleague that some star is classified such-and-such, she will know that this star resembles certain other stars, with whose properties she is already familiar. Hence, an essential first step in the study of stars is the setting up of convenient classification schemes.

Observations of a star rather straightforwardly yield two functions: the star's **spectrum**, or intensity of radiation as a function of wavelength, and its **light curve**, or intensity of radiation as a function of time. Since the great majority of stars have rather boring light curves, most stars are classified by their spectra rather than their light curves, and most of this section is devoted to spectral classification. A few important stellar classes are defined in terms of light curves, however. In §§5.1.10 and 5.2.3 we shall encounter several of these classes. And here we mention only two.

**Novae** A **classical nova** is a star that in a matter of hours brightens by up to 11 mag to an absolute magnitude in the range $-4.8 \gtrsim M_V \gtrsim -8.9$. The precursor object is usually observed to be a white dwarf. A typical nova has declined by $\sim 6$ mag six weeks after achieving peak brightness, but some novae fade much more slowly.

**Pulsars** A **pulsar** is an object that emits remarkably regular pulses of electromagnetic waves. A pulsar light curve usually has two peaks per period. The shapes of these peaks, which can be very different, vary with the frequency at which the light curve is measured, but the period at which the light curve repeats itself is the same at all frequencies, and is usually as stable as the periods of the of the best terrestrial clocks. Some pulsars have periods as short as a millisecond, while others have periods of tens of seconds.

Soon after their discovery by Hewish *et al.* (1968) it was established that pulsars are rotating **neutron stars**: the cores of dead stars that have been crushed by gravity to densities larger than those in ordinary atomic nuclei – a typical neutron star has a mass of $\sim 1.4 \, \mathcal{M}_\odot$ (Figure 5.6) and a radius of 10 km, so its mean density is an astonishing 38 million tonnes per cubic centimeter. Most of the volume of a neutron star is occupied by a fluid that consists primarily of neutrons but contains a sprinkling of protons and electrons.

The mechanism by which pulsars pulse is not fully understood, but is believed to involve a magnetic field that is anchored within the star in the same way that the Earth's magnetic field is anchored within the Earth, and rotates with the star. Two beams, or fans, of broad-band electromagnetic radiation sweep round as the star rotates, very much as two beams of light emerge from the rotating lantern of a light-house. Shklovskii (1978) gives a very readable account of the discovery and physics of neutron stars.

### 3.3.1 Classification of stellar spectra

There are two very different ways in which to set about extracting astrophysical information from complex stellar spectra. We shall categorize them as 'spectrum analysis' and 'spectral classification'. In this section, we shall deal mainly with the latter, but it is worthwhile to consider both in broad terms for the sake of orientation.

When we examine the light from a star with a spectrograph, we observe a bright continuum of radiation, upon which are superimposed absorption lines (and, occasionally, emission lines). The distribution of energy in the continuum and the profiles of the lines are determined by the physical conditions in the stellar atmosphere – its temperatures, densities, velocity fields, magnetic fields, element abundance and so on. Hence, by a suitable theoretical analysis of these features, we can derive a detailed physical picture of the structure and composition of the outer layers of stars.

The basic data required for a **spectrum analysis** are quantitative descriptions of line profiles, and descriptions of the frequency variation of the continuum. These data are obtained from **spectrophotometric** measurements. The continuum information can be obtained from color indices or from absolute energy distributions (see §2.3). The line data are obtained from high-dispersion spectra and are usually quantified by determining the **equivalent width** of each line: this is the width of a section of the continuum near the line which contains as much light as is either contributed by the line (for an emission line) or blocked by the line (for an absorption line) – see Figure 3.3.

The essence of the procedure for determining the physical conditions in the outer layers of a star is to describe the star's spectrum with a small number of numerical indices and then, by a theoretical calculation using established physical laws, to determine the physical conditions in the stellar atmosphere that are required to match these numerical values. The procedure is complicated and we shall not discuss it here [see Mihalas (1978)], although we shall use the results of such work in following chapters, particularly Chapter 5. In practice, these analyses are very time consuming and are restricted to relatively bright stars for which suitable spectra can be obtained.

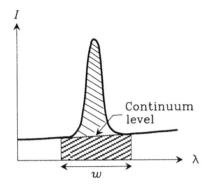

**Figure 3.3** The equivalent width of
the emission line shown here is the
width of the shaded trapezium; this
trapezium has the same area as the
diagonally hatched area.

In **spectral classification**, one simply groups together spectra that
have similar structure (morphology), without, in the first instance, seeking
to understand the physical significance of these features. In principle, one
could classify in terms of the whole spectrum, but in practice most classifications are based on only the part of the spectrum transmitted through the
Earth's atmosphere. One first chooses one or more sets of labels (dimensions) with which to categorize the spectrum. The standard system now in
use (the 'MK system' discussed below) employs two sets of labels; that is,
it is two-dimensional. For each choice of labels, one declares the name of
a standard star whose spectrum defines what is meant by that particular
classification. Successive choices of labels to which standards are assigned
establish classification 'boxes'. and the variation of the nature of the spectrum from box to box is fixed by the properties of the standards in those
boxes. Each classification box corresponds to a unique **spectral type** of the
system. When the whole system is defined, one classifies a star by finding the
standard whose spectrum it most closely matches (or interpolates between
standards, if necessary). In this way, each star is put into one of the boxes of
the system. We shall see later that similar systems are used for classifying
star clusters, galaxies, and clusters of galaxies.

If an astute choice of dimensions has been made, and if the classification process itself is done with care using well-defined criteria, then (i) a
group of stars having the same type will, in fact, be nearly identical to one
another in those physical properties that produce variations of the classification criteria (but not necessarily in all properties), and (ii) the groups with
different types will be distinguishable from one another in a significant way.
After the system has been defined and the classification carried out, one
then calibrates the system in terms of physical parameters, such as stellar
temperature luminosities, compositions, and so on. Quantitative estimates
of these parameters are derived for each spectral type by performing a detailed spectrum analysis on a typical member of that type. At that point one
can say that, if some star has a certain spectral type, then the temperature,

luminosity, or other property appropriate to that type can be assigned to it without further analysis. The spectral type thus gives a concise description of both the spectrum and the physical properties of a star.

In practice, classification is normally done by visual inspection of spectra of moderate dispersion, and can be carried out for large numbers of stars. Pioneering work in spectral classification was done in the 1860s by A. Secchi, who divided stars into four broad spectral classes. Parallel efforts were made at about the same time by W. Huggins and H.C. Vogel. The first great steps toward our present system were made at Harvard College Observatory in 1890. Under the direction of E.C. Pickering, Williamina P. Fleming published a catalog of 10,000 stars grouped into a system of spectral types denoted by the letters A, B, C, and so on. In 1888, Antonia C. Maury, without benefit of astrophysical data (which were almost nonexistent at the time), rearranged these spectral types into the order that has been used ever since, solely by studying the progression of line patterns observed in the spectra. Subsequently, Annie J. Cannon introduced decimal subdivisions of the spectral types, and, in the four years from 1911 she classified $\sim 225\,000$ stars on this system – these classifications were published as the classical *Henry Draper Catalog* [Cannon & Pickering (1918–1924), Cannon (1925–1936), Cannon & Mayall (1949) – see Appendix B].

From the observed variation of ratios of line strengths of successive ionization stages of the chemical elements, and from photometric data, this spectral sequence, O, B, A, F, G, K, M, was later recognized to be primarily a temperature sequence, listed here in order of decreasing temperature. At the cool end of the sequence, additional types R, N, and S were added to describe stars in the K-M temperature range that have markedly different compositions. Finally, with the advent of M. Saha's ionization theory in 1920 (Saha 1920–1921), quantitative analysis became possible, and in 1925 a comprehensive theoretical interpretation of the Harvard spectral sequence appeared in Cecilia Payne's book *Stellar Atmospheres* (Payne 1925).

With the work of E. Hertzsprung and H.N. Russell in the years from 1905 to 1913, it became evident that stars of a given spectral type could have vastly different luminosities (see §3.5), and this fact implied that they could have markedly different radii and atmospheric densities. The most common, fainter stars are called **dwarfs**; brighter, larger stars are called **giants**; and the brightest, largest stars are called **supergiants**. The effects of differing stellar sizes are reflected directly in changes in the spectrum, and therefore a second parameter is required to describe the spectrum completely. A start in this direction was made in the 1890s by Antonia Maury at Harvard, who added the symbols a, b, or c to some spectral types; we now know that her type c corresponds to supergiants and the others correspond to dwarfs.

**The MK system**     A fully two-dimensional system – the **MK system** – was developed by Morgan, Keenan & Kellerman (1943). In addition to a spectral class, they assigned stars to one of five **luminosity classes**, which were labeled by Roman numerals I – V. In physical terms, a star's luminosity class reflects the star's envelope size and atmospheric density and hence correlates with the star's surface-gravity. The luminosity class is closely correlated with stellar luminosity. The MK system, with subsequent revisions and extensions has become the standard classification system in use today.

The basic precepts of the MK system are:

1. It is empirical; only directly observable features of the spectrum are used to determine a star's classification.

2. It is based on homogeneous material. It uses spectra of a standard dispersion that is high enough to provide sensitive criteria but low enough to allow one to reach faint stars, and hence ones at large distances in the Galaxy.

3. It is defined by standards. Thus the classification system is autonomous, in the sense that it remains unchanged even when the interpretation of the types in terms of physical conditions in the stars changes, as models of stellar structure are refined. Furthermore, observers using different spectrographs and dispersions can classify on the same system simply by reobserving the standard stars with their own particular equipment.

The spectral types of the MK system are essentially those of the Harvard sequence, and some of the principal spectral features characterizing each of these types are listed in Table 3.1. The luminosity classes, and the stars to which they pertain, are listed in Table 3.2. The characteristics mentioned in these lists are only illustrative; the system is defined by standard stars. The complete spectral type is specified by both the spectral type and the luminosity class of a star as determined by comparison with the standards. In §3.5 we give detailed tables of the stellar physical properties (for example, temperatures and luminosities) that have been associated with MK spectral class through astrophysical calibrations.

Spectral types are subdivided into decimal subtypes, running from 0 at the hot end through 9 at the cool end: B0, B1, B2, ..., B9; A0, A1, A2, ..., A9; F0, F1, F2, ..., F9, and so on. The luminosity classes are usually not subdivided except for supergiants (Table 3.2). Examples of spectral types are: Sun (G2V), $\varepsilon$ Ori (B0Ia), $\alpha$ Lyr (A0V), $\alpha$ Tau (K5III). Stars hotter than the Sun (types O, B, A, F) are commonly called **early types**, and solar-type and cooler stars (types G, K, M) are called **late** types.[1] Although it is explicitly two-dimensional, the MK system implies the existence of, and the need for, formal consideration of at least one more dimension in localized regions of the system in order to describe, say, **weak-lined stars** (for example, the 'subdwarf' HD 140283 or the variable star RR Lyrae) or peculiar stars (see the following discussion).

---

[1] These designations are archaic remnants of an obsolete scheme of stellar evolution. Although devoid of physical significance, they are universally used by astronomers.

**Table 3.1**   Principal characteristics of spectral types

| Spectral type | Spectral features |
|:---:|:---|
| O | He II lines visible; lines from highly ionized species, for example, C III, N III, O III, Si IV; H lines relatively weak; strong UV continuum |
| B | He I lines strong, attain maximum at B2; He II lines absent; H lines stronger; lower-excitation ions, for example, CII, O II, Si III |
| A | H lines attain maximum strength at A0 and decrease towards later types; Mg II, Si II strong; Ca II weak and increasing in strength |
| F | H weaker, Ca II stronger; lines of neutral atoms and first ionization states of metals appear prominently |
| G | Solar-type spectra; Ca II lines extremely strong; neutral metals prominent, ions weaker; G band (CH) strong; H lines weakening |
| K | Neutral metallic lines dominate; H quite weak; molecular bands (CH, CN) developing; continuum weak in blue |
| M | Strong molecular bands, particularly TiO; some neutral lines, for example, CA I, quite strong; red continuum |
| C | Carbon stars; strong bands of carbon compounds $C_2$, CN, CO; TiO absent; temperatures in range types K and M |
| S | Heavy-element stars; bands of ZrO, YO, LaO; neutral atoms strong as in types K and M; overlaps these types in temperature range |

**Table 3.2**   MK luminosity-class designations

| | |
|:---|:---|
| Ia-0 | Most extreme supergiants |
| Ia | Luminous supergiants |
| Iab | Moderate supergiants |
| Ib | Less-luminous supergiants |
| II | Bright giants |
| III | Normal giants |
| IV | Subgiants |
| V | Dwarfs |

**Table 3.3** Distribution of stellar types in HD catalog ($V \leq 8.5$)

| Spectral type | Percent |
|:---:|:---:|
| O | 1 |
| B | 10 |
| A | 22 |
| F | 19 |
| G | 14 |
| K | 31 |
| M | 3 |

**Table 3.4** Distribution of MK classes of apparently bright stars (%)

| Spectral type | Luminosity class | | | | |
|:---:|:---:|:---:|:---:|:---:|:---:|
| | V | IV | III | II | I |
| O, B | 10 | 3 | 6 | 2 | 3 |
| A, F | 14 | 3 | 5 | 1 | 4 |
| G, K, M | 1 | 4 | 25 | 6 | 4 |

More than 90% of all stars can be classified with standard MK spectral classes. The relative numbers of stars in different spectral types in the HD Catalog are listed in Table 3.3, and a similar distribution for MK classes is given in Table 3.4. Both of these tables refer to apparently bright stars, and they do not reflect true space densities, which are discussed in §3.6.

Additional letters may be appended to spectral types to denote special characteristics; for example, p(peculiar), e(emission lines), f(He II and N III emission in O stars), and n(broad lines). Some stars have special notations. Some examples are **peculiar A stars** (Ap), which show strong lines of certain elements (Si, Mn, Cr, Sr, Eu) and have strong magnetic fields, and **metallic-line A stars** (Am), which show abnormally strong metal lines[2] and weak Ca II. The prefix w or D is often used to denote **white dwarfs** (wA, wG, DA, and so on), and the prefix sd denotes **subdwarfs**, extremely metal-poor dwarfs (for example, sdG). In some work, particularly at low dispersion, it is not possible to derive accurate MK luminosity classes, although it is possible nevertheless to distinguish among dwarfs, giants, and supergiants; this is of interest because we can then assign intrinsic brightnesses to the stars. In such cases, the prefixes d, g, and c are often used to denote these three groups (for example, dK, gK, cA).

---

[2] Astronomers often refer to all elements heavier than lithium as **metals**. In particular, carbon and oxygen are referred to as metals.

Additional spectral types are used for certain unusual stars: type C (which replaces Harvard types R and N) for carbon stars, whose spectra show strong bands of $C_2$ and CN; S for stars whose spectra lines of heavy elements such as ZrO are more prominent than the usual TiO lines; and WC and WN (or sometimes WR) for Wolf-Rayet stars.

Certain stars do not fit into the MK classification but nevertheless fall into classes. There is no systematic way of naming these classes; sometimes they are named after a prototypical star, sometimes after the authors of a catalog which identified early examples. Table 3.5 lists some of the more important such classes.

The standard stars that define the MK system are given in Morgan & Keenan (1973), Johnson & Morgan (1953), Abt *et al.* (1969), Morgan, Abt & Tapscott (1978), and Morgan & Keenan (1973). A fairly large number of stars have been classified on this system, and extensive lists can be found in the literature (see Appendix B). A reworking of the HD Catalog has for many years been under way at the University of Michigan Observatory, using high-quality objective prism plates (Houk 1975–). This survey will yield almost a full two-dimensional classification, very nearly on the MK system, for most stars in the HD catalog. When completed, it will have enormous value for studies of galactic structure.

# 3.4 Physical interpretation of stellar spectra

The structure of a stellar atmosphere, and therefore the spectral energy distribution emitted by a star, is determined essentially by three atmospheric parameters:

(1) The **effective temperature,**

$$T_{\text{eff}} \equiv \left( \frac{L}{4\pi R^2 \sigma} \right)^{1/4}, \qquad (3.15)$$

where $\sigma$ is the Stefan-Boltzmann constant, $L$ the total luminosity and $R$ the radius of the star. $T_{\text{eff}}$ specifies the amount of energy passing outward through unit area of the surface. Since the energy distributions of stars are very roughly Planckian, $T_{\text{eff}}$ is a measure of the temperature of the gas that forms the bulk of the stellar atmosphere. Table 3.7 gives values of $T_{\text{eff}}$ as a function of MK spectral class.

(2) The **surface gravity**

$$g \equiv \frac{G\mathcal{M}}{R^2}, \qquad (3.16)$$

where $G$ is the gravitational constant and $\mathcal{M}$ is the mass of the star. The surface gravity fixes the pressure gradient in the atmosphere and largely determines the densities at which spectral lines are formed.

**Table 3.5** Classes of unusual stars

| Class | Defining characteristic | Remarks |
|---|---|---|
| Cepheids | Massive, luminous, variable stars with characteristic asymmetric light curves, periods 1–10 days and F-type spectra. | Sometimes called $\delta$ Cepheids or classical cepheids to distinguish them from W Virginis stars (see below). Their periods, luminosities and colors satisfy a period-luminosity-color relationship that plays an important rôle in determining the cosmic distance scale see §§5.1.10 and 7.3.1. Their initial masses exceed $\sim 5\,M_\odot$ and their ages are $\lesssim 0.1$ Gyr. |
| Miras | Very red variable stars with periods 80–1000d and amplitudes up to 7 mag. Most show emission lines. | Stars near their deaths. Satisfy period-luminosity relation; typical $M_{bol} = -5$. Frequently surrounded by shells of gas detectable in the infrared and through mm emission by CO. |
| OH/IR stars | Stars with strong line emission at 1612 MHz from masing OH radicals. Also luminous mid-infrared emitters. | Luminous end of population of Miras. OH/IR emission comes from gas that has been ejected by the star. The ejection velocity of the gas can be measured and used to estimate the age of the star. |
| RR Lyrae | Variable stars with characteristic asymmetric light curves showing periods $\lesssim 1$ d, peak luminosity $M_V \sim 0$ and A-type spectra. | Stars of roughly solar mass. Occur in globular clusters and in other low-metallicity systems. |
| T-Tauri stars | Irregular variable stars with Balmer lines and Ca HK in emission. Frequently show Li I absorption. Spectral types F–M. | Pre-main-sequence stars (§5.1.8), perhaps with accretion disks. |
| Wolf-Rayet stars | Luminous O or B stars showing strong broad emission lines with P-Cygni profiles (Figure 3.4). Also lines of highly-ionized species. | Usually very massive, short-lived stars. WN stars show strong emission in lines of nitrogen, while carbon lines are prominent in the spectra of WC stars. |

**Table 3.5** (*continued*)

| Class | Defining characteristic | Remarks |
|---|---|---|
| W Virginis stars | Old variable stars with periods 7–60 d | Also called Type II Cepheids; they are $\gtrsim 1.5$ mag fainter than Cepheids at a given period. |
| Dwarf novae | Short-period binary systems subject to occasional eruptions that brighten the system by $\lesssim 6$ mag in $< 5$ d. SS Cyg is the prototype. Quiescent G-type spectra. | Systems contain a white dwarf and a late-type star. Erruptions caused by mass being transferred on to the white dwarf. |
| Novae | Star whose brightness suddenly increases by 7 – 16 mag | Close binary systems containing a cool red giant and a hot, less-massive companion that accretes material, which then feeds explosive nuclear burning. Novae expel gas at $\lesssim 3\,000\,\mathrm{km\,s^{-1}}$. |
| Supernovae | Violently exploding stars. Type I supernovae reach $M_V \sim -19$ and expel metal-rich gas at $\sim 10\,000\,\mathrm{km\,s^{-1}}$. Type II supernovae reach $M_V \sim -17$ and expel hydrogen-rich gas at $\sim 5\,000 - 20\,000\,\mathrm{km\,s^{-1}}$. | The most luminous of all stars, they are potentially important for determining the cosmic distance scale. The principal suppliers of heavy elements ($\sim 0.6\,\mathcal{M}_\odot$ of Fe per Type I supernova) and energy ($\sim 10^{42}$ J per supernova) to the interstellar medium. |
| Low-mass X-ray binaries | X-ray sources with hard spectra and low luminosities | Short-period systems consisting of a compact object (usually a neutron star) and a conventional low-mass star that fills its Roche lobe. The x-ray emission is powered by accretion on to the compact object of material from the conventional star. |
| High-mass X-ray binaries | X-ray sources with soft spectra and high luminosities | Longer-period systems consisting of a compact object (usually a neutron star) and an early-type star. The x-ray emission is powered by accretion on to the compact object of material from the star. |

**Table 3.6** $V - K$ versus $T_{\text{eff}}$ for dwarfs and late-type giants

| Dwarfs | | Giants | | |
|---|---|---|---|---|
| $V - K$ | $T_{\text{eff}}/\text{K}$ | Sp. type | $V - K$ | $T_{\text{eff}}/\text{K}$ |
| −0.97 | 47000 | G8 III | 2.20 | 4930 |
| −0.93 | 38000 | K0 III | 2.30 | 4790 |
| −0.85 | 30500 | K1 III | 2.48 | 4610 |
| −0.67 | 23000 | K2 III | 2.68 | 4450 |
| −0.57 | 18500 | K3 III | 2.96 | 4270 |
| −0.43 | 15000 | K4 III | 3.26 | 4095 |
| −0.30 | 13000 | K5 III | 3.52 | 3980 |
| −0.25 | 12000 | M0 III | 3.78 | 3895 |
| 0.00 | 95000 | M1 III | 4.02 | 3810 |
| 0.35 | 8300 | M2 III | 4.30 | 3730 |
| 0.79 | 7300 | M3 III | 4.64 | 3640 |
| 1.01 | 6600 | M4 III | 5.10 | 3560 |
| 1.22 | 5900 | M5 III | 5.96 | 3420 |
| 1.53 | 5600 | M6 III | 6.84 | 3250 |
| 1.75 | 5100 | | | |
| 2.75 | 4200 | | | |
| 3.25 | 3700 | | | |
| 4.50 | 3000 | | | |
| 5.25 | 2500 | | | |

SOURCE: Dwarfs: data published in Peletier (1989); giants: data published in Ridgway *et al.* (1980)

(3) The chemical composition of the atmosphere, which determines the frequency variation of the opacity of the stellar material, and the relative strengths of spectral lines of the different atomic species present. We shall see that, to a first approximation, the chemical composition of a star in the solar-neighborhood may be characterized by quoting for it the abundance relative to hydrogen of any heavy element, such as Fe, because stars rich in Fe tend to be rich in other elements heavier than helium. On account of this phenomenon, one often speaks of the **metallicity** of a star as a short-hand for the complete chemical composition.

These atmospheric parameters play key rôles in the theory of stellar structure summarized in Chapter 5, so their empirical determination is of considerable interest.

The relationship between these parameters and the structure of a stellar spectrum is complex and can only be determined by large model-atmosphere programs. In practice, trial values of the parameters are adopted and used to calculate a theoretical spectrum, and then these values are adjusted to achieve the optimum fit between theory and observation. Different properties

**Table 3.7**   The effective-temperature and bolometric-correction scales

| Spectral type | V | | III | | I | |
|---|---|---|---|---|---|---|
| | $T_{eff}/K$ | $BC_V$ | $T_{eff}/K$ | $BC_V$ | $T_{eff}/K$ | $BC_V$ |
| O3 | 52 500 | −4.75 | 50 000 | −4.58 | 47 300 | −4.41 |
| O5 | 44 500 | −4.40 | 42 500 | −4.05 | 40 300 | −3.87 |
| O7 | 38 000 | −3.68 | 37 000 | −3.58 | 35 700 | −3.48 |
| O9 | 33 000 | −3.33 | 32 000 | −3.13 | 32 600 | −3.18 |
| B0 | 30 000 | −3.16 | 29 000 | −2.88 | 26 500 | −2.49 |
| B2 | 22 000 | −2.35 | 20 300 | −2.02 | 18 500 | −1.58 |
| B3 | 18 700 | −1.94 | 17 100 | −1.60 | 16 200 | −1.26 |
| B5 | 15 400 | −1.46 | 15 000 | −1.30 | 13 600 | −0.95 |
| B7 | 13 000 | −1.02 | 13 200 | −0.97 | 12 200 | −0.78 |
| B8 | 11 900 | −0.80 | 12 400 | −0.82 | 11 200 | −0.66 |
| A0 | 9 520 | −0.30 | 10 100 | −0.42 | 9 730 | −0.41 |
| A5 | 8 200 | −0.15 | 8 100 | −0.14 | 8 510 | −0.13 |
| F0 | 7 200 | −0.09 | 7 150 | −0.11 | 7 700 | −0.01 |
| F5 | 6 440 | −0.14 | 6 470 | −0.14 | 6 900 | −0.03 |
| G0 | 6 030 | −0.18 | 5 850 | −0.20 | 5 550 | −0.15 |
| G2 | 5 860 | −0.20 | 5 450 | −0.27 | 5 200 | −0.21 |
| G5 | 5 770 | −0.21 | 5 150 | −0.34 | 4 850 | −0.33 |
| K0 | 5 250 | −0.31 | 4 750 | −0.50 | 4 420 | −0.50 |
| K5 | 4 350 | −0.72 | 3 950 | −1.02 | 3 850 | −1.01 |
| M0 | 3 850 | −1.28 | 3 800 | −1.25 | 3 650 | −1.29 |
| M5 | 3 240 | −2.73 | 3 330 | −2.48 | 2 800 | −3.47 |
| M8 | 2 640 | −4.1 | | | | |

SOURCE: From data published in Schmidt-Kaler (1982)

of a spectrum are sensitive to different atmospheric parameters. For example, the Balmer lines are particularly pressure sensitive and provide insight into the value of $g$. A good first estimate of $T_{eff}$ is obtainable from a color index; $V - K$ is the favored index for all but the hottest stars on account of its long baseline in frequency and its avoidance of the blue-ultraviolet part of the spectrum that contains many spectral lines and is therefore sensitive to metallicity. Table 3.6 lists the temperatures of dwarfs and giants as a function of $V - K$. Clearly, the relative abundances of elements is directly reflected in the relative strengths of their spectral lines.

Unfortunately, a spectrum at good dispersion with high signal-to-noise is required if reliable results are to be obtained by fitting observed and theoretical spectra in this way. Such spectra are only available for a small number of apparently bright stars. The best available spectra of the vast majority of astronomical objects are full of features that are pure noise, and a fit of a theoretical spectrum to such a noisy spectrum would be heavily influenced

**Table 3.8** Spectral indices defined by Faber *et al.* (1985)

| Index | Feature | C. Band (nm) | S. Bands (nm) |
|---|---|---|---|
| G | CH | 428.325 − 431.700 | 426.825 − 428.325 |
| | | | 432.075 − 433.575 |
| Mg $b$ | Mg $b$ | 516.200 − 519.325 | 514.450 − 516.200 |
| | | | 519.325 − 520.700 |
| $Fe_1$ | $Fe^0$, $Ca^0$ | 524.800 − 528.675 | 523.550 − 524.925 |
| | | | 528.800 − 531.925 |
| $Fe_2$ | $Fe^0$, $Cr^0$, | 531.475 − 535.350 | 530.725 − 531.725 |
| | $Ca^0$, $Ti^+$ | | 535.600 − 536.475 |
| Na | Na D | 587.920 − 591.050 | 586.300 − 587.675 |
| | | | 592.450 − 594.925 |
| $H\beta$ | $H\beta$ | 484.950 − 487.700 | 482.950 − 484.825 |
| | | | 487.825 − 489.200 |
| CN | CN | 414.400 − 417.775 | 408.200 − 411.825 |
| | | | 424.600 − 428.475 |
| $Mg_1$ | MgH | 507.100 − 513.475 | 489.700 − 495.825 |
| | | | 530.300 − 536.675 |
| $Mg_2$ | MgH, Mg$b$ | 515.600 − 519.750 | 489.700 − 495.825 |
| | | | 530.300 − 536.675 |
| $TiO_1$ | TiO | 593.900 − 599.525 | 581.900 − 585.025 |
| | | | 604.100 − 610.475 |
| $TiO_2$ | TiO | 619.200 − 627.325 | 606.900 − 614.275 |
| | | | 637.500 − 641.625 |

NOTES: Indices in the first group have the units of equivalent
widths – see Figure 3.3. Indices in the second group are measured in magnitudes.

by chance fits of spectral features to noise features. Since the results of such
a fit are of little value, a less direct procedure is adopted to the measurement
of the atmospheric parameters of faint objects.

The first step is to discover by experimenting with theoretical models
which relatively strong lines and line-ratios are sensitive to the atmospheric
parameters. Next, one designs an index that quantifies each such interesting spectral feature. In designing this index, one aims to obtain a number
which can be reliably estimated from a low-dispersion, high signal-to-noise
spectrum, but which is nevertheless sensitive to the structure of the underlying spectrum. Table 3.8 defines the widely employed **spectral indices** of
Faber *et al.* (1985), while several other important indices are defined in Rose
(1994) and Jones & Worthey (1995). Each index is defined by a central band
of width $\Delta_0$ and two side bands. The 'side-band level' $I_s$ is defined to be the
mean intensity over the two side bands, while the intensity of the central feature $I_c$ is the mean intensity in the central band. The values of the first group
of indices in Table 3.8 are reported as equivalent widths: $w = (1 - I_c/I_s)\Delta_0$

(Figure 3.3). The values of indices in the second group are reported in magnitudes: $-2.5 \log(1 - I_c/I_s)$. All of these indices are moderately sensitive to the temperature of the emitting star and the first step in the interpretation of a set of indices is to estimate the value that each index would have if the star had some standard $V - K$ color rather than the measured value of $V - K$. Also, the measured values of a star's indices depend slightly on the spectral resolution employed and should first be corrected to some standard spectral resolution. Some of the corrected indices are primarily sensitive to the star's surface gravity $g$, while others are sensitive to its metallicity. The primary metallicity indicator is

$$\langle \text{Fe} \rangle \equiv \tfrac{1}{2}(\text{Fe}_1 + \text{Fe}_2), \tag{3.17}$$

where the values on the right are corrected ones.

Next one determines the values of each index for a sample of apparently bright stars. A good-quality spectrum should be available for each of these stars so that its atmospheric parameters can be determined from a detailed spectral fit as described above, but the indices are measured for these stars using the same dispersions and equipment as are used to observe the faint program objects. This step enables one to assert that, say, the metallicity of this faint star is similar to that of this bright and well-studied star, because the two stars have similar values of all metallicity-sensitive spectral indices.

The final step is to estimate the atmospheric parameters of each of the bright reference stars from a detailed fit of a high-quality spectrum with a theoretical spectrum. As more complete and reliable atomic-physics data become available and model-atmosphere computer codes become more sophisticated, the best available values of the atmospheric parameters of bright stars are constantly being revised. The scheme we have just outlined, in which faint objects are linked to bright stars by means of spectral indices, makes it possible to update the atmospheric parameters of faint stars along with those of bright stars without having to re-observe them.

Hundreds of spectroscopic analyses of stellar spectra have been performed. The results are most reliable for stars of types A–K; the optical spectra of O and B stars have few strong metal lines because in their hot atmospheres the heavy elements are highly ionized, while the spectra of M stars show strong absorption by molecules, which is difficult to model theoretically. To give some idea of the accuracy that can be achieved, a spectrum of a G-dwarf at $\sim 2$ nm resolution typically yields uncertainties in $T_{\text{eff}}$, $g$ and metallicity of 20 K, 0.03 dex and 0.02, respectively;[3] if only Strömgren photometry is available, the uncertainties in each quantity are larger by a factor $\sim 2.5$ (Cayrel 1992).

In the Sun the abundances by *number* of various elements relative to hydrogen are roughly as follows: helium, 0.10; carbon, $3 \times 10^{-4}$; nitrogen,

---

[3] The notation dex indicates an increment in the logarithm to base 10.

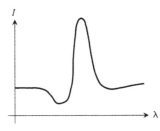

**Figure 3.4** A P-Cygni line profile is characterized by an absorption trough on the blue wing of an emission line.

$1 \times 10^{-4}$; oxygen, $6 \times 10^{-4}$; neon, $1 \times 10^{-4}$; magnesium, $3 \times 10^{-5}$; silicon, $3 \times 10^{-5}$; and iron, $4 \times 10^{-5}$. Element abundances in other stars are generally reported relative to these abundances. For example, the "iron abundance," denoted [Fe/H], is defined to be

$$[\mathrm{Fe/H}] \equiv \log_{10}\left(\frac{n(\mathrm{Fe})}{n(\mathrm{H})}\right)_{\mathrm{star}} - \log_{10}\left(\frac{n(\mathrm{Fe})}{n(\mathrm{H})}\right)_{\odot}. \qquad (3.18)$$

Similarly, the "relative abundance" of an element other than iron is often expressed in terms of a symbol such as [O/Fe], which is defined by

$$[\mathrm{O/Fe}] \equiv \log_{10}\left(\frac{n(\mathrm{O})}{n(\mathrm{Fe})}\right)_{\mathrm{star}} - \log_{10}\left(\frac{n(\mathrm{O})}{n(\mathrm{Fe})}\right)_{\odot}. \qquad (3.19)$$

[Fe/H] varies from as low as $\lesssim -3$ to $> 0$. Ratios such as [O/Fe] cover a much smaller range: that is, stars poor in Fe are generally also poor in other metals. Consequently it is useful to define parameters $(X, Y, Z)$, which give the fractional abundances by *weight* of hydrogen, helium, and everything else, respectively. For the solar mixture, $X = 0.70$, $Y = 0.28$, and $Z = 0.02$.

A small number of stars show anomalous line strengths that may be indicative of unusual atmospheric abundances. Examples include Ap and AM stars, carbon stars, S stars, Wolf-Rayet stars (§3.3). Some of these anomalous line strengths have been theoretically traced to credible abundance anomalies. Others remain puzzling and it is unclear whether they are simply due to anomalous abundances or will be eventually explained by more sophisticated model atmospheres (which include, for example, the effects of magnetic fields and stellar rotation) in terms of normal abundances.

The shape of a spectral line, in addition to its strength, can convey important astrophysical information. A case in point is that of lines that have a **P-Cygni profile**. P Cygni is a 5th-magnitude B star whose spectrum contains strong emission lines that go over into absorption at their blue edges (Figure 3.4). The physical interpretation of this phenomenon is that along the line of sight to the star, there is matter flowing away from the star and towards us. This material absorbs radiation in the same spectral line that the star emits, but this absorption is blueshifted relative to the star. An analogous phenomenon is observed in many stars, and is usually interpreted as evidence that the star is losing mass in a strong stellar wind.

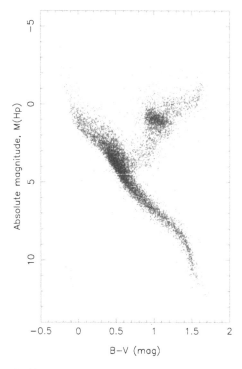

**Figure 3.5** The CM-diagram for 10 793 stars with good Hipparcos parallaxes. The great majority of stars fall along the MS that runs diagonally from bottom right to top left. The subgiant, red giant, and white-dwarf sequences are also apparent, as is the red clump. The MS and WD stars were selected to have parallaxes with errors smaller than 10%, while the giants were chosen to have parallaxes in error by < 20%. [After Perryman *et al.* (1995) courtesy of M. Perryman]

## 3.5 Color-magnitude diagrams

Unquestionably the most important correlation among stellar properties discovered to date is the **Hertzsprung–Russell diagram (HR-diagram)**, which was developed independently by the Dane E. Hertzsprung in 1911 and the American H. N. Russell in 1913. The HR-diagram has been of profound importance to the development of our understanding of stellar evolution; it provides both one of the most stringent tests of the theory of stellar evolution, and one of the most incisive tools for exploring the history of the Galaxy as a whole.

In its original form, the HR-diagram was a plot of absolute visual magnitude versus spectral type, but variants are now more commonly used in which the discrete spectral type is replaced by a continuous coordinate. Observationally, the most useful form is the **color-magnitude (CM) diagram**, which is a plot of a color versus a magnitude, either absolute (for stars of known distances) or apparent (if the stars are known to be all at the same distance). Figure 3.5, in which each point represents a nearby field star of known distance, is an example of this genre. Insofar as stars of a given MK spectral class have a unique color, a CM-diagram is equivalent to the classical HR-diagram.

From a theoretical point of view, the most convenient form of the HR-diagram is a plot of $T_{\rm eff}$ versus $\log(L/L_\odot)$; this is usually called the **theo-**

**retical CM-diagram**. Insofar as all stars of a given MK spectral class have the same luminosity and effective temperature, the theoretical CM-diagram is equivalent to both a classical HR-diagram and a CM-diagram. However, to make a quantitative correspondence between the theoretical CM-diagram and a CM-diagram, we must know the relations $T_{\text{eff}}$(MK spectral class) and BC(MK spectral class) with good accuracy. Establishing this mapping between the two diagrams remains a significant astrophysical problem (§5.1.1).

### 3.5.1 Observed CM-diagrams

Figure 3.5 shows the CM-diagram for nearly 11 000 stars with Hipparcos parallaxes. Clearly the stars are not scattered at random in this figure, but fall into several distinctive groups or **sequences**. The vast majority of all stars lie along the **main sequence (MS)**, which stretches from luminous, hot, blue, O stars to faint, cool, red, M stars; these stars are the dwarfs (MK luminosity class V). The **subgiant branch (SGB)** joins the MS at $B - V \simeq 0.7$ and $M_{\text{Hp}} \simeq 4$ and extends horizontally to $B - V \simeq 1$, where the populated region turns sharply upwards into the **red-giant branch (RGB)**. The RGB stretches from $(B - V) \simeq 1$, at a luminosity about thirty times (3.7 mag) brighter than MS stars of the same type, toward cooler and brighter stars. These stars are the red giants; they correspond to MK luminosity class III. The strong enhancement in the density of stars near $B - V = 1.1$ and $M_{\text{Hp}} = 1$ is known as the **red clump**. Centered on $B - V \simeq 0$, $M_{\text{Hp}} \simeq 12$ we find a few faint, hot stars known as **white dwarfs**. The region between the MS and the RGB at $M_{\text{Hp}} \sim 1$ that is almost devoid of stars is known as the **Hertzsprung gap**.

The difficult step in the construction of Figure 3.5 is the determination of the distance to each star that is plotted, in order to calculate the star's absolute magnitude $M_{\text{Hp}}$ from its apparent magnitude. Since even Hipparcos data only allow us to determine the distances to relatively nearby stars, it should not be surprising if some of the more unusual classes of star are unrepresented in this plot.

Our knowledge of the CM-diagram can be rounded out by studying star clusters. As we shall see in Chapter 6, the Galaxy contains many such agglomerations of stars, which include a rich variety of stellar types. In general, we will not know the distance to a cluster, so we cannot plot its members on an absolute-magnitude scale as a CM diagram such as Figure 3.5 requires. However, most clusters are sufficiently small relative to our distance from them that we can assume that all a cluster's stars lie at the same distance. Thus, the difference between absolute and apparent magnitude will be the same for every star, so the structures that we have traced in Figure 3.5 should be apparent in a plot of apparent magnitude versus color for cluster members. In fact, as we show in Chapter 6, features in the CM-diagrams of clusters tend to be even more sharply defined than those in Figure 3.5. The absence of errors due to uncertainties in the distances of

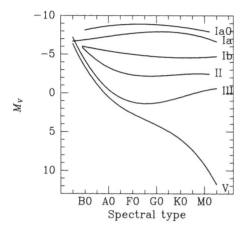

**Figure 3.6** Luminosity as a function of MK spectral class. The curves correspond to the luminosity classes marked at right.

stars reduces the scatter in the stellar sequences of a cluster compared to that in the sequences of nearby stars. Further, it turns out that stars in clusters have very homogeneous properties, which also helps squeeze cluster stars into narrowly defined regions of the CM-diagram.

### 3.5.2 Luminosity and color as functions of spectral class

As its name implies, the MK luminosity class that a star belongs to is an indication of the star's luminosity. The range of luminosities that are associated with a given luminosity class varies with spectral type, so the complete MK spectral classification of a star is required before one can predict the star's luminosity from its spectrum. Figure 3.6 gives an approximate indication of the combined dependence of luminosity on spectral type and luminosity class; at a given spectral type, stars of each luminosity class will mostly lie within a magnitude or two of the appropriate curve in the figure.

Since MK spectral type is tightly correlated with $B - V$ color, and MK luminosity class is connected with luminosity, the complete MK classification of a star gives one a fair idea of where the star lies in the CM diagram. In so far as stars are confined to narrow sequences in a CM diagram such as Figure 3.5, even a rough prediction of a star's location in the CM diagram enables us to predict its luminosity rather accurately. For example, Figure 3.5 indicates that a MS star with $B - V = 0$ is likely to have $M_{\rm Hp} \simeq 1.5$, and in §3.6.4 we shall see that the error on this estimate is smaller than 0.5 mag. Other cases may be more problematical. A giant star with $B-V = 1$ could lie on the SGB at $M_{\rm Hp} \simeq 3$, or in the red clump at $M_{\rm Hp} \simeq 1.$[4] It may or may not be possible to distinguish these cases on the basis of additional information, such as the star's MK luminosity class (III or II), its value of a gravity-sensitive spectral index (§3.4), or its apparent magnitude and likely distance.

---

[4] See Table 2.1 for the definition of Hipparcos magnitudes, $M_{\rm HP}$.

In other cases an approximate location in the CM diagram does not enable us to refine our original estimate of the star's luminosity because the star in question is not expected to lie on a narrow sequence in the CM diagram. Consider, for example, a supergiant (MK luminosity class I). Figure 3.6 shows that supergiants occur over the whole of the upper part of the CM diagram, so we cannot argue that a particular object is likely to lie on some narrow sequence in the CM diagram, and cannot improve on the estimate of its luminosity that follows from its MK spectral class. Hence, the accuracy with which luminosity can be inferred from MK class varies considerably with location in the CM diagram. Nonetheless, relations between absolute magnitude and MK class play an important part in astronomical research. Figure 3.6 gives approximate relations for most classes. Tables 3.9 to 3.11 repeat some of these data in tabular form, as well as specifying relations between MK spectral type and color for MK luminosity classes I, III and V. It should be borne in mind that Figure 3.6 and Tables 3.9 to 3.11, like Figure 3.5, refer to stars of approximately solar metallicity.

The data used in this section rely on several types of observations. In §3.6 we shall find that low-luminosity stars are relatively numerous. Consequently, spectral types fainter than $M_V \simeq 1$ usually have several nearby representatives for which accurate trigonometric parallaxes, and hence distances and luminosities, have been determined. For spectral types brighter than $M_V \simeq 1$, there is often no representative with a good trigonometric parallax. Determinations of the luminosities of such spectral types rely heavily on secular and statistical distances (§2.2). Clusters also play an important rôle in the determination of the luminosities of certain spectral-types because they contain both stars whose luminosities can be calibrated locally, for example from trigonometric parallaxes, with rarer stars that cannot be studied near the Sun. Since all the stars in a given cluster usually have very similar distances, knowledge of the luminosity of any star in the cluster in principle yields a good estimate of the luminosity of any other cluster star (Chapter 6). Moreover, in §2.2.2 we saw that there is an absolute method of determining the distances to, and hence luminosities of, stars in nearby clusters.

The main techniques used for the calibration of various stellar types can be summarized as follows:

1. supergiants: clusters;
2. O–A stars: clusters, secular and statistical parallaxes;
3. F–M dwarfs: trigonometric parallaxes, moving cluster method;
4. F–M giants: moving cluster method, clusters, secular and statistical parallaxes;
5. White dwarfs: trigonometric parallaxes, binaries, clusters.

Special calibration techniques may also be helpful in specific cases. For example, absolute magnitudes can be estimated for several supergiants which are components of noninteracting binaries containing another star of known

**Table 3.9**   The zero-age main sequence (ZAMS)

| $(B-V)_0$ | $(U-B)_0$ | $M_V$ | $(B-V)_0$ | $(U-B)_0$ | $M_V$ |
|-----------|-----------|-------|-----------|-----------|-------|
| −0.30 | −1.08 | −3.25 | 0.30 | 0.03 | 2.8 |
| −0.25 | −0.90 | −2.1 | 0.40 | −0.01 | 3.4 |
| −0.20 | −0.69 | −1.1 | 0.50 | 0.00 | 4.1 |
| −0.15 | −0.5 | −0.2 | 0.60 | 0.08 | 4.7 |
| −0.10 | −0.30 | 0.6 | 0.70 | 0.23 | 5.2 |
| −0.05 | −0.10 | 1.1 | 0.80 | 0.42 | 5.8 |
| 0.00 | 0.01 | 1.5 | 0.90 | 0.63 | 6.3 |
| 0.05 | 0.05 | 1.7 | 1.00 | 0.86 | 6.7 |
| 0.10 | 0.08 | 1.9 | 1.10 | 1.03 | 7.1 |
| 0.20 | 0.10 | 2.4 | 1.20 | 1.13 | 7.5 |
|  |  |  | 1.30 | 1.20 | 8.0 |

SOURCE: From data published in Schmidt-Kaler (1982)

absolute magnitude (for example, a dwarf), by taking the difference in apparent magnitudes of the two components to equal the difference in their absolute magnitudes. Implicit in this method is the assumption that both stars in the system are normal and unaffected by the presence of the companion.

The stars on the MS in Figure 3.5 do not all have the same age, and in Chapter 5 we shall see that in consequence they have a significant spread in absolute magnitude at a given color. The stars of a hypothetical population of zero age would form a narrower MS, the so-called **zero-age main sequence** or **ZAMS**. Table 3.9 lists $U - B$ and $M_V$ as a function of $B - V$ along the ZAMS. Compiling this table would be straightforward if observations of a nearby, populous and very young cluster were available. Sadly, no such cluster exists. Hence one has to make do with clusters of non-negligible age. In §6.2.1 we shall see that as a cluster ages, the most luminous stars move away from the ZAMS first, with the result that the older a cluster is, the fainter is the magnitude below which stars still lie near the ZAMS. Consequently, the standard strategy for determining the ZAMS has been to use nearby but old clusters at faint magnitudes, and more distant and younger clusters at bright magnitudes. Hence, data are used only for stars which lie near the ZAMS, and for such stars stellar-evolution theory (Chapter 5) can predict accurately by how much they lie above the ZAMS. The relative distances to the clusters employed is determined by requiring that between them the clusters define a single ZAMS. Recently, data from the Hipparcos satellite have suggested that this crucial assumption may be in error (Mermilliod *et al.* 1997) but it is too early to assess the implications of this finding.

Table 3.10 gives values of $M_V$ and several colors for dwarfs and for luminosity-class III giants where these lie on the RGB. As we saw above,

**Table 3.10**  Luminosities and colors of dwarfs and luminosity-class III giants

| MK class | $M_V$ | $(U-B)$ | $(B-V)$ | $(V-J)$ | $(J-H)$ | $(H-K)$ |
|---|---|---|---|---|---|---|
| | | | Dwarfs (luminosity class V) | | | |
| O3 | −6.0 | −1.15 | −0.33 | −0.73 | −0.19 | −0.05 |
| O5 | −5.7 | −1.15 | −0.33 | −0.73 | −0.19 | −0.05 |
| O8 | −4.9 | −1.14 | −0.32 | −0.72 | −0.17 | −0.05 |
| B0 | −4.0 | −1.08 | −0.30 | −0.68 | −0.12 | −0.05 |
| B3 | −1.6 | −0.71 | −0.20 | −0.46 | −0.08 | −0.03 |
| B5 | −1.2 | −0.60 | −0.17 | −0.38 | −0.06 | −0.02 |
| B8 | −0.2 | −0.38 | −0.11 | −0.24 | −0.03 | −0.02 |
| A0 | 0.6 | −0.06 | −0.02 | −0.05 | 0.00 | 0.00 |
| A5 | 1.9 | 0.11 | 0.15 | 0.28 | 0.08 | 0.01 |
| F0 | 2.7 | 0.06 | 0.30 | 0.59 | 0.16 | 0.02 |
| F5 | 3.5 | 0.01 | 0.44 | 0.79 | 0.20 | 0.03 |
| G0 | 4.4 | 0.10 | 0.58 | 0.92 | 0.24 | 0.04 |
| G5 | 5.1 | 0.23 | 0.68 | 1.19 | 0.31 | 0.05 |
| K0 | 5.9 | 0.45 | 0.81 | 1.37 | 0.38 | 0.06 |
| K5 | 7.4 | 1.03 | 1.15 | 2.08 | 0.57 | 0.10 |
| M0 | 8.8 | 1.25 | 1.40 | 2.79 | 0.65 | 0.15 |
| M5 | 12.3 | 1.22 | 1.64 | 4.48 | 0.52 | 0.28 |
| | | | Giants (luminosity class III) | | | |
| MK class | $M_V$ | $(U-B)$ | $(B-V)$ | $(V-J)$ | $(J-H)$ | $(H-K)$ |
| G5 | 0.9 | 0.50 | 0.86 | 1.49 | 0.44 | 0.04 |
| K0 | 0.7 | 0.82 | 1.00 | 1.72 | 0.50 | 0.07 |
| K5 | −0.2 | 1.68 | 1.50 | 2.83 | 0.74 | 0.16 |
| M0 | −0.4 | 1.81 | 1.56 | 2.95 | 0.76 | 0.17 |
| M2 | −0.6 | 1.84 | 1.59 | 3.09 | 0.76 | 0.18 |
| M5 | −0.3 | 1.88 | 1.63 | 3.42 | 0.78 | 0.19 |

SOURCE: From data published in Schmidt-Kaler (1982) and Peletier (1989)

it is for dwarfs and these giants that the correlation between MK class and luminosity is tightest. The absolute magnitudes tabulated for each spectral class are *averages* for stars of that class. Thus the MS given in the table includes an admixture of somewhat evolved stars, and it will therefore necessarily be brighter than the ZAMS for stars earlier than type G. For example, the ZAMS for O stars lies almost a magnitude below the MS given in Table 3.10.

Table 3.11 list values of $M_V$ and colors for luminosity class I supergiants. These stars are rare objects so they do not have nearby representatives for which absolute distances are readily obtained. Moreover, they are not con-

**Table 3.11**  Luminosities and colors of supergiants

| Spectral type | Ib $M_V$ | Ib $(U - B)$ | Ib $(B - V)$ | Ia $M_V$ | Ia $(U - B)$ | Ia $(B - V)$ | Ia0 $M_V$ |
|---|---|---|---|---|---|---|---|
| O5 |      | −1.17 | −0.32 | −6.8 | −1.17 | −0.31 |      |
| B0 | −6.1 | −1.07 | −0.24 | −6.9 | −1.05 | −0.23 | −8.2 |
| B5 | −5.4 | −0.70 | −0.10 | −7.0 | −0.76 | −0.08 | −8.4 |
| A0 | −5.2 | −0.33 | −0.01 | −7.1 | −0.44 | 0.02 | −8.5 |
| A5 | −5.1 | 0.00 | 0.09 | −7.4 | −0.10 | 0.09 | −8.8 |
| F0 | −5.1 | 0.15 | 0.19 | −8.0 | 0.15 | 0.17 | −9.0 |
| F5 | −5.1 | 0.27 | 0.33 | −8.0 | 0.27 | 0.31 | −9.0 |
| G0 | −5.0 | 0.50 | 0.76 | −8.0 | 0.52 | 0.75 | −8.9 |
| G5 | −4.6 | 0.81 | 1.00 | −7.9 | 0.82 | 1.03 | −8.6 |
| K0 | −4.3 | 1.15 | 1.20 | −7.7 | 1.18 | 1.25 | −8.5 |
| K5 | −4.4 | 1.79 | 1.59 | −7.5 | 1.60 | 1.60 |      |
| M0 | −4.5 | 1.90 | 1.64 | −7.0 | 1.90 | 1.67 | −8.0 |

SOURCE: From data published in Schmidt-Kaler (1982)

**Table 3.12**  Luminosities of white dwarfs

| Class | DB | DA | DF | DG | DM |
|---|---|---|---|---|---|
| $M_V$ | $10 - 11$ | $11 - 12$ | $13 - 14$ | $14 - 15$ | 15 |

fined to a narrow sequence in the CM diagram, but occur all over the upper part of the CM diagram. These facts make it inevitable that there is considerable scatter in the luminosities of these objects for any given spectral class, and the numbers in Table 3.11 should be used with more caution than the numbers in Table 3.10.

Table 3.11 shows that, for the hottest stars, both $(B - V)$ and $(U - B)$ reach limiting values. This is a result of the fact that, at high temperatures $(T \gtrsim 40,000\,\mathrm{K})$, the Planck function in the visible reduces to the Rayleigh-Jeans form $B_\nu(T) = (2kT\nu^2/hc^2)$. It follows from equation (2.33) that, in this limit, the color indices will be essentially independent of $T$.

Table 3.12 gives characteristic ranges of absolute magnitude for the various white-dwarf spectral types.

The data in Tables 3.10 and 3.11 can be used to construct **color–color diagrams** (or **two-color diagrams**), such as those shown in Figure 3.7. If adequate allowance can be made for interstellar reddening, stars of different spectral and luminosity classes can be distinguished on the basis of colors alone. Similar diagrams using other photometric indices can sometimes be used to carry out a two-dimensional photometric "spectral classification" within limited regions of the CM-diagram. This is valuable because we can then classify very faint stars.

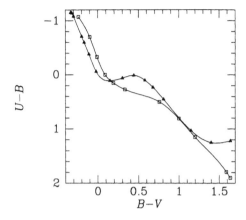

**Figure 3.7** Two-color diagram for MS stars (triangles) and luminosity-class I supergiants (squares). [From the data in Tables 3.10 and 3.11]

### 3.5.3 The physical properties of stars on the MS and RGB

The spectrum of a star is essentially determined by its effective temperature and surface gravity (see §3.4). Furthermore, because a stellar spectrum can be characterized by giving its MK spectral class, to each MK spectral class, there should correspond definite values of $T_{\rm eff}$, BC, and the surface gravity $g$. Table 3.7 gives $T_{\rm eff}$ and BC as functions of MK class. The most reliable values given are for main-sequence B–K stars. The accuracy of the estimates deteriorates for O and M stars and for giants and supergiants. Table 3.7 shows that MK spectral type correlates closely with $T_{\rm eff}$. The MK luminosity classes are primarily sensitive to the surface gravity $g$.

Naïvely, one would expect massive stars to be very luminous and stars with small masses to be faint. Broadly speaking, it is true that faint stars have small masses, but it turns out that luminous stars are not necessarily massive. As we shall see in Chapter 5, the luminosity of a star tends to increase as it evolves, and this increase is very pronounced for low-mass stars. Thus, toward the end of its life, even a low-mass star can become a luminous giant. On the MS, however, both a **mass–luminosity relation** and a **mass–radius relation** hold. Table 3.13 gives these relations (cf. §5.1.7). On the RGB there is quite a well-defined relation between spectral type and both radius and luminosity (Table 3.14).

## 3.6 The stellar luminosity function

The density of stars clearly varies from point to point within the Galaxy. Moreover, within any volume of space there will be both luminous and faint stars. Let the number $dN$ of stars with absolute magnitudes in $(M+dM, M)$ in the volume $d^3\mathbf{x}$ around the point $\mathbf{x}$ be

$$dN = \Phi(M, \mathbf{x})\, dM\, d^3\mathbf{x}. \tag{3.20}$$

**Table 3.13**  Physical properties of MS stars

| Spectral type | $\mathcal{M}/\mathcal{M}_\odot$ | $\log(L/L_\odot)$ | $M_{bol}$ | $M_V$ | $R/R_\odot$ | $\bar{\rho}/\bar{\rho}_\odot$ |
|---|---|---|---|---|---|---|
| O3 | 120 | 6.15 | −10.7 | −6.0 | 15 | 0.035 |
| O5 | 60 | 5.90 | −10.1 | −5.7 | 12 | 0.035 |
| O8 | 23 | 5.23 | −8.4 | −4.9 | 8.5 | 0.037 |
| B0 | 17.5 | 4.72 | −7.1 | −4.0 | 7.4 | 0.043 |
| B3 | 7.6 | 3.28 | −3.5 | −1.6 | 4.8 | 0.069 |
| B5 | 5.9 | 2.92 | −2.7 | −1.2 | 3.9 | 0.099 |
| B8 | 3.8 | 2.26 | −1.0 | −0.2 | 3.0 | 0.14 |
| A0 | 2.9 | 1.73 | 0.3 | 0.6 | 2.4 | 0.21 |
| A5 | 2.0 | 1.15 | 1.7 | 1.9 | 1.7 | 0.41 |
| F0 | 1.6 | 0.81 | 2.6 | 2.7 | 1.5 | 0.47 |
| F5 | 1.3 | 0.51 | 3.4 | 3.5 | 1.3 | 0.59 |
| G0 | 1.05 | 0.18 | 4.2 | 4.4 | 1.1 | 0.79 |
| G5 | 0.92 | −0.10 | 4.9 | 5.1 | 0.92 | 1.18 |
| K0 | 0.79 | −0.38 | 5.6 | 5.9 | 0.85 | 1.29 |
| K5 | 0.67 | −0.82 | 6.7 | 7.4 | 0.72 | 1.79 |
| M0 | 0.51 | −1.11 | 7.4 | 8.8 | 0.60 | 2.36 |
| M5 | 0.21 | −1.96 | 9.6 | 12.3 | 0.27 | 10.7 |
| M7 | 0.12 | −2.47 | 10.8 | 14.3 | 0.18 | 20.6 |
| M8 | 0.06 | −2.92 | 11.9 | 16.0 | 0.1 | 60 |

SOURCE: Data published in Schmidt-Kaler (1982)

**Table 3.14**  Radii and luminosities of red giants

|  | G0 | G5 | K0 | K5 | M0 | M5 |
|---|---|---|---|---|---|---|
| $\log(R/R_\odot)$ | 0.8 | 1.0 | 1.2 | 1.4 | 1.6 | 1.9 |
| $\log(L/L_\odot)$ | 1.5 | 1.7 | 1.9 | 2.3 | 2.6 | 3.0 |

SOURCE: From data published in Allen (1973)

To a first approximation it is useful to imagine that the *mix* of stars of different luminosities is the same everywhere. To express this idea mathematically, we approximate the function $\Phi(M, \mathbf{x})$ defined above by the product of two functions $\Phi(M)$ and $\nu(\mathbf{x})$. That is, we write

$$dN = [\Phi(M)\, dM][\nu(\mathbf{x})\, d^3\mathbf{x}]. \tag{3.21}$$

$\Phi(M)$ is called a **luminosity function**, which measures the relative fractions of stars of different luminosities, while $\nu(\mathbf{x})$ measures the total number density of stars at the point $\mathbf{x}$. In its simplest form, $\Phi$ gives the distribution over luminosity of stars irrespective of their spectral or physical types. In this case we call $\Phi$ the **general luminosity function**. Frequently we are interested in the distribution over luminosity of stars of a special type, for example MS stars, or, more specifically, say, stars of MK spectral class G5 V. Hence, we shall encounter many different stellar luminosity functions.

The concept of a luminosity function generalizes, moreover, to extragalactic studies. In particular, in equation (3.21) $\nu(\mathbf{x})$ could be the number density of normal galaxies within the Universe and $\Phi(M)$ could be the luminosity function of normal galaxies, which we shall study in §4.1.3. Similarly, $\nu(\mathbf{x})$ might be the number density of some species of active galaxies, such as Seyferts (see §4.6) and $\Phi(M)$ could be the luminosity function of Seyferts.

Equation (3.21) does not determine the normalization of $\Phi$ since we can multiply $\Phi$ by any constant provided we divide $\nu$ by the same constant. Conventionally, stellar luminosity functions for the solar neighborhood are normalized such that $\Phi(M)\,\mathrm{d}M$ gives the number of stars per cubic parsec near the Sun with magnitudes in the range $(M + \mathrm{d}M, M)$. With this normalization, $\nu(\mathbf{x})$ is the ratio of the stellar density at $\mathbf{x}$ to that near the Sun.

### 3.6.1 Malmquist bias

Most methods for determining a luminosity function involve counting the number $\mathrm{d}N/\mathrm{d}m$ of objects that have apparent magnitudes in the range $(m + \mathrm{d}m, m)$ and that lie within some given area of the sky. The star-count function

$$A(m) \equiv \frac{\mathrm{d}N}{\mathrm{d}m} \tag{3.22}$$

clearly depends on both the spatial distribution of the objects and on their luminosity function. Since it is impossible to determine $A(m)$ to arbitrarily faint magnitudes, there will be some **limiting magnitude** $m_l$ such that $A(m)$ is available only for $m < m_l$. The simplest sample of objects upon which $A(m)$ could be based is **magnitude-limited** in that it consists of all objects brighter than $m_l$ that lie within a specified area of the sky. It is not hard to see that the mean absolute magnitude of objects in such a sample will be brighter than the mean absolute magnitude of the population as a whole: the volume within which we can see the most luminous objects is larger than that within which we can also see the faintest objects. Consequently, luminous objects are over-represented in a magnitude-limited sample. This effect is called **Malmquist bias** after the Swedish astronomer K.G. Malmquist (Malmquist 1922, Malmquist 1936). It plays an important role in nearly every field of astronomy. We now evaluate the effect of Malmquist bias for the case in which the luminosity function of the population is a Gaussian. This analysis leads to a relationship between the population and sample means, equation (3.34), which is often used to correct for Malmquist bias. Unfortunately, this equation involves the variance $\sigma^2$ of the luminosity function, so it is necessary to derive equation (3.39), which relates the sample variance $\sigma_m^2$ to $\sigma^2$.

Let us calculate the number of stars in a magnitude-limited sample that have absolute magnitudes in $(M + dM, M)$ and apparent magnitudes in $(m + dm, m)$:

$$\frac{d^2 N}{dm dM} = \Phi(M) \frac{dn}{ds} \left( \frac{\partial s}{\partial m} \right)_M, \tag{3.23}$$

where $s(m, M)$ is the distance to a star of the given apparent and absolute magnitudes and $dn$ is the number of stars that lie within the surveyed region of the sky at distances in $(s + ds, s)$. If the survey covers a solid angle $\omega$, the volume at distances in $(s + ds, s)$ is $\omega s^2 ds$ and we have that

$$\frac{dn}{ds} = \omega s^2 \nu(s). \tag{3.24}$$

Inserting (3.24) into (3.23), multiplying both sides of the resulting equation by $dM$ and integrating, we find that

$$A(m) = \int_{-\infty}^{\infty} dM \frac{d^2 N}{dm dM} = \omega \int_{-\infty}^{\infty} dM \, \Phi(M) \left( \frac{\partial s}{\partial m} \right)_M s^2 \nu(s). \tag{3.25}$$

We now change the integration variable from $M$ to $s$, which is a function of $M$ through equation (2.36).[5] This operation yields

$$A(m) = \omega \int_0^{\infty} ds \, \Phi(M) s^2 \nu(s), \tag{3.26}$$

where we have taken advantage of the relation $(\partial M / \partial s)_m = -(\partial m / \partial s)_M$, which is an immediate consequence of equation (2.36).

Consider now the mean absolute magnitude of the stars in the sample that have apparent magnitude $m$:

$$\langle M \rangle_m = \frac{\displaystyle\int_{-\infty}^{\infty} dM \, M \frac{d^2 N}{dm dM}}{\displaystyle\int_{-\infty}^{\infty} dM \frac{d^2 N}{dm dM}}. \tag{3.27}$$

By the reasoning that carried us from equation (3.23) to equation (3.26), equation (3.27) is equivalent to

$$\langle M \rangle_m = \frac{\int_0^{\infty} ds \, M \Phi(M) s^2 \nu(s)}{\int_0^{\infty} ds \, \Phi(M) s^2 \nu(s)}. \tag{3.28}$$

Notice that the bottom of this equation is the integral that appears in equation (3.26) for $A(m)$. Moreover, in this integral $M$ is a function of $m$ through

---

[5] In equation (2.36) we must substitute $s$ for $d$.

equation (2.36) with $(\partial M/\partial m)_s = 1$, so on differentiating (3.26) with respect to $m$ we obtain

$$\frac{\mathrm{d}A}{\mathrm{d}m} = \omega \int_0^\infty \mathrm{d}s \frac{\mathrm{d}\Phi}{\mathrm{d}M} s^2 \nu(s). \tag{3.29}$$

Comparing this result with equation (3.28), we see that

$$\frac{1}{A}\frac{\mathrm{d}A}{\mathrm{d}m} = \left\langle \frac{1}{\Phi}\frac{\mathrm{d}\Phi}{\mathrm{d}M} \right\rangle_m. \tag{3.30}$$

Differentiating equation (3.29) with respect to $m$ we similarly find that

$$\frac{1}{A}\frac{\mathrm{d}^2 A}{\mathrm{d}m^2} = \left\langle \frac{1}{\Phi}\frac{\mathrm{d}^2\Phi}{\mathrm{d}M^2} \right\rangle_m. \tag{3.31}$$

To proceed further we must adopt a definite functional form for the luminosity function $\Phi(M)$. Consider the case in which $\Phi(M)$ is a Gaussian distribution:

$$\Phi(M) = \frac{1}{(2\pi\sigma^2)^{1/2}} \exp\left(-\frac{(M-M_0)^2}{2\sigma^2}\right), \tag{3.32}$$

where for simplicity we have set the local stellar density to unity. The constants $M_0$ and $\sigma$ appearing in equation (3.32) are, respectively, the mean absolute magnitude and the dispersion in absolute magnitude of a **volume-limited** sample of the stars – that is, a sample formed by all the stars that lie within some specified volume. When $\Phi$ is given by equation (3.32), equation (3.30) becomes

$$\frac{1}{A}\frac{\mathrm{d}A}{\mathrm{d}m} = -\left\langle \frac{M-M_0}{\sigma^2} \right\rangle_m, \tag{3.33}$$

so

$$\langle M \rangle_m - M_0 = -\sigma^2 \frac{\mathrm{d}\ln A}{\mathrm{d}m}. \tag{3.34}$$

Since the number of objects in a magnitude-limited survey almost invariably increases towards fainter apparent magnitudes, $\mathrm{d}A/\mathrm{d}m > 0$. Consequently, equation (3.34) implies that the objects in the survey that have any a given apparent magnitude $m$, will, on the average, have a higher luminosity than the mean luminosity of the population as a whole.

The quantity $\langle M \rangle_m$ is potentially measurable. For example, we could determine trigonometric parallaxes for all the stars of a magnitude-limited survey that have apparent magnitudes in some narrow, suitably bright range. Once the parallaxes of these stars were known, their absolute magnitudes could be calculated, and thus $\langle M \rangle_m$ determined. If the dispersion $\sigma$ of the luminosity function were known, equation (3.34) could be used to 'correct $\langle M \rangle_m$ for Malmquist bias', that is to infer $M_0$ from $\langle M \rangle_m$ and the observed form of $A(m)$. Unfortunately, $\sigma$ will, in general, have to be determined from the same survey, and, as we now show, its measurement is also complicated by Malmquist bias.

When we substitute equation (3.32) into equation (3.31), the latter becomes

$$\frac{1}{A}\frac{d^2 A}{dm^2} = \left\langle \left(\frac{M - M_0}{\sigma^2}\right)^2 - \frac{1}{\sigma^2} \right\rangle_m . \tag{3.35}$$

Hence

$$\frac{1}{\sigma^4}\left\langle (M - M_0)^2 \right\rangle_m - \frac{1}{\sigma^2} = \frac{1}{A}\frac{d^2 A}{dm^2}$$

$$= \frac{d^2 \ln A}{dm^2} + \left(\frac{1}{A}\frac{dA}{dm}\right)^2 \tag{3.36}$$

$$= \frac{d^2 \ln A}{dm^2} + \left(\frac{\langle M \rangle_m - M_0}{\sigma^2}\right)^2 ,$$

so

$$\sigma^4 \frac{d^2 \ln A}{dm^2} = \left\langle (M - M_0)^2 \right\rangle_m - (\langle M \rangle_m - M_0)^2 - \sigma^2$$

$$= \left\langle M^2 \right\rangle_m - \langle M \rangle_m^2 - \sigma^2. \tag{3.37}$$

Now the variance of the measured absolute magnitudes is

$$\sigma_m^2 \equiv \left\langle (M - \langle M \rangle_m)^2 \right\rangle_m = \left\langle M^2 \right\rangle_m - \langle M \rangle_m^2 , \tag{3.38}$$

so from equation (3.37) we have that

$$\sigma_m^2 - \sigma^2 = \sigma^4 \frac{d^2 \ln A}{dm^2}. \tag{3.39}$$

To assess the significance of this equation, consider the case of a spatially uniform population: $\nu(s) = $ constant. From equation (2.36) it follows that $(\partial s/\partial m)_M = (0.2 \ln 10)s$, and with a little algebra we can show that for constant $\nu$, equation (3.25) may be written

$$A(m) \propto e^{(0.6 \ln 10)m} \int_{-\infty}^{\infty} dM$$

$$\times \exp\left\{ -\left(\frac{M^2 + 2M\left[(0.6 \ln 10)\sigma^2 - M_0\right] + M_0^2}{2\sigma^2}\right) \right\} \tag{3.40}$$

$$\propto \exp\left[ 0.6 \ln 10(m - M_0) \right].$$

Hence in the case of a homogeneously distributed population,

$$\frac{d \log_{10} A}{dm} = \frac{d \ln A/ \ln 10}{dm} = 0.6, \qquad \frac{d^2 \ln A}{dm^2} = 0. \tag{3.41}$$

Consequently, by equations (3.34) and (3.39), neither $\langle M \rangle_m - M_0$, nor $\sigma_m^2 - \sigma^2$ depends upon $m$. This fact can be used as a test of the hypothesis that the luminosity function is Gaussian: one can divide the sample by apparent magnitude, and for each subsample evaluate the mean and variance of $M$. If these quantities vary significantly between subsamples, either the Gaussian hypothesis or the assumption of homogeneity must be rejected. Since equations (3.34) and (3.39) are only valid for a Gaussian luminosity function, they must not be used to correct for Malmquist bias if there is evidence that the luminosity function is non-Gaussian.

One case in which the assumptions of a Gaussian luminosity function and homogeneity are reasonably valid is that of MS stars with good Hipparcos parallaxes. As we will see in §3.6.4, in this case $\sigma_m \simeq 0.5\,\mathrm{mag}$, so by equations (3.34) and (3.41), $\langle M \rangle_m - M_0 \simeq 0.15\,\mathrm{mag}$.

Equations (3.39) and (3.41) show that in the case of a homogeneously distributed population, the sample variance $\sigma_m$ is an unbiased estimator of the population variance $\sigma$. However, it often happens that a population's density falls with distance from the Sun – perhaps because the surveyed field lies towards a Galactic pole. In this case, the star-count function $A(m)$ will increase with $m$ less rapidly at faint than at bright magnitudes, with the result that $\mathrm{d}^2 \ln A / \mathrm{d}m^2 < 0$ and equation (3.39) then implies that the sample variance will be smaller than the population variance.

### 3.6.2 Lutz-Kelker Bias

The underlying cause of the Malmquist bias – that, for a survey of fixed solid angle, the volume studied increases with distance – gives rise to a similar effect in the analysis of trigonometric parallaxes. This effect, known as the **Lutz-Kelker bias**, causes an observed parallax to be on average higher than its true value (Lutz & Kelker 1973). This overestimate translates into an underestimate of distance, and hence an underestimate of an object's luminosity as derived from its apparent brightness. Given the renewed interest in parallax determinations excited by Hipparcos, this bias and its eradication are of some importance.

To quantify the importance of this effect, we now calculate the probability $P(\varpi | \varpi')\mathrm{d}\varpi$ that the true parallax of a given star lies in $(\varpi, \varpi + \mathrm{d}\varpi)$ given that its measured parallax is $\varpi'$. Bayes theorem [equation (C.4)] tells us that this probability is given by

$$P(\varpi | \varpi') = \frac{P(\varpi' | \varpi) P(\varpi)}{P(\varpi')}, \qquad (3.42)$$

where $P(\varpi' | \varpi)\mathrm{d}\varpi'$ is the probability that observational errors will cause the parallax of a star that has true parallax $\varpi$ to be measured to be $\varpi'$, and $P(\varpi)\mathrm{d}\varpi$ and $P(\varpi')\mathrm{d}\varpi'$ are 'prior probabilities' which reflect the knowledge that we have of the star before we measure its parallax. Specifically, $P(\varpi)\mathrm{d}\varpi$

is the prior probability that the star has true parallax in $(\varpi, \varpi + d\varpi)$, and $P(\varpi')d\varpi'$ is the prior probability of measuring a parallax in $(\varpi', \varpi' + d\varpi')$. Since we are interested in the dependence of $P(\varpi|\varpi')$ on $\varpi$ at fixed $\varpi'$, we do not need to determine the prior probability in the denominator of equation (3.42). The prior probability in the numerator of equation (3.42), by contrast, is of fundamental importance.

The prior $P(\varpi)$ will reflect (i) the known value of the star's apparent magnitude $m$, and (ii) the probabilities that we assign to the various possible values of the star's absolute magnitude $M$. Quantitatively, let $P(s|m)ds$ be the probability of finding a star of magnitude $m$ at a distance in $(s, s + ds)$. By Bayes theorem (equation (C.4)) this is given by

$$P(s|m) = \frac{P(m|s)P(s)}{P(m)}, \tag{3.43}$$

where $P(m|s)dm$ is the probability that a star at distance $s$ has apparent magnitude $m$ and $P(s)ds$ is the probability of finding a star at distance $s$. From the definition of the luminosity function, equation (3.21), we have

$$P(m|s) \propto \Phi[M(s,m)] \quad \text{and} \quad P(s) \propto \nu(s)s^2, \tag{3.44}$$

while $P(m)$, the probability density of observing of star of apparent magnitude $m$, irrespective of distance, is proportional to the star-count function $A(m)$ calculated above. Equations (3.43) and (3.44) now yield that the prior probability that a star of apparent magnitude $m$ has true parallax in $(\varpi, \varpi + d\varpi)$ is

$$
\begin{aligned}
P(\varpi)d\varpi &= P(s|m)\left|\frac{\partial s}{\partial \varpi}\right|_m d\varpi \\
&\propto \Phi(M)\nu(s)s^2\left|\frac{\partial s}{\partial \varpi}\right|_m d\varpi,
\end{aligned}
\qquad \text{where} \quad
\begin{cases}
M = m + 5\log(\varpi/10) \\
s = 1/\varpi.
\end{cases}
\tag{3.45}
$$

When we evaluate the derivative in this equation and use the result to eliminate $P(\varpi)$ from equation (3.42), we obtain

$$P(\varpi|\varpi') \propto P(\varpi'|\varpi)\Phi(M)\nu(s)\varpi^{-4}. \tag{3.46}$$

To proceed further we must make assumptions regarding the distribution of measuring errors, $P(\varpi'|\varpi)$, the luminosity function $\Phi(M)$ and the space-density of stars $\nu(s)$. We assume that the probability of measuring a parallax $\varpi'$ for a star whose true parallax is $\varpi$ is given by the Gaussian

$$P(\varpi'|\varpi) = \frac{1}{\sqrt{2\pi}\sigma_\varpi}\exp\left[-\frac{(\varpi' - \varpi)^2}{2\sigma_\varpi{}^2}\right], \tag{3.47}$$

where $\sigma_\varpi$ is the standard deviation of the measurement errors. Since parallaxes can only be measured for nearby stars, a reasonable assumption regarding $\nu$ is that it is independent of $s$. With this assumption and equation (3.47), equation (3.46) becomes

$$P(\varpi|\varpi') \propto \Phi(M)\varpi^{-4}\exp\left[-\frac{(\varpi' - \varpi)^2}{2\sigma_\varpi{}^2}\right].  \tag{3.48}$$

where $M = m + 5\log(\varpi/10)$. The noteworthy feature of equation (3.48) is the factor $w^{-4}$, which strongly biases the probability $P(\varpi|\varpi')$ that the star has true parallax $\varpi$ towards small values of $\varpi$. This is Lutz–Kelker bias. For given $m$, very small values of $\varpi$ are associated with very bright absolute magnitudes $M$, at which the luminosity function $\Phi$ will be small. Hence Lutz–Kelker bias is moderated by the luminosity function.

The form of $\Phi(M)$ that we should use in equation (3.48) depends on what we know about the star before we measure its parallax. If we know the MK spectral class of the star, $\Phi$ should be the luminosity function for that MK class, which can be approximated by a Gaussian – see equation (3.57) below. At sufficiently bright magnitudes, $\Phi$ then declines extremely strongly and Lutz–Kelker bias is strongly moderated. If we only know the star's apparent magnitude, $\Phi$ should be the general luminosity function of nearby stars. We shall see below that at bright absolute magnitudes this can be approximated by a straight line in a plot of $\log\Phi$ versus $M$, with slope $\simeq 0.65$:

$$\begin{aligned}\Phi(M) &= \text{constant} \times 10^{\beta M}, \\ &= \text{constant} \times 10^{\beta[m+5\log(\varpi)]},\end{aligned} \quad \text{where} \quad \beta \simeq 0.65 .  \tag{3.49}$$

Substituting this simple form of $\Phi(M)$ into equation (3.48) we conclude that when we know only the star's apparent magnitude prior to measuring its parallax $\varpi'$, the probability distribution of its true parallax is

$$P(\varpi|\varpi') \propto \exp\left[-\frac{(\varpi' - \varpi)^2}{2\sigma_\varpi{}^2}\right]\varpi^{5\beta-4}.  \tag{3.50}$$

Figure 3.8 plots this probability distribution for three values of $\varpi'/\sigma_\varpi$ and $\beta = 0.65$, when the exponent $5\beta - 4 \simeq -0.75$. When $\varpi'/\sigma_\varpi = 2$, the peak in the probability density occurs at very small values of true parallax, $\varpi$. Physically, this result implies that unless a parallax measurement is more than a $2\sigma$ result, nothing useful can be inferred from it about the star's distance because it is at least as likely to have a negligible true parallax as a value near the measured value. The curve of $P$ for the case $\varpi'/\sigma_\varpi = 3$ peaks only once, near the measured parallax, but the peak occurs significantly to

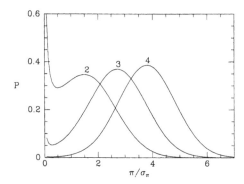

**Figure 3.8** The probability distribution $P(\varpi|\varpi')$ given by equation (3.50) for $\beta = 0.65$ and $\varpi'/\sigma_\varpi = 2, 3, 4$. Each curve is labeled by its value of $\varpi'/\sigma_\varpi$.

**Table 3.15**   The Lutz–Kelker bias in parallax and absolute magnitude and its 90% confidence limits

| $\varpi'/\sigma_\varpi$ | 40 | 20 | 15 | 10 | 8 | 5 | 3 |
|---|---|---|---|---|---|---|---|
| $(\varpi - \varpi')\sigma_\varpi$ | −0.016 | −0.033 | −0.043 | −0.065 | −0.082 | −0.134 | −0.235 |
| $\varpi_L/\sigma_\varpi$ | 38.7 | 18.7 | 13.7 | 8.6 | 6.6 | 3.5 | 1.2 |
| $\varpi_U/\sigma_\varpi$ | 41.3 | 21.2 | 16.2 | 11.2 | 9.2 | 6.1 | 4.0 |
| $M - M'$ | −0.001 | −0.004 | −0.006 | −0.014 | −0.022 | −0.059 | −0.177 |

the left of the measured parallax. In fact, by setting the derivative of equation (3.50) to zero it is easy to show that the peak occurs at

$$\frac{\varpi}{\sigma_\varpi} = \frac{1}{2}\left( \frac{\varpi'^2}{\sigma_\varpi^2} + \sqrt{\frac{\varpi'^2}{\sigma_\varpi^2} + 4\beta} \right). \tag{3.51}$$

The offset $\varpi - \varpi'$ between the location of this peak and the measured parallax, which is a measure of the seriousness of Lutz–Kelker bias, is given in Table 3.15, which also gives $\varpi_L$ and $\varpi_U$, the lower and upper 90% confidence limits of $\varpi$. For example, if $\varpi'/\sigma_\varpi = 5$, the most probable value of $\varpi$ is $\varpi' - 0.134$ and 90% of the probability lies between $\varpi = 3.5\sigma_\varpi$ and $\varpi = 6.1\sigma_\varpi$. On account of the non-Gaussian nature of the distribution $P(\varpi|\varpi')$, these confidence limits are asymmetric, and exceed those that would be returned by a naive propagation of the observational errors.

Since parallaxes are often used to calibrate the absolute magnitudes of objects, it is useful to define the correction that should be applied to the magnitude inferred from the raw parallax measurement, $M'$, into a better estimate of the true absolute magnitude $M = M' + 5\log(\varpi/\varpi')$ [see equations (2.7) and (2.36)]. A plausible form of this correction is obtained by setting $\varpi$ equal to the value given by equation (3.51). The last row of Table 3.15 gives the corresponding values of $M - M'$.

Notice that Lutz–Kelker bias leads to an essentially circular problem: to correct for this bias we need to know the luminosity function of stars $\Phi$, which is precisely what we are trying to determine when we first encounter Lutz–Kelker bias. In such circumstances we proceed iteratively: we use a guess at the form of $\Phi$ to correct for bias, use the corrected data to measure $\Phi$ and then repeat the analysis with refined corrections. The important thing is to be aware of the existence of the bias and to understand that its magnitude depends on the criteria upon which a given star was included in a parallax program.

### 3.6.3 The general luminosity function

**Cluster luminosity functions**  Conceptually the simplest method of determining $\Phi$ is to count stars within a cluster of known distance. If it were possible to resolve every star more luminous than some limiting absolute magnitude $M_{\max}$, then, for $M < M_{\max}$, $\Phi(M)$ would be proportional to the number of stars actually counted in $(M + dM, M)$. Actually, at faint magnitudes only some fraction $\phi(M)$ of the stars actually present will be detected because faint stars tend to be missed through 'confusion' – see §6.1.1. We defer discussion of how one corrects for confusion and the details of cluster luminosity functions to Chapter 6, but note here that observations with HST suggest that at magnitudes $M_V \gtrsim 6$, which correspond to stars with MS-lifetimes longer than the age of the Galaxy, $\Phi(M)$ has the same general form as it does in the field near the Sun. Specifically, for $M_I \lesssim 9$, $\Phi(M_I)$ increases with $M_I$, but at some characteristic magnitude $M_I^0 \sim 9$, which depends on the cluster's metallicity, $\Phi(M_I)$ flattens off and then turns down as $M_I$ becomes fainter than $M_I^0$ (Figure 6.16).

**Photometrically complete surveys**  Consider now the problem of determining $\Phi(M)$ from the mix of stars near the Sun. Clearly, this problem involves counting stars within a well defined volume for each of a series of intervals of absolute magnitude. The volume within which stars are counted will generally be different for every magnitude interval since, on the one hand, we want our samples to be as large as possible to minimize statistical errors, and, on the other hand, luminous stars are comparatively rare but can be seen to large distances. Hence it is helpful for the volume of our census to increase with the luminosity of the stars being counted. One convenient way of sampling volumes that increase with luminosity is to form a magnitude-limited sample. With this strategy, our sampling volume increases as $L^{3/2}$, where $L$ is the luminosity of the sampled stars.

To determine the star's absolute magnitude, we need an estimate of its distance $d$. Ideally this would be a trigonometric distance. Unfortunately, trigonometric parallaxes are available only for a tiny fraction of all stars – Hipparcos determined parallaxes for all stars brighter than $m = 8$ mag and for only a fraction of the fainter stars. The lowest-luminosity MS stars have $M_V \simeq 18$ mag, so such a star is guaranteed to have a Hipparcos parallax only if it lies closer than $d = 0.001$ pc! Although such very low-luminosity stars are amongst the most numerous stars within the Galaxy, there is a negligible chance that one of them will lie so close to the Sun. Hence we have no chance of obtaining a useful estimate of the faint end of the luminosity function simply by relying on the complete Hipparcos parallax catalog. Yet long before the Hipparcos catalog became available, reasonable estimates of $\Phi(M)$ down to $M_V \sim 17$ had been made using parallax data that were order-of magnitude worse than those in the Hipparcos catalog. How was this task be accomplished?

One important strategy involves the use of photometric distances (§2.1.3). From the colors, and perhaps a low-resolution spectrum, of each star in our magnitude-limited survey, we decide what kind of star it is, and hence what its luminosity is. Its distance then follows from equation (2.36). Notice that this procedure does not eliminate the need for trigonometric parallaxes, since the calibration of the relation between colors and luminosities will be based on trigonometric parallaxes. The key point is that for this calibrating step one requires parallaxes for only a few faint stars rather than a complete sample of faint stars – see §10.4.3 for details.

Since color data are required for the determination of each star's distance, color generally features in the original magnitude-limited sample. For example, one might define the sample to be all stars brighter than some limiting magnitude in the $V$ band and redder than some particular value of $V - I$. In the absence of interstellar reddening (which will be unimportant if the volume sampled is not large), whether a star satisfies the $V - I$ criterion does not depend on its distance, so the selection criteria will turn up all sufficiently red stars that lie within a luminosity-dependent volume. The great advantage of imposing a color criterion is that it enables us to focus on one particular type of object, for example low-luminosity MS stars or cool white dwarfs, which can only be studied at faint apparent magnitudes. Without the color criterion, any complete sample of faint stars would include large numbers of distant luminous objects, whose number density has already been determined at brighter apparent magnitudes.

**Proper-motion selected surveys**     Since a powerful strategy for identifying stars that lie close to the Sun is to look for stars with large proper motions, a classical procedure for determining $\Phi(M)$ relies upon a catalog of high-proper motion stars. This procedure is particularly important for the determination of the space density of halo stars (see §10.5.2) which are scarce near the Sun but are the dominant population more than $\sim 1 \, \text{kpc}$ from the plane. As these stars pass through the solar neighborhood, they have large space velocities (§10.5.2), so they have high probabilities of entering a proper-motion catalog.

The determination of $\Phi(M)$ from a proper-motion catalog involves handling difficulties that arise from the incompleteness of the observations and selection effects. Since similar difficulties arise in many branches of astronomy, such as the analysis of pulsars, X-ray sources, radio galaxies, and

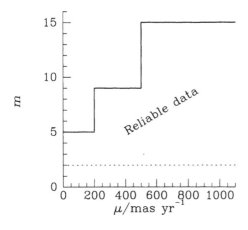

**Figure 3.9** The observed sample of stars is complete below the full curve. At the very bright magnitudes that lie below the dotted curve, there are too few stars to permit an adequate statistical discussion. [After Trumpler & Weaver (1953)]

quasars, we now describe the derivation of $\Phi(M)$ from a proper-motion catalog in some detail.

Consider the stars within a sphere of radius 100 pc centered on the Sun. This volume is large enough to contain a statistically useful sample of stars but still small enough that interstellar absorption can be neglected within it. Inside this sphere, stellar parallaxes can be measured accurately because they are $\geq 10$ mas.

We begin by counting the number of stars for a specified set of values of apparent magnitudes $m$ and proper motions $\mu$ in a proper-motion catalog. We restrict the region in the $(m, \mu)$ plane over which counts are made (see Figure 3.9) by fixing, for each $m$, a minimum proper motion $\mu_0$ chosen to be large enough to ensure that, at the apparent magnitude specified, most stars with $\mu \geq \mu_0$ do appear in the proper-motion catalogs. Representative choices for various ranges of $m$ are as follows: $(m, \mu_0) = (5, 0); (5 - 9, 0''.2); (9 - 15, 0''.5)$. Now we estimate from the catalogs the probability $P_{\varpi'}(m, \mu)$ that a star with magnitude $m$ and proper motion $\mu$ has a measured parallax $\varpi'$. We have

$$P_{\varpi'}(m, \mu) \simeq \frac{\text{number of stars with chosen } (m, \mu) \text{ and measured } \varpi'}{\text{number of stars with chosen } (m, \mu)}. \quad (3.52)$$

Those stars referred to in the numerator of equation (3.52) will have some distribution over parallax. If we can assume that all stars in the $(m, \mu)$ group have the same distribution, then $P_{\varpi'}^{-1}(m, \mu)$ is a valid estimate of the factor by which any star with some measured parallax within some $(m, \mu)$ group should be multiplied to give the total number that really have that value of the parallax in that $(m, \mu)$ group. Thus to correct for incompleteness of the parallax data, we count each star that has a measured parallax as $P_{\varpi'}^{-1}(m, \mu)$ stars. By summing over all values of $\mu$ at a set of fixed values of $(m, \varpi')$, we can then construct numerically the distribution function $G'(m, \varpi'|\mu \geq \mu_0)$, which is such that

$$dN = G'(m, \varpi'|\mu > \mu_0) \, dm d\varpi' \quad (3.53)$$

---

### Box 3.1:    Inverse problems

Equation (3.54) is an integral equation of the type

$$f(x) = \int d\xi K(x, \xi) g(\xi), \qquad (1)$$

where $f$ and $g$ are observed and theoretical distributions, respectively, and $K$ is a smooth, non-negative kernel. An equation of this type is said to pose an **inverse problem**. The key to dealing with such an equation is to avoid solving it, as we now explain.

Since we expect $g$ to be a smooth function, $f$ should be extra smooth. In reality, $f$ won't be so smooth because noise – from small-number statistics and observational errors – will add jaggedness. In fact, in many cases $f$ will be more jagged than *any* non-negative function $g$ would be after convolution by the smooth kernel $K$.

If you actually solve an equation such as (3.54), you find that with $f$ estimated from high-quality data, $g$ comes out implausibly bumpy, and, as the quality of the data deteriorates, $g$ soon has unphysical negative stretches. So solving (3.54) will produce garbage unless the quality of the input data is high. But no matter how good the data, it is *stupid* to solve the equation, because in doing so one is failing to exploit one's ability to distinguish signal from noise. It is actually an advantage that noise in $f$ quickly produces an unphysical distribution $g$, since this implies that the effects of noise can be limited by exploiting one's prior knowledge that $g$ is smooth and non-negative. A powerful general method for doing this is described in Appendix C.

---

gives the number of stars with proper motion in excess of $\mu_0$ that have $m$ in $(m + dm, m)$ and $\varpi'$ in $(\varpi', \varpi' + d\varpi')$.

We now consider the effects of errors in the measured parallaxes. If the errors of measurement are random, then it is reasonable to model the probability $P(\varpi'|\varpi)$ of measuring a parallax $\varpi'$ for a star of true parallax $\varpi$ by equation (3.47). In this model, the true distribution of stars in the $(m, \varpi)$ plane, $G(m, \varpi|\mu > \mu_0)$, is related to the observed distribution, $G'(m, \varpi'|\mu > \mu_0)$, by

$$G'(m, \varpi'|\mu > \mu_0) = \frac{1}{\sqrt{2\pi}\sigma_\varpi} \int_{-\infty}^{\infty} d\varpi\, G(m, \varpi|\mu > \mu_0) \exp\left[-\frac{(\varpi' - \varpi)^2}{2\sigma_\varpi^2}\right].$$
$$(3.54)$$

Mathematically, this is an integral equation for $G$ given $G'$, which is equal to $G$ convolved with the Gaussian kernel. Equations of this type frequently occur in astronomy. Appendix C discusses how they should be handled. Here we simply assume that a method such as that described in Appendix C has

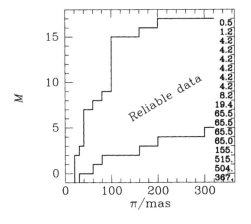

**Figure 3.10** For values of $\varpi$ and $M$ that lie between the two curves, equation equation (3.56) can be used to transform $F(m, \varphi)$ into $H(M, \varphi)$ with reasonable accuracy. The numbers on the extreme right give the volume, in $1000\,\mathrm{pc}^3$, within which stars of the given absolute magnitude are counted.

been used to estimate $G$ from $G'$.

Next, we allow for the fact that some stars within the 100 pc sphere have proper motions below our proper-motion completeness limit $\mu_0$; these stars are missing from the proper-motion catalogs, but they must be included in the determination of $\Phi(M)$. Given a kinematic model for the distribution of stellar random velocities, both in direction and in speed (see §10.5.2), we can estimate the probability $P(\mu > \mu_0|\varpi)$ that a star of parallax $\varpi$ will have a proper motion in excess of $\mu_0$. The quantity

$$F(m, \varpi)\,\mathrm{d}m\mathrm{d}\varpi \equiv \frac{G(m, \varpi|\mu > \mu_0)}{P(\mu > \mu_0|\varpi)}\,\mathrm{d}m\mathrm{d}\varpi \qquad (3.55)$$

is then the total number of stars of apparent magnitude $m$ and parallax $\varpi$ within the 100 pc sphere, irrespective of proper motion. For more detail see Trumpler & Weaver (1953), 375–377.

The final step in deriving $\Phi(M)$ is to convert $F(m, \varpi)$, the distribution over apparent magnitude $m$, into a distribution $H(M, \varpi)$ over absolute magnitude $M$ via the relation

$$M = m + 5 + 5\log\varpi. \qquad (3.56)$$

This conversion can be done reliably between the two curves of in Figure 3.10. For each value of $M$, we then sum the tabulated values of $H(M, \varpi)$ over all accessible values of $\varpi$ and divide the sum by the volume corresponding to that range of $\varpi$. We then have our estimate of $\Phi(M)$, the number of stars per unit absolute magnitude per cubic parsec in the solar neighborhood.

Table 3.16 presents the general luminosity function. For each absolute magnitude $M_V$ three numbers are given: (i) the number $\Phi(M_V)$ of stars with magnitudes in the range $(M_V + \frac{1}{2}, M_V - \frac{1}{2})$ per $10^4\,\mathrm{pc}^3$; (ii) the luminosity $\mathrm{d}L$ emitted by these stars; (iii) the mass $\mathrm{d}\mathcal{M}$ contained in these stars. Figure 3.11 plots as a function of $M_V$: (i) the cumulative luminosity density

**Table 3.16**   The general luminosity function (per $10^4 \, pc^3$)

| $M_V$ | $\Phi(M_V)$ | $\delta L/L_\odot$ | $\delta \mathcal{M}/\mathcal{M}_\odot$ | $M_V$ | $\Phi(M_V)$ | $\delta L/L_\odot$ | $\delta \mathcal{M}/\mathcal{M}_\odot$ |
|---|---|---|---|---|---|---|---|
| −6 | 0.0001 | 2.6 | 0.005 | 7 | 29 | 4.0 | 21.3 |
| −5 | 0.0006 | 5.1 | 0.020 | 8 | 33 | 1.8 | 21.8 |
| −4 | 0.0029 | 9.4 | 0.060 | 9 | 42 | 0.90 | 24.2 |
| −3 | 0.013 | 17.1 | 0.17 | 10 | 70 | 0.60 | 35.0 |
| −2 | 0.05 | 28.2 | 0.05 | 11 | 90 | 0.30 | 36.0 |
| −1 | 0.25 | 53.9 | 1.6 | 12 | 127 | 0.17 | 36.3 |
| 0 | 1 | 95.8 | 4.0 | 13 | 102 | 0.055 | 20.8 |
| 1 | 3 | 111 | 7.4 | 14 | 102 | 0.022 | 16.3 |
| 2 | 5 | 64 | 8.7 | 15 | 127 | 0.011 | 16.3 |
| 3 | 12 | 66 | 17.3 | 16 | 102 | 0.0035 | 10.5 |
| 4 | 17 | 36 | 19.4 | 17 | 51 | 0.0007 | 4.3 |
| 5 | 29 | 25 | 28.1 | 18 | 22 | 0.0001 | 1.6 |
| 6 | 30 | 10 | 24.7 | 19 | 13 | 0.0000 | 0.7 |
|   |   |   |   | Total | 1008 | 532 | 356 |

SOURCE: For $M_V \leq 0$ from data published in Allen (1973); for $M_V > 0$ from data published in Jahreiss & Wielen (1983) and Kroupa, Tout & Gilmore (1990)

$j_V$ from stars more luminous than $M_V$ (full curve); (ii) the cumulative mass density $\rho$ contributed by these stars (dotted curve); and (iii) the mass-to-light ratio $\Upsilon_V = \rho/j_V$ of the stars more luminous than $M_V$. Inspection of Table 3.16 reveals several noteworthy features.

1. *Most stars in the solar neighborhood are intrinsically faint.* The maximum in $\Phi(M)$ occurs near $M_V = 14$. These stars are typically K and M dwarfs.

2. *Essentially all the light emitted is contributed by intrinsically luminous stars* despite the fact that they are extremely rare. The peak emission comes from stars with $M \simeq 1$. These stars are typically A dwarfs and K and M giants. It is worth stressing just how strong this bias toward high luminosities really is. For example, stars at $M_V \simeq -4$, typically B stars, emit as much light per unit volume in the $V$ band as do the stars at $M_V \simeq +6$, mostly G and K dwarfs, even though the latter are 10 000 times as numerous and contain 400 times as much mass! The average $V$-band light output per unit volume is about $0.053 \, L_\odot \, pc^{-3}$

3. *Most of the stellar mass density in the solar neighborhood is contributed by the vast number of low-luminosity stars.* Figure 3.11 shows clearly that whereas nearly all the luminosity comes from stars brighter than the Sun ($M_v = 4.83$), the mass is fairly uniformly distributed in the magnitude range $3 < M_V < 15$. The large mass in faint stars has the unpleasant implication that the dynamics of the Galaxy are dominated by stars that are at best inconspicuous, while the very luminous objects that can be readily observed in our own galaxy and in other galaxies

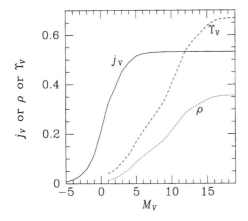

**Figure 3.11** The full curve shows the $V$-band luminosity density (in $10^{-1}L_\odot\,\mathrm{pc}^{-3}$) in stars brighter than $M_V$. The dotted curve shows the mass density (in $10^{-1}\mathcal{M}_\odot\,\mathrm{pc}^{-3}$) in these stars. The dashed curve shows the corresponding values of the mass-to-light ratio $\Upsilon_V$. [From data published in Jahreiss & Wielen (1983) and after Kroupa, Tout & Gilmore (1990) from data kindly provided by P. Kroupa]

contribute little to the gravitational forces in the system. Therefore, except insofar as the few high-luminosity stars can be used as tracers of the entire population, it is dangerous to assume that the luminosity density in galaxies is proportional to the mass density. The only safe procedure is to infer the mass distribution by studying the gravitational field of the whole system through the dynamics of the population of bright stars. The total mass density from the stars in Table 3.16 is about $0.036\,\mathcal{M}_\odot\,\mathrm{pc}^{-3}$. However, these data omit most of the white dwarfs, which are numerous, are intrinsically very faint ($M_V \simeq +13$ to $+16$), and yet have masses of the order of 0.5 to $1\,\mathcal{M}_\odot$. One finds (Weidemann 1990) that they contribute approximately an additional $0.003\,\mathcal{M}_\odot\,\mathrm{pc}^{-3}$ (see Tables 3.19 and 3.20). Hence the total stellar mass density is about $0.039\,\mathcal{M}_\odot\,\mathrm{pc}^{-3}$. This number gives a feeling for the relative emptiness of space in the Galactic disk in the solar neighborhood.

4. The average **mass-to-light ratio** – the mass per unit volume divided by the light emission per unit volume (in solar units) – is $\Upsilon_V \simeq 0.67\,\mathcal{M}_\odot/L_\odot$. This number is useful, for, insofar as it is typical for a galactic disk, it can be used to convert observed light distributions in other galaxies into first estimates of stellar mass distributions, thereby providing some input to dynamical analyses of these objects. On the other hand, note that even for the solar neighborhood the value of $\Upsilon_V$ just quoted is surely a lower limit, because many kinds of objects (for example, faint companions in multiple stars, dead white dwarfs and neutron stars, stars lost in dense interstellar clouds, and so on) would evade discovery. Cumulatively, such objects could add appreciably to the mass density while contributing little or nothing to the light.

We have mentioned that Malmquist bias causes a sample that is complete down to some limiting apparent magnitude $V_0$ to contain a higher proportion of luminous stars than a volume-limited sample. Table 3.17 quantifies this phenomenon by giving the expected absolute-magnitude distributions of

**Table 3.17**  Absolute-magnitude distributions of volume- and magnitude-limited samples of 100 stars

| $M_V$ | $-6$ | $-5$ | $-4$ | $-3$ | $-2$ | $-1$ | 0 | 1 | 2 | 3 | 4 | 5 |
|---|---|---|---|---|---|---|---|---|---|---|---|---|
| $N_{\rm vol}$ | 0.0 | 0.0 | 0.0 | 0.0 | 0.1 | 0.4 | 1.5 | 4.8 | 7 | 18 | 25 | 43 |
| $N_{V_0=2.5}$ | 4 | 7 | 10 | 12 | 12 | 16 | 16 | 14 | 5 | 3 | 1 | 0 |
| $N_{V_0=6}$ | 2 | 4 | 6 | 10 | 12 | 17 | 19 | 17 | 6 | 4 | 2 | 1 |

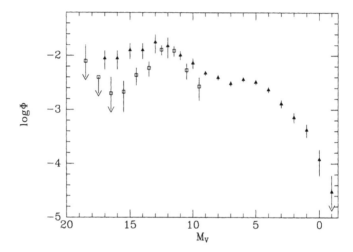

**Figure 3.12** The open squares show the luminosity function obtained by Reid & Gilmore (1982) from a color-selected survey. The filled triangles show the luminosity function obtained by Wielen, Jahreiss & Krüger (1983) from a proper-motion selected sample.

100 stars in a volume-limited sample, and in two magnitude-limited samples (see Problem 3.6 for details).

If any one point emerges from the preceding discussion, it is that the behavior of $\Phi(M)$ at faint magnitudes is of great importance to Galactic research. For this reason, tremendous effort has been lavished on its determination. Figure 3.12 shows results from two important studies of the faint-end luminosity function: the open points are from Reid & Gilmore (1982), who carried out a color-selected survey, while the filled points are from Wielen, Jahreiss & Krüger (1983), whose data were based on proper-motion surveys. Both studies show a peak in $\Phi(M)$ at $M_V \simeq 12.5$, rather than at $M_V = 14$ as Table 3.16 implies, and recent HST observations of faint red stars are consistent with $\Phi(M)$ peaking at $M_V \sim 12.5$ (Gould, Bahcall & Flynn 1996, Santiago, Gilmore & Elson 1996). In §5.1.9 we shall see that this peak reflects the rapid decline in the luminosity of a MS star as its mass $\mathcal{M}$ approaches the limit for hydrogen burning at $\mathcal{M} \simeq 0.08\,\mathcal{M}_\odot$. On the faint side of this peak, the data from the color-selected survey fall off more steeply than do those from the proper-motion surveys. Kroupa, Tout & Gilmore (1993) argue that a combination of photometric errors and the existence of unresolved binaries within a color-selected survey cause the resulting values of $\Phi(M)$ to fall below the true values. Hence the full symbols in Figure 3.12 probably more accurately reflect the truth. Nevertheless, the structure of $\Phi(M)$ for $M_V \gtrsim 12$ must be regarded as rather ill determined despite the tremendous amount of labor that has been lavished on it.[6]

### 3.6.4 The luminosity function of a given MK spectral class

The general luminosity function lumps together all spectral classes, and, at a given $M$, several radically different kinds of stars may contribute to $\Phi(M)$. It is therefore also interesting to break down the luminosity function by the stars' spectral classes, $S$.

To determine $\Phi(M, S)$, we proceed essentially as we did in the determination of $\Phi(M)$, but for one spectral class at a time. The same problems arise as before, one of the more troublesome being the scarcity of luminous stars. Consequently, $\Phi(M, S)$ is even less well determined than the general luminosity function $\Phi(M)$.

For each MK class $S$, $\Phi(M, S)$ peaks at some absolute magnitude $M_0(S)$ that is characteristic of the class. For some MK classes, for example those with luminosity classes I and II, the speed with which $\Phi(M, S)$ falls as one moves away from $M_0(S)$ is simply a measure of the sensitivity of MK system to luminosity. That is, $\Phi(M, S)$ falls as one moves away from $M_0(S)$ simply because as $|M - M_0(S)|$ increases at fixed spectral type, stars are more likely to be classified as belonging to a different luminosity class. Figure 3.6 implies that for classes $S$ with luminosity classes I or II, the full width at half maximum of the peak in $\Phi(M, S)$ is 2–3 mag and we cannot make a more precise statement about the structure of $\Phi(M, S)$ at this time.

The case of a MK class with luminosity class V (MS stars) is different, for now as one moves vertically in the CM diagram from $M_0(S)$, the density of stars falls. Thus the width of the peak in $\Phi(M, S)$ does not merely reflect the luminosity-sensitivity of the MK system, but a real, physical paucity of stars. This fact makes the peak in $\Phi(M, S)$ narrowest for luminosity class V. Moreover, MS stars are numerous, so Murray *et al.* (1997) and Houk *et al.* (1997) were able to determine the form of $\Phi(M, S)$ rather accurately for MK classes from late AV to early KV. Figure 3.13 shows four of their histograms that are, in effect, plots of $\Phi(M, S)$ as a function of $M$ at fixed $S$. In each panel a dotted vertical line shows the absolute magnitude brighter than which the sample should be volume-limited. Fainter than this line, both dotted and full histograms can be distinguished. The dotted histogram shows the stars actually counted, while the full histogram shows the result

---

[6] For an independent determination of $\Phi(M)$ at faint magnitudes, see Kirkpatrick *et al.* (1994).

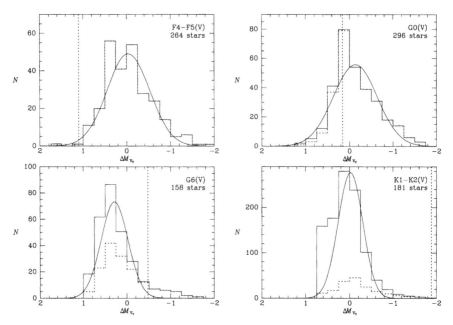

**Figure 3.13** Distributions over luminosity of MS stars with given spectral types. The dotted histograms show the stars actually counted, while the full histograms show these data after correction for Malmquist bias – the dotted vertical lines show the magnitude brighter than which the sample should be volume-limited. The curves are Gaussian fits to the full histograms. The ordinate, $\Delta M_V$ is centered on the median absolute magnitude of each sample. [After Houk *et al.* (1997)]

of correcting these counts for Malmquist bias to the counts expected from a volume-limited sample.

The histograms in Figure 3.13 have a characteristically skew shape, which rises more sharply at faint magnitudes than it falls at bright magnitudes. At least part of this asymmetry will be due to the unresolved binary stars in the sample – in such a case the MK classification is liable to be based on the brighter star, while both binary components contribute to the measured magnitude of the system.

When Gaussian distributions

$$\Phi(M, S) = \frac{\Phi_0}{\sqrt{2\pi}\sigma} \exp\left(-\frac{(M - M_0)^2}{2\sigma^2}\right), \tag{3.57}$$

are fitted to histograms like those shown in Figure 3.13, the resulting dispersions $\sigma$ lie in the range 0.3 to 0.5 mag. The stars studied by Houk *et al.* were chosen because they have trigonometric parallaxes with standard errors smaller than 10%. Since a 10% distance error gives rise to an error of 0.2 mag in absolute magnitude, when we correct for the effect of these distance errors, a dispersion in measured absolute magnitude of 0.4 mag corresponds to a dispersion in intrinsic absolute magnitude of order 0.35 mag.

**Table 3.18** Luminosity function of MS stars

| $M_V$ | 0 | 1 | 2 | 3 | 4 | 5 | 6 | 7 | 8 | 9 |
|---|---|---|---|---|---|---|---|---|---|---|
| $S$ | B8.5 | A1.5 | A5.6 | F1.9 | F7.8 | G4.3 | K0.3 | K3.7 | K7.1 | M0.3 |
| $\Phi_{\text{gen}}(M_V)$ | 1 | 3 | 5 | 12 | 17 | 29 | 30 | 29 | 33 | 42 |
| $\Phi_{\text{MS}}$ | 0.23 | 0.74 | 1.9 | 9.6 | 19 | 24 | 24 | 33 | 17 | 20 |
| $\pm$ | 0.07 | 0.12 | 0.2 | 0.4 | 0.6 | 0.9 | 1.7 | 4 | 6 | 12 |

SOURCE: From data published in Murray *et al.* (1997)

**Table 3.19** Number density of stars by spectral type

| Luminosity group | Spectral type | | | | | | | Totals |
|---|---|---|---|---|---|---|---|---|
| | O | B | A | F | G | K | M | |
| Giants and supergiants | | | | −0.3 | 0.2 | 0.6 | −0.6 | 0.8 |
| Main sequence | −3.6 | 0.0 | 0.7 | 1.4 | 1.8 | 2.0 | 2.8 | 2.9 |
| White dwarfs | | 1.8 | 2.0 | 1.7 | 1.7 | 1.4 | | 2.4 |

NOTES: Each entry is the logarithm of the number of stars brighter than $M_V = 16$ in $10^4 \, \text{pc}^3$.
SOURCE: From data published in Allen (1973)

Consequently, in favorable cases photometric distances to MS stars can have standard errors down to $\sim 17\%$. The non-Gaussian nature of the histograms shown in Figure 3.13 requires, however, that some caution be exercised in the analysis of errors in photometric distances. In particular, the long tail to bright magnitudes has the effect of significantly increasing the upper limits on photometric distances.

Whereas Houk *et al.* (1997) determined the dependence of $\Phi(M, S)$ at fixed $S$, Murray *et al.* (1997) used the same sample of stars with accurate parallaxes to determine the luminosity function, $\Phi_{\text{MS}}$ of MS stars, irrespective of spectral type. Table 3.18 gives these values alongside the corresponding values of the general luminosity function from Table 3.16. One sees that at spectral types later than $S = \text{F2}$ ($M_V \gtrsim 3$), stars of luminosity class V account for most of the general luminosity function. This is in accordance with what we would expect from inspection of Figure 3.5, which shows that the SGB joins the MS near $M_V = 3.5$ with the result that at fainter magnitudes nearly all stars are of luminosity class V. Table 3.18 suggests that at $M_V \lesssim 2$, the majority of stars near the Sun are giants of one sort or another.

Table 3.19 gives the number densities, down to $M_V = 16$, of giant stars, MS stars and white dwarfs in the solar neighbourhood, irrespective of luminosity. The total space density of white dwarfs in the solar neighborhood can be seen to be about 30% that of MS stars, while the total density of giants and supergiants is about 1% that of the MS.

The local *mass* density of these stellar types is given in Table 3.20, which again shows how the dynamical effects arising from stars in the disk

**Table 3.20**   Mass densities ( $\mathcal{M}_\odot/10^3\,\mathrm{pc}^3$ ) contributed by various objects to the solar neighborhood

| Object | Mass density | Object | Mass density |
|--------|--------------|--------|--------------|
| O–B | 0.9 | White dwarfs | 20 |
| A | 1 | Cepheids | 0.001 |
| F | 3 | Long-period variables | 0.001 |
| dG | 4 | Planetary nebulae | $5 \times 10^{-6}$ |
| dK | 9 | Galactic clusters | 0.04 |
| dM | 25 | | |
| gG | 0.8 | Subdwarfs | 0.15 |
| gK | 0.1 | RR Lyrae variables | $10^{-6}$ |
| gM | 0.01 | Globular clusters | 0.001 |

SOURCE: From data published in Allen (1973), Oort & Plaut (1975) and Schmidt (1965)

are dominated by the most inconspicuous stars – the K and M dwarfs and the white dwarfs. For completeness, the table also lists average mass densities from other distinctive disk objects. The last three entries give local densities of members of the spheroidal component – see §10.5.2 (that from globular clusters being averaged over a spherical shell of the appropriate radius). Of these objects, only the subdwarfs have an appreciable local mass density, about 3% of the total. As we shall see in §10.2, the situation is entirely different near the Galactic center, where the mass density of the spheroidal component dominates that of the disk.

### 3.6.5 Catalogs of the nearby stars

It is very interesting to draw up an inventory of all known stellar objects in the immediate vicinity of the Sun. One chooses a small volume so that even inherently faint objects can be detected. Early inventories of this type were drawn up by Hertzsprung, Luyten and Haas, who adopted limiting distances of 5 pc, 10 pc and 15 pc, respectively. In 1951 Gliese began assembling a list of all objects thought to lie within 20 pc of the Sun. In 1957 he released a list of 915 systems, containing 1094 components, with distances of 20 pc or less; this first edition of the *Catalogue of Nearby Stars* is usually called the first *Gliese Catalog*. Since 1957 Gliese and his colleagues at the Astronomisches Rechen-Institut in Heidelberg have continuously updated and extended the Gliese catalog by searching the literature for objects believed to lie within first 22 pc and most recently 25 pc. By 1993 the list included 3264 systems, containing 3803 components (Gliese 1969, Gliese & Jahreiss 1979, Jahreiss & Wielen 1997).

    Since the Gliese catalogs are based on many different astrometric and photometric programs, that have no common selection criterion, it is extremely hard to decide down to which magnitude the catalogs are reasonably complete, or to estimate the degree of incompleteness at faint magnitudes.

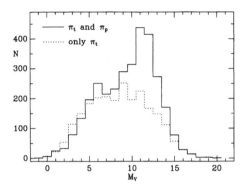

**Figure 3.14** The distribution over absolute magnitude of stars in the 1993 version of the Gliese catalog (full histogram). The dotted histogram shows the distribution of stars that one would obtain if one used only trigonometric parallaxes rather than a combination of trigonometric parallaxes and photometric distances. [From Jahreiss (1993) courtesy of H. Jahreiss]

The full histogram in Figure 3.14 gives some insight into these questions, however, by showing the number of stars of each absolute magnitude in the 1993 version of the catalog. Comparison of this histogram with the luminosity functions of Figure 3.12 makes it clear that the decline in numbers faintward of $M_V \sim 12$ is due to the catalog's incompleteness; within 25 pc of the Sun there must be great numbers of stars with $M_V \sim 15$ and $V \lesssim 16$ that have still to be discovered.

Within 10 pc of the Sun one finds a sample of 8 white dwarfs, 2 subgiants, and 245 MS stars down to $M_V = 18$. This sample includes no O–B stars and only 4 A stars. It contains no giants or supergiants, and at least 50% of the stars are members of binaries or multiple systems.

## 3.7 Interstellar dust

The space between the stars is not empty. It is filled with rarefied but exceedingly filthy gas; if this gas were compressed to the density of ordinary air (that is, by a factor of $\sim 10^{21}$) the density of smoke in it would be such that objects would disappear into the haze at a distance of much less than a meter.

Interstellar gas is so filthy because many stars are furnaces of the least environment-friendly type: vast quantities of hot gas stream out of every red giant, and smoke particles condense out of this gas as it streams away from the glowing surface of the star in exactly the same way that smoke particles form in flue gases as they come off a furnace. Some of these smoke particles will later gather together to form comets, asteroids and planets, but the vast majority hang around in interstellar space obscuring the view. For some reason astronomers call smoke in space, 'interstellar dust'. Whittet (1992) is a useful general reference on interstellar dust.

As we saw in Chapter 1, the realization by Trumpler in the 1920s, that dust prevents us seeing the Galaxy clearly, was one of the great milestones in astronomy. The key to his realization was the fact that dust not only

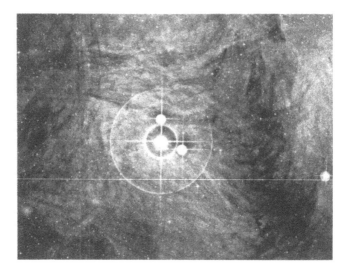

**Figure 3.15** The reflection nebula around the star $\rho$ Ophiuchi. [Courtesy of D. Malin]

dims the light from distant stars, but it also reddens it. Dust accomplishes these things in two ways. First, it scatters some photons out of the line of sight from a star to the Earth. Second, it absorbs other photons, converting their energy into heat. In regions in which bright stars lie close to dense gas clouds, light scattered by dust can be directly observed; clouds from which scattered light can be detected are called **reflection nebulae**. Figure 3.15 shows an example of such a reflection nebula. Evidence that the nebula's light is indeed scattered starlight is provided by its spectrum, which shows the characteristic spectral features of the early-type stars that illuminate the nebula.

The absorption of light by dust constitutes an important energy source for the interstellar medium and has the overall effect of making many galaxies emit most of their energy in the far infrared (at $\lambda \approx 200\,\mu$m). Thus dust transforms blue light into infrared light. The essential physics is as follows. Most of the energy radiated by stars emerges at wavelengths that can be absorbed by dust grains, and, moreover, young, luminous stars generally live out their lives in regions rich in dust – they are born in dense clouds, and in their brief lifetimes cannot get far away from the cloud of their birth. Hence a significant fraction of the energy radiated by the stars in the Galaxy is absorbed by dust grains.

Dust grains are effective absorbers of photons whose wavelength $\lambda$ is comparable to, or smaller than, the characteristic size $a$ of the dust grains, but poor absorbers of longer-wavelength photons. Since there are a great many dust grains larger than about 100 nm, interstellar dust absorbs UV radiation most effectively. But very few dust grains are tens of microns in size, so far-infrared radiation is able to pass through quite dense interstellar clouds. The absorption of starlight warms dust grains to $\approx 10\,\mathrm{K}$. At this temperature they radiate significantly at $\lambda \approx 200\,\mu\mathrm{m}$, and photons of this wavelength can escape the Galaxy.

### 3.7.1 Extinction and reddening

Both scattering and absorption dim the light from distant stars. Astronomers refer to this dimming as 'extinction'.[7] As we saw in §2.3.3, the **extinction** $A_X$ of a star in some waveband $X$ is defined to be the difference between the observed $X$-band magnitude $m(X)$ and the magnitude $m_0(X)$ that would be observed in the absence of dust:

$$A_X \equiv \left(m - m_0\right)_X. \tag{3.58}$$

The **reddening** or **color excess** $E(X-Y)$ in some color $X-Y$ is defined to be the difference between the observed color $m(X) - m(Y)$ and the intrinsic color $m_0(X) - m_0(Y)$:

$$E(X - Y) \equiv \left[m(X) - m(Y)\right] - \left[m_0(X) - m_0(Y)\right] = A_X - A_Y, \tag{3.59}$$

where the second equality follows from (3.58). The most often cited extinction is $A_V$, and the most often cited color excess is $E(B-V)$. Since colors are always defined such that the shorter waveband is on the left and the strength of interstellar extinction generally decreases from short to long wavelengths, color excesses are usually positive.

We have seen that it is sometimes possible to predict the absolute magnitude $M_X$ of a star from a knowledge of its MK spectral class. Fortunately, the latter may be determined irrespective of moderate extinction, since it follows from details of the star's spectrum which are not significantly modified by the presence of dust along the line of sight. Combining equations (2.36) and (3.58), we have

$$m_X = M_X + A_X + 5\log d - 5, \tag{3.60}$$

so, if the distance $d$ to a star of known spectral type is known, $A_X$ follows immediately from the measured value of $m_X$. Similarly, measurements of stars of known MK class and therefore of known intrinsic colors in two wavebands $X, Y$ yield $E(X - Y)$ directly from equation (3.59).

In practice, it is usually necessary to determine $A_V$ from observations of a star of unknown distance. In such a case, we assume that $A_X$ tends to

---

[7] Frequently the words extinction and absorption are used interchangeably, but in view of the clear distinction between absorption and scattering, this confusion of terms is to be avoided.

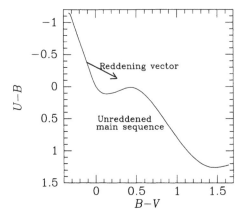

**Figure 3.16** Effects of interstellar reddening in the $UBV$-system two-color diagram.

zero at long wavelengths as $A_X = A_0 f(\lambda_X)$, where $A_0$ is a constant to be determined and $f$ is a theoretically-inspired function, and $\lambda_X$ is the effective wavelength of our wave-band.[8] Typically $f(\lambda) \propto \lambda^{-1.8}$ at the longest wavelengths for which observations are available. Once a form has been adopted for $f(\lambda)$, we seek the values of the two numbers $A_0$ and $\log d$ that minimize

$$\sum_X \left\{ m_X - \left[ M_X + A_0 f(\lambda_X) + 5 \log d - 5 \right] \right\}^2, \tag{3.61}$$

where the sum extends over several different wave-bands $X$.

The classic studies in the $UBV$ system are those by Hiltner & Johnson (1956) and Johnson & Morgan (1953). Many references to more recent work will be found in the reviews of Savage & Mathis (1979) and Whittet (1992). Most studies focus on O–B stars, which are ideal for determining extinction and reddening because (i) they are highly luminous and hence can be seen to large distances, over which the column density of dust can become significant; (ii) they are intrinsically blue and hence strongly susceptible to reddening; and (iii) they have distinctive spectra that can be classified with high precision. Figure 3.16 shows how stars of identical MK class are distributed along a **reddening line** in the two-color diagram. The bluest stars on this line are taken to be unreddened and these define the intrinsic two-color sequence given in Tables 3.10 and 3.11 and in Figure 3.7. More reddened stars move down the line to larger values of $(U - B)$ and $(B - V)$. This procedure can be applied to a group of stars of any spectral type to determine both the intrinsic color and the reddening line for the type.

The amount of extinction and reddening along a line of sight depends on the density of dust along it. But in §8.1.6 we shall see that if the *type* of dust

---

[8] In the notation of §2.3.4, the **effective frequency** of a filter $X$ when observing an object with absolute energy distribution $f_\nu$ is defined by $\nu_{\rm eff} \equiv \int d\nu\, \nu S_X(\nu) f_\nu / \int d\nu\, S_X(\nu) f_\nu$. The effective wavelength is $\lambda_{\rm eff} = c/\nu_{\rm eff}$.

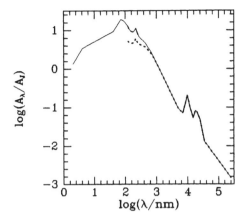

**Figure 3.17** The interstellar extinction curve for two representative lines of sight. The full curve is characteristic of lines of sight that pass only through the intercloud medium, while the dashed curve is for a line of sight that penetrates deep into a molecular cloud. [From data published in Mathis (1990)]

were everywhere the same, the ratio of the extinctions at any two frequencies would be the same for all lines of sight. The full curve in Figure 3.17 shows, for a representative line of sight, the ratio $A_\lambda/A_J$ of the extinction at wavelength $\lambda$ to that in the Johnson $J$-band. This curve is called the **extinction curve** or **extinction law** for that line of sight. $A_\lambda$ peaks in the far-UV, near 73 nm. Shorter wavelength radiation (X-rays) passes right through grains, while much longer wavelength radiation refracts around them.

At the longest wavelengths plotted ($100\,\mu\mathrm{m}$), $A_\lambda$ is decreasing with increasing $\lambda$ as $\sim \lambda^{-1}$. In most of the optical and UV regions of the spectrum, $A_\lambda$ is falling slightly more steeply with increasing $\lambda$. This steady fall is broken by three distinct bumps. One in the UV centred on 217 nm, and two in the IR, at $9.7\,\mu\mathrm{m}$ and $18\,\mu\mathrm{m}$. The origin of these bumps has not been certainly established, but it is thought that the UV bump is due to tiny pieces of graphite – these objects probably contain only $\sim 50$ atoms. The IR features are probably associated with Si–O bonds in silicate grains.

In the visual band and at shorter wavelengths, the extinction curve is found to vary from one line of sight to another. The dashed curve in Figure 3.17 indicates the sense of this variation: along some lines of sight the extinction does not increase into the UV as strongly as along 'standard' lines of sight. The origin of this effect is thought to be variations in the size-distribution of grains in different regions: the extinction rises strongly into the UV along lines of sight rich in very small grains, which have negligible effect on red and IR radiation. Some lines of sight appear to be deficient in such small grains and have less steeply rising extinction curves.[9]

The slope of the extinction curve near the $V$ band is $A_V/(A_J R_V)$, where

$$R_V \equiv \frac{A_V}{A_B - A_V} = \frac{A_V}{E(B-V)}. \tag{3.62}$$

---

[9] We discuss this phenomenon further in §8.1.6.

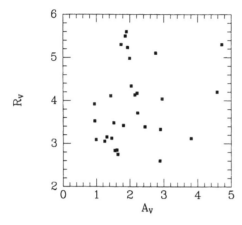

**Figure 3.18** Values of $R_V$ for 29 lines of sight to O and B stars. [From data published in Cardelli, Clayton & Mathis (1989)]

Consequently, $R_V$ quantifies whether the extinction curve is rising steeply into the UV or not: $R_V \simeq 3$ for a steeply-rising extinction curve, and $R_V \simeq 5$ for a slowly rising one. Figure 3.18 shows the distribution of $R_V$ for 29 lines of sight studied by Cardelli, Clayton & Mathis (1989). Smaller values of $R_V$ generally apply for lines of sight that do not penetrate deeply into a molecular cloud.

The classical studies of extinction (e.g., Schultz & Wiemer 1975, Snedden *et al.* 1978) derived

$$R_V \simeq 3.1 \qquad (3.63)$$

for lines of sight that do not pass through dense clouds. Equation (3.63) is widely used because it allows one to estimate $A_V$ from the much more readily measured quantity $E(B - V)$. Unfortunately, these studies did not derive $R_V$ for individual lines of sight but, for each of a series of wavebands $X$, found the slope of the best fitting line to the positions of the program stars in the $\big(E(V - X), E(B - V)\big)$ plane. These slopes were then plotted against $\lambda_X^{-1}$ and extrapolated to $\lambda = 0$ to yield the *mean value* of $R_V$. This procedure does not allow one to determine the scatter in values of $R_V$ for individual lines of sight and thus determine the likely error associated with equation (3.63). The scatter would certainly be smaller than that shown in Figure 3.18, which is for a sample of lines of sight chosen for their anomalous UV extinction curves, but it is probably not negligible.

Cardelli, Clayton & Mathis (1989) show that along any line of sight $A_\lambda/A_J$ can be well approximated by

$$\frac{A_\lambda}{A_J} = a(\lambda) + \frac{b(\lambda)}{R_V}, \qquad (3.64)$$

where $a$ and $b$ are polynomials in $\lambda^{-1}$.

As Figure 3.17 illustrates, the extinction curve is universal in the $R$ band and redward from it, but variable at $V$ and blueward of $V$. This fact implies

**Table 3.21**  The standard interstellar extinction law

| Band $X$ | $\dfrac{E(X - V)}{E(B - V)}$ | $\dfrac{A_X}{A_V}$ |
|:---:|:---:|:---:|
| $U$ | 1.64 | 1.531 |
| $B$ | 1.00 | 1.324 |
| $V$ | 0.00 | 1.000 |
| $R$ | −0.78 | 0.748 |
| $I$ | −1.60 | 0.482 |
| $J$ | −2.22 | 0.282 |
| $H$ | −2.55 | 0.175 |
| $K$ | −2.74 | 0.112 |
| $L$ | −2.91 | 0.058 |
| $M$ | −3.02 | 0.023 |
| $N$ | −2.93 | 0.052 |

SOURCE: From data published in Rieke & Lebofsky (1985)

that extinctions and reddenings should be normalized to a waveband that lies to the red of $V$, which has traditionally been used as the normalizing band.

In §8.1.6 we shall see that along any line of sight $E(B - V)$ is approximately proportional to the column density $N_H$ of interstellar hydrogen atoms, irrespective of whether the hydrogen atoms are in the form of HI or are locked up in $H_2$ molecules. Specifically, one has (Bohlin *et al.* 1978, Kent, Dame & Fazio 1991)

$$E(B - V) = \frac{N_H}{5.8 \times 10^{25}\ \text{m}^{-2}}. \tag{3.65}$$

$E(B - V)/N_H$ is called the **dust/gas ratio**. Its constancy down different lines of sight suggests that a fixed number and size distribution of dust grains is associated with a given mass of interstellar hydrogen. In §9.2 we shall find that a typical value for the number density of H near the Sun is $n_H = 10^6\ \text{m}^{-3}$, so along a line of sight of length $d$, the column density of H is $N_H \simeq 3.1 \times 10^{25}(d/\text{kpc})\ \text{m}^{-2}$. With equations (3.62), (3.63) and (3.65) it then follows that along a typical line of sight through the disk we have

$$E(B - V) \simeq 0.53\left(\frac{d}{\text{kpc}}\right) \quad \text{and} \quad A_V \simeq 1.6\left(\frac{d}{\text{kpc}}\right). \tag{3.66}$$

Table 3.21 expresses the extinction law for 'standard' lines of sight in terms of the standard $UBVRI$ wavebands. From the numbers in this table it may not be evident just how much more transparent the ISM is in the near infrared than in the blue. So consider the case of a star that is located at the Galactic center. For such a star we might have $A_M \simeq 0.6$, and therefore, according to Table 3.21, $A_B \simeq 34.5$. The probability that a $M$-band photon reaches us from such a star is $10^{-0.24} = 0.57$, while the probability of a $B$-band photon reaching us is only $10^{-13.8} = 1.6 \times 10^{-14}$.

Extinction is caused by a combination of absorption and scattering of radiation by dust. The fraction of extinction which is caused by scattering is called the **albedo**, $\gamma_\lambda$. For example, suppose a star that emits luminosity $L_\lambda$ at wavelengths in the range $(\lambda + d\lambda, \lambda)$ is observed through an optically thin cloud which generates extinction $A_\lambda$, and that the cloud as seen from the star subtends solid angle $\Omega \ll 4\pi$. The luminosity scattered by the cloud is

$$\gamma_\lambda \left(1 - 10^{-A_\lambda/2.5}\right) \left(\frac{\Omega}{4\pi}\right) L_\lambda, \qquad (3.67)$$

while the luminosity absorbed by the cloud is

$$(1 - \gamma_\lambda)\left(1 - 10^{-A_\lambda/2.5}\right) \left(\frac{\Omega}{4\pi}\right) L_\lambda. \qquad (3.68)$$

Dust grains are not expected to scatter radiation isotropically; more radiation will be scattered through small angles than large angles, so forward scattering predominates, especially in the case of large dust particles. In principle, one would like to determine the full angular dependence of scattering. In practice, the most that can be hoped for is a reliable estimate of the mean cosine of the scattering angle, which is generally denoted $g_\lambda$.

$\gamma_\lambda$ and $g_\lambda$ have been determined from studies of (i) reflection nebulae, and (ii) 'diffuse Galactic light'. It is relatively straightforward to determine the amount of light scattered by a reflection nebula, but very hard to determine the angle through which it is scattered because this angle depends sensitively on the location of the scattering dust relative to the illuminating star. 'Diffuse Galactic light' is light scattered by diffuse interstellar dust from the general interstellar radiation field of the Milky Way. It is of low surface brightness, so its intensity can be determined only by carefully correcting the surface brightness of an apparently blank piece of sky for the other contributions – see §4.2. The advantage of using diffuse Galactic light to measure $\gamma_\lambda$ and $g_\lambda$ is that the relative positions of the scattering grains and illuminating sources are then well determined. The night-sky brightness measured at 3 AU by the Pioneer 10 spacecraft lead to estimates $\gamma(440\,\text{nm}) = 0.61 \pm 0.07$ and $g(440\,\text{nm}) = 0.6 \pm 2$ (Witt 1988). Scattering should become more isotropic $(g_\lambda \to 0)$ as $\lambda$ increases, while $\gamma_\lambda$ is believed to decrease with $\lambda$.

**Reddening-free indices**    Thus far, we have assumed that the spectral types of the stars under study are known, so that we can determine color excesses by comparing the observed colors of each star with those of an unreddened star of the same (known) type. However, in many cases, we do not know the spectral types of the stars that we wish to study. Nevertheless,

**Table 3.22**  $Q$ versus spectral type for
early-type stars

| Spectral type | $Q$ | Spectral type | $Q$ |
|:---:|:---:|:---:|:---:|
| O5 | −0.93 | B3 | −0.57 |
| O6 | −0.93 | B5 | −0.44 |
| O8 | −0.93 | B6 | −0.37 |
| O9 | −0.90 | B7 | −0.32 |
| B0 | −0.90 | B8 | −0.27 |
| B0.5 | −0.85 | B9 | −0.13 |
| B1 | −0.78 | A0 | 0.00 |
| B2 | −0.70 | | |

SOURCE: From data published in Johnson &
Morgan (1953).

given the slope of the reddening line, it is possible to define a photometric
parameter that depends only on the spectral type of a star and is independent
of the amount of reddening. In the $UBV$ system, this parameter is (see
Problem 3.8)

$$Q \equiv (U - B) - \frac{E(U - B)}{E(B - V)}(B - V)$$

$$\simeq (U - B) - 0.72(B - V).$$

(3.69)

Values of $Q$ for stars of spectral types O through A0 are listed in Table 3.22.
The importance of this parameter is that, for early-type stars, it uniquely
determines a star's intrinsic color from photometric data alone without the
need for spectra. Furthermore, $Q$ can be measured easily for stars that are
much too faint for spectral classification. From the data given in Table 3.10
and Table 3.22, one finds that, to a good approximation,

$$(B - V)_0 = 0.332Q.$$

(3.70)

To summarize, we can determine $(U - B)_0$ and $(B - V)_0$ for early-type stars
from Q. Once $(B - V)_0$ is known, we find $E(B - V)$ from $(B - V)$, and, as
we shall see later, the visual extinction $A_V$, (or $A_X$, at any other waveband
$X$) follows from $E(B - V)$.

For spectral types later than A0, $Q$ is not a unique function of spectral
type [see Strand (1963), Chapter 13, for the full dependence of $Q$ on spectral
type], and hence it is no longer useful for estimating the amount of reddening
present. This failure occurs because the reddening line happens to have
almost the same slope as the unreddened MS curve for late-type stars, shown
in Figure 3.16, which means that it is nearly impossible to tell (from $UBV$
data alone) whether a given star has been reddened or is unreddened but of
a later type.

Finally, it is worth noting that reddening-free indices can be defined in other photometric systems as well. For example, in the Strömgren *uvby* system, we can define

$$
\begin{aligned}
[c_1] &\equiv c_1 - 0.2(b - y) \\
[m_1] &\equiv m_1 + 0.18(b - y) \\
[u - b] &\equiv (u - b) - 1.84(b - y) \\
&= [c_1] + 2[m_1],
\end{aligned}
\tag{3.71}
$$

from which intrinsic stellar properties can be inferred in certain ranges of spectral types.

**Polarization of starlight by dust**    Light from obscured stars tends to be slightly polarized, and the degree of this polarization tends to increase with $A_V$. The origin of this phenomenon is believed to be alignment of dust grains by a magnetic field **B** that permeates interstellar space (see §8.1.5 for details). The alignment of dust grains with **B** is believed to be caused by the **Davis–Greenstein effect**, which causes spinning dust grains to align their spin axes with **B**. When a grain's spin axis is not parallel to **B**, the direction of **B** relative to axes fixed in the grain changes as the grain rotates. Dissipative effects, such as hysteresis and eddy currents, then torque the grain in such a way that the grain's spin axis moves closer to **B**. Since grains tend to spin about the axis that has the greatest moment of inertia, the net result of the Davis–Greenstein effect is to preferentially align the grains perpendicular to **B**. When an electromagnetic wave passes a grain, the wave's electric field **E** will be in some general direction. Since the grain is much more readily polarized parallel to its long axis, the grain responds most readily to the component of **E** in this direction. Consequently, radiation scattered or absorbed by the grain tends to be polarized parallel to the grain's long axis and hence perpendicular to **B**. Conversely, transmitted radiation tends to be polarized parallel to **B** – for more details see §4.4 of Whittet (1992).

The Davis–Greenstein effect enables us to map out the direction of the interstellar magnetic field by measuring the polarizations of large numbers of stars. Axon & Ellis (1976) gathered from the literature polarization directions for 5064 stars with reliable distances and produced plots like those shown in Figure 3.19 of the directions of **B** as a function of Galactic longitude and latitude for several distance ranges. The more distant stars show that, averaged over kpc-scales, **B** usually points within the plane, while the polarization vectors of relatively nearby stars suggest that plumes of magnetic field rise tens of parsecs above the plane in a way reminiscent of magnetic loops observed in the chromosphere of the Sun.

### 3.7.2 Extinction of sight-lines out of the Galaxy

Burstein & Heiles (1982) have mapped $E(B - V)$ over most of the sky that lies outside the zone of avoidance ($|b| < 10°$) – see Figure 3.21. Their values

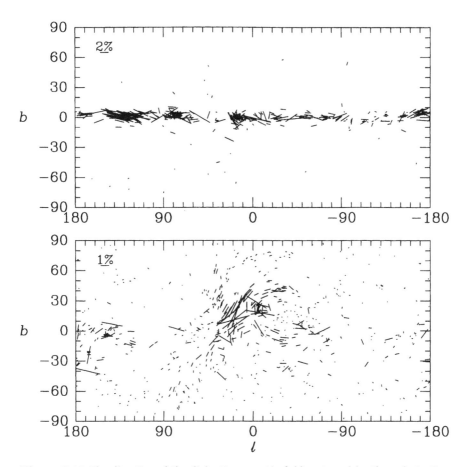

**Figure 3.19** The direction of the Galactic magnetic field as traced by the polarization of light from stars with distances between 100 pc and 200 pc (bottom) and between 1 and 1.5 kpc (top). The length of each line is proportional to the fractional polarization of the light from a star according to the scale indicated in the top-left corner of each panel. [After Axon & Ellis (1976)]

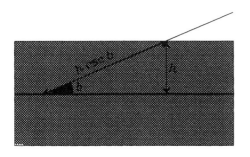

**Figure 3.20** Path length through a homogeneous layer of dust in the Galactic disk. Perpendicular to the Plane the layer has half-thickness $h$.

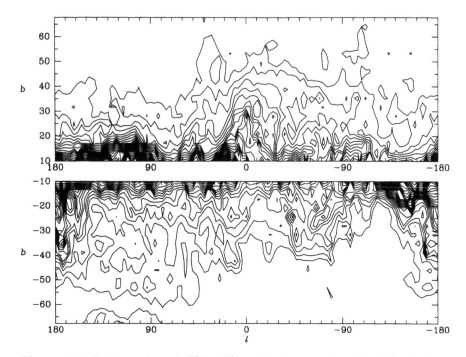

**Figure 3.21** Contours of equal $E(B - V)$ on the sky away from the Galactic plane ($|b| > 10°$). The contour levels are uniformly spaced in $E(B - V)$ from 0.025 to 0.75. Burstein & Heiles (1982) obtained these data by combining galaxy counts with column densities of neutral hydrogen measured from 21-cm line profiles. [After Burstein & Heiles (1982) from data kindly provided by D. Burstein]

of $E(B - V)$ were obtained by exploiting two phenomena: (i) extinction makes external galaxies fainter and therefore reduces the number of galaxies counted in any zone to some given limiting magnitude; (ii) extinction is often associated with neutral hydrogen that can be detected through its 21-cm line (see §8.1.4). Neither of these methods by itself provides a reliable estimate of the extinction along a given line of sight. One problem with the first method is that galaxies cluster, so an enhancement or deficit in the number of galaxies counted may be unconnected with extinction. The second method has the problem that, along a given line of sight, ionized and/or molecular gas may make significant contributions to the total column density of interstellar material, and these contributions are not detectable through the 21-cm line. However, Burstein & Heiles (1978) have shown that these weaknesses can be largely overcome by combining the two approaches. Comparison of their values of $E(B - V)$ with values derived from Strömgren *ubvy* photometry of stars indicates that their values are accurate to 0.01 mag or 10%, whichever is the larger. The reddening at the Galactic poles is $\sim 0.01$ mag, comparable to the overall uncertainty.

Figure 3.21 shows that, although the extinction along lines of sight out of

---

## Box 3.2: Galaxy Counts

Hubble (1936) modeled the distribution of interstellar dust by a slab of half-thickness $h$ – see Figure 3.20. In this model, a line of sight from the plane out of the Milky Way along latitude $b$ passes for a distance $h \csc b$ through the slab. Hubble assumed that $A_V$ is proportional to this distance [see equation(8.48)], so

$$A_V = h_1 \csc b, \tag{1}$$

where $h_1$ is some constant. Now imagine a population of galaxies of absolute magnitude $M$. All those that lie within distance $d$ will have apparent visual magnitudes brighter than $m$, where

$$m = M + 5 \log d - 5 + A_V \tag{2}$$

If these galaxies are uniformly distributed in space with number density $n_M$, there will be $\frac{1}{3}\delta\Omega d^3 n_M$ such galaxies in a cone that subtends solid angle $\delta\Omega$. The number $N(m)$ counted per unit solid angle to apparent magnitude $m$ thus satisfies

$$\log N = 3 \log d + \text{constant} = \tfrac{3}{5}(m - A_V - M) + \text{constant}$$
$$= 0.6(m - A_V) + \text{constant}. \tag{3}$$

Combining this with equation (1), we have

$$\log N = 0.6(m - h_1 \csc b). \tag{4}$$

Thus $h_1$ can be deduced from the slope of a plot of the observed number counts $\log N(m)$ against $\csc b$. From equation (1), $h_1$ is the $V$-band extinction at the Galactic pole.

---

the Milky Way generally diminishes as $|b|$ increases, it varies very irregularly over the sky. Away from the direction to the Galactic center, $E(B-V)$ tends to be small in the North at $b > 20°$, while in the South there is a great spur of high $E(B-V)$, which reaches to $b = -40°$ at $l = 170°$. The asymmetries between the maps for the northern and southern hemisphere arise, at least in part, because the Sun lies slightly above the Galactic plane (§10.1).

## Problems

**3.1** Prove that, if the orbital planes of binaries are oriented randomly with respect to the plane of the sky, then the average value of $\sin^3 i$ for the binaries is $\langle \sin^3 i \rangle = 0.59$.

**3.2** Show that, if the surface of a pulsar were at the same temperature as the photosphere of the Sun, the pulsar would have $M_V \simeq 30$.

**3.3** Show that a star's effective temperature, $T_{\text{eff}}$, can, in principle, be determined without knowing its distance by expressing $T_{\text{eff}}$ as a function of the star's bolometric magnitude, $m_{\text{bol}}$, and its angular diameter, $\theta$. This technique was used by Ridgway *et al.* (1980) to compile the data for giants listed in Table 3.6.

**3.4** Evaluate the difference between the mean luminosities of (i) a magnitude-limited sample and (ii) a volume-limited sample of objects with luminosity function given by equation (3.32). Assume that the population is homogeneously distributed in space.

**3.5** Show that, if the distribution of stars in space is uniform, the distribution of stars over absolute magnitude in a magnitude-limited survey is independent of the survey's limiting magnitude.

**3.6** The **effective volume** of a survey is the ratio of the number of objects reached by the survey to the local space-density of the objects. Consider the case in which the space density of stars is proportional to $\exp(-|z|/h)$, where $|z|$ is distance from the Galactic plane and $h$ is the 'scale-height' of the density distribution (cf. §4.4.3). Assuming that we are located in the Galactic plane, show that the effective volume of a magnitude-limited, all-sky survey is then

$$V_{\text{eff}} = 4\pi h^3 \left[ \tfrac{1}{2}x^2 + (x+1)e^{-x} - 1 \right], \qquad (3.72)$$

where $x$ is related to the absolute magnitude $M$ of the surveyed objects and the limiting magnitude of the survey, $m_{\text{lim}}$, by

$$x(h/\text{pc}) = 10^{0.2(m_{\text{lim}}-M)+1}. \qquad (3.73)$$

By evaluating $V_{\text{eff}}$ for $h = 300\,\text{pc}$, $M = -6, -5, \ldots$ and $m_{\text{lim}} = 2.5, 6$ reproduce the numbers in Table 3.17.

Explain why Malmquist bias is more pronounced for $m_{\text{lim}} = 2.5$ than for $m_{\text{lim}} = 6$.

**3.7** From a study of the red clump in Baade's window Stanek (1996) concludes that $A_V/E(V-I) = 2.49 \pm 0.02$. Is this datum compatible with the numbers cited in §3.7.1?

**3.8** Verify that the quantity $Q$ defined by equation (3.69) is reddening independent.

**3.9** Equation (1) of Box 3.1 defines the generic inverse problem. Appendix C describes the Richardson–Lucy algorithm for the solution of such equations. This algorithm requires that the functions $f, g$ and $K$ satisfy equation (C.1). In case $f, g$ and $K$ do not satisfy this equation, show that the functions

$$\widetilde{f}(x) \equiv \frac{f(x)}{f_0}, \quad \widetilde{g}(\xi) \equiv \frac{1}{f_0}g(\xi)k_0(\xi) \text{ and } \widetilde{K}(x,\xi) \equiv \frac{K(x,\xi)}{k_0(\xi)} \qquad (3.74)$$

do, where

$$f_0 = \int \mathrm{d}x\, f(x) \text{ and } k_0(\xi) = \int \mathrm{d}x\, K(x,\xi). \qquad (3.75)$$

# 4

# Morphology of Galaxies

In this chapter, we begin our study of external galaxies by discussing their shapes and something of what can be learned of their composition from the spectrum of the light they emit. In §4.1 we discuss the **morphological classification** of galaxies, that is, the classification of galaxies into types, either on the basis of their gross appearance, or on the basis of their integrated spectra. The morphological type of a galaxy turns out to be correlated with whether that galaxy resides in a high- or a low-density region of the Universe. In §4.1.2 we review the connection between a galaxy's environment and its morphology. Galaxies vary enormously in their sizes and luminosities. The least luminous galaxies can only be observed if they happen to lie very close to us. The Milky Way is the second most luminous member of a group of about 35 galaxies that is called the "Local Group". Most of the luminosity in the Universe is believed to come from galaxies which are members of similar galaxy groups. Consequently, studies of the Local Group are important and we review them in §4.1.4. It turns out that we can put much of the morphological classification of galaxies on a quantitative basis by measuring the surface-brightness profiles of galaxies. In §4.2, the techniques involved in such measurements are described. In §§4.3 and 4.4 we discuss the brightness distributions that seem to characterize the various morphological classes defined in the previous section. In particular, we present empirical laws which approximately describe the radial brightness distributions of the various components of galaxies, and we explore what information can be

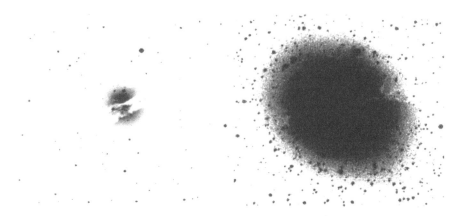

**Figure 4.1** An illustration of how the appearance of a galaxy can depend on how its image is displayed. Both panels depict a CCD frame of the peculiar galaxy NGC 5128, which was obtained in a 10 minute exposure on a 10-inch telescope. Both are shown as negative images, but the adopted grayscales are very different. [From data kindly provided by J. Torres (`http://www.dogtech.com/ccdast/`)]

gleaned about the three-dimensional forms of galaxies from their observed surface-brightness distributions. Most galaxies are attended by an entourage of globular clusters. In §4.5 we discuss these populations and seek correlations between the characteristics of a given galaxy's morphological type and its globular cluster population. A small fraction of all galaxies do not fit comfortably into the standard classification scheme introduced in §4.1. In §4.6 we discuss such galaxies, many of which are believed to be objects that have been more or less violently disturbed by an encounter with another galaxy.

# 4.1 Morphological classification of galaxies

When classifying galaxies on the basis of their images, one has constantly to bear in mind that galaxies are not hard-edged objects of the type that surround us in our daily lives. Indeed, galaxies fall far outside the range of our ordinary experience as regards both their self-illumination and their ghost-like insubstantiality. Figure 4.1 illustrates just how misleading pictures in books and journals can be. Here one sees two images of the same galaxy on the same scale and in the same orientation. Yet, because of the different depths of the two images, the appearance of the object is so different that one may feel a degree of disbelief that both illustrations are of the same system. Similar differences arise between two images in different wavebands, and between spectra of very different resolutions: a feature that is conspicuous in blue light may totally vanish at longer wavelengths, and a narrow spectral line that is prominent in a high-resolution spectrum may be undetectable in a lower-resolution spectrum. Therefore, when classifying objects one has to try to avoid being misled by superficial appearances that are little more than optical illusions, and one must seek quantitative confirmation of qualitative impressions wherever possible.

Imagine that you have collected a library of galaxy images in some way. If your collection is a large one, it will contain objects with a wide range of appearances. When confronted with a set of complex objects, a natural first step is to identify classes of similar objects. So you look carefully at each image and notice features that some galaxies have in common and others do not. Eventually, after much consideration you arrange the images in piles, one for each of your classes. Finally, you may decide that some classes are closely related to one another, while others are not. For example, the galaxies of classes a, b and c might each show spiral structure, while those of class e might not. Moreover, the objects of class b might be intermediate between those of classes a and c in that their spiral arms are intermediate between the open arms of class c objects and the tightly-wound arms characteristic of class a objects.

This example illustrates several fundamental considerations. First, different libraries of galaxy images are liable to lead to the identification of different classes. For example, if none of the images shows spiral structure, the final classification system is likely to focus on features that are apparent in spiral-free images, such as the degree of central concentration of the light, or ellipticity of the image. Similarly, if the library contains images of each object in several wavebands, the colors of an objects may play an important role in the classification scheme, whereas it clearly can play no role if the library contains only one image per object. The classifications schemes to be described below are based on a small number of catalogs[1] for which galaxies were selected on the basis of (i) their apparent magnitudes and (ii) their typical surface brightness in the $B$ band. It is now known that these catalogs exclude large numbers of objects, some of which may be important for the structure of the Universe (Impey *et al.* 1996).

The second fundamental consideration is that it makes no sense to compare, for example, a $B$-band image of one galaxy with a $R$-band image of another. Nor to compare a shallow image of one galaxy with a deep image of another. In short, it is essential that the images in one's library be homogeneous in the sense that a similar set of images is available for every object. If this criterion is not met, objects that are actually nearly identical are liable to be assigned to very different classes because, for example, one object was imaged in a band that enhances spiral structure and the other was imaged in a band that obscures it. This problem is particularly acute if one wishes to compare the morphology of a galaxy observed at large redshifts with that of a nearby galaxy: for this comparison one requires images in wavebands

---

[1] Principally the *Shapley Ames Catalogue of Bright Galaxies* (Sandage & Tammann 1981).

**Table 4.1**  General galaxy classification schemes

| System | Principal criteria | Symbols | Examples |
|---|---|---|---|
| Hubble–Sandage[1] | barrishness; openness of arms/disk-bulge ratio; degree of resolution of arms into stars | E, S0, S, SB, Irr a, b, c | M87=E1 M31=Sb M101=Sc LMC=Irr I |
| De Vaucouleurs[2] | barrishness; openness of arms/disk-bulge ratio; rings or s shapes | E, S0, S, SA, SB, I a, b, c, d, m (r), (s) | M87=E1P M31=SA(s)b M101=SAB(rs)cd LMC=SB(s)c |
| Yerkes[3] | central condensation of light; barrishness/smoothness | k, g, f, a E, R, D, S, B, I | M87=kE1 M31=kS5 M101=fS1 LMC=afI2 |
| DDO[4] | young-star richness of disk; barrishness; central condensation of light; quality and length of arms | E, S0, A, S, Ir B a, b, c I, II,.., V | M87=E1 M31=Sb I–II M101=Sc I LMC=Ir III–IV |

NOTES: [1]Sandage (1961–1995), [2]de Vaucouleurs (1959b), [3]Morgan (1958–1970), [4]van den Bergh (1960–1976) that are similar in the *rest* frames of the two galaxies. For example a $R$-band image of a galaxy at redshift $z = 1$ should be matched by a $U$-band image of the nearby galaxy. Few such images are currently available.

The third fundamental consideration is that the chosen classification scheme is likely to reflect one's views as to what is 'physically important' and what is 'merely superficial.' Since morphological classification is an early step taken towards understanding the objects in an image library, these prejudices are likely to be mere hunches; it would not be surprising if it later emerged that some features that were originally dismissed as 'superficial' are in fact fundamental, and vice versa.

The fourth fundamental consideration is that the criteria used to assign a galaxy to a class should permit a unique classification in each case. In particular, if more than one criterion is employed, it is important to be clear in what order the criteria are to be applied lest the different criteria lead one to place a given galaxy into more than one class.

Table 4.1 lists four galaxy-classification schemes that have been devised and more or less widely employed. These schemes have been developed from libraries of $B$- or photographic-band images. Hence colors do not play a defining role. The defining libraries are designed to include all galaxies brighter than some given apparent magnitude, or having isophotal diameters in excess of some given angle. We shall see in §4.1.3 that these criteria effectively ensure that the schemes are defined in terms of rather luminous galaxies, that are actually not at all typical of the generality of galaxies in the Universe. Most of these schemes employ more than one classification criterion in violation of one of the principles discussed above, with the result that the classification of a given galaxy is occasionally ambiguous.

Most published galaxy classifications are the result of a subjective assessment of a picture by a human being. Do different astronomers place a given object in the same class when they work from identical material? Naim *et al.* (1995) investigated this question by getting six astronomers to classify 831 galaxies. Much of the time two astronomers placed the same galaxy in different classes. However, in such cases the two classes were invariably very similar ones – adjacent ones when the classes are placed in an ordered linear sequence.[2] This result implies that, while the classification scheme used is fundamentally sound, one should not take a galaxy's exact classification too seriously.

### 4.1.1 The Hubble sequence

The most widely used classification scheme is that introduced by Hubble (1936) in his book *The Realm of the Nebulae*, and subsequently modified as his collection of large-scale plates of giant galaxies grew. Reproductions of many of these plates were published posthumously in 1961 by Sandage as

---

[2] The RMS dispersion in Hubble $T$ type defined by Table 4.2 was typically 1.8.

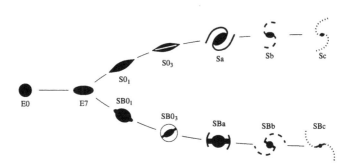

**Figure 4.2** The Hubble tuning-fork diagram. This figure differs
from Hubble's original diagram in that it shows stages of lenticular
galaxies interposed between the ellipticals and the spirals.

*The Hubble Atlas of Galaxies*, alongside an introduction by Sandage which
is generally regarded as the definitive exposition of Hubble's scheme. In
1994 the *Hubble Atlas* was superseded by the much more extensive *Carnegie
Atlas of Galaxies* (Sandage & Bedke 1994). Hubble classifications for all
the galaxies in the *Shapley–Ames Catalogue of Bright Galaxies* are given
by Sandage & Tammann (1981). Figure 4.2 illustrates the scheme in its
final form. Hubble suggested that galaxies evolved from the left-hand end of
this sequence to the right. This now discredited speculation lives on in the
convention that galaxies that lie towards the left-hand end of the sequence
are referred to as **early-type** galaxies, while those towards the right-hand
end are referred to as **late-type** galaxies.

On the left of Hubble's **tuning-fork diagram** are placed those galaxies
which appear smooth and structureless. These **elliptical galaxies** vary in
shape from round to fairly highly elongated in form. A galaxy of this type is
designated E$n$, where the number $n$ describes the apparent axial ratio $(b/a)$
by the formula $n = 10[1 - (b/a)]$. Thus, a galaxy which appears round on
the sky is designated E0, and one whose major axis is twice as long as its
minor axis is and E5 galaxy. Figure 4.3 shows examples of each type of
elliptical galaxy, ranging from the round E0 galaxy NGC 3379 through the
highly elongated (E6) system NGC 3377 – few, if any, ellipticals are more
elongated than E6.

The galaxies shown in Figure 4.3 are all highly luminous objects. Most
elliptical galaxies are much less luminous, and are designated dE for **dwarf
elliptical galaxy**. Three well-studied dE galaxies in the Local Group are
NGC 147 (dE3), NGC 205 (dE5) and M32 (dE2). Figure 4.4 shows a photograph of NGC 147. Notice the low surface-brightness envelope of the system
against which many foreground stars are conspicuous. The Local Group also
contains a large number of **dwarf spheroidal galaxies** (dSph). In these
systems the star density is everywhere so low that on images obtained with
even the largest ground-based telescopes, they appear as mere clusterings of
faint (but intrinsically moderately luminous) stars. No smooth background

**Figure 4.3** Four elliptical galaxies. From top left to bottom right NGC 3379 (E0), NGC 4261 (E3), NGC 4486 = M87 (E0), and NGC 3377 (E6). [Photographs from the *Carnegie Atlas* courtesy of A. Sandage]

of light produced by less luminous stars is visible. The Sculptor system, discovered by Shapley (1938), was the first dSph galaxy to be discovered, but more of these systems are discovered each decade. Ground-based telescopes cannot detect dSph galaxies that lie significantly outside the Local Group. In §4.3.4 we shall see that these systems may differ fundamentally from classical E galaxies but form the low-luminosity tail of a sequence whose brighter members include dE galaxies such as NGC 205 but not M32.

Globular clusters are not normally accorded the dignity of being termed elliptical galaxies, being considered mere appendages of more massive systems like our Galaxy. But from a structural point of view, it is sometimes useful to regard them as a species of extremely low-mass, elliptical galaxy. Indeed, one finds that the stellar content of globular clusters is frequently quite similar to what one might expect to find in a very low-mass elliptical galaxy, and their density profiles prove to be intermediate in type between those of giant ellipticals and of dwarf spheroidals. Many of the globular clusters that are located far out in the galactic halo, for example the *Palomar*

**Figure 4.4** The dwarf elliptical galaxy NGC 147. [Photograph from the *Carnegie Atlas* courtesy of A. Sandage]

*clusters* (Abell 1955), probably provide a link between the classical clusters, which tend to be tightly bound to giant galaxies, and the more massive but less centrally concentrated dwarf spheroidals. If one extends Hubble's original elliptical class in this way, one finds that elliptical galaxies span an enormous range (at least six decades) of intrinsic brightness (see §4.1.3).

After the elliptical galaxies, Hubble's diagram bifurcates into two branches – the 'normal' and the 'barred' galaxies. The word 'normal' should not mislead one into supposing these galaxies are more common than the barred species; the two types actually occur with similar frequencies. The reader should also beware of imagining that the division of galaxies into normal and barred type is clear and unambiguous. Most galaxies show some of the characteristics of barred galaxies, and it is only the more extreme examples that are classified as barred systems. Furthermore, barlike characteristics may be lost on small-scale plates, especially of nearly edge-on systems, so that misclassification in this respect is quite common.

In the middle of Hubble's tuning-fork diagram, at the junction of the elliptical and the spiral galaxies, comes a class of galaxies known as **lenticular galaxies**. These galaxies are designated as type S0 or type SB0 according to whether or not they are barred. The S0 galaxies are characterized by a smooth central brightness condensation (the bulge or spheroidal component) similar to an elliptical galaxy, surrounded by a large region of less steeply declining brightness. This latter component, which is generally rather structureless (although it may sometimes contain some dust), is believed to be intrinsically rather flat. Indeed, when systems of this type are seen edge-on, one sees clearly that they are very flat in their outermost parts. Between the inner, elliptical-like bulge region and the outer (**envelope**) parts, many lenticular galaxies possess smooth subsidiary features called 'lenses' – see below. The family of S0 galaxies is subdivided into three classes, $S0_1$, $S0_2$, and $S0_3$ galaxies, according to the strength of dust absorption within their disks. The $S0_1$ galaxies do not show any signs of absorption by dust, whereas the $S0_3$ galaxies have a complete dark band of dust absorption running within their disks around their elliptical-like components (see Figure 4.5). The

**Figure 4.5** The normal lenticular galaxy NGC 4976 (S0$_1$) at top left and the barred lenticular galaxy NGC 4371 (SB0) at bottom left. The top right panel shows a deep image of the normal lenticular galaxy NGC 6861 (SB0$_3$). In the lower right panel a much shallower photograph shows the dust lane near the center of this same galaxy. [Photographs from the *Carnegie Atlas* courtesy of A. Sandage]

strength of dust absorption in S0$_2$ galaxies is intermediate between these two extremes. The **barred lenticular galaxies** are also divided into three types, SB0$_1$, SB0$_2$, and SB0$_3$, but here the division is made according to the prominence of the bar rather than the presence of dust-absorption lanes. In SB0$_1$ galaxies, the bar shows only as two broad regions of slightly enhanced brightness on either side of the central bulge. In SB0$_3$ galaxies, the bar is narrow and well defined and extends completely across the lens. SB0$_2$ galaxies have bars of intermediate prominence.

In the Hubble sequence, the lenticulars are followed by the spiral galaxies. A **normal spiral galaxy** comprises a central brightness condensation, which resembles an elliptical, located at the center of a thin disk containing more or less conspicuous spirals of enhanced luminosity, the **spiral arms**. A barred spiral has, interior to the spiral arms, a **bar**, often containing dark lanes believed to be produced by absorption of light by dust. The spiral arms of barred spirals generally emanate from the ends of the bar.

Within each class of spiral, normal, or barred, a sequence of subtypes is identified by division according to a combination of three criteria (see Figures 4.6 and 4.7): (1) the relative importance of the central luminous bulge and

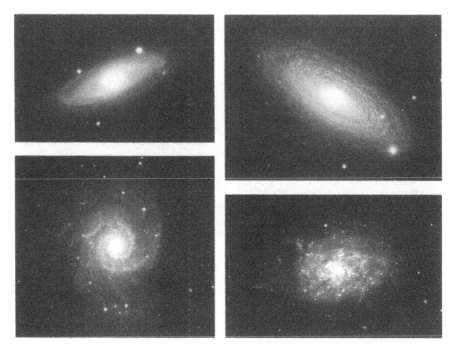

**Figure 4.6** Four normal spiral galaxies: at top left NGC 2811 (Sa); at top right NGC 2841 (Sb); at lower left NGC 628 (Sc); at lower right NGC 7793 (Sd). [Photographs from the *Carnegie Atlas* courtesy of A. Sandage]

the outlying disk in producing the overall light distribution of the galaxy; (2) the tightness with which the spiral arms are wound; and (3) the degree to which the spiral arms are resolved into stars and individual emission nebulae (HII regions). Properly speaking, this superposition of criteria is highly unsatisfactory, but it turns out that there is sufficient correlation among these three parameters – in the sense that galaxies with conspicuous central bulges tend to have tightly wound spiral arms that are not highly resolved into stars – that experienced observers independently classifying objects usually agree rather closely in their class assignments. Early-type spirals (that is, those more to the left in the Hubble diagram; see Figure 4.2) are those having conspicuous bulges and tightly wound, smooth arms. These spirals are designated Sa or SBa according to whether they are barred (SBa) or not (Sa).[3] Late-type spirals, having small central brightness condensations (bulges) and loosely wound, highly resolved arms, are designated Sc and SBc. In between are Sb and SBb galaxies. Intermediate stages of central condensation and tightness of winding of the arms are designated Sab, Sbc, and so on.

---

[3] Transition cases between lenticulars and spirals are designated S0/a.

**Figure 4.7** Four barred spiral galaxies: top left NGC 1291 (SBa); top right NGC 1300 (SBb); lower left NGC 7741 (SBc); lower right the LMC (SBm). [Photographs from the *Carnegie Atlas* courtesy of A. Sandage]

**Table 4.2** Hubble stage $T$

| Hubble | E | E–S0 | S0 | S0/a | Sa | Sa–b | Sb | Sb–c | Sc | Sc–Irr | Irr I |
|---|---|---|---|---|---|---|---|---|---|---|---|
| de Vaucouleurs | E | S0$^-$ | S0$^0$ | S0$^+$ | Sa | Sab | Sb | Sbc | Scd | Sdm | Im |
| $T$ | | -5 -3 | -2 | 0 | 1 | 2 | 3 | 4 | 6 | 8 | 10 |

Unsymmetrical galaxies were assigned by Hubble to two classes of **irregular galaxies**. **Irr I galaxies** are objects that lack symmetry or well defined spiral arms and display bright knots that contain O and B stars. **Irr II galaxies** are unsymmetrical objects that have rather smooth images. They frequently display dust lanes.

Later workers have generally considered Hubble's original scheme satisfactory in regard to the ellipticals, but many have said that Hubble's classification of the spirals is incomplete and that his treatment of irregular galaxies – of which more will be said presently – was quite inadequate. In particular, de Vaucouleurs (1959b) argued that Hubble's division of the spirals does not do justice to the great variety of morphologies of real galaxies, and proposed a more elaborate classification. Most importantly, de Vaucouleurs extended

**Figure 4.8** Images of galaxies containing: a) a nuclear ring (ESO 565–11); b) an inner ring (IC 5240); c) an outer ring (NGC 1543); and d) all three ring types (NGC 6782). Note that the images are displayed at different contrast levels to show the various features. [After Buta & Combes (1996) from data kindly provided by R. Buta]

the prongs of Hubble's tuning fork with additional classes Sd, Sm and Im. The Sd class overlaps Hubble's Sc class to some extent, but it also contains some more extreme objects which were classified as Irr I galaxies in Hubble's scheme. The Sm and Im classes contain the remaining galaxies of Hubble's Irr I class – the 'm' stands for 'Magellanic' because the Large Magellanic Cloud (LMC) is classified SBm. Only very irregular and loose objects like the Small Magellanic Cloud (SMC) and IC 1613 are placed in the Im stage by de Vaucouleurs.

De Vaucouleurs introduced the notation SA for an unbarred galaxy to match Hubble's designation SB for barred systems; an SAB galaxy is weakly barred. In this system a lenticular galaxy is designated S0 only if one cannot

tell whether it has a bar or not – for example because it is seen edge-on.
De Vaucouleurs and subsequently Kormendy (1979) have emphasized the
importance for morphological studies of the rings and lenses that often occur
in disk galaxies, especially barred ones. Figure 4.8 shows some examples of
these phenomena. In a barred galaxy a ring is classified as a **nuclear ring**,
an **inner ring**, or an **outer ring** according to the ratio of its radius $r$ to the
length $a$ of the bar's semimajor axis: a nuclear ring has $r/a < 1$, an inner
ring has $r/a \simeq 1$ and an outer ring has $r/a > 1$. Thus inner rings are touched
by the ends of the bar, while nuclear and outer rings are, respectively, smaller
and larger than the bar. A **lens** is an elliptical (but not bar-shaped) feature
that has a very shallow brightness gradient interior to a sharp outer edge,
beyond which the brightness gradient is steep. Thus a lens may be thought
of as a filled ring. Lenses frequently contain a bar, each end of which touches
the lens' perimeter. De Vaucouleurs uses the symbols (r) and (s) to indicate
systems with and without rings.

Table 4.2 indicates the correspondence between the Hubble and de Vaucouleurs schemes, and also defines the numerical Hubble stage $T$ that is
useful in some quantitative studies.

Elmegreen & Elmegreen (1982) divided galaxies into 12 **arm classes**
according to the morphology of their spiral arms. Galaxies in arm class 1
have chaotic, fragmented, unsymmetrical arms, while galaxies in arm class 12
have two long symmetric arms that dominate the optical disk. These classes
correlate with luminosity and, in fact, a simpler predecessor of this scheme
formed the basis for a now obsolete distance indicator (van den Bergh 1960).

### 4.1.2 Effects of environment

The wide range of galaxy morphologies raises the issue of what causes this
diversity. One clue to finding an answer to this question has come from
studies of the environments in which galaxies of different types are found. In
particular, as soon as reliable observations of galaxy types were made, it was
recognized that galaxies in clusters are much more likely to be ellipticals or
S0s than are those in the field (e.g. Hubble & Humason 1931). This finding
suggests that environmental factors play an important role in determining
the morphology of a galaxy.

Not all clusters are the same, however: Oemler (1974) found that as
many as 40% of the galaxies in some clusters are elliptical, while in others
the proportion of ellipticals is more like 15%. This **elliptical fraction**, $f(E)$
was found to correlate with the morphology of the cluster, in that a cluster
with a large value of $f(E)$ tends to have a regular, symmetric appearance,
often with a giant cD galaxy (see §4.3.1) at its center, while a cluster with a
small value of $f(E)$ generally has a ratty appearance.

Oemler's study also provided the first quantitative evidence that the
balance of morphological types varies within individual clusters. The pro-

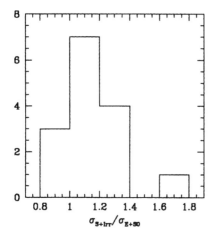

**Figure 4.9** The distribution of the ratio of the random velocities of spiral and irregular galaxies in a cluster, $\sigma_{S+Irr}$, to the random velocities of ellipticals and S0s, $\sigma_{E+S0}$. [From data published in Sodré *et al.* (1989)]

jected number density of galaxies in a centrally-concentrated cluster decreases monotonically with radius $R$, but Oemler discovered that the projected density of spirals actually *increases* with $R$ in the inner parts of such a cluster. Thus, the fraction of spirals, $f(\mathrm{Sp})$ in these systems increases with $R$ from very small values at the centers. This result is particularly striking because we are only measuring the *projected* density of galaxies, that is, the average density along lines of sight through the cluster. When proper allowance is made for this fact, the observations are consistent with there being essentially *no* spiral galaxies in the cores of regular clusters; all of those that are seen at small projected radii probably lie at large distances from the cluster's center. This discovery provided the first example of a **morphology–radius** relation in clusters. Similarly, for several clusters Melnick & Sargent (1977) compared the ratio of the density of spiral galaxies to that of S0 galaxies, and found that the S0s become increasingly dominant at small radii. Studies at other wavelengths show similar trends: radio studies of HI in spiral galaxies (see §8.2.1) have found that those few spirals that are seen projected close to the center of a cluster contain significantly less HI than do spirals at larger $R$, and the average HI deficiency is found to vary monotonically with $R$ (Magri *et al.* (1988), and references therein).

The spatial segregation of galaxy types in clusters should give rise to measurable differences between the kinematics of galaxies of different types. Indeed, the reason spiral galaxies lie further from the centers of clusters than do ellipticals must be that spirals follow more energetic orbits through the cluster. Hence, *at a given distance from the center of a cluster*, spiral galaxies should have larger random velocities than elliptical galaxies. Figure 4.9 suggests that this effect is evident in observed galaxy velocities. However, some caution is appropriate in the interpretation of these data since one would expect the velocity dispersion within cluster galaxies to decrease with projected radius $R$. Hence the mean random velocity of *all* the spirals in a

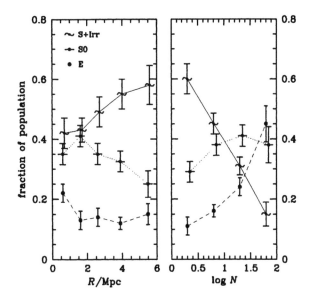

Figure 4.10 The fraction of galaxies of different types in a sample of 6 clusters plotted both as a function of radius and as a function of local projected galaxy density. [From data published in Dressler (1980)]

cluster should be smaller than the mean random velocity of *all* the cluster's ellipticals. In observed samples this effect is likely to be masked by the inclusion within the spiral sample of many galaxies that lie either in front of or behind the cluster and are not physically associated with it: the measured 'random velocities' of these foreground and background galaxies will reflect the Hubble flow rather than the velocity dispersion of the cluster.

The first largescale study of morphological segregation was made by Dressler (1980), who obtained the morphological types of $\sim 6000$ galaxies in 55 clusters. Dressler's analysis confirmed that $f(\text{Sp})$ increases with $R$ – see the left panel of Figure 4.10. In this larger data set, subtler effects are also evident. For example Figure 4.10 indicates that the fraction of S0s, $f(\text{S0})$, falls near the center and detailed analysis of these data shows that $f(\text{S0})$ decreases from $54 \pm 7\%$ at 0.25 Mpc to $30 \pm 5\%$ at 0.075 Mpc (Whitmore & Gilmore 1991). However, Dressler concluded that correlations involving $R$ are not the fundamental ones, and that galaxy type is really dictated by the local density of galaxies. That is, Dressler argued that the fundamental relations are **morphology–density relations** between the local galaxy density and $f(\text{Sp})$ and $f(\text{S0})$. The existence of such relations is illustrated in the right panel of Figure 4.10, which $f(\text{Sp})$ and $f(\text{S0})$ as a function of the projected number density of galaxies, $N$.

It is extremely difficult, if not impossible, to ascertain which of the relations shown in Figure 4.10 is the more fundamental. The distribution of galaxies in clusters is such that $N$ is a monotonic function of $R$: Beers & Tonry (1986) have shown that $N \propto R^{-1}$ over a large range of radii in many

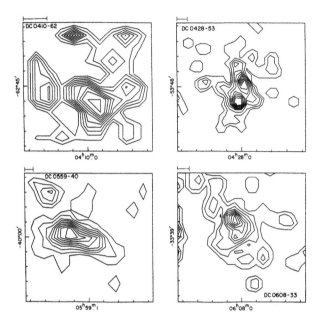

**Figure 4.11** Contour plots showing the projected number density of galaxies in several clusters. The bar in each diagram represents 0.24 Mpc at the cluster distance. Note the variety of morphologies, and the prevalence of substructure. [After Geller & Beers (1982)]

clusters, while Merrifield & Kent (1989) found that at small radii the relation is better described by an exponential, $N \propto \exp(-R/100\,\mathrm{kpc})$. This close relation between $N$ and $R$ implies that if morphology is correlated with $R$, it is necessarily also correlated with $N$, and vice-versa. Without invoking some underlying physical mechanism, it is thus essentially impossible to decide which correlation is the more fundamental.

In fact, the true picture is a little more complicated. Although the density of galaxies does broadly drop with radius in clusters, these systems also contain significant **substructure** in the form of local enhancements in the projected density of galaxies – see Figure 4.11. A simple argument shows that the existence of such substructure is a prerequisite for a physically-meaningful morphology–density relation. In the course of its orbit, any given galaxy will travel a large distance through its cluster, and will normally pass through regions of very different galaxy density. It is thus somewhat surprising that the morphology, which was presumably decided some time ago, is strongly correlated with the current surrounding density. The only way that the local density can dictate morphology is if each galaxy is accompanied by a large chunk of its local environment on its orbit. In other words, the cluster must contain a number of fairly coherent subgroups – there must be substructure.

Several studies have sought to establish the role that substructure plays in dictating the morphologies of galaxies. Sanromà & Salvador-Solé (1990) have carried out a simple test which seems to show that substructure cannot be the driving mechanism. They took the data from Dressler's study, and produced a new data set by randomizing the positions of each cluster's galaxies in azimuth whilst preserving their radial positions in the cluster. Thus, any small-scale substructure will be wiped out, while the large-scale radial variations in cluster properties are preserved. Repeating Dressler's analysis on the "shuffled" clusters, they found that the morphology–density relation was essentially unaltered, and hence that this correlation cannot be primarily caused by small-scale substructure.

Additional support for this view has come from a further analysis of Dressler's data by Whitmore, Gilmore & Jones (1993). They considered only the data from the innermost parts of clusters, and looked to see how the fraction of ellipticals from these comparable radii varied with the density of their surroundings. This study found that ellipticals always make up $\sim 50\%$ of the central population, irrespective of the density of galaxies around them. It thus appears that a galaxy's radius within a cluster is the primary factor that dictates its morphology.

The issue is not entirely clearcut, however. The fact that Oemler (1974) found that wimpier clusters contain lower global fractions of ellipticals implies that some factor other than radius dictates the mix of galaxy types. Postman & Geller (1984) took this analysis further by studying the morphology–density relation for even poorer groups of galaxies. This study showed that groups seem to obey the same relation as clusters: the inner parts of groups have comparable densities to the outer parts of clusters, and the distribution among the different morphological types in these two environments are indistinguishable. In this case, therefore, the principle correlation seems to be the morphology–density relation. Combining the results from the groups and clusters, Postman and Geller were able to define a single, apparently Universal, morphology–density relation which is obeyed over six orders of magnitude in density.

Even within rich clusters, where the morphology–radius relation appears to be more fundamental, there are phenomena that can only be attributed to local causes. Whitmore, Gilmore & Jones (1993) have found that where a galaxy has a nearby companion, it has a disproportionate probability of being elliptical. In the outer parts of clusters the elliptical fraction is only $\sim 15\%$, but close pairs of galaxies, separated by projected distances of less than $\sim 0.05\,\mathrm{Mpc}$, are $\sim 55\%$ elliptical even in the outer parts of clusters. This is clearly a very localized phenomenon, which falls under the category of a morphology–density relation.

It is apparent, and perhaps not surprising, that the morphology of a galaxy is dictated by a combination of factors, some of which relate to its immediate surroundings, while others reflect its broader environment. Trying to capture the complete physics of the issue in anything as crude as a morphology–radius or morphology–density relation is never going to do justice to this complex process, although it clearly provides a partial picture.

### 4.1.3 The galaxy luminosity function

Just as the distribution of stellar luminosities reflects the physics of star formation and stellar structure, we might hope to learn about galactic evolutionary processes by studying the distribution of galaxy luminosities. This distribution is quantified in an analogous manner to that of stars (§3.6), via the **galaxy luminosity function**, $\Phi(M)$. The quantity $\Phi(M)\,\mathrm{d}M$ is proportional to the number of galaxies that have absolute magnitudes in the range $(M, M + \mathrm{d}M)$. Conventionally, this function is normalized by setting

$$\int_{-\infty}^{\infty} \Phi(M)\,\mathrm{d}M = \nu, \qquad (4.1)$$

where $\nu$ is the total number of galaxies per unit volume, so that $\Phi\,\mathrm{d}M$ specifies the number density of galaxies in the magnitude range $(M, M+\mathrm{d}M)$.

**The field galaxy luminosity function**    In order to determine the generic behavior of the galaxy luminosity function, we need to look at the luminosities of galaxies in some unexceptional region of the Universe. The classical procedure for estimating $\Phi(M)$ for such "field" galaxies follows a similar series of steps to that involved in calculating the stellar luminosity function (§3.6). In its simplest form, it involves measuring the apparent magnitudes of all the galaxies in some representative sample. The individual brightnesses are converted to absolute magnitudes by estimating the galaxies' distances, usually by applying the Hubble law to their observed redshifts (see §7.1). If the redshifts are substantial, then it will also be necessary to apply a K-correction to the magnitude, so that magnitudes calculated for galaxies at different redshifts are all converted to a common waveband (see §2.3.3). Finally, one divides the number of galaxies in each absolute magnitude range $(M, M + \mathrm{d}M)$ by the volume of space that has been surveyed, $V$, to convert to the requisite galaxy number density. It should be recalled that $V$ is likely to vary with $M$: since more luminous objects can be seen to greater distances, $V$ is greater for intrinsically brighter galaxies in a survey that is complete down to some fixed limiting apparent magnitude.

In practice, this simple approach has a variety of shortcomings. First, corrections must be made for the Malmquist-type biases that hamper the analysis of magnitude limited surveys. As we saw in §3.6.1, basic properties of the luminosity function will be distorted whenever the function has a finite spread in luminosity. In fact, such distortions occur even if galaxies all have intrinsically identical luminosities, but a range of estimated absolute magnitudes due to errors in their adopted distances. Estimating the distances to galaxies via the Hubble law is an intrinsically approximate process, so errors of this kind are unavoidable. The problem is particularly acute for nearby galaxies, since the random motions of such systems can exceed their Hubble velocities (see §7.4.2), so the errors in their distance determinations can be very large. This shortcoming is unfortunate, since low-luminosity galaxies can only be observed if they are nearby, so the faint end of the field galaxy luminosity function is rather poorly determined.

A further complication to this analysis arises from the assumption implicit in the classical method that the galaxies in any absolute magnitude range $(M, M + dM)$ are uniformly distributed through the surveyed volume $V(M)$. We now know that this assumption is unfounded. Nearby in the Universe, there are large enhancements in galaxy density in some regions, such as the Virgo cluster. Even on larger scales, galaxies are arranged in large wall-like structures, with intervening voids which are almost empty of galaxies. Thus, for a given absolute magnitude $M$, the number of galaxies found within the volume $V(M)$ can depend as much on the spatial distribution of galaxies in the Universe as it does on $\Phi(M)$. To obtain an unbiased estimate for $\Phi(M)$, we must remove this spatial dependence. A variety of techniques for circumventing this problem are described by Binggeli, Sandage & Tammann (1988) and Efstathiou, Ellis & Peterson (1988).[4]

When these effects are allowed for in the analysis of galaxy surveys, the field galaxy luminosity function is found to take a reasonably simple form. The number of galaxies drops monotonically with increasing luminosity: at faint absolute magnitudes, $\Phi(M)$ decreases approximately exponentially with $|M|$, but brighter than some characteristic absolute magnitude, $M^*$, $\Phi(M)$ cuts off very sharply. A convenient parameterization which reproduces these properties is the **Schechter function**,

$$\Phi(M) = (0.4 \ln 10)\Phi^* 10^{0.4(\alpha+1)(M^*-M)} \exp\left(-10^{0.4(M^*-M)}\right), \qquad (4.2)$$

where $\Phi^*$, $M^*$ and $\alpha$ are chosen to fit the observations. This rather contrived-looking formula takes a simpler form if one works in luminosities rather than magnitudes: defining $\Phi(L)\,dL$ to be the number density of galaxies with luminosities in the range $(L, L + dL)$, the corresponding function is

$$\Phi(L) = (\Phi^*/L^*)(L/L^*)^\alpha \exp(-L/L^*), \qquad (4.3)$$

where $L^*$ is the luminosity corresponding to an absolute magnitude of $M^*$ (Schechter 1976). Thus, $\alpha$ sets the slope of the luminosity function at the faint end, $L^*$ or $M^*$ gives the characteristic luminosity above which the number of galaxies falls sharply, and $\Phi^*$ sets the over-all normalization of galaxy density. This formula was initially motivated by a simple model of galaxy formation (Press & Schechter 1974), but it has proved to have a wider range of application than originally envisaged.

The Schechter function provides a reasonable fit to the luminosity functions derived from a number of galaxy redshift surveys. Efstathiou, Ellis & Peterson (1988) analyzed the data from several such studies, and

---

[4] With a spatially-varying density of galaxies, $\nu$ in the normalization of equation (4.1) is taken to be the mean density of the volume surveyed.

found that almost all could be fitted by a Schechter function with $\Phi^* = (1.6\pm0.3)\times10^{-2}h^3\,\mathrm{Mpc}^{-3}$, $M_B^* = -19.7\pm0.1+5\log h$ and $\alpha = -1.07\pm0.07$ as the parameters of the model.[5] The photometric data in these various surveys were obtained through different optical bandpasses, so assumptions had to be made about the colors of the galaxies in order to infer the common $B$-band luminosity function whose parameters are listed above. The value of $M_B^*$ corresponds to a luminosity of $L_B^* = (1.2\pm0.1)h^{-2}\times10^{10}L_\odot$.

One drawback to studying the luminosity function using blue light is that, as we have seen in §3.6, the young stars that produce most of the blue light are a small minority of the stellar population of the Milky Way, and hence, presumably, other galaxies. Thus, the luminosity function of galaxies in blue light may not be very representative of the average stellar properties of these systems. More recent studies have therefore been made of the galaxy luminosity function at infrared wavelengths, where the light is dominated by more "typical" stars. The $K$-band luminosity function is also found to be well-represented by the Schechter function, with best-fit parameters of $\Phi^* = (1.6\pm0.2)\times10^{-2}h^3\,\mathrm{Mpc}^{-3}$, $M_K^* = -23.1\pm0.2+5\log h$ and $\alpha = -0.9\pm0.2$ (Gardner *et al.* 1997). Apart from the difference between $M_B^*$ and $M_K^*$ (which can largely be attributed to the intrinsic $B-K$ colors of typical galaxies), the parameters of the best-fit Schechter functions in the optical and infrared are indistinguishable. It would therefore appear that the optical data are reasonably representative of the galaxies' over-all stellar populations.

Since $\alpha$ seems to lie close to $-1$, there is a serious danger that integration of equation (4.3) over luminosity will imply that there are infinite numbers of low-luminosity galaxies. This divergence is not as implausible as it might seem: the total luminosity density contributed by the Schechter function is

$$l_{\rm tot} = \int_0^\infty L\Phi(L)\,dL = \Phi_0 L^*\Gamma(2+\alpha),  \qquad (4.4)$$

where $\Gamma$ is the gamma function. For $\alpha = -1$, we obtain $l_{\rm tot} = \Phi_0 L^*$. Thus, although the Schechter function predicts huge numbers of low luminosity galaxies, the integrated contribution from all galaxies to the total luminosity density of the Universe remains finite, as the observed darkness of the night sky requires. In reality, the divergent form of the galaxy luminosity function probably breaks down at some point. However, as we shall see in §4.1.4, in the one place where we can readily detect low-luminosity galaxies, we do, indeed, find that they exist in large numbers.

With larger, deeper surveys, the limitations of the simple Schechter function start to become apparent. Figure 4.12 shows the luminosity function

---

[5] The quantity $h$ parameterizes our ignorance as to the exact conversion from redshift to distance [see equation (7.3)]. It almost certainly has a value somewhere between 0.5 and 1.

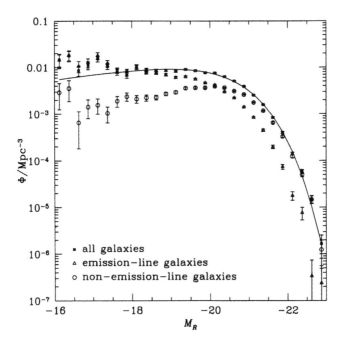

**Figure 4.12** Luminosity function as derived from the Las Campanas Redshift Survey, together with the Schechter function that best fits the data at points brighter than $M = -17.5$ (with parameters $\alpha = -0.70 \pm 0.05$ and $M_R^* = -20.29 \pm 0.02 + 5 \log h$.) The component luminosity functions of emission-line and non-emission-line galaxies in the survey are also shown. [After Lin *et al.* (1996) from data kindly provided by H. Lin]

derived from the Las Campanas Redshift Survey of some 19 000 galaxies. Although fitting a Schechter function to the bright end of this luminosity function reproduces the observations very well, it does not correctly predict the number of faint galaxies.

**The cluster galaxy luminosity function**     Clusters of galaxies provide a very different environment for galaxy formation and evolution. It is therefore interesting to see whether the same luminosity function is found in rich clusters as is found in the field. In some ways, the luminosity function of a cluster is easier to obtain than that of the field: cluster members lie within a small region of the sky, so their photometry can be obtained efficiently with few images. In addition, members are all at essentially the same distance, so one only needs to obtain the single distance to the cluster in order to calculate the galaxies' absolute magnitudes. The only real problem in studying cluster luminosity functions is that rich clusters are quite rare, so they are typically at large distances, making it hard to obtain photometry for their fainter members.

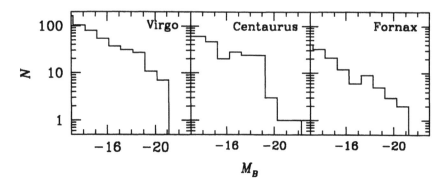

**Figure 4.13** Number of galaxies as a function of absolute magnitude [∝ $\Phi(M)$] found in the central regions of the Virgo, Centaurus and Fornax clusters [From data published in Jerjen & Tammann (1997)]

Studies of some of the nearer clusters have revealed that their luminosity functions are fairly similar to that of the field, and are reasonably well represented by Schechter functions. However, the parameters of the best fit do seem to differ from those of the field. Trivially, $\Phi^*$ will be larger due to the higher density of galaxies in a cluster, but other differences are also apparent. In a study of nine moderately-rich clusters, Lugger (1986) found that on average these systems have Schechter function parameters of $M_B^* = -19.5 \pm 0.1 + 5\log h$ and $\alpha = -1.27 \pm 0.04$. Thus, although the cut-off magnitude is indistinguishable from that of the field galaxy luminosity function, the slope of the faint end of the luminosity function is significantly steeper.

The galaxy luminosity function does not seem to vary greatly from cluster to cluster: Lugger (1986) found all the clusters in her sample to have a luminosity function that is reasonably well reproduced by the average Schechter function described above. However, she also noted that there is structure apparent in the luminosity functions of some clusters that is inconsistent with the smooth Schechter function, and that the form of this structure varies between clusters. Some departures from the Schechter function do occur in more than one cluster, though: quite frequently, there seems to be a dip in the luminosity function at $M_B \sim -16 + 5\log h$, and an excess of galaxies at the bright end of the function. In fact, the existence of an excess of bright galaxies in clusters was recognized in the original paper by Schechter (1976), where he pointed out that cD galaxies (see §4.3.1), with typical luminosities of $10L^*$, cannot fit satisfactorily within the simple form of equation (4.3). These localized features complicate attempts to resolve the question of whether all clusters share similar over-all luminosity functions, since the derived values of the Schechter function parameters can be grossly distorted if the magnitude range over which the fit is made happens to include such a small-scale distortion.

The same story also seems to hold when we look at poorer clusters. Figure 4.13 shows the luminosity functions of three closer, poorer systems. Since these clusters are too close for reliable distances to be obtained from the Hubble law (see §7.4.2), the calibration on to an absolute magnitude scale is not directly comparable to the above analyses. However, it is clear from Figure 4.13 that the over-all picture remains the same: the luminosity function declines steadily with luminosity before cutting off rather sharply at a maximum luminosity. There are also clear signs in these data of sub-structure in the luminosity function, which varies from cluster to cluster, and which cannot be described using the Schechter function.

**The luminosity function divided by morphological type.**     A clue to understanding the complexity of the luminosity function is offered by Figure 4.12. If field galaxies are divided into two sub-groups, depending on whether or not their spectra contain significant emission lines, one finds that the luminosity functions of the two sub-groups differ significantly. Since emission lines are associated with cool gas, which is only found in large quantities in spiral and irregular galaxies, these sub-groups essentially divide the late-type (Sa – Irr) from the early-type (E – S0) galaxies. Given the very different appearances of these sub-classes of galaxies (see §4.1.1), one might reasonably suppose that their members have followed disparate evolutionary paths, and so the difference in their luminosity functions is not unexpected. If anything, it is surprising that their luminosity functions are as similar as they appear: fitting Schechter functions to the individual subgroups, one finds values of $M_R^*$ that differ by only a few tenths of a magnitude. Whether this similarity tells us something about the common ancestry of these objects, or whether it is simply a coincidence, remains to be seen.

As Figure 4.14 shows, if one sub-divides the morphological types of galaxy more finely, one finds that different classes of galaxy have very different luminosity functions. In their decomposition of the luminosity function into its constituent parts, Jerjen & Tammann (1997) have found that spiral galaxies (Sa – Sc) scatter around a well-defined mean luminosity, and their luminosity function can be adequately described by a Gaussian with a mean magnitude of $\overline{M_B^{\rm Sp}} = -16.8 + 5 \log h$ and a dispersion of $\sigma_B^{\rm Sp} = 1.4$ magnitudes. S0 galaxies are found to follow a similar distribution, but are marginally brighter, with $\overline{M_B^{\rm S0}} = -17.5 + 5 \log h$ and $\sigma_B^{\rm S0} = 1.1$ magnitudes. Ellipticals show a slightly more complex distribution: although they are also distributed about a fairly well-defined characteristic luminosity of $\overline{M_B^{\rm E}} = -16.9 + 5 \log h$, their distribution is skewed toward an excess of bright galaxies. This distribution can be characterized by a modified Gaussian distribution,

$$\Phi_{\rm E}(M_B) \propto \exp\left(-[\overline{M_B^{\rm E}} - M_B]^2 / [2\{\sigma_B^E(M_B)\}^2]\right), \qquad (4.5)$$

where $\sigma_B^E(M_B) = 2.2$ if $M_B < \overline{M_B^{\rm E}}$, or 1.3 if $M_B > \overline{M_B^{\rm E}}$. Irregular galaxies are, on average, much fainter. In fact, as one looks to fainter magnitudes,

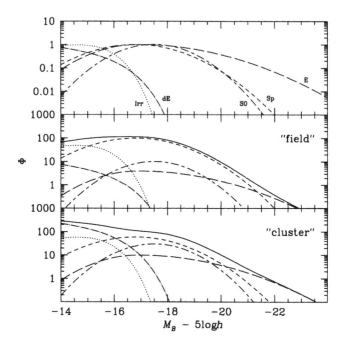

**Figure 4.14** Luminosity functions for galaxies of various morphological types. The top panel shows The separate functions at arbitrary normalization, while the lower panels show approximately how these components combine to produce the total luminosity function in the field and in clusters.

their numbers continue to grow, with no clear signs of a lower limit to their luminosity function. The faint nature of these systems makes their luminosity function hard to characterize, but it can be crudely described by a Schechter function with parameters $M_B^*(\text{Irr}) = -15 + 5\log h$ and $\alpha(\text{Irr}) = -0.3$. The dwarf elliptical galaxies show a similar open-ended luminosity function, which can be fitted by a Schechter function with $M_B^*(\text{dE}) = -16 + 5\log h$ and $\alpha(\text{dE}) = -1.3$.

The disparity in the luminosity functions of galaxies of different types goes a long way toward explaining the variations in the total luminosity function described above. As we saw in §4.1.2, the mix of galaxy types varies with the density of the surroundings. As Figure 4.14 shows, by adding together the mixtures of galaxies appropriate for field and cluster populations, the differences in their luminosity functions become self-explanatory. The high-luminosity cutoffs in these different environments occur at approximately the same magnitude because spiral galaxies and elliptical galaxies have comparable characteristic luminosities. The steeper slope at the faint end of the cluster luminosity function arises from the substantial population of dwarf ellipticals in these systems. We can even explain much of the small-scale structure, such as the dip in the cluster luminosity function at $M_B \sim -16$, which arises at the point where domination of the luminosity function transfers from dwarf to giant galaxies.

Although it is now clear that there is no "universal" total luminosity function, it has been argued that the shapes of the luminosity functions of the individual morphological types may be invariant, with just their amplitudes changing with environment (Binggeli, Sandage & Tammann 1988). To some extent, the truth of this argument must depend on the definitions of morphological types adopted. For example, cD galaxies are almost always found in regions of high galaxy density. If these galaxies are lumped in with the ellipticals, then the elliptical galaxy luminosity function varies with galaxy density and is not universal. If there do exist distinct categories of galaxies whose luminosity functions have universal shapes, then the forms of these function may contain clues as to the histories of these classes of galaxy. Further, a characteristic feature in such an invariant luminosity function might provide a suitable "standard candle" for estimating the distances to quite remote clusters (see §7.3.2). The existence (or non-existence) of universal luminosity functions for particular morphological types therefore remains an important issue in extragalactic astronomy.

### 4.1.4 The Local Group

As we have seen above, the number density of galaxies increases as one considers fainter systems. Thus, the large spiral and elliptical galaxies, which have traditionally caught the eye of astronomers, do not by any means represent a fair sample of the galaxies that populate the Universe. Further, we have seen in §4.1.2 that the mix of galaxies of different types depends quite strongly on environment. Therefore, if we are to ask questions about the "typical" properties of a population of galaxies, we must avoid looking in atypical regions such as rich clusters. Fortunately, the Milky Way is located in a remarkably dull corner of the Universe, so that we can assess a typical galaxy population simply by looking at our neighbors. By looking so nearby, we can also readily observe the large numbers of faint galaxies that are all-but undetectable at greater distances.

The loose collection of galaxies to which the Milky Way belongs is known as the **Local Group**. The criteria for membership of the Local Group are somewhat nebulous. One could simply consider all the galaxies within a distance of, say, 1.5 Mpc to be members, but such a choice affords the Milky Way the unwarranted distinction of being at the group's center. Alternatively, one could examine the radial velocities of nearby galaxies to determine which objects are gravitationally bound to the same system of galaxies as the Milky Way. Unfortunately, this criterion for membership is not entirely clear-cut either, since the uncertainties in the total mass of the Local Group and our ignorance of candidates' transverse velocities mean that we cannot be sure which galaxies are free to depart. We also run into the problem that the Local Group is almost certainly gravitationally bound to other nearby groups, so we have to consider whether these other systems are themselves simply sub-units of the Local Group. Finally, it should be borne in mind that gravitational interactions between galaxies will occasionally cause a galaxy to be flung out of the Local Group, or to be captured from a neighboring group, so membership is not an entirely static property.

Despite the absence of a clear-cut criterion, there is general agreement that the 35 galaxies listed in Table 4.3 belong to the Local Group. The brightest member of the group is the Andromeda Galaxy, M31, with the Milky Way and M33 making up the complement of spirals. As befits its poor environment, there are no particularly bright ellipticals in the group, but dwarf ellipticals and dwarf spheroidals make up more than half the total membership. The remaining galaxies, including our near neighbors the Large and Small Magellanic Clouds, are irregulars.

Figure 4.15 shows the untidy appearance of the Local Group. The system lacks any real symmetry, but rather divides into two main groupings centered on the dominant spiral galaxies, M31 and the Milky Way. There also seems to be a third sub-group loosely clustered around NGC 3109. Since the galaxies in each of these sub-groups tend to have similar line-of-sight velocities, it would appear that they represent physical associations rather than transient conjunctions. Although a lot of Local Group galaxies are associated with these three sub-groups, some members seem to follow more solitary orbits.

It is very unlikely that Table 4.3 provides a complete list of members of the Local Group. Figure 4.15 shows that very few faint members have been identified close to the Galactic plane, particularly in the direction of the Galactic Center. This deficit arises because dust obscures our view in these directions, coupled with the difficulty in identifying diffuse low surface-brightness objects against the bright foreground starlight of the Galactic plane. It was on account of this latter difficulty that the Milky Way's closest neighbor, the Sagittarius Dwarf, was only discovered in 1994 (Ibata, Gilmore & Irwin 1995). Although this system is relatively bright for a dwarf spheroidal, it eluded previous detection because it happens to lie almost directly behind the Galactic center. Its discovery was actually hampered by its proximity, since the large area of the sky over which its emission is spread ($\sim 8° \times 2°$) made it extremely difficult to pick out against the bright Galactic foreground light. Even away from the Galactic plane, new faint members of the Local Group continue to be identified: Whiting, Irwin & Hau (1997) obtained CCD observations of a faint smudge on sky survey plates in the constellation of Antlia, which confirmed that it was another dwarf spheroidal member of the Local Group. Thus, we still don't know the full complement of fainter Local Group galaxies; all we can say with reasonable confidence is that more than 90% of the galaxies in the Local Group have absolute magnitudes fainter than $M_V \approx -18$.

**Table 4.3** Local Group members

| Name | Alternate Name | Coordinates RA (1950) Dec | | | Type | Distance (kpc) | $M_V$ |
|---|---|---|---|---|---|---|---|
| M31 | NGC 224 | 00 40.0 | +40 | 59 | Sb | 725 | −21.1 |
| Milky Way | Galaxy | 17 42.4 | −28 | 55 | Sbc | 8 | −20.6 |
| M33 | NGC 598 | 01 31.1 | +30 | 24 | Sc | 795 | −18.9 |
| LMC | | 05 24.0 | −69 | 48 | Irr | 49 | −18.1 |
| IC 10 | | 00 17.7 | +59 | 01 | Irr | 1250 | −17.6 |
| NGC 6822 | DDO 209 | 19 42.1 | −14 | 56 | Irr | 540 | −16.4 |
| M32 | NGC 221 | 00 40.0 | +40 | 36 | dE2 | 725 | −16.4 |
| NGC 205 | | 00 37.6 | −41 | 25 | dE5 | 725 | −16.3 |
| SMC | | 00 51.0 | −73 | 06 | Irr | 58 | −16.2 |
| NGC 3109 | DDO 236 | 10 00.8 | −25 | 55 | Irr | 1260 | −15.8 |
| NGC 185 | | 00 36.2 | +48 | 04 | dE3 | 620 | −15.3 |
| IC 1613 | DDO 8 | 01 02.2 | +01 | 51 | Irr | 765 | −14.9 |
| NGC 147 | DDO 3 | 00 30.5 | +48 | 14 | dE4 | 589 | −14.8 |
| Sextans A | DDO 75 | 10 08.6 | −04 | 28 | Irr | 1450 | −14.4 |
| Sextans B | DDO 70 | 09 57.4 | +05 | 34 | Irr | 1300 | −14.3 |
| WLM | DDO 221 | 23 59.4 | −15 | 45 | Irr | 940 | −14.0 |
| Sagittarius | | 18 51.9 | −30 | 30 | dSph/E7 | 24 | −14.0 |
| Fornax | | 02 37.8 | −34 | 44 | dSph/E3 | 131 | −13.0 |
| Pegasus | DDO 216 | 23 26.1 | +14 | 28 | Irr | 759 | −12.7 |
| Leo I | DDO 74 | 10 05.8 | +12 | 33 | dSph/E3 | 270 | −12.0 |
| Leo A | DDO 69 | 09 56.5 | +30 | 59 | Irr | 692 | −11.7 |
| And II | | 01 13.5 | +33 | 09 | dSph/E3 | 587 | −11.7 |
| And I | | 00 43.0 | +37 | 44 | dSph/E0 | 790 | −11.7 |
| SagDIG | | 19 27.9 | −17 | 47 | Irr | 1150 | −11.0 |
| Antlia | | 10 01.8 | −27 | 05 | dSph/E3 | 1150 | −10.7 |
| Sculptor | | 00 57.6 | −33 | 58 | dSph/E3 | 78 | −10.7 |
| And III | | 00 32.6 | +36 | 12 | dSph/E6 | 790 | −10.2 |
| Leo II | DDO 93 | 11 10.8 | +22 | 26 | dSph/E0 | 230 | −10.2 |
| Sextans | | 10 10.6 | −01 | 24 | dSph/E4 | 90 | −10.0 |
| Phoenix | | 01 49.0 | −44 | 42 | Irr | 390 | − 9.9 |
| LGS 3 | | 01 01.2 | +21 | 37 | Irr | 760 | − 9.7 |
| Tucana | | 22 38.5 | −64 | 41 | dSph/E5 | 900 | − 9.6 |
| Carina | | 06 40.4 | −50 | 55 | dSph/E4 | 87 | − 9.2 |
| Ursa Minor | DDO 199 | 15 08.2 | +67 | 23 | dSph/E5 | 69 | − 8.9 |
| Draco | DDO 208 | 17 19.2 | +57 | 58 | dSph/E3 | 76 | − 8.6 |

SOURCE: From data kindly provided by M. Irwin.

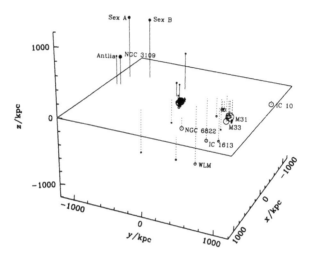

**Figure 4.15** Galactocentric coordinates of Local Group galaxies. The Milky Way lies at the origin, with the Galactic plane at $z = 0$, and the vector from the Sun to the Galactic center parallel to the $x$-axis. Galaxies at positive Galactic latitudes are shown as filled points, while those at negative latitudes are shown as open symbols. The projection on to the $z = 0$ plane is also indicated. The size of the symbol increases with the luminosity of the galaxy.

## 4.2 Surface Photometry of Galaxies

An essential step towards putting the morphology of galaxies on a quantitative basis is the measurement of galaxies' surface-brightness distributions.[6] This is a surprisingly difficult undertaking for two reasons. First the sky is never completely dark. Second, the atmosphere is not completely transparent. In the next section we discuss the way in which these phenomena limit our knowledge of the brightness distributions of galaxies. Then we describe the characteristic brightness distributions of galaxies of each Hubble type. More details of the techniques of surface photometry may be found in de Vaucouleurs (1984), Djorgovski & Dickinson (1989) or Morrison *et al.* (1997).

---

[6] For a quantitative classification of the Revised Shapely-Ames galaxies (Sandage & Tammann 1981) see Michard & Marchal (1994).

### 4.2.1 The night sky

In §2.3.6 we saw that the surface brightness of a given point in a galaxy is independent of the galaxy's distance from us. We saw also that optical astronomers measure surface brightnesses in units of magnitudes per square arcsecond, here abbreviated $\mu_B$, $\mu_R$, etc., depending on the waveband of the measurement. We shall find that half of the light in a typical giant elliptical galaxy comes from points at which the surface brightness is less that $\sim 22\,\mu_B$, while essentially all the luminosity of a typical spiral galaxy comes from points at which the surface brightness is less than $21.5\,\mu_B$. Unfortunately, the surface brightness of blank sky at a good site on a moonless night typically lies near $23\,\mu_B$ (see Table 4.4). To obtain a reasonable understanding of the light distribution of a galaxy, its brightness distribution must be followed down to below $26\mu_B$, that is, to less than 6% of the night sky brightness, and sometimes it is necessary to follow a galaxy down to less than 0.5% of the sky. Clearly, such measurements will be meaningless unless the brightness of the night sky can be *extremely* accurately determined.

The brightness of the moonless night sky is made up of four contributions (Allen 1973, Dube, Wickes & Wilkinson 1972):

1. *Air glow* produced by photochemical processes in the upper atmosphere. This component has a very irregular spectrum, but it is most prominent at long wavelengths. Its intensity varies from point to point on the sky in an irregular way, while, on the average, increasing from latitude 20° to latitude 70° by about a factor of two. It fluctuates in magnitude by about 20% on timescales that can be as short as ten minutes. At many observatories, this component is augmented by mercury and sodium line radiation from the street lamps of nearby cities.
2. *Zodiacal light*, which is sunlight scattered off particulate matter in the solar system.
3. *Faint and unresolved stars* in our Galaxy.
4. *Diffuse extragalactic light*, coming from distant, faint, unresolved galaxies.

The relative proportions of these various components, and the total intensity they produce when added together, vary from one observing site to another. They also vary with galactic and ecliptic longitude and latitude. Generally, zodiacal light is the greatest source of brightness, followed by air glow and diffuse galactic light. Diffuse extragalactic light usually makes the smallest contribution to the night sky. Together, these components shed more light onto the surface of the Earth than all the resolved stars, nebulae, and galaxies put together.

Table 4.4 gives typical values for the brightness (in magnitudes per square arcsec) of the night sky in five wavebands. Notice that the night

**Table 4.4**    Typical values brightness of the night sky

| $\mu_U$ | $\mu_B$ | $\mu_V$ | $\mu_R$ | $\mu_I$ |
|------|------|------|------|------|
| 22.0 | 22.7 | 21.8 | 20.9 | 19.9 |

SOURCE: NOAO newsletter **37**, 1994 March 1

sky is rather red; it has a color index near $(B - V) = 0.9$, similar to that of a fairly red galaxy. Seen from a space vehicle, the sky is darker by a factor that varies with waveband and distance from the ecliptic. In the $V$ band this factor can be as large as 1.5 mag, and still larger factors can apply at longer wavelengths. Moreover, space-based photometry is relatively easy to correct for the night-sky brightness, because airglow, which varies rapidly in time and space, does not contribute to space-based data.

The first steps in the reduction of photometric observations of a galaxy are: (i) to determine the sensitivity of each pixel of one's imaging device; and (ii) to measure the dimming of images towards the edge of the telescope's field of view – that is, the **vignetting** of the field. The process of correcting for these errors is called **flat-fielding**, and errors arising from this step usually dominate the final error budget, especially in the case of HST observations.

Once the imaging device has been flat-fielded, one has to subtract the contribution of the sky background from the flux measured in each region of the sky. It is apparent from the numbers given above that this contribution will, in general, be much greater than the contribution of the galaxy itself. Moreover, the sky brightness at the location of the galaxy cannot be directly measured. One must infer the sky brightness at the galaxy's location by extrapolation from neighboring portions of 'blank' sky. This last requirement makes it desirable to use a detector that covers the largest possible area of the sky, especially if the galaxy being observed has a large angular diameter.[7] Suppose that the errors in this combined extrapolation are comparable to 3% of the measured value. Then the uncertainty in the galaxy brightness becomes of order unity where the surface brightness is of order 26 $\mu_B$. Remarkably, surface-brightness data are commonly reported to be reliable to even below this value.

Figure 4.16 indicates how the derived brightness profile will depend upon the chosen sky-brightness by showing the profile of a galaxy, whose true profile follows the $R^{1/4}$ law, as one would observe it for three different sky-brightness levels. The left-hand panel shows the profile one obtains with too low a background level. The profile declines almost as a power law [a straight line in the $(\mu, \log R)$ plane] before flattening off at some brightness level $\mu_\infty$, which gives the error in the assumed background level. The center panel in Figure 4.16 shows the profile one obtains with an overestimated sky level. The brightness profile now steepens steadily and plunges towards zero at

---

[7] Inadequate sky coverage was for many years a problem with CCD observations of reasonably nearby galaxies and remains a serious problem with HST data.

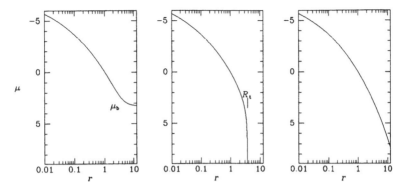

**Figure 4.16** The effects of errors in the subtracted sky background on the profile of a galaxy with a truncated brightness distribution. Left: the background has been underestimated so that the night sky contributes to the "galaxy" brightness at large radii. Center: the background has been overestimated with the consequence that the surface brightness appears to go to zero at the 'tidal radius' $R_t$. Right: with correctly estimated background level the galaxy is seen to follow the $R^{1/4}$ law [equation (4.18) below].

some finite 'tidal radius' $R_t$. Finally, the right-hand panel shows the shape of the profile one obtains with the correct background. This figure shows that an overestimated background level, in contrast to an underestimated one, gives rise to a an entirely plausible surface-brightness profile for the galaxy. Consequently, a profile like that of the center panel in Figure 4.16 cannot by itself be taken as evidence that the galaxy in question has no luminous material beyond $R_t$.

This discussion shows that small errors in the sky-brightness can result in large errors in the derived surface-brightness profile at large radii. Unfortunately, these errors will lead to even more dramatic errors in the derived colors of galaxies at large radii, since these have to be obtained by differencing two derived surface-brightness profiles.

Once one has decided how much to subtract from the measured surface brightness to obtain the galaxy's surface brightness, one can integrate the latter over the area of the image to obtain the galaxy's integrated light. Next one wishes to express this as an apparent magnitude, which is essentially the ratio of the galaxy's light to the light from a standard star. Ideally this is determined by using the same equipment to measure brightness of a standard star. Sometimes, however, the CCD photometry is calibrated by reference with a published apparent magnitude for the galaxy. This will usually have been determined by observing both the galaxy and a standard star with a non-imaging photoelectric detector. The weakness of this procedure is that the spectral responses of the CCD and the photoelectric detector will be different, as will the colors of the galaxy and the standard star. These differences oblige one to use necessarily approximate color transformations. One not infrequently encounters zero-point offsets of order 0.2 mag between the results of different observers, in part because determination of the pho-

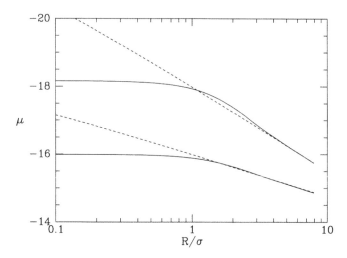

**Figure 4.17** The dashed curves show the true central surface-brightness profiles of circularly-symmetric galaxies in which $\mu \sim r^{-1}$ and $\mu \sim r^{-0.5}$ at small radii. The full curves show the profiles that would be measured for these galaxies if the PSF were a Gaussian of dispersion $\sigma$.

tometric zero-point involves comparing the brightnesses of objects that are widely separated on the sky. Since the properties of the sky vary very little on the angular scales of typical galaxies, the relative brightnesses of different points within a single image should be subject to errors that are very much smaller than 0.2 mag.

### 4.2.2 Effect of seeing

In §3.2.3 we encountered the phenomenon of 'seeing': the blurring of images by the atmosphere. We now investigate how seeing degrades measurements of the surface-brightnesses of galaxies.

The seeing is characterized by the **point-spread function** (PSF) $P(\mathbf{d})$ which gives the probability density that a photon will hit the imaging device at a point that is offset by the vector $\mathbf{d}$ from where it would have hit in the absence of seeing. If, in the absence of seeing, the surface brightness at position $\mathbf{R}'$ in the detector is $I_t(\mathbf{R}')$, then the measured brightness at location $\mathbf{R}$ will be

$$I_{\rm app}(\mathbf{R}) = \int {\rm d}^2\mathbf{R}'\, P(\mathbf{R} - \mathbf{R}')I_t(\mathbf{R}'). \tag{4.6}$$

The simplest case to consider is that of a circularly-symmetric surface-brightness distribution $I_t(r')$ and a Gaussian PSF. If the dispersion of the

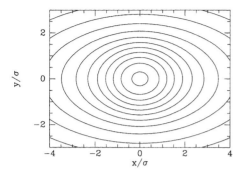

**Figure 4.18** Isophotes for a model E5 galaxy observed with a Gaussian PSF of dispersion $\sigma$. The galaxy's intrinsic surface brightness is proportional to $[x^2 + (y/2)^2]^{-1/2}$.

PSF is denoted $\sigma$ then we have

$$P(d) = \frac{1}{2\pi\sigma^2} \exp\left(-\frac{d^2}{2\sigma^2}\right) \tag{4.7}$$

and it is not hard to show[8] that equation (4.6) can be reduced to

$$I_{\text{app}}(R) = \int_0^\infty dR'\, R' I_{\text{t}}(R') I_0\left(\frac{RR'}{\sigma^2}\right), \tag{4.8}$$

where $I_0$ is a modified Bessel function of order zero. We shall find that the true surface-brightnesses of galaxies tend to diverge as $R \to 0$ as negative powers of $R$; that is for $R$ small we have $I_{\text{t}} \sim R^{-\gamma}$, with $0 < \gamma < 1$. If we plug this form of $I_{\text{t}}$ into equation (4.8), we obtain the apparent surface-brightness profiles shown in Figure 4.17. For $R \lesssim \sigma$ the measured surface brightness is smaller than the observed surface brightness, while for $R \gtrsim \sigma$ the measured surface brightness is larger than the true surface brightness since the light removed from the smallest radii has to emerge at slightly larger radii. The overall effect of seeing is to introduce into a featureless power-law profile an apparent **core**, that is, a central region of nearly constant apparent surface brightness. Since some intuitively attractive theoretical galaxy models possess similar cores, it was until recently widely believed that the cores in apparent surface-brightness profiles reflected cores in the true brightness profiles rather than artifacts introduced by seeing.

Now consider the effects of seeing on a galaxy image that is not circularly symmetric. Figure 4.18 shows isophotes of a model galaxy in which the intrinsic surface brightness is proportional to $[x^2 + (y/2)^2]^{-1/2}$ when the galaxy is observed with a Gaussian PSF of dispersion $\sigma$. In the absence of seeing all isophotes would have axis ratio 2:1, but it is clear that the central isophotes in Figure 4.18 are rounder than those further out. The full curve in Figure 4.19 quantifies this impression by plotting the apparent ellipticity

$$\epsilon \equiv 1 - b/a \tag{4.9}$$

---

[8] See Problem 4.1

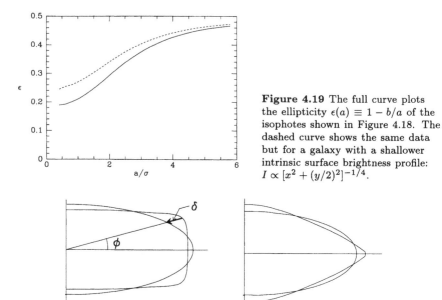

**Figure 4.19** The full curve plots the ellipticity $\epsilon(a) \equiv 1 - b/a$ of the isophotes shown in Figure 4.18. The dashed curve shows the same data but for a galaxy with a shallower intrinsic surface brightness profile: $I \propto [x^2 + (y/2)^2]^{-1/4}$.

**Figure 4.20** Left: an outer boxy isophote fitted by an ellipse with geometry of $a_4$ determination shown. Right: a disky isophote and its fitting ellipse.

of the isophotes as a function of the length $a$ their semimajor axes. Even the isophote with semimajor axis length $a = 4\sigma$ is made appreciably rounder by seeing. The dashed curve in Figure 4.19 shows that the effects of seeing on the measured ellipticity are less pronounced when the surface-brightness increases less steeply towards the center: the dashed curve shows the same quantities as the full curve except for an intrinsic surface brightness proportional to $[x^2 + (y/2)^2]^{-1/4}$.

Isophotes are generally not exactly elliptical. In particular, some isophotes are **boxy** like the outer isophote in Figure 4.20, while others, like the inner isophote in Figure 4.20 resemble a leaf – such isophotes are called **disky**. The boxiness or diskiness of isophotes is often quantified by measuring a quantity denoted $a_4$. Figure 4.20 illustrates this process. First the ellipse $R_e(\phi)$ is fitted to the isophote. Then for each angle $\phi$ one determines the distance $\delta(\phi) \equiv R_i(\phi) - R_e(\phi)$ between the radii of the corresponding points on the ellipse and on the isophote. Finally one expresses the function $\delta(\phi)$ as a Fourier series:

$$\delta(\phi) = \overline{\delta} + \sum_{n=1}^{\infty} a_n \cos n\phi + \sum_{n=1}^{\infty} b_n \sin n\phi. \qquad (4.10)$$

If the isophote is four-fold symmetric and the ellipse has been correctly fitted, $\overline{\delta}$, $a_1$, $a_2$, $a_3$ and all the $b_n$ should be small. If the isophote is disky, $a_4$ will

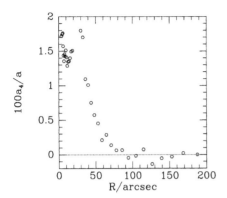 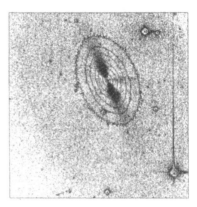

**Figure 4.21** The diskiness [equation (4.10)] of NGC 4697 is large from the center out to $R \simeq 40\,\text{arcsec}$, and from there out is effectively zero (left panel). The right panel demonstrates that this result reflects the existence of a disk at the center of NGC 4697 by showing an image of the galaxy from which a best-fitting elliptical model has been subtracted. [After Peletier *et al.* (1990) from data kindly supplied by R. Peletier]

be greater than zero, while $a_4$ is less than zero for a boxy isophote. The quantity $a_4/a$, where $a$ is the length of the isophote's semimajor axis, forms a dimensionless measure of the **diskiness** of the isophote.

The diskiness $a_4/a$ is even more strongly diminished by poor seeing than is $\epsilon$.

Figure 4.21 illustrates the power of $a_4/a$ to detect a buried disk by showing in the left panel the diskiness of NGC 4697 and in the right panel a picture of the galaxy after subtraction of the best-fitting elliptical model of the system.[9] The diskiness is large out to $R \simeq 30''$ and falls to zero beyond $R \simeq 50''$. The subtracted galaxy image clearly shows a band of luminosity along the major axis, which is almost certainly a disk.

### 4.2.3 Deprojecting galaxy images

What can we infer about the three-dimensional luminosity density $j(\mathbf{r})$ in a transparent galaxy from its projected surface-brightness distribution $I(\mathbf{R})$? If $I(\mathbf{R})$ is circularly-symmetric, it is possible that $j(\mathbf{r})$ is spherically symmetric. Figure 4.22 shows that in this case, $j$ is related to $I$ by

$$I(R) = \int_{-\infty}^{\infty} \mathrm{d}z\, j(r) = 2 \int_{R}^{\infty} \frac{j(r) r \mathrm{d}r}{\sqrt{r^2 - R^2}}. \tag{4.11}$$

---

[9] The technique of showing the difference between an image and a smooth model of the image is called **unsharp masking**. In early work the smooth model was often just an out-of-focus copy of the image itself. Now it is more likely based on either ellipse-fitting or digital smoothing of the original image.

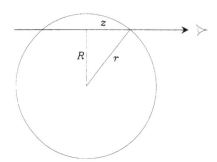

**Figure 4.22** Projecting a spherical luminosity density along the line of sight.

This is an Abel integral equation for $j$ as a functional of $I$, and its solution is [see §4.2 of BT for details]

$$j(r) = -\frac{1}{\pi} \int_r^\infty \frac{dI}{dR} \frac{dR}{\sqrt{R^2 - r^2}}. \qquad (4.12)$$

While equation (4.12) is useful when $I(R)$ is available as a smooth function, it should not be used when $I(R)$ is contaminated by noise, as it will be if it has been inferred from surface-brightness measurements. The Richardson–Lucy technique described in Appendix C provides an appropriate means of recovering $j(r)$ from a noisy surface-brightness profile – see Binney & Mamon (1982) for details.

Often it is desirable to fit to an observed surface-brightness profile a formula that corresponds to a simple analytical form of $j(r)$. Perhaps the simplest pair of formulae that are related by projection is

$$I(R) = \frac{I_0}{1 + (R/r_0)^2} \quad \leftrightarrow \quad j(r) = \frac{j_0}{[1 + (r/r_0)^2]^{3/2}}, \qquad (4.13)$$

where $I_0 = 2r_0 j_0$. The surface-brightness profile given by (4.13) is known as the **modified Hubble law** because of its similarity to Hubble's law introduced in Box 4.1 below. Notice from equations (4.13) that $I \sim R^{-2}$ for $R \gg r_0$, while $j \sim r^{-3}$ for $r \gg r_0$. It is, in fact, a general property of spherical models that if the luminosity density falls as a power law at large $r$, the surface-brightness falls as power law with an exponent that is smaller by unity than the exponent of the luminosity density.

A significant problem with the modified Hubble law is that it predicts that the total luminosity interior to projected radius $R$ diverges as $\log R$:

$$L(R) \equiv 2\pi \int_0^R I(R')R' \, dR' = \pi r_0^2 \ln \left[1 + (R/r_0)^2\right]. \qquad (4.14)$$

**Table 4.5**  Surface brightness profiles associated with equation (4.15)

| $\gamma$ | $I(R)/(L/\pi a^2)$ | $R_e/a$ |
|---|---|---|
| 0 | $\dfrac{-2 - 13s^2 + 3s^2(4 + s^2)|s^2 - 1|^{-1/2}c(s)}{4(s^2 - 1)^3}$ | 2.91 |
| 1 | $\dfrac{(2 + s^2)|s^2 - 1|^{-1/2}c(s) - 3}{2(s^2 - 1)^2}$ | 1.815 |
| $\frac{3}{2}$ | $\dfrac{s^2 - 5}{2(s^2 - 1)^2} + \dfrac{(s^2 - \frac{3}{2})F(\frac{\pi}{4}, s')}{\sqrt{s}(s + 1)^{5/2}(s - 1)^2} + \dfrac{\sqrt{s}(3 - s^2)E(\frac{\pi}{4}, s')}{(s + 1)^{3/2}(s - 1)^2}$ | 1.28 |
| 2 | $\dfrac{\pi}{4s} - \dfrac{1}{s^2 - 1}\left(\dfrac{1}{2} - \dfrac{(2 - s^2)c(s)}{|s^2 - 1|^{1/2}}\right)$ | 0.763 |

NOTES: $s \equiv R/a$; $s' \equiv \sqrt{\frac{2}{s+1}}$; $c(s) \equiv \begin{cases} \cosh^{-1}(s^{-1}) & \text{for } s \leq 1 \\ \cos^{-1}(s^{-1}) & \text{for } s \geq 1 \end{cases}$;
$E$ and $F$ are incomplete elliptic integrals.

Clearly, a real galaxy could not follow this law all the way to $R = \infty$.

A family of simple density profiles that does not suffer from this defect is given by (Dehnen 1993)

$$j(r) = \frac{3 - \gamma}{4\pi} \frac{La}{r^\gamma(r + a)^{4-\gamma}}. \tag{4.15}$$

Here $L$ is the total luminosity of the model and $a$ is a characteristic scale length. When $\gamma$ is either an integer or a half-integer, this form of $j$ yields an analytically tractable projected density distribution – see Table 4.5 for details. Three interesting special cases are noteworthy within this one-parameter family of **Dehnen models**. The model with $\gamma = 1$ is known as the **Hernquist model** (Hernquist 1990), the model with $\gamma = \frac{3}{2}$ yields the closest approximation to the '$R^{1/4}$ model' that is defined below [equation (4.18)], and the model with $\gamma = 2$ is known as the **Jaffe Model** (Jaffe 1983).

It is instructive to consider the surface brightness of a general Dehnen model. Inserting equation (4.15) into (4.11) and making the substitution $r = R \cosh \psi$, we find

$$I(R) = R^{1-\gamma} \int_0^\infty \frac{\sinh^{1-\gamma} \psi \, d\psi}{(1 + R \cosh \psi)^{4-\gamma}}. \tag{4.16}$$

When $\gamma > 1$, $\sinh \psi$ is raised to a negative power on the top of the integrand in (4.16). On account of this the dominant contributions to the integral comes from small values of $\psi$, at which the bottom of the integrand is approximately equal to unity. Hence, for $\gamma > 1$ the integral in (4.16) is approximately independent of $R$ for $R \ll 1$ and $I$ diverges as $R^{1-\gamma}$ at small

$R$. If, by contrast $\gamma < 1$, then $\sinh\psi$ is raised to a positive power in (4.16) and the dominant contributions to the integral come from those values of $\psi$ for which $R\cosh\psi \simeq 1$. In these circumstances the integral is sensitive to the value of $R$ and we no longer have $I \simeq R^{1-\gamma}$.[10] In fact, if $R \ll 1$, we can argue that for the dominating values of $\psi$, $\sinh\psi \simeq \cosh\psi \simeq R^{-1}$ and when we use this estimate to evaluate the integrand, we find that $I$ is approximately independent of $R$ for very small $R$. These approximate results are confirmed by the exact results listed in Table 4.5. Thus the case $\gamma = 1$ is a sort of watershed: luminosity densities that increase inwards more strongly than $r^{-1}$ give rise to cuspy surface brightness distributions, while less steeply rising luminosity densities give rise to surface brightness profiles that tend to a constant at small $R$.

If $I(\mathbf{R})$ is not circularly-symmetric, the galaxy cannot be spherically symmetric, but it might be axisymmetric. If we assume that this is the case, and that the line of sight to the galaxy lies within the galaxy's equatorial plane, then $I(\mathbf{R})$ and $j(\mathbf{r})$ are again related by a soluble integral equation. Binney, Davies & Illingworth (1990) give a Richardson–Lucy algorithm suitable for solving this equation for noisy data $I(\mathbf{R})$.

In general the line of sight will be inclined at an angle to an axisymmetric galaxy's equatorial plane. In this case Rybicki (1987) showed that the image contains insufficient information to reconstruct the three-dimensional luminosity density $j(\mathbf{r})$. Let the galaxy's **inclination** $i$ be the angle between the normal to the equatorial plane and the line of sight. Then the amount of missing information increases steadily from the edge-on case, $i = 90°$, in which $j$ can be uniquely inferred from $I$ to the face-on case $i = 0°$. In the face-on case it is easy to see that many different luminosity densities project to the same (circularly-symmetric) surface-brightness distribution: given one possible axisymmetric luminosity distribution, others can be generated by moving stars parallel to the line of sight. In the case $0 < i < 90°$ it is not so easy to see how to change $j(\mathbf{r})$ without changing $I(\mathbf{R})$. However, for any angle $i < 90°$ Gerhard & Binney (1996) display **konus densities**. These densities contain positive and negative regions arranged such that they are completely invisible at any inclination smaller than $i$. Consequently, a konus density can be added to any possible luminosity density $j(\mathbf{r})$ to generate another that projects to exactly the same surface-brightness distribution. It follows that whenever $i < 90°$ it is in principle not possible to determine a galaxy's true shape from photometry alone.

Notwithstanding the existence of konus densities, the Richardson–Lucy algorithm of Binney, Davies & Illingworth (1990) works for inclinations less than $90°$ in the sense that it recovers one of the infinitely many luminosity densities that are compatible with given $I(\mathbf{R})$. However, for $i \lesssim 60°$ the

---

[10] Thus the simple relationship between power-law luminosity densities and surface brightnesses that holds at large radii and was noted above does not necessarily hold at small radii.

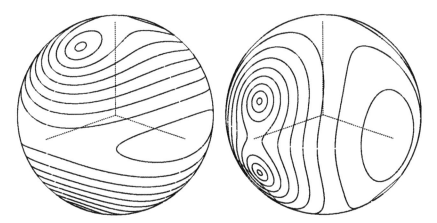

**Figure 4.23** Contours of equal apparent axial ratio on the sphere of viewing directions. At left for an oblate body with $\eta = 0.9$, $\zeta = 0.5$ and at right for a prolate body with $\eta = 0.6$, $\zeta = 0.5$. The dashed lines show the directions of the axes of the ellipsoid, with the shortest axis upward and the longest axis to the left. The small circular contours at the upper left of each figure surround the directions at which the body appears circularly symmetric and $q = 1$. The contours run from $q = 0.975$ to $q = \zeta + 0.025$ in intervals of $\delta q = 0.05$.

recovered luminosity density is liable to be sensitive to noise in the data and physically implausible.

We shall see below that many galaxies cannot be axisymmetric. So let us suppose that a given galaxy is triaxial in the sense that it has three orthogonal symmetry planes, and investigate the relationship between its luminosity density $j(\mathbf{r})$ and its surface-brightness distribution $I(\mathbf{R})$. In the interests of simplicity we additionally assume that $j$ is constant on ellipsoidal surfaces, that is, that $j$ is a function of the variable

$$a_v \equiv \sqrt{x^2 + \frac{y^2}{\eta^2} + \frac{z^2}{\zeta^2}}. \tag{4.17}$$

Here $\eta$ and $\zeta$ are the axis ratios of the ellipsoids of constant $a_v$. Stark (1977) showed that when such a galaxy is projected onto the sky, its isophotes are similar, coaxial ellipses. Both the axis-ratio $q$ and the orientation of these ellipses depend on the direction of the line of sight. Figure 4.23 shows $q$ on the sphere of possible viewing directions for two triaxial bodies. There are always lines of sight along which the body appears circularly symmetric. Figure 4.36 below gives the probability of observing the body to have a given value of $q$.

The orientation on the sky of the elliptical isophotes of a triaxial body depends not only on the orientation of the body, but also on the body's axis ratio. Figure 4.24 shows a prolate ellipsoid inside an oblate ellipsoid. Now, although the coordinate axes in the figure are the principal axes of both the

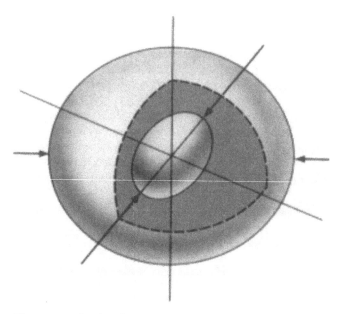

**Figure 4.24** Isophotal twist as a consequence of triaxiality. Two concentric, coaxial ellipsoids are shown. The dashed lines mark the intersections of the ellipsoids with the coordinate planes, while the solid lines show their outlines to the observer. The arrows mark the directions of their apparent principal axes.

inner and the outer ellipsoids, it is evident that the apparent major axis of the outer ellipsoid makes an angle of $50°$ with the apparent major axis of the inner body. This skew alignment of the apparent major axes arises from the fact that the axial ratios of the two bodies are different. The inner body has $a{:}b{:}c = 1{:}2{:}1$, whereas the outer body has $a{:}b{:}c = 1{:}1{:}0.5$. Imagine that these two bodies are two of the isodensity surfaces of an elliptical galaxy, and that the isodensity surfaces between these two have axial ratios such as $1{:}1.8{:}0.9$, $1{:}1.1{:}0.6$, and so on, which form a continuous sequence between the ratios of the two surfaces drawn in Figure 4.24. Then, if the isodensity surfaces do have axial ratios that vary with increasing mean radius, the isophotes will be seen (from the given observing position) to twist continuously through a substantial angle. Thus if galaxies are triaxial, we expect variation in intrinsic axial ratios to be associated with **isophote twist**. By contrast, the isophotes of an axisymmetric system must always be aligned with one another.

Two points are worth noting in connection with isophote twists. (1) If a galaxy happens to be aligned in space such that the line of sight is one of the principal axes, then its isophotes will not twist.[11] (2) Although changes

---

[11] There is another much more remarkable case in which no twist will be observed – see Franx (1988).

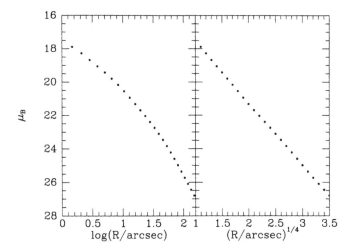

**Figure 4.25** The major-axis brightness profile of NGC 1700 plotted against (a) $\log r$ and (b) $R^{1/4}$. [From data published in Capaccioli, Piotto & Rampazzo (1988)]

with radius in the true axial ratios generate twists, it does not follow that, in any particular galaxy, an isophotal twist must be accompanied by a change in the apparent axial ratio. To specify a triaxial model, one has to determine the two true axial ratios, but fixing the apparent axial ratio at some definite value still leaves one degree of freedom, which allows the position angle of the observed isophotes to vary.

# 4.3 Photometry of Elliptical Galaxies

To a good approximation elliptical galaxies are transparent. Hence the luminosity of these systems may be considered to be the unattenuated light from myriads of individual stars.

### 4.3.1 Radial surface-brightness profiles of elliptical galaxies

Figure 4.25 shows the surface brightness as a function of distance $R$ along the major axis of the giant elliptical galaxy NGC 1700. In the left-hand panel $\mu_B$ is plotted against $\log R$, while in the right-hand panel the independent variable is $R^{1/4}$. If $\mu_B$ varied as a power of $R$, the points in panel (a) would lie on a straight line. At the smallest and largest radii the data can be crudely fitted by straight lines and from the slopes of these lines one infers $I \sim R^{-0.8}$ at small $R$ and $I \sim R^{-1.7}$ at large $R$. To a remarkably good approximation, the points in the right-hand panel lie on a single straight line. This indicates that $\mu_B = x - yR^{1/4}$, where $x$ and $y$ are positive constants, and therefore that

$I \propto 10^{-0.4\mu_B} \propto 10^{-0.4yR^{1/4}}$. Since this relationship between $I$ and $R^{1/4}$ was discovered by de Vaucouleurs (1948), it is known as the **de Vaucouleurs** $R^{1/4}$ **law**. It is conventionally written

$$I(R) = I_e 10^{\{-3.33[(R/R_e)^{1/4}-1]\}}$$
$$= I_e \exp\{-7.67[(R/R_e)^{1/4} - 1]\}. \tag{4.18}$$

The length scale $R_e$ is known as the **effective radius**, and the numerical factor 3.33 in (4.18) is chosen such that, if the galaxy image were circularly symmetric and if the formula were valid all radii, both large and small (and it must be remembered that its validity has been established only for a restricted range of radii), then one-half of the total light of the system would be emitted interior to $R_e$. Thus

$$2\int_0^{R_e} dR\, I(R) 2\pi R = \int_0^\infty dR\, I(R) 2\pi R$$
$$= \frac{8!\, \exp(7.67)}{(7.67)^8}(\pi R_e^2 I_e) = 7.22\pi R_e^2 I_e. \tag{4.19}$$

The parameter $I_e$ is clearly the surface brightness at $R = R_e$, and, according to the $R^{1/4}$ law, the central brightness of the galaxy is $10^{3.33} I_e \simeq 2000 I_e$. From equation (4.19) it follows that the mean surface brightness interior to $R_e$ is $\langle I \rangle_e = 3.61 I_e$.

Unless an elliptical galaxy is exactly circularly symmetric, different values of the effective radius will be obtained by fitting (4.18) to the galaxy's brightness profiles along its major and minor axes. Quoted effective radii are usually the geometric mean of these two values, say $a_e$ and $b_e$. Thus $R_e = (a_e b_e)^{1/2}$. With this definition, half the total light is predicted to be contained within the isophotal ellipse that has area $\pi R_e^2$.

Given that it contains only two free parameters, $I_e$ and $R_e$, the $R^{1/4}$ law provides a remarkably good fit to the surface-brightness profiles of many elliptical galaxies. However, as Figure 4.27 illustrates, the brightness profiles of elliptical galaxies show genuine individuality and some deviate significantly from the best-fitting $R^{1/4}$ profile. Box 4.1 describes some three-parameter fitting functions that have been employed to parameterize the brightness profiles of elliptical galaxies more completely than the $R^{1/4}$ law can, but none of these formulae is as important as the $R^{1/4}$ law.

To some extent the deviations of galaxies from the $R^{1/4}$ law depend systematically on the galaxies' luminosities. Specifically, when $\mu$ is plotted versus $R^{1/4}$, the surface-brightness profiles of low-luminosity galaxies tend to curve downwards at all values of $R^{1/4}$, whereas the surface-brightness profiles of the most luminous galaxies curve upwards over significant intervals in $R^{1/4}$. This phenomenon is illustrated by Figure 4.26, which shows characteristic luminosity profiles for galaxies whose luminosities inside 16 kpc lie within six ranges.

**cD galaxies**    Galaxies whose surface-brightness profiles turn strongly upwards in a $(\mu, R^{1/4})$ plot are called **cD galaxies**. Figure 4.28 shows the

## Box 4.1: Three-Parameter Fitting Functions

Fitting functions that contain more than two free parameters would be expected to yield good fits to a larger range of galaxies than can be obtained with the $R^{1/4}$ law. One possibility is to replace the $1/4$ in the $R^{1/4}$ law with $1/n$, where $n$ is fitted to the data. That is, the data are fitted by the **generalized de Vaucouleurs law**

$$I(R) = I_e 10^{\{-b_n[(R/R_e)^{1/n}-1]\}}. \qquad (1)$$

The constant $b_n$ is again chosen such that half the total luminosity predicted by the law comes from $R < R_e$ (Sersic 1968, Ciotti 1991). When $n < 4$, equation (1) predicts a brightness profile that curves downwards in a plot of $\mu$ versus $R^{1/4}$, whereas when $n > 4$ the model profile curves upwards in this plot. Hence when equation (1) is fitted to data for a low-luminosity elliptical galaxy, the best-fitting value of $n$ is likely to be smaller than 4. Conversely, when equation (1) is fitted to data for very luminous elliptical galaxies, particularly cD systems, values of $n$ in excess of 4 are usually found.

Another three-parameter fitting function that is sometimes used is the **Hubble–Oemler law**

$$I(R) = \frac{I_0 e^{-R^2/R_t^2}}{(1 + R/r_0)^2}. \qquad (2)$$

Here $I_0$ is the central surface brightness, and $r_0$ is the radius interior to which the surface brightness is approximately constant. For $R$ in the range $r_0 < R < R_t$ the surface brightness varies approximately as $I \sim R^{-2}$. Beyond $R_t$ the surface brightness falls off rapidly, with the consequence that the law predicts finite total luminosity. In the limit $R_t \to \infty$, equation (2) reduces to the **Hubble law**

$$I(R) = \frac{I_0}{(1 + R/r_0)^2}. \qquad (3)$$

This functional form was introduced by Reynolds (1913) and popularized by Hubble (1930). Note that in Hubble's law, the quantity $R/r_0$ is squared after it has been added to unity, rather than before, as in the modified Hubble law of equation (4.13). In consequence of this difference, the central gradient of the surface brightness, $dI/dR|_{R=0}$, is non-zero according to the Hubble law but zero according to the modified Hubble law. Both laws predict $I \sim R^{-2}$ for $R \gg r_0$ and therefore result in an unphysically-infinite value for the total luminosity of the galaxy being modeled.

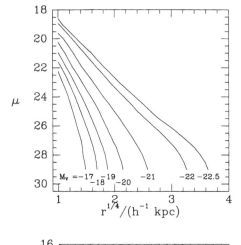

**Figure 4.26** Characteristic surface-brightness profiles for elliptical galaxies of different luminosities. The luminosities are expressed in terms of absolute $V$-band magnitudes and correspond to the portion of the galaxy that lies in projection interior to a circle of radius 16 kpc. These profiles are based on photometry of 261 elliptical galaxies. [After Schombert (1986) from data kindly provided by J. Schombert]

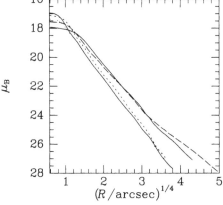

**Figure 4.27** Surface-brightness profiles of NGC 720, NGC 1199, NGC 1209, NGC 1395 and NGC 1426. [From data published in Capaccioli, Piotto & Rampazzo (1988)]

surface brightness profile of the cD galaxy at the center of the cluster Abell 1413. These galaxies are always luminous and tend to lie in regions of exceptionally high galaxy density, such as the center of a cluster of galaxies (Beers & Geller 1983) or of a compact group (Morgan, Kayser & White 1975). The picture of the Coma cluster shown in Figure 4.29 illustrates this phenomenon. Schombert (1986) finds that $I \propto R^{-1.6}$ in the halos of cD galaxies.

The displacement at large $R$ of the surface-brightness profiles of these galaxies above the $R^{1/4}$ law is interpreted as being due to the galaxy being embedded in an extensive luminous halo. It may be profitable to think of this halo as belonging to the surrounding cluster of galaxies rather than to the galaxy itself (Merritt 1985). In particular, in normal elliptical galaxies the velocity dispersion is constant or falling as one moves outwards (see §11.2.1), whilst in cD galaxies the velocity dispersion rises outwards into the halo (Dressler 1979, Carter *et al.* 1985). The increase implies that the velocity dispersion in the halo is comparable to that of the inner galaxy cluster as

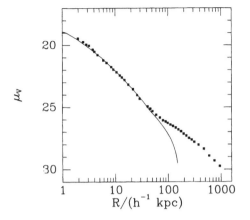

**Figure 4.28** The surface-brightness profile of the cD galaxy that lies at the center of the cluster Abell 1413 (points). The line shows the $R^{1/4}$-law that best fits the inner points. [From data kindly provided by J. Schombert based on the work of Oemler (1976).]

**Figure 4.29** The cD NGC 4881 is located near the center of the Coma cluster and is surrounded by a swarm of much less luminous galaxies. [Figure courtesy of STScI]

a whole. The ellipticity of cD galaxies generally increases outwards and the principal axes of the outer isophotes are always aligned with the principal axes of the **isopleths** (curves of constant number-density of galaxies) of the surrounding cluster (Carter & Metcalfe 1980, Rhee & Katgert 1987). This

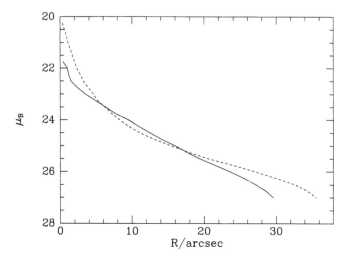

**Figure 4.30** The full curve shows the surface-brightness profile of a typical dE galaxy in the Virgo cluster from the photographic photometry of Ichikawa, Wakamatsu & Okamura (1986). Note that the profile is nearly straight in this log-linear plot because it is well fitted by the exponential law (4.20). The dashed curve shows the surface brightness profile of another galaxy from the same study. This galaxy approximately obeys the $R^{1/4}$ and is thought to be a background giant elliptical galaxy.

phenomenon again points to cD halos belonging to the cluster rather than the galaxy. Finally, and most crucially, Struble (1988) discovered in the galaxy cluster Abell 545 what appears to be a cD halo that lacks a central galaxy.

**Dwarf elliptical galaxies**     The majority of galaxies in the Local Group are low-luminosity elliptical galaxies of the dE or dSph classes and it is likely that such dwarf ellipticals are the commonest type of galaxy in every environment. Certainly they exist in great numbers in the nearby clusters of galaxies such as the Virgo and Fornax clusters (Bingelli *et al.* 1985, Wirth & Gallagher 1984). On account of their low luminosities, few are found in magnitude-limited samples and in consequence they have not been as extensively studied as the giant ellipticals with which we have been largely concerned above.

There is evidence that dwarf ellipticals fall into two classes, compact ones and diffuse ones (Kormendy & Djorgovski 1989). The only compact dwarf elliptical in the Local Group is M32, which lies very close to the Milky Way's giant companion, M31. However, numbers of compact dE galaxies that are not close companions of any giant galaxy have been found in nearby galaxy clusters. The radial surface-brightness profiles of these galaxies, which are reasonably well fitted by the $R^{1/4}$ law, fall at one end of the sequence of profiles for giant elliptical galaxies shown in Figure 4.26.

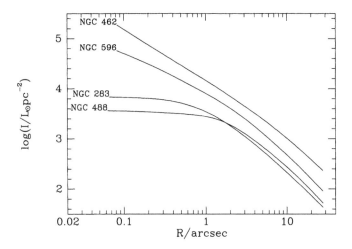

**Figure 4.31** The surface-brightness profiles of four early-type galaxies observed by the HST. [After Lauer *et al.* (1995) from data kindly provided by K. Gebhardt]

By contrast, the radial surface-brightness profiles of the other class of dE galaxies, the diffuse ones, are best fitted by the **exponential law**

$$I(R) = I_{\rm d} \exp(-R/R_{\rm d}). \tag{4.20}$$

Figure 4.30 illustrates this phenomenon. We shall see below that the exponential law is the hall-mark of galactic disks, with the result that its parameter $R_{\rm d}$ is called the **disk scale length**. However, an analysis, along the lines described in §4.3.3 below, of the distribution of the apparent ellipticities of diffuse dE galaxies indicates that these are not highly flattened objects (Ichikawa, Wakamatsu & Okamura 1986).

**Centers of elliptical galaxies**    Since it has long been suspected that massive black holes lurk at the centers of elliptical galaxies, an enormous amount of effort has been devoted to understanding these regions. Much of this work has been frustrating because the nearest elliptical galaxies (such as M32) are small systems, and the largest specimens lie far away. Consequently, the effects of seeing seriously degrade ground-based observations of the crucial central regions of these systems and completely convincing observations had to wait until the launch of the HST.

Lauer *et al.* (1995) and Byun *et al.* (1996) have analyzed the surface-brightness profiles that were measured by the HST (unfortunately prior to its refurbishment in March 1994) for 42 early-type galaxies that have neither an optically luminous AGN nor conspicuous absorption by nuclear dust. Each isophote was fitted by an ellipse with semi-axes $a$ and $b$, and then the isophote's surface-brightness was plotted against $r \equiv \sqrt{ab}$. Figure 4.31 shows some of these profiles, which extend to $10''$ in radius, with an effective

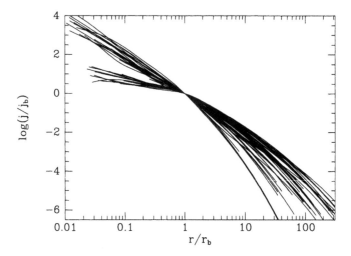

**Figure 4.32** The luminosity densities of 40 early-type galaxies with HST surface-photometry. The point of sharpest curvature on each profile has been shifted on this log–log plot to $r = r_b$, $j = 1$. Each curve is obtained by least-squares fitting a spline to the observed surface-brightness profiles and then applying equation (4.12) to the resulting spline. [After Gebhardt *et al.* (1996) from data kindly provided by K. Gebhardt]

spatial resolution of better than $0.1''$. This figure shows at a glance (i) that the central brightness profiles display considerable individuality, and (ii) that many brightness profiles continue to rise steeply at the smallest measured radii.

This last observation suggests that many of these galaxies have surface-brightness profiles that diverge as $I \sim R^{-\alpha}$ when $R \to 0$. From §4.2.3 we recall that divergent central surface-brightness profiles are only generated by luminosity densities that diverge more strongly than $j \sim r^{-1}$. Hence it seems that many of the galaxies observed by the HST have *strongly* divergent central luminosity densities. Moreover the galaxies that show no evidence of divergent surface-brightness profiles probably have luminosity densities that diverge as $r \to 0$, but do so sufficiently slowly for $I(R)$ to remain finite as $R \to 0$.

Gebhardt *et al.* (1996) have used equation (4.12) to examine the luminosity densities of the galaxies observed by Lauer *et al.* (1995) and Byun *et al.* (1996). Figure 4.32 shows their estimate of each galaxy's luminosity density, scaled in radius and luminosity so as to place the point on the curve that has the greatest curvature at $j = 1$, $r = 1$. At $r < 1$ the curves clearly fall into two groups. The curves of one group rise steeply towards $r = 0$, clustering around the power law $j = r^{-1.9}$. The curves of the second group cluster around $j = r^{-0.8}$. In each group most curves are reasonably straight, so that $j$ may be approximated by a power-law in $r$, although one curve in the second group bends steadily downwards as $r \to 0$, suggesting that in this galaxy $j$ may tend to a finite limit as $r \to 0$. At $r > 1$ all curves appear to belong to a single group that clusters around $j = r^{-2}$.

### 4.3.2 Color and line-strength gradients in elliptical galaxies

Elliptical galaxies are redder near their centers than further out. For example, Franx, Illingworth & Heckman (1989a) find that the 17 galaxies in their sample exhibit mean color gradients $\mathrm{d}(U - R)/\mathrm{d}\log_{10} R \simeq -0.25$ and $\mathrm{d}(B - R)/\mathrm{d}\log_{10} R = -0.1$[12] Peletier *et al.* (1990) find similar mean color gradients in a sample of 39 elliptical galaxies and conclude that contours of constant color coincide with isophotes. These color gradients are of great interest because they are almost certainly due to gradients in the mean ages and metallicities of the galaxies' stars. However, as we saw in §4.2, color gradients are exceptionally difficult to determine because they depend sensitively on the sky backgrounds estimated for each band. Fortunately, they should be associated with changes in the strengths of various spectral lines in the galaxies' spectra, and it is easier to measure accurately changes in spectrum than in color in the presence of an ill-determined sky background. Hence much recent work has probed gradients in the compositions of galaxies through spectral indices.

Davies, Sadler & Peletier (1993) measured several of the spectral indices defined in Table 3.8 out to $R \simeq R_e$ in 13 elliptical galaxies. They found that the indices tended to be constant on isophotes. H$\beta$ generally increases with radius while Mg$_1$, Mg$_2$ and $\langle$Fe$\rangle$ decrease with radius. Using the rather uncertain calibrations discussed in §3.4, they conclude from the Mg and Fe indices that

$$\frac{\mathrm{d}[\mathrm{Fe/H}]}{\mathrm{d}\log_{10} r} = -0.22 \pm 0.09 \ . \tag{4.21}$$

This gradient agrees with earlier work on index gradients and on color gradients. However, it must be stressed that the interpretation of indices in terms of metallicity is very uncertain. In particular, variations in the mean ages of stars at different radii can have a similar effect on indices to that of metallicity variations – see Figure 5.23.

The origin of the outwards increase in H$\beta$ is controversial. One possibility is that at small radii extended H$\beta$ emission is filling in the absorption lines in the stellar spectra, thus diminishing the H$\beta$ index. An alternative explanation is that the mean age of the stars increases from the center outwards, perhaps because accreted gas has caused new stars to form at small radii within the last few Gyr.

Davies, Sadler & Peletier (1993) found only weak correlations within their sample between the index-gradients and other measured quantities such

---

[12] See Figure 5.20 for other evidence of color gradients in ellipticals.

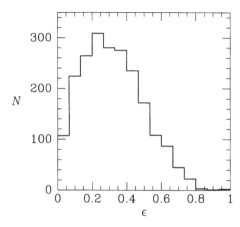

**Figure 4.33** Histogram of the ellipticities of 2135 elliptical galaxies. [After Lambas, Maddox & Loveday (1992) from data kindly supplied by S. Maddox.]

as central velocity dispersion. The best correlation is between the gradient in $Mg_2$ and the gradient in $U - R$. However, the measurements are subject to substantial errors, which are capable of degrading tight correlations between the variables into weak ones such as those observed.

### 4.3.3 Shapes of elliptical galaxies

**Ellipticity**    In §4.2 we saw that the elongation of isophotes is commonly characterized by the ellipticity $\epsilon$. When the effects of seeing discussed in §4.2.2 are discounted, the ellipticity sometimes increases from the center outwards (as in cD galaxies), sometimes decreases and sometimes varies non-monotonically as a function of radius. In short, no general rule seems to govern the variation of $\epsilon$ with $R$, and it is not clear what physical significance attaches to the observed variations. Consider therefore assigning a single characteristic ellipticity to each galaxy. Fasano & Vio (1991) have used CCD photometry to assign to each of 204 elliptical galaxies the ellipticity of isophotes near $R_e$ (which is where most of the galaxy's light in concentrated). Lambas, Maddox & Loveday (1992) have determined average ellipticities for 20 399 galaxies that were automatically measured on survey plates; 2135 of these galaxies were ellipticals. Figure 4.33 shows the frequency distribution of the Lambas *et al.* data for ellipticals. The distribution rises gradually from zero at $\epsilon = 0.8$ to a broad peak around $\epsilon = 0.25$ and then falls sharply as $\epsilon = 0$ is approached. The smaller sample of Fasano & Vio (1991) is consistent with this distribution.

What can we learn about the true shapes of elliptical galaxies from the data of Figure 4.33? Let us start by assuming that all elliptical galaxies are spheroidal (that is, either oblate or prolate figures of revolution). Consider then a spheroidal galaxy whose isodensity surfaces are described by the

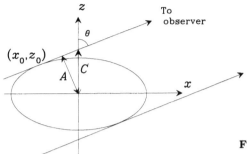

**Figure 4.34** Evaluating the apparent axial ratio of a rotationally symmetric galaxy.

equation

$$(\zeta x)^2 + (\zeta y)^2 + z^2 = c^2. \tag{4.22}$$

If $\zeta > 1$, the galaxy will be prolate, and if $\zeta < 1$, it will be oblate. In either case, $c$ is half the length of the symmetry axis of the isodensity surface that is defined by equation (4.22); the $x$ and $y$ semi-axes of this surface are each of length $(c/\zeta)$. Consider the appearance of the galaxy to an observer whose line of sight makes an angle $\theta$ with the $z$ axis (Figure 4.34 shows the geometry of this situation). Without loss of generality, we may take the $y$ axis to be the line of nodes,[13] so that one apparent semi-axis has length $(c/\zeta)$ and the other principal semi-axis has length $A$, as shown in Figure 4.34. In the notation of the figure,

$$(\zeta x_0)^2 + z_0^2 = c^2, \tag{4.23}$$

and, by differentiation of this equation, we find for the slope $s$ of the tangent at $(x_0, z_0)$

$$s = \cot \theta = -\frac{\zeta^2 x_0}{z_0}, \tag{4.24}$$

so

$$\cot^2 \theta = \frac{\zeta^4 x_0^2}{z_0^2} = \zeta^2 \left( \frac{c^2}{z_0^2} - 1 \right), \tag{4.25}$$

which implies

$$\frac{c^2}{z_0^2} = \frac{\cot^2 \theta}{\zeta^2} + 1. \tag{4.26}$$

But

$$C = \frac{A}{\sin \theta} = z_0 - x_0 s$$
$$= z_0 + \frac{\zeta^2 x_0^2}{z_0} = \frac{c^2}{z_0}. \tag{4.27}$$

---

[13] An object's **line of nodes** is the line through the center of the object that lies in both the plane of the sky and the object's equatorial plane.

Thus, using equation (4.26), we obtain

$$A^2 = c^2 \left( \frac{c^2}{z_0^2} \right) \sin^2 \theta = c^2 \left( \frac{\cos^2 \theta}{\zeta^2} + \sin^2 \theta \right). \tag{4.28}$$

In the oblate case, the apparent axial ratio (when defined to be less than 1) is $q = \zeta A/c$, and $q = c/\zeta A$ in the prolate case. Therefore we have

$$\zeta^2 \sin^2 \theta + \cos^2 \theta = \begin{cases} q^2 & \text{(oblate)}, \\ 1/q^2 & \text{(prolate)}. \end{cases} \tag{4.29}$$

Let us now use equation (4.29) to determine the distribution of apparent ellipticities that will arise from a large number of spheroidal galaxies whose symmetry axes are randomly oriented about the line of sight. If the symmetry axes of the galaxies are randomly oriented with respect to the line of sight, then, of the $n(\zeta) \, d\zeta$ galaxies under discussion, a fraction $\sin \theta \, d\theta$ will have their symmetry axes directed at angle $\theta$ to the line of sight. By equation (4.29), $q$ is a function $q(\theta)$ of $\theta$. To have $q$ between $q$ and $q + dq$, the symmetry axis of the galaxy has to be oriented between $\theta$ and $\theta + d\theta$, where $d\theta = dq/|dq/d\theta|$. Therefore the probability $P(q|\zeta)dq$ that a galaxy with true axis ratio $\zeta$ is observed to have apparent axis ratio in the range $(q, q + dq)$ is

$$P(q|\zeta)dq = \frac{\sin \theta \, dq}{|dq/d\theta|}. \tag{4.30}$$

Let there be $f(q) \, dq$ galaxies with axial ratios in the range $(q, q + dq)$. This number is equal to $\int d\zeta \, n(\zeta) P(q|\zeta) dq$. So equating these expressions and dividing out $dq$ we have

$$f(q) = \int d\zeta \, n(\zeta) \frac{\sin \theta}{|dq/d\theta|}. \tag{4.31}$$

From equation (4.29), one has for *oblate* geometry,

$$\left| \frac{dq}{d\theta} \right| = \frac{1}{q}[(1 - q^2)(q^2 - \zeta^2)]^{1/2}$$

$$\sin \theta = \left[ \frac{(1 - q^2)}{(1 - \zeta^2)} \right]^{1/2}, \tag{4.32}$$

and for *prolate* geometry,

$$\left| \frac{dq}{d\theta} \right| = q[(1 - q^2)(q^2\zeta^2 - 1)]^{1/2}$$

$$\sin \theta = \frac{1}{q} \left[ \frac{(1 - q^2)}{(\zeta^2 - 1)} \right]^{1/2}. \tag{4.33}$$

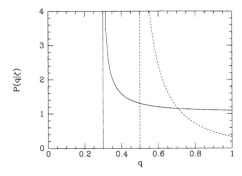

**Figure 4.35** Probability density of observing axis ratio $q$ for a spheroidal system of true axis ratio $\zeta$. The full curve corresponds to oblate geometry and $\zeta = 0.3$, while the dashed curve is for prolate geometry and $\zeta = 0.5$.

For prolate geometry, equation (4.31) therefore becomes

$$f(q) = \frac{1}{q^2} \int_{1/q}^{\infty} \frac{n(\zeta)\,d\zeta}{[(\zeta^2 - 1)(\zeta^2 q^2 - 1)]^{1/2}}. \tag{4.34}$$

In the prolate case, if the *true axial ratio* $\beta$ is considered to be a quantity less than 1, then one has $\beta = 1/\zeta$. Therefore, if we define $N(\beta)$ to be the number density of galaxies with true axial ratios near $\beta$, we have $N(\beta)\,d\beta = n(\zeta)\,d\zeta$, and

$$f(q) = \frac{1}{q^2} \int_0^q \frac{N(\beta)\beta^2\,d\beta}{[(1 - \beta^2)(q^2 - \beta^2)]^{1/2}}. \tag{4.35}$$

In the oblate case, $\beta = \zeta$ and equations (4.31) and (4.32) lead immediately to

$$f(q) = q \int_0^q \frac{N(\beta)\,d\beta}{[(1 - \beta^2)(q^2 - \beta^2)]^{1/2}}. \tag{4.36}$$

Equations (4.35) and (4.36) are integral equations relating the known (observed) frequency of galaxies of apparent axial ratio $q$ to the unknown frequency $N(\beta)$ of galaxies of true axial ratio $\beta$. The only difference between them is the extra factor of $(\beta^2/q^3)$ in the prolate case, which reduces the contribution of highly aspherical ($\beta \ll 1$) galaxies to the observed number of apparently round galaxies. Lambas, Maddox & Loveday (1992) solve these equations for polynomial fits to the data of Figure 4.33 and find that, for both prolate and oblate geometry, $N(\beta)$ turns negative near $\beta = 1$. Since it is logically impossible for a frequency function such a $N(\beta)$ to be negative, this result implies that something is seriously wrong with either the data or its analysis.

The nature of the problem is made clear by the full curve in Figure 4.35, which shows, for oblate geometry and $\zeta = 0.3$, the probability density $P(q|\zeta)$ that is given by substituting equations (4.32) into equation (4.30). This declines from very large values near $q = \zeta$ to a plateau that continues to $q = 1$. When an infinite number of such distributions (one for each value of $\zeta$) is added together, it is clear that the resulting summed distribution

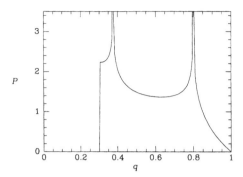

**Figure 4.36** $P(q|\eta, \zeta)$ for $\eta = 0.8$, $\zeta = 0.3$.

will be fairly flat near $q = 1$. By contrast, the observed distribution of Figure 4.33 shows a pronounced drop in the vicinity of $q = 1$. The result of Lambas, Maddox & Loveday (1992) is that this drop cannot be produced by any sum of distributions like the full curve in Figure 4.35. The dashed curve in Figure 4.35 shows that the apparent axis ratio $q$ of a prolate system is much more likely to lie near the true axis ratio $\zeta$ and less likely to lie near unity than in the oblate case. Nevertheless, $P(q|\zeta)$ remains non-zero in the neighborhood of $q = 1$, so the study of Lambas, Maddox & Loveday (1992) finds that the drop in the observed distribution of ellipticities near $q = 1$ cannot be reproduced by any distribution of prolate spheroidal systems.

Figure 4.36 shows the probability density $P(q|\eta, \zeta)$ of observing axis ratio $q$ for a triaxial system in which the luminosity density is a function $j(a_v)$ of the ellipsoidal radius that is defined by equation (4.17). For the distribution shown, $\eta = 0.8$ and $\zeta = 0.3$; the body in question is strongly flattened and only slightly elliptical in the $xy$-plane. The probability density diverges at two values of $q$, namely $q = \zeta/\eta$ and $q = \eta$, and falls steeply to zero at both $q = \zeta$ and $q = 1$. It is clear that the steep decline in $P(q|\eta, \zeta)$ as $q = 1$ is approached has the potential to fit the depression near $q = 1$ in the observed distribution of Figure 4.33, and Lambas, Maddox & Loveday (1992) do indeed find a distribution of triaxial bodies that can account for the data. In this distribution $\zeta$ is Gaussianly distributed with central value $\zeta_0 = 0.55$ and dispersion $\sigma_\zeta = 0.2$, while $\eta$ is likewise Gaussianly distributed with central value $\eta_0 = 0.95$ and dispersion $\sigma_\eta = 0.35$. (The Gaussian distributions are obviously truncated at 0 and 1 and are further constrained such that $\eta > \zeta$.)

The distribution derived by Lambas, Maddox & Loveday (1992) is not unique, but it seems likely that any acceptable distribution will be dominated by systems which are nearly oblate axisymmetric bodies, rather than nearly prolate axisymmetric objects.

This conclusion is strongly reinforced by the observation that galaxies of all types frequently exhibit the phenomenon of isophote twist that was discussed from a theoretical view-point in §4.2.3. As an example, Figure 4.37 shows the twisted isophotes of the E3 galaxy NGC 5831. In §4.2.3 we saw

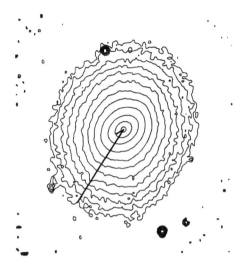

**Figure 4.37** Isophotes (at 0.5 $\mu_r$ intervals) of the E3 galaxy, NGC 5831. The heavier lines indicate the semi-major axes of the ellipses that best fit the isophotes at two radii. These axes have lengths $a = 4$ arcsec and $a = 40$ arcsec. [From data kindly provided by R. Bender]

that axisymmetric objects cannot have twisted isophotes, whereas triaxial systems are expected to exhibit them. Unfortunately, it has not proved possible to use observations of isophote twists to quantify the triaxiality of galaxies.

In §11.2.1 we shall see that kinematic data also lead to the conclusion that many galaxies are mildly triaxial.

**Deviations from ellipses**   In §4.2 we saw that deviations of isophotes from ellipses may be conveniently quantified by the Fourier coefficients $a_n$ defined by equation (4.10). Significantly non-zero values of $a_3/a$, $b_3/a$, $b_4/a$ and $a_4/a$ are frequently measured, but the higher Fourier coefficients in equation (4.10) are usually smaller than the measurement errors. Peletier *et al.* (1990) find that $a_3$ and $b_3$ are sensitive diagnostics for the presence of dust in elliptical galaxies. As we have seen, the diskiness $a_4/a$ is sensitive to the presence of an embedded disk.

Normal elliptical galaxies have values of $a_n/a$ that are smaller than a few per cent. Larger values are easily detectable by eye and an object with large $|a_n/a|$ has in the past been classified as either an S0 or an irregular object. The least round galaxies have the largest values of $|a_4/a|$. In fact, galaxies with $\epsilon \gtrsim 0.35$ almost always have $|a_4/a| > .005$ and are therefore either disky or boxy (Bender *et al.* 1989). This suggests that any elliptical galaxy will have non-elliptical isophotes when viewed edge-on, and apparently elliptical galaxies are those that are viewed at inclinations $i \lesssim 60°$. Indeed, we know from the deprojection studies mentioned in §4.2 that one cannot tell whether a galaxy is really disky or boxy when one views it at $i \lesssim 60°$.

Bender *et al.* (1989) discuss some intriguing correlations between the diskiness $a_4/a$ and other, apparently unrelated, properties of elliptical galaxies. Figure 4.38 shows plots against $a_4/a$ of (i) radio power at 1.4 GHz, and (ii) X-ray luminosity. The left panel shows that essentially all disky ellipti-

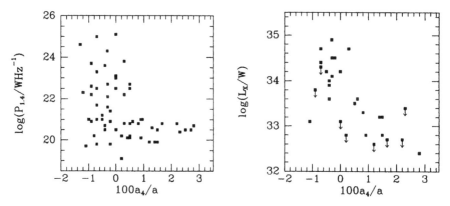

**Figure 4.38** Disky galaxies show neither much X-ray power nor much radio power. In the left-hand panel, X-ray luminosity is plotted against $a_4/a$, while the right-hand panel shows a plot of power at 1.4 GHz against $a_4/a$. There is a conspicuous absence of points in the upper-right quarter of each panel. [From data published in Bender *et al.* (1989)]

cals are only weak radio sources ($P_{1.4} < 10^{21}$ W Hz$^{-1}$), while boxy ellipticals are frequently up to a factor of 10 000 more powerful at 1.4 GHz. The division between the two groups is remarkably abrupt, at $a_4/a \simeq 0.005$. The right-hand panel in Figure 4.38 shows that all disky ellipticals in the sample have X-ray luminosities below $3 \times 10^{33}$ W, while most boxy ellipticals have larger X-ray luminosities. This correlation becomes even more striking if one decomposes the measured X-ray luminosities of the galaxies into the probable contributions from binary stars and the interstellar medium (see §8.3): this decomposition suggests that *only* boxy galaxies have X-ray luminous interstellar media.

The correlations shown in Figure 4.38 are the more striking because we know that $a_4/a$ depends on the inclination at which we view a galaxy, whereas $P_{1.4}$ and $L_X$ do not. Therefore even if there were a *perfect* correlation between the true shapes and luminosities of elliptical galaxies, there would still be an imperfect correlation between their apparent shapes and their luminosities.

Are the correlations shown in Figure 4.38 reflections of other correlations, or somehow intrinsic? It is known that both radio power and X-ray luminosity are strongly correlated with optical luminosity – see §8.3. Hence the correlations of Figure 4.38 would be expected if diskiness were correlated with optical luminosity. Bender *et al.* (1989) show that any such correlation is weak. Hence it seems that the correlation of diskiness with radio and X-ray luminosities is real. Its origin is not adequately understood.

A possible key to this puzzle is provided by examination of the connection between diskiness and the rate at which elliptical galaxies rotate. In §11.2.1 we define the indicative rotation rate $(v/\sigma)^*$, which is of order unity for systems that are flattened by rotation, and much less than unity if rotation is dynamically unimportant. The left-hand panel of Figure 4.39 shows

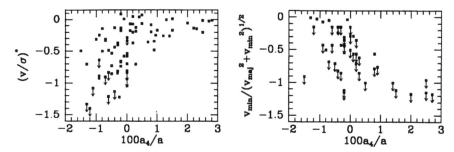

**Figure 4.39** Boxy galaxies frequently rotate very slowly, and then about an axis other than the apparent minor axis. The left panel shows that the indicative rotation rate $(v/\sigma)^*$ [equation (11.14)] is invariably of order unity for disky galaxies but sometimes very small for boxy galaxies. The right-hand panel shows that boxy galaxies frequently show a non-negligible velocity gradient along their minor axes. [From data kindly provided by R. Bender]

that $(v/\sigma)^*$ is invariably of order unity for disky galaxies, which suggests that these systems are indeed flattened by their spins. The right-hand panel of Figure 4.39 lends support to this conjecture by showing that in these galaxies the velocity gradient along the apparent major axis is invariable much larger than that along the apparent minor axis, just as we expect if these systems are axisymmetric bodies that rotate about their short axes. Since, by contrast, many boxy galaxies have small values of $(v/\sigma)^*$, rotation is often dynamically unimportant in these systems. Moreover, the right-hand panel of Figure 4.39 shows that there is often a significant velocity gradient along the apparent minor axis of a boxy galaxy. This may be because these objects are triaxial, in which case the apparent minor axis is not expected to be of any dynamical significance.

The correlations with $a_4$ that are shown in Figure 4.39 strongly suggest that many boxy galaxies are triaxial systems with little net rotation. We would not expect the ISM in such a system to have much net rotation either. Moreover, triaxiality of the galaxy's potential will facilitate the exchange of angular momentum between the stellar component and the ISM. Consequently, in a boxy galaxy interstellar gas may not be prevented from reaching the nucleus by a centrifugal 'barrier' of the type that operates in a disky galaxy. This absence of a centrifugal barrier may account for the propensity to powerful radio and X-ray emission in boxy galaxies that is demonstrated by Figure 4.38.

**Fine structure**    The three-parameter fitting functions discussed above provide a good description of the largescale properties of elliptical galaxies. However, detailed photometric studies have revealed that some galaxies contain significant small-scale departures from these smooth radial profiles. Sadler & Gerhard (1985) estimated that about 40% of ellipticals contain dust lanes – the fraction of systems in which dust lanes have been detected is significantly smaller because the visibility of a dust lane varies significantly

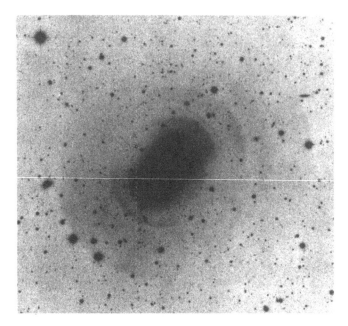

**Figure 4.40** The elliptical galaxy NGC 3923. This negative image
from the Anglo-Australian Telescope has been manipulated to enhance the appearance of sharp-edged features; this process reveals
the presence of a number of ripples or shells. [Courtesy of D. Malin]

with its orientation to the line of sight.

Between 10% and 20% of early-type galaxies are found to contain sharp
steps in their luminosity profiles like those apparent around NGC 3923 in
Figure 4.40 (Schweizer 1980, Malin & Carter 1983, Sadler & Gerhard 1985).
These features are called **ripples** or **shells**. In many, but by no means all
systems (Prieur 1990), ripples tend to form arcs of ellipses that are centered
on the galactic nucleus. In highly developed systems of ripples such as that
shown in Figure 4.40, the radii of ripples on opposite sides of the galaxy are
interleaved. That is, if on one side there are ripples at $r = a_1, a_2, \ldots$ and on
the opposite side ripples occur at $r = b_1, b_2, \ldots$, then $a_1 < b_1 < a_2 < b_2 < \cdots$.
The radii of ripples within a single galaxy can differ by a factor of 30 or more
– in the case of NGC 3923 the innermost ripple is at $r \simeq 3\,\mathrm{kpc}$ while the
outermost lies at $r \simeq 95\,\mathrm{kpc}$. Ripples generally involve 3–5% deviations
of the galaxy's surface-brightness profile from a smooth underlying law, and
their colors indicate that their light comes from the kind of stellar population
that one would expect to encounter in an early-type galaxy (Pence 1986).

Schweizer & Seitzer (1988) have shown that ripples occur in S0 and
Sa galaxies as well as in ellipticals. Since their detectability in early-type
systems depends upon the contrast between their sharp edges and the smooth
background brightness profiles of the underlying galaxies, ripples would be

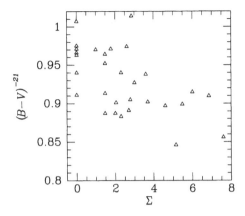

Figure 4.41 Galaxies with fine structure have bluer colors. Here we plot the correlation between the fine-structure parameter $\Sigma$ [equation (4.37)] and the color $(B-V)^{-21}$. This is the galaxy's color after correction for the color-magnitude effect (§4.3.4) to absolute magnitude $M_B = -21$. [After Schweizer & Seitzer (1992) from data kindly supplied by F. Schweizer]

very hard to detect in late-type systems, so we do not know how universal a phenomenon they are.

A stellar system can display a sharp edge only if some parts of its phase space (see BT §4.1) are very much more densely populated with stars than neighboring parts of phase space. In the classical dynamical model of an elliptical (BT §4.4), phase space is populated very smoothly. Therefore the existence of ripples directly challenges the classical picture of ellipticals. One likely possibility is that ellipticals acquire ripples late in life as a result of accreting material from a system within which there are relatively large gradients in phase-space density. Systems with large density gradients in phase space include disk galaxies and dwarf galaxies: in a thin disk the phase-space density of stars peaks strongly around the locations of circular orbits, while in a dwarf galaxy all stars move at approximately the systemic velocity, so that there is only a small spread in velocity space.

Numerical simulations suggest that ripples can indeed form when material is accreted from either a disk galaxy or a dwarf system – see Barnes & Hernquist (1992) for a review. Moreover, simulations have successfully reproduced the interleaved property of ripples described above. Despite these successes significant uncertainties still surround the ripple phenomenon because the available simulations have important limitations, and it is not clear how probable their initial conditions are.

Schweizer $et$ $al.$ (1990) defined an index $\Sigma$ that quantifies the amount of fine structure such as ripples that a galaxy possesses:

$$\Sigma \equiv S + \log(1 + n) + J + B + X. \qquad (4.37)$$

Here $S$ measures the strength of the most prominent ripple on a scale of 0 to 3; $n$ is the number of detected ripples; $J$ is the number of optical 'jets'; $B$ is a measure of the boxiness of the galaxy's isophotes on a scale of 0 to 4; $X$ is 0 or 1 depending on whether the galaxy's image shows an X structure. For a sample of 69 nearby early-type galaxies $\Sigma$ varies from 0 to 7.6. Notice that

$\Sigma$ is defined such that it does *not* include a term sensitive to the presence of dust – this definition is motivated by the observation that while ripples, jets, boxiness and X structures tend to occur together, no correlation between these features and dust lanes is apparent (Sadler & Gerhard 1985, Schweizer & Seitzer 1992). Schweizer *et al.* (1990) and Schweizer & Seitzer (1992) find that $\Sigma$ is correlated both with line-strengths and with broad-band colors in the sense that the larger $\Sigma$ is, the weaker are system's lines and the bluer its colors; Figure 4.41 shows the correlation of $\Sigma$ with $B - V$ for a sample of 32 elliptical galaxies. As we shall see in §5.4, the most natural interpretation of this finding is that galaxies that have significant fine structure contain larger numbers of relatively youthful stars than do galaxies that lack such structure. This is very much what we would expect if fine structure formed during accretion events since there is evidence that such events are associated with bursts of star formation – see §4.6.1.

### 4.3.4 Correlations among global parameters of elliptical galaxies

The basic structural parameters of an elliptical galaxy can be divided into two sets. One set consists of the shape parameters $\epsilon$, $a_4/a$ and $(v/\sigma)^*$ discussed above. The other set consists of the shape-independent parameters, namely the luminosities, $L_X$, in various bands $X$, effective radius, $R_e$ (measured in kpc), effective surface brightness, $I_e$, mean colors, $B - V$ etc, linestrength indices, $Mg_2$ etc., and central velocity dispersion $\sigma_0$. Just as there are correlations between the shape parameters, there must be correlations between the shape-independent parameters. For example, it is obvious that more luminous galaxies will be physically larger, so it would be very surprising if $L_B$ and $R_e$ were not positively correlated. These correlations are enormously important for both practical and theoretical cosmology. Their practical importance arises because they relate quantities such as $\sigma_0$ that can be determined without a knowledge of a galaxy's distance, to quantities such as $L_B$ that can be determined only for galaxies of known distance. Hence these correlations can be used to infer the distances to galaxies. Their importance for theoretical cosmology is that they provide much-needed constraints on theories of galaxy formation.

We saw above that the observed gradients within galaxies in colors such as $B - R$ and line-strength indices such as $Mg_2$ are consistent with one another. Similarly, Bender *et al.* (1992) report that the $B - V$ colors of E and S0 galaxies are tightly correlated with their measured $Mg_2$ values – see Figure 4.42. The tight relationship between colors and line strength in the left panel of Figure 4.42 is a non-trivial result because the two variables plotted relate to different volumes within each galaxy: the $Mg_2$ index depends only on the spectrum of light from the innermost few arcseconds of the image, while the measured value of $B - V$ is dominated by light from $R \simeq R_e$. Thus the tight correlation of these variables displayed in the left panel of

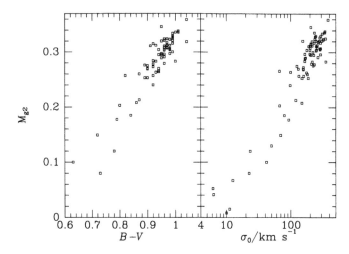

**Figure 4.42** Both the broad-band colors and the central velocity dispersions of elliptical galaxies are strongly correlated with line-strength indices such as Mg$_2$. [From data published in Bender *et al.* (1992)]

Figure 4.42 does not merely imply that line-strength is tightly correlated with broad-band color in a given spectrum. In addition it requires a degree of consistency in galactic color and line-strength gradients. We saw independent evidence for this consistency in §4.3.2 above.

Rather surprisingly, it turns out that the central velocity dispersion $\sigma_0$ of a galaxy is extremely tightly correlated with the galaxy's colors and line-strength indices – see the right panel of Figure 4.42. The origin of this correlation between the speeds with which a stars move and the mean spectrum of the light that they emit is not properly understood.

In §11.2.1 we shall see that in a typical elliptical galaxy, the velocity dispersion $\sigma$ varies with radius just as Mg$_2$ does (Figure 11.6). Davies, Sadler & Peletier (1993) find that the relationship $\sigma(\mathrm{Mg}_2)$ defined by the gradients in $\sigma$ and Mg$_2$ differs slightly from relationship $\sigma_0(\mathrm{Mg}_2)$ that is defined by the right panel of Figure 4.42.

Let us now turn to correlations that depend on the distances to galaxies. Kormendy (1977) demonstrated that a correlation exists between $R_e$ and $I_e$ in the sense that larger galaxies have fainter effective surface brightnesses. This **surface brightness–effective radius relation** is shown in the bottom left panel of Figure 4.43, where we plot $\langle I \rangle_e$, the mean surface brightness interior to $R_e$ in place of $I_e$, which is more susceptible to measurement error. The mean regression is $R_e \propto \langle I \rangle_e^{-0.83 \pm 0.08}$ (Djorgovski & Davis 1987). Let $L_e$ be the luminosity interior to $R_e$. Then since $L_e = \pi \langle I \rangle_e R_e^2$, it follows from the correlation of $R_e$ and $\langle I \rangle_e$ that $\langle I \rangle_e$ *decreases* with $L$; one finds $\langle I \rangle_e \sim L^{-3/2}$. More luminous ellipticals have lower surface brightnesses.

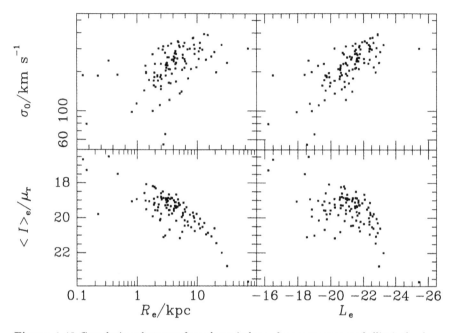

**Figure 4.43** Correlations between four shape-independent parameters of elliptical galaxies. The parameters are the effective radius $R_e$, the mean surface brightness within $R_e$, $\langle I \rangle_e$, the central velocity dispersion $\sigma_0$ and $L_e$, the luminosity in Djorgovski's $G$ band interior to $R_e$. The luminosity and the surface brightness are expressed in magnitudes and in magnitudes per square arcsecond, respectively. [From data published in Djorgovski & Davis (1987)]

The lower right panel of Figure 4.43 shows the correlation of which this is the mean regression.

More luminous ellipticals have larger central velocity dispersions. The upper right panel in Figure 4.43 illustrates this correlation, which is called the **Faber–Jackson relation** after its discoverers (Faber & Jackson 1976). Quantitatively one has

$$L_e \sim \sigma_0^4. \tag{4.38}$$

Since $\sigma_0$ is correlated with $L_e$ and $L_e$ is correlated with $R_e$ it follows that $\sigma_0$ must be correlated with $R_e$. The top left panel in Figure 4.43 displays this correlation.

Since $\sigma_0$ is strongly correlated with line-strengths and colors, the existence of a host of additional correlations involving $Mg_2$, $B - V$ etc is implied by the top two panels of Figure 4.43. One of the earliest of these to be discovered was the **color–magnitude relation**: Faber (1973) showed that more luminous elliptical galaxies have stronger absorption lines, and Visvanathan & Sandage (1977) showed that more luminous galaxies are redder.

In all the correlations of Figure 4.43 there is **cosmic scatter**; the scatter of the points about the mean relations is larger than can be accounted

---

### Box 4.2: Principal Component Analysis

Given $M$ points $\mathbf{x}^{(\alpha)}$ in a space of $d$ dimensions we wish to choose a $d$-dimensional unit vector $\mathbf{n}$ and a constant $p$ such that our points lie as nearly as possible on the plane $\mathbf{n} \cdot \mathbf{x} = p$. To this end we minimize with respect to $p$ and the components of $\mathbf{n}$ the quantity

$$S \equiv \sum_\alpha |\mathbf{n} \cdot \mathbf{x}^{(\alpha)} - p|^2$$

subject to the constraint $|\mathbf{n}|^2 - 1 = 0$. By the method of Lagrange multipliers this problem leads to the equations

$$0 = \sum_{\alpha=1}^{M} \left[ (\mathbf{n} \cdot \mathbf{x}^{(\alpha)} - p) x_i^{(\alpha)} \right] - \lambda n_i \quad (i = 1, .., d)$$

$$0 = \sum_{\alpha=1}^{M} (\mathbf{n} \cdot \mathbf{x}^{(\alpha)} - p),$$

where $\lambda$ is the undetermined multiplier. We solve these equations by using the last equation to eliminate $p$ from each of the first $d$ equations. Then we find that

$$0 = \sum_\alpha \left( \mathbf{n} \cdot \mathbf{x}^{(\alpha)} - \frac{1}{M} \sum_\beta \mathbf{n} \cdot \mathbf{x}^{(\beta)} \right) x_i^{(\alpha)} - \lambda n_i.$$

This equation can be rewritten $\mathbf{A} \cdot \mathbf{n} = \lambda \mathbf{n}$, where

$$A_{ij} \equiv \sum_\alpha \left( x_j^{(\alpha)} - \frac{1}{M} \sum_\beta x_j^{(\beta)} \right) x_i^{(\alpha)}. \tag{1}$$

Thus the required vector $\mathbf{n}$ is an eigenvector of the real-symmetric matrix $\mathbf{A}$. One may show (see Problem 4.7) that the eigenvalues $\lambda$ are the values taken by $S$ for $\mathbf{n}$ equal to the corresponding eigenvector. Hence the desired $\mathbf{n}$ is the eigenvector with the smallest eigenvalue.

In order to minimize distortion by measuring errors, the $x_i$ should have been scaled such that they have numerically equal measuring errors.

---

for by measurement errors alone. Interestingly, it turns out that the residuals between the positions of individuals objects and the mean relations are correlated. For example, galaxies which lie above the mean Faber–Jackson relation in the top right panel tend to lie above the mean surface-brightness–magnitude relation in the lower right panel. To understand the significance of both the basic correlations and these correlations between residuals, it is useful to imagine each galaxy to be represented by a point in a three-

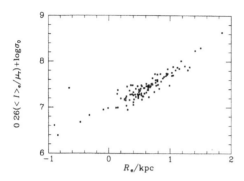

**Figure 4.44** An edge-on view of the fundamental plane as defined by the data of Figure 4.43. Note how much narrower is the distribution of points than in either of the left-hand panels of Figure 4.43.

dimensional space. The Cartesian coordinates $x_i$ of a point in this space are the three numbers of the set $(\log R_e, \langle I \rangle_e, \log \sigma_0)$, where $\langle I \rangle_e$ is in units of $\mu_B$.[14] If there were no correlations between the variables, the points of individual galaxies would be fairly uniformly distributed within a cuboidal region of our three-dimensional space. Correlations between the variables will confine the data points to a sub-volume of this cuboid. For example, if two of the variables are genuinely independent and the third dependent on these two, the points will be confined to a plane, while if there is only one independent variable, the points will lie on a line. Our first step in analyzing such data is to ask whether the points are nearly confined to a plane; if we can find a suitable plane, we can further enquire whether the points lie on a line within that plane. In statistics books this type of investigation is called **principal component analysis**. Box 4.2 explains what has to be done.

For the data in Figure 4.43 one finds that the points are as nearly confined to a plane as observational errors allow; in the absence of observational errors, they might lie *precisely* on a plane! If the position vector of a galaxy in our three-space is $\mathbf{g} = (\log R_e, \langle I \rangle_e, \log \sigma_0)$, where $R_e$ is measured in kpc, $\langle I \rangle_e$ in units of $\mu_B$ and $\sigma_0$ in km s$^{-1}$, then the equation of this **fundamental plane** is $\mathbf{n} \cdot \mathbf{g} = 1$ where $\mathbf{n} = (-0.65, 0.22, 0.86)$.

Naturally one wants to 'see' the fundamental plane. One way to do this is to choose an axis, for example the $R_e$-axis, as the horizontal axis, and rotate the space about this axis until the plane appears edge-on. In this orientation, the plane's normal, $\mathbf{n}$, lies in the plane of the projection, which is spanned by the unit vector, $\mathbf{e}_R$, that runs parallel to $R_e$-axis and some other, orthogonal, unit vector $\mathbf{e}$. Thus $\mathbf{n} = \alpha \mathbf{e}_R + \beta \mathbf{e}$, where $\alpha$ and $\beta$ are suitable numbers. Comparing this with the value of $\mathbf{n}$ given above we see that $\beta \mathbf{e} = 0.22 \mathbf{e}_I + 0.86 \mathbf{e}_\sigma$. Thus the fundamental plane will be seen edge-on if we plot $\log R_e$ against $0.22(\langle I \rangle_e / \mu_B) + 0.86 \log \sigma_0$. Actually, it is conventional to plot $0.26(\langle I \rangle_e / \mu_B) + \log \sigma_0$, which is simply $1/0.86$ times the linear combination we have derived. Figure 4.44 shows this plot. The

---

[14] $L_e$ need not be included in the set since it is related to $R_e$ and $\langle I \rangle_e$ by $L_e = \pi \langle I \rangle_e R_e^2$, where $L_e$ is measured in W, $I_e$ in W m$^{-2}$sterad$^{-1}$ and $R_e$ in m.

data points nearly lie on the line which is the edge-on projection of the fundamental plane. The equation of this line

$$\log R_e = 0.36(\langle I \rangle_e / \mu_B) + 1.4 \log \sigma_0 \qquad (4.39)$$

is also an equation for the fundamental plane, as one may verify by multiplying $\mathbf{n} \cdot \mathbf{g} = 1$ by $1/0.65$.

**The $D_n - \sigma_0$ correlation**   Dressler *et al.* (1987) have defined a readily measured photometric parameter that has a tight correlation with $\sigma_0$ by virtue of the fundamental plane. This parameter, $D_n$, is the diameter within which the mean surface brightness is $I_n \equiv 20.75 \, \mu_B$. Since $D_n$ is defined in terms of a fiducial surface brightness, $D_n/R_e$ is larger for galaxies of high surface brightness than for galaxies of low surface brightness but the same luminosity. To quantify this idea, let us suppose that all galaxies have the same surface-brightness profile, for example the $R^{1/4}$ law. Let this profile be $I(R) = I_e f(R/R_e)$. Then from the definition of $D_n$ we have

$$
\begin{aligned}
I_n &= \frac{2\pi I_e \int_0^{D_n/2} dR \, R f(R/R_e)}{\frac{1}{4}\pi D_n^2} \\
&= 8 I_e \left( \frac{R_e}{D_n} \right)^2 \int_0^{D_n/2R_e} dx \, x f(x).
\end{aligned}
\qquad (4.40)
$$

From Figure 4.25 we deduce that most of the light interior to $I = 20.75 \, \mu_B$ will come from radii at which $f(x) \sim x^{-\alpha}$ with $\alpha \simeq 1.2$. Using this approximation to evaluate the integral, we obtain

$$D_n \propto R_e I_e^{1/\alpha} \sim R_e I_e^{0.8}. \qquad (4.41)$$

When equation (4.39) is used to eliminate $R_e$ from (4.41), one obtains[15]

$$D_n \propto \sigma_0^{1.4} I_e^{0.07}. \qquad (4.42)$$

The weak dependence upon $I_e$ in this equation is responsible for the tight correlation between $D_n$ and $\sigma_0$ that was discovered by Dressler *et al.* (1987). If we adopt a distance of 16 Mpc for the Virgo cluster (§7.4.2), their mean relation becomes

$$\frac{D_n}{\text{kpc}} = 2.05 \left( \frac{\sigma}{100 \, \text{km s}^{-1}} \right)^{1.33} \qquad (4.43)$$

with a 15% scatter from galaxy to galaxy.

**Dwarf elliptical galaxies**   In §4.3.1 we saw that while the radial surface-brightness profiles of giant ellipticals and compact dwarf ellipticals obey the

---

[15] Recall that $\langle I \rangle_e / \mu_B \propto -2.5 \log(I_e)$.

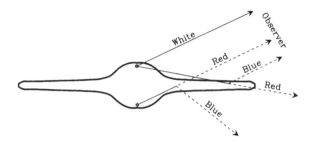

**Figure 4.45** Effects of scattering and absorption of light by dust on the images of disk galaxies. Light from the top of the bulge reaches the observer without obstruction by dust in the disk. Light from the lower portion of the bulge is partially absorbed by the disk. Some light is forward scattered by the disk into the path to the observer.

$R^{1/4}$ law fairly closely, the exponential law provides a better fit to the surface-brightness profiles of diffuse dwarf ellipticals. There are two other indications that diffuse dwarf ellipticals are fundamentally different from giant ellipticals and compact dwarf ellipticals:

1. Diffuse dwarf ellipticals do not lie on the fundamental plane (Kormendy 1987).
2. In §11.2.1 we shall see that, although the rotational kinetic energy of the most luminous ellipticals is dynamically negligible, lower-luminosity, higher surface-brightness ellipticals appear to be largely flattened by their rotational kinetic energy. By contrast, diffuse dwarfs of comparable or lower luminosity have dynamically insignificant rotational kinetic energy (Figure 11.7).

## 4.4 Photometry of Disk Galaxies

Although elliptical galaxies are the simplest systems observed, we have seen that, on closer inspection, even these display considerable variety and individuality of form. Now we turn to a discussion of the brightness distributions of galaxies which possess a conspicuous disk. The complexity of these systems can be very great, for in addition to a disk they often possess an elliptical-like bulge and/or a bar, and the disk may display pronounced spiral structure. Moreover, they usually contain significant quantities of dust, with the result that they are far from transparent in some wavebands. This lack of transparency severely complicates the task of deducing their three-dimensional shapes from images. Therefore it is appropriate to start our study with a discussion of the photometric effects of dust.

### 4.4.1 Photometric effects of dust

The apparent luminosity of a transparent galaxy is independent of the galaxy's orientation to the line of sight because the stars emit light isotropically. But if the galaxy is filled with dust, the galaxy will appear to be less luminous when seen at edge-on orientation than when viewed face-on. This is because at edge-on orientation starlight typically has to pass a longer column of the galaxy's dusty ISM to reach us, and is liable to be either absorbed or scattered out of the line of sight en route to us.[16]

Consider how dust affects a galaxy's measured surface brightness. The peak measured surface brightness of a highly flattened but transparent galaxy increases as the galaxy is tipped from face-on to edge-on, since at edge-on orientation the same luminosity comes from a smaller area of the sky than at face-on orientation. By contrast, all the light we receive from a very dusty galaxy will have been either emitted or last scattered in a thin layer around the galaxy; the thickness of this layer is one optical depth to scattering or absorption by dust (see §3.7). Hence a very dusty galaxy will resemble a glowing lump of metal in that its surface brightness will be independent of orientation, and therefore its apparent luminosity will be proportional to its projected area.

These effects are extremely important because they bias the galaxies that are included in catalogues. Consider for example a catalogue that includes all galaxies with isophotal major axes larger than some value.[17] If a galaxy is transparent, its isophotal major axis will lengthen as it is tipped towards edge-on, and its chances of being included in the catalogue will increase. By contrast, if the galaxy is dusty, its isophotal major axis and thus its chances of being included in the catalogue are essentially independent of orientation. It is easy to see that corresponding biases occur when galaxies that are brighter than some specified absolute magnitude are included in a catalogue. Thus it is impossible to compile a catalogue that is unbiased with respect to orientation until one has decided just how dusty galaxies are.

As we saw in §3.7, blue light is more strongly absorbed and scattered than red light, and there is a tendency for dust to scatter light through only a small angle. Figure 4.45 shows how these effects lead to color and brightness asymmetries between the near and far sides of disk galaxies. These asymmetries are important because they enable us to determine which is the nearer side of a galaxy, and therefore to conclude from line-of-sight velocity measurements in which sense a galaxy is rotating – this knowledge is crucial for studies of spiral structure, for example.

In Figure 4.45, the light received by the observer from the part of the bulge on the near side of the galaxy's major axis is more strongly absorbed

---

[16] See §3.7 for a discussion of absorption and scattering by dust.

[17] The UGC (Nilson 1973) is an important diameter-limited catalogue. However, its diameters are eye-estimates, and differ from photometric diameters in just the way required for them to be approximately inclination-independent (Huizinga & van Albada 1992, Giovanelli et al. 1995).

than the light from the far side, because much of the former has to pass right through the absorbing dust within the disk. The absorption is so strong in type S0$_3$ galaxies that it produces a crescent of blackness on their near sides. This differential absorption also makes the light from the near side of a galaxy appear redder than that from the far side. Scattering by dust within the disk has precisely the opposite effect. Light emitted by the upper half of the bulge in Figure 4.45 is strongly forward scattered off the dust in the disk and thus contributes extra blue light to the galaxy's image on the near side of the major axis, but little light is scattered toward the observer by the far side of the disk. In consequence the near side of the image tends to be bluer and brighter than the far side. The relative importance of these two competing effects varies with the angle of inclination $i$ of the galaxy's disk to the plane of the sky. At small inclinations, absorption dominates over forward scattering, and the near side appears dim and red. At intermediate inclination ($80° \gtrsim i \gtrsim 60°$), forward scattering can make the near side bluer and brighter than the far side. At very large inclinations, light from the near side is very heavily absorbed, and forward scattering cannot make up for the light lost by absorption (which can be so heavy that it produces dark lanes across the image of the galaxy). Extensive discussions of this problem will be found in Elvius (1956) and van Houten (1961).

Unfortunately, the degree to which photometry of disk galaxies is affected by dust remains controversial – see Giovanelli *et al.* (1995) or Davies & Burstein (1995) for recent discussions. The classical studies of Holmberg (1958) and de Vaucouleurs (1959a) argued that disks are fairly transparent by showing that the mean surface brightnesses of samples of galaxies that had been selected by apparent diameter, is positively correlated with inclination. Subsequently, a powerful case was developed that disks are by no means transparent (Disney, Davies & Phillipps 1989). In §9.3.1 we shall see that the Milky Way's dust reprocesses roughly half of the Milky Way's luminosity, which suggests that much of our disk is not transparent. On the other hand, the importance of dust for the dependence of photometric parameters, such as scale lengths and total luminosities, upon inclination angle depends not only on how much dust a disk contains, but also on how it is distributed. For example, if the dust forms a smooth, vertically extended layer, it will profoundly affect inclination-dependencies, while if it is confined to small clumps around luminous stars, it will have a negligible effect on inclination-dependencies.

We shall not submerge ourselves and the reader in these complex and unresolved questions, but only remark that in view of the possible importance of dust, it is desirable to work as much as possible in the near-IR, where absorption by dust is considerably less important than in the optical.

### 4.4.2 Overall shapes of disk galaxies

Figure 4.46 shows the distributions of the apparent axial ratios of a sample of $\sim 5000$ S0 galaxies (left panel) and $\sim 13\,000$ spiral galaxies (right panel) that

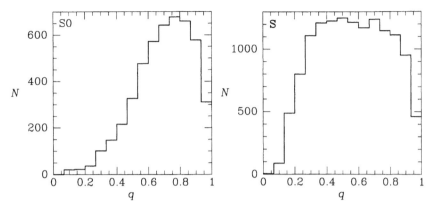

**Figure 4.46** The distribution of apparent axial ratios of the APM galaxies. The left-hand panel shows the data for 4782 S0 galaxies, the right-hand panel for 13482 spiral galaxies [After Lambas, Maddox & Loveday (1992) from data kindly supplied by S. Maddox]

have been automatically measured on survey plates. The two distributions are very different: whereas the frequency of S0 galaxies rises steadily from near zero at $q \simeq 0.2$ to a sharp peak at $q = 0.8$, the frequency of spirals is roughly constant from $q \simeq 0.23$ to $q \simeq 0.85$. If we hypothesize that spiral galaxies are axisymmetric oblate bodies, then from the flatness of the distribution in the right-hand panel of Figure 4.46 it follows that the distribution of true axial ratios is sharply peaked around some small value of the axial ratio $q \simeq 0.2$. In fact, from equation (4.36) one can see that, if $N(\beta)$ is non-zero only in the interval $0 \leq \beta \leq \beta_0$, where $\beta_0 \ll 1$, then $f(q)$ is approximately independent of $q$ for $q \gg \beta_0$. This result lends quantitative support to the subjective judgment that spiral galaxies are intrinsically quite thin.

By contrast, the long upward sloping portion of the left-hand histogram in Figure 4.46 implies that S0 galaxies are widely distributed in true axial ratio, from $q \simeq 0.25$ to $q \simeq 0.85$. This finding presumably reflects the fact S0 galaxies tend to have prominent bulges, and these are sometimes nearly spherical. Clearly, the apparent axial ratio of a galaxy that has a prominent, nearly spherical bulge will be close to unity no matter what the galaxy's inclination happens to be.

In common with the distribution of apparent ellipticities of elliptical galaxies that is shown in Figure 4.33, the distributions for both S0s and spirals shown in Figure 4.46 dips as it approaches axial ratio $q = 1$ ($\epsilon = 0$). Lambas, Maddox & Loveday (1992) show that this dip makes it impossible to interpret this distribution as the distribution of projected ellipticities of any group of axisymmetric oblate bodies. Evidently these objects are not axisymmetric, a conclusion reached earlier by Binney & de Vaucouleurs (1981) from a study of the distribution of apparent axial ratios of disk galaxies that appear in the Second Reference Catalogue. This sample is much smaller

than that which underlies Figure 4.46, but has the advantage that it may
be subdivided according to Hubble type. In addition to finding that many
disks are slightly elliptical they concluded that galaxies of type Sdm and later
are not particularly flat: their true axial ratios are approximately uniformly
distributed in the interval $0.2 - 0.8$

It is not clear whether the ellipticity of a typical disk reflects merely spiral structure or ellipticity of the dark halo that may contribute significantly
to the potential in which the disk rotates, especially at large radii (Binney
1978). In an attempt to resolve this question, Rix & Zaritsky (1995) analyzed
$K$-band photometry of 18 near face-on galaxies. By observing in the near-IR,
they hoped to maximize the contribution to the measured luminosity from
old stars relative to that of young stars. Galaxies were selected for study
on the basis of the widths of their 21-cm lines: this width is proportional to
$v_c \sin i$, where $v_c$ is a characteristic intrinsic circular speed (§8.2.4), so if $v_c$
is known, $i$ can be determined. Rix & Zaritsky estimated $v_c$ from the absolute magnitudes of their galaxies and equation (7.26). They found that their
galaxies were by no means axisymmetric. About one third were significantly
lop-sided and about half had two-armed spirals with arm–interarm contrasts
of order unity. They estimated the ellipticity $\epsilon_\Phi$ of the underlying potentials
and concluded that this peaks near $\epsilon_\Phi = 0.045$ and has a characteristic width
$\sim 0.025$. Thus few potentials are genuinely axisymmetric.

Even when a galaxy is not face-on, it is in principle possible to determine
$\epsilon_\Phi$ from an analysis of the shape and velocity field of any gas ring that can be
studied in sufficient detail (§8.2.4). From an analysis of kinematic data for
NGC 3198 (Figure 8.35), Schoenmakers, Franx & de Zeeuw (1997) tentatively
concluded that $\epsilon_\Phi = 0.019 \pm 0.003$.

### 4.4.3 Bulge–disk decomposition

The literature contains detailed surface photometry of several hundred nearby
disk galaxies. Simien & de Vaucouleurs (1985) summarize many results obtained photographically, while Kent (1984) and de Jong & van der Kruit
(1994) present a considerable body of CCD data. Other useful references are
Walterbos & Kennicutt (1987) and Wevers, van der Kruit & Allen (1986).

Figure 4.47 shows surface brightness plotted against radius for three spirals – these profiles represent average brightnesses around ellipses of similar
shape to the galaxies' outer isophotes. In this plot of $\log I$ versus $R$ the
outermost points for two of the galaxies tend to lie on straight lines; that is,
at large radii the surface brightness obeys the exponential law (4.20). Since
this brightness distribution fits the outer brightness profiles of the majority
of disk galaxies, it has, in fact, come to *define* the normal disk component of
all flat galaxies. Deviations from this profile, such as one sees in the profile
of NGC 5194 shown in Figure 4.47, are generally ascribed to contamination
of the profile by other components.

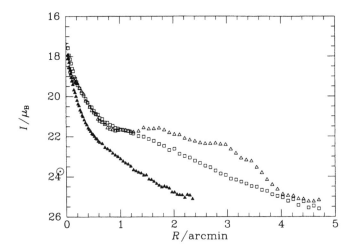

**Figure 4.47** Blue surface brightness versus elliptical radius for three spiral galaxies: NGC 2841 (open squares); NGC 3898 (filled triangles); NGC 5194 (open triangles). The solar symbol ⊙ indicates the estimated surface brightness of the Milky Way near the Sun. The outer parts of two of the profiles are fairly straight in accord with equation (4.20). [From data published in Boroson (1981)]

Figure 4.47 clearly shows that both NGC 2841 and NGC 3898 are significantly brighter near their centers than the exponential law predicts. Images of the galaxies indicate that in most cases the additional brightness is contributed by an elliptical-like bulge, which might well obey the $R^{1/4}$ law. Hence we next attempt to fit the observed profile of these galaxies with the sum of an $R^{1/4}$ law and an exponential law. Figure 4.48 shows that good fits can be obtained in this way.

The right-hand panel of Figure 4.48 shows that the slope of the derived disk component can be significantly smaller than the slope of the straight line that is tangent to the measured profile at large radii. This difference in slope arises because the contribution of the bulge component to the brightness of the galaxy is non-negligible even at large $R$. In fact, if the bulge were to obey the $R^{1/4}$ law at very large radii, there would be a radius beyond which the bulge's light once again dominated that of the disk; clearly, the exponential decrease of disk brightness predicted by equation (4.20) will ultimately lead to a lower brightness at large $R$ than the less steeply falling $R^{1/4}$ law. The bulge contribution drops below that of the disk at intermediate radii only because the scale length of the bulge is smaller than that of the disk. It is therefore dangerous to ignore the brightness of the bulge and fit a straight line to the large-$R$ behavior of the composite profile. Such a fit to the composite profile will inevitably be steeper than the line of the true disk contribution and thus yield a smaller scale length $R_d$ (from the steeper slope) and a higher

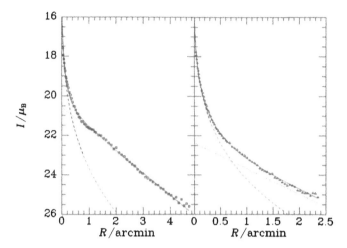

**Figure 4.48** Fits to the surface-brightness profiles of NGC 2841 (left) and NGC 3898 (right). The dotted curves show the exponential fits to the disks and the dashed curves show the $R^{1/4}$ fits to the bulges; the full curves show the sums of these components. [From data published in Boroson (1981)]

intercept $I_d$ than the true values for an exponential disk [equation (4.20)].

The largest residuals between disk–bulge decompositions and the data occur in the transition region between disk and bulge. Freeman (1970) defined type I systems to be those in which the data bob above the model in the transition region, and type II systems to be those, like NGC 3898 in Figure 4.48, in which the data dip below the model. It is not clear how significant this division is physically. Many type II galaxies may have holes in their disks. Moreover, the bumps in the profiles of some type I galaxies are caused by bars – see below. On the other hand, Prieto *et al.* (1992) caution that type II profiles can be caused by dust lanes, while type I profiles may be due to a narrow region of enhanced star formation just outside the bulge. In principle these various possibilities can be distinguished if accurate colors are available, but this task is complex because there is inevitably a steep color gradient in the transition region, as the red bulge gives way to the relatively blue disk.

It turns out that the parameters of the disk and bulge that one obtains from fits like that shown in Figure 4.48 depend to some extent on how the fitting is done. The only reliable way to choose between possible bulge-disk decompositions is to fit the models directly to the two-dimensional surface photometry rather than to surface brightness profiles such as those shown in Figure 4.47. Figure 4.49 shows an example of this process. The right-hand panel in the figure shows the difference between projection onto the sky of the best-fitting model (middle panel) and the data (left-hand panel). The quality of the fit is such that only the galaxy's spiral arms are visible in this

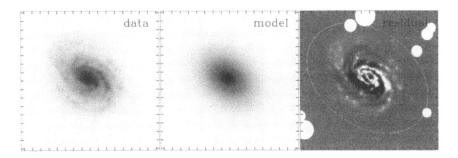

**Figure 4.49** Modeling photometry of a galaxy in two dimensions. The left panel shows a B-band CCD image of NGC 214; the middle panel shows the projection onto the sky of the best-fitting combination of an exponential disk and an $R^{1/4}$-law bulge; and the right panel shows the residuals after this model has been subtracted. The tick marks are at 10 arcsec intervals. The ellipse in the right panel is the region over which the fit was performed, and the empty circles are regions that were excluded because they are contaminated by light from other objects. [Data kindly provided by R. de Jong]

residual plot.

The superiority of the two-dimensional data is connected with the radial gradient in ellipticity that marks the transition from the bulge-dominated region to the disk-dominated region: since the bulge is a truly three-dimensional object, its ellipticity will be less than that of the thin disk.[18] When the sum of a bulge and a disk is fitted directly to the two-dimensional data, the residuals will be smallest when the ellipticity gradient in the model fits that in the data.

Studies of nearly edge-on galaxies are important for two reasons. First they enable us to test the conjecture that the surface-brightness profiles of bulges may be fitted with the $R^{1/4}$ law. Second, they enable us to study the vertical structures of disks. van der Kruit & Searle (1981b) analyzed photographic photometry of the edge-on galaxy NGC 891 and Wainscoat, Freeman & Hyland (1989) have analyzed optical and near-IR photometry of the edge-on galaxy IC 2531. The images of both galaxies are cut through the middle by a dust lane, so that obscuration by dust is a major problem for both studies. van der Kruit & Searle (1981b) tried to evade the obscuration by fitting model profiles to their data away from the dust lane. These profiles were line-of-sight projections of the model luminosity density[19]

$$j(R, z) = \begin{cases} j_0 e^{-R/R_d} \operatorname{sech}^2(z/2z_0) & \text{for } R < R_{\max} \\ 0 & \text{for } R > R_{\max}. \end{cases} \quad (4.44)$$

Here $(R, z)$ are cylindrical polar coordinates that define the symmetry axis of the model, and $j_0$, $R_d$, $z_0$ and $R_{\max}$ are constants to be fitted to the data. At

---

[18] If the bulge is triaxial, its ellipticity will be greater than that of the disk for sufficiently face-on orientations.

[19] Our definition of $z_0$ differs from that of van der Kruit & Searle (1981b) by a factor 2 in order that $j \propto e^{-z/z_0}$ at $z \gg z_0$.

face-on inclination this model obeys the exponential surface-brightness law at $R < R_{max}$. The $sech^2$ functional form in (4.44) was motivated by theory – see Problem 4-25 of BT. Remarkably, van der Kruit & Searle (1981b) were able to fit their data for NGC 891 without allowing any dependence of the parameter $z_0$ upon $R$ – the characteristic thickness of the disk appears to be independent of radius. They found that their data could be satisfactorily fitted by the model with $R_d = 4.9\,kpc$, $z_0 = 0.5\,kpc$, $R_{max} = 21\,kpc$. They concluded that the observed profile drops extremely steeply at $R_{max}$, where the surface brightness of the galaxy is about 4 mag below the estimated sky level. Barteldrees & Dettmar (1994) reach a similar conclusion on the basis of CCD photometry of a sample of 27 more distant edge-on galaxies. This cutoff is harder to see in face-on galaxies, but by going to surface-brightnesses as low as $26\,\mu_R$, Morrison, Boroson & Harding (1994) detected cutoffs in two galaxies. By contrast, de Jong & van der Kruit (1994) found no evidence of cutoffs in a larger sample of galaxies.

Because they possessed near-IR photometry, Wainscoat, Freeman & Hyland (1989) were able to model their data for IC 2531 right down to the middle of the dust lane. They concluded that a successful model has to contain three components: an old disk of $B - V$ color 0.78, a young disk of $B - V$ color $-0.04$, and a dust layer. The vertical density profiles of these components need to be more sharply peaked near $z = 0$ than the model of (4.43) allows. The data can be successfully modeled if the density profiles of all three components are assumed to be of the form

$$j(R, z) = \begin{cases} j_0 e^{-R/R_d - |z|/z_0} & \text{for } R < R_{max} \\ 0 & \text{for } R > R_{max}, \end{cases} \tag{4.45}$$

with $R_d = 6.4\,kpc$ and $R_{max} = 4R_d$. The old disk had $z_0 = R_d/12$, the young disk had $z_0 = R_d/96$, the dust layer had $z_0 = R_d/48$. The luminosity density $j_0$ was the same for both the old and the young stellar disks. For the dust layer $j_0 = 1.6\,mag\,kpc^{-1}$.

Burstein (1979) examined the distribution of light along lines that run perpendicular to the major axes of five highly inclined S0 galaxies. The crosses in Figure 4.50 show three representative profiles for one of these galaxies, NGC 4570: those along lines that cut the major axis at $R = 6''$, $42''$ and $60''$. The diamonds show the corresponding surface brightness profiles that are predicted by modeling the major-axis brightness profile with a sum of an infinitesimally thick exponential disk and a bulge that obeys the $R^{1/4}$ law. In each case the diamonds fit the crosses down to $I \simeq 23.5\,\mu_B$, confirming that the inclination of the disk and the flattening of the bulge have been correctly determined. Fainter than $I \simeq 23.5\,\mu_B$ the crosses lie systematically above the diamonds. From this observation Burstein concluded NGC 4570 has a **thick disk** in addition to a thin disk and an $R^{1/4}$ bulge.[20] The

---

[20] In her 1977 University of Texas thesis, Tsikoudi suggested that some S0 galaxies had thick disks.

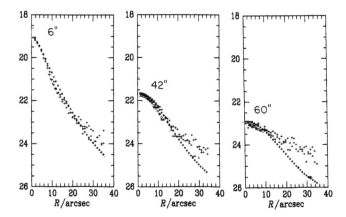

**Figure 4.50** Perpendicular profiles of NGC 4350 at major-axis intercepts $R = 6''$, $R = 42''$, $R = 60''$ [After Burstein (1979) courtesy of D. Burstein]

luminosity density of a thick disk is often modeled by equation (4.45) with $z_0 \simeq 1\,\mathrm{kpc}$. From the observation that in Figure 4.50 the thick disk starts to dominate the thin disk at $I \simeq 23.5\,\mu_B$, it follows that the radial scale-length $R_d$ must significantly exceed the radial scale length of the thin disk.

The phenomenon detected by Burstein (1979) in five S0 galaxies, has since been observed in many but not all disk galaxies. It is now thought that most S0 galaxies and many spirals have perpendicular profiles of Burstein's type. However, the correct physical interpretation of these profiles is unclear. One possibility is that the light Burstein attributes to the thick disk should be assigned to the bulge. Indeed, even if bulges were physically identical to low-luminosity elliptical galaxies, their luminosity profiles would have been distorted from those characteristic of elliptical galaxies by the strong gravitational fields of the embedded disks (Freeman 1978, Rowley 1988). Another possibility is that not all disks have exponential vertical profiles; there is no theoretical reason to expect an exponential dependence on $z$ – see Problem 4.10. In §10.4.3 we shall see that the Milky Way displays a similar phenomenon to that studied by Burstein, and important additional clues as to its physical origin are provided by studies the chemical composition of the stars as a function of their distance above the Galactic plane.

The minor-axis luminosity profiles of nearly edge-on galaxies enable us to test the conjecture that bulges can be adequately fitted with the $R^{1/4}$ law. Kent (1985) shows the minor-axis profiles of 105 disk galaxies of all morphological types, together with fits to the data by the sum of exponential and $R^{1/4}$ laws. The great majority, of these fits are of high quality, although there are significant deviations between model and observations in a few cases. Similarly, Burstein (1979) was able to obtain good fits to his minor-axis data with the sum of exponential and $R^{1/4}$ laws. However, the minor-

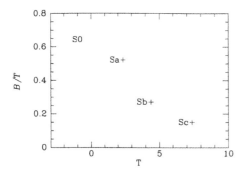

**Figure 4.51** The later the Hubble stage $T$ of a disk galaxy, the smaller is its bulge fraction $B/T$. The plotted values are means. Individual values scatter significant about these means. [From data published in Kent (1985)]

axis profiles of some galaxies are definitely not adequately fitted by this model and are better fitted by an exponential profile. The classic example is the near edge-on Sb galaxy NGC 4565 (Kormendy & Bruzual 1978, van der Kruit & Searle 1981a). Bahcall & Kylafis (1985) have shown that existing data are usually unable to distinguish between bulges that follow the $R^{1/4}$ law and ones that take the form of thick disks.

Once surface photometry of large numbers of disk galaxies has been decomposed into disk and bulge components, it is interesting to look for correlations between the values of the parameters that are determined by the fits, and also seek correlations with Hubble type (Simien & de Vaucouleurs 1985, Kent 1985). The fits involve four parameters: $I_e$, $R_e$ for the bulges and $I_d$ and $R_d$ for the disk – see equations (4.18) and (4.20). From these it is interesting to calculate a fifth parameter, the **bulge fraction**,

$$B/T \equiv \frac{R_e^2 I_e}{R_e^2 I_e + 0.28 R_d^2 I_d},\qquad(4.46)$$

which is the fraction of the total luminosity that is contributed by the bulge. $B/T$ is related to the **disk-to-bulge ratio**, which is frequently cited in the literature, by $D/B = (B/T)^{-1} - 1$. The advantage of $B/T$ is that it is confined to the interval $(0, 1)$.

Figure 4.51 shows that the bulge fraction $B/T$ is quite tightly correlated with Hubble type, falling from a mean value of 0.65 for lenticular galaxies, to a mean value of 0.15 for galaxies of type Sc and later. This correlation quantifies the basic idea of the Hubble's classification according to the prominence of the bulge. From the point of view of cosmology it is important because it argues against the possibility that all S0 galaxies are simply typical spiral galaxies that have somehow lost their gas and therefore ceased to display spiral structure or form new stars. Of course, a disk that ceases to form new stars will gradually fade relative to its associated bulge, but estimates of this effect (see §5.4) indicate that it is insufficiently powerful to account for the steepness of the trend displayed in Figure 4.51.

The left-hand panel in Figure 4.52 shows that the bulges of galaxies of Hubble type Sb and earlier have a rather similar distribution in the $(R_e, I_e)$

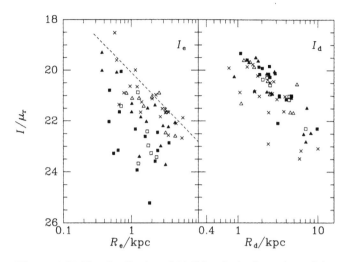

**Figure 4.52** The distribution of 66 disk galaxies in analogs of the lower left panel of Figure 4.43. The left panel shows the surface-brightnesses and effective radii of the bulges, while the right panel shows the corresponding disk parameters. Different symbols correspond to galaxies of different Hubble types: S0 (crosses); Sa–Sab (open triangles); Sb (filled triangles); Sbc (open squares); Sc–Scd (filled squares). If the ellipticals of Figure 4.43 were to be plotted in the left-hand panel, they would cluster around the dashed line. [From data published by Kent (1985), who adopts $H_0 = 100 \, \mathrm{km \, s^{-1} \, Mpc^{-1}}$.]

plane to that of elliptical galaxies – compare the bottom left-hand panel of Figure 4.43. In fact, if the elliptical galaxies of Figure 4.43 were to be plotted in the left-hand panel of Figure 4.52, they would cluster around the dashed line shown.[21] The bulges of Sc galaxies lie systematically lower in the $(R_e, I_e)$ plane.

The right-hand panel of Figure 4.52 shows that the corresponding parameters for the disk, $R_d$ and $\mu_d$, are tightly correlated in the same sense: physically larger systems have lower central surface brightnesses. Freeman (1970) noted that the measured central surface brightnesses of disk galaxies clustered around $I = 21.7 \, \mu_B$ – this result is often called **Freeman's law**. Subsequently, it turned out that this clustering was a selection effect: the apparently largest and brightest galaxies were the first to be measured carefully. In reality there exist significant numbers of intrinsically fainter galaxies, but one has to look carefully for such objects. Schombert *et al.* (1992) and Impey *et al.* (1996) have systematically searched for **low surface brightness (LSB)** galaxies, which may usefully be defined to be objects

---

[21] The dashed line in Figure 4.52 has been calculated by displacing the ridge-line of the distribution of Figure 4.43 to allow for the different distance scales and measures of $I$ employed.

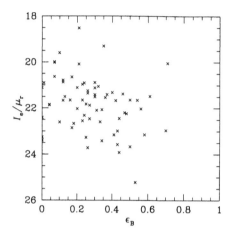

**Figure 4.53** Plot showing bulge surface brightness, $I_e$, versus bulge ellipticity, $\epsilon_B$, for the 66 disk galaxies shown in Figure 4.52. [From data published by Kent (1985)]

with $I(0)$ fainter than $23\,\mu_B$. Since for a bulge $23\,\mu_B \simeq 21.7\,\mu_r$, only a few LSB galaxies contribute to the right-hand panel in Figure 4.52.

It is worth noting that Peletier *et al.* (1994) have shown that near-infrared photometry generally yields smaller values of $R_d$ than $B$-band photometry. This phenomenon reflects the fact that disks tend to become bluer at larger radii – see §4.4.5.

### 4.4.4 Shapes of bulges

Ideally, we would like to investigate the three-dimensional shapes of bulges using the same sort of analysis that we applied to the apparent ellipticities of elliptical galaxies in §4.3.3. Unfortunately, the contribution of light from the other components in disk galaxies complicates matters. For example, if a disk galaxy contains a bulge that is highly flattened towards its galactic plane, it will be very difficult to disentangle the bulge light from that of the disk. The distinction between bulges and galactic bars is even harder to draw: a prolate spheroidal structure whose major axis lies in the plane of a disk galaxy could equally well be identified as a prolate bulge or a bar. It is therefore perhaps unsurprising that – as we shall see in §4.4.7– the properties of bulges and bars seem to be very closely related.

Despite these complexities, several studies have estimated the shapes of galactic bulges. In his photometric decomposition of a sample of 66 disk galaxies, Kent (1985) estimated the shapes of the bulges as well as their radial profiles, and found ellipticities spanning the range $0 < \epsilon_B < 0.7$. As Figure 4.53 illustrates, this sample shows no correlation between $\epsilon_B$ and the characteristic surface brightness, $I_e$, of the $R^{1/4}$ law that fits the bulge [equation (4.18)]. If the bulges were all identical oblate spheroids viewed from different directions, the more edge-on systems would have larger values of both $I_e$ and $\epsilon_B$. Even if the bulges were oblate spheroids with a range of flattenings, one would expect $I_e$ to be correlated with $\epsilon_B$ in a sample of galaxies with a range of inclinations. The absence of such a correlation might therefore be taken as evidence that bulges have more complex shapes than simple oblate spheroids. Alternatively, it might simply reflect the small size of the sample, systematic difficulties in disentangling the various components in disk galaxies, or the pervasive influence of obscuration by dust on measured surface brightnesses (see §4.4.1).

One piece of evidence which suggests that bulges *can* be represented by simple oblate spheroids, has come from studies of their kinematics. As we shall see in §11.3.1, analyses of the kinematics of bulges in edge-on disk galaxies show that flatter bulges rotate more rapidly. In fact, their ellipticities turn out to be very close to what one would expect for simple axisymmetric oblate spheroids that have been flattened by rotational motion around axes that are parallel to those of the surrounding disks.

As for elliptical galaxies, further information on the properties of bulges can be gleaned from any departures in their isophotes from perfect ellipses. Once again, the presence of the other components in disk galaxies complicates the analysis: if part of the disk's light is erroneously attributed to the bulge, then, not surprisingly, the bulge will appear to have disky isophotes with $a_4 > 0$ [see equation (4.10)]. Similarly, the presence of non-axisymmetric structure such as spiral arms and patchy obscuration by dust can distort the shapes of fitted isophotes. Studies of the shapes of bulge isophotes have therefore generally been restricted to disk galaxies that are close to edge-on: away from the plane of such a galaxy bulge light is relatively uncontaminated by other components. Since bulges tend to be flattened towards the plane of their surrounding disks, the exclusion of emission close to the plane makes it significantly harder to detect deviations of isophotes from ellipses – see Figure 4.20. Nevertheless, such analyses have revealed departures from elliptical isophotes in a significant fraction of bulges. de Souza & dos Anjos (1987) found that $\sim 25\%$ of a sample of edge-on galaxies contained bulges with dramatically boxy isophotes. In some cases, the isophotes have been squashed to the degree that they have a double-lobed structure, rather like an unshelled peanut. The image of NGC 5746 in Figure 4.60 provides an example of such a peanut-shaped bulge. As we shall see in §4.4.7, this phenomenon seems to be related to the presence of a bar in the galaxy.

### 4.4.5 Color and metallicity gradients in disk galaxies

Figure 4.54 shows three color indices as a function of radius in the Andromeda nebula, M31. Interior to $\sim 6\,\text{kpc}$ the light is dominated by the bulge and the colors are those of an elliptical galaxy. Slightly further out there is a steep drop in $U - B$ and a smaller drop in $B - V$ as young stars begin to make a significant contribution to the overall surface brightness. All three

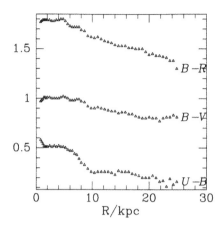

**Figure 4.54** Three colors as a function of radius in the giant Sb galaxy M31. These data represent average colors in elliptical annuli of given semimajor axis. They have not been corrected for Galactic extinction. [From data published in Walterbos & Kennicutt (1987)]

color indices show the disk to be getting steadily bluer from $\sim 9\,\mathrm{kpc}$ to at least $20\,\mathrm{kpc}$. However, the gradients are shallow and require highly smoothed data for their detection. Few galaxies have been as carefully observed and reduced as M31. Consequently the evidence for systematic color gradients in the disks of other galaxies is weak – see Prieto *et al.* (1992) for an indication of how varied major-axis color profiles typically are. It is reassuring that Peletier *et al.* (1994) have shown that near-infrared photometry generally yields smaller values of $R_\mathrm{d}$ than $B$-band photometry.

There are three potential causes of color gradients in disks: (i) gradients in the degree of internal extinction by dust, (ii) gradients in the mean ages of stars, and (iii) metallicity gradients – see §8.2.6. In practice all three factors are probably in competition in any given galaxy, with the result that many major-axis color profiles show no systematic trend.

### 4.4.6 Spiral structure in disk galaxies

Spiral structure can take many forms. Some galaxies display **grand-design** spiral structure; that is, in the $B$ band their disks are dominated by two great arms. M51, M81 and M100 are famous examples of grand-design spirals. Since these objects are photogenic, they tend to feature in coffee-table books and on posters. Grand-design spirals make up arm-class 12 in the Elmegreen & Elmegreen (1982) classification. Only $\sim 10\%$ of spiral galaxies fall into this class (Elmegreen & Elmegreen 1987); most spiral galaxies have more or less ratty or **flocculent** spiral structure.

A good way to study spiral structure is to analyze the difference between an image and the azimuthal average of that image. Figure 4.55 shows such difference images for the grand-design spiral M51. These images show two general properties of spiral structure: (i) spiral structure is present in both blue and red images but it has larger amplitude in blue images; (ii) spiral structure is smoother in red than in blue images. They also show a general

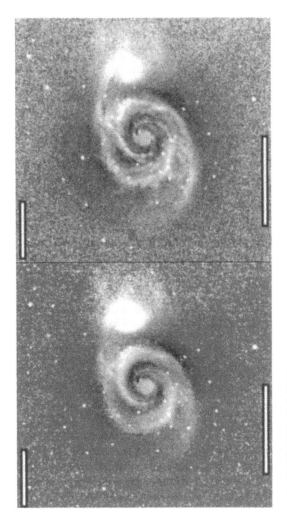

**Figure 4.55** Spiral structure in M51. These images are the differences between images of M51 and the azimuthal averages of those images. The top panel is for the $B$ band and the bottom image is for the $I$ band. [After Elmegreen, Elmegreen & Seiden (1989) by courtesy of B. Elmegreen]

characteristic of grand-design structure: the amplitude of the arms is modulated in a systematic way as a function of angle around the arm. Specifically, if in the lower panel of Figure 4.55 one follows the arm whose inner end is at 3 o'clock, one sees that is passes through a minimum in amplitude around 5 o'clock, and then another at 9 o'clock. Each of these minima is mirrored in the other spiral arm.

Figure 4.56 shows $U$, $B$ and $O$ azimuthal profiles for the giant Sb galaxy M81 at $r = 475''$, together with two of the associated color indices. Two points should be noted about this diagram. (i) The two spiral arms which feature so clearly in the top panel appear in the $U$ band to be made up of a fairly symmetrical broad wave, on top of which is superimposed a pair of narrower peaks. In the $B$ and $O$ bands, these narrower peaks are much less

---

### Box 4.3:    Pitch Angles

The spiral arms in Figure 4.55 are roughly logarithmic spirals, $\phi \propto u$, where $\phi$ is the usual azimuthal angle and $u \equiv \ln(R)$. This observation suggests that it might be useful to express the brightness $I$ of a disk as a Fourier sum of logarithmic spirals:

$$I(u, \phi) = \frac{1}{2\pi} \sum_{m=-\infty}^{\infty} \int_{-\infty}^{\infty} dp \, I_m(p) e^{i(pu+m\phi)}.$$

$I_m(p)$ is the amplitude of the $m$-armed spiral of **pitch angle** $\psi = -\arctan(m/p)$. For a grand design spiral $I_m(p)$ peaks strongly at $m = 2$ and some value $p_0$ of $p$, and one says that the galaxy has pitch angle $\psi_0 = -\arctan(2/p_0)$. Pitch angles have been estimated for many galaxies – see Garcia Gomez & Athanassoula (1993) for details. They increase with Hubble stage from $\sim 10°$ at Sa to $\sim 18°$ at Sc.

---

conspicuous. (ii) The bottom panel of Figure 4.56 shows that the $(B - O)$ color averages $\approx 0.8$ mag and changes by no more than $\pm 0.2$ mag across an arm.

In his classic study of six giant spirals Schweizer (1976) found that, in the $O$ (orange) passband, the arms typically contribute about 17% of the total brightness at 3 kpc, rising to 50% of the brightness at 15 kpc. The arms are generally 20% stronger in the $B$ passband and 50% stronger in the $U$ passband than they are in the $O$ band. The underlying disks prove to be very uniform in color and rather red. For the disks in his sample, $(B - V) \approx 0.75 \pm 0.05$, which is redder than many old galactic clusters, although it is a little over 0.1 mag bluer than the central parts of giant elliptical galaxies. As Figure 4.55 shows, the ridges of arms are bluer than a disk, so the growth in strength of the arms relative to the disk with increasing radius causes the overall colors of the galaxy to become slightly bluer at large radii.

If we anticipate results concerning the interstellar media of galaxies that will be discussed in §8.2.3, the observations described above are fairly easy to understand. First, in all cases in which it is possible to determine which is the nearer and which the further side of a spiral galaxy (see §4.4.1), velocity measurements show that the spiral arms are trailing.[22] That is, stars and gas enter the arms on their concave inner edges and leave them on their convex outer edges.

Second, surface brightness at the blue end of the spectrum is sensitive

---

[22] In the majority of cases it is not clear which is the nearer side of a galaxy. Then one determines this by *assuming* that the arms trail. NGC 4622 has both leading and trailing arms Buta, Crocker & Byrd (1992).

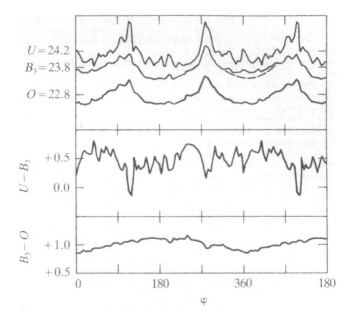

**Figure 4.56** Spiral structure in M81. $U$, $B$, and $O$ azimuthal profiles at $r = 475$ arcsec. The angle $\phi$ is measured from the apparent major axis in the direction of rotation. The profile has been continued periodically beyond $\phi = 360°$. Notice how the $U$ profile consists of narrow peaks superposed on a broader wave pattern which resembles the $B$ and $O$ profiles. Much of the small fluctuation in the $(B_3 - O)$ color can be accounted for by an instrumental malfunction. [After Schweizer (1976)]

to the presence of short-lived stars. These form on the inner edges of spiral arms as a response to shock-compression of interstellar gas, blaze briefly as very blue objects as they traverse the peak of the arm, and are then extinguished. Surface brightness in the red and infrared is determined by a wider range of stars, including objects that do not evolve significantly in the time it takes a star to drift across a spiral arm. Hence $O$- and $I$-band surface brightnesses give truer impressions of the enhancement in the mass density that underlies spiral structure. This is why in Figure 4.55 arms are fuzzier and less pronounced in $I$ than in $B$.

The uniformity and numerical magnitude of the color of the broad components in Figure 4.56 is of some importance since it very strongly suggests that old disk stars (of relatively late spectral types) contribute about 40% of the increased brightness found near the spiral arms.[23] If we accept this conclusion, it follows that spiral structure is not an entirely gas-dynamic

---

[23] In principle the observed uniformity of color could be consistent with all the arms' light coming from young stars if the latter were unusually heavily reddened – by $A_V \approx 0.5$ mag.

phenomenon, which leaves the bulk of the mass in the disk quite unmoved. Instead it must be associated with some sort of gravity wave that propagates through the stellar disk.

Finally, the minima in the amplitudes of the arms as a function of azimuth have been convincingly interpreted as nodes in the interference pattern of two spiral waves, a dominant trailing wave and a subordinate leading wave (Toomre 1981). That is, at the minima, a trough of the leading wave crosses the peak of the trailing wave, partially canceling it. In this picture the disk forms a resonant cavity within which energy is carried in and out by leading and trailing spiral waves.

It is thought that grand-design spiral structure is usually generated by either a rotating bar within the galaxy (see below), or by the tidal gravitational field of a companion galaxy. In the case of M51, the perturbing companion, NGC 5152 is clearly visible to the North in Figure 4.55.

The more common but less photogenic ratty spiral structure is probably caused by ephemeral gravitational instabilities within the disk (Sellwood & Kahn 1991). The dynamical details are complex and not fully understood but we know that in disks both stars and gas have small random velocities (see §8.1.4, §9.2.6 and §11.3.2) with the consequence that any initial irregularities are quickly amplified into short spiral arms. These arms convert some of the disk's ordered rotational kinetic energy into random kinetic energy, which inhibits the formation of further arms. Hence galaxies can continue to display spiral structure only if they have a mechanism for dissipating disordered kinetic energy. In gas-rich systems, interstellar gas provides this mechanism. In §8.2.5 we shall see that S0 galaxies have relatively little interstellar gas, and this may explain why they do not display conspicuous spiral structure.

### 4.4.7 Barred galaxies

Of order one half of all disk galaxies are barred. In particular the Milky Way is believed to be barred (see §10.1). Therefore the study of barred galaxies is an important field. It is, however, a difficult field and the effort that has so far been devoted to it is small compared with the effort that has been devoted to the study of elliptical galaxies, for example.

Fortunately, over the last decade advances in the techniques of surface photometry have opened up the field and the literature now contains high-quality photometry of a fair number of barred systems. Bars occur in the bright inner parts of galaxies, so study of them benefits considerably from the large dynamic range of CCD detectors relative to photographic plates. As we saw in §4.1, dust lanes are often prominent on blue images of barred galaxies. While a dust lane can be of considerable interest when modeling the flow of interstellar gas, it does make it hard to determine the luminosity distribution. Hence, near-IR photometry, in which dust lanes are much less prominent, is extremely valuable and is now becoming accessible.

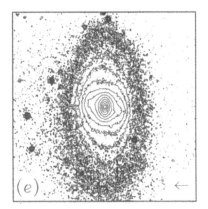 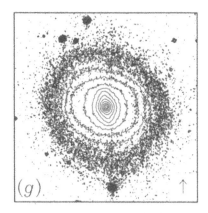

**Figure 4.57** $B$-band isophotes of two typical SB0 galaxies, NGC 4371 (left) and NGC 4477 (right). The fields are 5.3 arcmin on a side and the isophotes are 0.5 mag apart. [After Wozniak & Pierce (1991) courtesy of H. Wozniak]

Elmegreen & Elmegreen (1985) and Baumgart & Peterson (1986) between them published photographic photometry of 24 barred spiral galaxies of all types. Jarvis *et al.* (1988), Forbes & DuPoy (1992) and Wozniak & Pierce (1991) have published CCD photometry for over a dozen early-type barred galaxies. Figure 4.57 shows the $B$-band isophotes of two typical SB0 galaxies. At small radii nearly elliptical isophotes crowd closely together. Then the isophotes become suddenly more elongated and less densely spaced. This is the region of the bar. Slightly further out still there is an abrupt change in the direction of the isophotes' major axis as the bar gives way to an inclined disk. Figure 4.58 shows the results of fitting ellipses to the isophotes of these two galaxies by plotting the surface-brightness and ellipticity of each ellipse versus the length of its semimajor axis. In these plots each bar is associated with characteristic features: a slight bump in the luminosity profile and a sharp peak in the ellipticity profile.

Notice in Figure 4.57 that the isophote of NGC 4371 that most nearly delineates the bar is not at all elliptical; rather it resembles an oval tray with two handles sticking out of it. This morphology is characteristic of bars, although it is not always observed as the example of NGC 4477 shows. Athanassoula *et al.* (1990) show that good fits to such isophotes can be obtained with **generalized ellipses**, which are the curves defined by

$$(|x|/a)^c + (|y|/b)^c = 1, \tag{4.47}$$

where $a$, $b$ and $c$ are constants characterizing the curve. The larger $c$ is, the squarer is the generalized ellipse.

It is obviously important to determine the true shapes of bars. There are two difficulties. First one must determine the inclination of the system. This is usually done by assuming that the outermost isophotes, which should be

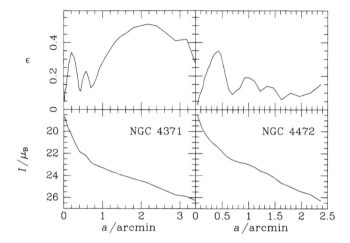

**Figure 4.58** The results of fitting ellipses to the isophotes of NGC 4371 and NGC 4477 shown in Figure 4.57. Here $a$ and $\epsilon$ are, respectively, the length of an ellipse's semimajor axis and its ellipticity $1 - b/a$. The bar in NGC 4371 produces a bump in the luminosity profile near $a = 0.5$ arcmin while that in NGC 4477 produces a bump at $a = 0.4$ arcmin. There are corresponding peaks in the ellipticity profiles. The bulge of NGC 4371 is triaxial and gives rise to a peak in the ellipticity profile at $a \simeq 0.2$ arcmin. [After Wozniak & Pierce (1991) from data kindly supplied by H. Wozniak]

generated by the outer disk, beyond the influence of the bar, are intrinsically round. The second and more difficult problem is making allowance for the presence of the bulge, which is obviously prominent in early-type galaxies such as those shown in Figure 4.57. Athanassoula *et al.* (1990) simply mask out the center of their galaxies and fit generalized ellipses to what remains of the isophotes. This procedure indicates that the ellipticity of the bar declines roughly linearly with radius, from values of order 0.7 at the smallest radius for which an estimate can be obtained. Thus they conclude that bars have axis ratios in the plane of 3:1 or greater.

The presence of a prominent bulge makes it hard to determine how surface brightness varies as a function of distance down the major axis of a bar. Nonetheless, Elmegreen & Elmegreen (1985) and Elmegreen *et al.* (1996) have estimated this quantity for about two dozen galaxies. Figure 4.59 shows results for two characteristically different galaxies. In each panel the full curve shows the surface-brightness profile of the galaxy along the bar's apparent major axis, while the dashed curve shows surface-brightness along the bar's minor axis. Double-headed arrows show the extent of each bar along the major axis. The left-hand panel is for the SBbc galaxy NGC 1300. In this galaxy the surface brightness is essentially independent of radius outside of a nuclear cusp. The right-hand panel in Figure 4.59 is for the SBcd galaxy NGC 7741, in which the surface brightness falls at least as steeply within

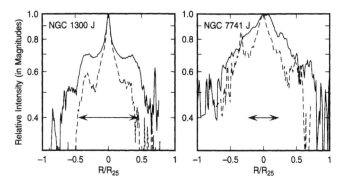

**Figure 4.59** *J*-band surface brightness profiles of two barred galaxies; full curves are for the major axes, dashed curves are for the minor axes. The arrows indicate the lengths of the bars. The major-axis profile of the SBbc galaxy NGC 1300 at left is characteristic of early-type galaxies. In it the surface brightness drops less steeply in the bar than outside it. The major-axis profile of the SBcd galaxy NGC 7741 at right is characteristic of late-type systems. [Adapted from Elmegreen *et al.* (1996) courtesy of B. Elmegreen.]

the bar as it does outside it. Elmegreen & Elmegreen (1985) suggested that in early-type systems the surface brightness declines more slowly with increasing radius within the bar than in the surrounding disk, as in the left-hand panel of Figure 4.59, while in late-type galaxies the surface brightness declines as steeply within the bar as outside it, as in the right-hand panel of Figure 4.59.

A key dynamical parameter of a stellar bar is its **pattern speed**, that is, the angular velocity with which its figure rotates. It is also useful to define the **corotation radius**, $R_{CR}$, which is the radius at which a star can move on a circular orbit at the same angular frequency as the bar's pattern – see §3.3.2 of BT. Elmegreen *et al.* (1996) summarize the available estimates of the ratio of $R_{CR}$ to the length of the bar's semi-major axis, and conclude that, with one exception, this ratio is $1.2 \pm 0.2$. Thus, bars extend to about 80% of the corotation radius.

**Vertical structure of bars** In order to study the structure of galactic bars in the third dimension, perpendicular to the planes of their host galaxies, we must obtain photometry of edge-on barred galaxies. Unfortunately, the structure that we identify as a bar is only discernible in a reasonably face-on galaxy, and so it is almost impossible to tell whether an edge-on galaxy contains a bar. The development of our understanding of the three-dimensional structure of bars is therefore largely dictated by the results of numerical simulations. The most important discovery from these calculations is that thin bars seem to be generically unstable structures which buckle about their midpoints out of the plane of the bar (Combes & Sanders 1981, Raha *et al.* 1991). The simulations show that the bar initially bends into an asymmetric banana-shape, but that it ultimately regains its symmetry about the midplane of the galaxy, forming a thickened double-lobed structure rather like an unshelled peanut. Viewed end-on such fattened bars appear round, but when viewed from the side they reveal their distinctive double-lobed appearance. These numerical results are intriguing, since, as we saw in §4.4.4, $\sim 25\%$ of edge-on bulges seem to have boxy or peanut-shaped isophotes. It is therefore tempting to suggest that these boxy bulges contain buckled bars that are viewed more-or-less edge-on, while the remaining $\sim 75\%$ of galaxies with elliptical bulges are made up from unbarred galaxies and barred galaxies where the bar happens to be oriented approximately end-on.

Circumstantial support for this hypothesis has come from a study of the boxy-bulged S0 galaxy NGC 1381. de Carvalho & da Costa (1987) have applied photometric decomposition techniques (§4.4.3) to images of the galaxy, and they find that this system appears to contain a third component in addition to the usual disk and bulge. This component, intermediate in size between the bulge and the disk, has an approximately uniform surface brightness that cuts off sharply at the same radius on either side of the galaxy. As we have seen above, bars in early-type disk galaxies have exactly this structure, and so these observations have been interpreted as implying that this boxy-bulged galaxy also contains a bar. A more extensive study of the photometry of edge-on galaxies by Dettmar & Barteldrees (1990) has confirmed that extra bar-like components in the light distribution are closely associated with boxy and peanut-shaped bulges. However, this evidence is not conclusive: the extra component could be an entirely axisymmetric distribution of light rather than a bar. Similarly, it is possible to construct axisymmetric bulge models with a toroidal structure which appear peanut-shaped in projection (Rowley 1988). Thus, the photometric observations do not provide direct evidence for non-axisymmetric structures in these edge-on galaxies.

A further probe of non-axisymmetric structure in edge-on galaxies has been proposed by Kuijken & Merrifield (1995). They point out that the orbits followed by material in a barred potential are more complicated than those in an axisymmetric system. They have therefore obtained optical spectra of edge-on galaxies in order to study the motions of the gas within these systems via their Doppler shifts. As Figure 4.60 illustrates, galaxies with elliptical bulges tend to have simple kinematic structures, with all the material on one side of the galaxy redshifted due to its motion away from us, whilst all the material on the other side is blueshifted due to its motion towards us. These simple kinematics imply that the material is following the circular orbits that we would expect in an axisymmetric galaxy. Galaxies with peanut-shaped bulges, on the other hand, reveal much more complex structure in the line-of-sight motions of their constituent gas, implying that this material follows the more complicated orbits associated with a barred potential. As we shall see in §9.4, similar reasoning has been applied to spectral observations of the Milky Way to infer that our own galaxy is also barred; it is therefore

NGC 1055                      NGC 5746

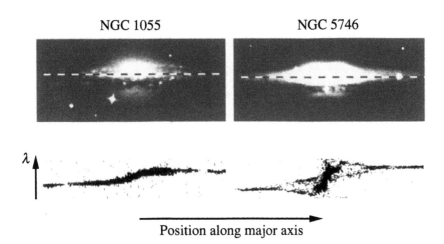

Position along major axis

**Figure 4.60** Illustration of the complex gas kinematics associated with peanut-shaped bulges. The top panels show images of a round-bulged galaxy (NGC 1055) and a galaxy with a peanut-shaped bulge (NGC 5746). The dust lanes which cut across the bulges of these galaxies indicate that they are close to edge-on. The lower panels show the 658.3 nm [NII] emission line as observed along the major axes of these systems (dashed lines in top panels). The plots show the observed wavelength $\lambda$ of the emission line (either redshifted or blueshifted due to the line-of-sight motion of the gas) as a function of position along the major axis.

interesting to note that the bulge of the Milky Way has a distinctly boxy appearance (see Figure 1.8). Thus, there is a clear kinematic link between peanut-shaped bulges and barred galaxies, which strongly suggests that the vertical structure of bars produces this double-lobed structure.

**Rings in SB galaxies**   In §4.1 we saw that barred galaxies often contain rings and/or lenses. Buta (1986) measured the diameters and orientations of these features in a sample of over 1200 objects. He found that $\sim 75\%$ of his galaxies have an inner ring. From the distribution of apparent axial ratios of these rings he concluded that the rings have intrinsic axial ratio $q \gtrsim 0.9$ in SA galaxies and $q \simeq 0.8$ in SB galaxies. The measured distribution of angles between the apparent major axes of the bar and the ring can be reproduced by a model in which the true major axis of the ring coincides with the true major axis of the bar.

Buta found outer rings in $\sim 50\%$ of his galaxies, almost all in clearly barred systems. These rings are on average slightly less elliptical than inner rings. He was unable to reproduce his distribution of angles between the apparent major axes of the bar and outer ring by assuming that the corresponding true axes are aligned; he required that some be mutually parallel and some mutually perpendicular, with the majority being perpendicular.

Some galaxies have more than one ring and it is interesting to inspect the

**Figure 4.61** The Small Magellanic Cloud. Notice the egg-shaped bar which is clearly displaced from the center of the surrounding luminosity. [Photograph from the *Carnegie Atlas* courtesy of A. Sandage]

distribution of the ratio of the diameters of these rings. This distribution is strongly peaked around 2, independent of Hubble type, but the width of the peak in the distribution does seem to exceed that expected from measuring errors alone.

Since rings are visible in near-IR frames as well as in $B$-band frames, they cannot be exclusively hydrodynamical phenomena that influence only the distribution of young stars. The mechanism responsible for rings must alter the distribution of older, redder stars.

**Dust lanes in SB galaxies**    Many barred galaxies display characteristic dust lanes. These often run almost but not quite radially from the ends of the bar; they keep toward the side of the bar which leads in the rotation of the whole galaxy. An excellent example of this phenomenon can be seen in the photograph of NGC 1300 in Figure 4.7. In galaxies later than the SBb galaxy NGC 1300, the dust lanes run along lines further removed from the principal axis of the bar, and through the less-well-illuminated disk. In very late-type specimens, the absorption lanes may be seen to curve in toward the nucleus as they reach the bar's minor axis, thus partially obscuring the center of the galaxy. In this event, a contour map of the galaxy's brightness will appear double peaked. These dust lanes are produced by the streaming of gas *along* the length of the bar.

**Lop-sidedness in SB galaxies**    Barred galaxies are commonly lopsided.

This phenomenon is most clearly seen in the Magellanic-type galaxies, which usually have rather egg-shaped bars that are often not situated at the center of the surrounding light distribution (de Vaucouleurs & Freeman 1970). The photograph of the Small Magellanic Cloud shown in Figure 4.61 illustrates this phenomenon, which may also be detected in giant SB systems – see, for example, the photograph of NGC 4548 in the *Hubble Atlas*. This phenomenon is very ill-understood. It may be related to lop-sidedness in unbarred galaxies, which is discussed in §8.2.3.

## 4.5 Globular cluster systems

The Milky Way possesses at least 160 globular clusters. These objects, which have been intensively studied for more than a century and have played a prominent rôle in the development of our understanding of the Universe, are the subject of Chapter 6. Most external galaxies also possess significant numbers of globular clusters and the study of the **globular cluster systems** of external galaxies has gradually become an important branch of galactic astronomy.

In Chapter 6 we shall see that a typical Galactic globular cluster has absolute magnitude $M_V \simeq -7.3$ and half-light radius $R_e \simeq 5\,\mathrm{pc}$. At the distance of M31 ($700\,\mathrm{kpc}$) a similar cluster would have $V = 16.9$ and $R_e = 1.5''$. As such it would be identifiable from the ground as a moderately faint, slightly non-stellar object. The HST would be able to discern its non-stellar nature at distances of order $10\,\mathrm{Mpc}$, but at greater distances it would be merely a faint stellar object. Nevertheless, since the pioneering work of Haynes (1977) and Strom *et al.* (1981) globular clusters have been studied in galaxies as distant as $\sim 100\,\mathrm{Mpc}$. The technique by which this is done is as follows.

A helpful first step is to subtract from a deep image of the galaxy under study a model of galaxy's own luminosity distribution. This step is most satisfactory if the galaxy is an elliptical and therefore has a smooth luminosity distribution. Globular clusters will manifest themselves in the galaxy-subtracted image by an enhancement in the density of faint objects towards the position of the galaxy's nucleus. Some of these faint objects will be distant galaxies, but these can be identified as such by the non-stellar nature of their images. Some of the remaining faint objects will be foreground stars in the Milky Way, but most of these will be redder than any globular cluster. When non-stellar objects and red objects are eliminated, spectroscopic observations have shown that in the case of an elliptical galaxy, of order 90% of the remaining objects will be globular clusters (Zepf, Geisler & Ashman 1994). If colors are not available, it will be necessary to estimate the number of foreground stars in the surviving sample from a model of the Milky Way (§10.6), and then make a statistical correction to obtain an estimate of the number of clusters in the sample. This correction will be large

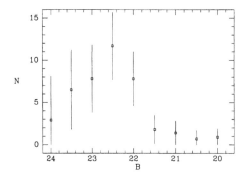

**Figure 4.62** Histogram of the numbers of globular clusters around NGC 3379. The numbers of clusters are non-integer because statistical corrections have been made for the inclusion within the sample of foreground stars. [From data published in Pritchet & van den Bergh (1985)]

and uncertain for the faintest magnitude bins and will dominate the error in the number of clusters present.

In general the detection of globular clusters in a spiral galaxy is very much harder than in a similarly distant elliptical galaxy, because it will be impossible to model accurately the uneven luminosity distribution of the disk. Moreover, even when globular clusters have been successfully identified, it is hard to determine the appropriate corrections to their magnitudes for extinction by dust in the spiral's disk. These problems are minimized if the galaxy is edge-on.

### 4.5.1 Globular cluster luminosity function

Once the apparent magnitudes of the globular clusters around a galaxy have been determined, it is natural to plot a histogram of the numbers of clusters in each of several magnitude intervals. It has been found that such a histogram is invariably well represented by a Gaussian that has been clipped at the faint end by virtue of the impossibility of detecting apparently faint clusters. That is, if the limiting magnitude of the survey is $m_{\text{lim}}$, then number of clusters with magnitudes in $(m + dm, m)$ is

$$\frac{\mathrm{d}N}{\mathrm{d}m} = A \times \begin{cases} \mathrm{e}^{-(m-m_0)^2/2\sigma_m^2} & \text{for } m < m_{\text{lim}}, \\ 0 & \text{otherwise.} \end{cases} \tag{4.48}$$

Figure 4.62 shows the luminosity distribution of the clusters in the E0 galaxy NGC 3379, which is sufficiently close ($cz = 630\,\text{km s}^{-1}$) for $m_{\text{lim}}$ for ground-based counts ($B \simeq 24$) to be significantly fainter than the peak magnitude $m_0$ in (4.48). The best-fitting Gaussian to the data shown in Figure 4.62 has $B_0 = 22.67$ and $\sigma_m = 1.2\,\text{mag}$.

If $m_{\text{lim}}$ is too bright, the counts will not turn over and one cannot simultaneously determine both $\sigma_m$ and $m_0$ in (4.48). That is, satisfactory fits to the counts will be obtainable with a one-parameter family of Gaussians along which $\sigma_m$ will increase as $m_0$ increases. In these circumstances the value of $\sigma_m$ is usually set to $\sigma_m = 1.25\,\text{mag}$ on the basis that this value fits the data for nearby galaxies (including the Milky Way – see §10.5.1)

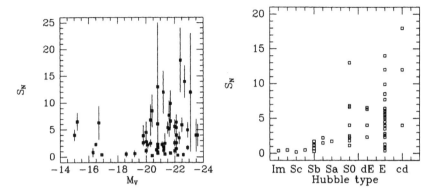

**Figure 4.63** The specific frequency of globular clusters in 56 galaxies plotted against $M_V$ (left-hand panel) and against Hubble type (right-hand panel). [From data published in Harris (1991) and Zepf, Geisler & Ashman (1994).]

and then one determines the value of $m_0$ that provides the best fit of (4.48) to the data (Hanes & Whittaker 1987). When this is done, one finds that $m_0$ varies with the distance $d$ to a galaxy in a way that is consistent with $M_0 \equiv m_0 - 5 \log d$ being a universal constant. This suggests that there is a universal luminosity function for globular clusters of the form

$$\Phi(M) = \text{constant} \times e^{-(M-M_0)^2/2\sigma_m^2}. \tag{4.49}$$

Harris (1991) finds $M_0(V) = -7.27 \pm 0.23$. This result gives rise to one method for determining the distances to galaxies – see §7.3.2.

### 4.5.2 Specific frequency of globular clusters

A quantity that is of considerable significance for theories of galaxy formation is the number of globular clusters that galaxies of different types and sizes possess. Since by default we expect more luminous galaxies to contain more globular clusters than less luminous galaxies, we define the **specific frequency of globular clusters** in a galaxy of absolute magnitude $M_V$ to be

$$S_N \equiv N_t 10^{-0.4(M_V+15)}, \tag{4.50}$$

where $N_t$ is the number of globular clusters that the galaxy is estimated to possess. Clearly, $S_N = N_t$ for a galaxy of absolute magnitude $M_V = -15$ (for example the Local-Group dwarf elliptical NGC 147). Since it is usually possible to detect only the more luminous globular clusters, $N_t$ is determined by fitting (4.48) to the observed number counts and then setting $N_t = \sqrt{2\pi}\sigma_m A$. Figure 4.63 plots $S_N$ versus absolute magnitude $M_V$ and versus Hubble type. Several conclusions can be drawn from these plots: (i) at both high and low luminosity $S_N$ varies by at least an order of magnitude; (ii) spiral and Magellanic galaxies never have large values of $S_N$; (iii) the largest values of $S_N$ occur in luminous early-type galaxies. There is one striking exception to this last rule: the galaxy with the largest value of $S_N$ is the dwarf elliptical Fornax. The corresponding point ($S_N = 73 \pm 12$) lies way off the top of the scales in Figure 4.63. Actually Fornax only possess five globular clusters, but for this galaxy the second factor on the right of equation (4.50) is large because Fornax's absolute magnitude is a piffling $M_V = -12.3$.

It has been argued that the lower values of $S_N$ measured for spiral galaxies arise because globular clusters are associated with spheroidal components and not disks, so the specific frequency should be defined such that $M_V$ in (4.50) is the luminosity of a galaxy's spheroidal components (which is essentially the total luminosity in the case of an elliptical). This procedure would clearly increase the $S_N$ values for spirals. However, there are two objections to defining $S_N$ in this way. First reliable absolute magnitudes for spheroidal components are in some cases hard to obtain because the spheroidal components are so faint – the Local-Group galaxy M33 is such a case. Second, in §10.5.1 we shall see that a large fraction of the Milky Way's globular clusters seem to be more closely associated with the disk than the halo, and, moreover, globular clusters are seen to be forming in the disk of the LMC (Mould, Xystus & Da Costa 1993, Elson & Fall 1988). Thus the conventional definition of $S_N$ makes perfect sense and a successful theory of galaxy formation must explain why some luminous early-type galaxies contain very large numbers of globular clusters.

A classic example of such a well-endowed galaxy[24] is M87, which sits at the middle of the Virgo cluster. This galaxy is estimated to possess $\sim 16\,000$ globular clusters, of which $\sim 6000$ have been identified to magnitude $B = 24.2$. The total luminosity of these clusters is $L_t \simeq 2.6 \times 10^9\,L_\odot$, which may be compared with the total luminosity of the galaxy $L_{tg} = 10^{0.4(M_V(\odot) - M_V)} = 1.0 \times 10^{11}\,L_\odot$. Thus even in this unusually well endowed system, globular clusters contribute only a couple of percent of the system's light.

### 4.5.3 Radial density profiles and shapes

The number of globular clusters per unit area of sky, $N(R)$, can only be readily studied as function of projected radius $R$ in well-endowed elliptical galaxies since (i) we have seen that it is hard to detect globular clusters against the irregular background of a late-type galaxy, and (ii) the surface density in several annuli can be accurately determined only if the galaxy possesses many globular clusters. The available studies of $N(R)$ show that $N$ falls with $R$ less rapidly than does the galaxy's luminosity profile $I(R)$. There are two aspects to this difference. First, in a plot of $\log N$ versus $\log R$ the slope at large $R$ is generally slightly shallower than the corresponding slope in a plot of $\log I$ versus $\log R$ (Harris 1987). Second, the surface density of globular clusters flattens off near the center in a way that $I$ does not. For

---

[24] M87 = NGC 4486

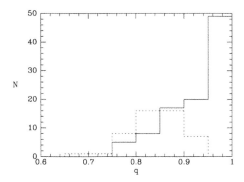

**Figure 4.64** The Milky Way's globular clusters are significantly rounder on the sky than are the globular clusters of the LMC. Here histograms are plotted of apparent axis ratio $q$ for globular clusters in the Milky Way (full lines) and in the LMC (dashed lines). [From data published in Han & Ryden (1994)]

example, $N(R)$ for M87 rises very little inside $R \sim 50\,\mathrm{arcsec}$, whereas the stellar light rises steeply in to a radius $\lesssim 7\,\mathrm{arcsec}$ (Lauer & Kormendy 1986). Although the Virgo elliptical[25] M49 has only a third of the number of globular clusters that M87 possesses, the shapes of the plots of $N(R)$ for these two galaxies are indistinguishable – the curve for M49 simply lies below that of M87. This observation clearly constrains any theory of why some galaxies are much better endowed with globular clusters than other, otherwise similar, galaxies.

The shapes of globular clusters can only be studied in the nearest galaxies since such a study requires well-resolved images. Han & Ryden (1994) describe the available data, which are effectively confined to the globular cluster systems of the Milky Way, M31 and the Magellanic Clouds. Figure 4.64 illustrates the major result of these studies: while the globular clusters of the Milky Way and M31 are very nearly round, those of the LMC and SMC are significantly elongated. Specifically, half of the Milky Way's globular clusters have undetectable ellipticity, while the modal ellipticity of the LMC's clusters is $\epsilon \equiv 1 - q \simeq 0.85$. It is likely that this difference in the shapes of globular clusters reflects differences in their ages – dynamical processes tend to make globular clusters more nearly spherical over time and most of the LMC's globular clusters appear to be less than $\sim 3\,\mathrm{Gyr}$ old (Da Costa 1990) while those in the Milky Way are more like $15\,\mathrm{Gyr}$ old (§6.1.4).

### 4.5.4 Color distributions

The ages and metallicities of globular clusters in external galaxies are clearly of considerable interest. In §5.4 we shall see that high-resolution spectra of individual stars are required if one is to distinguish observationally between the effects of changes in the age and the metallicity of an old stellar population. On account of the faintness of the globular clusters of all galaxies outside the Local Group, such spectra are not available and the physical interpretation of the available color and line-strength data is inevitably ambiguous. Nevertheless such interpretations are routinely given in the literature in that

---

[25] M49 = NGC 4472

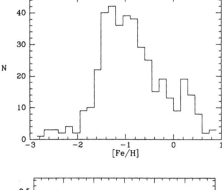

Figure 4.65 The globular clusters of M87 vary widely in color. Here the distribution of Washington colors of 407 M87 globular clusters are interpreted in terms of metallicities. [From data published in Lee & Geisler (1993)]

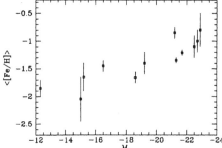

Figure 4.66 The mean metallicity of a galaxy's globular clusters seems to increase with the luminosity of the galaxy. [From data published in Harris (1991).]

measured values of [Fe/H] are reported. These values are derived by (i) measuring colors or line-strength indices, and then (ii) assigning to the observed globular clusters the value of [Fe/H] of a Galactic globular cluster that has the same colors or line-strength indices. The implicit assumption employed is that the ages of all observed globular clusters are similar to those of the Milky Way. In some cases this is certainly untrue: young globular clusters are known in both the LMC and in the giant E/S0 galaxy NGC 1275 that sits at the center of the Perseus cluster (Zepf *et al.* 1995).

Lee & Geisler (1993) used Washington photometry (Table 2.1) to determine [Fe/H] for over 400 globular clusters in M87 and give references to much earlier work on metallicity determinations. A major result of these studies is that globular clusters are systematically bluer by $\sim 0.25$ in $B - V$ than the light from the underlying galaxy. If this color difference arises from a metallicity difference, the globular-cluster stars have to be on average more metal-poor by $\sim 0.5$ in [Fe/H].

The colors of individual globular clusters within a given galaxy span a wide range. Figure 4.65 illustrates this result in the case of M87, by showing the metallicity distribution that Lee & Geisler derive from their colors. As regards shape, this is not dissimilar to metallicity distribution Galactic globular clusters shown in Figure 10.32, but (i) the steep rise in $N([Fe/H])$ on the left of the plot occurs at values of [Fe/H] that are larger for M87 by $\sim 0.5$ than they are for the Milky Way, and (ii) on the right of the plot the distribution for M87 extends to values of [Fe/H] that are larger by $\sim 1$. Thus M87 seems to possess significantly more metal-rich globular clusters than does the Milky Way, although the metal-poorest globular clusters in M87 would not be thought remarkable if they lay in the Milky Way. Figure 4.66 suggests that the higher metallicities of the M87 globular clusters is connected with the large luminosity of M87 itself by showing that there is a general correlation between the mean metallicity of globular clusters $\langle[\text{Fe/H}]\rangle$ and the luminosity of the parent galaxy.

We have seen that the integrated light from elliptical galaxies becomes bluer with increasing galactocentric distance $R$. There is now fairly convincing evidence that the mean color of the globular clusters at $R$ also becomes bluer with increasing $R$ – see Lee & Geisler (1993) for details.

## 4.6 Abnormal galaxies

One goal of any classification scheme is to discover which galaxies are abnormal and therefore merit further investigation. The Hubble classification system and its modifications and derivatives refer principally to normal galaxies – Hubble labeled all abnormal galaxies as Type II Irregulars (Irr II galaxies). In this section we take a look at the objects in this 'rag-bag' bin.

There are broadly two very different ways in which a galaxy can stand out from the crowd and be classified as 'abnormal': (i) it can have an unusual spectral distribution, such as being anomalously powerful at radio wavelengths, or showing strong, broad optical emission lines; (ii) it can look 'funny' in optical images perhaps by having a patchy, irregular brightness distribution, displaying a long luminous structure of some sort, or by possessing an anomalously bright nucleus. We shall find that funny-looking galaxies frequently have abnormal spectra, so studies of abnormal galaxies frequently focus on the same objects irrespective of whether their starting point is spectroscopy or imaging.

### 4.6.1 Starbursting systems

Several astronomers have published catalogs of funny-looking galaxies (Arp 1966, Arp, Madore & Roberton 1987, Vorontsov-Velyaminov et al. 1962–1974, Zwicky & Zwicky 1971). As Figure 4.67 shows, these systems often involve jets, tails, or ringlike structures, but are extremely varied in form; what they have in common is that they don't conform to any set pattern. An important clue to the nature of funny-looking galaxies was provided by Larson & Tinsley (1978), who showed that they are more widely distributed in the $(B - V, U - B)$ plane (Figure 7-19 of BT). In §5.4.2 we shall see that these galaxies tend to be blue and interpret this as a evidence that these systems have formed significant numbers of stars within the last several Gyr.

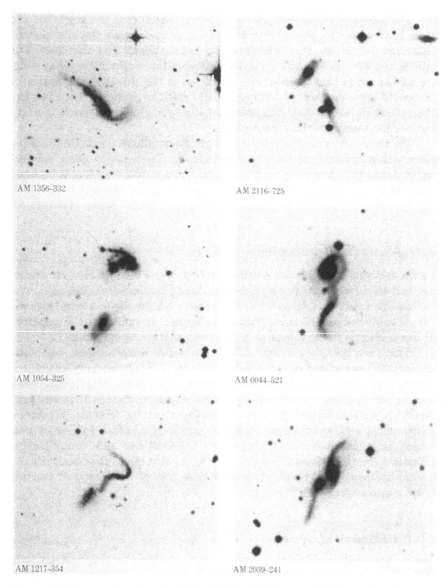

**Figure 4.67** A page from the atlas of Arp, Madore & Roberton (1987) reproduced courtesy of H. Arp.

Many of the features such as tails and ringlike structures visible in Figure 4.67 can be convincingly explained in terms of colliding and/or merging galaxies – see Schweizer (1990) and Barnes & Hernquist (1992). It is likely that many galaxies have, at one time or another, suffered a catastrophic collision of this type (Toomre & Toomre 1972), so developing an understanding

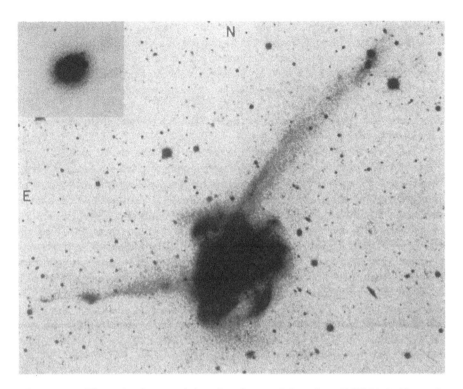

**Figure 4.68** The main photograph is a deep image of the galaxy NGC 7252. Two tails are conspicuous and the main body of the galaxy is seen to be highly irregular. The insert at top left shows a much shallower image of the galaxy on the same scale. The central part of the galaxy appears as smooth as an elliptical galaxy. [After Schweizer (1982) courtesy of F. Schweizer]

of what happens when galaxies collide and merge is a central rather than a peripheral branch of galactic astronomy.

A convincing demonstration that when two large spiral galaxies merge an elliptical is formed was provided by Schweizer (1982), who studied the peculiar galaxy NGC 7252 shown in Figure 4.68. Two enormous tails of luminosity stick out of the main body of the galaxy. Dynamical simulations clearly demonstrate that each of these has formed by tidal stretching of a galactic disk (e.g. Barnes & Hernquist 1992). So we can be pretty sure that NGC 7252 represents the wreckage of two large disk galaxies. In Figure 4.68 the main body of the galaxy is clearly highly disturbed, but Schweizer showed that near the center of this blob the radial luminosity profile obeys the $R^{1/4}$ law rather accurately. So it looks as if NGC 7252 will in time settle down to being a normal elliptical galaxy.

A rather general dynamical argument indicates that an elliptical galaxy cannot form from the merger of two purely stellar disks. The essence of this argument is that in a giant elliptical galaxy stars are more tightly packed together than in a spiral, and the tightness of packing of the stars can only decrease during a merger. Hence the formation of a giant elliptical out of two disk galaxies requires the participation of gas as well as stars – gas can pack down during a merger. In fact, simulations indicate that during a merger a large fraction of the progenitor galaxies' gas rushes headlong to the center of the merger remnant. The simulations cannot tell us what it does when it gets there, but three observations indicate that a substantial fraction of it is turned into stars.

First, there is Schweizer's demonstration that the center of NGC 7252 obeys the $R^{1/4}$ law, as well it might if sufficient new stars can form. Second, the spectra of suspected merger remnants frequently display Balmer absorption, which is characteristic of A stars (Table 3.1). A stars have masses in the range $2 - 3\,\mathcal{M}_\odot$ (Table 3.13) and have lifetimes $\lesssim 1\,\mathrm{Gyr}$ (Table 5.2), so galaxies with Balmer absorption have formed stars recently. Third, in the 1980s the 'IRAS' satellite – see §8.1.6 for details – discovered that some galaxies have extremely large infrared luminosities. The infrared radiation observed from these IR-luminous systems is believed to be energy originally radiated by young stars that has been absorbed and then reradiated by dust grains (see §8.1.6). The most luminous of these objects must have converted much of their interstellar gas into luminous stars fairly recently, and will fade markedly within a Gyr as these stars burn out. Since it is found that galaxies that appear to be interacting are likely to be IR-luminous (Joseph 1990), it is natural to conclude that bursts of star formation are induced by interactions.

### 4.6.2 Systems with active galactic nuclei

Many abnormal galaxies contain peculiar point-like sources at their centers. These sources are called **active galactic nuclei** or **AGN** for short. An AGN can be so bright that it outshines the entire surrounding galaxy. Yet AGN must be physically small because their huge luminosities frequently change dramatically in less than a year: even if the luminosity density throughout an extended object of size $a$ were to double simultaneously, the observed brightness of the object would only adjust to its new level over a time that is comparable to the time $\tau \simeq a/c$ that it takes light to pass from the back to the front of the source. Thus, the observed rapid variations in the brightness of AGN imply that their emitting regions are significantly less than a parsec across. These bizarre objects, which can pack an entire galaxy's worth of luminosity into less than the space between the sun and its nearest stellar neighbor, provide a fascinating field of astronomical study. The details of our understanding of their physics lies beyond the scope of this text, and the interested reader is referred to the book by Robson (1996) for an introduction to the subject, the AGN section of Osterbrock (1989) for a more quantitative discussion, and the paper by Antonucci (1993) for a thorough review. Here, we describe only the bare bones of what has become known as the **AGN paradigm** and its associated taxonomy.

Carl Seyfert (1943) discovered the first class of systems with AGN. A **Seyfert galaxy** is a galaxy whose nucleus shows strong, fairly broad emission lines of high excitation. There are two types of Seyfert galaxy. The spectra of **type 2 Seyferts** contain both permitted and forbidden lines (see Box 8.1) that appear to be broadened by Doppler velocities of the order of $500\,\mathrm{km\,s^{-1}}$. **Type 1 Seyferts** differ from type 2 Seyferts in that their permitted spectral lines have very broad wings, suggesting that they are Doppler broadened by velocities of $1000 - 5000\,\mathrm{km\,s^{-1}}$.[26] Both the forbidden lines and the cores of the permitted lines in the spectra of type 1 Seyferts are of similar width to the lines in the spectra of type 2 Seyferts. About one-third of the Seyferts are of type 2. Seyferts of both types usually show strong, variable X-ray emission and emit strongly in the infrared.

Since Seyfert's pioneering work, a whole zoo of AGN types have been classified. As with the normal galaxies, the classification schemes fall short of the ideal, since they are often non-unique and somewhat subjective. For example, Markarian *et al.* (1978) made a wide-area survey which picked out galaxies which emitted more strongly in the ultraviolet than would be expected from their optical emission if their luminosity were dominated by stars. We now know that the ultraviolet excesses of many of these **Markarian galaxies** is emitted by an AGN so it is not surprising that 10% of Markarian galaxies are also classified as Seyfert galaxies.

Many galaxies have nuclei that resemble those of type 2 Seyferts except that their forbidden lines tend to arise in less highly ionized atoms. These **low-ionization nuclear emission-line regions**, or **LINERs**, are found at the centers of some 80% of Sa and Sb galaxies, and in smaller proportions of Sc galaxies and ellipticals. LINERs were originally treated as a distinct phenomenon, but they are now thought to represent the low-luminosity tail of the Seyfert phenomenon.

Radio telescopes have proved powerful detectors of AGN. The angular resolution of early radio telescopes – which had single-dish radio antennae – was low with the consequence that they could not give accurate positions for the sources they detected. The nomenclature adopted for the sources detected in early studies reflects the limited spatial resolution of the observations: for example, the first discrete source of radio emission detected outside the solar system was initially only localized to somewhere within the constellation Cygnus, and so it was designated Cygnus A. With the development of radio interferometers the angular resolution of radio observations improved dramatically. In particular a position for Cygnus A became available that allowed Baade & Minkowski (1954) to identify the source with

---

[26] Seyferts which display only very weak broad components are sometimes given such fractional designations as Seyfert 1.5 – see Osterbrock (1989) for a thorough discussion of these type designations.

an optically-peculiar elliptical galaxy. As radio telescopes gained sensitivity, more and more extragalactic radio sources were discovered: the most famous study was the third survey carried out at Cambridge [designated the **3C catalog**; Edge *et al.* (1959)], which covered the northern hemisphere at a wavelength of 177 MHz and turned up some 471 radio sources brighter than 9 Jy; its successor, the **4C catalog**, contains almost 5000 sources (Pilkington & Scott 1965, Gower, Scott & Wills 1967).[27] Over time almost all of the objects detected in the 3C catalog were unambiguously identified with optical sources. Some are associated with Galactic sources, particularly supernova remnants, but the majority are extragalactic in origin, and a large fraction are at redshifts in excess of $z = 0.5$. Some two-thirds of the bright extragalactic radio sources are associated with what appear to be more-or-less normal elliptical galaxies (Spinrad *et al.* 1985), and these objects are known generically as **radio galaxies**.

Radio interferometers have been able to resolve structure in most radio galaxies. In addition to a nuclear compact source, nearly all radio galaxies possess two amorphous regions of radio brightness, or **radio lobes**. These lobes are often placed roughly symmetrically on opposite sides of the nucleus and hundreds to millions of parsecs from it. At low radio frequencies ($\lesssim 1.4\,\mathrm{GHz}$) the lobes are most prominent and the central source is not always detected. Conversely, at high frequencies the nuclear source stands out prominently. High-quality radio images reveal that the nuclear source is often connected to one or both lobes by a thin structure that is called a **radio jet**. In favorable cases jets can be traced over several decades in radius and remain narrow and remarkably straight through most of their length (see Figure 4.70). Occasionally, for example in the giant E0 galaxy M87 (which is identified with the radio source 3C 274), a jet is also visible at optical frequencies.

Within a lobe the surface brightness usually peaks at a well defined **hot spot**. On the basis of the relative location of these hot spots, Fanaroff & Riley (1974) divided radio galaxies into **FR I** and **FR II** sources. The distance between the hot spots of FR I sources is less than a half of the maximum diameter of the source, while those of FR II sources are separated by more than half the maximum diameter of the source. Very roughly, the hot spots of a FR II sources lie on the outside of the system, while those of a FR I source do not. Fanaroff & Riley observed that FR I sources tend to have $P_{1.4} \lesssim 10^{24.5}\,\mathrm{W\,Hz^{-1}}$, where $P_{1.4}$ is radio power at 1.4 GHz, while FR II sources tend to have $P_{1.4} \gtrsim 10^{24.5}\,\mathrm{W\,Hz^{-1}}$. Owen & Ledlow (1994) showed that FR I and FR II sources occupy distinct regions of the $(L_R, P_{1.4})$ plane; FR I sources occupy the region that corresponds to large $R$-band luminosities and relatively small radio powers, and the FR II the other half of the plane

---

[27] Somewhat confusingly, the 3C Catalog is arranged in right ascension order, so that the source 3C 279 lies to the east of 3C 278, while the 4C catalog is sorted by declination, so the source 4C 34.16 is the sixteenth source at a declination of 34 degrees.

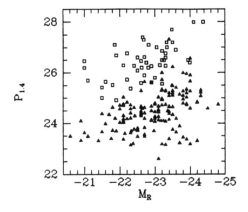

Figure 4.69 The Fanaroff & Riley (1974) morphological classification of radio sources leads to a sharp division of these sources in a plot of radio luminosity at 1.4 GHz against $R$-band luminosity. FR I sources are plotted as filled triangles and FR II sources as open squares. [After Ledlow & Owen (1996) from data kindly supplied by M. Ledlow]

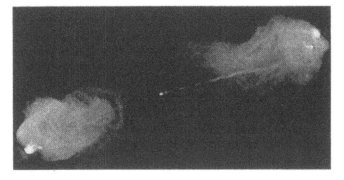

Figure 4.70 Image showing the emission from the radio galaxy Cygnus A at 0.06 m as mapped by the VLA. Note the narrow, straight jet joining the central unresolved emission to the largescale radio lobes. [After Carilli & Barthel (1996) using data kindly provided by C. Carilli and R. Perley]

(see Figure 4.69). Like Seyferts, radio galaxies can be divided on the basis of their optical spectra into **narrow-line radio galaxies** which emit only the narrow emission lines characteristic of type 2 Seyferts, and **broad-line radio galaxies** which also emit the broad lines seen in the spectra of type 1 Seyferts

The huge power of the radio lobes is believed to flow from the AGN along the narrow channels that are defined by the jets. Within the jets, energy is stored as the ordered kinetic energy of bulk flow. The observed radio waves are emitted by the synchrotron process (see §8.1.4) after this energy has been converted by shocks into the random motions of electrons and ions. This crucial disordering of the kinetic energy that is emitted by the AGN occurs in part within the jets but is believed to be concentrated in the hot spots. Models of radio lobes show that enormous quantities of random kinetic energy must be stored within the lobes in plasma that has past through the hot spots.

Although most radio galaxies appear to be normal ellipticals when observed at optical wavelengths, approximately 10% contain exceptionally-bright central nuclei (Spinrad *et al.* 1985). Galaxies that contain such bright central nuclei (which also include Seyferts) have been given the generic title of **N galaxies** (Morgan 1972).

At optical wavelengths the locations of some radio sources are marked by an unresolved point source rather than an elliptical galaxy. It was initially believed that these **quasi-stellar radio sources**, or **quasars** for short, were some strange new kind of star within the Milky Way. Their spectra, however, contain broad, strong emission features, totally unlike the spectral lines of normal stars. Schmidt (1963) conclusively demonstrated that quasars are extragalactic by identifying such features in the spectrum of 3C 273 with emission lines, similar to those seen in type 1 Seyferts, but redshifted by $z = \Delta\lambda/\lambda = 0.16$. Once this breakthrough was made, the spectra of other quasars were recognized as highly redshifted. The huge Doppler velocities implied by these redshifts strongly suggests that quasars do not lie within the Galaxy since they are moving much too fast to be confined by the Galaxy's gravitational field. Furthermore, if they were being shot out from the Galactic center, we would expect to see some blueshifted spectra and none has ever been found.

The similarity between the spectra of quasars and those of type 1 Seyferts suggested a more natural interpretation of quasars: they are distant active galaxies whose large redshifts simply reflect the cosmological expansion of the Universe (see §7.1). They are so distant that their optical angular sizes are comparable to the regions over which the light of their nuclei is spread by seeing (§4.2.2). Consequently, the brilliance of their nuclei completely swamps their stellar light and from the ground they appear to be unresolved point sources.[28]

Radio astronomy has also played a key role in identifying a further class of AGN, 'BL Lac objects', or 'BL Lacs'. BL Lacertae is an unresolved optical point source which can vary in luminosity by an order of magnitude in less than a month. It was originally believed that this object was some new kind of extremely variable star – hence its stellar name. Early spectra of BL Lacertae showed a non-thermal power-law spectrum that was devoid of spectral lines in either absorption or emission. Hence the heliocentric velocity of the object could not be deduced from these spectra. Radio observations, however, suggested that it is an AGN by showing that BL Lacertae is a strange compact radio source which varies rapidly and is strongly linearly polarized. This linear polarization is associated with large Faraday rotation (§8.1.5), and this suggests that BL Lacertae lies outside the Milky Way. Adams (1974) finally proved that BL Lac is indeed an AGN by showing that

---

[28] Actually 3C 273 has an extended optical image, but the extended structure is due to a jet similar to that seen in M87. High-resolution radio observations also show this jet.

the bright point source is surrounded by a faint nebulosity whose brightness distribution and color are consistent with its being an elliptical galaxy. The spectrum of this nebulosity is that of a normal elliptical galaxy at a redshift of $\sim 0.07$ (Miller, French & Hawley 1978). AGN which like BL Lacertae combine rapid variability, a non-thermal optical spectrum and strong radio emission are called **BL Lacs**.

BL Lacs are now frequently bracketed with another class of AGN, the **optically violently variable quasars**, or **OVVs** for short. These objects share many of the properties of BL Lacs: they are strong radio emitters; they can vary in brightness by large factors on timescales of weeks; and their radio and optical emissions are strongly polarized. The distinguishing factor between OVVs and BL Lacs is that OVVs possess the broad optical emission lines that are characteristic of quasars. However, the distinction between the two classes is made less than clear-cut by the variability of all these objects: an object's emission lines may stand out clearly when the underlying continuum is faint, but be almost invisible when the continuum is bright. Hence at different times a single object may be classified as an OVV or as a BL Lac. The two types of objects are therefore often grouped into a single category known as **blazars**.

A further class of powerful AGN comprises the **quasi-stellar objects** or **QSOs**, which have the same observable properties as quasars except that they are not strong radio sources.[29] These radio-quiet AGN are very similar to type 1 Seyferts, except that their nuclei are optically so bright that they outshine the surrounding galaxy. Thus, the distinction between these two classes is somewhat artificial, but it is customary to describe objects with absolute magnitudes brighter than $M_V \sim -23$ as QSOs, and fainter radio-quiet AGN as Seyferts.

Since QSOs have a stellar appearance and no radio signature, distinguishing them from stars is difficult. However, the non-thermal continua of AGN do not cut off in the ultraviolet as stellar spectra do, so QSOs stand out against stars by their very blue colors.[30] Further, the broad emission lines of QSOs are apparent in even poor-quality spectra, so QSOs can be identified in **objective prism surveys**, which disperse the light of each object in a field into a low-dispersion spectrum. Finally, QSOs, like quasars, are strong emitters at X-ray wavelengths, so surveys carried out with X-ray satellites such as ROSAT have proved a fruitful method for identifying QSOs. By gathering the sources detected by these various techniques, large catalogs of QSOs have been produced: Hewitt & Burbidge (1993) published a list of 7315 objects detected by one or more of these methods. Such studies have shown that most powerful AGN are not strong radio sources: for every

---

[29] There is a certain amount of confusion in the nomenclature of these objects: some authors refer to *all* bright AGN as quasars. The original acronym definition of quasars is then ignored, and such oxymoronic terms as "radio-quiet quasars" appear.

[30] At redshifts $z \gtrsim 2.5$, quasars become faint in the $U$ band on account of Ly$\alpha$ and Lyman continuum absorption. Hence high-redshift quasars do not have blue optical colors.

quasar, there are perhaps twenty QSOs. Surveys also confirm the rarity of BL Lacs: only $\sim 1\%$ of the objects in the Hewitt & Burbidge (1993) catalog are classified as a BL Lac. However, some care must be taken in interpreting such statistics, since BL Lacs are even harder to identify than QSOs, so they may be under-represented in catalogs of identified objects.

### 4.6.3 Host galaxies of AGN

Since this book is primarily concerned with the large-scale structure of galaxies, we now turn to the morphologies of the galaxies that host AGN. In the case of nearby low-power AGN such as LINERs and Seyfert nuclei, it is straightforward to produce images of the surrounding galaxies. Although some lower-powered LINERs reside in the centers of elliptical galaxies, the majority of these systems appear to be fairly normal spiral galaxies, with the usual complement of bulge, disk, and, in some cases, a bar. Radio galaxies are also relatively straightforward to image, and almost without exception they are found to be elliptical galaxies, although in many cases their morphologies are somewhat disturbed.

The host galaxy of a quasar or a QSO is much harder to image, since it is obscured by the glare of the central point source. However, several technological advances have enabled astronomers to image the host galaxies of even powerful AGN – these advances include the introduction of CCDs and infrared arrays, the limitation of the degradation of ground-based observations by seeing, and especially the refurbishment of the HST. Quasars all seem to be hosted by giant elliptical galaxies. Consequently they represent the luminous end of the radio-galaxy phenomenon. The morphologies of the QSO host galaxies is less clear-cut: although some of these galaxies are giant ellipticals, others are probably disk galaxies since their luminosity profiles are best described by the exponential law (4.20). Thus it is not clear whether QSOs are more closely related to quasars or Seyferts. There are indications that the host galaxies of some luminous AGN have been disturbed by interactions with other galaxies: imaging with HST has revealed multiple nuclei, close companions and tidal tails in several systems (see Figure 4.71).

A reasonably consistent picture of AGN host galaxies is beginning to emerge, but the number of objects studied is still small and by no means all issues are resolved. To every apparent rule there are exceptions: for example, the vast majority of BL Lacs, like other radio-bright AGN, appear to be hosted by elliptical galaxies, but an HST image of one BL Lac, PKS1413+135, reveals that it is located in an edge-on spiral galaxy (McHardy *et al.* 1994). Even more unexpectedly, a $V$-band study of eight nearby bright quasars and QSOs with HST revealed the host galaxies of only three of these objects – no stellar light could be detected in the other five systems (Bahcall, Kirhakos & Schneider 1995). These observations of apparently "naked" AGN imply that any host galaxies must be significantly

**Figure 4.71** A HST image of the quasar PKS2349−014. Note the strongly disturbed morphology of the diffuse host galaxy light surrounding the bright point-like quasar, and the large number of nearby galaxies. [Figure courtesy of STScI]

fainter than average field galaxies. A subsequent study by Disney *et al.* (1995) based on *R*-band HST imaging was more successful in detecting QSO host galaxies, and concluded that these objects are found in elliptical galaxies of average brightness. Since the objects observed were amongst the brightest nearby AGN, however, yet they are housed in galaxies of no more than average brightness, it seems that there is no direct correspondence between the luminosity of an AGN and the luminosity of its host galaxy.

### 4.6.4 The unified model of AGN

While it is in principle possible that the different types of AGN that we have encountered are physically unrelated objects, it is now generally accepted that they are physically very similar objects and that the differences in their observational properties may be largely explained by three factors: (i) some objects are significantly more luminous than others; (ii) some objects produce jets, while others don't; (iii) a given AGN can have very different observational properties depending on the direction from which one views it. This last chameleon-like property of AGN reflects their strongly anisotropic radiation patterns.

Direct evidence that the radiation pattern of AGNs is anisotropic comes from the discovery near some active galaxies of wedge-shaped regions of photo-excited gas (see Figure 4.72). The sharp edges of these wedges are interpreted as the boundaries between gas that is exposed to ionizing radiation and gas that is shielded from it. The wedge-shaped geometry suggests that the ionizing radiation is being beamed out from an AGN in a cone.

Figure 4.73 illustrates the geometry inferred from these observations. A powerful energy source, probably a supermassive black hole, lies at the center of a torus of gas and dust. The source's radiation is absorbed by the torus in all directions except along a cone around the torus's symmetry

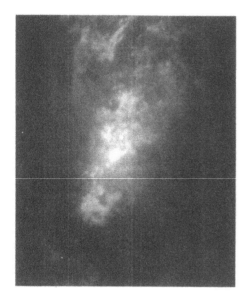

**Figure 4.72** HST image showing the emission from ionized material in the core of the Seyfert 2 galaxy NGC 1068. [Figure courtesy of STScI]

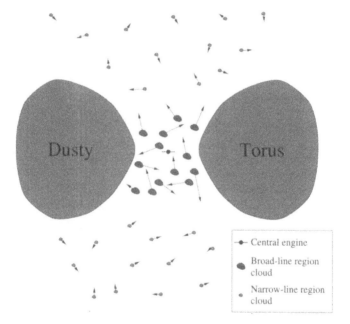

**Figure 4.73** Schematic cross-section of the geometry invoked by the unified AGN model. The high velocity, high density clouds in the broad-line region are obscured unless viewed from close to the axis of symmetry. The low velocity, low density clouds in the narrow-line region, which are ionized by the radiation from the central engine, are visible from all directions.

axis. Physically, this cone is probably a vortex tube akin to the one that often forms within the plug-hole when a bath-tub is emptied. When we can see down the vortex tube, we classify the system as a type 1 Seyfert or a QSO depending on luminosity because we see dense, fast-moving gas that is located in the **broad-line region** close to the AGN – this gas is responsible for the broad permitted lines in the object's spectrum. When we cannot see down the vortex tube, we classify the object as a type 2 Seyfert because in this case the spectrum is dominated by emission from low-density, relatively slowly moving, photo-excited gas that lies in the **narrow-line region** at some distance from the central energy source, as shown in Figure 4.73.

This picture gained enormously in credibility when Miller & Antonucci (1983) obtained a spectrum for the component of the light of NGC 1068 which is polarized with its electric vector perpendicular to the long-axis of the cone of emission. They found that the permitted emission lines in this polarized spectrum have broad wings similar to those seen in a type 1 Seyfert. The significance of this result is the following: since NGC 1068 is a type 2 Seyfert, it should be oriented such that we cannot see the region within which the wings of broad emission lines are formed. However, photons emitted deep in the vortex can pass up it and then be scattered into our line of sight by free electrons that lie within our field of view. In order to be scattered in this way, light must be polarized with its electric vector perpendicular to the plane containing our line of sight and the direction to the original point of emission. Thus, by measuring the spectrum of the component of the radiation of NGC 1068 that is polarized in this direction, Miller & Antonucci were picking out precisely light that had been scattered up from deep in the vortex. The unified model therefore successfully predicted that this spectrum should contain the emission from the broad-line region that is usually hidden in type 2 Seyferts.

Unification in terms of orientation also accounts successfully for the phenomenology of radio-loud AGN. We imagine radio jets emerging along the vortex tube shown in Figure 4.73. When a jet is pointing towards us we classify the system as a blazar because we then see to very small radii, deep in the core of the AGN, and the observed brightness can vary rapidly. Near the central engine jets move at relativistic speeds. In a blazar this motion is almost straight toward us with the result that synchrotron emission is Doppler shifted to markedly higher frequencies. Domination of the optical spectrum by such Doppler-boosted synchrotron radiation explains the featureless optical continua of BL Lacs. When our line of sight does not pass right down the jet, our picture of the galaxy is dominated by the radio lobes and we classify the system as a radio galaxy. The comparative rarity of blazars is explained by the small fraction of lines of sight that pass right down a jet. Narrow-line radio galaxies are interpreted as systems that are seen rather nearly edge-on so we do not see far down the vortex tube, while broad-line radio galaxies are systems that are seen sufficiently pole-on that we see deep

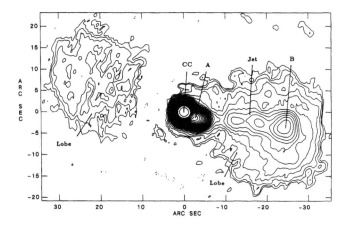

**Figure 4.74** Very high dynamic range radio map of the blazar
3C 371 obtained using the VLA. Note the faint extended emission,
and its limited angular extent (which amounts to only ~ 40 kpc
at the distance of this source). [Reproduced from Wrobel & Lind
(1990)]

down the vortex tube, although not so precisely pole on that we look right
along the jet.

We can test the unified model of radio-load AGN observationally: if
blazars are nearly end-on radio galaxies, they should have radio lobes but
in projection these should lie closer to the nucleus than they do in normal
radio galaxies. In fact, in many cases one would expect a projected lobe to
surround the nucleus. Since the diffuse radiation from lobes should come
from plasma that is approximately stationary in the galaxy's rest frame, it
should be fairly isotropic. Hence the lobes of blazars should be of comparable
luminosity to those of normal radio galaxies. Observationally, the detection
of diffuse radio emission against the glare of the nucleus is far from trivial,
however, just as it is difficult to image QSO host galaxies optically. None the
less, Figure 4.74 shows that the blazar 3C 371 does possess diffuse emission of
limited angular extent, just as the unified model predicts. Moreover, in this
and other blazars the total power associated with diffuse emission is comparable to that observed from the lobes of normal radio galaxies. Thus the
data are consistent with the distinction between blazars and radio galaxies
being purely one of orientation.

The unified model predicts that the mid-infrared and hard X-ray emissions of an AGN should be rather isotropic, because the jet should not
dominate such radiation and the accretion torus, which obscures the central source at optical wavelengths, should be transparent to hard X-rays and
mid-infrared radiation. Since hard X-ray and mid-infrared luminosities differ
by orders of magnitude between the weakest and strongest AGN, it is clear
that AGN differ in intrinsic luminosity as well as in orientation.

The power emitted by an AGN is believed to be derived from the accretion of matter on to a black hole. For a black hole of mass $\mathcal{M}_{bh}$ an approximate upper limit on this power is given by the **Eddington luminosity**. This is (see Problem 4.12)

$$L_{\text{Edd}} = 3 \times 10^4 \left( \frac{\mathcal{M}_{bh}}{\mathcal{M}_\odot} \right) L_\odot. \tag{4.51}$$

For example, a bright QSO with a luminosity of $\sim 10^{12} \, L_\odot$ requires a black hole of mass $\gtrsim 3 \times 10^7 \, \mathcal{M}_\odot$. One possible explanation for the range of AGN luminosities is that AGN are powered by black holes which span a wide range of masses and are radiating at close to their respective Eddington limits. However, since the luminosity of an individual AGN can vary by orders of magnitude on timescales of days to months, and there is no mechanism by which the responsible black hole can vary its mass in a corresponding manner, the mass of the black hole cannot be the sole factor which dictates the luminosity of the AGN.

The primary factor which affects the luminosity produced by a given black hole is the rate at which it accretes material. Once a black hole has gobbled up everything that lies within its sphere of influence, it must shut down for lack of fuel. If its host galaxy now undergoes a close encounter with a companion galaxy, the distribution of gas and stars around the black hole may be radically restructured, and fresh material may be enter the black hole's sphere of influence. Thus, there may be a link between the brightness of AGN and the length of time since it was last gravitationally disturbed. Support for such a connection comes from the large fraction of QSO host galaxies with disturbed morphologies mentioned in §4.6.3.

Many other questions about AGN are also as yet unresolved. Perhaps the biggest gap in our understanding at present is that we have no good explanation as to why some AGN produce radio jets while others do not. The fact that radio-loud AGN are generally located in elliptical galaxies while radio-quiet AGN tend to be found in spirals must be relevant. The causal factor may be the very different structures of the interstellar media of ellipticals and spirals (see Chapter 8). Another possibility is that the black holes in ellipticals have different spin properties from those in spirals and that this difference renders them more liable to produce jets.

## Problems

**4.1** For a Gaussian PSF, and an image that is circularly-symmetric about the point $R = 0$, show that equation (4.6) can be written

$$I_{\text{app}}(R) = \int_0^\infty dR' \, R' I_t(R') \exp\left( -\frac{R^2 + R'^2}{2\sigma^2} \right) \int d\theta' \exp\left( -\frac{RR' \cos\theta'}{\sigma^2} \right).$$

Hence obtain equation (4.8).

**4.2** If the coefficient $a_3$ in equation (4.10) is greater than zero, what does this tell us about the shape of the isophote?

**4.3** One way of fitting an ellipse to an isophote is to perform a least-squares fit of the polar equation for an ellipse, $R^{-2} = C - A\cos 2\phi - B\sin 2\phi$, to the points $(R_i, \phi_i)$ that lie on the isophote. Express the semi-major axis length, $a$, the ellipticity, $\epsilon$, and the position-angle of the major axis, $\phi_0$, in terms of $A$, $B$ and $C$.

**4.4** Peletier *et al.* (1990) measure the diskiness of an isophote by fitting an ellipse to it, and then writing the surface brightness as a function of angle $\phi$ around this ellipse as a Fourier series,

$$I(\phi) = I_0(a)\left(1 + \sum_{n=1}^{4} C_n \cos(n\phi) + \sum_{n=1}^{4} S_n \sin(n\phi)\right), \qquad (4.52)$$

where $a$ is the semi-major axis of the ellipse. Assuming that all the $C_n$ and $S_n$ are small, show that

$$\frac{a_4}{a} \simeq \frac{C_4}{0.4\ln 10(\mathrm{d}\mu/\mathrm{d}a)}, \qquad (4.53)$$

where $\mu(a)$ is the galaxy's surface brightness (measured on a magnitude scale) at distance $a$ down the major axis, and $a_4/a$ is the diskiness parameter as defined in equation (4.10).

**4.5** Using equation (4.12), show that if the luminosity density follows a power law $j \sim r^{-\alpha}$ then the surface brightness $I$ varies with projected radius $R$ as $I \sim R^{-(1+\alpha)}$.

**4.6** Show that the luminosity $L(R)$ that is predicted by the $R^{1/4}$ law to lie inside projected radius $R$ is

$$L(R) = L(\infty)\left[1 - e^{-y}\left(1 + y + \frac{y^2}{2!} + \frac{y^3}{3!} + \frac{y^4}{4!} + \frac{y^5}{5!} + \frac{y^6}{6!} + \frac{y^7}{7!}\right)\right], \qquad (4.54)$$

where $y = 7.67(R/R_e)^{1/4}$. Hence verify equation (4.19), and show that the mean surface brightness inside $R = R_e$ is $\langle I \rangle_e = 3.61 I_e$.

**4.7** Show that the matrix $\mathbf{A}$ defined in Box 4.2 is $M$ times the covariance matrix of the observations. Hence show that its eigenvalues, $\lambda$, are the values taken by the quantity $S$ for each eigenvector $\mathbf{n}$.

**4.8** An edge-on view of the fundamental plane is required with $\log \sigma_0$ plotted horizontally. What linear combination of $\log R_e$ and $\langle I \rangle$ should be plotted vertically?

**4.9** Obtain an expression for the total luminosity $L$ of the model defined by equations (4.44). Show that the edge-on surface brightness of this model is

$$I(R, z) = I(0,0)\frac{R}{R_\mathrm{d}}K_1(R/R_\mathrm{d})\operatorname{sech}^2(z/2z_0),\qquad(4.55)$$

where $K_1$ is a modified Bessel function. Determine the dependence of $I(0,0)$ upon $L$ and the model's scale lengths.

**4.10** In an isothermal distribution of particles, the particle density varies as $\exp(-\beta\Phi)$, where $\beta$ is the inverse temperature of the system, and $\Phi$ is the gravitational potential. Investigate the isothermal particle density, $n(z)$, when $\Phi(z)$ is generated by: (a) an infinitesimally-thin uniform-density sheet occupying the plane $z = 0$; and (b) an infinite medium of uniform density.

**4.11** Consider the globular cluster population of a galaxy at a distance $d$ from us. Let $\Psi(m)\mathrm{d}m$ be the luminosity emitted by clusters with apparent magnitudes in the range $(m + \mathrm{d}m, m)$. Given that the luminosity function of the globular clusters satisfies equation (4.49), show that $\Psi(m)$ is a Gaussian distribution that peaks at $m_1 \equiv m_0 - 0.92\sigma_m^2$. Hence, show that the total luminosity of $N_\mathrm{t}$ globular clusters is

$$L_\mathrm{t} = N_\mathrm{t}\exp\left(\frac{m_1^2 - m_0^2}{2\sigma_m^2}\right) \times 10^{2\log(d/\mathrm{pc})}.\qquad(4.56)$$

Suggest how one might observationally determine $m_1$ for a globular cluster system so distant that $m_0$ is fainter than the limiting magnitude of cluster counts.

**4.12** Consider a fully ionized and optically thick cloud of material that lies near a point source of mass $\mathcal{M}$ and luminosity $L$. Given that the dominant interaction of the cloud with the radiation field is Thomson scattering by free electrons (§8.1.4), show that radiation pressure on the cloud exactly balances gravity when $L$ equals the Eddington luminosity

$$L_\mathrm{Edd} \equiv \frac{4\pi G\mathcal{M}c}{\sigma_T}\frac{\Sigma}{N_\mathrm{e}},\qquad(4.57)$$

where $\Sigma/N_\mathrm{e}$ is the mass per free electron in the cloud. Hence derive equation (4.51).

# 5
# Evolution of Stars and Stellar Populations

The theory of stellar structure and evolution is fundamental for our understanding of almost every aspect of star clusters and galaxies. We start this chapter by reviewing the understanding of the color-magnitude (CM) diagram that has emerged from the theory. Then we summarize its results concerning the origin of the chemical elements – where and how they were synthesized, and in what quantities. Finally, after describing practical techniques for determining the physical and chemical characteristics of stars from observations, we use our results to predict the observational characteristics of entire populations of stars within which the star-formation rate has some simple time dependence. These models will assist greatly in the interpretation of observations of both the Milky Way and external galaxies.

Since we only use the results of the theory of stellar structure as interpretive tools, we merely state them without detailed justification. The subject's principles and detailed results are ably expounded in the monographs of Kippenhahn & Weigert (1990), Shapiro & Teukolski (1983), Clayton (1968) and Cox & Guili (1968), while valuable review articles include Iben (1967), Iben (1974) and Chiosi, Bertelli & Bressan (1992).

## 5.1 Stellar evolution and the CM diagram

In §3.5 we defined the CM diagram and saw that stars congregate near particular points and lines of its plane. The goal of the theory of stellar evolution is to explain the arrangement of stars in the CM diagram and to reproduce certain other regularities, such as the main-sequence (MS) mass-luminosity relation (Table 3.9). A basic theoretical picture of stellar evolution can be developed by assuming that stars are spherically symmetric and in hydrostatic equilibrium, and by writing a set of four nonlinear partial differential equations that describe the structure of a star as a function of radius from the star's center and as a function of time. One of these equations describes the increase with radius $r$ of the mass interior to $r$, another describes the increase with $r$ of the luminosity passing through the sphere of radius $r$, and the other two equations describe the decrease with $r$ of the pressure and temperature within the star. When these equations are supplemented with an equation of state for the star's gas and the assumption that the luminosity is generated at the center, one can show (as was done by Russell, Vogt, Eddington, and others) that the structure of the star is determined by its mass and its chemical composition.

One of the key factors affecting the course of stellar evolution is the nature of the energy-production mechanism. Early in the 20th century, it was believed that the energy radiated by a star was produced by the conversion of gravitational energy into thermal energy by a (relatively) slow contraction of the star. But, when it was discovered, on the basis of geophysical evidence, that the Earth is at least 1 Gyr old and that therefore the Sun must be at least this old, it was realized that the time interval over which gravitational energy release could sustain the Sun's luminosity is two orders of magnitude too short to match the geophysical scale. The problem was solved when it was recognized that the required energy could be provided by thermonuclear fusion of light elements into heavier elements at the centers of stars. In particular, the MS can be identified with an episode of conversion of hydrogen (cosmically the most abundant element) into helium.

There are two important channels for the conversion of hydrogen to helium. The **p-p chain** starts with this pair of nuclear reactions:[1]

$$
\begin{aligned}
p + p &\rightarrow {}^2H + e^+ + \nu \\
{}^2H + p &\rightarrow {}^3He + \gamma.
\end{aligned}
\tag{5.1}
$$

In the final stage of the p-p chain the $^3$He created by the reactions above is destroyed either following the collision of two $^3$He nuclei or through the collision of a $^3$He nucleus with a $^4$He nucleus:

$$
\text{either} \quad {}^3He + {}^3He \rightarrow {}^4He + 2p \quad \text{or} \quad
\left(
\begin{aligned}
{}^3He + {}^4He &\rightarrow {}^7Be + \gamma \\
{}^7Be + e^- &\rightarrow {}^7Li + \nu \\
{}^7Li + p &\rightarrow 2{}^4He.
\end{aligned}
\right)
\tag{5.2}
$$

---

[1] See §5.2.1 for an explanation of the notation used for nuclear species.

The other important channel for the conversion of hydrogen to helium is the **CNO-cycle**, which involves the reactions[2]

$$^{12}C + p \rightarrow {}^{13}N + \gamma$$
$$^{13}N \rightarrow {}^{13}C + e^+ + \nu$$
$$^{13}C + p \rightarrow {}^{14}N + \gamma$$
$$^{14}N + p \rightarrow {}^{15}O + \gamma \tag{5.3}$$
$$^{15}O \rightarrow {}^{15}N + e^+ + \nu$$
$$^{15}N + p \rightarrow {}^{12}C + {}^4He.$$

The net effect of these reactions is to convert four protons into a helium nucleus; the C, N, and O nuclei catalyze the reaction without themselves being consumed.

The speed with which both the p-p chain and the CNO cycle convert hydrogen to helium is proportional to the square of the density and increases rapidly with the temperature within the star. The CNO cycle is the more temperature sensitive of the two processes, with the result that it dominates in the hottest stellar cores. The p-p chain is the dominant process in the Sun.

When the equations of stellar structure are solved, zero-age, chemically homogeneous, H-burning stars of different masses are found to be arranged along the MS as observed; by changing the assumed chemical composition of the material, families of theoretical main sequences can be generated. While a star is on the MS, its luminosity is supported by the slow conversion of hydrogen into helium. When an appreciable fraction of the original hydrogen in a star's core has been consumed, the star moves away from the MS in the first phase of a series of structural changes. These reflect changes in its chemical composition that are produced by successive episodes of thermonuclear burning. The structural changes in stars as they evolve are manifested in the CM diagram. Hence, by analysis of CM diagrams, we are able to infer significant information about stellar evolution.

When using the results of stellar-evolution calculations, it is worthwhile bearing in mind that the errors and uncertainties in these calculations accumulate from early to late evolutionary stages. The largest sources of uncertainty lie in the treatment of convection and mass loss. Convection is assumed to occur at some radius $r$ within a star when the **Schwarzschild criterion** is satisfied at $r$. The Schwarzschild criterion is that the temperature gradient that is required at $r$ for the entire heat flux to be carried by the outward diffusion of photons is greater than the temperature gradient that one would have at $r$ if the specific entropy of the gas was constant in

---

[2] The last reaction in the sequence (5.3) gives only the most important process for the recovery of $^{12}C$ from $^{15}N$; the CNO cycle can have other endings.

the neighborhood of $r$. In these circumstances cells of hot gas move upwards and are replaced by cells of cooler gas that fall downwards. This interchange of gas transports energy extremely efficiently, so evolutionary calculations assume that in a convective region the temperature gradient is the minimum gradient at which convection will work, namely the adiabatic gradient.

In the standard phenomenological model of convection, which is known as **mixing length theory**, all rising and falling convective cells travel one **mixing length** $l$ before dissolving into the ambient medium. This length-scale is reduced to a dimensionless parameter $\alpha \equiv l/H$ by dividing it by the local scaleheight[3] of the star's atmosphere, $H$. Stellar-evolution theory is sensitive to the assumed value of $\alpha$ because this parameter determines the speed with which convection erodes gradients in chemical composition.

Convective regions usually have a top and a bottom within the star, and these boundaries often coincide with a region of exceptionally strong chemical inhomogeneity. For example, the lower boundary may coincide with a shell within which nuclear burning is taking place, with the consequence that a particular nuclear fuel is abundant above the shell and exhausted below it. Now a convective zone cannot have a sharp boundary, since turbulence generated by rising and falling cells is bound to spill over into adjacent convectively (marginally) stable regions. There is now considerable evidence that such **convective overshoot** plays an important role in the evolution of stars and the results of stellar-evolution calculations are sensitive to the size, $\beta$, of the overshoot region (measured in units of $H$) that is assumed to exist.

The standard strategy is to determine appropriate values for $\alpha$ and $\beta$ by tuning stellar structure calculations so that they yield the best possible match to the observational characteristics of the Sun and other well-observed stars. On account of this empirical calibration, we should not to read too much into the ability of mixing-length theory to model the properties of other stars. It is also not obvious that the appropriate values of the theory's parameters are the same for stars of every type; by picking a single value we may be distorting our predictions in a systematic way.

Except at the extreme ends of the sequence, MS stars only contain small convective regions, so the uncertainties in their structure are small. By contrast, stars that have evolved on to the subgiant branch and beyond have large convective envelopes and correspondingly large uncertainties in their structure.

Matter flows away from most stars in the form of a **stellar wind** – a roughly spherically symmetric flow of plasma from the surface of the star out into the interstellar medium. This mass loss, like convection, becomes a more important effect as the star evolves away from the MS, where mass loss tends to be very slow. At the present time it is extremely hard to predict

---

[3] The scaleheight is the distance $H = (\mathrm{d}\ln P/\mathrm{d}r)^{-1}$ over which the pressure $P$ declines by a factor $\mathrm{e} = 2.7\ldots$

the rate of mass loss theoretically, and it is conventional to subject a stellar model to a mass-loss rate that has been estimated from observations of stars that are at a similar observational stage.

Next in importance to uncertainties introduced by convection and mass loss, is uncertainty in the rate at which elements heavier than H settle to-wards the centers of stars: in thermal equilibrium a star would be stratified with the heaviest elements at the center and the lightest, namely H, floating on the surface. To a large extent nuclear burning establishes this sort of stratification naturally, but most of the He in any star is primordial, and starts out homogeneously mixed with H. The detailed results of evolutionary calculations depend on the assumed rate at which He sinks towards the cen-ter. The downward motion of He (**helium diffusion**) reduces the fraction of H in the core and brings forward in time the moment at which nuclear burning exhausts the supply of H in the core and a star turns off the MS.

Any error that is introduced into a model at one evolutionary stage when either convection, mass-loss or diffusion is inadequately described, clearly undermines the validity of the model at all subsequent evolutionary stages, even if no further error is introduced.

### 5.1.1 Placing models in the CM diagram

The final difficulty in comparing stellar-evolution models to observed CM di-agrams is the conversion of the temperature and bolometric luminosity of a model into a color and an absolute magnitude. There are two fundamentally different approaches to this task. In one approach, we assemble a sample of nearby stars for which effective temperatures, absolute magnitudes, metal-licities and colors have been observationally determined. Then we assign to our theoretical model the spectral properties of the star within the sample that has the nearest values of $M_{\rm bol}$ and $T_{\rm eff}$. In fact, by interpolating in a table of values of $T_{\rm eff}$ and colors (such as Table 3.7) that has been assembled from the observational results, we can estimate the colors of our theoretical star rather accurately even if our observational sample does not contain a precisely identical star.

Clearly, the procedure just outlined for determining the colors of the-oretical stars is viable only if a sufficient sample of well-observed stars of similar metallicity can be assembled. Unfortunately, this requirement effec-tively restricts the method to stars that do not differ too much in metallicity from the Sun. In particular, very metal-rich or metal-poor stars, stars that have markedly different abundance ratios from those in the Sun, and very old stars are very scarce near the Sun, although they may dominate the light from a distant object. Since reliable bolometric magnitudes and effective temperatures are not available for large samples of such stars, the spectral properties of their theoretical counterparts cannot be obtained by interpola-tion in an appropriate table of observational values. An alternative procedure

involves tuning a stellar-atmosphere code to the star's effective temperature
and bolometric magnitude, and then determining the star's colors by 'observing' the resulting theoretical spectrum in the computer (Tripicco & Bell
1995). Since the spectra of many stars contain enormous numbers of spectral
lines, this procedure requires both a sophisticated stellar-atmosphere code
and a mass of reliable atomic and molecular data. The cooler and more
metal-rich the star is, the greater the number of lines, and hence the more
formidable the theoretical problem that must be solved if accurate colors are
to be determined in this way.

### 5.1.2 Features in the CM diagram

Let us now indicate what, according to the theory of stellar structure, is
happening in key areas of the CM diagram (§3.5). These areas are marked
in Figure 5.1 and summarized in Table 5.1.

The lives of stars are governed by the same law that drives star clusters
and galaxies: *let heat flow from the inside to the outside, so that the inside
may become denser and hotter.*[4] As the inside of a star becomes denser and
hotter, progressively heavier nuclei form – the nuclei at the center change
successively from those of H to He to a mixture of C and O to Si and so on
eventually up to Fe if the star is sufficiently massive. The energy released
in the formation of a given nuclear species can for a while fulfill the law's
requirement for an outward heat flux without further compression of the interior material. When the raw material for this process is exhausted the period
of stasis is followed by further central contraction. Just as traffic builds up
near road-works on a highway, so each period of steady nuclear burning is
associated with a concentration of stars in the CM diagram. Quantitatively,
the number of stars at each burning stage varies roughly as $\tau$, the time it
takes to exhaust the nuclear fuel burnt at that stage.

Of the concentrations of stars in the CM diagram that are associated
with these static periods, the most important is the MS – see Figure 5.1
and §3.5. Here stars are burning H to He in their cores. This is the longest
period of a star's life because it is a period of relative fuel-economy: more
nuclear energy will be released after leaving the MS than on it, but after
leaving the MS a star rapidly increases in luminosity with the result that it
rather quickly exhausts its fuel – see Figure 5.4. The position occupied by
a star on the MS depends on its mass: more massive stars lie up and to the
left of less massive ones – see Figure 5.2. The locus of the MS depends on
metallicity: the MS of a metal-poor population lies below and to the left of
the MS of a metal-rich one.

When a massive star leaves the MS it moves to the right in the CM diagram and not long afterwards dies in a supernova explosion. Running nearly

---

[4] In stellar dynamics, this law leads to the 'gravithermal catastrophe' – see §8.2 of BT
or Lynden-Bell & Wood (1968).

**Table 5.1**   Features in the CM diagram

| Feature | Physical significance | Remarks |
|---------|----------------------|---------|
| Main sequence (MS) | Core H burning | |
| Subgiant branch (SGB) | Transition from core to shell H burning | Prominent in globular-cluster CM diagrams |
| Red giant branch (RGB) | Shell burning of H | For lower-mass stars terminated by He flash |
| Horizontal branch (HB) | Core He burning | Has characteristic luminosity; color sensitive to metallicity |
| Red clump (RC) | Stubby red HB formed by more metal-rich stars | Prominent in disk CM diagrams |
| Asymptotic giant branch (AGB) | Shell He burning | Associated with significant and increasing mass loss; stars often irregular variables |
| Instability strip | $He^+$ ionization zone gives rise to regular variables | RR Lyrae and W Virginis stars lie at intersection with HB; Cepheids are massive stars that lie in strip |
| White-dwarf sequence | Cooling electron-degenerate stars | Blue and faint |

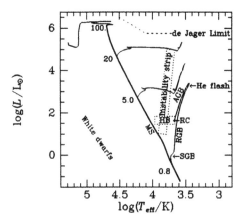

**Figure 5.1** The principal features in a CM diagram. The MS and evolutionary tracks are for chemical composition $Y = 0.28$, $Z = 0.2$. [After Chiosi, Bertelli & Bressan (1992) from data published in Bressan *et al.* (1993)]

vertically through the CM diagram is a strip within which stars pulsate. This **instability strip** is the subject of §5.1.10 below. The evolutionary tracks of higher-mass stars intersect the instability strip, and at this intersection are found **Cepheid variables**. These pulsating stars are discussed in detail in §5.1.10 below because they are important for the problem of establishing the cosmic distance scale (§7.3.1).

After leaving the MS, stars of mass $\mathcal{M} \lesssim 8\,\mathcal{M}_\odot$ spend some years moving up the approximately vertical red-giant branch (RGB) – see Figure 5.1 and §3.5. In the CM diagrams of globular clusters, the base of the RGB is joined to the end of the MS by the subgiant branch (**SGB**) which is short and slopes from lower left to upper right in the CM diagram. [In the CM diagram of the solar-neighborhood (Figure 3.5) this region is rather sparsely populated and called the Hertzsprung gap – see §3.5.] The SGB is occupied by stars that are burning H to He in a shell rather than in their cores, but have not yet developed fully convective envelopes. The RGB is occupied by shell-burning stars that possess fully convective envelopes.

An important concentration of stars is the **horizontal branch (HB)**, on which stars burn He in their cores just as on the MS they burn H in their cores. The locus of stars of different masses that have just started core He-burning is called the **zero-age horizontal branch (ZAHB)** by analogy with the ZAMS. During a star's time on the HB, its luminosity slowly increases – see the upper panel of Figure 5.5. Consequently the HB has a non-zero vertical extent with the ZAHB bounding it below. The stars of metal-rich clusters and the solar neighborhood congregate at the red end of the horizontal branch, so that CM diagrams for such stars contain HBs that resemble blobs rather than lines. The HB is then usually called the **red clump (RC)**.

The HBs of metal-poor clusters intersect the instability strip, and at this intersection are found **RR Lyrae variables**. Like Cepheid variables, RR Lyrae stars are important for the problem of establishing the cosmic distance scale (§7.3.1) and are described in §5.1.10 below.

Once a star has run out of He in its core, it moves rather quickly away from the HB up the **asymptotic giant branch (AGB)**, which runs parallel to the RGB, but slightly to the blue (Figure 5.3). Here a star burns He in a shell and possibly C in its core. As it moves up the AGB, a star becomes more and more subject to variability and mass loss. Probably all stars eject material in the form of a stellar wind, but the rate of mass loss is usually small early in a star's life. As it gets older and more luminous, a star loses mass faster and faster, with the result that many stars return most of their mass to the interstellar medium (Chapter 8) before they die. As a star approaches the top of the AGB, its rate of mass loss increases and oscillations develop within the star. The cause of these oscillations is a kind of duet between a He-burning shell at small radii and a H-burning shell around it (Schwarzschild & Härm 1965). For 90% of the time the H-burning shell is the more important source of the star's energy, but occasionally it is

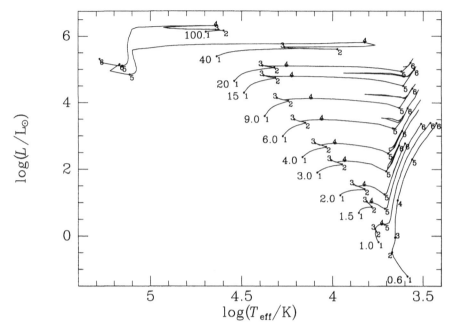

**Figure 5.2** Evolutionary tracks for solar-metallicity stars $(Y, Z) = (0.28, 0.02)$ with initial masses from $0.6\,\mathcal{M}_\odot$ to $100\,\mathcal{M}_\odot$. On each track several points are marked and numbered. Table 5.2 gives the time it takes a star to reach each of these points starting from point 1. To avoid confusion tracks for $\mathcal{M} \leq 2\,\mathcal{M}_\odot$ terminate at the He flash – see Figure 5.3 for the further tracks of these stars. All models assume convective overshoot. [From data published in Bressan *et al.* (1993)]

extinguished by bursts of activity on the part of the He-burning shell. This duet generates a complex pattern of convection which may bring C nuclei from below the He-burning shell to the surface. In this way a **carbon star** is made out of an M giant. The movement of C nuclei through the He-burning shell into the star's envelope is one example of a general phenomenon known as **dredge-up**. It seems that such mixing into the envelope of material that has been processed by nuclear reactions deep in the interior of a star is not at all uncommon. In §5.2.2 we shall see that dredge-up is of considerable importance for galactic astronomy.

By the time a star has reached the HB it is thought to have lost about 20% of its mass (§6.1.3). Observations indicate that the rate of mass loss increases markedly as a star approaches the top of the AGB, although there is as yet inadequate theoretical understanding of the causes of this increased mass loss. A likely consequence of this enhanced mass loss is the formation of OH/IR stars – see Table 3.5. Both carbon stars and OH/IR stars have come to play important roles in Galactic structure research, because they are readily identified objects of high luminosity and can therefore be studied throughout the Milky Way. Stars that lie near the top of the AGB usually

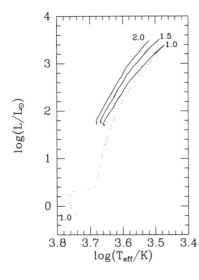

**Figure 5.3** The full curves show the evolutionary tracks of solar-metallicity stars of mass 1, 1.5 and 2 $\mathcal{M}_\odot$ from the ZAHB to the tip of the AGB. The dotted curve shows the evolution of a 1 $\mathcal{M}_\odot$ star prior to the He flash. Figure 5.5 shows how long these stars linger near the ZAHB before moving on up the AGB. [From data published in Bressan *et al.* (1993)]

have appreciable infrared luminosities, which are thought to derive from an envelope of dusty gas that has been thrown off by the star and is now drifting away from it in a roughly spherical configuration.

Soon after reaching the top of the AGB a star will shed essentially all of its remaining H. Stripped of its blanketing layer of H, the star becomes extremely blue, making a violent lurch leftwards in the CM diagram. The intense ionizing radiation from the star may cause the ejected H envelope to fluoresce brightly as a **planetary nebula**. Gradually He burning ceases in the star and it starts to cool, moving down the **white-dwarf cooling sequence** at the extreme left of the CM diagram. Its mass now probably lies in the range $0.55 - 0.6\,\mathcal{M}_\odot$ regardless of its initial mass, although some white dwarfs have masses as low as $0.4\,\mathcal{M}_\odot$. Its density is now so great that it is sustained against gravitational collapse by the minimum motion required of its electrons if they are to satisfy the Pauli exclusion principle, just as an ordinary atom is sustained against electrostatic collapse by the minimum motion of its electrons. Now that the star is degenerate, it can finally evade the law that has hitherto required it to lose heat and contract.

### 5.1.3 Characteristic initial masses

Let us now return to the MS and note some key masses. Figures 5.2 and 5.3 show how stars of solar metallicity and various initial masses move in the CM diagram when they move off the MS. Table 5.2 gives the times that elapse between the numbered points on the evolutionary tracks plotted in Figure 5.2, while Figures 5.4 and 5.5 plot the luminosities of some of these models versus time.

The least massive stars have $\mathcal{M}_H \simeq 0.08\,\mathcal{M}_\odot$. Objects less massive than

## Box 5.1:   Degenerate Objects

Consider a box filled with $N$ non-interacting Fermions. Since no two Fermions can occupy the same state, $N$ different single-particle states must be occupied. At zero temperature, these are the $N$ lowest-lying energy eigenstates. If the linear scale $L$ of the box is changed, the momentum $p$ of each state scales as $p \propto \hbar/L \propto \rho^{1/3}$, where $\rho$ is the density of the material in the box. The pressure $P$ exerted by the material in the box is proportional to the energy density, so $P \propto E/L^3 \propto E\rho$. If the particles are moving non-relativistically, $E \propto p^2 \propto \rho^{2/3}$ and we conclude that $P \propto \rho^{5/3}$. This is the relation that is satisfied during adiabatic compression of a gas that has ratio of principal specific heats $\gamma = \frac{5}{3}$. If the fermions are moving relativistically, we have $E \propto p$ and similarly deduce $P \propto \rho^{4/3}$. Thus for a gas of relativistic fermions $\gamma$ takes the value $\frac{4}{3}$. In Problem 5.4 we will see that this value is critical for gravitational collapse.

If one calculates the mass $\mathcal{M}$ of a fully relativistic degenerate object as a function of the object's central density, $\rho_0$, one finds that as $\rho_0 \to \infty$, $\mathcal{M}$ tends to a finite value, $\mathcal{M}_{\text{Chan}}$, called the **Chandrasekhar limiting mass**. Numerically, $\mathcal{M}_{\text{Chan}} = 5.87\mu^{-2}\,\mathcal{M}_\odot$, where $\mu$ is the number of atomic mass units per relativistic fermion (Weinberg 1972). For a He white dwarf, $\mu = 2$, so $\mathcal{M}_{\text{Chan}} = 1.47\,\mathcal{M}_\odot$.

In many stars, atoms are crushed together so tightly that the volume per atom is substantially less than the volume of an isolated atom, and most electrons are no longer bound to individual nuclei. If for some reason the star contracts, the wavefunctions of the electrons shrink, and their energies are driven upwards as calculated above. Hence our results are valid even though the electrons have non-negligible electrostatic interactions with the nuclei and thus do not conform to our simple picture of a non-interacting electron gas. Similarly, in a neutron star, the neutrons interact strongly with one another, but our results will be valid so long as their wavefunctions scale with the overall size of the star.

$\mathcal{M}_{\text{H}}$ never get hot enough in their cores to burn H; they are supported by electron-degeneracy pressure and become **black dwarfs**. Jupiter is a low mass ($\sim 0.001\,\mathcal{M}_\odot$) object of this class. Although they never burn H, they may burn light elements such as D and Li, and during this period they are known as **brown dwarfs**.

The next significant mass on the MS is $\mathcal{M}_{\text{conv}} \simeq 1.1\,\mathcal{M}_\odot$. Stars more massive than $\mathcal{M}_{\text{conv}}$ have convective cores, whereas the cores of less massive stars are entirely radiative. (Actually, the size of convective regions grows gradually, starting at $\mathcal{M}_{\text{conv}}$ and engulfs the core only at $\simeq 1.5\,\mathcal{M}_\odot$.) The cores of massive stars are convective because at their higher temperatures,

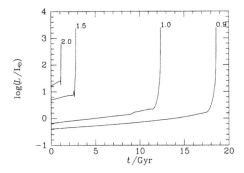

**Figure 5.4** Luminosity as a function of time for stars evolving from the MS to the He flash. Curves are labelled by initial masses ($\mathcal{M}_\odot$). [From data published in Bressan *et al.* (1993)]

the CNO cycle [equation (5.3)] is a more important source of heat than the p-p chain [equation (5.1)], and the strong temperature-sensitivity of the CNO cycle concentrates energy production at the center of the star.

A further important mass on the MS is $\mathcal{M}_{\text{HeF}} \simeq 1.8 - 2.2\,\mathcal{M}_\odot$ depending on the chemical composition. A star that starts out less massive than $\mathcal{M}_{\text{HeF}}$ experiences a **helium flash**. This is the explosive ignition of its helium core as the star reaches the top of the RGB – see Box 5.2. Stars that are initially more massive than $\mathcal{M}_{\text{HeF}}$ ignite helium quiescently because, unlike the cores of lower-mass stars, at the point of ignition their cores are not electron degenerate. From our point of view, the importance of $\mathcal{M}_{\text{HeF}}$ is that initially more massive stars do not burn their He on the HB or in the red clump. The luminosity of HB stars is almost independent of mass (Figure 5.3). The reason for this is that He ignites explosively when a core has accumulated a characteristic mass ($\simeq 0.45\,\mathcal{M}_\odot$) of He and the HB luminosity is largely determined by the mass of the He core. By contrast, at the point of He ignition, the mass of the He core of a star of initial mass $\mathcal{M}_i > \mathcal{M}_{\text{HeF}}$ increases with $\mathcal{M}_i$, with the consequence that the luminosity of the star during core He burning increases with $\mathcal{M}_i$. Hence stars with $\mathcal{M}_i > \mathcal{M}_{\text{HeF}}$ lie above the HB during core He burning by an amount that increases with $\mathcal{M}_i$. In Figure 5.2 the point of He ignition on tracks for $\mathcal{M}/\mathcal{M}_\odot \leq 20$ is marked by the digit 6 and for $\mathcal{M} \geq 3\,\mathcal{M}_\odot$ most He is burnt near the bottom of the track that runs first down from the 6 and then up to the tip of the AGB above the 6. Notice that a star of mass $\gtrsim 6\,\mathcal{M}_\odot$ makes a significant horizontal excursion as it burns He in its core. This excursion is caused by the cessation of convection in the star's outer layers.

In stars with initial mass $\mathcal{M}_i > \mathcal{M}_{\text{up}} \simeq 8\,\mathcal{M}_\odot$, C ignites quiescently in the core. In all but the most massive stars ($\mathcal{M} \gtrsim 50\,\mathcal{M}_\odot$) nuclear burning then proceeds until the core consists of iron-peak nuclei (see §5.2.1 below). Since these are the most tightly bound nuclei, further energy cannot be released by burning them to still more massive nuclei. Consequently, the core tends to contract, and as this contraction drives up the Fermi energy of the core's electrons and heats the core, two processes lead to the core's catastrophic gravitational collapse: the first process is the capture of electrons

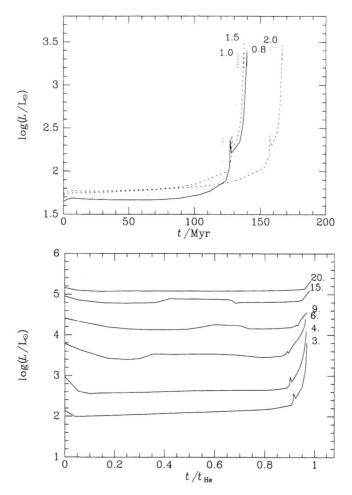

**Figure 5.5** Luminosity as a function of time from the onset of core He burning. The top panel is for stars that experience the He flash and the lower panel is for more massive stars. For these stars the time axis has been scaled by the time $t_{He}$ given in the last column of Table 5.2. [From data published in Bressan *et al.* (1993)]

by nuclei and the second is the shattering of nuclei into $\alpha$-particles by energetic photons. Both processes diminish the ambient pressure by removing energy from the photon and electron fields, which sustain the core against the inward pull of gravity. This leads to a runaway contraction because the smaller the core gets, the higher the Fermi energy becomes, and the faster electron capture and the photo-dissociation of nuclei proceed.

As the core implodes, some of the energy released may lead to a spectacular display as a supernova – see §5.2.3 below. Depending on the mass

**Table 5.2** Stellar-evolution times for $Z = 0.02$

| $\mathcal{M}/\mathcal{M}_\odot$ | Times (Myr) between points in Figure 5.2 | | | | | $t_{He}$ |
|---|---|---|---|---|---|---|
| | $1-2$ | $2-3$ | $3-4$ | $4-5$ | $5-6$ | |
| 0.6 | 78245 | 4724 | 1591 | 175.3 | 17.70 | 143.7 |
| 1.0 | 8899 | 376.3 | 1995 | 178.3 | 907.5 | 133.0 |
| 1.5 | 2471 | 53.76 | 5.05 | 63.42 | 188.9 | 137.5 |
| 2.0 | 1030 | 20.26 | 1.700 | 13.82 | 44.75 | 166.7 |
| 3.0 | 347.4 | 6.785 | 0.492 | 2.199 | 2.558 | 107.1 |
| 4.0 | 171.1 | 3.119 | 0.254 | 0.551 | 0.683 | 30.14 |
| 6.0 | 67.92 | 1.057 | 0.070 | 0.088 | 0.147 | 5.94 |
| 9.0 | 29.45 | 0.414 | 0.023 | 0.020 | 0.041 | 1.94 |
| 15.0 | 12.37 | 0.136 | 0.008 | 0.008 | 0.011 | 0.82 |
| 20.0 | 8.46 | 0.085 | 0.005 | 0.006 | 0.006 | 0.61 |
| 40.0 | 4.43 | 0.138 | 0.007 | 0.541 | 0.032 | – |
| 100.0 | 1.27 | 1.079 | 0.619 | 0.415 | 0.023 | – |

NOTES: The final column gives the time from point 6 until the star's death.

SOURCE: From data published in Bressan *et al.* (1993)

of the core, the implosion leads to·the formation of either a neutron star or a black hole. Neutron stars, like white dwarfs, cannot have masses in excess of some limiting value (Chandrasekhar 1931, Oppenheimer & Volkoff 1939). This value depends upon the equation of state of the star's material, which is uncertain in the case of neutron stars. Observationally, neutron star masses cluster around $1.4\,\mathcal{M}_\odot$ – see Figure 5.6 – which suggests that the limiting mass is $\simeq 1.4\,\mathcal{M}_\odot$, and that a core that is significantly more massive will form a black hole when it implodes.

Stars initially more massive than $\mathcal{M}_{ee} \simeq 60\,\mathcal{M}_\odot$ become unstable before forming iron-peak elements in their cores – at the point of instability, their cores consist largely of O. When the core temperature reaches $T_e \equiv 2m_e c^2/k \simeq 2 \times 10^7$ K, the thermal density of positrons becomes significant and the **pair-production instability** sets in. The rapid increase with temperature $T$ in the number density of $e^- e^+$ pairs implies that the specific heat of the vacuum becomes unusually large, and the ratio of principal specific heats $\gamma$ in the star drops below the value $\gamma = \frac{4}{3}$.[5] This destabilizes the star – see Problem 5.4. What happens next is uncertain. Some calculations suggest that when the star's mass is less than $\sim 120\,\mathcal{M}_\odot$, a few violent pulsations occur in which episodes of O-burning temporarily reverse gravitational collapse. Finally the core collapses and the overlying envelope may be ejected (Woosley & Weaver 1986). Similar calculations suggest that when

---

[5] Recall that the principal specific heats always satisfy $c_P - c_V = R$, the gas constant. Hence increasing $c_V$ lowers $\gamma \equiv c_P/c_V$.

## Box 5.2: Ignition in a Degenerate Object

As Box 5.1 explains, at zero temperature, the $N$ electrons of a body occupy the $N$ lowest-lying quantum states. The energy of the highest-energy occupied state is called the body's **Fermi energy**, $E_F^e$. An object whose temperature is significantly less than $E_F^e/k$ is said to be **electron degenerate** because its electrons are still largely confined to the lowest-lying $N$ quantum states. In such a body the thermal energy of the electrons is almost independent of temperature $T$ because there are few quantum states into which an electron could be excited by the provision of energy of order $kT$.

Mestel (1952) pointed out that nuclear fuels are liable to ignite explosively in an electron-degenerate object. The pressure in such an object is provided by electrons, most of which have energies comparable to the electron Fermi energy $E_F^e$. When the object is heated, the energies of these electrons change appreciably only when the temperature approaches $T_F \equiv E_F^e/k$ and the object ceases to be completely degenerate. Since in a non-relativistic system, $E_F$ is inversely proportional to particle mass, the nuclei become non-degenerate at temperatures thousands of times lower than $T_F$. Hence over a large range of temperatures, the velocity distribution of the nuclei is Maxwellian while that of the electrons, which govern the pressure, is essentially independent of temperature. As nuclear reactions heat the object, the rate at which they occur increases because the number of fast nuclei that are capable of penetrating Coulomb barriers increases in accordance with the Maxwell–Boltzmann law. However, little extra pressure is generated, with the result that there is negligible expansion of the system and thus the reaction rate, which is proportional to the square of the particle density, is scarcely moderated by expansion. Hence the reaction rate increases almost exponentially, and an explosion ensues.

$120 < (\mathcal{M}_i/\mathcal{M}_\odot) < 300$, O-burning completely disrupts the star, while an even more massive star collapses to a black hole without a single pulsation. Before they become unstable it is likely that all stars of mass $\mathcal{M}_i > 60\,\mathcal{M}_\odot$ will have largely ejected their H envelopes, with the result that in any subsequent supernova event H-lines will not be prominent. This will cause them to be classified as type I supernovae – see §5.2.3 below.

A relationship that is of particular importance for galactic astronomy is the **remnant-mass–initial-mass relation**, $\mathcal{M}_r(\mathcal{M}_i)$, which gives the mass $\mathcal{M}_r$ of the remnant to which a star of initial mass $\mathcal{M}_i$ evolves. This relation is sketched in Figure 5.7. It is of interest only for $\mathcal{M}_i \gtrsim 0.8\,\mathcal{M}_\odot$ because no lower-mass stars have had time to complete their evolution. At low values of $\mathcal{M}_i$, $\mathcal{M}_r$ climbs slowly through the range of possible white-dwarf masses.

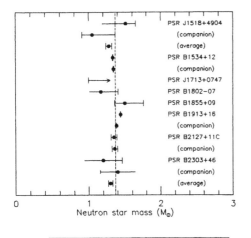

PSR J1518+4904
(companion)
(average)
PSR B1534+12
(companion)
PSR J1713+0747
PSR B1802−07
PSR B1855+09
PSR B1913+16
(companion)
PSR B2127+11C
(companion)
PSR B2303+46
(companion)
(average)

0        1        2        3
Neutron star mass ($M_\odot$)

**Figure 5.6** The masses of neutron stars cluster around $1.4\,M_\odot$. [From data kindly supplied by J. Taylor]

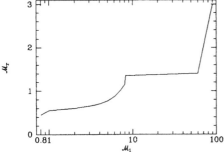

**Figure 5.7** The remnant-mass–initial-mass relation. In the range $1 < M/M_\odot < 8$ the curve follows data published in Weidemann (1990) for solar-neighborhood white dwarfs. For larger masses the curve is more uncertain.

At $\mathcal{M}_i = \mathcal{M}_{\rm up} \simeq 8\,M_\odot$ the remnant switches from being a white dwarf to a neutron star, and since, as Figure 5.6 shows, the latter are observed to have masses that cluster tightly around $1.4\,M_\odot$, we adopt this remnant mass for values of $\mathcal{M}_i$ up to $\mathcal{M}_i = 60\,M_\odot$. Some less massive stars may form black holes, but stellar evolution calculations suggest that when they implode, the cores of these stars are not much more massive than $1.4\,M_\odot$. Significantly more massive black holes may be produced by stars of mass $\mathcal{M}_i > 60\,M_\odot$: X-ray astronomy has revealed remnants with masses $\gtrsim 3\,M_\odot$. In Figure 5.7 we arbitrarily assume that the mass of the black hole remnant rises linearly with $\log \mathcal{M}_i$ from $1.4\,M_\odot$ at $\mathcal{M}_i = 60\,M_\odot$ to $3\,M_\odot$ at $\mathcal{M}_i = 90\,M_\odot$.

The $\mathcal{M}_r(\mathcal{M}_i)$ relation plotted in Figure 5.7 is very approximate. Not only is it subject to the uncertainties at the high-mass end just discussed, but even at the low-mass end it is observationally clear that the very concept of a remnant-mass–initial-mass relation is incorrect in detail because stars of identical initial mass may evolve to white dwarfs of differing mass for at least two reasons. First differences in chemical composition and/or rotation may lead to differing amounts of mass loss on the RGB and AGB. Second, most stars are in binary systems and a binary companion can either accelerate mass loss through a stellar wind or add material to the surface of a

---

## Box 5.3:   The Hayashi Line

Imagine a furnace that floats freely in space and is enveloped in a massive cloud of gas. You are in control of the fuel supply to the furnace. At the lowest setting, the heat it produces is radiatively conducted away through the gas cloud and radiated from its surface. Then at some critical setting, some part of the cloud becomes convective – in places cells of hot gas rise, just as hot air rises from a radiator, and are replaced by downward moving cells of cold gas. As you turn up the fuel supply more, the convectively stirred region grows steadily in radial extent. Eventually, the cloud becomes fully convective.

Consider the location, at each stage in the experiment, of the cloud in a plot of luminosity $L$ versus surface temperature $T$. Initially there is not much heat to radiate, so both $L$ and $T$ are small. Then as $L$ rises, $T$ tends to rise also. But it turns out that once the cloud becomes fully convective, $T$ becomes very nearly independent of $L$; at this point the cloud swells sufficiently rapidly with increasing $L$ that the additional luminosity can be radiated from the surface without an appreciable change in temperature. Hayashi (1961) discovered this surprising behavior of fully convective stars, and their locus in the $(T, L)$ plane is called the Hayashi line in his honor. As Figure 5.10 below shows, the Hayashi lines of stars of different masses lie close to one another, with more massive stars on the left (at higher $T$).

---

star. Notwithstanding these caveats, the general form of the curve shown in Figure 5.7 is probably sound and of considerable astrophysical importance. In particular, it makes clear that *stars eventually return most of their mass to the interstellar medium.*

### 5.1.4 Bounding curves in the CM diagram

The dashed lines in Figure 5.1 bound a region within which important pulsating stars lie. The mechanism driving the pulsation of these stars will be discussed in §5.1.10 below. The most important stars within this instability strip are the RR Lyrae variables, which are fairly metal-poor HB stars, and the Cepheid variables, which are supergiants – see Table 3.5. The strip also contains W Virginis stars (Table 3.5), $\delta$ Scuti stars (pulsating MS stars) and pulsating white dwarfs.

In the uppermost region of the CM diagram only variable stars are found (de Jager 1984), and the curve that forms the upper edge of the region occupied by non-variable stars is called the **de Jager limit**. This limit slopes downwards from left to right, so that the most luminous blue stars are about a factor of six brighter than the most luminous red stars. This is

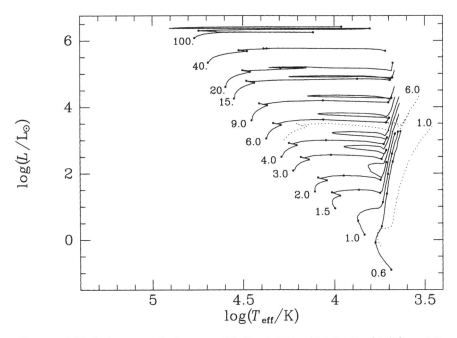

**Figure 5.8** Evolutionary tracks for stars with $Z = 0.0004$, which implies [Fe/H] $= -1.7$. These tracks should be compared to those of solar-metallicity stars shown in Figure 5.2; to facilitate this comparison the dotted curves show those tracks for $\mathcal{M} = 1$ and $6\,\mathcal{M}_\odot$. For clarity the tracks for $\mathcal{M} = 1.5$, 1 and $0.6\,\mathcal{M}_\odot$ terminate at the He-flash – see Figure 5.9 for the subsequent evolution of these stars. Table 5.3 gives the times required to get between the marked points. [From data published in Fagotto *et al.* (1994)]

interesting because massive stars evolve from the MS at essentially constant luminosity. Hence the evolution of the most massive stars carries them across the de Jager limit. It is observed that the largest mass-loss rates occur in the vicinity of the de Jager limit (de Jager, Nieuwenhuijzen & van der Hucht 1988). One possible explanation for the absence of stars to the right of the de Jager limit is that mass loss arrests the redward movement of these stars over this boundary. An alternative explanation is that the stars evolve so quickly once they reach the de Jager limit that we stand little chance of observing them in the region beyond the limit.

To the right of the populated part of the CM diagram lies the **forbidden region**; this region is forbidden in the sense that an equilibrium star cannot lie there. The boundary between the forbidden and populated regions of the CM diagram is formed by the **Hayashi line** of the lowest-mass stars – see Box 5.3 for the definition of a star's Hayashi line, which is essentially coincident with the RGB on to which the star evolves from the MS.

## 5.1.5 Dependence of CM diagrams upon metallicity

While the gross structure of the CM diagram is the same for stars of any chemical composition, the precise locations of the various branches do depend on the composition parameters $(X, Y, Z)$ that were defined in §3.4. Figure 5.8 plots evolutionary tracks for moderately metal-poor stars, for which appropriate values are $Y = 0.23$ and $Z = 0.0004$ – as we shall see below (§5.1.6), while most He was produced in the Big Bang, significant quantities of He are produced alongside heavy elements, and it is appropriate to decrease both $Y$ and $Z$ together.

Careful comparison of these tracks of Figure 5.8 with the ones for $Y = 0.28$ and $Z = 0.02$ plotted in Figure 5.2 reveals that lowering $Z$ by 1.7 dex modifies individual stellar tracks in several important ways that we now discuss. To facilitate discussion, the tracks of solar-metallicity stars of mass $1 \mathcal{M}_\odot$ and $6 \mathcal{M}_\odot$ are shown as dotted curves in Figure 5.8.

Lowering $Z$ causes individual stars to become both brighter and hotter. Above $4 \mathcal{M}_\odot$ the main effect of lowering $Z$ is to make stars hotter: on the ZAMS, a $0.6 \mathcal{M}_\odot$ star is brighter by a factor 2.3 and hotter by a factor 1.26 at the lower metallicity, while at the lower metallicity a $4 \mathcal{M}_\odot$ star is brighter by a factor 1.3 and hotter by a factor 1.33. Although lowering $Z$ makes individual stars brighter, the overall effect of lowering $Z$ by 1.7 dex is to move the ZAMS *down* by just under 1 mag at a given color.[6] In §3.3 we saw that extremely metal-poor MS stars are called subdwarfs. These stars owe their name to the fact that they lie below the ZAMS of the Hyades cluster, which is the standard for stars of approximately solar metallicity. Box 5.4 describes an important technique for identifying subdwarfs.

Lower-metallicity stars make longer excursions towards higher temperatures and bluer colors during He burning. In Figure 5.2 only stars of $6 \mathcal{M}_\odot$ and above make such excursions, and these are of smaller extent than the corresponding excursions in Figure 5.8. Moreover, at $Z = 0.0004$ blueward excursions are significant for all stars with masses of $2 \mathcal{M}_\odot$ and above. These excursions are of particular importance because they are liable to carry the star into the instability strip, when it will become an easily identified variable star (§5.1.10).

Table 5.3 lists the times required to evolve between the points marked in Figure 5.8. Since the luminosity of a star generally increases with decreasing metallicity, it evolves more quickly, and the times in Table 5.3 are generally shorter than those in Table 5.2.

Comparison of Figures 5.2 and 5.8 shows that at low metallicity the RGB is much more nearly vertical in the CM diagram, and has a color that depends less strongly on the masses of the stars that lie on it, than is the case for solar metallicity. Observationally, these differences are apparent in Figure 6.3.

---

[6] At a given luminosity lowering $Z$ by 1.7 dex shifts the ZAMS to the left by $\sim 0.06$ in $\log T_{\rm eff}$, which corresponds to about $-0.32$ in $V - K$ color (Table 3.6).

## Box 5.4: Ultraviolet Excess

An important way of identifying subdwarfs of unknown distance and therefore unknown location in the CM diagram is to measure the **ultraviolet excess**, $\delta(U - B)$, which is the difference between $U - B$ for a star and that of the Hyades MS at the star's value of $B - V$. (Both $U - B$ and $B - V$ should have been corrected for any reddening.) At fixed metallicity, $\delta(U - B)$ is a decreasing function of $B - V$ – it tends to be largest for blue stars because line blanketing occurs mostly in the blue region of the spectrum (§2.3.2).

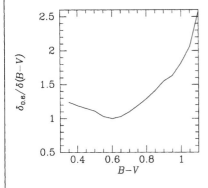

In an attempt to define a color index that is an unambiguous metallicity indicator for any dwarf, Sandage (1969) defined $\delta_{0.6}$ to be $\delta(U - B)$ multiplied by the correction factor that is plotted at left: this correction factor is designed to make $\delta_{0.6}$ independent of $B - V$ at fixed metallicity. Laird, Carney & Latham (1988) found that the relationship between $\delta_{0.6}$ and [Fe/H] is

$$\delta_{0.6} \simeq -0.0776 + \left(0.01191 - 0.05353[\text{Fe/H}]\right)^{1/2} \tag{1}$$

and that the vertical offset $\delta M_V$ between the Hyades ZAMS and that formed by subdwarfs of given $\delta_{0.6}$ is

$$\delta M_V = 0.862\left(-0.6888\delta_{0.6} + 53.14\delta_{0.6}^2 - 97.004\delta_{0.6}^3\right). \tag{2}$$

Figures 5.3 and 5.9 show the evolution of low-mass stars from the HB up the AGB. At low $Z$ the AGB, like the RGB, runs more nearly vertically in the CM diagram and lies blueward of the AGB for solar-metallicity stars; a star half way up the AGB for $Z = 0.0004$ is hotter than a solar-metallicity star that is half way up its AGB by $\sim 0.1$ dex.

Careful comparison of the locations of the bottoms of the tracks in Figures 5.3 and 5.9 reveals that for low $Z$ the HB lies slightly lower than it does for solar metallicity. Calculations of Lee, Demarque & Zinn (1990) yield for $Y = 0.23$

$$M_V(\text{HB}) = 0.17[\text{Fe/H}] + 0.82 . \tag{5.4}$$

but this dependence is unfortunately still controversial (Bounanno, Corsi & Fusi Pecci 1989, Chaboyer, Demarque & Sarajedini 1996).

**Table 5.3**   Stellar-evolution times for $Z = 0.0004$

| $\mathcal{M}/\mathcal{M}_\odot$ | Times (Myr) between points in Figure 5.8 | | | | | $t_{He}$ |
| | $1-2$ | $2-3$ | $3-4$ | $4-5$ | $5-6$ | |
|---|---|---|---|---|---|---|
| 0.6 | 41939 | 2616 | 782 | 111 | 17.40 | 137.5 |
| 1.0 | 5095 | 1423 | 140 | 36.5 | 10.57 | 114.0 |
| 1.5 | 1516 | 38.7 | 30.1 | 17.1 | 93.31 | 112.2 |
| 2.0 | 662 | 17.2 | 7.17 | 4.69 | 8.291 | 158.2 |
| 3.0 | 253 | 6.38 | 1.72 | 0.931 | 0.657 | 43.67 |
| 4.0 | 139 | 3.38 | 0.759 | 0.317 | 0.176 | 18.61 |
| 6.0 | 64.6 | 1.120 | 0.229 | 0.070 | 0.047 | 5.91 |
| 9.0 | 31.6 | 0.431 | 0.069 | 0.017 | 0.016 | 2.21 |
| 15.0 | 14.31 | 0.160 | 0.028 | 0.004 | 0.006 | 0.89 |
| 20.0 | 9.87 | 0.095 | 0.040 | 0.030 | 0.597 | − |
| 40.0 | 5.02 | 0.035 | 0.021 | 0.016 | 0.367 | − |
| 100.0 | 2.90 | 0.148 | 0.012 | 0.001 | 0.258 | − |

NOTES: The points are numbered according to the scheme used in Figure 5.2. The final column gives the time from point 6 until the star's death.

SOURCE: From data published in Fagotto *et al.* (1994)

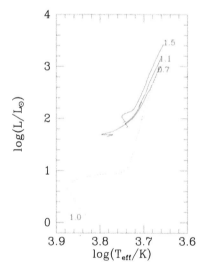

**Figure 5.9** The full curves show the evolutionary tracks of low-metallicity stars ($Z = 0.0004$) of mass 0.7, 1.1 and 1.5 $\mathcal{M}_\odot$ from the ZAHB to the tip of the AGB. The dotted curve shows the evolution of a 1 $\mathcal{M}_\odot$ star prior to the He flash. [From data published in Fagotto *et al.* (1994)]

In Figure 5.9 it can be seen that on the HB 0.7 $\mathcal{M}_\odot$ stars lie significantly to the left of 1.1 $\mathcal{M}_\odot$ stars, which themselves lie slightly below 1.5 $\mathcal{M}_\odot$ stars. In Figure 5.3, by contrast, the less massive a solar-metallicity star is, the further to the right it sits on the HB. For solar metallicity the total extent of the HB is small, however, so that a red clump at the bottom end of the

AGB is all that can be seen of the HB in metal-rich clusters.

### 5.1.6 The cosmic helium abundance

Since the discovery of the cosmic background radiation by Penzias & Wilson (1965), there has been little doubt that most of the He in the Universe was synthesized in the first few minutes of the Big Bang.[7] Hence even the oldest and most metal-poor stars should have a substantial He abundance, and we would very much like to measure this abundance since this would tell us how much He was made in the Big Bang. Unfortunately, the only old stars that are hot enough to excite spectral lines of He are evolved stars that have themselves synthesized additional He. Clearly, it is necessary to measure $Y$ in an unevolved star. One approach is to observe hot, young stars in dwarf, metal-poor galaxies, and then to plot the values of $Y$ that one obtains as a function of the metallicity $Z$ in these stars: the primordial value of $Y$ is obtained by extrapolating $Y(Z)$ back to $Z = 0$. Another technique involves a similar extrapolation of the values of $Y$ and $Z$ that are measured for the interstellar media of metal-poor galaxies. These techniques lead to the value $Y = 0.228 \pm 0.005$ (Pagel *et al.* 1992), which does not differ significantly from the standard value $Y = 0.235$, which is assumed in many stellar-evolution calculations.

Lowering $Y$ at fixed $Z$, raises the ZAMS. For astrophysically plausible values of $Y, Z$, Lebreton (1998) finds that $(\partial M_{\rm bol}/\partial Y)_{Z,T_{\rm eff}} \simeq 3$. Hence, fixing $Z = 0.02$ and lowering $Y$ from 0.28 to 0.23 raises the ZAMS at constant $T_{\rm eff}$ by about 0.15 mag. The connection between $Y, Z$ and the location of the ZAMS provides a method for directly determining the He content of the oldest, most metal-poor stars. Cayrel (1968) analyzed several cool subdwarfs with accurately known distances, particularly Groombridge 1830, and concluded that the ZAMS of these stars lies about $0.75 \pm 0.3$ mag below the Hyades MS in the theoretical CM diagram. Eggen (1973) confirmed this conclusion by extending the photometry to much cooler subdwarfs and showed that, in a plot of $M_V$ versus $(R - I)$, the subdwarfs lie 0.75 mag below the Hyades MS. Eggen concluded that $Y \approx 0.23$ in the subdwarfs. This figure is to be compared to $Y \approx 0.28$ for stars of solar composition. In number ratios, $N({\rm He})/N({\rm H}) \approx 0.075$ for subdwarfs and 0.10 in solar-composition stars.

### 5.1.7 Simple numerical relations

The theory of stellar structure relies heavily on the results of large-scale computations. We now describe a few numerical results that are simple enough to be remembered and powerful enough to be valuable when making

---

[7] In fact Gamow (1948) predicted the existence and temperature of the background radiation precisely from the consideration that most He should have been synthesized in the Big Bang.

order-of-magnitude estimates of the way in which a stellar system should evolve.

The models of Bressan *et al.* (1993) imply that on the ZAMS a star of mass $\mathcal{M}$ has luminosity[8]

$$\frac{L_{\rm MS}}{L_\odot} \propto \begin{cases} 81(\mathcal{M}/\mathcal{M}_\odot)^{2.14} & \text{for } \mathcal{M} \gtrsim 20\,\mathcal{M}_\odot, \\ 1.78(\mathcal{M}/\mathcal{M}_\odot)^{3.5} & \text{for } 2\,\mathcal{M}_\odot < \mathcal{M} \lesssim 20\,\mathcal{M}_\odot, \\ 0.75(\mathcal{M}/\mathcal{M}_\odot)^{4.8} & \text{for } \mathcal{M} \lesssim 2\,\mathcal{M}_\odot. \end{cases} \qquad (5.5)$$

Figure 5.4 shows that as it sits on the MS a star's luminosity steadily increases from this ZAMS value.

The **MS lifetime**, $\tau_{\rm MS}$, of a star is fixed by the length of time that its luminosity can be supported by thermonuclear conversion of H to He. The fusion of four protons into one He nucleus releases an energy of 26.7 MeV, of which about 25 MeV is injected into the burning region rather than radiated in neutrinos.[9] Thus the conversion of a mass $\Delta\mathcal{M}$ of hydrogen releases $E = 0.0067\Delta\mathcal{M}c^2$ of useful energy. If a fraction $\alpha$ of the total mass of a star can be converted, then $\tau_{\rm MS} = (0.0067\alpha\mathcal{M}c^2/L)$. Detailed computations show that, when about one-tenth of the stellar mass has been converted, the star evolves rapidly away from the MS. Hence, to order of magnitude, we find

$$\tau_{\rm MS} \sim 10\left(\frac{\mathcal{M}}{\mathcal{M}_\odot}\right)\left(\frac{L}{L_\odot}\right)^{-1} \text{Gyr}. \qquad (5.6)$$

Stars of initial mass $\mathcal{M}_{\rm i} < \mathcal{M}_{\rm HeF}$ have bolometric luminosity $L_{\rm HB} \simeq 50\,L_\odot$ while on the HB (Figure 5.5), and they remain there until they have burnt their He cores, which have mass $\mathcal{M}_{\rm He} = 0.45\,\mathcal{M}_\odot$. Typically half of the He will be converted to C and most of the other half to O. Converting a mass $\Delta\mathcal{M}$ of He to half C and half O releases $7.2 \times 10^{-4}\Delta\mathcal{M}c^2$ of energy. Hence the time these stars spend on the HB is of order (cf. Figure 5.5)

$$\tau_{\rm HB} \simeq \frac{7.2 \times 10^{-4} \times 0.45\,\mathcal{M}_\odot c^2}{50\,L_\odot} \sim 0.1\,\text{Gyr}. \qquad (5.7)$$

A star of initial mass $\mathcal{M}_{\rm i} > \mathcal{M}_{\rm up}$ converts $\sim 1.4\,\mathcal{M}_\odot$ to Fe before exploding as a supernova. Conversion of a mass $\Delta\mathcal{M}$ of H to Fe releases about $0.0085\Delta\mathcal{M}c^2$ of useful energy. This can sustain the star at luminosity $L = L_3 \times 10^3\,L_\odot$ for a time

$$\tau_{\rm SG} = \frac{0.0085 \times 1.4\,\mathcal{M}_\odot c^2}{1000 L_3\,L_\odot} \simeq 0.18 L_3^{-1}\,\text{Gyr}. \qquad (5.8)$$

---

[8] Tout *et al.* (1996) give more elaborate and accurate analytic fits to the function $L(\mathcal{M}, Z)$

[9] The fraction of energy carried off by neutrinos depends in detail on the temperature and density in the burning region.

More precise results are given in Table 5.2.

Since most of the support of a white dwarf derives from electron-degeneracy pressure, which is independent of temperature, the radius of a white dwarf is approximately constant as it cools. Hence the white-dwarf cooling line in the CM diagram is given by

$$L \propto T_{\text{eff}}^4 \quad \text{or} \quad M_{\text{bol}} = -10 \log T_{\text{eff}} + \text{const.} \tag{5.9}$$

These constant-radius lines run roughly parallel to the MS. The observed positions of white dwarfs in the CM diagram – see Figure 6.17 – are in good agreement with theory (Cool, Piotto & King 1996).

### 5.1.8 Star formation

The earliest stages in the life of a star lie beyond the scope of the conventional theory of stellar structure, which assumes that the star is at all times in hydrostatic equilibrium.[10] This is one of the reasons why the theory of star formation is much less well developed than that of quasi-static stars. Other factors that make it hard to develop a convincing theory of star formation include the complex nature of the interstellar medium from which stars form (Chapter 8), which makes it hard to know what initial conditions to use, and the likely importance of magnetic fields for the formation process, which make it essential to treat the problem as one in magnetohydrodynamics and to calculate accurately the degree of residual ionization at each point in a collapsing cloud. Moreover, stars are observed to form as members of binaries, which are likely themselves to be members of an association or a cluster, and a star's interactions with its neighbors is likely to be crucial for its early development.

For all these reasons we cannot present an authoritative account of the formation of stars. Nonetheless, given the importance of star formation for galactic astronomy it would be inappropriate to pass over the topic in complete silence. We therefore now sketch in broad outline how stars probably form, and draw attention to the key physical processes involved.

Clusters of young stars are invariably associated with dense interstellar clouds (see §6.2), and, as observations of external galaxies show (see §5.1.8), they are often also associated with spiral arms. From this coincidence we conclude that stars form from dense interstellar gas, probably after the gas has been compressed in a large-scale shock front.

The internal dynamics of dense interstellar clouds are not well understood although spectacular advances are currently being made in the observational study of these complex systems. Gravity pulls them together, but some combination of turbulent energy and magnetic energy resists gravity

---

[10] In the theory of pulsating stars, hydrodynamic effects are taken into account, but the configurations considered are near-equilibrium ones.

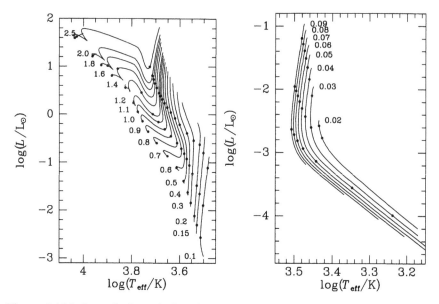

**Figure 5.10** Left panel: the evolution of stars massive enough to burn H on the MS before they reach the MS. Each curve is labeled with the object's mass in solar units. At first these objects are fully convective and descend almost vertically along the Hayashi track. Stars with masses $\mathcal{M} \gtrsim 0.3\,\mathcal{M}_\odot$ eventually cease to be fully convective and move sharply leftwards towards the ZAMS. The positions reached by stars after 1, 10, and 100 Myr from birth are marked with dots. Right panel: the equivalent figure for low-mass objects. At first the descent of these objects is slowed by the combustion of Li and D. Once an object has exhausted its D, it moves more rapidly towards lower temperatures and luminosities. [From data published in D'Antona & Mazzitelli (1994)]

with surprising effectiveness. At the densest points within clouds the ionization level of the material, which is governed by the fluxes of ionizing radiation and cosmic-ray particles (see §8.1.4), can become small, and the coupling between the magnetic field and the gas weakens. This decoupling allows the magnetic energy, which plays a large role in supporting the material against gravitational collapse, to dissipate. As magnetic energy is dissipated, the material becomes more dense, the ionization level falls still further and the magnetic energy ebbs away even faster. A **protostar** has formed.

Imagine a protostar in essentially free fall. As the protostar collapses, gravitational energy is released and must be radiated away. Because the material is initially nearly transparent to infrared radiation, radiative energy loss is efficient, the luminosity rises rapidly, and the temperature remains quite low (10 K to 20 K). Eventually the material becomes opaque in a dense core; this opacity traps the radiation and allows the core to heat. As the core heats, pressure builds and arrests the collapse. The core continues to heat until its temperature rises to about 2000 K, at which point molecular hydrogen dissociates into atomic hydrogen. This dissociation provides an energy sink and triggers another free-fall collapse of the core, which continues to heat until the hydrogen not only dissociates but also ionizes. The newly freed electrons trap radiation rather efficiently. Pressure balance can again be achieved, and core collapse halts. At this point the luminosity of the star is sufficiently large for it to be fully convective – see Box 5.3. Hence in the CM diagram it lies upon its Hayashi line – see Figure 5.10.

Once a star has reached its Hayashi line, it can achieve a state of hydrostatic equilibrium. It then moves down the Hayashi line on a much longer timescale – this downward motion is further slowed by nuclear burning of such light isotopes as deuterium and lithium. In stars more massive than $\mathcal{M} \sim 0.3\,\mathcal{M}_\odot$, convection eventually peters out in the core, and, as the size of the radiative core grows, the track turns sharply to the left – see Figure 5.10. Subsequent evolution of these more massive objects occurs with the star contracting at nearly constant luminosity and ever-increasing effective temperature. Ultimately, the temperature in the core rises to the point at which thermonuclear fusion of hydrogen into He becomes possible, and the star settles on to the MS. The pre-MS evolution of a star proceeds rather rapidly – for a $1\,\mathcal{M}_\odot$ object it takes $\sim 75\,\mathrm{Myr}$.

Stars with masses in the interval $(0.08\,\mathcal{M}_\odot, 0.3\,\mathcal{M}_\odot)$ settle on to the MS while still fully convective, so they continue to move vertically in the CM diagram until they reach the ZAMS. Stars with masses $\mathcal{M} \lesssim 0.08\,\mathcal{M}_\odot$ run out of fuel once their D has been burnt, and then slide down and to the right in the MS as Figure 5.10 shows.

Many young stars are observed to be surrounded by differentially rotating disks of gas and dust. In fact, the solar system is presumed to have formed from such a disk, and T-Tauri stars (Table 3.5) possess disks. Gravitational energy released as the material of the disk accretes on to the central star can make a significant contribution to the luminosity of such systems. Some of this energy is channeled into the formation of jets of material that is ejected along the rotation axis of the disk. The origin and astrophysical significance of these jets are not at all well understood, but jets seem to be a generic by-product of accretion (see, for example, §4.6).

### 5.1.9 The initial mass function

A fundamental quantity for studies of galaxy formation and evolution is the **initial mass function (IMF)**, $\xi(\mathcal{M})$, that specifies the distribution in mass of a freshly formed stellar population. Specifically, just after a burst of star formation, let there be

$$dN = N_0 \xi(\mathcal{M})\, d\mathcal{M} \qquad (5.10)$$

stars with masses in $(\mathcal{M}, \mathcal{M} + d\mathcal{M})$. The normalizing constant $N_0$ depends on both the magnitude of the burst and the way that $\xi$ is normalized. Since very large numbers of inconspicuous low-mass stars are likely to be formed, it makes more sense to normalize $\xi$ in terms of the total mass of the new-born stars rather than their total number. Therefore we normalize $\xi$ such that

$$\int \mathrm{d}\mathcal{M}\,\mathcal{M}\xi(\mathcal{M}) = \mathcal{M}_\odot. \tag{5.11}$$

With this normalization, $N_0$ is the number of solar masses contained in the star-burst.

We have no *a priori* reason to suppose that $\xi$ is a universal function that applies to all starbursts. However, the available data are consistent with $\xi$ having been the same at $\mathcal{M} \lesssim \mathcal{M}_\odot$ for starbursts that have occurred under remarkably different circumstances over the entire age of the Galaxy.

The procedure for determining $\xi$ is as follows. First we determine the luminosity function, $\Phi(M)$, for MS stars in a cluster or the solar neighborhood – see §3.6. Then we correct this luminosity function for the effects of stellar evolution. If the population under study is coeval (i.e., all its stars were born at the same time), there is no correction to make. If, by contrast, the star-formation rate within the population has been roughly constant, as is thought to be the case in the solar neighborhood, we estimate the **initial luminosity function** $\Phi_0(M)$ to be

$$\Phi_0(M) = \Phi(M) \times \begin{cases} t/\tau_{\mathrm{MS}}(M) & \text{for } \tau_{\mathrm{MS}}(M) < t \\ 1 & \text{otherwise.} \end{cases} \tag{5.12}$$

Here $t$ is the time since the population started to form and $\tau_{\mathrm{MS}}(M)$ is the MS lifetime of a star of absolute magnitude $M$. The factor $t/\tau_{\mathrm{MS}}(M)$ corrects for the fact that the only stars of magnitude $M$ that we see are those that formed in the last fraction $\tau_{\mathrm{MS}}(M)/t$ of the population's life.[11] Once the initial luminosity function $\Phi_0(M)$ has been determined, we have

$$\xi(\mathcal{M}) = \frac{\mathrm{d}M}{\mathrm{d}\mathcal{M}}\Phi_0[M(\mathcal{M})], \tag{5.13}$$

where $M(\mathcal{M})$ specifies the relationship between mass and absolute magnitude for stars of the appropriate type.

The function $M(\mathcal{M})$ can in principle be determined from either theory or observation. Theoretically, we obtain $M(\mathcal{M})$ by coupling stellar-atmosphere models to models of a MS stars of various masses and appropriate metallicities. This procedure leads to reliable results for most stellar masses, but becomes difficult and unreliable for masses $\lesssim 0.5\,\mathcal{M}_\odot$ – see D'Antona & Mazzitelli (1994), D'Antona & Mazzitelli (1996) for details. One problem is that low-mass stars are slow to settle on to the MS, so one may need to include models that lie significantly above the MS. A more serious problem

---

[11] See Problem 5.8 for the correction factor appropriate to an exponentially declining star-formation rate.

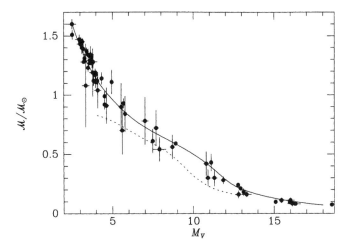

**Figure 5.11** The data points show the masses of MS binary components with the given absolute magnitudes $M_V$. The full curve was derived by a complex procedure that takes into account both these data and the solar-neighborhood luminosity function shown in Figure 3.12 (Kroupa, Tout & Gilmore 1993). The dashed curve shows the $\mathcal{M}(M_V)$ relation for metal-weak stars. [After Kroupa, Tout & Gilmore (1993) with additional data from Henry & McCarthy (1993) and Brewer *et al.* (1993)]

is that the atmospheres of cool stars have extremely rich spectra that deviate significantly from that of a black body – Bessell (1991) shows sample spectra. Simulating the complexities of these spectra, which depend not only on the details of chemical abundances but also on the abundances of molecules such as TiO and $H_2O$, is a highly complex business. Finally, for $\mathcal{M} \lesssim 0.3\,\mathcal{M}_\odot$ the star is fully convective, and this adds to the complexity of the modeling.

In view of these complexities in the theoretical determination of $M(\mathcal{M})$, this relation is often obtained empirically from binary stars. Figure 5.11 shows the empirical data, which are sparse and subject to significant observational uncertainty.

These uncertainties are unfortunate since we see from equation (5.13) that $\xi$ and $\Phi_0$ are related by the *derivative* of the $M(\mathcal{M})$ relation. One cannot hope to constrain this derivative strongly with data of the quality shown in Figure 5.11, especially as theory predicts that the $M(\mathcal{M})$ relation should reflect complex physics and therefore be itself complex. Kroupa, Tout & Gilmore (1990) argue that the natural procedure in these circumstances is to assume that $\xi(\mathcal{M})$ is a simple function and to choose $M(\mathcal{M})$ such that (i) it is compatible with the empirical data shown in Figure 5.11, and (ii) it reproduces $\Phi_0(M)$, wiggles and all. The full curve in Figure 5.11 shows the $M_V(\mathcal{M})$ relation that emerges from this procedure (Kroupa, Tout & Gilmore 1993).

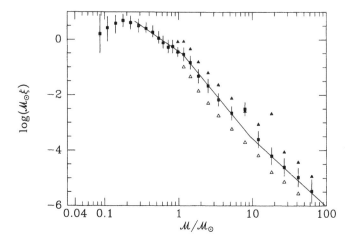

**Figure 5.12** The IMF from Scalo (1986). For masses $\mathcal{M} > \mathcal{M}_\odot$ three sets of points are shown, each set being for a different assumption about the ratio of the current rate of star formation to its average over the lifetime of the solar neighborhood. The curve defined by the points that are based on the assumption of a constant star-formation rate (squares) can be approximated by the three power-law segments defined in equation (5.16).

Why should $\xi(\mathcal{M})$ be a simple function and $\Phi_0(M)$ a complicated one? The argument is that star formation is a chaotic process that proceeds through densities and temperatures that vary by orders of magnitude. It is unlikely that such a scrambling process bears the imprint of any particular length or energy scale, so $\xi(\mathcal{M})$ should be rather a featureless function. $\Phi_0(M)$ is, in effect, this same function reflected in the distorting mirror of the function $M_V(\mathcal{M})$. As we have seen, this last function is rich in structure, and we expect this structure to be reflected in $\Phi_0(M)$.

The prototypical simple function is a power law, so $\xi$ is frequently assumed to be a simple power of $\mathcal{M}$. In a classic study Salpeter (1955) concluded that the then-available evidence pointed to $\xi$ having the form

$$\xi(\mathcal{M}) \propto \mathcal{M}^{-2.35} \quad \text{(Salpeter IMF)}. \tag{5.14}$$

Since for a power law,

$$\xi \propto \mathcal{M}^{-\alpha}, \tag{5.15}$$

equation (5.11) predicts that the total mass involved in a starburst diverges at low masses for $\alpha > 2$ and at high masses for $\alpha < 2$, it is not surprising that recent studies imply that $\xi$ tends to steepen with increasing mass, so that the effective value of $\alpha$ is less than 2 at small masses and greater than 2 at large masses. Figure 5.12 shows the IMF derived by Scalo (1986). For masses $\mathcal{M} \gtrsim 0.2\,\mathcal{M}_\odot$, Scalo's data can be adequately fitted by three power-law segments:

$$\xi(\mathcal{M}) \propto \begin{cases} \mathcal{M}^{-2.45} & \text{for } \mathcal{M} > 10\,\mathcal{M}_\odot, \\ \mathcal{M}^{-3.27} & \text{for } 1\,\mathcal{M}_\odot < \mathcal{M} < 10\,\mathcal{M}_\odot, \quad \text{(Scalo IMF).} \\ \mathcal{M}^{-1.83} & \text{for } \mathcal{M} < 0.2\,\mathcal{M}_\odot \end{cases} \tag{5.16}$$

Kroupa, Tout & Gilmore (1993) likewise advocated a three-power form for $\xi$, but one that falls off much more steeply at the high-mass end:

$$\xi(\mathcal{M}) \propto \begin{cases} \mathcal{M}^{-4.5} & \text{for } \mathcal{M} > 1\,\mathcal{M}_\odot, \\ \mathcal{M}^{-2.2} & \text{for } 0.5\,\mathcal{M}_\odot < \mathcal{M} < 1\,\mathcal{M}_\odot, \\ \mathcal{M}^{-1.2} & \text{for } \mathcal{M} < 0.5\,\mathcal{M}_\odot. \end{cases} \tag{5.17}$$

In §8.2.8 we shall encounter evidence that the *average* IMF within external galaxies has $\alpha \lesssim 2.5$ even at the highest masses. This result is not necessarily incompatible with equation (5.17), however, because the solar neighborhood could be deficient in massive stars relative to the Galaxy as a whole.

Note that $\xi(\mathcal{M})$ continues to increase with declining mass down to the smallest masses for which reliable data are available; the peak in $\Phi(M)$ at faint magnitudes seen in Figure 3.12 arises because the luminosity of a MS star of mass $\mathcal{M}$ falls precipitately as the minimum mass for hydrogen burning, $\mathcal{M} \sim 0.08\,\mathcal{M}_\odot$ is approached, with the result that the stars with masses in a small range near $0.1\,\mathcal{M}_\odot$ are strung out over a huge interval in absolute magnitude $M$. Unfortunately, neither observation nor theory is able to determine with any accuracy the function $M(\mathcal{M})$ that describes this distribution, so it is currently impossible to make any definitive statement about the form of $\xi(\mathcal{M})$ for $\mathcal{M} \lesssim 0.2\,\mathcal{M}_\odot$.

### 5.1.10 Pulsating stars

All stars are to a greater or lesser degree variable because stars tend to oscillate, and these oscillations give rise to variations in luminosity. The amplitude of a star's oscillations is usually very small, so that the associated variations in luminosity are tiny – solar-type stars vary in luminosity by of order a micro-magnitude. The study of the small-amplitude stellar oscillations, or **asteroseismology**, has become an important branch of astronomy because the oscillations involve the entire star, so from them one can glean information about the deep stellar interior. However, we focus here on a minority of stars that oscillate with much larger amplitudes, since the study of these stars has had a profound influence on our understanding of Galactic structure and cosmology. Readers who wish to learn about asteroseismology are referred to Brown & Gilliand (1994).

Like a bell, a star is capable of oscillating at a number of discrete frequencies – the frequencies of its normal modes. The normal modes of a star fall into two broad classes, **p-modes** and **g-modes**. For our purposes it suffices to focus on spherical p-modes. A p-mode involves the star's atmosphere breathing in and out, and is associated with substantial pressure fluctuations, which is the origin of the 'p' in the name of this class of modes.[12] When a star oscillates in its fundamental p-mode, all points in the atmosphere move in and out together and there is no point at which the pressure fluctuations vanish throughout an oscillation. By contrast, when the star oscillates in its **first overtone mode**, some points in the atmosphere move inwards while others are moving outwards, and there is a radius within the star at which the pressure fluctuations vanish throughout an oscillation. Very crudely, the fundamental p-mode is a radially propagating sound wave that is confined within the star with half its wavelength equal to the stellar diameter, while the first overtone mode is a similar sound wave that has three halves of its wavelength equal to the diameter of the star.[13] The light curves of stars that pulsate in the first overtone mode tend to be rather sinusoidal, while those of fundamental-mode pulsators are frequently of larger amplitude and highly asymmetrical in form.

Whereas low-amplitude stellar oscillations such as those observed in the Sun are excited by the random pressure and velocity fluctuations that are associated with convection, most large-amplitude pulsations of stars are excited by an instability, the **$\kappa$-mechanism**, which was identified by Eddington. The physics underlying this mechanism is that the opacity, $\kappa$, of material can increase rather than decrease when the material is heated. As the following argument shows, this behavior enables the atmosphere of a star to function as a heat engine.

At one point in the cycle a layer of material loses support against the star's gravity and falls downwards. This downward motion tends to compress the layer, with the result that it heats up and becomes more opaque to radiation. Since radiation now diffuses more slowly through the layer, heat builds up below it. In consequence of this build-up of heat, the pressure rises below the layer and pushes the layer back outwards. As it moves upwards, the layer expands, cools, and becomes more transparent to radiation. This increased transparency allows energy to escape freely from below the layer, with the result that the layer loses pressure support. The layer now falls back downwards and the cycle repeats. In effect, the stellar envelope behaves like a steam engine, with radiation playing the rôle of steam, the layer the rôle of the piston, and the anomalous behavior of the opacity the rôle of the valve.

Using an order-of-magnitude analysis, which in essence argues that the pulsation period should be of the order of the time required for a sound wave to move through the star (see Problem 5.6), one can show that a relationship

---

[12] The 'g' of g-mode stands for 'gravity'.

[13] The pressure fluctuations must vanish at the stellar edge and be maximal at the centre, where there must be zero displacement.

of the form

$$P \left( \frac{\bar{\rho}}{\bar{\rho}_\odot} \right)^{1/2} = Q \tag{5.18}$$

exists. Here, $P$ is the period, $\bar{\rho}$ denotes the mean density of the star, and $Q$, the **pulsation constant**, is a number of order $(G\bar{\rho}_\odot)^{-1/2} \simeq 1\,\mathrm{hr}$. A simple analysis shows that at constant effective temperature, $T_{\mathrm{eff}}$, the mean density of a star decreases with increasing luminosity – see Problem 5.7. Hence, if we apply equation (5.18) to stars of similar $T_{\mathrm{eff}}$, we predict that they should obey a **period-luminosity relation** (**PL relation**) in the sense that more luminous stars should have longer pulsation periods. Henrietta Leavitt (1912) discovered that the Cepheids obey just such a relation. This was a major breakthrough because if the PL relation could be calibrated properly – see Box 5.5 – one could deduce the distance to any Cepheid simply by determining its period, inferring its luminosity from the PL relation and comparing this with its apparent brightness.

As Figure 5.1 shows, there is an entire strip of the CM diagram – the instability strip – within which stars pulsate. The $\kappa$ mechanism drives pulsations in stars that lie within this strip because in these stars the ionization of $\mathrm{He}^+$ to $\mathrm{He}^{2+}$ causes $\kappa$ to increase with temperature at radii that are strategically placed for the excitation of the fundamental and/or first overtone modes. Stars in the instability strip play a central rôle in astronomy because their very distinctive light variations allow them to be identified unambiguously even in a crowded field of stars, and their absolute magnitudes can be determined from their periods. A great deal of information regarding both the physical principles and applications of pulsating stars may be found in Cacciari & Clementini (1990) and Gautschy & Saio (1995), (1996). Here we summarize the properties of the most important stars that lie within this instability strip. We work from the top of the strip downwards.

**Classical Cepheid variables**   Stars with $\mathcal{M} \gtrsim 3\,\mathcal{M}_\odot$ make excursions to the left in the CM diagram after reaching the RGB – see Figure 5.2. More massive stars move further to the left, and a star with $\mathcal{M} \gtrsim 5\,\mathcal{M}_\odot$ is liable to pass into the classical instability strip on such an excursion and become a classical Cepheid variable. The lines in Figure 5.13 indicate the extent of the instability strip, as predicted by the models of Chiosi, Wood & Capitanio (1993). Notice that the strip is predicted to become wider with brightening absolute magnitude $M_V$. The dots in Figure 5.13 give the location in the CM diagram of the 98 Galactic Cepheids with well determined reddening. All fall in or near the theoretically predicted instability strip.

Strictly, the sobriquet 'classical' is necessary to distinguish classical Cepheids from W Virginis stars, which are often called 'Population II Cepheids' and are discussed below.[14] We shall often drop the 'classical' on the understanding that a plain Cepheid variable is a classical Cepheid. Since

---

[14] Classical Cepheids are also called $\delta$ Cephei stars.

---

### Box 5.5: How the Universe Doubled in Size During 1952

The Cepheid PL relation discovered by Leavitt (1912) was a linear correlation between $\log P$ and the apparent magnitudes of Cepheids in the Small Magellanic Cloud. Before this relation could be exploited for distance estimation, it had to be expressed in terms of absolute magnitudes by determining the zero-point. This problem has an interesting history – see Struve & Zebergs (1962), Chapter 15.

No Cepheid was close enough to the Sun to have a measurable trigonometric parallax, so for the calibration Shapley used the statistical parallaxes (see §2.2.4) of Cepheids. Unfortunately, interstellar absorption, which is always significant along lines of sight to Cepheids, was ignored, so the estimated Cepheid luminosities were much too small. This problem was compounded by the assumption that Cepheids have the same properties as W Virginis stars in globular clusters. With this assumption it was found that the RR Lyrae stars and Cepheids all fell along a single apparently well-defined period-luminosity law. This harmony lulled astronomers into a false sense of security about the accuracy of the results.

The problem was not clearly resolved until 1952 when Baade discovered that 200-inch telescope photographs of M31 failed to reveal RR Lyrae variables in the globular clusters at the magnitude predicted by the then-adopted distance scale, but instead showed only the brightest cluster giants. Inasmuch as the absolute magnitudes of the RR Lyrae stars had been determined independently from statistical parallaxes and cluster CM diagrams, he realized that the error lay in the absolute magnitudes of the Cepheids, which had been set about 1.5 mag too faint. Making this revision doubled the estimated distance to, and the size of, M31 (making it comparable in size to our Galaxy), and the revision brought its globular-cluster luminosities into line with those in our Galaxy. Furthermore, it doubled the estimated size of the Universe, because the extragalactic distance scale had been calibrated using Cepheids – see Chapter 7.

---

Cepheids are massive, they are young stars and tend to lie in spiral arms within the Galactic plane (§10.4.5).

Classical Cepheid light curves are strictly periodic. In the $B$ band they usually show an abrupt rise by $\sim 1$ mag (in about 20% of the period) followed by a slower decline (over about 50% of the period). At infrared wavelengths their light curves show smaller-amplitude and more nearly sinusoidal variations. At maximum light (phase $= 0.0$), the star has its highest temperature, earliest spectral type, and greatest outward radial velocity; at minimum light

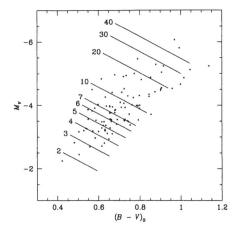

**Figure 5.13** The Cepheid instability strip in the CM diagram. The points represent the positions of individual Galactic Cepheids from Fernie (1990). The lines indicate the extent of the instability strip as calculated from stellar pulsation models (Chiosi, Wood & Capitanio 1993). Each line marks the locus of Cepheids with a particular period $P$, which is given in days.

(phase $\approx 0.75$), it has its lowest temperature, latest spectral type, and greatest inward velocity. The star achieves its maximum radius on the descending part of the light curve (phase $\approx 0.4$) and its minimum radius shortly after the next light-rise begins. The absolute magnitudes of Cepheids lie in the range $-6 \lesssim M_V \lesssim -2$ and their periods $P$ range from several days to some hundreds of days.

In many Cepheids both the fundamental and the first overtone mode are excited. Two independent mass estimates can be made for these stars: stellar-evolution theory provides an estimate from the star's location within the CM diagram, while pulsation theory provides an independent estimate from the frequencies at which the star oscillates. For many years these two estimates tended to conflict, but the introduction of more accurate opacities in the early 1990s resolved these discrepancies.

As we saw above, to a first approximation Cepheids satisfy a period-luminosity (PL) relation that reflects the variation with mass $\mathcal{M}$ of the frequency of the fundamental mode. Since the detailed structure of a star depends on its metallicity, it is evident that the PL relation should depend on metallicity. Moreover, since the color of a star of given mass varies with both temperature and metallicity, it follows that in a more precise treatment the periods, luminosities and colors of Cepheids should obey a period-luminosity-color, or PLC, relation. The non-zero slope of lines of constant $P$ in Figure 5.13 quantifies this expectation that $P$ should be a function of color as well as luminosity.

The empirical determination of the PLC relation is problematic, however, because Cepheids occur in regions of significant interstellar absorption. Hence non-negligible reddening corrections to colors are required, and it is difficult to make these with sufficient accuracy. Moreover, the full PLC relation proves to be a non-linear function of $P$, $M_V$ and $(B - V)_0$ – see Caldwell & Coulson (1986) for details.

On account of the complexity of the PLC relation, recent work has concentrated on PL relations at infrared wavelengths. Observations of Cepheids at longer wavelengths have several advantages. The two most important of these are:

1. In the IR reddening corrections are small, and their uncertainties are negligible.
2. The PL relation is both tighter and steeper in the IR than in $B$. Hence observations in the $K$ band lead to very much smaller errors in $L$ for a given error in $P$ than do observations in the $B$ band.

Since Cepheid light curves have smaller-amplitude variations in the IR than at optical wavelengths, IR observations typically lead to larger errors in $P$. On the other hand, the smallness of the fluctuations in an IR band such as $K$ reduces the uncertainty in the estimate of the period-average of the apparent magnitude that one obtains from a small number of randomly phased measurements. Optimal results are obtained by measuring in $B$ or $V$ as well as in $K$ so that the period can be accurately determined in $B$ or $V$ while the apparent magnitude is determined in $K$. A detailed discussion of the relative benefits of optical and infrared observations of Cepheids can be found in §3 of Jacoby *et al.* (1992).

Feast & Walker (1987) have collated the results from a large number of PL determinations, and they find a best estimate for the $V$-band absolute magnitude of a Cepheid of period $P$ to be given by

$$\langle M_V \rangle = -2.78 \log(P/10\,\mathrm{d}) - 4.13 \quad \text{with scatter} \quad \sim 0.3\,\mathrm{mag}, \qquad (5.19)$$

where angled brackets again indicate the magnitude obtained by averaging the flux over a period. From a study of Cepheids in the Magellanic Clouds, Madore & Freedman (1991) find these PL relations:

$$\langle M_B \rangle = -(2.43 \pm 0.14) \log(P/10\,\mathrm{d}) - 3.50 \pm 0.06 \text{ with scatter } \sim 0.36\,\mathrm{mag}$$
$$\langle M_V \rangle = -(2.76 \pm 0.11) \log(P/10\,\mathrm{d}) - 4.16 \pm 0.05 \text{ with scatter } \sim 0.27\,\mathrm{mag}$$
$$\langle M_R \rangle = -(2.94 \pm 0.09) \log(P/10\,\mathrm{d}) - 4.52 \pm 0.04 \text{ with scatter } \sim 0.22\,\mathrm{mag}$$
$$\langle M_I \rangle = -(3.06 \pm 0.07) \log(P/10\,\mathrm{d}) - 4.87 \pm 0.03 \text{ with scatter } \sim 0.18\,\mathrm{mag}$$
$$\langle M_K \rangle = -(3.42 \pm 0.09) \log(P/10\,\mathrm{d}) - 5.70 \pm 0.04 \text{ with scatter } \sim 0.13\,\mathrm{mag}.$$
$$(5.20)$$

**Mira variables**     As they approach the top of the AGB, many stars become Mira variables – see Table 3.5. Although Miras lie a little to the red of the instability strip, their long-period ($\gtrsim 80\,\mathrm{d}$), large-amplitude pulsations are thought to be driven by the $\kappa$ mechanism within zones of H and $He^+$ ionization just as Cepheid pulsations are. However, their pulsations are not well understood because their atmospheres are highly convective, and there is no definitive theory of pulsations in the presence of convection. In particular, it is unclear whether it is the fundamental or first overtone which is the excited mode in these stars (Gautschy & Saio 1996).

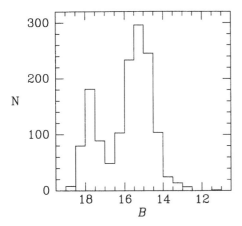

**Figure 5.14** The distribution with respect to apparent magnitude of RR Lyrae variables in a field just south of the Galactic center. The main peak is associated with RR Lyraes in the Galactic bulge, and the subsidiary peak at fainter magnitudes is associated with stars in the Sagittarius Dwarf Galaxy, that lies behind the Galactic center. [After Alard (1996) from data kindly supplied by C. Alard]

The periods of Miras increase with luminosity; Feast *et al.* (1989) have determined the PL relation for Miras in the LMC. For a distance to the LMC of 50 kpc they find

$$\langle M_K \rangle = -(3.57 \pm 0.16) \log(P/1\,\mathrm{d}) + 1.21 \pm 0.39 \,. \tag{5.21}$$

Within the Galaxy, the periods of Miras are correlated with their kinematics in the sense that longer-period Miras are more strongly confined to the plane (Jura & Kleinmann 1992). Since hundreds of millions of years are required for a star to change its orbit significantly, and Miras certainly do not live so long, this correlation implies that any change in the period of a given Mira during its lifetime must be small.

**W Virginis stars**    As a low-mass, metal-poor star evolves away from the HB on the exhaustion of He in its core, it may enter the instability strip and become a **Population II Cepheid**. The prototype object of this class is W Virginis, so Population II Cepheids are frequently called **W Virginis** stars – we shall use this name to avoid confusion with the astronomically more important classical Cepheids. Since they are halo stars (see §10.5), they are found at high Galactic latitudes and in globular clusters, and they are concentrated toward the Galactic center.

These stars are physically much more closely related to RR Lyrae stars than to classical Cepheids. However, their periods, $0.8\,\mathrm{d} \lesssim P \lesssim 30\,\mathrm{d}$, are similar to those of classical Cepheids. They do not satisfy as tight a PL relation as do the classical Cepheids. Gautschy & Saio (1996) discuss W Virginis stars in some detail and give many references to the literature.

**RR Lyrae stars**    As we noted in §5.1.2, RR Lyraes are HB stars that happen to lie in the instability strip; their spectral types are centered around A. They are always old, relatively metal-poor stars with periods $P \simeq 0.5\,\mathrm{d}$. Their current masses are thought to be $\sim 0.8\,\mathcal{M}_\odot$, while their original MS-masses must have been $\sim 1\,\mathcal{M}_\odot$.

RR Lyrae stars with periods $\gtrsim 0.5\,\mathrm{d}$ and $\Delta S \gtrsim 5$ [see equation (5.22) below] are commonly found in globular clusters, in the Galactic halo, and concentrated toward the Galactic center. Hence they are halo stars. RR Lyrae stars with periods $\lesssim 0.4\,\mathrm{d}$ and $\Delta S \lesssim 5$ are concentrated to the Galactic plane and are old disk stars.

Although RR Lyrae stars are much fainter than Cepheids, in regions of low interstellar absorption ('Baade's windows' – see §10.2) they can be seen all the way to the Galactic center. In fact, surveys designed to identify gravitational microlensing events (§2.4) have recently detected large numbers of RR Lyraes along with many other types of variable star (Alcock *et al.* 1995a). The numbers of RR Lyraes detected in windows on the Galactic center peak at a definite apparent magnitude and then fall off – see Figure 5.14. These observations can be used to estimate the Sun's distance from the Galactic center (see §10.6). Moreover, because RR Lyrae stars are so readily identified, they can be sampled in a given field to a high level of completeness, and hence they can be used in stellar-density analyses of the spheroidal component of our Galaxy (see §10.5.2). Finally, RR Lyrae stars are typically only 2 to 3 mag fainter than the brightest stars in globular clusters. Thus they are relatively easy to observe, in addition to being easily identified. Because they have well-defined absolute magnitudes, they provide good distance indicators for clusters, and they allow us to avoid having to observe down to the cluster MS (which is 5 mag fainter).

RR Lyrae stars are classified into **Bailey types** $a$, $b$, $c$, and $d$ according to the shapes of their light curves. Classes $a$ and $b$ have been amalgamated into one class, $RR_{ab}$. These stars have asymmetric (Cepheid-like) light curves with periods $P_{ab} \gtrsim 0.4\,\mathrm{d}$ and an average period $\langle P_{ab} \rangle \approx 0.55\,\mathrm{d}$. In general, these stars have large amplitude luminosity variations, $0.5\,\mathrm{mag} \lesssim \Delta m \lesssim 1.5\,\mathrm{mag}$, and pulsate in the fundamental mode. The $RR_c$ variables have symmetric (practically sinusoidal) light curves with smaller amplitudes, $\Delta m \lesssim 0.5\,\mathrm{mag}$, and shorter periods, $P_c \lesssim 0.4\,\mathrm{d}$ (average $\langle P_c \rangle \approx 0.3\,\mathrm{d}$). These stars pulsate in the first-overtone mode. In $RR_d$ variables both the fundamental and the first overtone modes are excited. A substantial majority of RR Lyrae stars are $RR_{ab}$ with the rest divided about equally between $RR_c$ and $RR_d$. The $RR_{ab}$ and most $RR_d$ variables are halo stars, while the $RR_c$ are old disk stars (see also §10.5).

Because most RR Lyrae stars are metal poor, their metallic line spectral types (classified by the standard criteria) tend to be earlier than the spectral types appropriate to their hydrogen lines. Preston (1959) introduced the parameter

$$\Delta S \equiv 10[\text{spectral type (H lines)} - \text{spectral type (Ca II K lines)}], \quad (5.22)$$

where the spectral types are measured at minimum light in tenths of a spectral class. This parameter correlates closely with the metal content of a star; $\Delta S \approx 0$ for stars of near solar abundances and rises to $\Delta S \approx 10$ to 12 for the most extreme metal-poor stars. $\Delta S$ has been calibrated in terms of [Fe/H] by means of curve-of-growth analyses of field RR Lyrae stars (Butler 1975). A very tight correlation,

$$[\text{Fe/H}] \approx -0.16\Delta S - 0.23, \tag{5.23}$$

is found. This relation can be applied to cluster variables to estimate metal abundances in globular clusters.

There are essentially five independent methods of estimating $M_V$ for RR Lyraes: (1) from the statistical parallaxes (see §2.2.4) of field RR Lyrae stars (Layden et al. 1996); (2) by the calibration of globular-cluster CM diagrams by MS fitting; (3) by combining the apparent magnitudes of RR Lyrae stars in the Magellanic Clouds with the known distance to the Clouds (§7.4.1); (4) the Baade–Wesselink method that is discussed in §7.2.1; and (5) by matching observations to the predictions of complete stellar models. Each method has advantages and disadvantages; the statistical parallaxes suffer from any errors present in the proper motions and from the intrinsic spread produced by variations in metal content among the stars. The problems with MS fitting of cluster CM diagrams are discussed in §6.1.2. Clearly, calibration from RR Lyraes in the Clouds is reliable only if (i) the RR Lyraes in the Clouds are identical to their counterparts in our galaxy, and (ii) the distance to the Clouds can be reliably determined without reference to Galactic RR Lyraes. Both the Baade–Wesselink method and the use of complete stellar models can be compromised by errors in stellar-structure calculations, including poor values for important opacities.

RR Lyrae stars have absolute magnitudes in the range $0.5 \lesssim M_V \lesssim 1$, with a population-weighted mean value of $\sim 0.6$. The observed absolute magnitudes correlate quite closely with metallicity [Fe/H]. Jones et al. (1992) find

$$\langle M_V \rangle = (0.16 \pm 0.03)[\text{Fe/H}] + 1.02 \pm 0.03, \tag{5.24}$$

where angled brackets again indicate the magnitude obtained by averaging the flux over a period. Thus, if the metallicity of an RR Lyrae can be determined from the star's spectrum, the star's absolute magnitude can be predicted to within $\sim 0.15$ mag, and its distance then follows from its apparent magnitude to better than 10%.

As with Cepheids, the dependence of the PL relation on metallicity becomes less important at infrared wavelengths. Longmore et al. (1990) showed that the $K$-band magnitudes of RR Lyraes in a given globular cluster are extremely tightly correlated with $P$. Moreover, the form of this correlation was the same in all the clusters studied, independent of their metallicities. This suggests that RR Lyraes satisfy a universal $K$-band PL relation. Calibration of the relation using the Baade–Wesselink method (§7.2.1) yields (Jones et al. 1992)

$$\langle M_K \rangle = -(2.3 \pm 0.2)\log(P/1\,\text{d}) - 0.88 \pm 0.06 \ . \tag{5.25}$$

This relationship offers a very useful distance measure: the pulsation period is straightforward to measure, and the observed $K$-band flux is almost unaffected by obscuration. However, even with current sensitive infrared detectors, RR Lyrae stars are hard to detect in external galaxies, and they have yet to be used as distance estimators beyond the Local Group.

Equations (5.24) and (5.25) show that the luminosities of RR-Lyraes are correlated with both period and metallicity. These two-variable correlations are projections of the full three-variable correlation (Jones *et al.* 1992)

$$\langle M_K \rangle = -(2.0 \pm 0.3) \log(P/1\,\mathrm{d}) + (0.06 \pm 0.04)[\mathrm{Fe/H}] - 0.7 \pm 0.1. \quad (5.26)$$

## 5.2 Synthesis of the chemical elements

One of the most notable triumphs of theoretical astrophysics has been the provision of a convincing explanation of the origin of the chemical elements. Hydrogen is primordial, and the majority of the He was synthesized during the first three minutes of the Big Bang, but nearly everything else has been manufactured by stars. Hence the chemical compositions of stars and the interstellar medium provide important information about how many stars of different types have been active during the life of the Milky Way and other galaxies. In this section we review our present understanding about the amounts of heavy elements that stars of different types return to the ISM. It is helpful to start by recalling some basic nuclear physics.

### 5.2.1 Basic nuclear physics

An atomic nucleus may be considered to be a bound state of a certain number $Z$ of protons and a certain number $N$ of neutrons. Hence each nucleus occupies a certain point in the $(Z, N)$ plane, and each element is associated with column of fixed $Z$ in this plane. Since it is sometimes useful to distinguish between points in the $(Z, N)$ plane and a nucleus as a physical entity, we shall refer to a point in the $(Z, N)$ plane as a **nuclide**, reserving 'nucleus' for a physical object. The **atomic number** of a nuclide is $A \equiv Z + N$ and one conventionally specifies a particular nuclide by preceding the chemical symbol associated with the $Z^{\mathrm{th}}$ element in the periodic table of the elements with a superscript $A$: for example $^{16}\mathrm{O}$ has $A = 16$ and $Z = 8$ and therefore lies at the point $(8, 8)$ in the $(Z, N)$ plane.

Only a few of the points in the $(Z, N)$ plane are associated with bound nuclear states, and these lie in a band, the **stability band**, that runs roughly parallel to, but centered just above, the line $N = Z$. Many of these bound states are liable to spontaneous radioactive decay into other states – Figure 5.15 shows the diagonal motions within the $(Z, N)$ plane of a nucleus

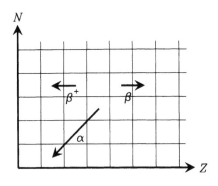

**Figure 5.15** Motion in the $(Z, N)$ plane due to radioactive decay.

that experiences either $\beta^{\pm}$-decay (emission of an electron or positron) or $\alpha$-decay (emission of an $\alpha$-particle; i.e., a He nucleus).

The binding energy per proton or neutron of nuclides, $\epsilon$, varies systematically within the $(Z, N)$ plane. As one moves across the stability band along a line $A = Z + N = \text{constant}$, the absolute value of $\epsilon$ first increases and then decreases. Thus if we imagine each nuclide to lie a distance $|\epsilon|$ below the $(Z, N)$ plane, the stability band takes the form of a valley, and hence is often called the **stability valley**. The bottom of this valley slopes from each end towards the middle. The deepest point in the valley, $^{56}$Fe, lies 8.79 MeV below H at the valley's bottom left end, and 1.21 MeV below $^{238}$U at the valley's top right end. Stars derive much of the energy that they radiate by shifting their protons and neutrons down the valley from H and He towards Fe.

It proves useful to identify a number of groups of nuclides within the $(Z, N)$ plane. Since it is generally only possible to distinguish different isotopes of the same element within the solar system, elements are frequently assigned to the same group as their commonest isotope.

$^{20}$Ne, $^{24}$Mg, $^{28}$Si, $^{32}$S, $^{36}$A and $^{40}$Ca comprise the $\alpha$ **nuclides**. They owe their name to the fact that they can be formed by adding $2,3,\ldots\alpha$ particles to $^{16}$O.[15] These nuclides form both during C- and O-burning by reactions such as

$$^{12}\text{C} + {}^{12}\text{C} \rightarrow {}^{20}\text{Ne} + {}^{4}\text{He}$$
$$^{16}\text{O} + {}^{16}\text{O} \rightarrow {}^{28}\text{Si} + {}^{4}\text{He} \tag{5.27}$$

and through the capture of $\alpha$ particles in reactions such as

$$^{20}\text{Ne} + {}^{4}\text{He} \rightarrow {}^{24}\text{Mg} + \gamma. \tag{5.28}$$

They are **primary nuclides** in the sense that they can be made in a star that starts out with nothing but hydrogen and He, and, moreover, the amount

---

[15] Sometimes the next stable nuclide in this sequence, $^{48}$Ti, is considered to be an $\alpha$ nuclide, but by our definition below it belongs to the iron-peak group.

produced in any given star is essentially independent of the star's initial metallicity. By contrast, a **secondary nuclide** is one that requires for its synthesis the presence of a primary nuclide that was synthesized in an earlier generation of stars.

The very direct routes of equations (5.27) and (5.28) to the formation of the $\alpha$ nuclides are reflected in the abundances of these nuclides: they are all more abundant than their immediate neighbors in the $(Z, N)$ plane. Moreover, the abundances of these nuclides decrease smoothly with increasing atomic number.

$^{23}$Na and $^{27}$Al are the sole stable isotopes of these two odd-$Z$ elements. They are related to the $\alpha$ nuclides in as much as they are products of C and Ne burning in hot stars – see equation (5.29) – and if this process were entirely responsible for their synthesis, they would be primary elements. However, they are also synthesized when He is burned in the presence of $^{16}$O, so they are not primary elements.

The **iron peak nuclides** are those with atomic number in the range $40 < A < 65$ – this range corresponds to the elements Sc, Ti, V, Cr, Mn, Fe, Co, Ni and Cu. The floor of the stability valley is here rather flat, with the consequence that the most stable isotopes of the elements that lie in the iron peak have similar binding energies. The iron-peak nuclides are primary nuclides that form late in the evolution of a star when the core is exceedingly hot. In fact, the temperature is usually so great that nuclei are constantly forming and then splitting apart, with the result that the system is able to come into thermal equilibrium. This fact greatly simplifies the calculation of the relative abundances of the various nuclides, since the probability of each possible configuration is simply proportional to the Boltzmann factor $e^{-E/kT}$, where $E$ is the configuration's energy. The flatness of the valley floor around the iron peak has the consequence that in thermal equilibrium many different nuclides are present in significant numbers.

The **s-process nuclides** lie along the valley floor up towards higher $A$ from the iron peak. The 's' stands for 'slow' for a reason that will emerge shortly. Abundant s-process nuclides include $^{88}$Sr, $^{89}$Y, $^{90}$Zr, $^{138}$Ba, $^{139}$La, $^{140}$Ce, $^{141}$Pr, $^{208}$Pb, $^{209}$Bi.

Certain reactions that take place during nuclear burning release free neutrons. For example, during C-burning a small fraction of colliding C nuclei undergo the reaction

$$^{12}C + {}^{12}C \rightarrow {}^{23}Na + n. \tag{5.29}$$

Because it experiences no Coulomb repulsion on approaching a nucleus, the free neutron will quickly be absorbed by a nucleus, to form a heavier isotope of the same element. For example, it could convert $^{56}$Fe to $^{57}$Fe, which is also stable. Subsequent absorption of free neutrons might convert the $^{57}$Fe nucleus to $^{58}$Fe and then to $^{59}$Fe. Now $^{59}$Fe is unstable (half-life 45 d) and soon $\beta^-$-decays to $^{59}$Co. Absorption of yet another neutron now produces

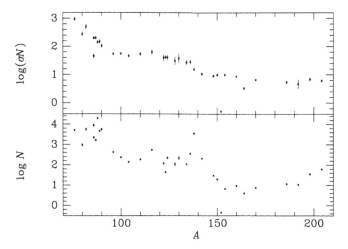

**Figure 5.16** Lower panel: the solar abundances (on an arbitrary scale) of 34 pure s-process nuclides versus atomic number. Upper panel: each abundance multiplied by the corresponding neutron-absorption cross section. [From data published in Käppeler, Beer & Wisshak (1989)]

$^{60}$Co, which is unstable (half-life 330 d) and $\beta^-$-decays to $^{60}$Ni. Proceeding in this way our original $^{56}$Fe nucleus moves ever further up the periodic table. At any given time within the nuclear-burning regions of a star there will be nuclei at all different points in this chain of transmutations, and since all nuclei are exposed to the same flux of neutrons, the time any given stable nucleus spends at a given point in the chain will be inversely proportional to its current cross-section for the absorption of a neutron. Hence when the neutron-generating reactions cease, and with them motion up the chain comes to a halt, the number of nuclei at each point in the chain is inversely proportional to the local neutron-absorption cross-section. Figure 5.16 shows the abundances of the s-process elements before and after multiplication by their neutron-absorption cross-sections. The post-multiplication values are not actually independent of $A$ as our naive analysis suggested would be the case, but they do decline smoothly with increasing $A$, which the raw abundances do not.

The s-process produces secondary elements in the sense that it requires iron-peak elements to be present in a C- or O-burning zone. Since a star synthesizes iron-peak elements only at the very end of its life, after it has largely ceased burning C and O, these iron-peak elements must have been present in the star at its birth, having been synthesized in some other star. Naively one would expect the abundances of secondary elements to be proportional roughly to the square of the abundance of primary elements. For example, suppose the rate for formation of primary elements is constant in time:

$$\frac{da_{\text{prim}}}{dt} = c_1. \tag{5.30}$$

Then given that the rate of formation of secondary elements is given by

$$\frac{da_{\text{sec}}}{dt} = c_2 a_{\text{prim}}, \tag{5.31}$$

where $c_2$ is a second constant, integration of equations (5.30) and (5.31) with respect to time reveals that

$$a_{\text{sec}}(t) = \tfrac{1}{2}c_1 c_2 t^2 = \frac{c_1}{2c_2}a_{\text{prim}}^2(t). \tag{5.32}$$

In §§8.2.6 and 10.4 we shall see that the observed slope of the mean relation between the logarithmic abundances of secondary and primary elements is significantly smaller than two. The likely reason is that the density of 'neutron poisons' increases with metallicity – a **neutron poison** is a nuclide such as $^3$He or $^{14}$N that readily absorbs neutrons. The increasing density of neutron poisons diminishes the neutron flux, and the rate of formation of s-process nuclides is obviously proportional to the product of the neutron flux and the density of heavy seed nuclei.

The **r-process nuclides** lie on the neutron-rich side of the valley floor just above the s-process nuclides – abundant representatives include $^{80}$Se, $^{81}$Br, $^{84}$Kr, $^{128,130}$Te, $^{127}$I, $^{192}$Os, $^{193}$Ir, $^{196,198}$Pt. The 'r' stands for 'rapid' because these nuclides, like the s-process nuclides, form by neutron capture but by *rapid* rather than *slow* neutron capture. The key issue is this: does the first unstable nuclide that is encountered on moving vertically up the $(Z, N)$ plane from the bottom of the stability valley have time to $\beta$-decay to a stable nuclide of higher $Z$ before a further neutron is absorbed? If the answer is 'yes', the nucleus shuffles along the chain of s-process nuclides, which lie near the bottom of the stability valley. If the answer to the question is 'no', the nucleus is driven up the side of the valley, where nuclides are more unstable, until the local half-life to $\beta$-decay becomes comparable to the mean time between neutron captures, and then jogs along the length of the valley. When the neutron-generating reactions cease, each nucleus slides down towards the valley floor by $\beta$-decay, coming to rest at an r-process nuclide. Notice that the abundances of these final nuclides are *not* expected to be inversely proportional to their neutron-capture cross-sections.

The distinction between r- and s-process nuclides is not a sharp one because many nuclides can be synthesized by both processes. The distinction is sharpest when several stable nuclides lie on the neutron-rich side of the valley of stability and are isolated from the floor itself by unstable nuclides. Figure 5.17 illustrates this phenomenon by showing the valley in the region of Sb, where $^{124}$Sn and $^{128}$Te form such islands and respectively ensure that

| Xe 124 | Xe 125 | Xe 126 | Xe 127 | Xe 128 | Xe 129 | Xe 130 | Xe 131 | Xe 132 |
|--------|--------|--------|--------|--------|--------|--------|--------|--------|
|        |        |        |        | s      |        | s      |        | rs     |
| I 123  | I 124  | I 125  | I 126  | I 127  | I 128  | I 129  | I 130  | I 131  |
|        |        |        |        | rs     |        |        |        |        |
| Te 122 | Te 123 | Te 124 | Te 125 | Te 126 | Te 127 | Te 128 | Te 129 | Te 130 |
| s      |        | s      |        | rs     |        | r      |        | r      |
| Sb 121 | Sb 122 | Sb 123 | Sb 124 | Sb 125 | Sb 126 | Sb 127 | Sb 128 | Sb 129 |
| rs     |        | rs     |        |        |        |        |        |        |
| Sn 120 | Sn 121 | Sn 122 | Sn 123 | Sn 124 | Sn 125 | Sn 126 | Sn 127 | Sn 128 |
| rs     |        | r      |        | r      |        |        |        |        |

**Figure 5.17** The $(Z, N)$ plane in the neighborhood of Sb. Stable nuclides have shaded squares. An 'r' indicates that a nuclide can be formed only by the r-process, while an 's' indicates a nuclide that can be formed only by the s-process. Stable nuclides that can be formed by either process are marked 'rs'.

$^{124}$Te and $^{128}$Xe are pure s-process nuclides. These isolated nuclides can be synthesized only by the r-process, while those on the valley floor are synthesized only by the s-process. Nearly all elements heavier than Fe have isotopes that can be synthesized by the s-process, but the s-process sometimes leads to small abundances of these isotopes because they have large neutron-absorption cross-sections.

It is at this stage unclear whether elements produced by the r-process are primary or secondary, because we have not established at what stage in a star's evolution material might be exposed to the large flux of neutrons that is required for the r-process.

The discussion above indicates that four processes contribute to the production of metals: (i) straightforward He- C- and O-burning produces not only $^{12}$C, $^{16}$O but also the $\alpha$ nuclides; (ii) the iron-peak nuclides are formed when nuclei achieve approximate thermal equilibrium in the hot cores of evolved stars; (iii) s-process elements are produced by slow neutron-irradiation of heavy nuclides that were synthesized in an earlier generation of stars; (iv) r-process elements are produced by rapid neutron-irradiation of heavy nuclides that may or may not have been synthesized in an earlier generation of stars. Now we have to understand (a) at what stages of stellar evolution the r-process occurs, and (b) how metals, once formed, are injected into the ISM.

### 5.2.2 Metal production at $\mathcal{M}_i < \mathcal{M}_{up}$

We saw in §5.1.3 that the remnants of stars are less massive, often much less massive, than their progenitor stars. Our interest focuses on the chemical composition of the material that a star returns to the ISM rather than locks up in its remnant. Stars of mass $\mathcal{M}_i < \mathcal{M}_{up}$ form white dwarfs, and these must contain most of the C, O and $\alpha$ elements that the star synthesizes. Thus, in a naive picture, these stars enrich the ISM only with He. However, the phenomenon of carbon stars (see §5.1.2) makes it clear that this picture is over-simplified. At points in their evolutions, stars are liable to dredge the products of nuclear processing up from their interiors into their envelopes, and this material will pass to the ISM when the envelope is ejected.

Spectroscopic analyses of planetary nebulae in the Milky Way and the Magellanic Clouds yield estimates of the abundances of He, N, O, and sometimes C in the ejecta. To determine from these data the mass of each element that was synthesized in the precursor star, we clearly need to know the *difference* between abundances of the ejecta and those characteristic of the ISM in the general vicinity of the planetaries. Unfortunately this difference is small, and correspondingly hard to determine. The difference is largest and easiest to measure in systems with a metal-poor ISM, such as the SMC. Monk, Barlow & Clegg (1988) find that in the SMC [N/H] is larger in planetaries than in HII regions by $\sim 1\,\mathrm{dex}$; in LMC and Galactic planetaries this difference falls to $\sim 0.8$ and $\sim 0.4$, respectively. The enhancements in the abundances of other elements are smaller and uncertain. For example, the enhancement in [O/H] for planetaries is $\sim 0.3$ in the SMC and unmeasurably small in more metal-rich systems. The enhanced production of N to which these observations point is interesting because much of the N that these stars produce is thought to be made when C nuclei absorb free neutrons. The neutrons that are responsible for this transmutation of C into N will at the same time be forming s-process elements. For this reason planetary nebulae are believed to be an important source of s-process elements.

In summary, heavy-element production by stars with $\mathcal{M}_i < \mathcal{M}_{up}$ probably plays an important rôle in the chemical evolution of galaxies, especially early on when the ISM is relatively metal-poor. Unfortunately, the sheer quantity of gas that these stars eject makes it difficult to determine observationally the rate at which these stars pump freshly made metals into the ISM. This difficulty is unfortunate because models of stellar evolution cannot reliably predict the mass of heavy elements that is typically carried into a stellar envelope and then ejected.

### 5.2.3 Supernovae

Supernovae are believed to be the prime sources of both iron-peak and r-process nuclei. They are classified according to their spectra near maximum brightness – see Harkness & Wheeler (1990). The spectra of **type II supernovae** contain hydrogen lines while those of **type I supernovae** do not. Type I supernovae whose spectra show absorption due to $Si^+$ are called **type Ia supernovae**, while all other type I supernovae are classified as **type Ib supernovae**. The spectra of type Ib supernovae usually show He lines and,

as we shall see, they are probably physically more closely related to type
II supernovae than to type Ia supernovae.[16]  In §7.3.3 we shall see that the
peak luminosities of type Ia supernovae are quite tightly bunched around
$L_B = 9.6 \times 10^9 L_\odot$. Type II supernovae have peak luminosities that span
a broader range: $0.4 \times 10^9 \lesssim (L_B/L_\odot) \lesssim 4 \times 10^9$ (Tammann & Schröder
1990).

Type II supernovae do not occur in early-type galaxies, while type Ia
supernovae occur in all types of galaxy. Moreover, in spiral galaxies type
II supernovae, like OB stars, tend to occur in spiral arms, while type Ia
supernovae show no such preference. Since spiral arms are delineated by
short-lived massive stars and early-type galaxies do not appear to contain
such objects, these observations suggest that the progenitors of type II supernovae are massive stars, while those of type Ia supernovae are not. Since
a distinction has been drawn between type Ia and type Ib supernovae only
since the mid 1980s, the statistics on the occurrence of type Ib supernovae
are poor. However, the well observed members of the Ib class have occurred
in spiral galaxies near spiral arms just as type II supernovae do. In the Milky
Way the supernova rate is thought to be of order one per 40 years, and only
$\sim 15\%$ of supernovae are of type Ia (Tammann, Löffler & Schröder 1994).

The physical interpretation of the different kinds of supernova is thought
to be the following (Woosley 1990). Both type Ib and type II supernovae
occur when a high-mass star suffers core collapse – for this reason we shall refer to them collectively as **core-collapse supernovae**. Type Ib supernovae
probably attend the deaths of the most massive stars, which have lost their
hydrogen envelopes, either to a stellar wind or a binary companion, before
suffering core collapse. Hence at the moment of core collapse, these objects
have He rather than hydrogen envelopes, and thus fail to display hydrogen
lines. Type Ia supernovae are thought to result from the thermonuclear explosion of a C/O white dwarf. This explosion is probably triggered by the
accretion of material from a binary companion.

**Metal production by core-collapse supernovae**    It is relatively simple
to calculate the structure of a massive star at the moment of core collapse.
A much harder task is to determine (i) how much of the star will be blown
away by the imploding core, and (ii) what changes will occur in its chemical
composition as it is blown away. What makes the problem extremely difficult
is that 99% of the energy released emerges in a vast blast of neutrinos. It
is extremely hard to determine accurately which tiny ($\sim 1\%$) fraction of
the energy of implosion is transferred to the stellar envelope, and thus to
determine whether the envelope is ejected and a supernova produced. In
fact, some implosions may not produce readily detectable electromagnetic
radiation and thus go undetected.

---

[16] Some authors actually require type Ib supernovae to show He lines, and classify type
supernovae that show neither Si nor He lines as type Ic supernovae.

**Table 5.4**　Chemical compositions of stars outside the mass-cut at core collapse

| $\mathcal{M}_\alpha$ | $\mathcal{M}_i$ | He | C | O | Ne | Mg | Si+Ca | Fe |
|---|---|---|---|---|---|---|---|---|
| 2.7 | 10 | 0.90 | 0.048 | 0.004 | 0 | 0 | 0 | 0.02 |
| 3 | 12 | 1.52 | 0.048 | 0.004 | 0 | 0 | 0 | 0.03 |
| 4 | 15 | 2.04 | 0.192 | 0.167 | 0.040 | 0.066 | 0.101 | 0.05 |
| 6 | 20 | 2.58 | 0.288 | 0.774 | 0.426 | 0.254 | 0.265 | 0.07 |
| 8 | 25 | 3.10 | 0.568 | 1.660 | 0.767 | 0.270 | 0.339 | 0.10 |
| 12 | 31 | 3.59 | 1.220 | 3.840 | 1.040 | 0.311 | 0.424 | 0.14 |
| 16 | 39 | 4.06 | 1.620 | 6.160 | 1.420 | 0.405 | 0.979 | 0.19 |
| 24 | 55 | 5.02 | 2.300 | 11.700 | 1.980 | 0.624 | 0.934 | 0.24 |
| 32 | 85 | 8.62 | 2.050 | 16.600 | 2.350 | 0.970 | 1.190 | 0.30 |

SOURCE: Adapted from Arnett (1991)

In view of the difficulty of modeling the violent ejection of a massive star's envelope, in the 1970s theorists developed the following rough-and-ready treatment of mass ejection in massive supernovae. One defines a radius within the pre-collapse star such that everything outside this radius is ejected and everything interior to it falls in to form the remnant. The shock wave that is responsible for ejecting the material that lies just outside this **mass-cut** induces a violent burst of nuclear reactions in the departing material. In particular, a substantial mass of Si is burnt to iron-peak nuclides, including the radioactive nuclide $^{56}$Ni. During this nuclear flash, iron-peak elements are subject to intense neutron bombardment, and many are converted to r-process elements. Further out, layers of material that had long ago been processed by the star to C, O, $\alpha$ nuclides and s-process nuclides are ejected by the shock. Hence core-collapse supernovae contribute every kind of nuclide to the ISM – iron-peak, H, He, C, O, $\alpha$, s- and r-process nuclides.

Table 5.4 lists for several stars the masses (in solar masses) of the various elements that they are expected to eject when their cores implode. These numbers have been obtained by calculating the composition of each star immediately prior to core collapse and choosing a mass-cut. Each element is approximately confined to a spherical shell, with the lightest element on the outside and the heaviest in a shell around the core. The first column in Table 5.4 gives the mass of the He core at the onset of He-burning – this is effectively what determines the subsequent evolution of the star. The second column gives the initial mass that would lead to such a He core in the absence of mass loss; the actual initial mass will be larger, probably by a significant factor in the cases of the most massive stars listed. The subsequent columns give the masses of the various elements that lie outside the adopted mass cut.

The fundamental correctness of this picture of the working of massive supernovae was confirmed by studies of the type II supernova 1987A, which was observed to explode in the LMC on 1987 February 23. Since it was so close to us ($\sim$ 50 kpc) it could be observed in incomparably greater detail than any other supernova has been, or indeed is likely to be for decades, perhaps centuries, to come. In particular, the spectacular burst of neutrinos that carries away most of the energy released by a supernova was observed. Such a burst had long been predicted by theory but never before been detected, and, because it lasted only $\sim$ 10 s, we could determine the precise origin in time of all other observations. Arnett *et al.* (1989) and McCray (1993) review this and many other aspects of SN 1987A.

Models of SN 1987A suggest that in its progenitor star He ignited when the He core contained $6 \mathcal{M}_\odot$. If the progenitor has been correctly identified on archival plates, it was a blue rather than a red giant (MK type B3 I), presumably because it had suffered significant mass loss prior to the explosion. From Table 5.4 it follows that its initial mass exceeded $20 \mathcal{M}_\odot$. Models of neutrino emission by the collapsed core suggest that the latter had a mass $\simeq 1.6 \mathcal{M}_\odot$. Of particular importance is evidence that from about 100 d after the explosion, the supernova's light emission was largely powered by the radioactive decay of $0.069 \mathcal{M}_\odot$ of $^{56}$Ni (mean-life 8.8 d) to first $^{56}$Co and then $^{56}$Fe. Specifically, in the interval $100 - 300$ d, the shape of the light curve is a declining exponential with a time-constant that perfectly fits the 111.3 d mean-life of $^{56}$Co, and $0.069 \mathcal{M}_\odot$ of $^{56}$Ni is required to generate the observed luminosity. (The mass cut upon which Table 5.4 is based has been chosen to yield this quantity of Ni, and hence after its complete decay of Fe.) After 300 d the light curve declined faster than exponentially because the ejecta had become too rarefied to degrade into optical photons all the $\gamma$ rays that carry much of the energy released by the decaying Co. Indeed, from about 175 d the 837 keV and 1240 keV $\gamma$-ray lines of $^{56}$Co were directly detected.

The observed spectra of SN 1987A imply a total kinetic energy $\sim 1.5 \times 10^{44}$ J in the ejecta. P-Cygni profiles of various spectral lines (Figure 3.4) indicated that the outermost layers of the star's envelope were ejected at $\gtrsim 20\,000$ km s$^{-1}$. This material was soon slowed to $\sim 2500$ km s$^{-1}$, however. The structure of the envelope was made highly complex by the development of hydrodynamical instabilities soon after the passage of the blast wave. The complexity of the envelope has defeated efforts to obtain from observed spectra accurate estimates of the amounts of the various metals that were thrown off by SN 1987A – see McCray (1993) for a review of this topic.

**Metal production by type Ia supernovae**   The progenitors of type Ia supernovae are believed to be C/O white dwarfs that by accretion from a binary companion achieve a mass $1.38 \mathcal{M}_\odot$, just barely smaller than Chandrasekhar's limiting mass for a C/O white dwarf. The electrons near the center of such a star are highly degenerate, so that the star's nuclear fuel is liable to ignite explosively – see Box 5.2. The details of how this ignition takes place are not fully understood because the event proves hard to model with sufficient accuracy, but there is a consensus that a wave of burning sweeps through the star in which about half of the star's C and O are converted to iron-group elements in about a second. The predicted light-curve and mix of nuclides produced by this event depend sensitively on the speed with which the wave sweeps through the star. The best agreement with observation is achieved when the wave is a **deflagration wave**, which propagates nearly as fast as the sound speed, but not as fast or faster – see Woosley (1990) for a discussion of the dynamics of type Ia supernovae.

The explosion of a C/O white dwarf releases only 1% of the energy that is released in a type II supernova, but a large fraction of it is channeled into the kinetic energy of the ejecta. Consequently, the mechanical energy of a type Ia supernova is, to within the uncertainties, identical with the mechanical energy of a core-collapse supernova: $\sim 10^{44}$ J. Since this energy is concentrated in $1.4\,\mathcal{M}_\odot$ rather than over $10\,\mathcal{M}_\odot$ of ejecta, the characteristic ejection velocity is larger: $(10-20) \times 10^3$ km s$^{-1}$. The star leaves no remnant, being completely disrupted by the event. The entire UV and optical display of a type Ia supernova derives from radioactive decay, overwhelmingly of $^{56}$Ni to $^{56}$Fe. A type Ia supernova produces roughly 5–10 times as much $^{56}$Fe as SN 1987A did: $\sim 0.7\,\mathcal{M}_\odot$, of which $\sim 0.45\,\mathcal{M}_\odot$ forms by the decay of $^{56}$Ni. Unfortunately, there are significant uncertainties in most of these numbers. Clearly, type Ia supernovae are significant sources of r-process elements.

# 5.3 Models of chemical enrichment

We now study models of the enrichment of stellar systems that are valuable for the interpretation of both observations of external galaxies (§8.2.6) and of stars within the Milky Way (§10.4). The three models that we consider differ from one another in the nature of their interaction with their surroundings: the first model is a completely isolated system; the second model vents metal-rich gas while the third model accretes metal-free gas from its surroundings.

### 5.3.1 The closed-box model

The closed-box model was introduced by Talbot & Arnett (1971) as the simplest possible model for the chemical evolution of a portion of a galaxy such as the solar neighborhood. It focuses on a narrow annulus of galactocentric radius, and assumes that in the period under study no material either enters or leaves this region. Initially, the material is entirely gaseous and free of heavy elements. As time goes on, stars are formed from the interstellar gas and massive stars return hydrogen, helium, and heavy elements to the interstellar medium. We assume that turbulent motions keep the gas well stirred and therefore homogeneous. The supply of interstellar gas is gradually consumed, and the remaining gas becomes steadily more polluted with heavy elements.

Suppose that at any time there is a mass $\mathcal{M}_h$ of heavy elements in the interstellar gas, which itself has mass $\mathcal{M}_g$. Then the **metallicity** of the interstellar gas is

$$Z \equiv \frac{\mathcal{M}_h}{\mathcal{M}_g} \qquad (Z_\odot \simeq 0.02). \qquad (5.33)$$

We suppose that at this time the stars have total mass $\mathcal{M}_s$ and consider the effect of forming a mass $\delta' \mathcal{M}_s$ of new stars. For simplicity we adopt the **instantaneous recycling approximation**. In this approximation we neglect the delay between the formation of a generation of stars and the ejection of heavy elements by that generation's massive stars. Since the progenitors of core-collapse supernovae have lifetimes that are very much shorter than the Hubble time, this approximation is reasonable. It is, however, not valid as regards the contributions to metal enrichment of either type Ia supernovae or of stars of a few solar masses. The theory can be refined to take into account the time delay between star formation and the ejection into the ISM of heavy elements – see Pagel (1997).

Let the mass of the stars in $\delta' \mathcal{M}_s$ that remain after the massive stars have died be $\delta \mathcal{M}_s$, and let the mass of the heavy elements produced by this stellar generation be $p\delta \mathcal{M}_s$, where $p$ is the **yield** of that generation of stars. Then the total change in the heavy element content of the interstellar gas that arises from these new stars is

$$\delta \mathcal{M}_h = p\delta \mathcal{M}_s - Z\delta \mathcal{M}_s = (p - Z)\delta \mathcal{M}_s. \qquad (5.34)$$

The metallicity of the interstellar gas changes by an amount

$$\delta Z = \delta \left( \frac{\mathcal{M}_h}{\mathcal{M}_g} \right) = \frac{\delta \mathcal{M}_h}{\mathcal{M}_g} - \frac{\mathcal{M}_h}{\mathcal{M}_g^2}\delta \mathcal{M}_g = \frac{1}{\mathcal{M}_g}(\delta \mathcal{M}_h - Z\delta \mathcal{M}_g). \qquad (5.35)$$

By conservation of mass $\delta \mathcal{M}_s = -\delta \mathcal{M}_g$, so, combining equations (5.34) and (5.35), we obtain

$$\delta Z = -p\frac{\delta \mathcal{M}_g}{\mathcal{M}_g}. \qquad (5.36)$$

If the yield $p$ of each generation of stars is the same, we may integrate equation (5.36) to obtain the metallicity at time $t$ as

$$Z(t) = -p\ln\left[\frac{\mathcal{M}_g(t)}{\mathcal{M}_g(0)}\right]. \qquad (5.37)$$

Here we have employed our assumption that the gas is initially free of metals.

Since $\mathcal{M}_g(0)$ is the total mass of the annulus, equation (5.37) makes a clear prediction: if we plot the observed metallicity $Z$ at different points within galactic disks against the logarithm of the fraction of the local mass density that is in gaseous form, we will obtain a straight line. The absolute value of the slope of this straight line will be the yield $p$. In §8.2.6 we shall review observational evidence for a relationship of this type.

In addition to predicting a relationship between the metallicity of the ISM and the fraction of the system's mass that is still in the ISM, the closed-box model also predicts the metallicity-distribution of stars. Specifically, by equation (5.37), the mass of the stars that have metallicity less than $Z(t)$ is

$$
\begin{aligned}
\mathcal{M}_s[< Z(t)] = \mathcal{M}_s(t) &= \mathcal{M}_g(0) - \mathcal{M}_g(t) \\
&= \mathcal{M}_g(0)[1 - e^{-Z(t)/p}].
\end{aligned}
\tag{5.38}
$$

This equation predicts that stars will be widely distributed in metallicity. In particular, a significant fraction of all stars will have metallicities smaller than, say, a third of the metallicity of currently-forming stars. Quantitatively, at time $t$ the fraction of stars with $Z < \alpha Z(t)$ is

$$
\begin{aligned}
\frac{\mathcal{M}_s[< \alpha Z(t)]}{\mathcal{M}_s[< Z(t)]} &= \frac{1 - e^{-\alpha Z(t)/p}}{1 - e^{-Z(t)/p}} \\
&= \frac{1 - x^\alpha}{1 - x}
\end{aligned}
\tag{5.39}
$$

where equation (5.37) for $Z/p$ has been used and $x \equiv \mathcal{M}_g(t)/\mathcal{M}_g(0)$ is the remaining gas fraction. A reasonable estimate for $x$ in the solar neighborhood is $x \simeq 0.1$. When this is substituted into equation (5.39) one finds $\mathcal{M}_s(\frac{1}{3}Z) = 0.51\mathcal{M}_s(Z)$. That is, if the solar neighborhood were a closed box and had started from zero initial metallicity, fully a half of all stars would have less than a third of the metallicity of the most metal-rich stars. In Chapter 10 we shall see that there are in reality many fewer metal-poor stars near the Sun.

Unfortunately, to determine $\mathcal{M}_s(Z)$ observationally we have to be able to resolve individual stars, and this requirement in practice confines our knowledge of $\mathcal{M}_s(Z)$ to the stellar populations of the Milky Way.

### 5.3.2 The leaky-box model

In the Milky Way stars form in molecular cloud complexes, and these systems are observed to be by no means closed boxes. For reasons that are inadequately understood, only a small fraction of the gas in a molecular cloud has been turned into stars by the time stars formed early in the cloud's life begin to explode as core-collapse supernovae. These supernovae, together with fast winds from massive stars that have not yet completed their lives, drive strong shocks into the cloud's remaining gas. The cumulative effect of these shocks is to blast the remaining gas out of the cloud. It is likely that something similar happened on a larger scale during the formation of globular clusters and elliptical galaxies. Hence we now investigate the chemical evolution of a **leaky box**; that is, one from which gas is driven by stars that form within the box.

We suppose that supernovae drive gas out of the box at a rate that is proportional to the star-formation rate:

$$\frac{d\mathcal{M}_t}{dt} = -c\frac{d\mathcal{M}_s}{dt}, \tag{5.40}$$

where $\mathcal{M}_t$ is the mass of stars and gas in the box and $c$ is a constant. Integrating this equation we find that $\mathcal{M}_t(t) = \mathcal{M}_t(0) - c\mathcal{M}_s(t)$, and thus by conservation of mass we have for the gas mass at any time,

$$\begin{aligned} \mathcal{M}_g(t) &= \mathcal{M}_t(t) - \mathcal{M}_s(t) \\ &= \mathcal{M}_t(0) - (1+c)\mathcal{M}_s(t). \end{aligned} \tag{5.41}$$

An analysis similar to that which leads to equation (5.36) shows that if the gas expelled had the same metallicity as the rest of the interstellar medium,

$$\frac{dZ}{d\mathcal{M}_s} = \frac{p}{\mathcal{M}_g} = \frac{p}{\mathcal{M}_t(0) - (1+c)\mathcal{M}_s}. \tag{5.42}$$

Integrating this differential equation subject to the initial condition $\mathcal{M}_s(Z = 0) = 0$ yields

$$\mathcal{M}_s(< Z) = \frac{\mathcal{M}_t(0)}{1+c}\left\{1 - \exp\left[-\frac{(1+c)Z}{p}\right]\right\}. \tag{5.43}$$

Comparing this expression with equation (5.38), we see that the only effect of the steady outflow of interstellar gas is to reduce the effective yield to $p/(1+c)$.

Hartwick (1976) formulated the leaky-box model to reproduce the observed metallicity distribution within the Milky Way's halo. We shall discuss these results in §10.7 and here concentrate on an entirely different application of the model. In §4.3.4 we saw that the line-strengths of elliptical galaxies are tightly correlated with their central velocity dispersions (see Figure 4.42). Consider the mean stellar metallicity that is predicted by the leaky-box model. In this model star formation ceased when the gas was exhausted. By equation (5.41) this occurred when $\mathcal{M}_s = \mathcal{M}_t(0)/(1+c)$. Thus the present mean stellar metallicity should be

$$\overline{Z}_s = \frac{1+c}{\mathcal{M}_t(0)}\int_0^{\mathcal{M}_t(0)/(1+c)} Z\,d\mathcal{M}_s. \tag{5.44}$$

When we use equation (5.43) to evaluate the integral in this equation, we obtain

$$\overline{Z}_s = \frac{p}{1+c}. \tag{5.45}$$

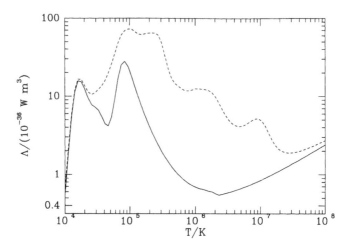

**Figure 5.18** Full curve: the cooling function $\Lambda$ of a hydrogen-He mixture in collisional equilibrium with 10% He by number. Dashed curve: the same quantity for gas with solar abundances. The radiative energy loss is $Q = \Lambda n_e^2 \, \mathrm{W\,m^{-3}}$, where $n_e$ is the electron density. [From data kindly supplied by R. Edgar]

If the system turns all its gas into stars, $c = 0$ and $Z_s = p$, but if supernovae are able to drive most of the original gas out of the system, $c \gg 1$ and $\overline{Z}_s \simeq p/c$. Thus the effective yield is expected to decline in step with the ease with which the hot, metal-rich component of the interstellar medium could flow out of the galaxy.

Figure 5.18 shows as functions of temperature the cooling rates of both primordial gas (full curve) and of gas that contains solar abundances of heavy elements (dashed curve).[17] In both cases the cooling rates rise from very small values at $T < 1000\,\mathrm{K}$ to peaks at $T \simeq 2 \times 10^4\,\mathrm{K}$ that are associated with the ionization of hydrogen. A second peak at $T \simeq 10^5\,\mathrm{K}$ is associated with the ionization of He. Beyond this peak the cooling rate of primordial gas declines steadily to a shallow minimum near $T = 2 \times 10^6\,\mathrm{K}$. The cooling rate of gas that contains heavy elements shows several subsidiary peaks beyond that associated with He ionization, but nonetheless declines fairly steadily from $T \simeq 3 \times 10^5\,\mathrm{K}$ to $T \simeq 2.5 \times 10^7\,\mathrm{K}$. The significance of these cooling rates for the present investigation is the following.

Equilibrium between heating and cooling processes is unlikely to be established at temperatures for which the relevant cooling curve falls with rising $T$. Indeed, at such temperatures, any small excess of heating over cooling would cause the temperature to rise and the cooling rate to fall, thus exacerbating the original small disequilibrium. Moreover, in the region

---

[17] The cooling rates in Figure 5.18 assume that photons created by the gas can escape freely and that the gas is the only source of radiation.

$10^4$ K $\lesssim T \lesssim 10^5$ K, cooling is so effective that, in the absence of strong heating, gas in this region generally cools to $T \lesssim 10^4$ K. Thus, under equilibrium conditions we expect the temperature of any given parcel of gas to satisfy either $T \lesssim 10^4$ K or $T \gtrsim 3 \times 10^6$ K.

In general we expect any given stellar system to contain both cool ($T \lesssim 10^4$ K) and hot ($T \gtrsim 3 \times 10^6$ K) parcels of gas – that is, its ISM will consist of two phases, one hot and one cold. The one-dimensional velocity dispersion that corresponds to temperature $T$ is

$$\sigma_{\text{gas}} = \left(\frac{kT}{\mu m_{\text{p}}}\right)^{1/2} \simeq 116\left(\frac{T}{10^6 \text{ K}}\right)^{1/2} \text{km s}^{-1}. \tag{5.46}$$

Hence the velocity dispersion associated with the hot component of the ISM is $\sigma_{\text{h}} \simeq 200$ km s$^{-1}$. For comparison, stars at the center of a globular cluster or dwarf spheroidal galaxy have dispersion $\sigma_* \simeq 5 - 10$ km s$^{-1}$ (see §6.1.11), while at the center of a giant elliptical galaxy they have $\sigma_* \gtrsim 200$ km s$^{-1}$ (cf. Figure 4.42). Hence in a globular cluster or a dwarf spheroidal galaxy, $\sigma_{\text{h}} \gg \sigma_*$ and the hot phase of the ISM is likely to flow freely out of the system, with the consequence that the system experiences little self-enrichment. By contrast, in a giant elliptical galaxy $\sigma_{\text{h}} \simeq \sigma_*$ and we expect the system's gravitational field to be able to confine the hot component of the ISM approximately as well as it confines the system's stars,[18] with the consequence that giant elliptical galaxies should experience substantial self-enrichment.

To obtain a quantitative relation between $\sigma_*$ and $Z$ we estimate the rate of ejection of gas from a giant elliptical galaxy by equating the rate at which the ejected gas gains potential energy to the rate of injection of energy into the ISM by supernovae. Thus with equation (5.40) we write

$$\left|\Phi\frac{\text{d}\mathcal{M}_{\text{g}}}{\text{d}t}\right| \simeq f\sigma_*^2 c\frac{\text{d}\mathcal{M}_{\text{s}}}{\text{d}t} = e_{\text{SN}}\frac{\text{d}\mathcal{M}_{\text{s}}}{\text{d}t}. \tag{5.47}$$

Here $\Phi \simeq -f\sigma_*^2$ is the typical gravitational potential from which the ejected gas is lifted, $f$ is a number of order a few, and $e_{\text{SN}}$ is the energy injected into the ISM per unit mass converted into stars. For the IMF characteristic of the solar neighborhood (see §5.1.9), one core-collapse supernova explodes per $\sim 200\,\mathcal{M}_\odot$ of fresh star formation. Given that each supernova injects $\sim 10^{44}$ J, $e_{\text{SN}} \simeq 10^{44}/200\,\mathcal{M}_\odot \simeq 2.5 \times 10^{11}$ J kg$^{-1}$. Inserting this estimate into equation (5.47) we find

$$c = \frac{0.6}{f}\left(\frac{200 \text{ km s}^{-1}}{\sigma_*}\right)^2. \tag{5.48}$$

This equation suggests that $c$ will increase from a value smaller than unity for giant elliptical galaxies to one of order a few for dwarf ellipticals. Since

---

[18] In §8.3 we shall encounter direct observational evidence for this conjecture.

we have assumed that none of mechanical energy output of supernovae is radiated rather than used to lift gas from the galaxies, our estimate of $c$ is probably on the low side. (But see Problem 5.12 for an indication that cooling cannot be very effective.) Hence it is probably sensible to treat the factor $0.6/f$ in (5.48) as a free parameter to be fitted to the data and to focus on the predicted dependence $c \propto \sigma_*^{-2}$. Inserting this dependence into equation (5.45) for the final mean metal-abundance of a leaky box's stars, we obtain the prediction

$$\overline{Z}_s = \frac{p}{1 + (\sigma_Z/\sigma_*)^2}, \tag{5.49}$$

where $\sigma_Z$ is a parameter to be determined from the observations. While Figure 4.42 shows this prediction to be qualitatively correct, quantitative confrontation of theory and observation must await determination of the relationship $\overline{Z}_s(Mg_2)$ between the mean metallicity of a stellar population and its $Mg_2$ line-strength index. At the moment all one can say is that there is no indication in Figure 4.42 that the observed $Mg_2(\sigma_*)$ relation flattens at $\sigma_* \gtrsim \sigma_Z$ as equation (5.49) predicts the corresponding relation for $\overline{Z}_s(\sigma_*)$ does.

Clearly, the analysis just given should apply with minor modifications to the radial line-strength gradients in elliptical galaxies that were discussed in §4.3.2. Differentiating equation (5.45) we find

$$\frac{d\ln \overline{Z}_s}{d\ln r} = -\frac{r}{1+c}\frac{dc}{dr}. \tag{5.50}$$

Now from equation (5.47) we have that

$$c = \frac{e_{SN}}{|\Phi(r)|}, \tag{5.51}$$

and on differentiating this and substituting the result into (5.50) we conclude that

$$\begin{aligned}
\frac{d\ln \overline{Z}_s}{d\ln r} &= \frac{c}{1+c}\frac{v_c^2}{\Phi} \\
&= \frac{e_{SN}}{e_{SN}-\Phi}\frac{v_c^2}{\Phi},
\end{aligned} \tag{5.52}$$

where $v_c = \sqrt{r\,d\Phi/dr}$ is the circular speed. For plausible values of $\Phi$ and $e_{SN}$, equation (5.52) predicts gradients of the same order $(-0.2)$ as those observed [see equation (4.21)]. Unfortunately, the appropriate values of $\Phi$ and $e_{SN}$ are uncertain, because the extent of dark halos around ellipticals is unclear, as is the importance of radiative cooling for the early dynamics of the hot phase of an elliptical's ISM.

### 5.3.3 The accreting-box model

In §8.4 we shall encounter evidence that galaxies such as our own are accreting gas. We now investigate an important consequence of such accretion, namely a tendency to make metal-poor stars much rarer than they would otherwise be.

Suppose that initially the total mass of gas and dust in our box is very small, and imagine that during some initial period the chemical evolution proceeds according to the standard closed-box model. At the end of an appropriate period, 90% of the gas will have been converted to stars and, as we saw at the end of §5.3.1, roughly half of all stars will be more metal-poor than a third of the current interstellar abundance. Now feed metal-free gas into the box at precisely the rate at which the box turns gas into stars. Imagine a small additional mass $\delta M$ of primordial gas joining the box. In a steady state, an exactly equal mass becomes locked up in stellar remnants, and a mass $p\delta M$ of freshly manufactured heavy elements is returned to the interstellar medium. Thus the overall effect on the interstellar gas is to remove mass $\delta M$ at metallicity $Z$ and to return the same mass at metallicity $p$. Consequently, if we continue to feed the box gas at exactly the rate at which it is locking material up in stars, the metallicity of the interstellar medium will eventually settle to the value $Z = p$. After a sufficiently long time, most of the stars in the box will have metallicity $Z \simeq p$, and the fraction of low-metallicity stars will be negligible.

This qualitative picture can be formulated mathematically. If the total mass of material $M_t$ in the box varies, equations (5.34) and (5.35) remain valid. However, we now have

$$0 \neq \delta M_t = \delta M_s + \delta M_g. \tag{5.53}$$

If we eliminate $\delta M_h$ and $\delta M_s$ between equations (5.34) and (5.35), we obtain

$$\delta Z = \frac{1}{M_g}[(p - Z)\delta M_t - p\delta M_g]. \tag{5.54}$$

When we divide this equation by $\delta M_t$, we obtain a differential equation for the evolution of $Z$:

$$\frac{dZ}{dM_t} = \frac{1}{M_g}\left[p - Z - p\frac{dM_g}{dM_t}\right]. \tag{5.55}$$

A change of variable enables us to write this equation in a particularly simple form:

$$\frac{dZ}{du} + Z = p\left(1 - \frac{d\ln M_g}{du}\right) \quad \text{where} \quad u \equiv \int \frac{dM_t}{M_g}. \tag{5.56}$$

The general solution is

$$Z = p\left(1 - Ce^{-u} - e^{-u}\int_0^u e^{u'}\frac{d\ln M_g}{du'}du'\right). \tag{5.57}$$

The simplest solution describes the case in which the gas mass $\mathcal{M}_g$ is constant and the initial metallicity of the gas is zero. Then

$$Z = p\left[1 - \exp\left(1 - \frac{\mathcal{M}_t}{\mathcal{M}_g}\right)\right] \qquad (\mathcal{M}_g = \text{constant}). \qquad (5.58)$$

Thus in this model, $Z \simeq p$ once $\mathcal{M}_t \gg \mathcal{M}_g$. Clearly, the mass contained in stars that are more metal-poor than $Z$ is $\mathcal{M}_s(< Z) = \mathcal{M}_t(Z) - \mathcal{M}_g$. By equation (5.58) this is predicted to be

$$\mathcal{M}_s(< Z) = -\mathcal{M}_g \ln\left(1 - \frac{Z}{p}\right) \qquad (\mathcal{M}_g = \text{constant}). \qquad (5.59)$$

The yield $p$ can be determined from the current gas fraction $\mathcal{M}_g/\mathcal{M}_t(t_0)$ and the current metallicity $Z(t_0) \simeq Z_\odot$ of the interstellar gas through equation (5.58). For example, in the solar neighborhood, $\mathcal{M}_t(t_0) \simeq 10\mathcal{M}_g$ and $Z(t_0) \simeq Z_\odot$, so by equation (5.58) we have $p \simeq Z_\odot$. Equation (5.59) then yields

$$\mathcal{M}_s(< 0.25Z_\odot) \simeq 0.3\mathcal{M}_g \simeq 0.03\mathcal{M}_t \qquad (\mathcal{M}_g = \text{constant}). \qquad (5.60)$$

Thus this simple model predicts that about 3% of all dwarf stars should be more metal poor than $0.25Z_\odot$, which is only slightly in excess of the fraction of such stars found in the solar neighborhood (see §10.7.2). 

Various modifications of this simple accretion model are possible. Suppose, for example, that only a fraction $1 - q$ of the infalling gas is locked up in star formation, so that $\delta\mathcal{M}_g = q\delta\mathcal{M}_t$. Then one can show (see Problem 5.13) that the number of low-metallicity stars is reduced below the value given in equation (5.60).

## 5.4 Evolution of stellar populations

Within a galaxy there are stars of every mass and frequently stars of widely differing ages. In this section we use the theory of stellar evolution to predict the observable properties of such stellar agglomerates. Even with the HST, colors and spectra can be obtained for individual stars only if they lie in the nearest galaxies; for most galaxies we can only analyze samples of radiation to which thousands or millions of stars make significant contributions. The question therefore arises, what can we learn from the colors or the spectrum of such integrated light about the nature of the stars in a galaxy? Presumably we will not be able to reconstruct the galaxy's CM diagram from a single spectrum, but we might hope to estimate the mean metallicity of the system and perhaps constrain its age. Although a great deal of effort has gone into this problem over two decades, it is still not solved. Nevertheless, solid progress has been made, and several important results are available.

The simplest case is that in which all the system's stars formed at the same time. Not only can we derive some interesting analytical results for this case of coeval stars, but, as we shall see in Chapter 6, it describes rather accurately the status of globular clusters. Moreover, we can regard a population with any history of star formation as a superposition of coeval populations of different ages, so once we know how a coeval population evolves it is a straightforward matter to determine the evolution of any other population.

### 5.4.1 Analytical results

We start by deriving some approximate analytical results that will tell us what to expect from detailed numerical models. From Table 5.2 we see that stars that are more massive than $1.25\,\mathcal{M}_\odot$ live less than 5 Gyr. Thus any coeval stellar population older than $\sim 5$ Gyr will contain only stars of initial mass $\mathcal{M}_i \lesssim 1.25\,\mathcal{M}_\odot$. Table 5.2 and Figure 5.2 show that these low-mass stars emit most of their light during the 1 Gyr or so that they are on the RGB, HB and AGB. In view of this result, it is not surprising that most of the integrated light from an old coeval population comes from evolved stars – see Problem 5.9. Some important conclusions follow easily from this fact.

The first conclusion depends on the fact that the effective temperature of the RGB, HB and AGB change only slowly with the masses of the stars on them for $\mathcal{M}_i \lesssim 2\,\mathcal{M}_\odot$. Consequently, the intrinsic colors of an old coeval population should not have evolved strongly since the epoch, about 1.5 Gyr after its formation, when all stars with $\mathcal{M}_i \gtrsim 2\,\mathcal{M}_\odot$ had died.

By contrast, the luminosity of a coeval population will still be evolving (Tinsley 1972). To see this, let there be $dN(\mathcal{M})$ stars in the population with mass in the range $\mathcal{M}$ to $\mathcal{M}+d\mathcal{M}$, and let the stars that turn off the MS at age $t$ be of mass $\mathcal{M}_{GB}(t)$. Table 5.2 shows that for $t \gtrsim 1.5$ Gyr, $\mathcal{M}_{GB} \lesssim 1.5\,\mathcal{M}_\odot$, and that the time $t_{GB}$ that the star spends on the giant branch is much less than $t$. Hence if stars of mass $\mathcal{M}$ emit total energy $E_{GB}(\mathcal{M})$ when they are on the giant branch, the luminosity of the population may be estimated as

$$L \approx \left( E_{GB} \frac{dN}{d\mathcal{M}} \right)_{\mathcal{M}_{GB}} \left| \frac{d\mathcal{M}_{GB}}{dt} \right|. \tag{5.61}$$

For the masses of interest, the MS lifetime of a star, $\tau_{MS}$, is related to its mass by

$$\frac{\tau_{MS}}{10\,\text{Gyr}} \simeq \left( \frac{\mathcal{M}}{\mathcal{M}_\odot} \right)^{-2.5}, \text{ which implies } \frac{\mathcal{M}_{GB}(t)}{\mathcal{M}_\odot} \simeq \left( \frac{t}{10\,\text{Gyr}} \right)^{-0.4}. \tag{5.62}$$

Hence

$$\frac{d\mathcal{M}_{GB}}{dt} \simeq -0.4 \left( \frac{\mathcal{M}_{GB}}{\mathcal{M}_\odot} \right)^{3.5} \left( \frac{\mathcal{M}_\odot}{10\,\text{Gyr}} \right). \tag{5.63}$$

We may fit the IMF (§5.1.9) $\xi = dN/d\mathcal{M}$ in the neighborhood of $\mathcal{M}_{GB}$ by a power law

$$\frac{dN}{d\mathcal{M}} \simeq K \left( \frac{\mathcal{M}}{\mathcal{M}_\odot} \right)^{-\alpha}. \tag{5.64}$$

In the solar neighborhood $\alpha \simeq 2.5$ for $\mathcal{M} \lesssim \mathcal{M}_\odot$. Substituting equations (5.63) and (5.64) into equation (5.61), we find

$$L \simeq \frac{K \mathcal{M}_\odot E_{GB}(\mathcal{M}_{GB})}{25 \, \text{Gyr}} \left( \frac{\mathcal{M}_{GB}}{\mathcal{M}_\odot} \right)^{3.5-\alpha}. \tag{5.65}$$

Differentiating this expression yields

$$\begin{aligned}
\frac{d \ln L}{d \ln t} &= \left[ \frac{d \ln E_{GB}}{d \mathcal{M}_{GB}} + (3.5 - \alpha) \right] \frac{d \ln \mathcal{M}_{GB}}{d \ln t} \\
&= 0.4\alpha - \left( 1.4 + 0.4 \frac{d \ln E_{GB}}{d \ln \mathcal{M}_{GB}} \right).
\end{aligned} \tag{5.66}$$

$E_{GB}$ probably depends only weakly on $\mathcal{M}_{GB}$ ($0 < d \ln E_{GB}/d \ln \mathcal{M}_{GB} < 1$), so unless $\alpha > 3.5$, the luminosity of an old coeval population is expected to be a decreasing function of time. In fact, detailed models by Tinsley & Gunn (1976) yield $(d \ln L/d \ln t) \simeq 0.3\alpha - 1.6$, in good agreement with equation (5.66). Thus we can reliably predict the luminosity evolution of a coeval population if we can accurately determine $\alpha$.

Ostriker & Thuan (1979) pointed out that the mass of heavy elements that is produced by a coeval population is related in a simple way to the present luminosity $L$ of the population and the slope of the IMF $\alpha$ defined by equation (5.64). Most of the present luminosity of the population comes from stars whose original masses lay in a narrow range around the characteristic mass $\mathcal{M}_{GB}$ of stars on the present giant branch, while the heavy elements generated in these systems were all contributed by stars more massive than some mass $\mathcal{M}_Z > \mathcal{M}_{GB}$. Suppose these massive stars return a fraction $f$ of their original mass to the interstellar medium in the form of heavy elements. Then the mass of heavy elements produced is

$$\mathcal{M}_h = \int_{\mathcal{M}_Z}^\infty f \mathcal{M} \frac{dN}{d\mathcal{M}} \, d\mathcal{M}. \tag{5.67}$$

If we take the IMF, $\xi = dN/d\mathcal{M}$, to be of the form of equation (5.64) for $\mathcal{M} < \mathcal{M}_{max}$ and zero for $\mathcal{M} > \mathcal{M}_{max}$, we have

$$\begin{aligned}
\mathcal{M}_h &= f K \mathcal{M}_\odot^2 \int_{\mathcal{M}_Z/\mathcal{M}_\odot}^{\mathcal{M}_{max}/\mathcal{M}_\odot} \mathcal{M}^{(1-\alpha)} \, d\mathcal{M} \\
&= \frac{f K \mathcal{M}_\odot^2}{\alpha - 2} \left[ \left( \frac{\mathcal{M}_\odot}{\mathcal{M}_Z} \right)^{\alpha-2} - \left( \frac{\mathcal{M}_\odot}{\mathcal{M}_{max}} \right)^{\alpha-2} \right].
\end{aligned} \tag{5.68}$$

If $\alpha > 2$ and $\mathcal{M}_{\text{max}} \gg \mathcal{M}_Z$, we may neglect the second term in the square bracket of (5.68). Then, dividing (5.68) by equation (5.65), we obtain the mass of heavy elements produced per unit present luminosity as

$$\frac{\mathcal{M}_{\text{h}}}{L} = \frac{f\,\mathcal{M}_\odot}{\alpha - 2}\left[\frac{25\,\text{Gyr}}{E_{\text{GB}}(\mathcal{M}_{\text{GB}})}\right]\left(\frac{\mathcal{M}_Z}{\mathcal{M}_\odot}\right)^{2-\alpha}\left(\frac{\mathcal{M}_{\text{GB}}}{\mathcal{M}_\odot}\right)^{\alpha-3.5}. \tag{5.69}$$

Inserting into this equation reasonable values of the parameters $f \simeq 0.1$, $\alpha \simeq 2.4$, $\mathcal{M}_{\text{GB}} \simeq 0.85\,\mathcal{M}_\odot$ and $E_{\text{GB}} \simeq 2.9 \times 10^{10}\,L_\odot$ yr, we have

$$\frac{\mathcal{M}_{\text{h}}}{L} \simeq 0.26\left(\frac{\mathcal{M}_Z}{\mathcal{M}_\odot}\right)^{2-\alpha}\mathcal{M}_\odot\,L_\odot^{-1}. \tag{5.70}$$

It is less clear what value we should adopt for $\mathcal{M}_Z$: if we are interested in elements primarily produced in core-collapse supernovae, $\mathcal{M}_Z = 8\,\mathcal{M}_\odot$ is appropriate; if we are concerned with the abundances of elements such as C, N, and O, a smaller value $\mathcal{M}_Z \simeq 2\,\mathcal{M}_\odot$ is appropriate. Fortunately, for $\alpha \simeq 2.4$ equation (5.70) is not sensitive to $\mathcal{M}_Z$: with $\mathcal{M}_Z = 8\,\mathcal{M}_\odot$ we find $(\mathcal{M}_{\text{h}}/L) = 0.11\,\mathcal{M}_\odot\,L_\odot^{-1}$, and with $\mathcal{M}_Z = 2\,\mathcal{M}_\odot$ we find $(\mathcal{M}_{\text{h}}/L) = 0.20\,\mathcal{M}_\odot\,L_\odot^{-1}$.

### 5.4.2 Numerical models of population evolution

A tremendous quantity of observational data for stellar systems of different types, observed at different stages in their evolution, is now available. Adequate exploitation of these data requires much more precise predictions than the simple analytical results that were derived in the last subsection. Since Beatrice Tinsley's pioneering work in the 1970s, several groups have worked hard on numerical models of population evolution that use the best current knowledge of stellar evolution to predict the observable properties of a coeval population. We now take a look at the predictions of these models and their limitations. We start by considering predictions for broad-band colors.

In §5.1.1 we described how apparent magnitudes and broad-band colors can be assigned to model stars. The same techniques can also be used to assign spectral indices (§3.4). Given these data, it is a straightforward matter to derive integrated magnitudes, colors and spectral indices by adding together the contributions of all the model stars in a theoretical CM diagram.

Charlot, Worthey & Bressan (1996) give a clear insight into the strengths and limitations of model populations through detailed comparison of results obtained with three independent modeling codes. They find that $B - V$ colors are typically uncertain by $\sim 0.03$ mag, while $V - K$ colors are uncertain by 0.1 to 0.14 mag. They find that the prime causes of these uncertainties are differences in the assumptions made regarding the evolution of stars once they leave the MS. Smaller but significant causes of uncertainty are

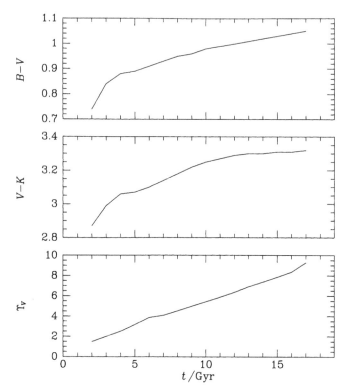

**Figure 5.19** The evolution of $B - V$, $V - K$, and $\Upsilon_V$ with time of a coeval population of solar metallicity that has a Salpeter IMF ($\alpha = 2.35$). [From data published in Charlot, Worthey & Bressan (1996)]

differences in the way colors are assigned to stars of specified position in the theoretical CM diagram.

Figure 5.19 shows the predictions of Charlot, Worthey & Bressan (1996) for the evolution of $B - V$, $V - K$ and the mass-to-light ratio $\Upsilon_V$ for a coeval population of solar metallicity. The population becomes steadily redder, but, in accordance with our analytical approximations, the colors change fairly slowly after $\sim 5\,\text{Gyr}$. Meanwhile, $\Upsilon_V$ increases at a roughly constant rate, just as our analytic work predicted.[19] 

The full curve in Figure 5.20 shows the evolution within the $(V - K, B - V)$ plane of the model population whose colors and mass-to-light ratio are plotted in Figure 5.19. Also shown in this figure are the colors of 12 elliptical galaxies from the study of Peletier (1989); for each galaxy a filled triangle marks the central color, while an open triangle shows the color at $r \sim 0.75 R_e$. The open triangles are associated with substantial errors – the

---

[19] The absolute value of $\Upsilon_V$ in Figure 5.19 is determined by the arbitrary truncation of the IMF $\xi(\mathcal{M}) \propto \mathcal{M}^{-2.35}$ at $\mathcal{M} = 0.15\,\mathcal{M}_\odot$.

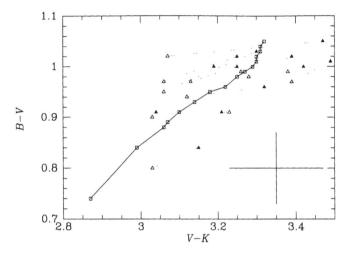

**Figure 5.20** The curve with squares shows the evolution of a coeval population of solar metallicity in the $(V - K, B - V)$ plane. After 2 Gyr the population has reached the lowest left square, and with each additional Gyr it moves up the curve by a further square; the last point corresponds to 17 Gyr. The triangles show central (filled) and peripheral (open) colors of 12 elliptical galaxies – the two points of each galaxy are joined by a line. The cross at bottom right indicates the typical error in the positions of the open triangles. [The curve is from data published in Charlot, Worthey & Bressan (1996) and the galaxy data are from Peletier (1989)]

cross at the bottom right of the figure shows typical error in the locations of these triangles. The error in the location of any full triangle should be smaller and correlated with the error in the location of the associated open triangle. When these errors and those in the location of the full curve are borne in mind, it is apparent that the galaxy data are consistent with the models. Moreover, there is a strong suggestion that the galaxies have a non-negligible spread along the full curve. Does this imply that there is a spread of several billion years in the ages of these galaxies' stellar populations?

Notice that the full curve in Figure 5.20 implies that a spread in age of $\pm 3$ Gyr around an age of 13 Gyr would be only barely detectable with data of the quality plotted in the diagram. Thus, if the spread of the triangles in Figure 5.20 does arise from age differences, the latter have to be substantial and, by the lower panel of Figure 5.19, be associated with variations in $\Upsilon_V$ of at least a factor 2.

Figure 5.21 suggests that age differences do contribute significantly to the spread in galaxy colors by showing a two-color distribution for 34 elliptical galaxies. The color indices plotted, $(V - B)^{-21}$ and $(U - B)^{-21}$, are not measured colors, but colors corrected for the color-magnitude effect (§4.3.4) to absolute magnitude $M_B = -21$. This correction has the effect of producing a highly linear distribution of points within the two-color plot.

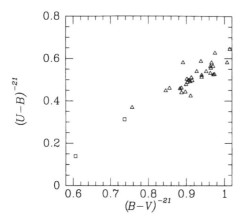

**Figure 5.21** A two-color plot for 34 elliptical galaxies. The plotted colors have been corrected for the color-magnitude effect. The two bluest galaxies (squares) show clear morphological signs of having recently emerged from violent mergers – one of them, NGC 7252 is shown in Figure 4.68. [After Schweizer & Seitzer (1992) from data kindly supplied by F. Schweizer]

The key fact about Figure 5.21 is that the two bluest galaxies are systems that show clear morphological evidence of having recently emerged from a violent merger event – the points belonging to these systems, NGC 3921 and NGC 7252, are marked with squares in Figure 5.21. As we saw in §4.6.1, there can be no doubt that within these systems there are many stars that are $\lesssim 1\,$Gyr old, so the blue colors of these galaxies surely reflect a relatively young mean age of the systems. In view of this finding, it is natural to suppose that galaxies that fall just above these merger remnants are also blue because they are comparatively young. Schweizer & Seitzer (1992) use this approach to estimate the epochs in the past when a sample of elliptical and S0 galaxies emerged from a major merger.

Unfortunately, the conclusion that we reached above, namely that anomalously blue colors imply youth, is insecure because the colors of an young population can be accurately mimicked by a somewhat older but less metal-rich population. Quantitatively, Worthey (1994) finds that

$$\left(\frac{\partial \ln t}{\partial \ln Z}\right)_{\mathrm{colors}} \simeq -\tfrac{3}{2}, \tag{5.71}$$

with the result that a 20% increase in $Z$ can be compensated for by a $\sim 30\%$ decrease in $t$. Therefore from colors alone one cannot hope to determine the age or metallicity of a stellar population.

Can we do better if we have a good spectrum of the population rather than just its broad-band colors? Consider first galaxies with measurable emission lines. In §8.1.3 we shall see that the rate at which the stars of a disk galaxy emit ionizing photons can be determined from the strength $I(\mathrm{H}\alpha)$ of the H$\alpha$ emission line at 657 nm in the galaxy's spectrum. Ionizing photons are produced only by hot stars, and most hot, luminous stars are short-lived. Hence, a measurement of $I(\mathrm{H}\alpha)$ allows one to estimate the rate at which massive stars are forming. The continuum radiation against

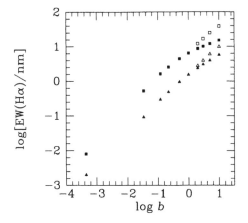

**Figure 5.22** The equivalent width of the Hα line versus the birthrate parameter $b$ defined by equation (5.72). Squares are for models with a Salpeter IMF while triangles for models with the Scalo IMF. [From data published in Kennicutt, Tamblyn & Congdon (1994)]

which the Hα line is observed receives important contributions from longer-lived stars, so its intensity is a measure of how many stars there are in the galaxy and is insensitive to the galaxy's star-formation history. From these considerations it follows that the equivalent width (§3.3.1) EW(Hα) of the Hα line is sensitive to both the IMF, which determines what fraction of each generation of stars will produce ionizing radiation, and the present star-formation rate, SFR. The difference between galaxies in which star-formation is still an important process and ones in which it has effectively ceased, is usually quantified by the **birthrate parameter** $b$

$$b \equiv \frac{\text{SFR}}{\langle \text{SFR} \rangle}, \qquad (5.72)$$

where the angle brackets imply an average over the galaxy's lifetime. Figure 5.22 shows EW(Hα) as a function of $b$ for various models. The squares are for models that have a Salpeter IMF [equation (5.14)], while triangles are for models with the Scalo IMF [equation (5.16)]. The impact of the larger numbers of massive stars predicted by the Salpeter IMF is apparent. All models with $b \leq 1$ have star-formation rates that decline exponentially with time. Two types of models with $b > 1$ are plotted: the open points are for models in which a current burst of star formation is superimposed on a constant background star-formation rate, while the filled points are for models in which the star-formation rate rises exponentially towards the present epoch. While the relation between $b$ and the equivalent width of Hα is not independent of star-formation history, it depends most sensitively on the adopted IMF.

Consider now what can be learned from the absorption-line spectrum that a typical stellar population produces. In practice integrated galaxy spectra never have high spectral resolution because the random velocities of stars within the population give rise to significant spectral broadening that washes out all narrow spectral features – see Chapter 11. In §3.4 we defined

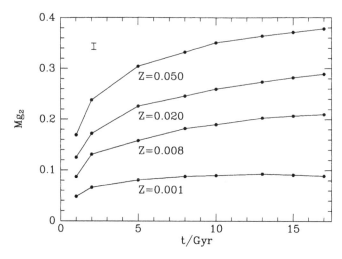

**Figure 5.23** The metallicity $Z$ implied by a given value of $Mg_2$ depends upon the age of the stellar population. Here $Mg_2$ is plotted against age for several values of $Z$. The error bar at upper left indicates the uncertainty in this calibration that is introduced by allowing $Y$ to vary from 0.26 to 0.32. [From data published in Casuo *et al.* (1996)]

several spectral indices that are designed to measure prominent spectral features in medium- and low-resolution spectra (Table 3.8). The first step in an analysis of an integrated spectrum is to determine from it the values of up to a dozen such spectral indices. The measured values of these indices depend on the stellar velocity dispersion within the galaxy and must be corrected to zero velocity dispersion before being compared with the measured indices of nearby stars of known chemical composition.

We constrain the nature of a galaxy's stellar population by comparing the measured index values with values determined for various model populations. The procedure for predicting the values of spectral indices for a stellar population is very similar to the procedure described above for predicting the population's colors: at each point in the theoretical CM diagram we identify a star in our library and add its measured spectral indices into the model's indices with an appropriate weighting. Clearly the final predicted spectral indices will be subject to uncertainties similar to those which plague predicted colors.

Figure 5.23 shows the dependence of one of the most widely studied spectral indices, $Mg_2$, upon $Z$ and the age $t$ of a coeval stellar population. Evidently, a measured value of $Mg_2$ does not uniquely determine either $Z$ or $t$. How much better can we do if we measure several different indices? The **$Z$-sensitivity** of an index or color $I$ is defined to be the value of the derivative $(\partial \log t / \partial \log Z)_I$ at constant $I$. For example, equation (5.72) states that the $Z$-sensitivity of colors is of order $-1.5$. Table 5.5 indicates the princi-

**Table 5.5** Sensitivities of spectral indices defined by
Faber *et al.* (1985)

| Index | Dominant | Others | $(\partial \log t / \partial \log Z)_{\text{index}}$ |
|---|---|---|---|
| G | CH | (O) | −1.0 |
| Mg $b$ | Mg | (C) | −1.7 |
| Fe$_1$ | Fe | C, (Mg), Ca | −2.3 |
| Fe$_2$ | Fe | (C), (Mg), Cr | −2.8 |
| Na | Na | C | −2.1 |
| H$\beta$ | H | (Mg), (Cr), C | −0.6 |
| CN | CN | C, N, (O) | −2.0 |
| Mg$_1$ | C | Mg | −1.8 |
| Mg$_2$ | Mg | C | −1.8 |
| TiO$_1$ | TiO | Ti, (Fe) | −1.5 |
| TiO$_2$ | TiO | V, Sc, Ti | −2.5 |

NOTES: Elements with brackets around them in the
"others" column contribute negatively to the index.
SOURCE: Worthey (1996)

pal sensitivities of the indices that are defined in Table 3.8. Each index is
sensitive both to certain element abundances and to the population's age
$t$. All indices have significant dependence on more than one element, and
this dependence is frequently in the sense that increasing the abundance of
the element decreases the value of the index – in Table 5.5 this condition is
indicated by brackets around the element's symbol. In every case, increasing
the overall abundance level affects the index in the same sense as decreasing the population's age – this fact accounts for the negative signs in the
last column of the table, which gives values for the derivative (5.71) at constant index. Several of these $Z$-sensitivities are similar to the $Z$-sensitivity
of broad-band colors [equation (5.71)]. These indices, which unfortunately
include the widely measured Mg indices, cannot help one separate the effects
of varying age and metallicity. Some other indices do have $Z$-sensitivities
significantly different from $-1.5$ and in principle we may determine $Z$ and $t$
independently by combining these indices with either broad-band colors or
indices with $Z$-sensitivities $\sim -1.5$.

The simultaneous determination of $Z$ and $t$ from quantities that all
depend on both $t$ and $Z$ is, in practice, fraught with difficulty. There are
two basic problems. First, the smaller the range that is covered by the $Z$-
sensitivities of the quantities one employs, the more accurately each quantity
must be determined for given final uncertainties in $t$ and $Z$. A second and
more fundamental difficulty is that, as we have seen, it is unlikely that a
stellar population can be characterized by a single metallicity $Z$ – see §§6.1.5
and 10.4.1. Rather, we expect the abundances of different elements relative to
H to vary. Therefore the number of quantities to be determined is in reality

greater than two: at the very least we should aim to determine separate abundances for $\alpha$ and iron-peak elements in addition to the age $t$, and ideally we would test the predictions of the theory of stellar evolution by determining independent abundances for every significant element. Clearly, if it is difficult to determine unique values for two quantities from the available data, we have no hope of determining from those data unique values of three or more quantities.

The ideal resolution of this difficulty would be to find an index that depended only on $t$ – the $Z$-sensitivity of such an index would be zero because, formally, an infinite change in $Z$ would be needed to compensate for an infinitesimal change in $t$. If an index of zero $Z$-sensitivity were available, we could at the outset determine the age of the population and then hold $t$ fixed while determining the relative abundances of the elements from other indices. The only spectral lines that have any prospect of being insensitive to metal abundances are lines of hydrogen. Table 5.5 shows that while H$\beta$ *is* less sensitive to metallicity than any other spectral index, it does have a significantly non-zero $Z$-sensitivity. Jones & Worthey (1995) show that indices that measure lower-frequency transitions in the Balmer series, such as H$\gamma$ and H$\delta$, can have negligible $Z$-sensitivity. Unfortunately, to measure these indices higher-resolution spectra are required than those on which the system of Faber *et al.* (1985) is based, with the result that few observational values of these indices are currently available.

## Problems

**5.1** Use Figure 5.7 to calculate the fraction of the mass of a stellar population that is returned to the ISM within 10 Gyr. Assume that the population forms with a Salpeter IMF, and adopt a lower mass cutoff of $0.1 \mathcal{M}_\odot$.

**5.2** Show that the mean molecular weight of a fully ionized H/He plasma lies near $\mu = 0.58$ for a primordial abundance of He, and near 0.61 for the solar abundance of He.

**5.3** Use the virial theorem to estimate the thermal energy in the Sun (assuming that the Sun has uniform density). Hence estimate the Sun's **Helmholtz time** – the time it would take to cool by a factor two in temperature if the nuclear energy release in it were to cease.

**5.4** Show that when a self-gravitating body of polytropic gas shrinks homologously, its thermal energy scales with the stellar radius $R$ as $E_{\text{th}} \propto R^{3(1-\gamma)}$, where $\gamma$ is the ratio of principal specific heats. Hence show that a polytropic star is unstable to gravitational collapse if $\gamma < \frac{4}{3}$.

**5.5** Check the crude estimate of main-sequence lifetimes given by equation (5.6) against the data plotted in Figure 5.4.

**5.6** Use the virial theorem to estimate a typical value for the speed of sound deep inside a star of mass $\mathcal{M}$ and radius $R$. Using this estimate, show that the time required for sound to cross the star is $P \sim (G\bar{\rho})^{-1/2}$. Hence show that, for the fundamental p-mode of a star, the pulsation constant $Q$ that is defined by equation (5.18) takes a value of order 1 hr.

**5.7** Given that the luminosity $L$ of a massive H-burning star varies with mass roughly as $L \propto \mathcal{M}^3$, show that, at fixed effective temperature, the fundamental pulsation period of such stars scales as $P \propto L^{7/12}$.

**5.8** Suppose the star-formation rate in a population varies with time $t$ as $e^{\alpha t}$, and let the oldest star in the population have formed at $t = 0$. Show that the analog of equation (5.12) for this case is

$$\Phi_0(M) = \frac{t \alpha e^{t\alpha}}{e^{\tau_{\rm MS}\alpha} - 1} \Phi(M). \tag{5.73}$$

**5.9** For a population forming at time $t = 0$ with a Salpeter IMF with upper mass cutoff at $\mathcal{M} = 2\mathcal{M}_\odot$, estimate the time after which most light comes from post-main-sequence stars.

**5.10** Suppose that the yield $p$ varies with metallicity $Z$ as $p(Z) = 0.002 + 0.6Z$. Show that the 'effective yield,' defined by equation (8.62), is then given by

$$p_{\rm eff} = \frac{0.6Z}{\ln(1 + 300Z)}. \tag{5.74}$$

**5.11** Consider a protogalactic cloud of gas that contains as much H and He as the present Milky Way and that occupies a similar volume. Assume that the cloud is homogeneous. Show that the cloud is optically thin to Thomson scattering. Estimate the supernova rate that would be required to maintain the cloud at its virial temperature.

**5.12** (a) By assuming that the square of the gas density within a forming elliptical galaxy is everywhere proportional to the star density ($\rho_{\rm g}^2 = K\rho_*$ – see §8.3), express the rate of radiative cooling $Q_{\rm c}$ of this phase in terms of an integral over $\rho_*$. (b) By assuming that the presently observed stars formed at a uniform rate over a period of $\tau$ Gyr, and that over this period one Type II supernova exploded per $200\,\mathcal{M}_\odot$ of newly-formed stars, obtain an expression for the rate of heating $Q_{\rm h}$ of the ISM by supernovae. (c) Show that if the ISM is isothermal, then $Q_{\rm c}/Q_{\rm h} \propto K$. (d) Consider the instant at which the gas mass is a fraction $\alpha$ of the stellar mass. Use the results of §4.3.4 to show that at this instant $Q_{\rm c}/Q_{\rm h} \propto R_{\rm e}^{-2.2}$. (e) Explain within the context of the leaky-box model why this result, taken with the metal-poverty of low-luminosity ellipticals, implies that radiative cooling cannot have been important for the early evolution of massive ellipticals.

**5.13** In this problem we investigate a more elaborate version of the accretion model presented in §5.3.3. We assume that initially all of the disk mass is in metal-free gas, and that a constant fraction $(1 - q)$ of each infalling gas parcel $\delta M_t$ is locked up by star formation. Thus the corresponding change in gas mass is $\delta M_g = q \delta M_t$.
(a) Using equation (5.57), show that a parametric solution for $Z(M_g)$ is

$$Z = p(1 - q)(1 - e^{-u}) \quad ; \quad M_g = M_{g0} e^{qu}, \tag{5.75}$$

where $M_{g0}$ is the initial gas mass.
(b) Show that the ratio of the stellar mass at $t_1$ to the mass in gas at the present time $t_0$ is

$$\frac{M_s(u_1)}{M_g(u_0)} = \frac{1 - q}{q}(e^{qu_1} - 1)e^{-qu_0}, \tag{5.76}$$

where $u_i$ is the value of the parameter $u$ at time $t_i$. Explain why this relation together with observations of our galactic disk imply $q \ll 1$.
(c) Now consider the case $u_0 \gg 1$, and let $u_1$ be an epoch at which the metallicity $Z_1$ was substantially lower than $Z_0$. Show (i) that the present metallicity $Z_0 \simeq p(1 - q)$; (ii) that $u_1 \simeq -\ln(1 - Z_1/Z_0) \ll 1$; (iii) that

$$\frac{M_s(u_1)}{M_g(u_0)} \simeq -\ln\left(1 - \frac{Z_1}{Z_0}\right)e^{-qu_0}. \tag{5.77}$$

This formula differs from that obtained for the simple accretion model of §5.3.3 [equation (5.59)] only by the presence of the factor $e^{-qu_0}$. Since this factor can take any value from 0 to 1 depending on $q$ and $u_0$, it follows that in this modified accretion model the fraction of low-metallicity stars can be made arbitrarily small.

**5.14** Figure 5.19 shows the evolution of $\Upsilon_V$ for a population with a Salpeter IMF with lower mass cutoff $\mathcal{M} = 0.15\,\mathcal{M}_\odot$. How should the $\Upsilon_V$ scale be changed to make the figure appropriate for mass cutoff $\mathcal{M} = 0.015\,\mathcal{M}_\odot$?

**5.15** Every entry in the last column of Table 5.5 is negative. Hence, show that the spectral indices of a coeval population all increase as the population ages. [Hint: you may wish to use the formula $(\frac{\partial y}{\partial x})_z = -(\frac{\partial y}{\partial z})_x/(\frac{\partial x}{\partial z})_y$, where $z(x, y)$.]

# 6

# Star clusters

In addition to a smooth distribution of stars, the Galaxy contains numerous agglomerations of stars. The study of these **star clusters** has played a key role in the development of our understanding of the Universe. Clusters vary enormously in the compactness of their structure, their luminosities and their metallicities. The more compact, luminous and metal-poor objects are classified as **globular clusters**, while the less compact, less luminous and more metal-rich objects are called **open clusters**. In this chapter, we present the wealth of information that has been gleaned from these systems.

## 6.1 Globular clusters

Globular clusters are roughly spherical agglomerations of stars found through-out the Milky Way (and, indeed, in other galaxies; §4.5) at positions ranging from near the Galactic center out to remote regions in the halo (§10.5.1). The study of these systems has played a key role in the development of our understanding of the Universe. William Herschel's resolution of these diffuse objects into individual stars led him to speculate that more distant nebulae might also be made up of many unresolved stars, and, as we saw in Chapter 1, Shapley's study of the distribution of globular clusters produced a revolution in our understanding of Galactic structure. Even today, as we will see below, studies of globular clusters have a bearing on such fundamental questions as the age of the Universe and the manner in which the Milky Way formed. Since they are some of the simplest-looking structures in the Universe, globular clusters also provide a useful test-bed for theories of stellar dynamics: if we cannot explain the properties of these objects, what hope do we have of understanding any of the more complex structures in the Cosmos?

Quite a number of the Milky Way's globular clusters are naked-eye objects which have been known since ancient times. Twenty-nine found their way into Messier's catalog of objects that should not be confused with comets, and still more are listed in the NGC and IC catalogs.[1] These catalogs were compiled by eye, and so do not include the faintest and lowest surface-brightness clusters. During the twentieth century, some of these fainter systems were discovered serendipitously on photographic plates, while others were discovered in the course of systematic, large-area photographic sky surveys. The most famous such survey was conducted by Abell (1955), who discovered 13 low surface-brightness globular clusters from the Palomar sky survey (hence their designations Pal 1...Pal 13). Combining all these catalogs, we find the current census of globular clusters in the Milky Way numbers $\sim 150$ objects (Monella 1985, Harris 1996). The only clusters which will have slipped through this net either are very low surface-brightness objects, or ones that lie very close to the plane of the Milky Way so that obscuration and confusion with field stars prevents their identification. Apart from these objects, the catalog of Galactic globular clusters is believed to be complete. Table 6.1 lists the physical properties of a number of the more widely-studied systems.

Figure 6.1 gives an indication of the range of properties that we see in the Milky Way's globular clusters. These clusters all lie at distances of $\sim 10\,\mathrm{kpc}$, so differences between the images mainly reflect intrinsic variations from cluster to cluster. Despite their clear differences, these systems all satisfy the defining requirements of a globular cluster: (i) they appear approximately circular in projection (and are hence presumably roughly spherical); and (ii) they seem to be purely stellar systems since no diffuse gas or dust is apparent in them.

Although close to circular in appearance, their images are slightly elliptical. The cluster M19, which is shown in the bottom left panel of Figure 6.1, is amongst the most elongated of the Milky Way's globular clusters with a minor-to-major axis ratio of $b/a \sim 0.73$. A study of 99 globular clusters by White & Shawl (1987) found that they have a mean axis ratio of $b/a = 0.93$, with only 5% more elongated than $b/a = 0.8$. As we shall see in §6.1.11, this modest degree of elongation seems to be caused by mild rotation of the systems.

---

[1] It is slightly unfortunate that the two brightest globular clusters in the sky, 47 Tucanae (47 Tuc) and $\omega$ Centauri ($\omega$ Cen), are both located in the southern hemisphere; their strange designations reflect the fact that northern hemisphere astronomers, who could only observe these objects close to the horizon, mistook them for stars.

**Table 6.1** Characteristics of selected globular clusters

| Name NGC | Other | $R_{gal}$ (kpc) | $M_V$ | $r_h$ (arcmin) | $\log (r_t/r_c)$ | Spec. type | [Fe/H] | $HB_{col}$ | Comments |
|---|---|---|---|---|---|---|---|---|---|
| 104 | 47 Tuc | 7.3 | −9.26 | 2.79 | 2.04 | G4 | −0.76 | −0.99 | Typical metal-rich cluster |
| 5272 | M3 | 11.6 | −8.77 | 1.12 | 1.85 | F6 | −1.57 | +0.08 | Typical intermediate-metallicity cluster |
| 7078 | M15 | 10.2 | −9.07 | 1.06 | 2.50 | F3 | −2.22 | +0.67 | Typical metal-poor cluster. Collapsed core |
| 288 | | 11.4 | −6.54 | 2.22 | 0.96 | F8 | −1.24 | +0.98 | Same [Fe/H] as NGC 362, but has a blue HB |
| 362 | | 9.0 | −8.26 | 0.81 | 1.94 | F9 | −1.16 | −0.87 | Same [Fe/H] as NGC 288, but has a red HB |
| 1851 | | 16.3 | −8.26 | 0.52 | 2.24 | F7 | −1.26 | −0.36 | Same [Fe/H] as NGC 288,362, but bimodal HB |
| 5139 | ω Cen | 6.3 | −10.16 | 4.18 | 1.24 | F5 | −2...−1 | +0.90 | Most luminous. Chemically inhomogeneous |
| | AM 4 | 24.2 | −1.50 | 0.42 | 0.50 | | −2.00 | +0.98 | Least luminous. Least concentrated core |
| 6121 | M4 | 6.1 | −7.06 | 3.65 | 1.59 | F8 | −1.18 | −0.06 | Closest cluster ($d = 2.1$ kpc) |
| | AM 1, E 1 | 117.2 | −4.60 | 0.50 | 1.23 | | −1.80 | −0.93 | Most distant cluster |
| | Liller 1 | 2.3 | −7.42 | 0.45 | 2.30 | | +0.22 | −1.00 | Most metal-rich. Collapsed core |
| 5053 | | 16.5 | −6.64 | 3.50 | 0.82 | | −2.41 | +0.52 | Most metal-poor |

NOTES: $R_{gal}$ is the radius in the Galaxy at which the cluster lies; $r_h$ is the half-light radius; other quantities are as defined in the text.

SOURCE: From data collected in Harris (1996).

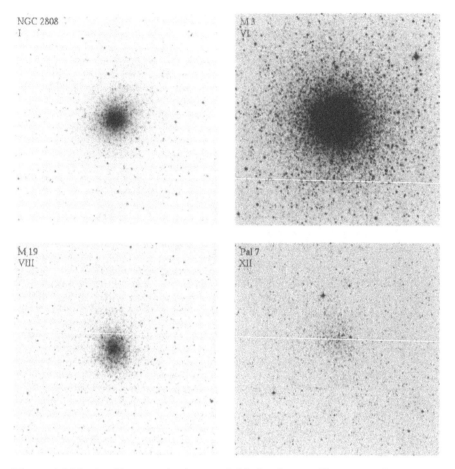

**Figure 6.1** Mosaic of four negative images of globular clusters. The name and concentration class of each cluster is marked. Each image is 15 arcmin on a side, which corresponds to ∼ 40 pc at the distance of these clusters (∼ 10 kpc). [DSS images from the Palomar/National Geographic Society Sky Survey and the SERC/AAO Southern Sky Survey, reproduced by permission]

It is clear from Figure 6.1 that globular clusters vary significantly in their luminosities. The brightest cluster in our galaxy, $\omega$ Cen, has an absolute magnitude of $M_V = -10.4$, and star counts reveal that this luminosity arises from some $\sim 10^6$ stars brighter than $V \sim 20$. At the other end of the spectrum, the faintest clusters detected contain only $\sim 10^3$ stars with corresponding absolute magnitudes of $\sim -3.0$. As is described in detail in §4.5 and §10.5.1, the number of globular clusters at absolute magnitudes between these extremes follows an approximately Gaussian distribution with the most common luminosity corresponding to $M_V \sim -7$.

Figure 6.1 also illustrates the variations in the radial distribution of stars from cluster to cluster. Some clusters (like NGC 2808) are strongly centrally-condensed distributions of stars, while others (like Pal 7) are much more diffuse. These variations were first quantified by Shapley & Sawyer (1927), who defined twelve **concentration classes** designated by Roman numerals I...XII. In this classification, class I clusters show the highest degree of central concentration and class XII clusters the lowest; the categories were defined such that similar numbers of clusters are found in each class. This crude and somewhat subjective measure of the structure of globular clusters has largely been superseded by more detailed analysis of the radial distribution of stars in these systems (see §6.1.10).

More can be learned about the properties of globular clusters from their spectra. Early spectral studies of the integrated light from globular clusters revealed that they have spectra similar to those of individual stars, except that the spectral lines are slightly broadened by differences between the Doppler shifts of the cluster members (see §6.1.11). The integrated spectra differ in line strength from cluster to cluster, covering a range of spectral types from about F3 to G5 (Monella 1985). Since the integrated light from a globular cluster is made up from the contributions of many individual stars of different types, the interpretation of the resulting composite spectrum is not straightforward (Morgan 1959). From studies of the spectra and colors of individual giant stars in clusters, it is now recognized that the primary cause of the variations in spectral type is the difference between the metal abundances in clusters. Measurements show that globular clusters have metallicities which run all the way from very metal poor ([Fe/H] $\sim -2.5$; Geisler, Minniti & Claria 1992) up to solar or even greater (Frogel, Kuchinski & Tiede 1995).

Although [Fe/H] provides a useful measure of metallicity, the use of a single parameter to quantify a globular cluster's heavy element abundances is an oversimplification. Comparisons between high resolution stellar spectra and the predictions of model atmosphere calculations have allowed estimates to be made for many elements in globular cluster stars. It is found that individual abundances, such as [O/H], [Mg/H], etc., vary independently of [Fe/H] – for a review, see Wheeler, Sneden & Truran (1989). The most dramatic example of this phenomenon is provided by oxygen: in a typical globular cluster, the abundance of iron may be $\sim 4\%$ of the solar value, while the abundance of oxygen is more like $\sim 10\%$. In the notation of equation (3.19), these stars have an oxygen enhancement [O/Fe] $\sim 0.4$. Other $\alpha$ nuclides (§5.2.1) are also found to be enhanced. These enhancements are believed to arise because the $\alpha$ nuclides are produced in core-collapse supernovae from short-lived high-mass progenitors (see §5.2.3). As we shall see below, globular clusters formed early in the evolution of the Milky Way, and so the material from which they formed could only have been enriched with metals produced by such short-lived stars.

### 6.1.1 Globular cluster stellar photometry

One of the simplest properties of a globular cluster that we can measure is the distribution of the brightnesses of its component stars. As we have seen in §3.5, the absolute brightnesses and colors of stars are directly related to their masses and ages. Thus, photometry of the stars which make up globular clusters contains important clues to the processes by which the stars in these systems formed as well as their subsequent evolution.

In principle, it is straightforward to obtain photometry of stars in globular clusters. Since globular clusters are compact systems, many stars fit within the field of view provided by a single CCD frame, which makes it much simpler to obtain the requisite data than is the case when measuring the magnitudes of more diffusely distributed field stars. Further, since all the stars in a cluster are at essentially the same distance from us (a globular cluster is typically a few parsecs across at a distance of a few kiloparsecs), we need no distance information to measure the relative luminosities of different cluster stars, and we only need to determine a single distance to calculate the absolute magnitudes of all the cluster members.

There are, however, some practical difficulties. In particular, although most well-studied globular clusters are located away from the Galactic plane, they are still subject to some interstellar extinction. Since globular clusters subtend only small angles on the sky, and are themselves almost entirely devoid of dust and gas, all the stars will be viewed through comparable amounts of absorbing material, and so the amount of absorption can be characterized by a single visual extinction. Fortunately, as we will see below, the color–magnitude diagrams of the stars in globular clusters contain a great deal of structure. By relating this structure to the predicted intrinsic colors and absolute magnitudes of features in the CM diagram, we can solve for both the distance to the cluster and the amount of reddening.

A rather less tractable problem in the study of clusters' stellar photometry arises from their compact nature. As we have seen in §3.2.3, the phenomenon of seeing degrades ground-based images of stars so that their light is spread over a finite disk on the sky. The projected number densities of stars in globular clusters can be so high that light from adjacent stars blends together. In fact, in the central regions of clusters, the stars are so tightly crowded together that individual stars cannot be detected at all. Even outside the region within which individual stars are undetectable, crowding can produce two serious problems for studies of the photometric properties of cluster stars. First, we may still miss some stars entirely because they happen to lie very close together on the sky. Moreover, the difficulty of deciding whether a given patch of light is due to one star or two increases not only with the noise level in the data but also with the magnitude difference between the two stars. Hence crowding causes more faint stars to go undetected than bright ones, especially if the data are noisy. Second, if we do misidentify two or more adjacent stars as a single source, the measured luminosity of such a "star" will be larger than that of any of the real stars of which it is composed. These effects of stellar crowding, known generically as **source confusion**, distort the observed distribution of stellar luminosities in globular clusters. As we shall see below, there are many more faint stars than bright ones in globular clusters, so the impact of source confusion becomes more acute as we look to the properties of cluster stars at fainter magnitudes.

One obvious solution is to restrict our attention to the outer regions of clusters, where the density of stars is sufficiently low that confusion ceases to be an issue. However, at large radii contamination of one's sample by field stars can become a serious problem, and, as we will see in §6.1.10, the photometric properties of stars in the outer parts of a cluster may not be representative of the cluster as a whole. Fortunately, there have been a number of developments which have alleviated the problem of confusion in studies of the more densely-populated inner parts of globular clusters. With the introduction of high-quality digital data from CCDs and the massive expansion of available computing power, it has become possible to implement sophisticated algorithms for disentangling observations of crowded star fields. Software packages such as DAOPHOT (Stetson 1987) have been developed which measure the point-spread function (§4.2.2) from the stars in an image, and then measure the locations and brightnesses of all the objects in the field that have light distributions that match the point-spread function. The contributions that the detected objects make to the total light from the cluster can then be subtracted from the original image to search for residual contributions from fainter stars which were imperceptible in the raw data. Although this procedure recovers the photometry of a crowded field much more efficiently than could be achieved by traditional techniques, some sources will still be lost. Software has therefore also been developed for adding artificial stars to the observed images. By seeing how efficiently the photometry package recovers the existence and brightnesses of such artificial stars as a function of their magnitude, it is possible to estimate the distorting effects of crowding on the observed distribution of stellar luminosities (e.g. McClure *et al.* 1985). More recently, computational power has increased to a point where it is practical to simulate complete images of globular clusters with any intrinsic distribution of stellar properties that we choose. These images can be analyzed in exactly the same way as the real data to see how well the derived photometric properties match up with those derived from real data.

The second major weapon in astronomers' battle against confusion has come from improvements in image quality. Refinements in telescope technology have dramatically reduced the typical seeing (§3.2.3) at optical telescopes, so reliable photometry can be obtained closer to the centers of globular clusters. Even more dramatically, the repaired HST produces images of stars that are a mere 0.1 arcsec across, so much more densely populated regions can be studied before confusion becomes an issue. Combining the imaging capabilities of HST with the crowded-field analysis techniques de-

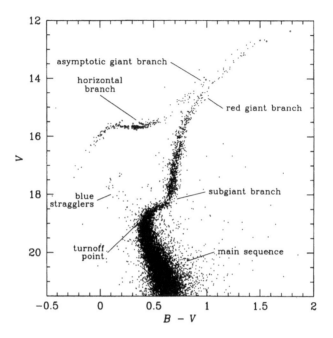

**Figure 6.2** The color-magnitude diagram for the globular cluster
M3. Known variable stars are shown as open circles, and the principal sequences are annotated. [From data published in Buonanno
*et al.* (1994)]

scribed above now allows us to probe stellar photometry in even the densest
regions of globular clusters (e.g. Guhathakurta *et al.* 1996).

### 6.1.2 Color-magnitude diagrams

Once we have calculated the apparent magnitudes for the stars in a cluster
in more than one color, it is straightforward to plot the CM diagram for the
cluster stars. Figure 6.2 shows the CM diagram of the well-studied globular
cluster M3. Comparison of Figure 6.2 with the CM diagram of the solar
neighborhood (Figure 3.5) shows that while most stars are confined to well
defined sequences in both figures, the two diagrams look very different. By
contrast, the CM diagrams of several globular clusters shown schematically
in Figure 6.3 all contain the sequences that are indicated in Figure 6.2: a
main sequence (MS); a subgiant branch (SGB); a red giant branch (RGB)
and a horizontal branch (HB). In addition many CM diagrams show an
asymptotic giant branch (AGB) that is clearly distinct from the RGB. Below,
we discuss the importance of the similarities and differences between these

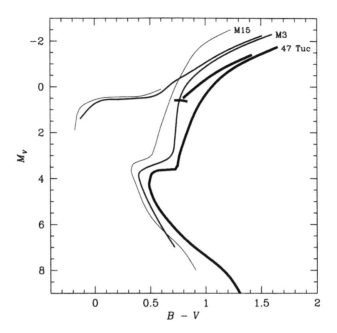

**Figure 6.3** Schematic illustration of the principal sequences in the CM diagrams for three globular clusters. The systems shown are a metal-rich cluster (47 Tuc), an intermediate-metallicity cluster (M3), and a metal-poor cluster (M15). [Sequences from Hesser *et al.* (1987) and Buonanno *et al.* (1994)]

various sequences for our understanding of the properties of globular clusters.

**The main sequence and subgiant branch** One striking difference between a globular-cluster CM diagram and that for the solar neighborhood is that in the former the MS stops well short of the luminosity of the HB,[2] while in the latter it continues up to include some of the brightest stars in the Galaxy. In §5.1 we saw that these bright, upper-MS stars have short lifetimes. Hence one possible explanation for this difference is that, while the solar neighborhood contains young stars, all the stars in a globular cluster are old and therefore none of them lie on the upper MS.

As Figure 6.2 illustrates, the SGB of a globular cluster is extremely narrow, and abuts the MS at a well defined **turnoff point**.[3] This structure indicates that all the stars that originally lay on the MS just above the turnoff point are currently evolving off the MS en masse. This implies that

---

[2] A few stars, termed **blue stragglers**, are sometimes found above the turnoff point in the CM diagram – see §6.1.8.

[3] The precise definition of the turnoff point is the point at which the ridge-line of the MS/SGB runs vertically up the CM diagram.

all these stars have essentially identical ages, namely their MS lifetime. In view of this fact, it is natural to assume that all stars in the globular cluster have this same age. Certainly, the absence of significant numbers of stars evolving away from the MS below the turnoff point indicates that the cluster does not contain stars much older than the MS lifetime of the turnoff stars. The sharpness of the observed sequences places strong limits on the length of time over which clusters formed. If the stars formed over an extended period, then the observed SGB and RGB would be made up from superposition of isochrones of different ages. Since the locations of these calculated sequences, particularly the SGB, shift significantly with age, the narrowness of these features in the CM diagram of clusters constrains the period of stellar formation in these systems to no more than $\sim 2\%$ of the clusters' age (Stetson 1993).

As discussed in §5.1.5, the locations of both the MS and SGB depend on the metallicity of the cluster: Figure 6.3 illustrates how these sequences in the CM diagram are observed to shift to the right, and how the SGB becomes flatter as we go from metal-poor to metal-rich globular clusters.

Deep observations make it possible to trace the MS clearly over a number of magnitudes of brightness (see Figure 6.2[4]). If we can accurately determine the location of the MS in the CM diagram, we can use this knowledge to measure the distance to a cluster by a technique termed **main sequence fitting**. The essence of the method is very simple: from the MS of nearby stars with known distances, we can establish a unique relationship between the color of a MS star and its absolute magnitude; using this relationship to predict the absolute magnitudes of MS stars in a cluster on the basis of their colors, and comparing this intrinsic brightness to the apparent magnitude of the stars then yields the distance to the cluster via equation (2.36). If we carry out this procedure for a whole sample of cluster stars, we are effectively shifting the MS in an observed CM diagram like Figure 6.2 up and down the vertical axis until it matches up with the MS in the CM diagram of nearby stars (Figure 3.5). The amount by which the scale has to be shifted is the distance modulus of the cluster.

In practice, a number of factors complicate the MS fitting procedure. The most important of these factors is the low metallicity of many globular clusters. Since the location of the MS in the CM diagram depends quite strongly on the metallicity of the stellar population, the calibration of globular cluster star magnitudes is usually tied to the absolute magnitudes of the metal-poor subdwarfs for which we have trigonometric parallaxes, and hence good distance determinations. Even with the vastly improved data available from Hipparcos, subdwarfs are sufficiently rare that reliable parallaxes are

---

[4] The apparent broadening of the sequence at the faint end is a purely instrumental effect: the uncertainties in magnitudes of the faintest stars detected become quite large, and so the error in the color $B - V$, the small difference between two large ill-determined numbers, becomes very large.

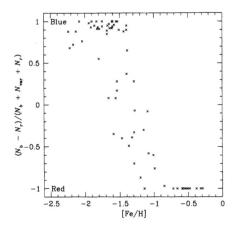

**Figure 6.4** Plot showing how the color of the HB in a globular cluster CM diagram varies with the cluster's metallicity. [After Lee (1990) from data kindly provided by Y.-W. Lee]

only available for $\sim 20$ suitable objects (Reid 1997), so the calibration is not all that firmly established. Also, as we shall see in §6.1.3, stars start to evolve to the right in the CM diagram before the turnoff point comes down to them. We must therefore be rather careful when determining distances by MS fitting to allow for possible age differences between the populations being compared.

**The horizontal branch**    The horizontal branches in the CM diagrams of globular clusters also differ dramatically from the equivalent feature for nearby stars. With no additional information, we would be hard pressed to conclude that the extensive HB apparent in Figure 6.2 is in any way related to the red clump in Figure 3.5. In fact, the differences between these features arise purely from the very different metallicities of M3 and the solar neighborhood: as Figure 6.3 illustrates, the HB becomes shorter and moves to the right in the CM diagram as we look to higher metallicity systems, so the metal-rich solar neighborhood contains only the vestigial red clump as its HB.

The HB is also where the RR Lyrae variables lie in the CM diagram (§5.1.10). As was first noted by Oosterhoff (1939), there appear to be two distinct types of globular cluster RR Lyrae population: clusters with red HBs contain mostly RR Lyrae stars of Bailey type $ab$ (§5.1.10) with periods of $P \sim 0.55$ days; while those with blue HBs contain more $c$-type RR Lyrae stars, and the $ab$-type stars that they do contain have longer periods, $P \sim 0.65$ days. These two types of cluster are designated **Oosterhoff class** I and II, respectively. This correlation between HB color and RR Lyrae period arises from the underlying dependence of both phenomena on cluster metallicity.

Since the instability strip populated by the RR Lyrae stars lies roughly in the middle of the region spanned by the HB, its position provides a useful fiducial mark for defining the color of the HB: if a globular cluster's CM diagram has a HB which contains $N_{\mathrm{var}}$ RR Lyrae variables, $N_r$ stars on the

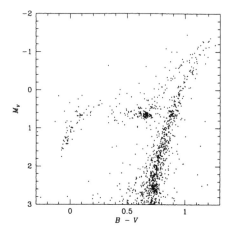

**Figure 6.5** Section of the CM diagram for the globular cluster NGC 1851. Note the concentrations of stars at both the blue and red ends of the horizontal branch. [From data published in Walker (1992b)]

red side of the variables, and $N_b$ stars on the blue side, then a measure of the overall color of the HB is provided by $HB_{col} = (N_b - N_r)/(N_b + N_{var} + N_r)$ (Lee 1990). Figure 6.4 shows the expected general trend in this quantity with metallicity: the average HB color varies from very blue ($HB_{col} \sim 1$) for the most metal-poor clusters to very red ($HB_{col} \sim -1$) for metal-rich clusters. In fact, the highest metallicity clusters have such red HBs that they contain essentially no stars on the blue side of the RR Lyrae stars, so $HB_{col}$ saturates as a color indicator at its minimum value of $-1$.

There are indications that the relation between $HB_{col}$ and metallicity is not monotonic, in that the most metal-poor clusters ($[Fe/H] < -2$) are somewhat redder than slightly metal-richer clusters – an effect first noted by Renzini (1983). As Sandage (1990) has pointed out, this non-monotonic behavior explains the dichotomy between the two Oosterhoff classes: a cluster with $HB_{col} \sim 1$ has such a blue HB that it contains almost no RR Lyrae stars. This absence of RR Lyrae stars means that clusters with metallicities in the range $-2 \lesssim [Fe/H] \lesssim -1.75$ cannot be classified using Oosterhoff's scheme, so there is a gap in the range of values of $[Fe/H]$ of clusters with Oosterhoff classifications. Since RR Lyrae period increases monotonically with the star's metallicity [equation (5.26)], this gap also shows up as a minimum in the period distribution; clusters on the metal-rich side of the gap form a group whose RR Lyrae stars have short periods, while those on the metal-poor side of the gap contain only long-period RR Lyrae stars.

The trend in HB color with metallicity in Figure 6.4 is clear, but there is also a spread in colors at any given metallicity. The width of the relation is greater than the uncertainties in metallicity and color determinations, and thus there are intrinsic variations between the colors of globular cluster HBs at any given metallicity. For example, the clusters NGC 288 and NGC 362 have essentially identical metallicities ($[Fe/H] = -1.2 \pm 0.1$ in each case), yet NGC 288 has a very blue HB ($HB_{col} = 0.98$), while NGC 362 has a very red

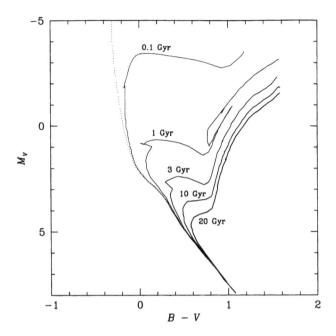

**Figure 6.6** Theoretically calculated isochrones showing how a stellar population with $Z = 0.004$, $Y = 0.24$ evolves away from the ZAMS (dotted line) in the CM diagram. Each isochrones is labeled by its age. [From the calculations of Bertelli *et al.* (1994)]

HB ($HB_{col} = -0.87$). Also, it has long been known (e.g. Racine 1971) that there are cases in which the HB cannot be described by a single characteristic color because the population along the HB displays a bimodal distribution, with concentrations of stars at both the red and blue ends of the sequence – see Figure 6.5.

### 6.1.3 Comparison with Theoretical CM diagrams

In order to see what can be learned from the observed CM diagrams of globular clusters, we must compare them with the predictions of the stellar population calculations that were introduced in §5.4. As we have seen above, the morphology of globular clusters' CM diagrams leads us to believe that all the stars in these systems formed in a single epoch, so the procedure adopted for calculating an appropriate model CM diagram is as follows:

(1) Choose initial abundances for the chemical elements in the stellar population. For heavy elements, the abundances can be matched to the observed line strengths in the spectra of stars in the cluster to be modeled. The relative overabundance of oxygen mentioned above is doubly impor-

tant in these stellar-structure calculations, since it affects the hydrogen burning efficiency of the star via the CNO cycle as well as altering the opacity of the stellar material. For the initial helium abundance, the primordial value of $Y = 0.23$ (§5.1.6) is usually adopted. Once these abundances have been fixed, we can calculate the initial locations of the stellar population on the ZAMS.

(2) Evolve the population forward in time, by repeatedly solving the stellar structure equations for each star, and keeping track of the chemical evolution which occurs as elements are mutated by fusion reactions.

(3) For each time step, calculate the luminosities and colors for all the stars in the population. The curve connecting all the stars in the color-magnitude diagram is called an **isochrone** from the Greek for "same time", since it represents a snapshot of the properties of the stellar population when they are all at an identical age.

If our assumed initial chemical composition is correct, and we have performed the stellar structure calculations accurately, then the calculated isochrone of the right age should trace the populated sequences in the observed CM diagram. Comparing isochrones to observed sequences thus provides a powerful tool not only for measuring the ages of globular clusters, but also for testing our understanding of the basic physics of stellar structure. Families of isochrones have therefore been calculated for a wide range of different metallicities, ages, and physical assumptions by researchers around the world, including vandenBerg & Bell (1985), Bergbusch & VandenBerg (1992), Chieffi & Straniero (1989), Bertelli *et al.* (1994) and the Yale group (Green, Demarque & King 1987; Chaboyer *et al.* 1995). Figure 6.6 shows the isochrones produced by one such calculation. Stars peel away from the MS one after the other, starting with the most luminous. Comparing Figures 6.3 and 6.6, it is clear that isochrones for ages somewhere between $\sim 10\,\mathrm{Gyr}$ and $\sim 20\,\mathrm{Gyr}$ can approximately reproduce a globular cluster's MS and SGB.

We can also determine the effects of varying the element abundances on the isochrones. Figure 6.7 shows what happens to the calculated isochrone for a system of age $14\,\mathrm{Gyr}$ as $Y$ and $Z$ are varied. Changing $Y$ from $Y = 0.2$ (comparable to the primordial value – see §5.1.6) to $Y = 0.3$ (approximately the value appropriate for stars with solar abundances) shifts the MS downwards by $\sim 0.2$ magnitudes and makes the SGB somewhat steeper. Increasing $Z$ from $Z = 0.0001$ (appropriate for a metal-poor globular cluster such as M15) to $Z = 0.006$ (appropriate for a metal-rich cluster such as 47 Tuc) shifts the isochrones to the right by $\sim 0.2$ in $B - V$, and significantly changes the shape of the SGB. Comparing Figures 6.3 and 6.7, we can see that the changes in the isochrones with $Z$ match the observed differences between the MSs and SGBs of globular clusters with a range of metallicities. The shift of the sequences to the right in the CM diagram with increasing $Z$ arises principally from the line blanketing effects of heavy elements, which suppress the emission of blue light by metal-rich stars (§2.3.2).

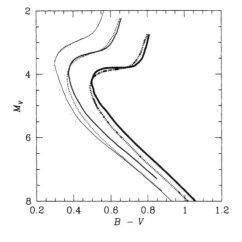

**Figure 6.7** Illustration of the effects of varying $Y$ and $Z$ on the shape and position predicted for the 14 Gyr isochrone. The line width indicates the metallicity of the isochrone: the heaviest lines are for $Z = 0.006$; the intermediate lines are for $Z = 0.001$; and the lightest lines are for $Z = 0.0001$. The solid lines are for a Helium abundance of $Y = 0.2$, and the dotted lines are for $Y = 0.3$. [From the calculations of vandenBerg & Bell (1985)]

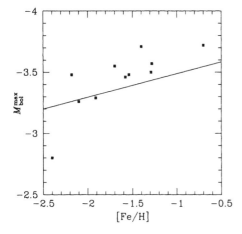

**Figure 6.8** Absolute bolometric magnitude of the brightest giant star as a function of the host globular cluster's metallicity (Sarajedini & Milone 1995). The line indicates the relation predicted by stellar evolution calculations (Da Costa & Armandroff 1990).

Ideally, we would carry out similar comparisons between the results from calculations of the later stages of stellar evolution and other features in the CM diagram such as the RGB and the HB. Unfortunately, as we have seen in §5.1, we are unable to model accurately the deep convective layers and mass loss in giant stars. Thus, the uncertainties in the calculations become large when trying to follow these later stages of stellar evolution, and the match between calculation and observation becomes rather poor. Calculations do predict the observed tendency for the color of the almost vertical RGB to depend on metallicity in the sense that the most metal-poor clusters have the bluest RGBs (see Figure 6.3). This dependence is once again attributable to the effects of line blanketing. However, the fit between theory and observation is poorer for the RGB than it is for the MS and SGB (Da Costa & Armandroff 1990).

As we look to still-later stages of stellar evolution further up the RGB, the discrepancies grow. Figure 6.8 compares the observed absolute magnitude of the brightest star in a number of globular clusters of different metallicities with the predictions of stellar-evolution calculations. It is clear that the match between theory and observation for these evolved stars is far from perfect. In particular, it is noteworthy that in most cases the brightest star is brighter than the theoretically-predicted maximum luminosity, whereas a simple argument leads us to expect the reverse: given the small number of stars near the top of any RGB and the speed of their evolution through this stage, we cannot expect to catch even one star right at the tip of the RGB. We should not be surprised at the inability of calculations to reproduce the luminosity of the tip of the RGB: the rapidity with which the luminosity $L$ of a star evolves at this point implies that $L$ is a very sensitive function of the star's internal structure. Consequently, small errors in the assumed internal constitution of a star must lead to large errors in the predicted luminosity. A good understanding of the luminosities of these very bright stars would enable us to use them as standard candles for estimating the distances to other galaxies at quite large distances (vandenBerg & Durrell 1990).

When a star in a globular cluster evolves to the tip of the RGB, it undergoes the helium flash (§5.1.3). This explosive phenomenon causes an almost instantaneous mass loss and a rearrangement of the structure of the star, which we have no hope of modeling in detail. It is thus not possible to follow the evolution of a star from the RGB on to the HB where it settles down to core helium burning. This transition is therefore treated in a semi-empirical manner. For a given cluster, stellar structure calculations allow us to calculate the approximate mass of a star at the very tip of the RGB, $\mathcal{M}_{RG}$, and the mass of the hydrogen-depleted helium core that it has acquired by this point, $\mathcal{M}_c$. A star marginally more massive than this will have already undergone the helium flash, after which it is assumed to have lost a mass $\Delta\mathcal{M}$ from its atmosphere, to end up with a total mass $\mathcal{M}_{HB} = \mathcal{M}_{RG} - \Delta\mathcal{M}$. Its helium core is unaffected by the helium flash and still contains a mass $\mathcal{M}_c$. After the helium flash, a star is burning helium in its core, and hence soon arrives on the ZAHB (§5.1.2). At this point, the rate at which its structure changes slows to a point where we can again use stellar structure calculations to estimate the observable properties of the star and follow its subsequent evolution.

As we have seen in §5.1.7, stars do not spend a very large fraction of their lives on the HB, so all the stars currently in this region of the CM diagram must have undergone the helium flash recently and hence have evolved from stars with very similar masses, $\mathcal{M}_{RG}$. Hence if all stars currently on the HB had lost the same amount of mass $\Delta\mathcal{M}$, they would all arrive at the same point on the ZAHB. Moreover, stars do not move far along the HB during core helium burning, so the HB would be very short. The finite extent of cluster HBs (Figure 6.2) therefore implies that at fixed $\mathcal{M}_{RG}$ there must be a spread in values of $\Delta\mathcal{M}$ and therefore of $\mathcal{M}_{HB}$. Rood (1973) modeled the

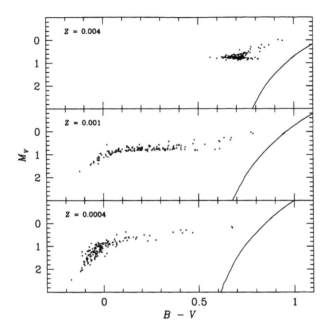

**Figure 6.9** The HB morphologies for three simulated clusters with the metallicities indicated. All three clusters have the same age (13 Gyr) and helium abundance ($Y = 0.22$). Individual points show the predicted locations of stars which have undergone different amounts of mass loss as described in the text, and are currently evolving through their core helium burning phase. RR Lyrae stars are shown as open points. The line shows the location of the RGB from the isochrones in Green, Demarque & King (1987). [After Lee, Demarque & Zinn (1990) from data kindly supplied by Y.-W. Lee]

probability distribution of $\mathcal{M}_{\mathrm{HB}}$ by

$$
P(\mathcal{M}_{\mathrm{HB}}) \propto
\begin{cases}
(\mathcal{M}_{\mathrm{HB}} - \mathcal{M}_c)(\mathcal{M}_{\mathrm{RG}} - \mathcal{M}_{\mathrm{HB}}) \\
\quad \times \exp\left[ -\frac{(\langle\mathcal{M}_{\mathrm{HB}}\rangle - \mathcal{M}_{\mathrm{HB}})^2}{2\sigma_{\mathcal{M}}^2} \right] & \mathcal{M}_c < \mathcal{M}_{\mathrm{HB}} < \mathcal{M}_{\mathrm{RG}}, \\
0 & \text{otherwise.}
\end{cases}
$$

$$(6.1)$$

This complex functional form is designed to ensure that stars neither gain mass nor lose part of their helium cores as they pass from the RGB to the HB. The quantity $\langle\mathcal{M}_{\mathrm{HB}}\rangle$ is a measure of the average mass of the stars arriving on the HB, while $\sigma_{\mathcal{M}}$ describes the dispersion in $\mathcal{M}_{\mathrm{HB}}$ (see Problem 6.1). By comparing Figures 6.3 and 6.9, we see that such a spread in $\mathcal{M}_{\mathrm{HB}}$ enables theory to reproduce the observed structure of the HB and its variation with metallicity. Typically, one requires $\mathcal{M}_{\mathrm{RG}} - \langle\mathcal{M}_{\mathrm{HB}}\rangle \sim 0.2\,\mathcal{M}_\odot$ and $\sigma_{\mathcal{M}} \sim$

$0.02 \, \mathcal{M}_\odot$ (e.g., Lee, Demarque & Zinn 1994).

These HB calculations explain the variation in HB color with cluster metallicity illustrated in Figure 6.4: as for the MS and SGB, most of the trend is caused by line blanketing in metal-rich stars (§2.3.2). The simulations also follow the evolution of HB stars as they move away from the ZAHB, and show how the stars begin to populate the AGB(see Figure 6.9).

Although this semi-empirical approach allows us to understand many of the features of the HB, its rather ad hoc formulation is unsatisfying. Why do some stars lose more mass than others in moving to the HB? The simple mass-loss model also fails to explain the differences in HB color seen in clusters with identical metallicities (see Figure 6.4), and does not reproduce bimodal HBs like the one shown in Figure 6.5. Clearly, we are still missing some of the physics that dictates the late stages of evolution in globular cluster stars.

### 6.1.4 Globular cluster ages

The discussion above shows that we understand the physics of stellar structure to the extent that we can use it to explain many of the features observed in the CM diagrams of globular clusters. That being the case, we can use the comparison between models and observation to tie down some of the basic physical parameters of the cluster. Of these parameters, perhaps the most interesting is age. As some of the oldest objects in the Milky Way, globular clusters place one of the strongest lower limits that we have on the age of the Universe, while possible differences between the ages of clusters at different locations in the Galaxy may provide important clues as to how the Milky Way formed (§10.7). Intense interest has therefore been focused on using CM diagrams as a diagnostic of cluster ages.

**Turnoff point ages**     As we have discussed, the presence of a sharp turnoff point from the MS provides evidence that the stars in a globular cluster formed in a single epoch. Since the location of the turnoff point works its way down the MS as a cluster ages (see Figure 6.6), the absolute magnitude of the turnoff point, $M_V(\mathrm{TO})$, provides a direct measure of a cluster's age, which can be calibrated using isochrone calculations. Quantitatively, Bergbusch & VandenBerg (1992) have found that $M_V(\mathrm{TO})$ is related to cluster age, $t$, and metallicity, [Fe/H], by

$$M_V(\mathrm{TO}) = 2.70 \log(t/\mathrm{Gyr}) + 0.30[\mathrm{Fe/H}] + 1.41. \qquad (6.2)$$

To extract a cluster's age from this equation, we first determine the apparent magnitude of the turnoff point. Then we find the cluster's distance, either by MS-fitting (§6.1.2) or by using the cluster RR Lyrae stars as standard candles (§5.1.10), and use it to calculate the absolute magnitude of the turnoff point. Finally, we measure the cluster's metallicity from spectra. Crudely, we can see from Figure 6.3 that clusters typically have a turnoff point at

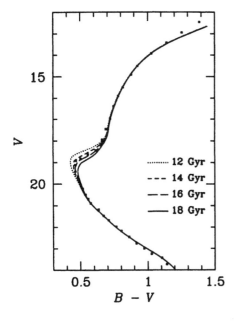

**Figure 6.10** A family of isochrones with ages as marked fitted to the MS and SGB in the CM diagram of M15 (points). The adopted distance modulus in this fit is $(m - M) = 15.4$, and the color excess is $E(B - V) = 0.10$. A residual color correction of $\delta(B - V) = 0.015$ has also been applied. [After Durrell & Harris (1993)]

$M_V(\mathrm{TO}) \approx 4$; for a typical globular cluster of metallicity $[\mathrm{Fe/H}] = -1.5$ (see Figure 10.32), this implies that $t \sim 13\,\mathrm{Gyr}$.

**Isochrone fitting** The vertical nature of the isochrones at the turnoff point means that the exact value of $M_V(\mathrm{TO})$ is not easy to measure observationally, and we should be able to obtain more accurate results by matching cluster sequences to theoretical isochrones throughout the transition from the ZAMS to the RGB – see Figure 6.6. Given the uncertainties in models of stars on the RGB, we focus on the upper end of the MS and the SGB.

The most straightforward way to exploit the shape of the combined MS and SGB as an age indicator is to fit the locations of both sequences directly to the predictions of isochrone calculations. The isochrone which minimizes the discrepancies between the observed and calculated sequences gives the estimated cluster age. By leaving the distance modulus as a free parameter in this fitting process, we can also solve simultaneously for the distance to the cluster. Figure 6.10 shows a typical example of this procedure, which compares the mean locus of stars on the MS, SGB and RGB of the globular cluster M15 to a family of isochrone. It is clear from this figure that a good match is achieved for an age of close to 15 Gyr.

We must take some care when interpreting the apparent success of this fitting procedure. As we have seen in §5.1, there are still considerable shortcomings in our understanding of stellar structure, particularly with regard to the treatment of convection. There are also still non-negligible uncertainties that arise from the conversion of model isochrones, for which we calculate luminosities and temperatures, into the observed CM diagram (§5.1.1). It is

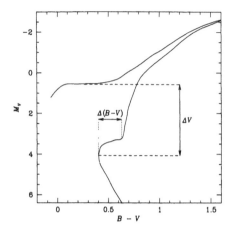

**Figure 6.11** Illustration of the distance-independent age diagnostics, $\Delta V$ and $\Delta(B - V)$, as calculated from the principal sequences in the CM diagram for a typical globular cluster.

therefore not surprising that the calculated isochrones usually fail to match the observed CM diagrams of globular clusters exactly. For this reason it is customary to apply a correction to the isochrones in the form of a small color shift, $\delta(B-V)$, in order to obtain a good fit between theory and observation. With good isochrone models such as those illustrated in Figure 6.10, these shifts are now typically only $\delta(B - V) \sim 0.02$. Nevertheless, since they are non-zero we must still be missing at least some of the detailed physics of these systems. Moreover, there is no reason why this residual error should take the form of a simple color shift for all stars, so we must be careful not to over-interpret the success or failure of different models to match the observations at this level. It is also worth noting that the color at which the turnoff occurs provides one of the major constraints in the isochrone fit procedure for tying down a cluster's age, and so the systematic uncertainty in color can represent a significant systematic error in the estimated age – see Problem 6.2.

**The $\Delta V$ method**    One of the biggest problems in the procedures described above is that the derived age is strongly coupled to the calculated distance modulus: if an isochrone has a turnoff point that lies above the observed value for a cluster, we could improve the fit either by using an older isochrone or by increasing our estimate for the distance to the cluster. A 10% uncertainty in the distance to a cluster corresponds to an error of $\sim 0.2$ in its distance modulus and hence the absolute magnitude scale for its CM diagram. Such an uncertainty in $M_V(\text{TO})$ yields a 20% error in the age inferred from equation (6.2). Similarly, a small uncertainty in the correction for extinction can lead to a sizeable error in the derived age.

We can circumvent these problems by instead basing our age estimate on the quantity $\Delta V = M_V(\text{TO}) - M_V(\text{HB})$, the difference in magnitude between the turnoff point and the horizontal branch in a cluster's CM diagram (see Figure 6.11). The brightness of the horizontal branch depends little on the mass of stars which populate it (and hence the age of the cluster) while the brightness of the turnoff point decreases quite rapidly with time, and so the value of $\Delta V$ should provide a useful measure of cluster age. Since this parameter depends only on the relative brightnesses of these features, it is independent of the adopted distance modulus. The clear definition of the turnoff point and horizontal branch in the CM diagram means that they can be identified without calibrating the absolute color of the cluster, so the measurement of $\Delta V$ is not compromised by uncertainties in absolute colors due to reddening. Further, since these features have comparable intrinsic colors, any reddening will affect the absolute magnitude of both by the same amount, and so we do not have to correct $\Delta V$ for any differential reddening effects. Clearly, $\Delta V$ provides a very robust diagnostic for investigating the ages of globular clusters; its only negative characteristic is the difficulty in specifying $M_V(\text{TO})$ mentioned above. Combining equations (5.4) and (6.2), we find that

$$\Delta V = 2.70 \log(t/\text{Gyr}) + 0.13[\text{Fe}/\text{H}] + 0.59. \tag{6.3}$$

Hence $\Delta V$ depends strongly on the age of the cluster, and only weakly on metallicity[5] (see Problem 6.3).

The values of $\Delta V$ derived for globular clusters are all remarkably similar: a compilation of data for 43 clusters by Chaboyer, Demarque & Sarajedini (1996) found that $\Delta V$ has a mean value of 3.55, and a standard deviation from cluster to cluster of only $\sim 0.15$. The small standard deviation is close to what we would expect from the measurement errors alone, so there is very little intrinsic variation in this quantity between clusters (but see also §6.1.6). Since the globular clusters in this sample have metallicities in the range $-2.5 < [\text{Fe}/\text{H}] < -0.5$, equation (6.3) implies that their ages are $\sim 15 \pm 1\,\text{Gyr}$.

**The $\Delta(B - V)$ method**    Other distance-independent properties of the CM diagram have also been employed as cluster age estimators. Sarajedini & Demarque (1990) advocate the use of the quantity $\Delta(B-V) = (B-V)_{\text{SGB}} - (B - V)_{\text{TO}}$, the difference between the color of the point at which the SGB turns up toward the RGB and the color of the turnoff point (see Figure 6.11). Like the $\Delta V$ method, this approach does not depend on the adopted distance modulus or reddening correction. It has the further advantage that $\Delta(B-V)$ turns out to be almost independent of metallicity, since both $(B-V)_{\text{SGB}}$ and $(B - V)_{\text{TO}}$ vary with heavy-element abundance in almost the same way. A similar technique has been developed by vandenBerg, Bolte & Stetson (1990), who measure the color and magnitude of the base of the giant branch relative to both the turnoff point and the MS of a cluster's CM diagram.

---

[5] The amount of variation in $\Delta V$ with metallicity is dictated by the still-disputed metallicity dependence of $M_V(\text{HB})$ (Bounanno, Corsi & Fusi Pecci 1989, Chaboyer, Demarque & Sarajedini 1996).

The principal disadvantage of methods that compare the colors of features in the CM diagram is that uncertainties in the physics of stellar structure calculations affect colors far more than absolute magnitudes. The luminosity of a star is fixed by the rate at which energy is generated by nuclear fusion in its core, which is a relatively well-understood process. The stellar color, on the other hand, depends on the effective temperature of the star, which in turn is set by its radius [see equation (3.15)]. The radius of the star is dictated by the outer structure of the star, which depends on the uncertain physics of convection and atomic opacities. The $\Delta(B - V)$ method is therefore more appropriate for comparing the relative properties of clusters to see if they differ (see §6.1.6) rather than for estimating absolute ages. It is nonetheless heartening that the observed values of $\Delta(B - V)$ are quite consistent with those calculated from isochrone models with ages of $\sim 15\,\mathrm{Gyr}$ (Sarajedini & Demarque 1990).

**Comparison with the age of the Universe**    The techniques described above, together with various hybrid combinations of these methods, all produce a fairly consistent picture showing that globular clusters formed some $\sim 15\,\mathrm{Gyr}$ ago. Quite how early globular clusters formed in the evolution of the Universe is an open question. However, the presence of quasars at redshifts as high as $z \approx 5$ implies that massive, well-ordered structures formed early on, and the presence of carbon, nitrogen and oxygen emission lines in these quasars show that the stars which synthesize such elements must already have formed. If we allow $\sim 1\,\mathrm{Gyr}$ for the formation of globular clusters (Sandage 1993a), we thus find that the Universe is $\sim 16\,\mathrm{Gyr}$ old. This value conflicts with the age of the Universe as predicted by its cosmological expansion. As we shall see in Chapter 7, most estimates of cosmological parameters favor an age significantly less than 16 Gyr; it is only by adopting one of the most extreme acceptable cosmological models that we can obtain an age for the Universe as large as 16 Gyr. Thus, either we must be prepared to adopt these most extreme acceptable cosmological models, or we must reconsider the absolute globular cluster age determinations.

The traditional view is that the weakness probably lies in the measurements of cosmological parameters, since there are still a number of internal inconsistencies in their derivation, while the age estimates of globular provide a more coherent picture. However, the degree of consistency in globular cluster age estimates may be deceptive. The absolute calibration of all of these estimates is obtained from isochrone calculations. Thus, any error in these calculations will alter all the age estimates in a correlated manner. Uncertainties in basic physics such as the opacities of stellar atmospheres affect the properties of stars at all stages in their evolution, so age estimates that use different features in the CM diagram will all be affected by one incorrect physical assumption. Errors in different stages of the stellar structure calculations are also strongly correlated because of the evolutionary nature of the problem. If, for example, we had missed some important piece of physics which shortens the MS lifetime of all stars by $\Delta t$, then stars would reach the MS turnoff and all subsequent points in the CM diagram $\Delta t$ sooner than our calculations had predicted, but the locations of the sequences would be unaffected. Thus, all the methods for comparing data and observation would consistently indicate that the two matched, but the derived ages would all be out by $\Delta t$. We cannot anticipate what pieces of physics may be missing from the existing stellar structure calculations, nor how large an affect they may have on globular-cluster age estimates. However, the recent discussion of the effects of helium diffusion (§5.1), which may lower MS lifetimes by as much as $\Delta t \sim 2\,\mathrm{Gyr}$ (Proffitt & Michaud 1991, Chaboyer, Sarajedini & Demarque 1992), should give some impression of the possible impact of phenomena which have yet to be incorporated. It is still quite possible that future changes in the modeling of globular clusters will lower the estimates of their ages to a point which is compatible with less extreme values of the cosmological parameters.

**Variations in age** We have seen that globular clusters are all old, but are they all equally old? Since the uncertainty in individual cluster ages is at least 2 Gyr, this is not an easy question to answer. Fortunately, however, much of the uncertainty in age determinations takes the form of an unknown systematic offset in absolute age, so it is possible to estimate age *differences* between clusters with very much greater accuracy [see Stetson, VandenBerg & Bolte (1996) for a review of relative age determinations for globular clusters]. Since the uncertainties in isochrone calculations depend on errors in our treatment of such features as convection theory and opacities, which, in turn, depend on the metallicity of the system, we can further minimize the errors in calculating relative ages by comparing systems with similar metallicities. For example, in the case of the equal-metallicity pair NGC 288 and NGC 362 (see page 338), the $\Delta(B-V)$-like methods imply that NGC 362 is $3 \pm 1\,\mathrm{Gyr}$ younger than NGC 288 (Sarajedini & Demarque 1990, vandenBerg, Bolte & Stetson 1990). Analysis of the cluster Pal 12, which has comparable metallicity ([Fe/H] $= -1.1 \pm 0.2$), imply that it is more youthful still, some $\sim 4\,\mathrm{Gyr}$ younger than NGC 362 (vandenBerg, Bolte & Stetson 1990). Similarly, M3 ([Fe/H] $= -1.7 \pm 0.1$) appears to be $\sim 1\,\mathrm{Gyr}$ older than M13 ([Fe/H] $= -1.7 \pm 0.1$), although this measurement lies right at the lower limit of what can be determined from the relative colors of the turnoff and giant branch (vandenBerg, Bolte & Stetson 1990). It is notable that age differences seem to become smaller as we look to lower metallicity clusters.

The $\Delta V$ method has also been applied to the measurement of relative cluster ages. The points in Figure 6.12 show the values of $\Delta V$ for a sample of globular clusters as a function of metallicity. The lines in this figure show the predictions of equation (6.3) for four cluster ages. Much of the scatter in the observed values can be attributed to observational uncertainties, but for given metallicity there does appear to be a real spread in $\Delta V$ consistent with variations in cluster ages of $\sim 6\,\mathrm{Gyr}$ (Chaboyer, Demarque & Sarajedini 1996). In Figure 6.12 the points do not scatter about any one of the sloping lines of constant age; if anything, $\Delta V$ decreases rather than increases with

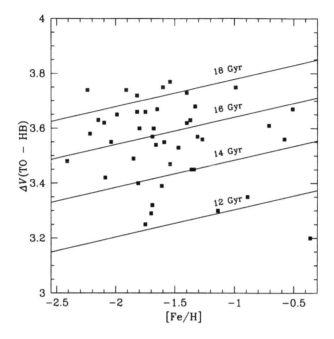

**Figure 6.12** Difference $\Delta V$ in $V$ magnitude between the horizontal branch and the turnoff point as a function of metallicity for a
sample of globular clusters. The lines show the values of $\Delta V$ as
predicted for the ages shown. [From data published in Chaboyer,
Demarque & Sarajedini (1996)]

metallicity, so the most metal-rich globular clusters are, on average, several
billion years younger than the metal poor clusters.

This correlation could reflect a general trend in the cluster population,
but it could also be explained if the Milky Way has tidally captured a few
young, metal-rich globular clusters from the Large Magellanic Cloud – such
clusters would populate the lower right-hand corner of Figure 6.12 (Lin &
Richer 1992).

### 6.1.5 Metallicities of globular clusters

As we have seen above, most of the variations between the shapes of sequences in globular cluster CM diagrams can be attributed to differences
between the metallicities of clusters. Once again, the narrowness of the sequences in a particular cluster imply that the cluster must be chemically
homogeneous, as a range in metallicity at a particular evolutionary point
would broaden the sequence dramatically.

Direct observations of line strengths in stellar spectra generally confirm this conclusion, but there are a few elements that provide notable exceptions. In particular, the fraction of O relative to Fe in red giant stars has been found to vary by more than a factor of ten within a single cluster (e.g. Kraft *et al.* 1992), and these abundance variations correlate with location on the RGB in the sense that stars higher up the sequence have lower O abundances (Kraft *et al.* 1993). Oxygen's role in the CNO hydrogen burning cycle [equation (5.3)] can cause its abundance to be depleted in the cores of stars. Thus, one credible explanation for the low abundances of O in the atmospheres of some stars is that the CNO cycle has reduced the O content of their cores, and some mixing process has then dredged this depleted material up to the stars' surfaces. In this picture the tendency for O depletion to increase as one ascends the RGB arises because stars further up the RGB are at more advanced evolutionary stages than those lower down the sequence, so their convective envelopes will have had longer to redistribute their oxygen-depleted cores into their envelopes. A test of this hypothesis is possible because the CNO cycle conserves the total amount of C, N and O: all O and C lost in the process is converted into N. Comparisons between data and model atmosphere calculations do, indeed, imply that although [O/Fe] varies significantly from star to star, [C+N+O/Fe] is approximately constant for all the stars in a single cluster (e.g. Dickens *et al.* 1991, Brown, Wallerstein & Oke 1991). However, the situation is not clear-cut, since there is considerable scatter in this relation, with stars from the same cluster at the same location in the CM diagram containing very different O abundances. This apparently random element in the depletion of O in red giant atmospheres is not yet understood.

Other observed element anomalies are also puzzling. Cohen (1978) found that the abundance of Na varies by an order of magnitude in the giant stars of M3. On the basis of a very small sample of stars, Peterson (1980b) suggested that the Na abundance was anti-correlated with the O abundance, and this result was subsequently confirmed by Sneden *et al.* (1992). Similarly, Al abundance is also found to be anti-correlated with O abundance (e.g. Norris & Da Costa 1995). It is believed that proton capture reactions can significantly enhance the abundances of Na and Al in stellar cores (Langer, Hoffman & Sneden 1993), and so the same processes which dredge up oxygen-deficient material may also bring the sodium- and aluminum-enhanced material to the surface of the star, explaining the anti-correlation between these abundances. However, we should stress at this point that the process by which this mixing occurs is not understood: the convective zones predicted by stellar structure calculations do not reach deep enough into the stars' interiors to drag material with modified O, Na and Al abundances to their surfaces. Some additional, as yet unknown, mixing process must be hypothesized if the observed chemical abundances are to be explained by nuclear processes within the present generation of stars.

$\omega$ **Cen** Other element abundance variations have proved undetectably small in all globular clusters bar one – $\omega$ Cen, the most luminous cluster in the Milky Way. In this cluster, the RGB has a large intrinsic width rather than forming the usual narrow sequence in the CM diagram, and it was early on suggested that this width might arise from a spread in chemical composition amongst its red giant stars (Dickens & Woolley 1967). Direct confirmation of this hypothesis came from spectral observations of giant stars, which showed that those on the blue side of the giant branch have Fe abundances a factor of ten lower than those on the red side (Mallia & Pagel 1981). Subsequently, high-resolution spectra of stars in this system have shown that the abundances of a wide range of elements vary from star to star (Norris & Da Costa 1995). Although the abundances of heavy elements all track one another to some extent, the relationships are not always linear. For example, while the Fe abundance varies by a factor of ten within $\omega$ Cen, the abundance of Ba varies by a factor of almost a hundred. These variations cannot be explained by the self-enrichment and mixing scenario described above, since many of the heavier elements which are observed to be enhanced in the red giants of $\omega$ Cen are not produced in such low-mass stars. However, we also see the same correlations between C, N, O, Na and Al abundances that have been observed in other clusters. It therefore seems likely that the chemical mix in $\omega$ Cen results from a combination of a strange primordial distribution of abundances and self-enrichment within the current generation of stars. Models can be cooked up to produce the required primordial abundances using an earlier generation of massive stars (e.g. Norris & Da Costa 1995), but these ad hoc explanations beg the question of why $\omega$ Cen is so different from the other globular clusters in the Galaxy. Perhaps the explanation lies in this cluster's other unusual properties such as its large size and high rotation rate (§6.1.11).

### 6.1.6 The third parameter problem

We have seen that to a large extent we can explain the variations between cluster CM diagrams in terms of variations in metallicity [Fe/H] and age $t$, with metallicity the dominant factor. It has, however, proved impossible to account fully for variations in CM diagrams using only [Fe/H] and $t$. For example Catelan & de Freitas Pacheco (1993) have investigated the difference in age $\Delta t$ between the equal-metallicity clusters NGC 288 and NGC 362 (see page 338). From the sequences near the turnoff they deduce $\Delta t = 3\,\text{Gyr}$, while the HB morphologies are compatible with $\Delta t$ being as small as this only if the clusters are both implausibly young ($t \lesssim 10\,\text{Gyr}$). Similarly, no combination of metallicity and age in standard stellar-structure calculations can reproduce the bimodal HBs seen in some clusters – see Figure 6.5. Either there is some fundamental failing in our understanding of stellar evolution, or there must be some other factor beside metallicity and age which dictates the properties of globular clusters. In the literature this unknown factor is generally referred to as the **second parameter** because it was mooted before variations in cluster ages had been established.

Various candidate third parameters have been proposed, which we now discuss.

**Variations in helium abundance**    The earliest suggestion was that the initial helium abundance varies from cluster to cluster, as such variations can reproduce the anomalous horizontal branches seen in some metal-poor clusters (Sandage & Wildey 1967). Unfortunately, the helium abundances of globular clusters are not amenable to direct spectroscopic measurement since helium lines are only excited at high temperatures, whereas globular clusters are populated by old, cool stars. Therefore, less direct measurements of helium abundance must be employed. For example, stellar-evolution calculations have shown that the ratio between the time that a star spends on the horizontal branch of the CM diagram and the time it spends on the red giant branch, $R = \tau_{\rm HB}/\tau_{\rm RGB}$, depends sensitively on the assumed helium abundance, but very little on the metallicity, age, or any other property of the star (Iben & Rood 1969). The number of stars in a cluster that we see at stage $i$ of their evolution is $N_i = \dot{n}_i \tau_i$, where $\dot{n}_i$ is the rate at which stars enter and leave that stage. Since stars leaving the red giant branch evolve directly on to the horizontal branch, $\dot{n}_{\rm RGB} = \dot{n}_{\rm HB}$, and so $R = N_{\rm HB}/N_{\rm RGB}$. Thus, by measuring the ratio of the number of stars on the horizontal branch to the number on the red giant branch of a cluster's CM diagram, we have a method for determining the helium abundance of the cluster. Application of this method to a number of clusters shows that they have a mean helium abundance of $Y = 0.23$, with a dispersion from cluster to cluster of $\sigma_Y = 0.02$ (Buzzoni *et al.* 1983). This dispersion is directly comparable to the errors inherent in the method, so the observations are entirely consistent with there being no real variations at all in helium abundance. The failure of such techniques to detect variations in helium abundance have led people to look elsewhere for the second parameter. However, it should be borne in mind that these indirect methods do not provide very accurate measurements of $Y$, so it is still possible that as-yet unobservable variations in helium abundance may have significant impact on the properties of globular clusters (Shi 1995).

**Variations in other element abundances**    More recently, attention has shifted to the abundances of the heavier trace elements. Particular attention has focused on C, N and O. These elements form the catalyst in the CNO cycle, which provides one of the channels by which H is burned into He in MS stars [equation (5.3)]. Increasing the abundance of these elements thus increases the efficiency of hydrogen burning, which decreases the time that a star spends on the MS. Further, the mixture of these trace heavy elements in a star's atmosphere also affects its opacity, which in turn dictates the star's structure. Thus, varying the abundances of C, N and O relative to Fe and other heavy elements can dramatically alter the appearance of a cluster's CM diagram, and it has been suggested that the relative abundance of these elements to iron, [C+N+O/Fe], may vary from cluster to cluster. An ideal test of the hypothesis that this quantity is the second parameter is provided by the equal-metallicity pair NGC 288 and NGC 362, since, as we have seen, their third parameters must differ. However, spectroscopic observations by Dickens *et al.* (1991) have shown that the two clusters have indistinguishable values of [C+N+O/Fe], and so the third parameter phenomenon cannot be explained by these abundances.

**Other candidates**   The third parameter is something that affects HB morphology. We know from §6.1.3 that the amount of mass, $\Delta\mathcal{M}$, lost between the RGB and the HB is a primary factor in determining HB morphology. Does the distribution of the stochastic variable $\Delta\mathcal{M}$ [see equation (6.1)] vary systematically from cluster to cluster? Clearly the cause of any such variation would be a third parameter of globular cluster CM diagrams.

It is possible that a dynamical factor like rotation could explain the differences: the envelopes of faster-rotating stars will be less tightly bound than their slower-rotating equivalents, so we might expect that they would be more prone to mass loss. If the initial conditions from which a cluster formed favored the production of fast-rotating stars, we might therefore expect such a cluster to have a blue horizontal branch (Suntzeff 1981). However, the absence of any correlation between HB morphology and the measured rotation rates of stars which populate the HB casts serious doubt on this explanation (Peterson, Rood & Crocker 1995). An alternative dynamical explanation is suggested by the analysis of Fusi Pecci *et al.* (1993), which showed that the HB morphology is correlated with the spatial distribution of stars in the cluster (see §6.1.10) in the sense that more centrally-concentrated clusters have bluer HBs. Perhaps, therefore, the blue colors of some horizontal branches are related to stellar encounters in clusters of higher-than-average density.

As more high-quality observations of globular clusters have been made, the constraints on evolutionary models of these systems have become tighter and tighter. It is therefore not entirely surprising that the number of parameters required to describe their properties satisfactorily has also increased. Metallicity and (to a lesser extent) age are established as quantities that vary from cluster to cluster, but there must also be other internal factors that affect their properties. As we shall see in §10.5.1, cluster properties also depend on external factors such as their location within the Galaxy. The full story of globular-cluster CM diagrams has yet to be unraveled.

### 6.1.7 Luminosity functions

As well as examining the regions of the CM diagram that are populated by globular cluster stars, it is also instructive to see how many stars occupy each region: the number of stars of different types that we see today tells us about the efficiency with which the different stellar types were produced and the way in which these stars subsequently mutate as they evolve. The simplest observational measure of the frequencies of different stellar types is the luminosity function of the cluster stars (§3.6.3). Measuring the luminosity function is a more difficult task than the construction of a useful CM diagram since source confusion plays a more damaging role: if we were to fail to detect some fraction of the stars shown in Figure 6.2, we would still be able to pick out the principal sequences in the diagram; however, we would obtain a totally meaningless luminosity function from these data if their incompleteness varied with magnitude. It is only by carefully quantifying any incompleteness due to crowding that we can calculate the luminosity function.

Early analyses were based on photographic plates without the benefit of sophisticated crowded-field photometric techniques. They therefore concentrated on measuring the luminosity function of only the relatively-scarce bright stars for which source confusion is not a serious problem. Since the most luminous stars are also the ones that evolve most rapidly, we would expect this bright end of the luminosity function to be the part of the distribution of luminosities most heavily affected by stellar evolution. This expectation is borne out by the observations: Figure 6.13 shows the significant difference between the luminosity functions for bright stars in a typical globular cluster and bright stars in the disk of the galaxy (§3.6.3). The globular-cluster luminosity function shows a deficit of stars in the range $1 \lesssim M_V \lesssim 4$. This deficit occurs because stars in this luminosity range in globular clusters are subgiants and giants; these stars pass through this evolutionary stage very rapidly compared to the MS stars of comparable luminosity, so our "snapshot" view of a single point in the life of a globular cluster catches relatively few stars with these luminosities. The excess in the globular cluster luminosity function in the range $-3 \lesssim M_V \lesssim 0$ reflects the well-developed RGBs of clusters and the small number of MS field stars at such high luminosities. The bump at $M_V \approx 0$ in Figure 6.13 is produced by the globular-cluster stars on the HB, which places a lot of evolved stars at very similar magnitudes (§6.1.2). Clearly, the shape of the bright end of the globular-cluster luminosity function is dictated by the processes of stellar evolution.

As we saw in §6.1.2, at magnitudes fainter than $M_V \approx 4$, most of the observed stars in globular clusters have yet to evolve away from the MS, so their properties more closely reflect the initial stellar content of the cluster. These stars are fainter and more abundant than the evolved stars discussed above, so the analysis of their luminosity function has had to await the improved techniques for dealing with source confusion discussed in §6.1.1. During the 1980s, deep CCD images were obtained for a number of nearby clusters, and these data were analyzed with crowded-field photometry packages in order to calculate the luminosity functions of these systems down to magnitudes of $M_V \approx 9$ and beyond – well into the part of the function made up from MS stars. The luminosity functions for a sample of clusters is presented in Figure 6.14. It is apparent from this figure that the MS portion of globular-cluster luminosity functions is not universal, but varies systematically with

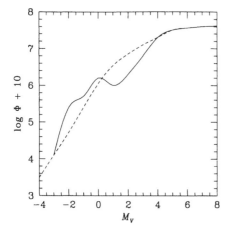

**Figure 6.13** Comparison of the general luminosity function for bright disk stars (dashed line) with that for globular cluster stars (solid curve). The functions have been arbitrarily normalized to match beyond $M_V = 5$. [From data published in Allen (1973)]

cluster metallicity.

The dependence of the luminosity function on metallicity is not surprising: as we saw in §6.1.2, the luminosity of a star depends on the star's heavy element abundance as well as on its mass, so it would be an amazing coincidence if the distribution of luminosities were identical for clusters with very different metallicities. To examine this issue more quantitatively, recall that the IMF is globally quite well-described by the Salpeter form of equation (5.14). Let us assume that the stars in a globular cluster formed with a similar power-law mass function such that the number of stars with masses between $\mathcal{M}$ and $\mathcal{M} + d\mathcal{M}$ was initially

$$\xi(\mathcal{M})d\mathcal{M} \propto \mathcal{M}^{-\alpha}d\mathcal{M}, \tag{6.4}$$

where the index $\alpha$ is as yet unknown. Since stars only lose significant amounts of mass when they evolve on to the subgiant branch and beyond, this function will also describe the mass distribution of MS stars. The stellar evolution calculations, which have been so successful in reproducing the CM diagrams of globular clusters (§6.1.2), also tell us how the absolute magnitude of a MS star in the cluster is related to its mass, $M(\mathcal{M})$. Since this quantity varies monotonically for MS stars, we can invert it to obtain $\mathcal{M}(M)$; thus, we can use equation (5.13) to calculate the luminosity function $\Phi(M)$ of MS stars which corresponds to the mass function of equation (6.4).

The results of such a calculation are presented in Figure 6.15, which shows the luminosity functions obtained for a range of indices $\alpha$ and two different values of metallicity. Comparing Figures 6.14 and 6.15, it is clear that the differences between the luminosity functions for different clusters can largely be explained by their different metallicities. Moreover, the shapes of the observed luminosity functions are well-reproduced by the simple power-law mass function. Hence, as in §5.1.9, we conclude that the structure in the

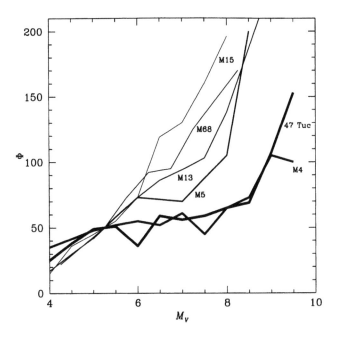

**Figure 6.14** The luminosity functions of globular clusters at magnitudes dominated by MS stars, normalized to a common value at $M_V = 5.25$. The names of the individual clusters are shown. The thickness of the lines increases with the metallicity of the clusters. [After McClure *et al.* (1986)]

luminosity function arises from the details of the relation between luminosity and mass rather than any complexity in the distribution of stellar masses.

Although the luminosity functions in Figure 6.14 can all be adequately reproduced by a power-law mass function, the required power-law index $\alpha$ varies from cluster to cluster – there is no universal mass function for globular cluster stars. McClure *et al.* (1986) find that $\alpha$ is smaller for metal-rich clusters than for metal-poor ones. This result implies that metal-rich cluster contain a larger fraction of high-mass stars. This correlation suggests that the extra high-mass stars in some clusters might have chemically enriched their environment early in the lifetime of the cluster, raising the metallicity of other stars that form (see Problem 6.4). It is also possible that the observed mass function is the result of nurture rather than nature in the sense that clusters all formed with a common IMF, but subsequent evolutionary effects have then altered their mass functions in different ways. For example, Capaccioli, Piotto & Stiavelli (1993) have suggested that clusters which form in the inner, high-metallicity parts of the Galaxy will have had more of their low-mass stars stripped away through interactions with the Galactic disk, resulting in

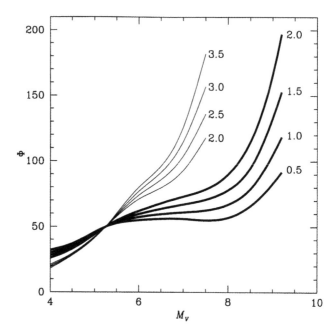

**Figure 6.15** The predicted luminosity functions of globular clusters for a power-law mass function, normalized to a common value at $M_V = 5.25$. The heavy lines were calculated assuming a metal abundance of $Z = 0.006$ (comparable to that of 47 Tuc), and the light lines for $Z = 0.0001$ (comparable to that of M15). Each line is annotated by the adopted mass function index, $\alpha$. These curves were calculated using the stellar evolution model of vandenBerg & Bell (1985), assuming an age of 16 Gyr. [After McClure *et al.* (1986)]

flatter present-day mass functions in these systems. Strong evidence that environment plays a key role in dictating the properties of globular clusters has come from the discovery that the slope of the mass function is more strongly correlated with a cluster's location in the Milky Way than it is with the system's metallicity (Djorgovski, Piotto & Capaccioli 1993).

If $\alpha \geq 1$, then integrating equation (6.4) over mass we find that there would have to be infinitely many low mass stars in the cluster. This result may seem implausible, but there is no observational evidence against it. With its superior imaging capabilities, HST has traced globular-cluster luminosity functions to very faint magnitudes. Figure 6.16 shows the results of several such studies. In every case, the luminosity function falls at the faint end. However, as we have discussed in §5.1.9, this drop is entirely consistent with a power-law mass function with $\alpha > 1$, and merely reflects the steep nature of the function $M(\mathcal{M})$ at low masses. The power-law form of $\xi(\mathcal{M})$ *could* break

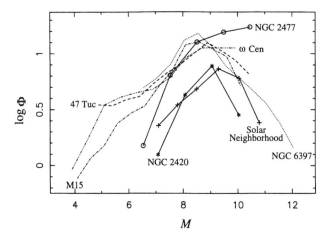

**Figure 6.16** HST I-band measurements of the faint end of the MS luminosity function for the globular clusters $\omega$ Cen, 47 Tuc, M15 and NGC 6397; the open clusters NGC 2420 and NGC 2477; and (for comparison) the solar neighborhood. The normalizations of the different functions are arbitrary. [After von Hippel *et al.* (1996)]

down for $\mathcal{M} \lesssim 0.2 \mathcal{M}_\odot$, but the large uncertainties in the form of $M(\mathcal{M})$ are such that the data currently provide no evidence for such a feature (Elson *et al.* 1995).

### 6.1.8 Binary stars

Binary stars play an important role in the evolution of globular clusters. The presence of even a small fraction of binary stars can affect the way in which the whole structure of the cluster evolves with time (see §6.1.10), and many of the oddities that we find in clusters such as millisecond pulsars (see §6.1.9) and blue stragglers may be connected with binary-star progenitors. Up to 50% of field stars are found to be members of binaries, but there is no reason to suppose that the fraction in clusters is similar since the environment is so different. Binaries which formed with large separations will tend to be torn apart by tidal forces, either due to the cluster as a whole or due to encounters with single passing stars. Conversely, in the high density environment of the cluster, there will be occasional very close encounters between single stars which can then be tidally captured into binaries. It is not obvious where the equilibrium between these processes lies, but, on the average, we would expect binaries in globular clusters to occur at a smaller separations than those in the field.

The observational techniques to search for binary stars in globular clusters have been extensively reviewed by Hut *et al.* (1992). Perhaps the simplest way to detect binaries is from the periodic Doppler shifts in their spectral lines (see §3.1.2). Since the requisite high quality spectra can only be obtained for the brightest globular cluster stars, this technique is only useful for searching for giant-star binaries. Spectroscopic studies uncovered very few binaries in globular clusters: for example, Pryor, Latham & Hazen (1988) found only one spectroscopic binary in 110 giant stars in M3. However, these studies will not detect any of the following: binaries with periods shorter than $\sim 2$ months or longer than $\sim 10$ years (since data have not been obtained which sample spectral variations on the appropriate timescale); binaries where the components have very unequal masses (since the Doppler motions of the brighter, more massive star decrease with increasing mass difference); and binaries with very eccentric orbits (since these stars spend most of their time at large separations moving at undetectably-low velocities). Once corrections have been made for these large numbers of undetectable binaries, the inferred fraction of binary stars in globular clusters is almost as large as that seen in the field.

We can extend the search for binaries to MS stars by careful studies of stellar photometry. Photometric surveys of stellar variability have uncovered a number of eclipsing binaries (see §3.2.5) in globular clusters. These studies preferentially detect very short-period binaries ($P \lesssim 1\,\mathrm{d}$), since only stars very close to each other have a reasonable probability of being eclipsed. Although the statistics are not yet very good, the fraction of eclipsing binaries detected (something like 1 in 1000 stars) is again comparable to what we see in the field.

Binary stars should also show up photometrically in the CM diagram. Imagine, for example, that a cluster contains a number of close binary systems that are made up from pairs of identical MS stars. Such pairs would not be resolved in photometric observations, so we would misidentify the binary as a single object with the same color as the MS stars, but brighter by a factor of two. A population of such binaries would thus show up on the CM diagram as a second sequence brighter than the MS by $2.5 \log 2$ magnitudes. If the stars are not identical pairs, they will not all lie along such a single line, but Romani & Weinberg (1991) have demonstrated that even unequal-brightness binaries will always produce a maximum along this second sequence in the CM diagram. In general, no second sequences are discernible in globular clusters, so this method only puts an upper limit on the fraction of MS stars in binaries, typically implying that $\lesssim 3\%$ are in such pairs. The clearest detection of a second sequence is in NGC 288, where 4% of the stars observed in the vicinity of the MS are offset by $\sim 0.75$ magnitudes on to a second sequence (Bolte 1992). However, the interpretation of such measurements is complicated because unrelated optical doubles (which are quite common in crowded globular cluster fields) will also lie on the second sequence. Further, very unequal-brightness pairings will lie very close to the normal MS in the CM diagram, so such binaries will not be detectable by this method.

One last feature in the CM diagram that may be related to binary stars is the collection of blue stragglers, which are sometimes found on the MS beyond the turnoff point (see Figure 6.2). Their location implies that they are more massive than the bulk of MS stars, and such massive stars should have evolved away from the MS unless they are younger than the rest of the cluster. Since we believe that all the stars in the cluster formed in a single epoch, such objects are something of a mystery. However, the properties of a blue straggler can be understood if it formed relatively recently from the merger of two less massive stars. Such mergers will occur most readily when the two stars are already bound in a binary; the mass may be transferred slowly from one star to the other (McCrea 1964), or the two may collide when the binary loses energy through an encounter with a third star (Hills & Day 1975). Thus, the presence of blue stragglers provides further indirect evidence that globular clusters contain a population of close binary stars.

### 6.1.9 Stellar remnants

**White dwarfs**     As we have seen in §5.1.3, stars end their days as compact stellar remnants. Single stars with initial masses $\mathcal{M}_i \lesssim 8\,\mathcal{M}_\odot$ evolve into white dwarfs, so all the single stars in globular clusters for which $8\,\mathcal{M}_\odot \gtrsim \mathcal{M}_i \gtrsim 1\,\mathcal{M}_\odot$ (the lower limit being set by the requirement that they have had time to complete their nuclear burning processes) will have become white dwarfs. For a typical globular cluster, white dwarfs make up a far-from-negligible $\sim 10\%$ of the total number of stars (see Problem 6.5).

Since white dwarfs lack an internal energy source, they cool and dim over time, so most of the cluster white dwarfs will be unobservably faint. Even the most luminous white dwarfs are very faint when viewed from the distance to a typical globular cluster, and direct observations of the white dwarfs in crowded cluster fields have proved unattainable from the ground. However, the imaging capabilities of HST allow us to observe down to the requisite faint limit: Figure 6.17 shows the distinct sequence of white dwarfs in the CM diagram of the globular cluster M4. Since white dwarfs fade in luminosity quite rapidly, those that we see must all have reached this evolutionary stage recently and have formed from stars that lie in a narrow mass range, a little above the current turnoff mass. Hence their own masses must lie in a narrow range, and, unlike the MS stars also apparent in Figure 6.17, the white dwarfs are not stretched out into a sequence by differences in mass; instead they are stretched out by differences in ages, with those that have had longest to cool lying at the bottom right-hand end of the sequence.

One possibility raised by observations of the white-dwarf cooling sequence in a cluster's CM diagram is that it might be used to measure the cluster's distance more accurately. By comparing the location of the sequence to the absolute magnitudes and colors of nearby white dwarfs with parallax distances, we can obtain the distance modulus to the cluster just as

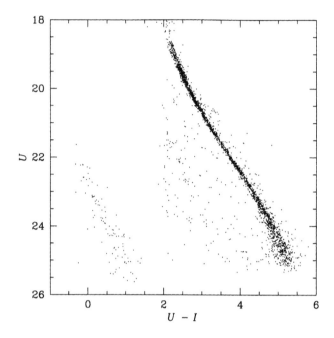

**Figure 6.17** CM diagram of the globular cluster M4 to very faint
limits derived from the HST observations of Richer *et al.* (1995).
The sequence of cooling white dwarfs is seen as the bluest stars
in the cluster, stretching from $U \sim 22$ down to $U \sim 26$. The
scattering of stars between the MS and white dwarf sequence is
due to field-star contamination. [From data kindly provided by
H. Richer]

in the classical MS fitting technique (§6.1.2). Moreover, the atmospheres of
white dwarfs are virtually metal free, so the metallicity corrections that must
be made for MS fitting can safely be neglected, and the absence of convection in these objects means that the physics which dictates their structure
is better understood. Finally, there are more local white dwarfs than subdwarfs, so the calibration of white dwarf absolute magnitudes is much better
constrained than that for the subdwarfs used in MS fitting. Renzini *et al.*
(1996) have shown that this approach can be used to reduce the uncertainty
in the distance modulus of the cluster NGC 6752 from $\delta(m-M) \sim 0.25$ to
$\delta(m-M) \sim 0.1$.

**Neutron stars**    Stars with initial masses $\mathcal{M}_i \gtrsim 8\,\mathcal{M}_\odot$ will have long-since
passed through their nuclear burning lifetimes and become compact rem-
nants, the vast majority of which will be neutron stars. Such old remnants
will have cooled and dimmed to the point of unobservability. However, as
we have seen in §6.1.8, some fraction of globular cluster stars live in binaries, and interactions between neutron stars in binaries and their companions

can render the neutron star luminous: if the components of the binary are close enough, material is transferred from the companion star to the neutron star, generating X-rays in the process. It is therefore noteworthy that globular clusters contain 12 of the 100 brightest X-ray sources in the Milky Way (while they only contain $\sim 10^{-4}$ of the optical luminosity of the Galaxy). Those X-ray sources which have been optically identified have been shown to be spectroscopic binary systems, while many of the others show periodic variations which strongly suggest that they lie in binary systems [see Hut *et al.* (1992) for a review of the evidence]. The observations thus imply that these X-ray sources are binary stars in which one of the components is an accreting neutron star.

In such a binary system, the companion transfers angular momentum as well as mass to the neutron star. This transfer increases the rate at which the neutron star spins. Rapidly-spinning neutron stars have been detected from their radio emission, which is modulated by the rotation of the neutron star with periods as short as $\sim 2\,\mathrm{ms}$. More than 30 of these **millisecond pulsars** have now been found in globular clusters, with 11 in 47 Tuc alone (Robinson *et al.* 1995). Doppler shifts in the arrival time of their pulses have shown that more than a third of these pulsars are in binary systems, as we would expect if accretion is responsible for their rapid spins.

One serious problem with this scenario of X-ray binaries evolving into millisecond pulsars is that there are estimated to be $\sim 100$ times as many millisecond pulsars as X-ray binaries in globular clusters.[6] A possible explanation for this imbalance is that each system spends 100 times longer as a pulsar than as an X-ray binary, but theory indicates that the lifetimes of both stages should be $\sim 1\,\mathrm{Gyr}$ (Bhattacharya & van den Heuvel 1991). Alternatively, some entirely different mechanism may be generating most of the millisecond pulsars: for example, Michel (1987) has suggested that white dwarfs in binary systems could accrete sufficient mass and angular momentum to turn into millisecond pulsars without passing through a lengthy X-ray binary stage.

### 6.1.10 Radial profiles

Having discussed the properties of the stars that make up a globular cluster, we now turn to their arrangement within the cluster. Since these systems appear remarkably round, the distribution of light in a cluster is almost completely specified by the radial brightness profile. Since we can resolve the stars in nearby globular clusters, there are two distinct approaches that can be used to measure their brightness profiles: either we can measure the brightnesses and locations of individual stars and use these data to calculate

---

[6] We have detected essentially all the $\sim 10$ X-ray binaries in Milky Way globular clusters, but only a few percent of the $\sim 10^3$ millisecond pulsars are detectable because most do not beam their emission towards us.

the number density of stars as a function of radius, or we can measure the integrated starlight as a function of radius using the techniques of integrated surface photometry that we discussed in the context of measuring galaxy brightness profiles (§4.2).

The Milky Way's globular clusters are sufficiently close that in their outer parts we can easily resolve their individual stars with large amounts of empty space in between cluster members. It is thus very inefficient to measure their outer brightness profiles using integrated starlight since most of the area over which one integrates contains no contribution from the cluster stars, but the whole area contributes towards the noise in the data via the intrinsic brightness of the night sky, unrelated background galaxies, etc. Thus counting individual stars is the method of choice for calculating clusters' outer brightness profiles. The only sources of noise in these data then arise from the Poisson statistics from the finite number of stars observed and confusion from Galactic stars along the line of sight to the cluster. The contribution from unrelated Galactic stars can be estimated by observing other nearby fields, but it is also possible to use the colors (§6.1.2) and velocities (§6.1.11) of individual stars to assess their membership of a cluster and thus reduce contamination from non-members to negligible proportions.

At small radii in a cluster, the density of stars can become very large. It then becomes difficult to resolve individual stars, so the surface-brightness profile is most straightforwardly calculated using the integrated light of the cluster. Fortunately, the relative contribution from other sources along the line of sight is generally negligible in the bright central region of the cluster. The surface brightness profile derived by this method can then be normalized to the star counts in an intermediate region of overlap, giving the brightness profile at all radii. This approach has been somewhat superseded with the advent of the Hubble Space Telescope, which has made it possible to resolve individual stars all the way into the very cores of clusters. These new data also reduce the degradation of the very centrally-concentrated cluster brightness profiles which occurs due to ground-based seeing (§4.2.2). Even with the resolving power of HST, however, there is still significant confusion in the cluster core, as the high surface density of stars means that adjacent stars will sometimes fail to be resolved, and hence will be misidentified as single sources. The number counts must then be corrected for the fraction of sources which have been missed, as discussed in §6.1.1.

It is also worth noting that the finite number of stars in a globular cluster imposes a fundamental limitation in the calculation of its brightness profile. At large radii, the low density of stars implies that the surface brightness can vary significantly just from the Poisson statistics unless we average over a significant range in radii. Even at small radii, the light is dominated by relatively few bright giant stars, so the concept of a smooth symmetric distribution of light is an oversimplification: there is, for example, no reason why the brightest point in the cluster should coincide exactly with the centroid of the positions of all the stars, and so the "exact" center of

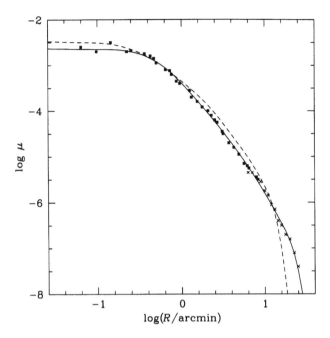

**Figure 6.18** plot of surface brightness, $\mu$, versus projected radius
for the globular cluster M3. The profile at small radii (solid symbols) is derived from integrated photometry, while that at large
radii (crosses) is derived from star counts. The dashed line shows
the King model which best fits these data, and the solid line shows
a more sophisticated multi-mass King model fit. [After Da Costa
& Freeman (1976)]

the cluster is not a well-defined quantity. It makes no sense to analyze the
radial brightness profile on scales smaller than that set by this fundamental
uncertainty, which, even in the most carefully-studied systems, may still be
as large as an arcsecond (Picard & Johnston 1994).

**Large-scale properties**    Figure 6.18 shows the complete brightness profile
of the nearby cluster M3 derived by combining star counts at large radii with
integrated photometry at small radii. The profile extends over 5 orders of
magnitude in surface brightness. This figure also illustrates the use of such
data in trying to understand the dynamical properties of globular clusters.
The dashed line shows the best fit that can be obtained by fitting a King
model to the observed brightness profile. As discussed in BT §4.4, the King
model makes the physically-reasonable assumption that the most energetic
stars in the cluster, which venture to large distances from its center, will have
been removed entirely by the tidal influence of the Milky Way. These losses
result in a cluster which is truncated at a finite **tidal radius**, $r_t$. The only
other variable parameters in this model are the central surface brightness

of the cluster and the **core radius**, $r_c$, which is defined to be the radius at which the surface brightness has fallen to half of its central value.[7] The **concentration** $c$ of the cluster, defined by $c = \log(r_t/r_c)$, then provides a useful distance-independent measure of the cluster's morphology with a more physical interpretation than the concentration class. As can be seen from Figure 6.18, although the simple King model predicts some of the properties that we see in the surface-brightness profile of M3, the best fit is far from perfect; it is only because we are able to probe the surface density of the cluster over such a large dynamic range that the shortcomings of this simple model are apparent.

One shortcoming of the simple King model is that it assumes that all the stars in the cluster have identical masses. As we have seen in §6.1.7, stars in globular clusters have a range of masses. It is most unlikely that stars of different masses will have identical kinematics: globular clusters are sufficiently old that encounters between stars of different masses will have significantly re-distributed their initial energies, driving them towards equipartition. On average, we therefore expect high-mass stars to be traveling more slowly than low-mass stars, so that their kinetic energies are approximately equal (BT §8.4.3). The solid line in Figure 6.18 shows the results of fitting a model which allows for the different kinematics of stars of different masses. This **multi-mass King model** divides the stellar population of the cluster into a number of different mass classes. Each class has a central velocity dispersion chosen such that at the center all stars are in energy equipartition. The fraction of stars in each class is determined by the observed distribution of stellar luminosities, with unobserved faint white dwarfs and neutron stars (§6.1.9) providing extra contributors to the mass of the system (but not its light). A dynamical model can then be generated which defines the kinematics of the individual classes in the gravitational potential arising from the mass distribution of all the stars. Adding the luminosity contributions from the different classes of star then gives the total surface-brightness profile indicated in Figure 6.18, which provides an excellent fit to the data.

Given the large number of parameters that define a multi-mass King model, it is not surprising that a model of this class provides a good fit to the photometry of M3. Other models are also possible: the assumption of full equipartition of energy between all stellar components can be relaxed, or we could consider more complex arrangements of stellar orbits. In fact, many different models can reproduce the observed brightness profile of a cluster, and it is only by examining other properties of the cluster such as its stellar kinematics (§6.1.11) that we can begin to distinguish between these various

---

[7] A natural parameter of a King model is the **King radius**, $r_0$, that is defined by BT equation (4-124b). In the limit of large $r_t/r_c$, the surface brightness at $r_0$ has fallen to 0.5013 of its central value, so $r_0 \simeq r_c$. In view of this result, $r_0$ is frequently called the core radius. This practice is misleading, however, because $r_c$ is significantly smaller than $r_0$ when $r_t/r_c$ is not very large – see Problem 6.6.

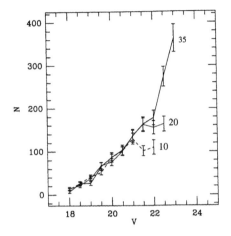

**Figure 6.19** The luminosity function of the globular cluster M30 derived from fields at three different radii. Each function is annotated by the radius at which it was obtained (measured in cluster core radii, $r_c$). [After Bolte (1989)]

globular cluster models.

**Luminosity segregation**   One conceptually-simple test of models of cluster evolution is to search for luminosity segregation. If equipartition of energy is important, as multi-mass King models assume, slower-moving more massive stars should be more concentrated toward the center of the cluster than the less massive stars. Observational support for this idea comes from the distributions of X-ray sources and blue stragglers in clusters. These objects are associated with binary systems (§6.1.8), which are more massive than single stars. Hence the finding that they are more centrally concentrated than normal stars (Grindlay *et al.* 1984, Nemec & Harris 1987) constitutes clear evidence of mass segregation.

We would also expect mass segregation to show up as a variation in the normal stellar luminosity function with radius. Bright, massive stars will make up a smaller fraction of the total stellar population at large radii than they do close to the center of the cluster, so the observed luminosity function should become increasingly steep with radius. In practice, this effect has proved difficult to detect. Unless observations are obtained over a wide range of radii, the variations in the luminosity function are rather small. Unfortunately, data obtained over a wide range in radii are subject to a variety of systematic errors: the contribution from possible background stellar contamination becomes large at large radii where the number density of cluster members is low; and the effects of stellar crowding become acute at small radii where the density is high. We therefore have to be certain that these systematic effects are well-understood before an unequivocal measurement of luminosity segregation can be made.

Figure 6.19 shows the luminosity function of the cluster M30 derived at a number of radii. These data imply that the fraction of faint stars in this cluster increases significantly at large radii. Unfortunately, these ground-based data may have been compromised by the crowding near the center. For

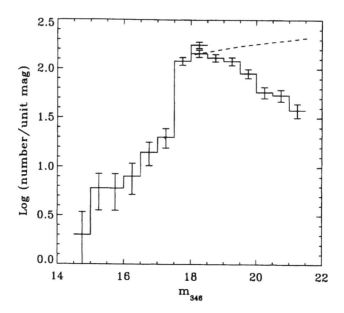

**Figure 6.20** The luminosity function of the globular cluster 47 Tuc derived from HST data within the central 4 arcsec of the cluster. The line indicates the luminosity function for faint stars at large radii derived from ground-based data by Hesser *et al.* (1987). [After Paresce, de Marchi & Jedrzejewski (1995)]

the innermost field presented in Figure 6.19, the star counts at the faintest magnitudes have been corrected by more than a factor of three; this inner field is so crowded that more than two-thirds of the abundant faint stars will have been missed due to their proximity to other stars. The credibility of the apparent difference between the luminosity functions shown in Figure 6.19 is therefore dependent on the reliability with which these large correction factors can be determined.

Figure 6.20 shows the luminosity function derived from HST observations of the innermost four arcseconds of 47 Tuc. Even at these small radii and at the faintest magnitudes detected, the post-repair HST images are so sharp that fewer than 10% of the stars are lost due to crowding, so the corrections to the luminosity function are small. This figure also shows the luminosity function derived from ground-based data at large distances from the center of the cluster, where the effects of crowding are similarly small. Clearly, with the small potential for systematic uncertainties due to crowding in these data, the difference between the luminosity functions at small and large radii in this system must be real, and luminosity segregation is occurring.

One consequence of the variation with radius in globular-cluster luminosity functions is that we must be rather careful when comparing luminosity

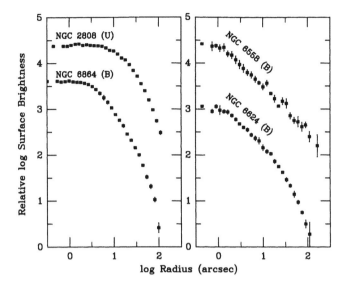

**Figure 6.21** surface-brightness profiles of globular clusters con-
trasting those with flat cores (left panel) to those with central cusps
(right panel). The identifications of the clusters and photometric
bands of the data are as indicated. [After Djorgovski (1988) courtesy of S. Djorgovski]

functions which have been derived from different data sets. If these analyses
are based on observations made at different radii in the clusters, then even
intrinsically identical globular clusters will appear to have different luminosity functions. Since crowding becomes a more serious problem in systems
at greater distances, luminosity functions of more distant clusters are likely
to be derived from stars at larger radii than those used in the analysis of
nearer clusters. Hence, if we are not careful we will end up with a distance-
dependent bias in the luminosity function. If we adopt a particular dynamical
model for each cluster, then the effects of mass segregation can be estimated,
and the observed luminosity function at a particular radius can be corrected
to give the global luminosity function for the cluster. However, the accuracy
of luminosity functions derived in this way depends on the validity of the
underlying dynamical models.

**Central cusps**    A generic property of the King-type models discussed
above is that inside the core radius, $r_c$, the surface density of stars tends
towards a constant value. Many clusters do display such flat cores in their
surface-brightness profiles, but there are also exceptions. In the case of M15,
for example, the surface brightness continues to rise all the way to its center.
When we factor in the effects of seeing, which will tend to smear out any
bright central cusp in the brightness distribution (§4.2.2), it is apparent that
M15 must have a remarkably sharp central peak in its brightness profile.

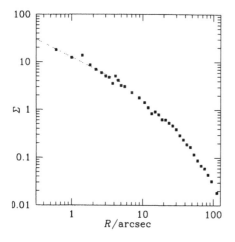

**Figure 6.22** The surface density of stars brighter than $m_V = 19$ as a function of projected radius in the globular cluster M15, as derived from a HST image (Guhathakurta *et al.* 1996). The dotted line shows a surface-brightness profile with a power law of index $-0.75$, as would be expected for the if the cluster contained a central massive black hole. [From data kindly provided by R. Guhathakurta]

The central cusp in M15 has been recognized for quite some time (e.g. King 1975), but it was treated as something of an oddity amongst globular clusters. However, the rarity of such central cusps simply reflected the fact that no-one had made any systematic search for the phenomenon. Interest in these systems began to grow in the early 1980s when a number of numerical studies of spherical self-gravitating bodies showed that, even if such systems initially contain a flat core, the central density increases dramatically over time, ultimately producing a central power-law cusp in the radial profile. This phenomenon of **core collapse**, discussed in BT §8.4.2, occurs generically in any stellar system that is old enough for two-body encounters to have redistributed the energy of the system between stars. This runaway process is only halted when binary stars at the center of the cluster provide a source of energy to halt the collapse (BT §8.4.6). Since globular clusters are all believed to be old systems, it is hard to see how they could have avoided collapsing to form central cusps similar to that in M15. Djorgovski & King (1986) therefore undertook a systematic search for collapsed cores by obtaining CCD photometry for a sample of 113 globular clusters, and found that clusters with central cusps are not all that unusual (see Figure 6.21). In fact, they discovered that around 20% of the clusters in their sample have surface-brightness profiles which cannot be described adequately by a model containing a flat central region, suggesting that these systems are examples of core-collapsed clusters. However, all the clusters they observed were old enough to have undergone core collapse, so the absence of cusps at the centers of 80% of the systems implies that there must be an efficient dynamical mechanism for re-inflating collapsed cores.

The repaired HST has enabled us to investigate the central regions of clusters in unprecedented detail. Figure 6.22 shows the surface density of stars brighter than $m_V = 19$ as derived from an HST image of M15. Although the steepness of the logarithmic gradient clearly drops as we look to smaller and smaller radii in this system, there is no evidence that it flattens out entirely even on the small scales probed by these data. The largest core radius consistent with this observation is only $r_c \sim 1$ arcsec. Moreover, the location of the center has an inherent uncertainty of $\sim 0.3$ arcsec on account of the finite number of stars in the system. Hence there is little chance of there being even a very small homogeneous core in M15.

It is worth noting that core collapse is not the only possible mechanism for generating central cusps in the surface brightness profile. One may show that if a globular cluster harbors a massive black hole in its core, the black hole will gather a cusp of stars around it (Bahcall & Wolf 1976). In fact, stars in such a system will produce a surface-brightness profile at small radii which follows the power law, $I(R) \propto r^{-0.75}$; as Figure 6.22 shows, such a power law matches the observed surface-brightness profile of M15 out to several arcseconds remarkably well. However, such a power-law slope is also quite consistent with the predictions of core collapse models. As with the larger-scale properties of globular clusters, we need more information than just the observed surface brightness in order to discriminate between these different explanations for central cusps. For example, we could measure the stellar kinematics of this central region. A central black hole would reveal its presence by an increase in the random velocities of stars in its immediate vicinity, whereas no such increase would be apparent if the central cusp were the result of core collapse.

### 6.1.11 Kinematics

**Velocities of individual stars** Since we can resolve individual stars in globular clusters, their individual line-of-sight velocities can be measured from the Doppler shifts in their spectral lines. As we shall see, the velocities of stars in a globular cluster typically differ by only $\sim 5\,\mathrm{km\,s^{-1}}$, so one requires optical spectra of sufficiently high quality to permit the measurement of shifts in spectral lines of $\lesssim 0.01$ nm. Only giant stars are bright enough for spectra of the required quality to be obtained, so only these objects can be used as kinematic tracers of the dynamics of globular clusters. Furthermore, large numbers of velocities are necessary: in a pioneering study, Gunn & Griffin (1979) obtained spectra of 111 giant stars in M3, and one of their principal conclusions was that this sample was too small to discriminate between many different dynamical models. Subsequently, observers obtained spectra for bigger samples of stars in a number of globular clusters. One of the largest such surveys consisted of line-of-sight velocities for 318 stars in $\omega$ Cen (Meylan & Mayor 1986). These data were derived by laboriously obtaining spectra for each star in the sample, with repeated observations made to identify the stars that are in binary systems – the velocities of such stars are confused by the orbital motion of the binary. These projects therefore required large investments of telescope time. The development of multi-object spectrographs (in which the light from many stars is channeled by fiber optic cables to the spectrograph) made it possible to determine globular cluster kinematics much more efficiently: Côté et al. (1995) have made use of such a spectrograph to measure the velocities of 399 stars in NGC 3201, and even larger samples lie within reach.

The simplest issue we can address with such data is whether globular clusters are rotationally flattened. As we shall see in §11.2.1, rotational motion in a stellar system tends to flatten it. In the case of globular clusters, the degree of flattening is usually rather small, so the amount of rotation required is also minimal. For typical cluster ellipticities of $\epsilon < 0.1$, the expected ratio of rotational to random motion is $v/\sigma < 0.3$. In general, the observed kinematics and ellipticities of globular clusters are quite consistent with their being rotationally-flattened systems, although the rotation rates and ellipticities of these systems are sufficiently small that they are subject to significant observational errors.

The small rotation rates and circular shapes of globular clusters mean that we can treat them as spherical systems to a good approximation. Kinematic observations therefore provide a data set that consists of the line-of-sight velocity relative to the cluster mean, $v_{los}$, and projected separation from the cluster center, $R$, for each of the giant stars observed. As an illustration of such a data set, Figure 6.23 shows the line-of-sight velocities that Peterson, Seitzer & Cudworth (1989) derived from spectra of 120 stars in M15 as a function of their projected distance from the cluster center. It is clear from this figure that the amplitude of random stellar velocities drops with radius: calculations of the velocity dispersion in these data show that it decreases from $\sim 15\,\mathrm{km\,s}^{-1}$ at small radii to $\sim 5\,\mathrm{km\,s}^{-1}$ in the outer parts of the cluster. Such a decrease in dispersion with radius is a generic property of any stellar system of finite extent which is populated by stars on isotropic orbits (like the King model discussed above). To see that this is the case, imagine a cluster wholly contained within a radius $r_t$. Any star currently lying at $r_t$ must have a radial velocity of zero, since any non-zero value would imply that the star follows an orbit which would take it beyond $r_t$. By definition, the velocity distribution of stars at any point in an isotropic system is equal in the tangential and radial directions, so the tangential velocities of stars in the outermost part of the cluster must also tend to zero. Thus, the observed line-of-sight velocities must drop toward zero as we near the edge of the cluster.

This description provides a qualitative explanation for the drop in dispersion with radius, but it often fails to match the observed kinematics in detail. In particular, the multi-mass King models that fit the photometric profiles of clusters generally predict a slower drop in the line-of-sight velocity dispersion with radius than is measured. This difference can be explained if the orbits in globular clusters are not isotropic. At large radii in a spherical system, the radial component of a star's velocity is mostly transverse to the line of sight, so its line-of-sight velocity is primarily made up from the star's

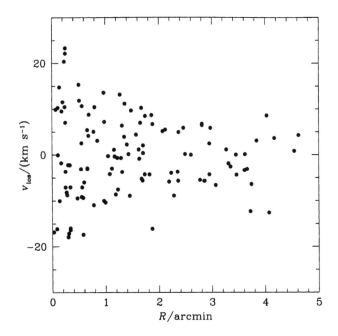

**Figure 6.23** Line-of-sight velocities of giant stars in M15 as a function of their projected radii. [From data published in Peterson, Seitzer & Cudworth (1989)]

tangential motion (see BT Figure 4-4). Thus, the observed rapid drop in the line-of-sight velocity dispersion can be explained if the stellar orbits at large radii are primarily radial with little tangential motion. Such an arrangement of orbits has some physical motivation: interactions between stars in the core place numbers of stars on highly eccentric orbits that reach from the core to large radii in the cluster. Such an eccentric orbit can be circularized only by an encounter with a star at a large radius; since the stellar density at large radii is small, such a circularizing encounter is improbable. Hence at large radii we now see an accumulated population of stars on deeply plunging orbits whose radial velocities generally exceed their tangential ones.

A generalization of the King model which allows for radial anisotropy in the orbits at large radii is provided by the **Michie model** (BT 4.4.4). This model adds a new parameter, the **anisotropy radius** $r_a$, a kinematic length-scale which supplements the photometric lengthscales, $r_c$ and $r_t$. Within $r_a$, the distribution of orbits is close to isotropic, but beyond $r_a$ the orbits rapidly become radial. As for the isotropic King model, the Michie model can be generalized for systems with more than one stellar species as a **multi-mass Michie model**, which has been successfully applied to the modeling of both the photometry and kinematics of globular clusters. The radius at which anisotropy becomes significant varies from cluster to cluster: in the case of $\omega$ Cen, for example, Meylan (1987) has shown that only models with $r_a \lesssim 3r_c$ can reproduce the observed photometry and kinematics satisfactorily, while Côté *et al.* (1995) have found that NGC 3201 can be adequately described by a model in which the stellar orbits remain isotropic all the way to the edge of the cluster ($r_a \gtrsim r_t$).

Although the generalized Michie models provide a good fit to the observed kinematics of globular clusters, these models are by no means unique. Even with the extra constraints provided by the line-of-sight velocities of stars, there is still a wide range of models which fit the observations. As with the kinematics of galaxies (§11.2.1), there is a trade-off between the distribution of stellar orbits and the form of the gravitational potential which enables us to reproduce the observed properties of a globular cluster with many different dynamical models.

Even though the gravitational potential derived from such dynamical studies is model-dependent, it is still often used to determine an estimate for the mass of the cluster, and hence the global mass-to-light ratio for the system, $\Upsilon = (\mathcal{M}/\mathcal{M}_\odot)/(L/L_\odot)$. In a study of 56 globular clusters, Pryor & Meylan (1993) found that globular clusters have a mean mass-to-V-band-light ratio of $\Upsilon_V = 2.3$. The calculated mass-to-light ratio varies significantly from cluster to cluster, and the dispersion in the values of $\Upsilon_V$ in this sample is 1.1. There are few clues as to why the mass-to-light ratio varies between clusters: its value does not correlate with the metallicity of the cluster, its location in the Galaxy, or its degree of central concentration. Pryor & Meylan (1993) did find a significant positive correlation between $\Upsilon_V$ and the total mass of the cluster, but since the mass is used to calculate $\Upsilon_V$, random errors in the calculated mass would naturally give rise to such a correlation, so the reality of the effect remains in doubt.

**Integrated-light kinematics**    At small radii in a cluster, crowding can make it impossible to obtain the spectra of individual stars. This problem is particularly acute in clusters with central density cusps, which is unfortunate, since it is for these systems that the central kinematics offers an important observational constraint on the nature of the central cusp (§6.1.10). For such clusters, we must resort to the integrated light spectral techniques that are used to study the stellar kinematics of galaxies (§11.1). Peterson, Seitzer & Cudworth (1989) obtained spectra of the central arcsecond of M15, the archetypal cusped cluster, and used cross-correlation analysis to obtain a velocity dispersion of $\sim 25\,\mathrm{km\,s^{-1}}$ for the stars in this region. This velocity dispersion appears to be significantly larger than the $\sim 15\,\mathrm{km\,s^{-1}}$ measured from individual stars slightly further out in the cluster (see Figure 6.23). Peterson, Seitzer & Cudworth (1989) have suggested that both the luminosity cusp and the high central velocity dispersion could result from the presence of a central black hole of mass $\sim 10^3\,\mathcal{M}_\odot$. However, as Dubath, Meylan & Mayor (1994) have pointed out, the HST images of the core of M15 have shown that the central arcsecond is dominated by the light of just three giant stars. The observed central velocity dispersion is thus at the mercy of small number statistics. If the three stars were drawn from a population with an intrinsic velocity dispersion of $15 \, \mathrm{km \, s^{-1}}$, their measured velocity dispersion could plausibly lie anywhere between $\sim 5 \, \mathrm{km \, s^{-1}}$ and $\sim 25 \, \mathrm{km \, s^{-1}}$ (see Problem 6.7). Thus, the large value for the central velocity dispersion in M15 as derived from the integrated light is consistent with a velocity dispersion which tends to a constant $15 \, \mathrm{km \, s^{-1}}$ at small radii, and so there is no need to invoke any strange behavior in the kinematics at small radii. Small number statistics present a fundamental limitation to studies of stellar kinematics in the very central regions of clusters.

**Proper motions**    We can also investigate the internal kinematics of globular clusters through the proper motions of their stars (§2.1.4). For a nearby globular cluster at a distance of $\sim 5 \, \mathrm{kpc}$, a typical transverse stellar velocity within the cluster of $\sim 5 \, \mathrm{km \, s^{-1}}$ will result in a proper motion of only $\sim 0.2 \, \mathrm{mas \, yr^{-1}}$. In order to detect such small shifts, studies must use archival photographic plates often reaching back as far as the beginning of the century. Even with such a long temporal baseline, the shifts in stellar positions are still very small, so careful astrometric measurements must be made in order to determine proper motions. These measurements are by no means simple, since telescopic images – particularly those recorded on the older photographic plates – can be significantly distorted. The analysis procedure must therefore solve for these largescale distortions which will displace the apparent locations of stars as well as the intrinsic stellar proper motions (e.g. Cudworth 1985).

As an example of the results of such an analysis, Figure 6.24 shows the proper motions of stars in M5 as measured by Rees (1993) from plates obtained between 1900 and 1982. The left panel shows the angular velocity of each star on the plane of the sky, $(\mu_x, \mu_y)$, relative to the cluster as a whole. The cluster members are concentrated quite tightly about the origin in this plot, reflecting the system's small internal velocity dispersion, while contaminating field stars have a much wider spread in proper motions, with a different mean velocity. Clearly, proper motions provide an efficient tool for identifying cluster members with very little ambiguity. Such measurements are thus very useful for cleaning up CM diagrams by removing interloping field stars.

Although the cluster members are tightly concentrated about the origin in this plot, they do show some scatter. This scatter can be quantified by fitting a Gaussian distribution to the arrangement of cluster members in the $(\mu_x, \mu_y)$ plane, and measuring its dispersion, $\sigma_{tot}$, which for the data in Figure 6.24 is $\sim 0.4 \, \mathrm{mas \, yr^{-1}}$. Part of this dispersion can be attributed to the finite uncertainties in the derived proper motions; even the careful analysis applied to these data results in RMS proper motion uncertainties of $\sigma_{err} \sim 0.1 \, \mathrm{mas \, yr^{-1}}$. However, $\sigma_{tot}$ is significantly greater than $\sigma_{err}$, so it is clear that data of this quality are detecting the intrinsic motions of the stars, which contribute a characteristic RMS proper motion $\sigma_{int}$. If both the errors

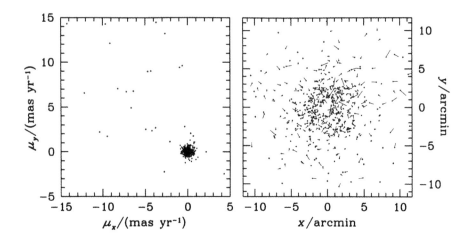

**Figure 6.24** Proper motions of stars in the field containing the globular cluster M5. The left panel shows the angular velocity for each star, $\{\mu_x, \mu_y\}$; the stars identified as cluster members are shown as solid points, while field stars are marked by open circles. The right panel shows the locations of these stars on the sky, with lines marking where the proper motions predict that the cluster members will have moved to in $5 \times 10^4$ years. Proper motions have not been determined at small radii due to crowding. [From data published in Rees (1993)]

and the intrinsic motions have Gaussian distributions in proper motion, then we can sum their contributions in quadrature so that $\sigma_{\rm tot}^2 = \sigma_{\rm err}^2 + \sigma_{\rm int}^2$.

The simplest use for this kinematic information is in determining the distance to the cluster via statistical parallax (see §2.2.4): if we equate the motions on the plane of the sky to the line-of-sight velocity dispersion, $\sigma_{\rm los}$, as measured from the Doppler shifts of cluster members, then we find the distance to be $D = \sigma_{\rm los}/\sigma_{\rm int}$ (where the angular shifts in $\sigma_{\rm int}$ are measured in radians). In practice, this formulation is too simplistic since, as we have seen above, the distribution of stellar velocities is not isotropic, and so the transverse and line-of-sight velocities will not in general be equal. This complication can be overcome by using the radial velocity data alone to calculate a full dynamical model for the stellar system. This model then allows us to predict the distribution of transverse velocities that we expect to observe at the locations in the cluster where stellar proper motions have been measured. The distance to the cluster can then be determined by reconciling the observed angular velocities to the predicted distributions of spatial velocities. Ultimately, this method offers an independent method for determining the distances to globular clusters which can be used to calibrate other important phenomena such as the absolute magnitudes of the MS and RR Lyrae stars. The main limitation at present is in uncertainties in the size of $\sigma_{\rm err}$, since a mistake in estimating this contribution to the observed proper motion will cause a corresponding error in the estimate for the intrinsic proper motion. If this problem can be overcome, proper motions can be used to derive distances to globular clusters to better than 10% (Lupton, Gunn & Griffin 1987).

Ultimately, there is a wealth of information that might be derived from samples of proper motions like those shown in the right panel of Figure 6.24. Leonard & Merritt (1989) have demonstrated that from a large sample of accurately-measured proper motions at all radii in a cluster it is possible to determine not only the shape and amplitude of the velocity ellipsoid (Box 10.2) but also the underlying gravitational potential of the cluster. Although existing proper motion data sets have neither the quality nor the quantity of data for such a complete analysis, some conclusions can already be drawn. For example, Lupton, Gunn & Griffin (1987) have shown that proper motions of stars at large radii in M13 are preferentially oriented toward (or away from) the cluster center, which implies that the stellar orbits are radially-dominated in the outer parts of this cluster, as predicted by the analysis of line-of-sight velocities. With improving imaging technology making it possible to measure ever smaller shifts in the positions of ever fainter stars, the gathering of large samples of high-quality proper motion measurements in globular clusters will soon greatly expand our understanding of the kinematics of these systems.

## 6.2 Open clusters

Not all the agglomerations of stars in the Milky Way have the tidy structure of globular clusters. As Figure 6.25 illustrates, many star clusters show little symmetry in their appearance, and lack the central concentration characteristic of globular clusters. Because of their more diffuse morphology, these systems are known as **open clusters**. Unlike the globular clusters, open clusters are found to be strongly concentrated toward the plane of the Milky Way (see Table 6.2), and hence they are also sometimes referred to as **Galactic clusters**.

On account of their rag-tag appearance and low central concentration, open clusters are less readily picked out by the eye than the symmetric globular clusters. Moreover, they occur near the Galactic plane, where they tend to be heavily obscured by dust and can easily be lost amongst the high density of field stars. Consequently, the Messier catalog lists only 27 open clusters. Nevertheless, they are intrinsically plentiful systems: the database compiled by Lynga (1987) lists some 1200 open clusters, and even this catalog covers only a small fraction of the Milky Way. Typical properties for a selection of open clusters are presented in Table 6.2.

Open clusters cover a broad spectrum in size, luminosity and morphology, ranging from groupings with a dozen or so members with integrated magnitudes of $M_V \simeq -3$ up to systems of many thousands of stars with

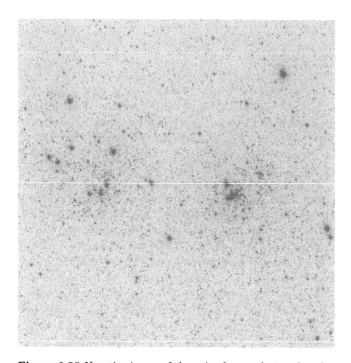

**Figure 6.25** Negative image of the pair of open clusters $h$ and $\chi$
Persei. The area shown in this figure is approximately $1\,\mathrm{deg}^2$. [DSS
image from the Palomar/National Geographic Society Sky Survey,
reproduced by permission]

integrated magnitudes as high as $M_V \simeq -9$. Their diameters range up
to a few parsecs. Since they are fairly large objects, and we can usually
only detect relatively nearby examples, open clusters frequently cover large
areas on the sky, subtending angles of several degrees. The densities of
these systems range from only marginally higher than the field-star density
($\sim 0.1\,\mathrm{stars\,pc}^{-3}$) up to $\sim 10^3\,\mathrm{stars\,pc}^{-3}$ at the centers of the richest clusters.
The lowest density systems, also termed **associations**, can frequently only
be distinguished from the field because they contain a high fraction of stars
of an unusual type, leading to a noticeable enhancement in the projected
density of these stars in the direction of the cluster. **OB associations**, for
example, are aggregations of stars containing a large fraction of O and B
stars, while **T associations** are rich in T Tauri stars.

The properties of an open cluster can be usefully summarized by its
**Trumpler classification** (Trumpler 1930), which specifies:
1) its degree of concentration as a roman numeral I...IV, where lower
   numbers indicate higher degrees of central concentration;
2) the range in brightness of its stars measured on a scale 1...3, where
   lower numbers indicate a smaller range of stellar brightnesses;

**Table 6.2** Characteristics of selected open clusters

| NGC | Name Other | $D$ (kpc) | $(R,z)_{\rm gal}$ (kpc) | $M_V$ | Diameter (arcmin) | Trumpler class | $(B-V)_{\rm TO}$ | $\log(t/{\rm yr})$ | [Fe/H] | Comments |
|---|---|---|---|---|---|---|---|---|---|---|
| 2264 | – | 0.79 | (9.23, 0.03) | −5.4 | 40 | III,3,p,n | −0.25 | 6.5 – 7.0 | −0.15 | Typical young cluster |
| 6705 | M11 | 1.72 | (6.96, −0.08) | −5.4 | 13 | I,2,r,- | −0.05 | 8.4 | +0.05 | Typical intermediate-age cluster |
| 188 | – | 1.55 | (9.35, 0.58) | −2.9 | 15 | II,2,r,- | +0.58 | 9.8 | −0.16 | Typical old cluster |
| 6791 | – | 4.20 | (8.12, 0.80) | −3.6 | 10 | I,2,r,- | +0.60 | 10.0 | +0.15 | Very old metal-rich cluster |
| 7261 | – | 2.12 | (9.23, 0.03) | −3.2 | 6 | III,1,p,- | −0.25 | 7.6 | −0.46 | Young metal-poor cluster |
| – | Berkeley 17 | 2.40 | (10.89, −0.15) | – | 8 | III,1,r,- | +0.58 | 10.1 | −0.29 | Oldest known cluster |
| – | Berkeley 20 | 8.14 | (16.12, −2.42) | – | 3 | I,3,p,- | – | 9.7 | −0.75 | Metal-poorest known cluster |
| – | Hyades | 0.05 | (8.55, −0.02) | −2.5 | 329 | II,3,m,- | +0.12 | 8.8 | +0.19 | Nearby cluster, classically used to establish the distance scale |
| – | Coma, Mel 111 | 0.08 | (8.56, 0.07) | −2.9 | 275 | III,3,p,- | +0.05 | 8.6 | −0.03 | Nearby poor cluster |
| – | Pleiades, M45 | 0.13 | (8.63, −0.05) | −3.7 | 109 | I,3,r,n | −0.11 | 8.0 | +0.11 | Nearby well-studied cluster |
| 2632 | Praesepe, M44 | 0.16 | (8.64, 0.08) | −2.9 | 95 | II,3,m,- | +0.15 | 8.8 | +0.19 | Nearby well-studied cluster |
| 869 | h Persei | 2.23 | (10.19, −0.14) | −7.4 | 29 | I,3,r,- | −0.25 | 6.7 | −0.05 | Rich young cluster, twin with χ Persei |
| 884 | χ Persei | 2.22 | (10.18, −0.14) | −7.3 | 29 | I,3,r,- | −0.25 | 6.7 | −0.05 | Rich young cluster, twin with h Persei |

NOTES: $D$ is the distance to the cluster; $(R,z)_{\rm gal}$ are its cylindrical coordinates relative to the Galactic center; $(B - V)_{\rm TO}$ is the color of its main sequence turn-off point; $t$ is its estimated age; other quantities are as defined in the chapter.

SOURCE: From the data collected in Lyngå (1987), Meynet, Mermilliod & Maeder (1993) and Friel (1995).

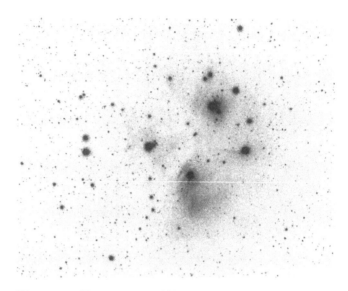

**Figure 6.26** Negative image of M45, the Pleiades open star cluster.
Note the nebulosity surrounding the stars [Image kindly provided
by J. Ware (`http://www.galaxyphoto.com/`)]

3) its richness, specified as p, m, or r depending on whether the system is
   poor (containing less than 50 stars), moderate (50 – 100 stars), or rich
   (more than 100 stars).

An 'n' is appended to the classification if there is nebulous, diffuse emission
around the cluster's stars. For example, the Pleiades (see Figure 6.26) is
designated I,3,r,n in this scheme.

The presence of diffuse emission from around the stars in many open
clusters provides a striking contrast with the purely stellar emission from
globular clusters. This nebulosity arises from starlight which has been reflected by dust grains in the cluster. As is usual in astrophysical systems, this
dust is accompanied by gaseous material, and so, unlike globular clusters,
open clusters contain a significant ISM. This observation ties in closely with
a further difference between the two classes of cluster: unlike globular clusters, many open clusters contain bright, blue stars. As we discussed above,
we believe globular clusters are old because they lack such stars; conversely,
the presence of bright blue stars in many open clusters implies that these systems formed much more recently. As we have seen in §5.3.2, star formation
is an intrinsically inefficient process, and the gas in open clusters is material
that has not been converted into stars. In the youngest star clusters, we can
observe directly the dense gaseous cores from which stars form and T Tauri
stars, which are believed to be stars in the final stages of formation (§5.1.8).
Hence star formation is an on-going process in these systems.

Further support for the recent origin of most open clusters comes from observations of their metallicities. Spectral studies of open-cluster stars yield heavy element abundances which scatter around the solar value, with metallicities lying in the range $-0.75 \lesssim [\mathrm{Fe/H}] \lesssim 0.25$ (see Table 6.2). Such high metallicities are to be expected in young objects which have formed from material in the Galactic disk that has already been chemically enriched by previous stellar generations. Since there are no very metal-poor open clusters, metallicity variations are a less important factor in interpreting the properties of these systems than is the case for studies of globular clusters.

Like their globular kin, open clusters have played a key role in the development of our understanding of galactic astronomy. Since they contain bright, blue stars, open clusters are testing grounds for investigating the properties of high-mass stars. The spatial distribution of open clusters yields a useful probe of the structure of the Galactic disk: the youngest systems delineate the spiral structure of the Galaxy (§10.4.5) while the older systems trace the kinematics of the outer Galaxy (§9.2.3). Since it is believed that many of the stars in the disk of the Milky Way originated in open clusters, the properties of these systems must dictate many of the properties of the stellar disk as a whole. Finally, as we see in §2.2.2 and Chapter 7, the nearby Hyades cluster has played an anchoring role in establishing the size of the entire Universe. We therefore now apply some of the techniques developed in the previous section to the study of open clusters.

### 6.2.1 Color-magnitude diagrams

As for globular clusters, the CM diagram offers a powerful diagnostic of the evolutionary state of open clusters. Although generally closer to us than globular clusters, the depth of an open cluster along the line of sight is usually still very much less than the cluster's distance, so we expect the sequences in its CM diagram to be sharply defined. In one respect, the CM diagrams of open clusters are easier to construct than those of globular clusters: the more diffuse nature of these systems means that the confusion which hampers stellar photometry in globular clusters (§6.1.1) is seldom an issue in studies of open clusters. However, the locations of open clusters close to the plane of the Galaxy mean that their CM diagrams are liable to be heavily contaminated by unrelated field stars, and some care must be taken to minimize this contamination by selecting stars on the basis of their kinematics, or by selecting only those stars with colors that are consistent with objects that have been reddened by the column of dust between us and the cluster. Since open clusters lie near the Galactic plane, extinction can be a large effect, which requires careful correction. An additional complication is that there is dust within these systems, so that stars within a single cluster suffer different amounts of extinction.

As Figure 6.27 illustrates, once these difficulties have been overcome, the MS shows up clearly in the CM diagrams of open clusters. In fact,

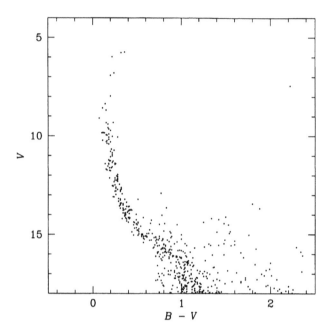

**Figure 6.27** CM diagram of the young open cluster NGC 4755.
Note the extensive MS, the scattering of stars to its right at faint
magnitudes, and the single evolved supergiant star at the top right
of the diagram. [From data published in Sagar & Cannon (1995)
and Dachs & Kaiser (1984)]

this sequence is usually much more extensive than in the CM diagram of a
globular cluster, stretching all the way up to the bright, blue stars at the top
end of the sequence. The presence of such massive, short-lived stars implies
that at least some of the stars in these clusters are very young. Further, the
sharp delineation of the sequence implies that all the stars are of very similar
age, since none of the less massive stars with longer MS lifetimes have yet
started to turn off toward the giant branch.

Towards the bottom of the MS in an open cluster, the sequence frequently becomes broader, with a broad scattering of stars on the red side
(see Figure 6.27). The width of this sequence is too large to be attributed to
observational errors. It also cannot be the result of patchy extinction, since
we would expect such random reddening to affect the colors of some of the
brighter stars as well, yet their part of the MS is narrow. The broadening of
the MS at low luminosities is, in fact, a consequence of the system's youth:
as we have seen in §5.1.8, before they reach the MS, stars evolve from the
Hayashi limit on the extreme right of the CM diagram over to the MS, with
lower mass stars taking longer to make this transition. For the low mass

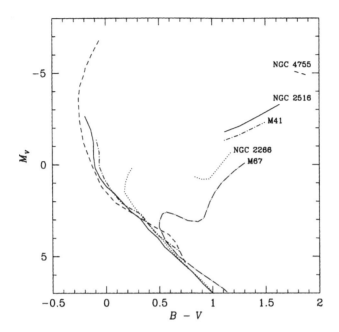

**Figure 6.28** Sequences found in the CM diagrams of a number
of open clusters, shifted to a common absolute-magnitude scale.
[From data kindly provided by J.-C. Mermilliod]

stars, the timescale to reach the MS is longer than the lifetime of the cluster,
and so we observe them following their Hayashi tracks down to the MS (see
Figure 5.10). Notice that the scattering of these stars in the CM diagram implies that they cannot all have formed simultaneously, as different stars with
the same masses have evolved to different points in their pre-main-sequence
evolution.

At the upper end of the MS, the stars have started to evolve away from
the MS, giving the sequence more curvature than the ZAMS. Subsequently,
massive stars evolve very rapidly across the CM diagram, have very brief
lives as supergiant stars, and explode as supernovae. The short timescale on
which these stars pass through the giant stages of their evolution means that
the late-evolution sequences in the CM diagram are very sparsely populated
with a pronounced Hertzsprung gap (§3.5).

To compare the CM diagrams of different clusters, we must place them
on a common absolute-magnitude scale by MS fitting. The calibration of
this scale is better established than for the globular clusters (§6.1.2) since
there are a large number of nearby stars with measured trigonometric parallaxes that have similar metallicities to the open cluster members. The
more extensive nature of the MS in open clusters also assists in the fitting

process, since it is possible to fix the calibration over a wide range of colors and magnitudes. Further, the Hipparcos satellite has measured trigonometric parallaxes out to distances that encompass a number of open clusters, so the absolute calibration of the CM diagrams for these nearby clusters can be obtained directly. Figure 6.28 shows schematically the sequences for a selection of open clusters which have been calibrated to an absolute magnitude scale. Although some of the clusters reveal very similar sequences, the most striking feature in Figure 6.28 is the wide range of MS turnoff points found in different clusters (see also Table 6.2). This variety of turnoff points, some of which take the form of an abrupt termination of the MS, contrasts markedly with the common location of a smooth turnoff point in the CM diagrams of globular clusters. As we shall discuss below, most of the differences between the sequences can be attributed to a wide range of ages amongst open clusters. There are, however, indications that some of the differences between the locations of sequences cannot be attributed to age or metallicity variations (Mermilliod *et al.* 1997) – like those of globular clusters (§6.1.6), open clusters CM diagrams seem to be under the influence of an as-yet unidentified third parameter.

### 6.2.2 The ages and demise of open clusters

Comparing Figures 6.28 and 6.6, it is apparent that the differences between the CM diagrams of different open clusters can be readily explained by differences in their ages. If we crudely apply equation (6.2) to the systems in Figure 6.28 (which, like most open clusters, have approximately solar metallicities) we find that the observed spread in their turnoff points corresponds to ages that range from $\sim 1\,\mathrm{Myr}$ all the way up to $\sim 10\,\mathrm{Gyr}$ (see Table 6.2).

More detailed calculations confirm this spread in cluster ages. The current record for the oldest reliable age determination for an open cluster is held by Berkeley 17, which has a turnoff point that places its age at $12 \pm 2\,\mathrm{Gyr}$. [See Friel (1995) for a discussion of the properties of this and other old open clusters.]

The slow approach to the MS of low-mass stars enables us to assign ages to very young clusters. For example, by comparing the locations of low-mass stars in the CM diagram to the predictions of pre-main-sequence contraction calculations (§5.1.8), Adams, Strom & Strom (1983) found that in the open cluster NGC 2264 there are low-mass stars with ages as high as $\tau_c \simeq 10\,\mathrm{Myr}$. This estimate conflicts with the cluster age one obtains by identifying the turnoff from the MS (§6.1.4). Walker (1956) found that NGC 2264 has a sharply defined turnoff at spectral type O7. From the MS nuclear burning lifetime of an O7 star, we derive a cluster age $\tau_n \simeq 3\,\mathrm{Myr}$. In fact, it is generally found that cluster ages determined from low-mass stars are greater than those determined from high-mass stars. As Herbig (1962) first pointed out, this ordering implies that low-mass stars form earlier than high-mass

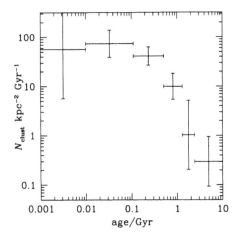

**Figure 6.29** The density of open clusters in the solar neighborhood as a function of cluster age. The horizontal error bars indicate the range in ages of each bin, while the vertical error bars show the uncertainty in the density due to Poisson statistics. The data are from the Mermilliod (1980) list of nearby open clusters, with the ages as determined by van den Bergh & McClure (1980).

stars in these systems, although the reasons for this difference are lost in the details of the ill-understood process of star formation. It is, however, just as well that the high-mass stars form last, since these stars pass through their lifetimes very rapidly, and when a number of such stars explode as supernovae they will disperse the remaining ISM in the cluster, bringing star formation to an abrupt end. If these stars had been created first, low-mass stars would never have had the chance to form.

As for the globular clusters, the absolute ages of individual open clusters are subject to significant uncertainties: using a different isochrone calculation, different conversion from theoretical stellar structure calculations to observed colors and magnitudes, or different reddening or distance modulus can all alter the derived age considerably. For example, estimates of the age of the old open cluster NGC 188 have ranged from 6.5 Gyr (Twarog & Anthony-Twarog 1989) up to 10 Gyr (vandenBerg 1985), depending on the adopted reddening and distance modulus. Similarly, age estimates based on pre-main-sequence evolutionary tracks are also found to be significantly model-dependent (D'Antona & Mazzitelli 1994). However, the differences between the properties of different open clusters are so extreme that there can be no doubt that they span a very wide range of ages. There also seems to be remarkably little correlation between clusters' ages and metallicities (see Table 6.2).

The spread in the ages of open clusters implies that they are continuously forming in the disk of the Galaxy. Mermilliod (1980) compiled a reasonably complete database of open clusters within 750 pc of the sun, which listed 14 open clusters younger than 0.1 Gyr. This density of young open clusters implies that they are forming at a rate of $\sim 80\,\mathrm{kpc^{-2}\,Gyr^{-1}}$. However, as Figure 6.29 shows, the density of older clusters is very much lower, dropping to less than $1\,\mathrm{kpc^{-2}\,Gyr^{-1}}$ for the oldest systems.

This underabundance of old open clusters has been known about for a considerable time (Oort 1958). It could be interpreted as meaning that the efficiency with which open clusters form has increased dramatically over the last billion years, but such a model would be contrived. In fact, it has also long been realized that open clusters are not very robust structures: tidal encounters with massive molecular clouds in the Galactic disk can rip them apart (Spitzer 1958). The deficit of old clusters can thus be simply understood if most of them have been destroyed by such encounters. Evidence in favor of this idea has come from the observation that old clusters are found preferentially at large distances from the Galactic center and far from the Galactic plane, where these encounters will be least frequent [van den Bergh & McClure (1980); see also Table 6.2].

Additional support for this hypothesis has come from studies of the morphological properties of open clusters as a function of age. Janes & Phelps (1994) have shown that while 49% of the old open clusters ($t \gtrsim 0.8\,\mathrm{Gyr}$) in the Lynga (1987) database are of Trumpler richness class 'r,' only 18% of the younger clusters ($t \lesssim 0.8\,\mathrm{Gyr}$) fall in this richest class. Further, 91% of the older clusters lie in Trumpler's highest two concentration classes (I and II), while only 62% of the younger clusters have such high degrees of central concentration. Thus, it would appear that because the poorer, less centrally concentrated clusters are less strongly gravitationally bound, they are less able to withstand tidal encounters with molecular clouds, and so they do not survive to old age.

We can even witness the final demise of some open clusters after they have been disrupted. When the cluster dissolves, its former members still "remember" the direction in which the cluster was moving at the time of its destruction, and continue to follow similar orbits in the Galaxy even though they are no longer gravitationally bound together. As we shall see in §10.3.2, it is possible to pick out such **moving groups** of stars which have similar velocities and similar metallicities even after they have spread out over large areas of the sky.

### 6.2.3 Structure and kinematics

We now turn to the internal structure of open clusters. The main difficulties in studying the dynamics of open clusters is that these objects are not such neat, regular systems as globular clusters. There is therefore less justification for modeling open clusters as spherically-symmetric systems in equilibrium. Nevertheless, simple King models (BT §4.4, and §6.1.10 above) have been shown to provide a reasonable description of the radial distribution of stars in a number of open clusters (e.g. King 1962, Leonard 1988). These model fits confirm the visual impression that open clusters are more diffuse systems than globular clusters: typically, they have core radii of $r_c \sim 1-2\,\mathrm{pc}$, and tidal radii of $r_t \sim 10-20\,\mathrm{pc}$, giving low concentration parameters of $c \lesssim 1$ (c.f. Table 6.1). Note, however, that the untidy appearance of many open

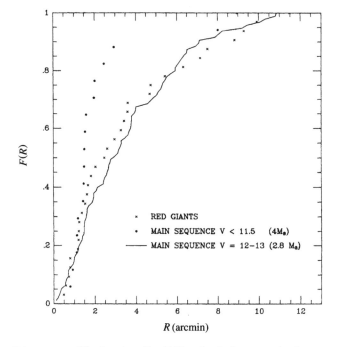

**Figure 6.30** The fraction, $F$, of MS and red-giant stars in the open cluster M11 that lie within a projected radius $R$ of the cluster's center. [After Mathieu (1984)]

clusters, combined with the large amount of contamination from field stars due to their low Galactic latitudes, mean that $r_t$ is usually poorly constrained by observations. The relatively low quality of the radial profile of a typical open cluster is such that the profile does not place very strong constraints on the cluster's dynamical state. In particular, it is not possible to distinguish between the simple and multi-mass King models (see §6.1.10) from such data.

Multi-mass models should be more appropriate, however, since the youth of open clusters implies that they contain stars with a wide range of masses. Further, an open cluster will evolve to a system in which different masses have different kinematics on a fairly short timescale. Equipartition of energy between stars of different masses will occur on a timescale characterized by the **relaxation time** of the cluster, $t_{\text{relax}}$. For a gravitating system made up of $N$ bodies, $t_{\text{relax}} \propto N/\ln(N)$ (BT §4.0.1), and so the relatively small numbers of stars in open clusters means that these systems should relax quite rapidly toward equipartition, with the low-mass stars travelling faster than their high-mass associates.

As we have discussed in §6.1.10, one of the observable consequences of equipartition is that the more massive stars become more centrally concentrated than the less massive stars. This phenomenon is apparent in a number of older open clusters, where, presumably, equipartition has had time to become established. For example, Figure 6.30 illustrates the effects of mass segregation in the intermediate-age cluster M11. It shows the fraction of the total population of different types of cluster star that lie within a projected radius $R$ as a function of $R$. It is clear that on average the fainter MS stars (with typical masses of $\sim 2.8\,\mathcal{M}_\odot$) lie at larger radii than the brightest MS stars (with typical masses of $\sim 4\,\mathcal{M}_\odot$). What is less well understood is why the red giants in the cluster seem also to be amongst the least centrally-concentrated cluster stars. Since these stars have recently evolved from the bright end of the MS, we would expect their masses to be slightly in excess of $4\,\mathcal{M}_\odot$, and as such one would naively predict that they should be the most centrally-concentrated population. The number of stars involved is quite small, but the extended nature of the giant population does not appear to be a statistical fluke, since the same phenomenon has been witnessed in a number of clusters (Hawarden 1975). The effect may be related to the mass loss that stars undergo in their giant phase (see §6.1.3), since the post-mass-loss stars should relax into a more extended distribution appropriate to their reduced masses. If this explanation is valid, then the amount of mass loss must be large. In Figure 6.30, for example, the red-giant distribution mimics that of the $\sim 2.8\,\mathcal{M}_\odot$ MS stars; if the two distributions are similar because the masses of the objects are similar, then the giant stars must have lost more than $1\,\mathcal{M}_\odot$ from their original mass. Such losses significantly exceed the $\sim 0.2\,\mathcal{M}_\odot$ believed to be shed by globular cluster giant stars (§6.1.3). Careful analysis of the CM diagrams of open clusters has uncovered evidence that the giant stars in these systems do, indeed, lose more mass than those in globular clusters (Tripicco, Dorman & Bell 1993). Since mass-loss is driven by luminosity, which increases strongly with mass [see equation (5.5)], it is entirely natural that the amount of mass a star loses in passing from the MS to the RGB should increase strongly with its initial mass. However, even if enough mass is lost by higher-mass stars, it is not clear that red-giant lifetimes are long enough for the spatial distribution of giants to have evolved significantly from that of their MS progenitors. Thus the extended distributions of red giants in open clusters remains problematical.

We might hope that direct studies of the motions of stars in open clusters will clarify the significance of mass segregation. The velocity dispersions of these low mass systems are typically only $\sim 3\,\mathrm{km\,s^{-1}}$, so it is quite difficult to measure their line-of-sight motions. Moreover, approximately a third of the stars in these systems turn out to be spectroscopic binaries (Liu, Janes & Bania 1989), and the velocity of a binary star's motion (typically $\sim 20\,\mathrm{km\,s^{-1}}$) completely masks the random motion of the star within the cluster unless velocities are averaged over the full binary period. A more profitable method for studying the kinematics of open clusters is provided by proper motions: since these systems are typically less than a kiloparsec away, the proper motions that their internal velocities induce can reach the readily-measurable level of $\sim 1\,\mathrm{mas\,yr^{-1}}$. In a study of proper motions in eight open clusters, Sagar & Bhatt (1989) found only one case (IC 2391) in which there was convincing evidence that the brighter (more massive) MS stars have smaller characteristic velocities than the fainter (less massive) MS stars, as we would expect if energy equipartition is occurring. In the other clusters in this study, there is no discernible trend in the velocity dispersion with stellar luminosity. These observations suggest that the stellar kinematics of most open clusters reflect the initial conditions under which the stars formed rather than the effects of relaxation. It should be noted, however, that this conclusion runs contrary to the observed luminosity segregation in the cluster radial profiles; it is not yet clear how these two observations can be reconciled.

As we have discussed in §6.1.11, the measurement of proper motions provides a valuable tool for studying the detailed orbital structure within a star cluster. Generally, open clusters seem to have rather simple kinematics: none of the eight clusters studied by Sagar & Bhatt (1989) show any significant change in the amplitude of the stellar random motions with projected distance from the cluster center, and so there is no evidence for the drop in velocity dispersion with radius that we see in globular clusters (see Figure 6.23). In one of the clusters in this sample (NGC 2516), the proper motions in the radial direction exceed those in the tangential direction at large radii, implying that the orbits in this system become radially-oriented at large radii. For the majority of clusters, however, there is no discernible difference between radial and tangential proper motions, and so the orbits seem to be more-or-less isotropic. Ultimately, limitations on such dynamical studies of open clusters arise from the relatively small numbers of stars in these systems, which are such that the Poisson uncertainty in kinematic measurements will always be quite large. Further, the lack of symmetry in open clusters means that simple spherical models are probably inappropriate. Finally, since the relaxation time of the poorer open clusters is not much longer than the time it takes a star to cross the system, the overall structure of these systems will be evolving quite rapidly; the sophisticated modeling required to include the effects of this secular evolution is beyond the scope of what can be achieved with the limited observational data.

### 6.2.4 Luminosity function

Some fraction of the stars in the disk of the Milky Way began their lives in the systems that we identify as open clusters (see Problem 6.8), and some further fraction must have been born in associations that would have been too poor to be identified as clusters. The distribution of disk-star masses is therefore closely coupled to the initial mass functions of open clusters and associations. There is no a priori reason to assume that all open clusters should have the same IMF: the mass function could vary wildly from cluster to cluster, with the mass function of the disk reflecting the sum of all the cluster and association IMFs from which it was produced. We therefore now consider the form of the open cluster IMF, whether it varies between systems, and how closely it matches the IMF of field stars.

As we have already seen in the case of globular clusters, we can only study the IMFs of clusters indirectly via their present-day luminosity functions. The situation is somewhat simpler for the open clusters, however, since we can look at young open clusters, where the effects of both stellar evolution and dynamical evolution are relatively small, so the observed luminosity function is closely related to the initial luminosity function and hence the IMF. In fact, the age window within which we can study the IMF directly is quite small. As we have seen in §6.2.2, high-mass stars form after the low-mass stars in open clusters. In order to study the complete IMF, we must therefore look at a cluster which has formed its O and B stars, so that the star formation process is essentially complete. However, if we are to study the high-mass end of the IMF, we must also catch the cluster before these O and B stars have exploded as supernovae and disappeared from the luminosity function.

Studies of the high-mass end of the luminosity function have therefore concentrated on OB clusters and associations which, by there very nature, fulfill the above requirements. A study of 11 OB associations by Massey, Johnson & DeGioia-Eastwood (1995) found that, to within the errors due to finite populations, the stars in these systems with masses greater than $7\,\mathcal{M}_\odot$ all seem to be drawn from a common power-law IMF with index $\alpha = 2.1\pm0.1$ [see equation (6.4)]. Studies of OB associations in the Magellanic Clouds, which differ in heavy-element abundances from those in the Milky Way by up to a factor of ten, reveal identical high mass IMFs (Massey *et al.* 1995), and so this generic cluster IMF does not seem to vary with metallicity. Since the lifetimes of massive stars are very much less than the timescales on which open clusters dissolve, we do not expect many massive stars to escape into the field. Studies of the IMF of field stars do, indeed, show that it has a much steeper power law $\alpha \simeq 4.5$ [see equation (5.17)]: there are many fewer very massive stars in the field than there are in open clusters.

Studies of the lower mass parts of the IMFs of open clusters are not easy. The already-faint low-mass stars are obscured by dust within the disk, and the high density of field stars can seriously contaminate the luminosity function. These difficulties can be minimized by looking at nearby clusters, for which obscuration will be smallest, and whose members can be selected on the basis of their large proper motions. However, nearby open clusters subtend large angles on the sky (see Table 6.2). Since luminosity segregation causes low-luminosity stars to be found preferentially at large radii (see §6.2.3), observations must be made across the entire diameter of the cluster in order to obtain an unbiased sample of stars. For example, a study of the Pleiades by Meusinger, Schilbach & Souchay (1996) required plates that covered 16.5 square degrees and involved measuring proper motions for some 49 000 stars. Of these stars, only 628 turned out to be cluster members. The

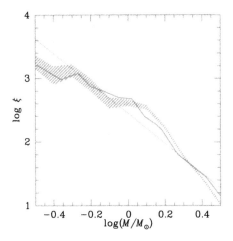

**Figure 6.31** The initial mass function for the Pleiades, as derived by Meusinger, Schilbach & Souchay (1996) (solid line) compared to the simple Salpeter IMF of equation (5.14) (dotted line) and the IMF derived by Basu & Rana (1992) for the solar neighborhood (shaded region). The IMFs have been normalized to agree at $2\,\mathcal{M}_\odot$. The uncertainty in the solar neighborhood IMF arises from the uncertain history of star formation in the Galaxy.

conclusion of this painstaking analysis, presented in Figure 6.31, is that the IMF of the Pleiades is indistinguishable from that of the field.

The comparison between the field and cluster IMFs is somewhat uncertain because the present-day distribution of stellar luminosities in the field reflects the star formation over an extended period – stars with masses $\lesssim 0.8\,\mathcal{M}_\odot$ have lifetimes longer than the Galactic disk and so could have formed at any time during its existence. Since we have little idea of how the star formation rate has changed over this period, there is a significant uncertainty here: the plethora of low mass stars in the Galaxy could have formed recently, in which case the IMF is very steep, or they could have formed much earlier, in which case fewer such stars are forming now and the corresponding IMF must be much flatter. This uncertainty notwithstanding, there is a respectable agreement between the lower-mass IMF of young open clusters like the Pleiades and that inferred for the field.

Using HST, it has become possible to probe the luminosity functions of open clusters to even fainter limits. Figure 6.16 shows the luminosity functions of the open clusters, NGC 2420 and NGC 2477, both of which are fairly old ($\sim 3\,\mathrm{Gyr}$ and $\sim 1\,\mathrm{Gyr}$ respectively). These clusters show evidence for a turn-over in the luminosity function at faint magnitudes, similar to that observed in globular clusters and in the solar neighborhood. However, as we discussed in §6.1.7, this feature reflects the rapid variation in stellar luminosity with mass for such faint stars rather than a turn-over in the mass function itself. It is notable that the location of the peak in $\Phi(M)$ varies from system to system in this plot. von Hippel *et al.* (1996) pointed out that these systems have very different metallicities, so the conversion from luminosity function to mass function will be different in each case. Intriguingly, the peaks in $\Phi(M)$ all occur at stellar luminosities that correspond to masses of $\sim 0.27\,\mathcal{M}_\odot$, with the shifts between systems entirely explicable by the differences in metallicity.

The agreement between field and open-cluster IMFs at masses below $\sim 3\,\mathcal{M}_\odot$ adds weight to the hypothesis that many, if not all, of the lower mass field stars were born in clusters and associations. Further support for this idea has come from studies of the luminosity functions of old clusters. As we have seen in §6.2.3, there is evidence that mass segregation occurs as an open cluster grows older, with the least massive stars ending up in an extended halo. These low-mass stars are the most likely to achieve escape velocity, so they should be the first cluster stars to join the field population. We would therefore expect the luminosity functions of old clusters to be depleted at the low-luminosity end. Such depletion is, indeed, observed, with the luminosity functions of many old clusters turning over and decreasing at faint magnitudes. The turn-over occurs at $\sim 1\,\mathcal{M}_\odot$, well above the $\sim 0.2\,\mathcal{M}_\odot$ where it might be attributed to the effects of a steep mass–luminosity relation (as discussed above), so it must represent a real deficit of low-mass stars in these systems. This phenomenon, which has been found to be widespread amongst old open clusters, is reviewed by Friel (1995). She also points to the extreme case of NGC 3680: a study of this old open cluster by Nordström, Andersen & Andersen (1996) found that only 13 of the 120 identified cluster members are single MS stars, with the remainder all more massive evolved objects. Apparently, while the higher-mass MS cluster stars have left the MS due to stellar evolution, the lower-mass MS stars have evaporated from the cluster to join the field. These two processes have almost met in the middle of NGC 3680's MS, so this cluster is in the final stages of dissolution.

## Problems

**6.1** The color of a HB star can depend quite sensitively on the amount of mass $\Delta\mathcal{M}$ it has lost. Given that the typical dispersion in mass lost is only $\sim 0.02\,\mathcal{M}_\odot$, estimate the dependence, $\partial(B - V)/\partial(\Delta\mathcal{M}/\mathcal{M}_\odot)$, for M3 (see Figure 6.2).

**6.2** Use Figure 6.6 to estimate how rapidly the turn-off point of an evolving globular cluster shifts in $B - V$ with time. Hence, show that typical residual color errors in isochrone fits can produce age estimates that are systematically wrong by $\sim 2$ Gyr.

**6.3** By how much would the adopted value of [Fe/H] have to be wrong to produce a 10% error in the age of a globular cluster as estimated using the $\Delta V$ method, assuming that equation (6.3) is correct?

**6.4** Using Figures 6.14 and 6.15 and Table 6.1, estimate the variation in metallicity with IMF power-law index, $d[\text{Fe/H}]/d\alpha$, in globular clusters. Assume that high-mass stars $(\mathcal{M} > 8\mathcal{M}_\odot)$ drawn from the appropriate IMF formed first in these systems, and all the material produced in the supernovae of such stars is recycled at solar metallicity into the lower mass stars $(0.08 < \mathcal{M} < 8\mathcal{M}_\odot)$, modifying the metallicity that we see today. Show that such recycling could go a long way toward explaining the observed variation of [Fe/H] with $\alpha$. Why is such efficient recycling difficult to envisage for globular clusters?

**6.5** Consider a globular cluster populated with stars drawn from an IMF of the Salpeter form [equation (5.14)] with an appropriate power law index. Allowing for the evolution that a globular cluster $\sim 15\,\mathrm{Gyr}$ old will have undergone, show that $\sim 10\%$ of the original stellar population will be white dwarfs.

**6.6** Use BT Figures 4-9 and 4-10 to show how the core radius, $r_c$, compares to the King radius, $r_0$, for King models with different concentrations $c = \log(r_t/r_c)$.

**6.7** If $n$ points are drawn from a Gaussian distribution of dispersion $\sigma$, then the standard deviation of the sample, $S$, will not be exactly equal to $\sigma$. In fact, it will be distributed such that $nS^2/\sigma^2$ follows a $\chi^2$ distribution with $n$ degrees of freedom. Show that the velocity dispersion as inferred from 3 globular cluster stars drawn from a Gaussian velocity distribution with $\sigma = 15\,\mathrm{km\,s}^{-1}$ will lie between $5\,\mathrm{km\,s}^{-1}$ and $25\,\mathrm{km\,s}^{-1}$ 90% of the time. How many stars would one have to observe to determine $\sigma$ to within 10% at 90% confidence?

**6.8** Assuming that the drop in the number of open clusters with age in Figure 6.29 arises from clusters dispersing into the field, and that each cluster initially contained $\sim 10^3$ stars, estimate the total number of stars per square parsec that these clusters have contributed to the field. If these stars end up distributed in a layer $\sim 100\,\mathrm{pc}$ thick (comparable to the typical distances of open clusters from the Galactic plane), estimate the number density of these stars in the solar neighborhood. What fraction of nearby stars were born in such clusters? Why might this calculation significantly under-represent the total contribution made by clusters and associations to the field?

# 7

# The Cosmic Distance Scale

For several centuries, telescopes have provided us with images of the cosmos showing a wealth of patterns and structures. However, the absence of depth information in these two-dimensional images has proved a major limitation on our ability to understand these structures. As we saw in Chapter 1, the argument over the nature of "spiral nebulae" was only finally resolved when the large distances to some of these systems were measured. Similarly, the discovery of a class of faint radio-emitting star-like objects only assumed its true significance when it was realized that these quasars lie at vast distances from us (see §4.6.2). If we are to understand the physics behind an astronomical phenomenon, we need to quantify the basic properties of the system under observation, and many of these properties cannot be determined unless we know the distance to the object in question.

Although we are not be able to determine many of the absolute properties of astronomical phenomena without distance information, we can compare their relative properties. For example, the stars in any galaxy outside our own are all at essentially the same distance from us. Thus, if observations of a nearby galaxy reveal that all the stars of a particular type have the same apparent brightness, then we know that this class of objects all have the same intrinsic luminosity. Furthermore, we can use this property to measure the relative distances to other galaxies: if we can identify the same type of star in these galaxies and measure their apparent brightnesses, then the ratio of distances follows directly from the ratio of apparent fluxes. In terms of the apparent magnitudes, $m_1$ and $m_2$, of the sources in two different galaxies, the distance ratio to the two galaxies, $D_1/D_2$, is given by

$$\log(D_1/D_2) = \frac{1}{5}\left[(m_1 - m_2) - (A_1 - A_2) - (K_1 - K_2)\right], \qquad (7.1)$$

where $A_1 - A_2$ corrects for possibly differing amounts of absorption towards the two sources, and $K_1 - K_2$ allows for possible differences in their Doppler shifts (see §2.3.3). Sources with identical intrinsic luminosities which can be used in this way are known as **standard candles**. Similarly, if a certain class of objects all have the same intrinsic size, then we can estimate the relative distances to objects containing these **standard rulers** by comparing their apparent angular sizes.

The classical procedure for estimating distances in the Universe typically uses a series of such **relative distance estimators**: the distance to a nearby galaxy might be determined by comparing the apparent brightnesses of its individual stars to those of similar stars in the Milky Way; the distance to a more remote galaxy whose individual stars cannot be resolved can then be measured by comparing the properties of the galaxy as a whole to the properties of the nearby galaxies, and so on. This bootstrapping approach to measuring distances has been dubbed the **cosmic distance ladder**, where each rung up the ladder takes us to greater distances. Notice, however, that this sequence of relative distance estimates must be calibrated by measuring at least one absolute distance using an **absolute distance estimator**. This absolute measurement provides the reference point that allows the series of relative measurements to be converted into absolute distances.

Great efforts have been expended in making the wide range of observations necessary to calibrate the distance ladder. Unfortunately, there are discrepancies in the derived distances: detailed reviews by Jacoby *et al.* (1992), van den Bergh (1992), Sandage (1993b) and Fukugita, Hogan & Peebles (1993) show that absolute distance determinations can still have an uncertainty of up to a factor of two. In this chapter, we discuss the various techniques and observations that have been used to define the cosmic distance scale. These methods combine the absolute distance measurements that are required to tie down the overall calibration of the distance scale (§7.2), and the relative distance measurements that provide most of the rungs from which the cosmic distance ladder is made up (§7.3). In §7.4, we show how the methods can be combined to obtain a reasonably coherent extragalactic distance scale, and where discrepancies between different analyses remain.

The cosmic distance ladder allows us to measure distances out to a significant fraction of the size of the observable Universe. In order to make sense of observations that probe these vast scales, we must first present the cosmological framework within which such data can be interpreted.

## 7.1 An introduction to cosmology

One crucial cosmological result, which was originally obtained without good absolute distance calibration, is the **Hubble law**. Following on from the discovery that the characteristic lines in the spectra of almost all galaxies are shifted systematically toward the red (Slipher 1914), Hubble (1929) investigated the relationship between the distance, $D$, to a galaxy and the amount $\Delta\lambda$ that a line of intrinsic wavelength $\lambda$ is shifted in its spectrum. He found that these two quantities are proportional to one another:

$$D = \frac{cz}{H_0},\qquad (7.2)$$

where $c$ is the speed of light and $z$ is the redshift of the galaxy (§2.2.1). Without good absolute distance measurements, the **Hubble constant**, $H_0$, is an ill-determined constant of proportionality. As we shall see below, the Hubble constant is believed to have a value of somewhere between 50 and $100\,\mathrm{km\,s^{-1}\,Mpc^{-1}}$, and so it is frequently written in the form

$$H_0 = 100\, h\,\mathrm{km\,s^{-1}\,Mpc^{-1}},\qquad (7.3)$$

where the dimensionless quantity $h$ parameterizes our ignorance of the Hubble constant, and probably has a true value somewhere between 0.5 and 1.0.[1]

The simplest physical interpretation of the Hubble law is that the entire Universe is expanding, and the recession of the galaxies gives rise to their redshifts via the usual Doppler formula $v = cz$ ($v \ll c$). Running the clock backwards, we see that one **Hubble time**,

$$t_0 = D/v = H_0^{-1},\qquad (7.4)$$

ago all the galaxies must have started from the same place. Thus, the discovery of the Hubble law revolutionized cosmology by introducing a **Big Bang** at a finite time in the past when the entire Universe was concentrated in a tiny volume. The fact that we can estimate the age of the Universe itself by determining the Hubble constant provides a powerful motivation for obtaining a reliable absolute calibration for the extragalactic distance scale.

The geometry of a general-relativistic expanding Universe is intrinsically different from Euclidean space. We therefore have to be a little careful to state our assumptions about the largescale geometry of the Universe, and even exactly what we mean when we talk about the "distance" to an object. A complete discussion of the geometry of the Universe lies beyond the scope of this book; here we simply sketch out the results that we will need to interpret cosmological observations. More detailed treatments can be found in Weinberg (1972), Peebles (1993), or Longair (1992).

---

[1] For SI purists, $H_0 = 3.2 \times 10^{-18} h\,\mathrm{s}^{-1} = 3.2h\,\mathrm{aHz}$.

The solutions to Einstein's general relativistic field equations depend on the distribution of mass in the system under consideration. Clearly, attempting to solve these equations for the entire Universe is a formidable undertaking. Light emitted in the Universe is strongly inhomogeneous, in that it mostly comes from the small regions that we call galaxies. Galaxies, in turn, are arranged into clusters and still larger structures. If this distribution of light-emitting material reflects the distribution of mass, then the detailed geometry of the Universe is extremely complex. However, on the largest scales the Universe does appear to be reasonably homogeneous. We might therefore hope to be able to describe its gross properties by solving Einstein's equations assuming a homogeneous mass distribution. Since there are no observations that seem to pick out preferred directions in the Universe on these largest scales, we might also assume that the Universe is expanding isotropically. Under these assumptions, the geometry of the Universe is greatly simplified, and can be described by the **Robertson-Walker metric**. The simplest solutions to the Einstein equations for such a geometry are known as **Friedmann cosmological models**.

These simple homogeneous cosmological models are fully defined by specifying the current expansion rate of the Universe (as quantified by the Hubble constant) and the current mass density of the Universe, $\rho_0$. The future fate of the Universe hangs upon the value of $\rho_0$. The simplest solutions to the field equations[2] imply that if $\rho_0$ exceeds the **critical density**,

$$\rho_c \equiv \frac{3H_0^2}{8\pi G} = 1.88 \times 10^{-26}\, h^2\, \mathrm{kg\, m^{-3}}, \qquad (7.5)$$

then there is sufficient mass in the Universe for its gravitational influence to ultimately halt the expansion, and for the Universe to collapse back towards a final "Big Crunch." Such a system is called a **closed** or **bound model**. If, on the other hand, $\rho_0 < \rho_c$, there is insufficient mass to ever halt the expansion, and we refer to such an eternally-expanding system as an **open** or **unbound model**. If $\rho$ is exactly equal to $\rho_c$, the Universe will just expand forever, but it's expansion velocity will tend asymptotically toward zero; such a system is termed a **critical** or **Einstein-de Sitter model**.

There are two dimensionless forms in which the present density of the Universe, and hence its ultimate fate, are often cast. The **cosmic density parameter**,

$$\Omega_0 \equiv \frac{\rho_0}{\rho_c} = \frac{8\pi G \rho_0}{3H_0^2}, \qquad (7.6)$$

simply measures the density in units of the critical value. Thus, the fate of the Universe depends on whether or not $\Omega_0$ exceeds unity. The second,

---

[2] These simple models assume that there is no pressure helping to support the Universe against collapse. Cosmologists sometimes refer to them as "dust" models to make the point that both gas and radiation produce pressure, so a Universe containing anything besides pressureless dust is not strictly represented by such a model.

slightly more abstract, way in which the density of the Universe can be quantified is the **cosmic deceleration parameter**, $q_0$. This parameter specifies in dimensionless form the rate at which the expansion of the Universe is currently decelerating. Clearly, this quantity is closely tied to the value of the mass density responsible for this retardation. In fact, for the Friedmann cosmological models considered here, the relation takes the exceptionally simple form $q_0 = \Omega_0/2$, and so the critical value for the deceleration parameter is $q_0 = 0.5$.

Once a particular cosmological model has been adopted by specifying values for $H_0$ and $q_0$, we can define what we mean by "distance." If, for example, we define the distance to an object of intrinsic luminosity $L$ as the quantity $D$ such that the observed flux from the source is $f = L/(4\pi D^2)$, then in the standard Robertson-Walker metric this **luminosity distance** for a source at redshift $z$ is given by

$$D = \frac{cz}{H_0} \left[ 1 + \frac{z(1 - q_0)}{(1 + 2q_0 z)^{1/2} + 1 + q_0 z} \right]. \tag{7.7}$$

Note that the luminosity referred to here is the bolometric value. Unfortunately, what we actually measure when we observe some standard candle at high redshift is its flux within a fixed bandpass. There is a serious danger that uncertainties in the K-correction needed to compensate for the object's redshift (§2.3.3) will render its luminosity distance unmeasurable.

Alternatively, we can define the **angular diameter distance**, $\mathcal{D}$, such that a source of intrinsic size $d$ subtends an angle $\delta\theta = \tan^{-1}(d/\mathcal{D}) \approx d/\mathcal{D}$ ($\mathcal{D} \gg d$). We then obtain a different distance measure,

$$\mathcal{D} = (1 + z)^{-2}D. \tag{7.8}$$

One counter-intuitive implication of equation (7.8) is that it does not vary monotonically with $z$: if we imagine moving an object to progressively higher redshifts, its angular size at first decreases as we would expect, but ultimately it starts to *increase* with redshift! This magnification at high redshift occurs because light travelling from the source to us follows a bent path in the non-Euclidean geometry of the Universe, distorting the image of the source – it is, in fact, a grand-scale example of gravitational lensing (Appendix A). The departures in the various distance measures from the simple linear relation between $D$ and $z$ of equation (7.2) mean that we can use the observed relation between redshift and distance as estimated by apparent luminosity or angular size to measure $q_0$ (see §7.4.4). In principle, we can therefore use standard candles or rulers to derive the large-scale mass density of the Universe. Note, however, that equations (7.7) and (7.8) are only strictly valid for the Friedmann models in which mass density is distributed uniformly through space. In reality, we know that at least some fraction of the material in the Universe is concentrated into galaxies and other discrete structures. As we will see in §7.2.4, these discrete mass concentrations also bend the path travelled by light, distorting the brightnesses and sizes of background objects. If a significant fraction of the mass of the Universe is non-uniformly distributed, then the distorting effects of discrete objects may dominate the departures from the simple Hubble law caused by the large-scale geometry of the Universe, complicating the calculation of the total mass density.

## 7.2 Absolute distance estimators

We have already discussed absolute distance measurements for nearby stars using simple geometric techniques – the available method include trigonometric parallax (§2.1.3), the moving cluster method (§2.2.2), secular parallax (§2.2.3) and statistical parallax (§2.2.4). The distances to these objects provide an absolute calibration to the relative distance estimators in the cosmic distance ladder, but the number of steps required to get from these local stellar measurements to extragalactic distances is undesirably large. In this section, we present a number of methods for calculating absolute distances which can be applied to objects at much greater distances. The successful application of these methods would allow us to tie down the absolute calibration of extragalactic distances on a variety of scales. In principle, these measurements offer a much more robust framework of distance determination than is available when the cosmic distance ladder is anchored solely by absolute distances to nearby stars.

As we will see, the major drawback to these absolute methods is that they suffer from a variety of potentially large systematic errors. If the distance scale calibration is to be effective, we must characterize these systematic effects as far as possible, and compare the results from all the different methods (which have different possible sources of systematic error) to check for consistency.

### 7.2.1 The Baade–Wesselink method

If we know a star's radius, $R$, then it is straightforward to estimate its distance from us. From its optical color or spectral shape, we can measure the effective temperature of the star (§3.4); its intrinsic luminosity is then given by

$$L = 4\pi R^2 \sigma T_{\text{eff}}^4 \tag{7.9}$$

[see equation (3.15)], or in terms of the star's bolometric absolute magnitude,

$$M_{\text{bol}} = -10 \log T_{\text{eff}} - 5 \log R + C, \tag{7.10}$$

where $C$ is a known constant. By comparing the apparent magnitude of the star to its absolute magnitude, we have a direct measurement of the star's distance from us (§2.3.3). Unfortunately, as we saw in §3.2, it is not usually possible to measure stellar radii directly, and those measurements which have been made give the radii of nearby stars which are not much help in setting the cosmic distance scale.[3]

This problem can be overcome through studies of variable stars such as Cepheids and RR Lyrae stars (§5.1.10). Lines in the spectrum of such a variable star show Doppler shifts that vary cyclically with the same period as the star's brightness variations. As we have seen in §5.1.10, both the brightness and velocity variations arise because the atmosphere of the star is unstable and undergoes global expansion and contraction, changing the size of the star. The amount by which the size of the star changes between times $t_0$ and $t_1$ can be measured by simply integrating the line-of-sight velocity of the atmosphere (as measured from Doppler shifts in its spectra) over this time:

$$\Delta r_1 = -p \int_{t_0}^{t_1} v_{\text{los}}(t) \mathrm{d}t. \qquad (7.11)$$

The negative sign occurs because $\Delta r_1$ and $v_{\text{los}}$ have opposite signs: an expanding atmosphere will blueshift the lines in the spectra resulting in a negative value for $v_{\text{los}}$. The factor of $p$ arises because the expansion of the star's atmosphere does not all occur along the line of sight – at the outer limb of the star, for example, the expansion velocity is transverse to the line of sight. The observed mean line-of-sight velocity is therefore less than the true expansion velocity of the atmosphere (see Problem 7.2).

Thus, if the star starts out with (unknown) radius $r_0$ at time $t_0$, then by time $t_1$ its size will have changed by a directly measurable amount, and its radius will now be $r_0 + \Delta r_1$. From equations (7.10) and (2.36), the change in the apparent magnitude of the source over this time will be

$$m_{\text{bol1}} - m_{\text{bol0}} = M_{\text{bol1}} - M_{\text{bol0}}$$
$$= -5 \left[ \log(r_0 + \Delta r_1) - \log r_0 \right] - 10 \left[ \log T_{\text{eff1}} - \log T_{\text{eff0}} \right]. \qquad (7.12)$$

If we can measure the effective temperature of the source at times $t_0$ and $t_1$ as well as the change in its apparent bolometric magnitude over this interval, then we can solve for $r_0$, the only unknown in equation (7.12). A particularly simple solution can be obtained if we select the time $t_1$ for our second observation to be a moment at which the star's effective temperature is the same as it was at time $t_0$. In this case, we find

$$r_0 = \frac{H \Delta r_1}{1 - H}, \quad \text{where } H \equiv 10^{(m_{\text{bol1}} - m_{\text{bol0}})/5}. \qquad (7.13)$$

---

[3] One method for measuring stellar radii which can, in principle, be extended to large distances comes from observations of eclipsing binaries (see §3.2.5). Such systems are rare and rather faint when they lie at interestingly-large distances, but the microlensing studies monitoring large numbers of stars in nearby galaxies (§2.4) are now turning up these objects in usable quantities.

A method for obtaining the distance to a pulsating star is now clear. By obtaining spectra throughout its period of oscillation and measuring the Doppler shifts in the spectral lines, we can derive $v_{los}(t)$. By also making a series of measurements of the apparent magnitude of the star in several bandpasses, we can estimate the variation in apparent bolometric magnitude with time (§2.3.4) once we have corrected the observations for the effects of extinction (§3.7.1). For a star of a particular type (dwarf, giant, etc), effective temperature varies monotonically with stellar color (see, for example, Table 3.6). Thus, by selecting data from two parts of the oscillation cycle at which the star has identical colors, we can pick out pairs of observations with the same effective temperature (without having to know anything about the details of the relationship between color and effective temperature). Calculating values for $\Delta r$ for each such pair using equation (7.11), and substituting these values together with the bolometric magnitudes into equation (7.13) allows us to obtain a series of estimates for $r_0$. Combining the best estimate for $r_0$ with equation (7.11) then tells us the radius of the star at any point during its oscillation cycle. Since we now know the radius of the star, it is straightforward to use the method outlined at the beginning of this section to calculate its distance from us. This approach, suggested by Baade (1926), was developed in the form described above by Wesselink (1946). All similar techniques for measuring distances based on variations in radial velocity and apparent magnitude are known generically as **Baade–Wesselink methods**; a review of this approach to measuring distances can be found in Gautschy (1987).

The classical Baade–Wesselink method described above has a number of shortcomings. First, real data will have observational errors associated with them, and it is not clear how the different estimates for $r_0$ obtained from different pairs of observations should be weighted when combining them to derive an overall "best" estimate (Balona 1977). More fundamentally, some of the assumptions that went into deriving the method may not hold. Specifically:

(1) We have assumed that the velocity derived from spectral line shifts can be used to characterize the overall expansion of the star. A real stellar atmosphere is not a single opaque surface, and since its opacity varies with wavelength, the depth into the atmosphere from which we observe most of the continuum emission may be different from the depth at which the spectral lines are formed. A non-uniform expansion in the atmosphere would mean that these two regions could be moving with different radial velocities. Since the structure of the atmosphere changes over the star's oscillation cycle, it is unlikely that there is a simple relation between these two velocities.

(2) The stellar oscillations may not be purely radial. Many types of variable star are believed to undergo oscillations which involve their shapes departing from spherical symmetry. The application of the Baade–Wesselink method to such stars is complicated because they can no

longer be described by a single radius and a single radial expansion velocity at any given time. The Baade–Wesselink method can only be
extended to allow for these non-radial effects if the shape of the stellar
oscillation is known (Balona & Stobie 1979).

(3) Finally, we have assumed that stellar color is a reliable measure of effective temperature. In fact, observations of variable stars using a range of
optical filters show that color indices centered on different wavelengths
(e.g. $U - B$, $V - R$) do not all vary in the same way with time. Hence
there is clearly not a universal one-to-one relationship between all these
optical colors and effective temperature, and it is not obvious which
measure of color best specifies the effective temperature. Given the
non-standard colors of these variable stars, it is also unclear how one
converts the apparent magnitudes in different wavebands into the requisite bolometric magnitude.

Although these problems complicate the Baade–Wesselink analysis, they
are not insurmountable. Recent applications of the method have concentrated on modeling the atmospheres of pulsating stars in some detail. From
these models, we can predict the observable photometric and spectral properties of the star in order to quantify the systematic errors that are introduced
into the Baade–Wesselink method by the detailed physics of variable stars.
Ultimately, directly fitting fully time-dependent models of variable star atmospheres to the observed spectra and photometry of these systems offers
the most complete approach to calculating distances to variable stars. However, the physical reason why such methods work is captured in the classical
Baade–Wesselink analysis: Doppler velocities allow us to measure a change
in the linear size of the source, and measurements of colors and apparent
magnitude provide us with a measure of the fractional change in its size;
putting these two constraints together yields the linear size of the source,
and hence its absolute luminosity and distance.

**Application to supernovae**    Thus far, we have discussed the Baade–
Wesselink method in terms of its application to variable stars. The same
technique can be applied to any radiating system whose size is changing
with time, and supernovae are obvious candidates for such studies. These
exploding stars (§5.2.3) are enormously bright and so can be observed out
to very large distances; obtaining a reliable measurement of their distances
would thus provide a giant leap toward establishing the extragalactic distance
scale.

The colors of a supernova vary monotonically as it expands and cools,
and so we cannot employ the Wesselink (1946) trick of selecting pairs of observations at different times which have the same colors. Instead, we must
assume a particular relation between color and effective temperature in order
to use pairs of observations to solve equation (7.12) for $r_0$ and hence the distance to the supernova. We might, for example, assume that the expanding
supernova radiates as a black body, and hence relate measurements of its

color directly to its temperature. By adopting this simple model, Branch & Patchett (1973) and Kirshner & Kwan (1974) made the first direct estimates for the distances to reasonably remote galaxies. Subsequent analyses used more complicated models for the expanding supernova atmospheres, and also started to include some of the physics of the supernova explosion via hydrodynamical models (Schurmann, Arnett & Falk 1979), and detailed model-atmosphere calculations (Schmidt, Kirshner & Eastman 1992).

The detonation of Supernova 1987A in the Large Magellanic Cloud (LMC) provided an ideal opportunity for testing the Baade–Wesselink method, since the distance to the LMC is already well-calibrated by other methods (§7.4.1). Further, the quality of data available for such a nearby supernova is far superior to that for more distant supernovae, and we have access to information – such as the exact time of detonation and the type of the progenitor star – which is usually unobtainable. Approximating the atmosphere of the supernova by a black body, Branch (1987) used the Baade–Wesselink method to obtain a distance to the LMC of $55 \pm 5$ kpc, in good agreement with other estimates. However, Schmutz *et al.* (1990) have made a detailed model for this supernova explosion, and they find that this agreement is somewhat fortuitous, due to the chance cancellation of several systematic errors. They also discovered that if the progenitor star had been a red supergiant rather than a blue supergiant, then this cancellation would not have occurred, and the simple Baade–Wesselink analysis would have produced a distance estimator that was wrong by a factor of two. Since we are very unlikely to be able to identify the progenitors of more distant supernovae, these calculations cast some doubt on the suitability of the simplest Baade–Wesselink analyses for determining the extragalactic distance scale.

### 7.2.2 The Sunyaev–Zel'dovich effect

The observed cosmic microwave background arises from the highly-redshifted thermal emission from the time when, not long after the Big Bang, the Universe became optically thin. Most observed background photons have travelled undisturbed since this time, and so we receive a very uniform illumination by microwave radiation from all directions. However, a small fraction of this light will have interacted with other material in the Universe between its creation at very high redshift and its arrival on Earth. In particular, photons that have passed through rich clusters of galaxies may have been Compton-scattered by electrons in the hot gas that pervades these systems. Since the electrons are more energetic than the photons, the photons will usually have their energies increased by the interaction, and so the light will on average be shifted to higher frequencies. Thus, the location of the peak in the black body spectrum of the undisturbed background radiation will be shifted to higher frequencies. Observations of the background radiation seen through a rich cluster of galaxies should therefore reveal an excess of photons at frequencies higher than the peak in the undistorted background spectrum, and a deficit at lower frequencies (Zel'dovich & Sunyaev 1969). This distortion in the microwave background is known as the **Sunyaev–Zel'dovich** or **SZ effect**. The size of the distortion depends only on the probability that a given photon is Compton scattered, $P_{SZ} = 1 - e^{-\tau_{SZ}}$, where

$$\tau_{SZ} = \sigma_T \int_S ds\, n_e(s) \qquad (7.14)$$

is the optical depth for scattering as the photon moves along the path $S$ through the cluster, $n_e$ is the electron density in the cluster's plasma, and $\sigma_T$ is the Thomson scattering cross-section. Since the distance to the cluster does not appear in equation (7.14), the distortion of the microwave background enables us to obtain a distance-independent constraint on the distribution of electrons in the intervening cluster.

We can exploit this phenomenon to measure the distance to the cluster by comparing the distance-independent observation to a distance-dependent measure of the electron density. Such a measure is provided by X-ray observations: as electrons scatter off protons in the hot cluster plasma, they emit X-rays via the Bremsstrahlung process (§8.1.2). The Bremsstrahlung emissivity of the gas at frequency $\nu$ is

$$\varepsilon(\nu) = A n_e^2 T_X^{-1/2} e^{-h\nu/kT_X}, \qquad (7.15)$$

where $T_X$ is the temperature of the plasma, and $A$ is a constant (see §8.1.2). By observing the X-ray emission at a range of frequencies, we can measure the spectrum produced by the hot gas, and thus obtain its temperature. The observed X-ray flux then depends only on the electron density and our distance from the cluster. Thus, combining this observation with the distance-independent measure of electron density provided by the Sunyaev–Zel'dovich effect, we can solve for the only remaining unknown quantity, the distance to the cluster.

To illustrate this method, consider a simple model for a cluster in which the cluster gas is isothermal and fills a uniform-density sphere of radius $r_c$ lying a distance $D$ from us. The optical depth measured by the SZ effect through the middle of such a cluster is $\tau_{SZ} = 2\sigma_T r_c n_e$. The observed X-ray flux is simply $f_X(\nu) = \frac{4}{3}\pi r_c^3 \varepsilon(\nu)/(4\pi D^2)$, and with an imaging X-ray telescope we can also measure the angular extent of the X-ray emission, $\delta\theta_X = 2r_c/\mathcal{D}$ radians. Combining these observable quantities with equation (7.15) and using the definition of angular diameter distance [equation (7.8)], we find that a cluster at redshift $z$ lies at a luminosity distance

$$D = \frac{A}{24\sigma_T^2} T_X^{-1/2} e^{-h\nu/kT_X} \tau_{SZ}^2 \frac{\delta\theta_X}{f_X(\nu)} (1+z)^{-2}. \qquad (7.16)$$

Thus, once the redshift has been measured using optical spectra of galaxies in the cluster, we can combine the X-ray observations and the Sunyaev–Zel'dovich measurements to solve directly for the cluster distance. In reality, of course, the electron density and temperature will vary with position in the cluster, but Silk & White (1978) have shown how detailed spatially-resolved X-ray observations can be used to allow for this greater degree of generality, subject only to the assumption that the cluster is spherically symmetric.

Here, then, is a simple method that enables us to calculate extragalactic distances directly. It relies on relatively straightforward observations and their interpretation using well-understood physics. The only dubious step in the process is the assumption of spherical symmetry for the cluster gas. This assumption is central to the method, since the distance is essentially determined by equating the length through the cluster (as measured by the SZ effect) to the angular scale across the cluster (as measured from X-ray images). Unfortunately, X-ray images of rich clusters reveal that they are far from spherically-symmetric: some have no clearly defined shape, while others appear in projection as reasonably regular ellipses. Where an elliptical shape does provide a reasonable fit, the mean ratio between the long and the short axes is $a/b \sim 1.5$ (Struble & Ftaclas 1994). If such an elliptical cluster were viewed from different angles, the path length through the cluster would change, leading to values of $\tau_{SZ}$ that also vary by a factor of $a/b$. If we were to assume incorrectly that the cluster under scrutiny was spherically symmetric, then the distance estimate derived from the above analysis could also be off by a factor of up to $a/b$ (See Problem 7.3), and so there is considerable uncertainty inherent in this method for determining distances.

It is tempting to suggest that this difficulty could be overcome by observing a random sample of clusters. Although the individual distance determinations would have large uncertainties due to the unknown asphericity of the cluster, it should be possible to obtain the correct answer on average, and thus calibrate the extragalactic distance scale. However, observational bias makes it difficult to select an appropriate random sample. For a sample of clusters of a given intrinsic richness and ellipticity, those that happen to lie with their longest axis along the line of sight will produce the most detectable X-ray emission, since their gas will appear concentrated in the smallest region. Such end-on clusters will also have the largest SZ effects due to the long path length through these systems. When analyzing samples of clusters, great care must be taken to ensure that we are not preferentially selecting these end-on systems for which the estimated distances will be systematically too large. In this context, it is interesting to note that Sunyaev–Zel'dovich distance estimates tend to be larger than the values obtained using other methods (e.g. McHardy *et al.* 1990).

### 7.2.3 Distances from time delays

If an astrophysical source's light output changes with time, then we may be able to estimate its distance by measuring a delay in the arrival of the time-varying signal. The details of the method depend on the physical reason for

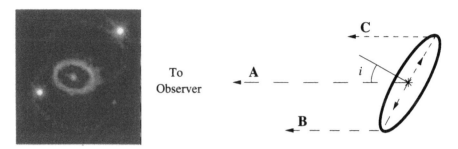

**Figure 7.1** Diagram showing the time delay in the signal received from SN 1987A due to reprocessing by material located in a ring of radius $r_{\text{ring}}$ at an inclination $i$ to the line of sight. The Hubble Space Telescope direct image of the inclined ring is shown on the left of the figure; note also the two fainter rings which imply that the full geometry of the system is more complex. [HST image reproduced with permission from AURA/STScI]

the time delay, but essentially all such methods use the observed delay, $\Delta t$, in the arrival of a signal traveling at the speed of light to obtain a measure of a physical dimension of the source, $d \sim c\Delta t$. If we can also measure the same characteristic scale in angular coordinates, $\theta$, derived from the size of the source on the sky, then we have a direct measure of the distance to the source, $\mathcal{D} \sim c\Delta t/\theta$, where $\theta$ is measured in radians.

**The ring around Supernova 1987A**   A simple example of distance determination via time delay is furnished by Supernova 1987A. Observations using the International Ultraviolet Explorer (IUE) satellite revealed narrow emission lines from highly ionized atoms in the spectrum of the supernova. These lines were only detected at a time $t_0 \approx 90$ days after the initial explosion, and their intensities rose to a maximum after $t_{\text{max}} \approx 400$ days, before decaying slowly back down to very low flux levels. The simplest explanation for this delayed light curve is that the lines arise from radiation that has been reprocessed by material at some distance from the supernova, with the delay arising from the extra light travel time involved in this detour. Support for this hypothesis comes from a Hubble Space Telescope image of the supernova obtained through a narrow-band filter which admits light produced in a line of ionized oxygen (see Figure 7.1). This image shows that the emission from this ion is concentrated in several rings, the brightest of which forms an ellipse centered on the supernova. An inclined circular ring of material around the supernova would have this elliptical appearance and might also explain the delayed light curve.

The geometry of the proposed tilted ring is illustrated in Figure 7.1. The first reprocessed light to reach us would travel via path B, which, by simple geometry, would be delayed as compared to the direct path A by an amount

$$t_0 = \frac{r_{\text{ring}}}{c}(1 - \sin i), \tag{7.17}$$

where $i$ is the inclination of the ring, and $r_{\text{ring}}$ is its radius. The strength of the line emission would then continue to rise until the entire ring is contributing to the reprocessed light output. The last point to "light up" around the ring would be the point which is illuminated only via the longest path, C. The extra time delay to this point is

$$t_{\max} = \frac{r_{\text{ring}}}{c}(1 + \sin i). \qquad (7.18)$$

Combining equations (7.17) and (7.18) with the observed values of $t_0$ and $t_{\max}$, we can solve simultaneously for $i$ and $r_{\text{ring}}$; this analysis yields $i = (42 \pm 5)°$ and $r_{\text{ring}} = (0.42 \pm 0.03)$ pc (Panagia *et al.* 1991). An entirely independent measure of the ring's inclination comes from its apparent ellipticity in the HST image, and these data yield an inclination of $i = (43 \pm 3)°$ (Jakobsen *et al.* 1991); the close agreement between these two results provides some confidence that our assumed geometry of a simple circular ring is correct. The major axis of the elliptical ring in the HST image has an angular size of $\theta = 1.66 \pm 0.03$ arcsec; equating this value to the derived linear diameter of the ring, $r_{\text{ring}}$, implies that the distance to the supernova is $52 \pm 3$ kpc in good agreement with other estimates (see §7.4.1).

**Gravitational lens time delays**    The measurement of distances by time delay can be extended to very large scales by observing gravitationally-lensed quasars. As we have discussed in §2.4, when light passes a massive body, its path is bent by gravity. If the light in question comes from a distant quasar, and the mass is a galaxy or even a whole cluster of galaxies, then the effects of this gravitational lensing can be quite dramatic. Such large masses are sufficient to separate lens-generated multiple images by several arcseconds, allowing the individual images to be resolved (see Figure 7.2). Further, the more complicated geometry of a galaxy's mass distribution (as compared to the point-mass lens discussed in §2.4) means that multiple images containing four components can also be produced.[4] To-date, around ten quasars have been found to be split into such double or quadruple images; a review of these and other cosmological gravitational lensing phenomena can be found in Blandford & Narayan (1992), and the entire subject is presented in great detail by Schneider, Ehlers & Falco (1992).

   The bending of the light path by the lens increases the distance that the light has to travel to reach us, so there is a time delay in the signals that we receive. If the image of a background source at redshift $z_s$ appears shifted by an angle $\alpha_i$ due to gravitational lensing by a deflecting mass at redshift $z_d$, then the extra delay can be shown to be

$$t_{\text{lens}}^{(i)} = (1 + z_d) \left[ \frac{1}{2c} \frac{\mathcal{D}_d \mathcal{D}_s}{\mathcal{D}_{ds}} \alpha_i^2 - \frac{2}{c^3} \int_{\gamma_i} ds\, \Phi(s) \right] \qquad (7.19)$$

---

[4] In fact, such quadruple images contain a fifth component, but this last, central image is heavily demagnified, so it is extremely hard to detect it.

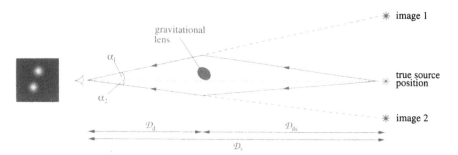

**Figure 7.2** The basic ray geometry of a gravitational lens system, showing how the bending of light rays by an intervening mass can produce multiple images of a background object which are offset from the object's true location. The inset image shows an example of this phenomenon, the "Twin Quasar" QSO 0957+561. [Image obtained by K. Kuijken and M. Merrifield]

[Schneider, Ehlers & Falco (1992); see also Appendix A]. The quantities $\mathcal{D}_d$ and $\mathcal{D}_s$ are the angular diameter distances to the deflecting lens and the source respectively, and $\mathcal{D}_{ds}$ is the angular diameter distance to the source as measured from the lens (see Figure 7.2). The integral of the Newtonian gravitational potential $\Phi$ is evaluated along the path $\gamma_i$ traversed by the light that produces this image. The first term on the right-hand side of equation (7.19) is simply the extra travel time due to the change in geometric distance that the gravitationally deflected light covers in getting from source to observer (see Figure 7.2); the second term arises from the general-relativistic time dilation caused by the distortion of space-time in the presence of a gravitational field.

Since we cannot remove the gravitational lens to measure the undeflected light path, $t_{\text{lens}}^{(i)}$ is not an observable quantity. However, if the quasar has been split into two images, then each of the images will be formed by light that has taken a different path and will have its own time delay, $t_{\text{lens}}^{(i)}$ (see Figure 7.2). The intrinsic light output from quasars varies with time, so we can obtain the difference between the time delays for two different quasar images,

$$\Delta t_{\text{lens}}^{(1,2)} = t_{\text{lens}}^{(1)} - t_{\text{lens}}^{(2)} \tag{7.20}$$

by observing the brightnesses of the images as a function of time and measuring the delay by which one of these functions lags behind the other. The time by which one image lags the other is typically a number of years, so a lensed quasar must be monitored over a long period if this delay is to be measured.

If we wish to use such relative time delays to obtain a quantitative estimate of distances, we require a detailed model for the distribution of mass in the gravitational lens. In general, this model will be quite complicated, since there is no reason why the lensing mass should be symmetrically arranged, and there may be several components (such as a single galaxy and a surrounding cluster of galaxies) which all contribute to the total mass distribution of the lens. Fortunately, we also have a number of observational constraints on the mass distribution: different mass models would produce very different arrangements on the sky for the multiple image locations and their relative brightnesses, so the range of possible lenses that could produce a given multiply-imaged quasar is relatively small (e.g. Falco, Gorenstein & Shapiro 1991). We may also be able to observe the lensing system directly. If the lens is a single galaxy, for example, then we can constrain our model by measuring its light distribution, and by using absorption line spectra to measure its kinematic properties (Chapter 11).

Once we have a model for the lens, we can derive values for the angles by which the quasar images have been shifted, $\alpha_i$, and the path integral in equation (7.19). If we have also obtained a measurement of the time delay by monitoring the light output of two quasar images, then the only remaining unknowns in equation (7.20) are the angular diameter distances to the source and deflecting gravitational lens, $\mathcal{D}_s$ and $\mathcal{D}_d$. Since we can measure the redshifts of the source and lens, the only remaining unknowns in these distances are the Hubble constant, $H_0$, and the deceleration parameter, $q_0$ [see equations (7.7) and (7.8)]. Thus by assuming a density model for the Universe (and hence fixing $q_0$), we can use the observed properties of the lens to solve uniquely for $H_0$. Alternatively, if similar observations were made of a second gravitationally lensed system then we could pool the results to solve simultaneously for both $H_0$ and $q_0$.

In practice, of course, life is never so simple. Even obtaining the basic measurement on which the method depends – the difference in time delay between images – is not straightforward. For example, in the case of the best-studied multiply-imaged quasar, QSO 0957+561, which has been monitored fairly closely over a period of more than ten years, there has been a long-running debate as to the time delay between the two images. Estimates vary between $410 \pm 20$ days (Pelt et al. 1994) and $540 \pm 10$ days (Press, Rybicki & Hewitt 1992). These discrepant estimates arise because the two images do not vary in the same way: although many of the variations observed in one image are also seen in the second image $\sim 500$ days later, other features occur in only one of the images. Clearly, these non-repeated features are not consistent with the simple gravitational lens model for such systems. One plausible explanation for the differences between the two images is that individual stars in the lensing galaxy occasionally pass through the light path from one or other of the quasar images, producing an additional amplification in the brightness of the image due to the microlensing influence of the star (Kayser, Refsdal & Stabell (1986); see §2.4). Fortunately, continuing observations of this source have now turned up a sharp spike in the luminosity of the secondary image 415 days after an identical feature occurred in the primary image (Kundić et al. 1997). The absence of any similar feature in the secondary image with a time lag of anywhere near 540 days all but rules out the longer time delay. From the amount of effort invested in this single measurement, it is clear that obtaining gravitational lens time delays is a difficult problem for even the simplest, best-studied systems.

A further complication is that just measuring the positions and relative brightnesses of the images is not, in general, sufficient to determine the lens model uniquely. Kochaneck (1991) has shown that plausible mass models for the QSO 0957+561 lens, which all satisfactorily reproduce the basic image properties, yield values of $H_0$ that are uncertain by a factor of ten. It is also quite possible that small-scale inhomogeneities in the cosmological distribution of mass along the line-of-sight to the quasar will produce additional gravitational lensing effects which are difficult to characterize in a simple lens model. Fortunately, additional constraints on the lens mass distribution have come from subtler properties of the images: Garrett et al. (1994) have obtained very high resolution radio observations of the QSO 0957+561 system, which resolve features in the individual images of the background quasar, showing a radio core and jet structure. By requiring the lens to reproduce the details of the individual distorted images, the range of possible mass distributions is dramatically reduced. A similar increase in the number of constraints on the lens mass can come from studying more complex gravitationally-lensed systems, such as the quadruply-imaged "Einstein Cross", QSO 2237+031 (Yee 1988). The positions and relative brightness of four images provide many more constraints on the lens model than does a two-component image. Moreover, in such a quadruple system there are three independent relative time delays that might be measured.

Clearly, difficult observations and modeling are required in order to use gravitational lens time delays to measure distances. However, the underlying physics – the gravitational bending of light rays – is both simple and well-understood. Furthermore, these systems can be seen right across the Universe, and so they provide direct distance determinations on the largest possible scales. These potential benefits more than offset the difficulty in making the requisite measurements, so the investment of effort will be well rewarded.

### 7.2.4 Water-maser proper-motions by VLBI

In §2.2.4, we studied the method of statistical parallax in which the distance to a set of objects is estimated by comparing their average line-of-sight velocities with their average proper motions. In the stellar case, the difficulty of obtaining accurate proper motions restricts the application of this technique to those stars which lie closer than $\sim 500\,\mathrm{pc}$. To extend this method to extragalactic distances, we have to meet the following requirements: we must study sources that are bright enough to be readily detectable at large distances; their line-of-sight velocities must be determinable with reasonable accuracy; and we must observe them using a technique that allows us to measure very small changes in their angular positions.

All of these criteria can be met by studying $H_2O$ maser spots. These sources, which occur preferentially in regions of active star formation, emit intense line radiation at a rest wavelength of $\lambda = 1.35$ cm. This emission arises from a molecular transition in water molecules, which is excited by nearby newly-formed stars. The excitation produces an inversion in the population of the $H_2O$ molecular energy states. Any spontaneously-emitted photon from this material will encounter other excited molecules in its path, causing them to emit further photons at this wavelength; these photons, in turn, induce further emission in the same direction. The resulting cascade of coherent mono-energetic photons – a **cosmic maser** – produces a cone of tightly-beamed radio emission which is sufficiently bright to be observed at very large distances. Since the maser emission occurs in a narrow line, we can use the observed Doppler shift to measure the line-of-sight velocity of the emitting region with an accuracy of better than $1\,\mathrm{km\,s^{-1}}$. Finally, because the emission occurs at radio wavelengths, we can use the technique of very long baseline interferometry (VLBI) whereby simultaneous observations from radio telescopes across the earth (of diameter $D_{\mathrm{earth}} \sim 10^7$ m) are combined to produce images with an angular resolution of $\sim \lambda/D_{\mathrm{earth}}$ radians $\sim 10^{-4}$ arcseconds. Images produced using this technique show that the maser emission occurs in discrete spots, and the positions of the brighter spots can be obtained to an accuracy of $\sim 10^{-5}$ arcseconds by measuring the centroid of the unresolved emission.

As we shall see in §7.4.1, studies of the statistical parallax of $H_2O$ masers in star forming regions near the Galactic center provide the only absolute determination of the Sun's radius in the Galaxy. Attempts have also been made to push this technique to extragalactic distances. Greenhill *et al.* (1993) have observed $H_2O$ masers in M33 at two epochs separated by $t = 479$ days. Some maser spots were detected in the data from both epochs, and they had moved on average by a RMS distance of $\delta\theta \sim 8 \times 10^{-6}$ arcseconds. The Doppler velocities of the spots within each star-forming region imply that they have a line-of-sight velocity dispersion of $\sigma_{\mathrm{los}} \sim 17\,\mathrm{km\,s^{-1}}$. Equating the transverse RMS motion to the line-of-sight velocity dispersion, we obtain a distance estimate for M33 of $D = \sigma_{\mathrm{los}}/[\delta\theta(\mathrm{radians})/t] \sim 600\,\mathrm{kpc}$. This value is lower than the accepted distance of $\sim 740\,\mathrm{kpc}$ (M33 is a companion galaxy to M31, whose distance has been investigated extensively; see §7.4.1). However, as Greenhill *et al.* (1993) point out, there are a large number of uncertainties in this analysis. If the orbits of the masing clouds were not entirely random, then the average transverse and line-of-sight velocities would not be equal, and so the simple statistical parallax argument would break down. It is also quite plausible in a star-forming region that the emitting clouds might have a net outward streaming velocity, which would again invalidate the statistical parallax technique. By fitting a complete dynamical model to the observed motion, we can in principle correct for these systematic effects – see Genzel *et al.* (1981) for the application of such a model to a Milky Way $H_2O$ maser source. Unfortunately, the very limited quantity of data available from M33

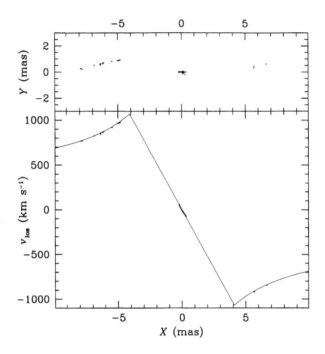

**Figure 7.3** The properties of the $H_2O$ masers in the core of NGC 4258, showing their spatial distribution (upper panel), and their line-of-sight velocities as a function of position along the $X$ axis (lower panel). The line indicates the line-of-sight velocities expected for maser spots in a simple Keplerian disk which is only populated at radii greater than 4.1 mas. [After Miyoshi *et al.* (1995) using data kindly provided by L. Greenhill]

does not justify this degree of complexity.

VLBI observations of the core of the nearby spiral galaxy NGC 4258 (Miyoshi *et al.* 1995) reveal that it contains a distribution of $H_2O$ maser spots with a remarkably simple geometry (see Figure 7.3). The emission spots are distributed in an almost-linear structure passing through the projected center of the galaxy, with the features on one side of the center redshifted relative to the galaxy's systemic velocity, while those on the other side appear blueshifted. The natural explanation for these observations is that the maser spots lie in an edge-on disk orbiting around a large central mass in the galaxy. Radio jets emanate from the core of NGC 4258 which also shows up as an X-ray source, so this galaxy clearly contains an active galactic nucleus (see §4.6.2). The massive black hole believed to cause these phenomena explains the large derived central mass, and it also provides the energy source required to power the maser spots in the surrounding disk.

Confirmation of this model has come from the measurements of the Doppler shifts of the individual maser spots. As Figure 7.3 shows, the line-of-sight velocity of the maser spots at large radii drops off with the distance from the center of rotation as $r^{-1/2}$, exactly as one would expect for Keplerian motion around a central point mass. At small projected radii, the line-of-sight velocities vary linearly with radius; this linear dependence can only occur if these maser spots all lie on a single circular orbit. We therefore end up with a picture of a disk around the central mass which is filled with masing material outside a ring of maser spots that defines the disk's inner edge.[5] We can infer the radius of this inner edge by measuring the point at which the extrapolation of the linear variation in line-of-sight velocity meets up with the $r^{-1/2}$ dependence at large radii; as Figure 7.3 shows, this intersection places the inner edge of the accretion disk at a radius of $\theta_{in} = 4.1$ mas. The orbital velocity of material at this inner radius is then simply the amplitude of the line-of-sight velocity at the intersection point, $v_{in} = 1080 \, \mathrm{km \, s^{-1}}$ (see Figure 7.3).

The individual maser spots follow their circular orbits because they are centripetally accelerated by the central mass. We can detect the line-of-sight component of this acceleration by measuring changes in the Doppler shifts of individual features with time. Since the acceleration is always directed toward the center of the disk, the largest line-of-sight accelerations should occur for maser spots that lie directly in line with the center of the accretion disk, where the full centripetal acceleration projects into the line of sight. VLBI observations spread over several years confirm that the velocities of spots which appear projected close to the center of the disk drift at a rate $\dot{v} = 9.5 \pm 1.1 \, \mathrm{km \, s^{-1} \, yr^{-1}}$ (Greenhill *et al.* 1995). As we have seen above, the maser spots that appear close to the center of the system are all emitted from close to the inner radius of the accretion disk, $r_{in}$. We can therefore equate the observed rate of drift in velocities to the centripetal acceleration, $\dot{v} = v_{in}^2/r_{in}$. Substituting the observed values for $\dot{v}$ and $v_{in}$, we find that $r_{in} = 0.12 \pm 0.01 \, \mathrm{pc}$. Comparing this radius with the angular size of the inner edge of the disk, $\theta_{in} = 4.1$ mas, we find that NGC 4258 must lie at a distance of $\mathcal{D} = 6.2 \pm 0.7 \, \mathrm{Mpc}$. Because of the simple geometry in this system, these observations provide an elegant technique for measuring absolute distances with very little ambiguity. The observed angular scales in NGC 4258 are well above the limits achievable by VLBI; assuming that similar bright $H_2O$ maser systems can be found in the nuclei of other galaxies, the same method could be applied to systems at much higher redshifts in order to provide a relatively error-free absolute calibration to the cosmic distance scale.

---

[5] This disk geometry, and the requirement that maser emission can only occur along lines of sight where all the material has similar line-of-sight velocities, also explains the marked absence of detectable maser spots at intermediate radii – see Problem 7.4.

## 7.3 Relative distance estimators

As we have seen in the last section, there exist a variety of methods for measuring absolute distances to particular classes of objects. We can estimate the distances to these objects directly because they possess unusually simple geometries, and so we can interpret observations of them on a straightforward physical basis. Unfortunately, most astronomical phenomena are the result of much more complicated processes, so no such simple modeling is possible when we wish to estimate their distances. However, many of these more complex systems do seem to follow simple empirical relations between their intrinsic properties, and these relations can be employed as distance estimators once they have been calibrated.

We have already discussed one good example of such a calibrated distance indicator for star clusters in §3.5.2: by comparing features in the CM diagrams of clusters at different distances, it is possible to derive the amount by which the apparent magnitudes of stars in the clusters differ, and thus derive their relative distances [equation (7.1)]. Once we have obtained an independent measure of the distance to one cluster (such as the Hyades, by using the moving cluster method; §2.2.2), we can convert all the relative measurements to an absolute scale.

This example illustrates many of the features typical of relative distance estimators. First, the distance estimate depends on combining a distance-independent property of objects (the stars' colors) with a property that does vary with distance (the stars' apparent magnitudes) in order to estimate their distances. Second, it is empirically based: without having to understand the complex physics of stellar structure which dictates stars' luminosities and colors, we can use the features that occur in the CM diagram to measure relative distances to clusters. Finally, since this distance estimator makes no use of the underlying physics, we have no *a priori* knowledge of the absolute calibration of the measure, and so we must obtain an independent measurement of the distance to at least one cluster in order to calibrate the relation.

Once calibrated on to an absolute scale, the relative distance estimators hold a number of advantages over the absolute methods. First, relative estimators depend on empirically established properties of objects. They do not require that we make assumptions about the geometry or physics of the source, so they are not subject to the systematic uncertainties that arise if our assumptions are incorrect. Second, relative distances can be estimated using common sources. They can therefore be used to measure the distances to many more objects, allowing us to estimate distances over the wide range that we require if we are to calibrate the complete extragalactic distance scale. They are also useful for measuring subtle effects such as the dependence of measured distance on $q_0$ [equations (7.7) and (7.8)], where large numbers of distances must be obtained in order to reduce the uncertainties which arise from random errors. A wide-ranging review of many of the relative distance estimators used in defining the cosmic distance scale can be found in Jacoby *et al.* (1992).

### 7.3.1 Luminosities of variable stars

In §5.1.10 we saw that the luminosities of certain pulsating stars can be determined fairly accurately from their periods. Clearly, the distance modulus of such a star can be estimated as the difference between its apparent magnitude and its predicted absolute magnitude.

The two most important classes of pulsating stars are Cepheid variables and RR Lyrae stars. Cepheids have absolute magnitudes $M_V \sim -3$, so they can be studied to distance moduli $m - M \sim 25$ from the ground and $m - M \sim 28$ with HST. RR Lyraes are significantly fainter than Cepheids, having absolute magnitudes $M_V \sim 0.6$. Hence they can be studied only to distance moduli $m - M \sim 22$ from the ground or $m - M \sim 25$ with HST. Consequently, whereas Cepheids can be studied in objects that lie near the center of the Virgo Cluster, even with HST it is impossible to find RR Lyrae stars outside the Local Group. Under optimal conditions, photometric distances to Cepheids and RR Lyrae stars can have errors that are as small as 10%.

Cepheid variables have played a central role in defining the extragalactic distance scale. Their widespread usefulness arises from their powerful combination of properties:

(1) They obey very well-defined relations such as equation (5.19), and so make excellent standard candles.
(2) Their characteristic variability means that they can be identified unambiguously.
(3) They are bright, and thus can be observed to reasonably large distances.
(4) They are plentiful, and so are found both near the Sun where their distances can be calibrated quite straightforwardly, and in moderately distant galaxies where they can be used as distance indicators.

As we have discussed in Chapter 1, Cepheids have the historical honor of providing Hubble with the evidence needed to demonstrate that an extragalactic distance scale existed. As we shall see in §7.4, even today these stars play a central role in refining the calibration of that scale.

### 7.3.2 Luminosity functions

Objects that have a large spread, $\sigma_M$, in the distribution of their absolute magnitudes are not individually useful as standard candles. However, if we observe a sample of $N$ such objects all at the same distance, the average magnitude of the sample will have a smaller scatter of $\sigma_M/\sqrt{N}$. If we observe a large enough sample, then the uncertainty in the average magnitude will become small enough for this quantity to serve as a standard candle. The success of this approach rests on the assumption that the distribution of the sources' luminosities – their luminosity function – does not vary from system to system, so that comparing the characteristic apparent luminosities of the distributions in different systems provides a direct measure of the relative distances to the systems.

**Globular clusters**    Shapley (1953) first noted that the average luminosities of globular clusters in the Milky Way, the Large Magellanic Cloud, and M31 all appear very similar, suggesting that this quantity might be used as a standard candle. It is very difficult to detect globular clusters in other more distant spiral galaxies: unless the galaxy is very close to edge-on, it is hard to pick out the globular clusters against the uneven background of the emission from galaxy's disk. In addition, the corrections to the globular cluster magnitudes due to extinction from dust in spiral galaxies are hard to characterize. However, it is straightforward to observe the globular clusters around a giant elliptical galaxy: these systems frequently have very large populations of globular clusters; the clusters are easy to detect against the smooth surface brightness distribution of their host elliptical; and the galaxy has little internal extinction. We are therefore faced with the unfortunate situation where the one system for which we have accurate measurements of the absolute magnitudes of globular clusters – the Milky Way – is very different from the giant elliptical galaxies for which we can obtain the best globular cluster data. Since there is no a priori reason why luminosity functions for the globular clusters in spiral and elliptical galaxies should be the same, there remains some doubt as to the absolute calibration of this distance estimator.

Despite this caveat, the globular cluster luminosity functions (GCLFs) for other galaxies look remarkably similar. The number of clusters per unit apparent magnitude is well-characterized by a simple Gaussian luminosity function,

$$\phi_{GC}(m) \propto e^{-(m - \overline{m}_{GC})^2 / 2\sigma_{GC}^2}. \tag{7.21}$$

By fitting this functional form to the observed luminosity distributions of globular clusters in elliptical galaxies, we can obtain a robust estimate for the apparent magnitude of the mean of the luminosity function, $\overline{m}_{GC}$. Assuming that there is a universal intrinsic distribution of luminosities, comparisons between the values of $\overline{m}_{GC}$ obtained for different cluster systems provides a direct measure of their relative distances. In those cases where there are independent measures of the relative distances to these galaxies, these estimates agree well with the GCLF distances, suggesting that the assumption of a universal luminosity function is valid. A more complete discussion of the application of the GCLF distance estimator can be found in Jacoby *et al.* (1992) §4.

The above fitting process works effectively as long as the GCLF has been measured to magnitudes fainter than $\overline{m}_{GC}$ so that the peak in the function is well-defined. To-date, the faintest globular cluster magnitudes that have been measured come from Hubble Space Telescope images of NGC 4881 in the Coma Cluster. These observations have derived the luminosity function of this galaxy's globular cluster population down to $m_V = 27.6$ (Baum *et al.* 1994). For the Milky Way, the peak of the GCLF is at a $V$-band absolute magnitude of $\overline{M}_{GC} = -7.4 \pm 0.1$ (Abraham & van den Bergh 1995). If we assume that the peak in the luminosity function of giant ellipticals lies at a similar absolute magnitude, then the limiting magnitude of 27.6 means that we should be able to see the peak in the GCLF for galaxies out to distances of $\sim 100$ Mpc.

**Planetary nebulae**    Young planetary nebulae (§5.1.2) provide a second population of objects whose luminosity function is well-suited for use as a standard candle. Since the emission from these systems occurs in lines, planetary nebulae can be detected even against the bright stellar continuum emission of a galaxy by obtaining images through narrow bandpass filters centered on these lines. Further, planetary nebulae occur in profusion in all types of galaxies, so their luminosity function can be accurately determined by measuring the luminosities of large samples in individual systems.

Observations of nearby galaxies through narrow bandpass filters centered on the 500.7 nm line of ionized oxygen show that the planetary nebulae luminosity function (PNLF) varies exponentially with magnitude in this bandpass. However, this relationship breaks down at high luminosities, and there are essentially no planetary nebulae brighter than a cut-off apparent magnitude, $m_{PN}^{cut}$. Ciardullo *et al.* (1989) have parameterized this abruptly-curtailed luminosity function in the form

$$\phi_{PN}(m) \propto e^{0.307m} \left( 1 - e^{3(m_{PN}^{cut} - m)} \right). \tag{7.22}$$

The 500.7 nm line lies within the bandpass covered by the Johnson $V$ filter, and the zero-point of the line emission intensity scale is customarily defined so that the derived magnitude is the same as would have been obtained if the line had been observed in the $V$-band. Numerically, if the flux collected from the line is $f_{500.7}$ W m$^{-2}$ then

$$m = -2.5 \log \left[ f_{500.7} \right] - 21.24. \tag{7.23}$$

Observations of the planetary nebula populations in a variety of types of galaxy reveal that all can be well-described by equation (7.22), suggesting that the PNLF is a universal function, and can hence be used as a standard candle. The particular functional form of the PNLF is also ideally suited for standard candle analysis: the sharp cut-off provides a distinctive feature in the luminosity function, and, even with only limited data, the cut-off magnitude can be measured quite accurately. Since this feature occurs at the bright end of the luminosity function, we only need observe the brightest planetary nebulae in a galaxy to measure this characteristic cut-off. We can

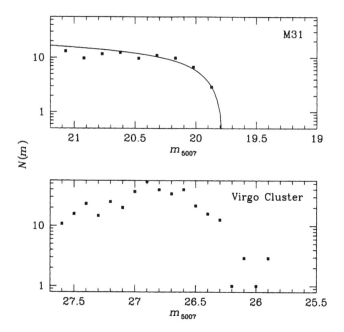

**Figure 7.4** The planetary nebula luminosity functions of M31 and of a composite of six galaxies in the Virgo Cluster. The 500.7 nm apparent magnitudes, $m_{500.7}$, have been corrected for extinction. The line shows the adopted functional from of equation (7.22). The cut-off at bright fluxes is apparent in both the M31 and Virgo Cluster data. [After Jacoby *et al.* (1992)]

therefore use the PNLF as a standard candle out to large distances where only the brightest planetary nebulae are detectable. For example, Figure 7.4 shows the observed planetary nebula luminosity functions for M31 and for galaxies in the Virgo Cluster. The sharp cut-off at bright magnitudes is apparent even in the distant Virgo Cluster data, and comparison between the values of $m_{cut}^{PN}$ yields an immediate estimate of the relative distances to these two systems (see §7.4.2).

Calibration of the absolute value of the cut-off luminosity is generally referred to observations of planetary nebulae in the bulge of M31. After correction for extinction, the maximum luminosity in the PNLF for this system occurs at a magnitude of $m_{PN}^{cut} = 19.77 \pm 0.04$ (Ciardullo *et al.* 1989). If we adopt a distance to this system of $740 \pm 40$ kpc (§7.4.1), we obtain an absolute magnitude for the cut-off of $M_{PN}^{cut} = -4.6 \pm 0.1$. With this absolute calibration, the distance moduli of other galaxies can be readily estimated.

### 7.3.3 Novae and supernovae

Novae, and especially supernovae, have large intrinsic luminosities. They have therefore always proved most attractive as possible standard candles for establishing the extragalactic distance scale. Indeed, part of the controversy in the original Shapley-Curtis debate on the existence of an extragalactic distance scale (§1.2.2) arose because supernovae and novae had been confused. Now that this confusion has been resolved, both novae and supernovae provide very useful distance estimators. A thorough review of the use of these systems for distance estimation can be found in Jacoby *et al.* (1992) §5 (novae) and §6 (supernovae).

**Novae** These transient systems, which result from the explosive ignition of material accreted on to the surfaces of white dwarf stars, have a number of desirable properties as standard candles. They are intrinsically bright ($M_V \sim -7$ at maximum brightness), and so can be observed out to large distances. They are produced by old stellar populations, and thus occur in elliptical galaxies as well as spirals. Since ellipticals do not contain intrinsically-bright objects such as early-type stars, novae are relatively easy to pick out in these systems. Further, since ellipticals do not suffer much from internal extinction, it is straightforward to measure the photometric properties of the novae that they host. Finally, observations indicate that the brightnesses of novae do not depend significantly on the metallicity of their surroundings (van den Bergh & Pritchet 1986), and so novae from environments with very different heavy element abundances can be compared directly.

For novae within the Milky Way, we can also determine distances by the simple absolute method of **expansion parallax**. The expanding shell of gas ejected by the explosion is thrown off with a typical velocity of $v_{\exp} \sim 1000 \, \mathrm{km \, s^{-1}}$; this quantity can be measured directly from the Doppler shifts in spectral lines. After a time $t$, the shell will show up in images as a structure of angular size $\delta\theta$, which we know corresponds to a physical dimension of $v_{\exp}t$. We thus know that the source lies at a distance of $\mathcal{D} = v_{\exp}t/\delta\theta$.

The angular diameters of extragalactic novae are too small for this method to be applicable, so we must return to the idea of using them as standard candles in order to determine their distances. Unfortunately, the intrinsic luminosities derived for the Galactic novae with measured distances show that these systems are not standard candles: their absolute magnitudes range from $M_V = -4.8$ to $M_V = -8.9$ (Cohen 1985). However, novae are not identical in other properties, either: Zwicky (1936) noticed that the light output from faint novae decays more slowly than that from bright novae. Since the measured rate of decline does not depend on the distance to the nova, we can use this measurement to predict the intrinsic brightness of the nova. Cohen (1985) quantified this relation using the 11 Milky Way novae with well-determined distances. She discovered that they obey a relationship

$$M_V(\mathrm{max}) = -10.7 + 2.3 \log(t_2/\mathrm{day}), \tag{7.24}$$

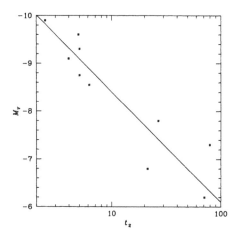

**Figure 7.5** Plot showing the absolute magnitudes of Milky Way novae at maximum brightness, $M_V(\text{max})$, as a function of the time they took to fade from this maximum by two magnitudes, $t_2$. The line shows the adopted linear relation of equation (7.24). [After Cohen (1985)]

where $M_V(\text{max})$ is the absolute magnitude of the nova at its maximum, and $t_2$ is the time that the nova took to decline in brightness by two magnitudes from this maximum (see Figure 7.5). The dispersion about this best fit is $\sim 0.5$ magnitudes, and so by measuring $t_2$ for a nova, we can predict its distance to within a factor of $\sim 25\%$. By observing many novae in a single galaxy, this statistical uncertainty can be reduced significantly.

Some caveats to this result come from the analysis of novae in M31 by Capaccioli *et al.* (1989). They showed that the brightest novae in M31 are fainter than predicted by equation (7.24). Since it is these brightest novae that are most likely to be detected in more distant galaxies, the application of this relation at large distances may not be appropriate. In addition, a few "anomalous" novae lie more than a magnitude brighter than equation (7.24). These anomalous systems are easy enough to spot in M31 where they are well-separated from the majority of the data. However, in more distant systems where fewer novae have been observed, they would be much harder to exclude.

As an alternative distance estimator based on novae in well-observed systems, we might use some characteristic of their luminosity function as a standard candle (see §7.3.2). The luminosity function of novae at maximum brightness appears to be a double-peaked function with a reasonably sharp central minimum (Capaccioli *et al.* 1989). It has therefore been suggested that the location of this minimum in the luminosity function might provide a useful standard candle.

**Type Ia supernovae**    It has been known for a long time that many of the brightest supernovae have very similar properties (Kowal 1968). In particular, type Ia supernovae (see §5.2.3) follow remarkably similar evolution in their light output with time. Since it is believed that these supernovae all originated from essentially identical systems, this similarity is not surprising. We might therefore also predict that the luminosities of type Ia supernovæ might also all be very similar, so these extremely bright sources should provide ideal standard candles which can be observed at very large distances.

A thorough review of the suitability of type Ia supernovæ as standard candles can be found in Branch & Tammann (1992). One particularly impressive piece of evidence which supports their use comes from studies of the few galaxies in which two type Ia supernovæ have been observed. In the three systems for which observations of two separate supernovæ are available, the apparent $V$-band magnitudes of the supernovæ at maximum brightness differ by only 0.1 mag on average. Since the separate supernovæ are likely to be viewed through different amounts of extinction in their host galaxies, the intrinsic scatter in maximum brightness is probably even smaller.

The luminosities of type Ia supernovæ have been calibrated against a number of other distance estimators. Most recently, Hubble Space Telescope images have been used to detect Cepheid variables in the galaxies that have hosted nearby type Ia supernovæ (Sandage *et al.* 1992, Saha *et al.* 1995). Adopting the Cepheid distance scale of §7.3.1, these observations imply that the supernovæ have brightest absolute magnitudes of $M_V = -19.52 \pm 0.07$, and $M_B = -19.48 \pm 0.07$.

A further technique for directly calibrating well-observed type Ia supernovæ is provided by the Baade–Wesselink method (§7.2.1). Application of this method seems to produce somewhat brighter estimates for peak absolute magnitudes (Branch, Drucker & Jeffery 1988), but given all the uncertainties in the Baade–Wesselink analysis, the difference may not be significant.

A final check on the absolute magnitudes of these supernovæ comes from direct modeling of the explosion. The physics of this process is sufficiently simple that it should be possible to model the explosion at the level required to predict the absolute magnitude of the resulting supernova. At present, there remain a number of uncertainties in the modeling process, but the values for unknown quantities can be derived by leaving them as free parameters and fitting to the observed detailed properties of the supernova. These properties include the exact spectrum of the supernova, and the variation of its observed brightness with time (or "light curve") in the various photometric bands. This fitting procedure yields a best fit which predicts an absolute magnitude of $M_B = -19.4 \pm 0.3$ (Branch 1992), consistent with the empirical calibrations.

Despite this reasonably coherent picture, some doubts have now been cast on the status of type Ia supernovæ as standard candles. For example, Maza *et al.* (1994) have studied two apparently normal type Ia supernovæ, one in a galaxy at a redshift of 0.020 and the other at a redshift of 0.019. If these supernovæ were identical in intrinsic luminosity and lay at the distances implied by the Hubble law for their redshifts, then their apparent brightnesses at maximum luminosity should differ by only 0.1 magnitudes. In fact, they differ by $0.8 \pm 0.2$ mag in the $B$ band and $0.4 \pm 0.2$ mag in the $I$ band.

Calibration of nine well-studied type Ia supernovae against other distance indicators has led Phillips (1993) to suggest that the colors of these systems vary systematically with the rate at which their luminosities decline. Work by Riess, Press & Kirshner (1995) has quantified this correction by fitting template light curves to the observed photometry of supernovae in order to measure the rate at which each declined. As for the novae discussed above, measurements of the shapes of light curves provide a mechanism for predicting accurately the intrinsic brightness of a supernova, and hence producing a distance estimator with a small scatter.

Supernova simulations are also beginning to come to grips with the variations between explosions: attempts have been made to explain the differences in peak luminosity and light curve shape by allowing the details of the explosion to vary from system to system (e.g. Höflich & Khokhlov 1996). Thus, as an alternative to the purely empirical luminosity calibration based on light curve shapes, one can fit complete theoretical models of type Ia supernova explosions to an observed light curve. An estimate of the intrinsic luminosity of the explosion is then provided by that of the best-fit model.

### 7.3.4 Distances from galaxy kinematics

Once we look beyond the Milky Way, the clearest structures that we see in the Universe are other galaxies. One obvious method for measuring the distance to these structures is by comparing their properties to those of nearby galaxies whose distances we can estimate using the methods discussed above. Unfortunately, galaxies are demonstrably not standard candles or standard rulers. Images of clusters of galaxies reveal that the members, which all lie at approximately the same distance, have an enormous range of apparent magnitudes and angular diameters. Clearly, we will have to look to more subtle properties if we are to use galaxies themselves as distance indicators.

One promising line of investigation is to use measurements of the internal kinematics of a galaxy in order to estimate its intrinsic luminosity. Very broad arguments based around the virial theorem (BT §4.3) imply that the average speed at which material orbits in a gravitating system increases with the mass of the system. We might also make the reasonable – but unproven – assumption that the amount of luminous material in a galaxy increases with its mass. Taken together, these properties imply that the intrinsic luminosity of a galaxy should increase with the amplitude of the orbital velocities of its constituents. We can obtain a distance-independent measure of the orbital velocities in a galaxy by looking at the Doppler shifts of its spectral lines relative to the galaxy's systemic velocity, and by measuring the broadening of the lines (see §8.2.4 and Chapter 11). We might then hope to use such kinematic observations to estimate the intrinsic luminosity of the galaxy, and thus, by comparison to its apparent magnitude, its distance.

**Spiral galaxies**    The first broadly successful implementation of this approach was presented by Tully & Fisher (1977). For a sample of ten nearby spiral galaxies, they showed that the galaxies' rotation velocities (measured using Doppler broadening in 21 cm emission) correlates quite tightly with their absolute magnitudes (measured from apparent magnitudes and distances calculated from Cepheid variables). This correlation is known as the **Tully–Fisher relation**.

The single dish radio telescope observations used in this analysis do not spatially resolve the HI disk in each system, and so the shape of spectrum around 21 cm provides us with a picture of the Doppler shifts in the HI throughout the galaxy. As we shall see in §8.2.4, most of the HI in the disks of spiral galaxies follows approximately circular orbits at the same speed, $v_c$. However, not all of this rotational motion will lie along the line of sight, so different parts of the galaxy will have their HI emission Doppler shifted by different amounts. If the galaxy is inclined at an angle $i$ to the line of sight, then the observed line-of-sight velocity for the HI will lie between $-v_c \sin i$ and $+v_c \sin i$ relative to the galaxy's systemic velocity. If one makes a spatially unresolved observation of such a disk, one finds that the 21 cm emission line is Doppler-broadened into a characteristic "two-horned" structure, with sharp peaks at frequencies corresponding to Doppler shifts of $\pm v_c \sin i$. Thus, measuring the line width gives a direct measure of $v_c$. Conventionally, this width is measured by finding the points in the line's wings at which the intensity has dropped to 20% of the peak value in the two horns. The difference in Doppler shift between these points in the high- and low-frequency wings, $W_{20}$, then provides the measure of the galaxy's kinematics.

To relate $W_{20}$ to the intrinsic rotation velocity, $v_c$, we need to correct for two effects. First, as we shall see in §§8.1.4 and 9.2.6, HI does not follow exactly circular orbits, but has a modest degree of random motion superimposed. These random motions broaden the line profile slightly, so $W_{20}$ overestimates the Doppler broadening due to rotational motion. This factor can be corrected for by subtracting an estimate of the contribution to $W_{20}$ arising from random motions: $W_R = W_{20} - W_{\mathrm{rand}}$.[6] For gas with a Gaussian random velocity distribution with dispersion $\sigma$, the correction to $W_{20}$ is $W_{\mathrm{rand}} = 3.6\sigma$ (Bottinelli *et al.* 1983). Since $\sigma \sim 10\,\mathrm{km\,s}^{-1}$, and $W_{20}$ is only $\sim 100\,\mathrm{km\,s}^{-1}$ for faint galaxies, this correction factor can be highly significant. The second effect we must correct for is inclination. Since the velocity width that we measure is essentially $2v_c \sin i$, the effects of inclination can be removed by dividing $W_R$ by $\sin i$. Thus,

$$W_R^i = (W_{20} - W_{\mathrm{rand}})/\sin i \qquad (7.25)$$

provides an inclination-independent measure of the rotation rates of spiral galaxies.

---

[6] More sophisticated formulations of this correction factor exist [e.g. Tully & Fouqué (1985)], but they do not produce significantly different answers.

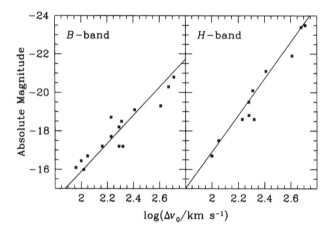

**Figure 7.6** Plot of absolute magnitude in $B$- and $H$- bands as a function of velocity width for galaxies with independently determined distances. [From the data published in Pierce & Tully (1992)]

Corrections must also be applied to the absolute magnitudes before they can be correlated with $W_R^i$. Unfortunately, a galaxy that contains sufficient HI for a profile to be obtained also contains large amounts of dust, so some of the galaxy's luminosity will be internally absorbed. Since the dust is concentrated in the disk of the galaxy, the amount of absorption will depend on the inclination of the galaxy. Absorption will be maximized when the galaxy is edge-on, as in this orientation the starlight from the disk of the galaxy has to pass through the longest path length of obscuring dust before leaving the galaxy. A number of methods have been developed, which correct for this factor by modeling the properties of the dust layer [e.g. Tully & Fouqué (1985)].[7] Alternatively, one can minimize the impact of the correction by obtaining the requisite photometry at infrared wavelengths, where the effects of extinction are small – an approach pioneered by Aaronson, Huchra & Mould (1979). The infrared holds the further advantage that light in this waveband is dominated by old stars rather than the young stars that dominate bluer wavelengths. Thus, the infrared flux provides a measure of stellar content that is less dependent on current star formation rates, so it better reflects the overall stellar population of the galaxy. The total stellar content is, presumably, closely tied to the total mass of the galaxy and hence its kinematics. We would therefore expect $W_R^i$ to be most closely correlated with the infrared luminosity of the galaxy.

Once the necessary corrections have been applied, the inclination corrected absolute magnitudes, $M^i$, can be determined for nearby "calibrator" galaxies with good Cepheid distances. The Tully–Fisher relation between

---

[7] The photometry must also, of course, be corrected for Galactic extinction – see §3.7.1.

$\log W_R^i$ and $M^i$ is remarkably tight (see Figure 7.6). As well as reflecting an interesting underlying piece of astrophysics, the tightness of the relation gives us some confidence in the correction factors that have been applied: since galaxies with comparable values of $W_R^i$ also have comparable values of $M_B^i$ even when their inclinations differ, it would seem that the correction factors have been chosen appropriately.

Pierce & Tully (1992) have calibrated Tully–Fisher relations based on photometry in $B$-, $R$-, $I$- and $H$-bands, using data from six calibrator galaxies. They discovered that the Tully–Fisher relations in these bands are

$$
\begin{aligned}
M_B^i &= -7.48(\log W_R^i - 2.50) - 19.55 + \Delta_B \pm 0.14, \\
M_R^i &= -8.23(\log W_R^i - 2.50) - 20.46 + \Delta_R \pm 0.10, \\
M_I^i &= -8.72(\log W_R^i - 2.50) - 20.94 \pm 0.10, \\
M_H^i &= -9.50(\log W_R^i - 2.50) - 21.67 \pm 0.08.
\end{aligned}
\tag{7.26}
$$

As predicted above, the relations in the infrared show a somewhat smaller scatter than those in the optical. The correction factors, $\Delta_B$ and $\Delta_R$, have been introduced because Pierce & Tully (1992) found that different bands produce inconsistent results. Using the local calibrations (for which $\Delta_B$ and $\Delta_R$ are zero), they calculated the distance to the Ursa Major Cluster by comparing members' absolute magnitudes as predicted by these relations to their apparent magnitudes. The distance modulus of this cluster was found to be systematically larger in the $B$- and $R$-bands than in the infrared. Arbitrary offsets of $\Delta_B = 0.25$ and $\Delta_R = 0.06$ were therefore introduced for cluster galaxies (but not the calibrating field galaxies) to explain the discrepancy – it would appear that cluster spiral galaxies are systematically redder than field galaxies. The ad hoc nature of the applied correction means that distances derived using these optical bands may well be subject to residual systematic errors; future users of the Tully–Fisher relation are likely to concentrate on the infrared bands where this uncertainty does not arise.

**Elliptical galaxies**    The Tully–Fisher relation is only applicable to spiral galaxies, so one obvious question is whether an analogous relationship holds for elliptical galaxies. Ellipticals are free of obscuration, so the existence of such a relation would be very useful for determining accurate distances. Further, since most rich clusters of galaxies contain very few spirals, the relation would be particularly useful for estimating the distances to these systems.

Happily, as we have already seen in §4.3.4, analogous relations hold for elliptical galaxies. Specifically, as Figure 4.43 illustrates, the luminosity inside the effective radius of an elliptical galaxy, $L_e$, is quite tightly correlated with its central velocity dispersion, $\sigma_0$ [the Faber–Jackson relation – see equation (4.38)]. Thus, we can predict the absolute magnitude of an elliptical galaxy on the basis of its central velocity dispersion, and derive its distance by comparing this quantity to the apparent magnitude of the galaxy.

The scatter in the relation is considerable, and so its ability to predict accurate distances is rather poor. As we also saw in §4.3.4, much of this scatter arises because the properties of elliptical galaxies also depend on their effective radii, $R_e$, and if we calculate $\log R_e$, $\langle I \rangle_e$ (the mean surface brightness inside $R_e$), and $\log \sigma_0$, we find that galaxies lie in a single plane in these coordinates [equation (4.39)]. Since both $\langle I \rangle_e$ and $\sigma_0$ are independent of the distance to a galaxy, we can use the observed values of these quantities to predict the intrinsic effective radius of the galaxy via equation (4.39), and by comparing this size-scale to the observed angular effective radius of the galaxy, we can predict its distance.

In practice, it is more common to use the approximation to the fundamental plane provided by the $D_n$–$\sigma_0$ relation [equation (4.43)] to estimate distances. This formulation has the advantage that the parameter $D_n$ – the radius within which the mean surface brightness is $20.75\mu_B$ – can be easily and robustly estimated from quite poor quality photometry. By comparing the observed angular size of $D_n$ to the scale predicted by equation (4.43), one obtains the distance to a galaxy.

It should be borne in mind that the distances to individual galaxies derived in this way are not very good: the 15% scatter about the $D_n$–$\sigma_0$ relation of equation (4.43), for example, means that the distance to any single galaxy has an intrinsic uncertainty of 15%. However, by observing many galaxies within a single cluster, the uncertainty in the distance to the cluster can be beaten down to insignificance. Similarly, if one's goal is to calibrate the Hubble constant (§7.4.3) or measure large-scale streaming (§7.4.2) by measuring distances to large samples of galaxies, then the large uncertainties in individual measurements become unimportant.

### 7.3.5 Surface brightness fluctuations

As we have already seen, if a galaxy is sufficiently close for its individual stars to be resolved, then we have access to a wealth of information that can be used to help determine its distance. Even in more distant galaxies where individual stars cannot be resolved, the discrete nature of the stars that make up the galaxy's light still has observable consequences.

As a simple illustration, consider a galaxy at a distance $D$ which is made up from identical stars of luminosity $L$ distributed such that there are $n$ stars per unit area across the observed face of the galaxy. If we make an image of this galaxy with an angular resolution of $\delta\theta$, then the image will contain an average of $\overline{N} = n(D\delta\theta)^2$ unresolved stars in each $\delta\theta \times \delta\theta$ resolution element. The flux that we observe from each star is $f = L/(4\pi D^2)$, so the average total flux from the stars in each resolution element will be

$$F = \overline{N}f = nL\delta\theta^2/(4\pi). \tag{7.27}$$

This average brightness is independent of $D$, and so clearly cannot be used as a distance estimator. However not all resolution elements will contain exactly the same number of stars: Poisson fluctuations mean that this number will have a dispersion of $\overline{N}^{1/2}$. The corresponding root mean square fluctuations in the observed flux per resolution element will be

$$\sigma_F = \overline{N}^{1/2} f = \frac{n^{1/2}\delta\theta L}{4\pi} D^{-1}. \tag{7.28}$$

Thus, the fluctuations in the surface brightness scale inversely with distance: a galaxy at twice the distance would appear twice as smooth. Combining equations (7.28) and (7.27), we find

$$\frac{\sigma_F^2}{F} = f = \frac{L}{4\pi D^2}; \tag{7.29}$$

by measuring the mean brightness per resolution element and the RMS fluctuations between different elements, we can infer the flux from each individual star. If two galaxies are made up from intrinsically identical stars, then comparing the values of $f$ obtained by this method will immediately yield their relative distances.

In fact, of course, galaxies are not made up from stars with identical luminosities. More realistically, a galaxy can be modeled as containing stars from a variety of species with different intrinsic luminosities $L_i$. If there are an average of $\overline{N}_i$ stars per resolution element from species $i$, and each of these stars contributes $f_i = L_i/(4\pi D^2)$ to the total brightness, then equation (7.29) generalizes to

$$\frac{\sigma_F^2}{F} = \frac{\sum_{\text{species}} \overline{N}_i f_i^2}{\sum_{\text{species}} \overline{N}_i f_i} = \frac{\langle L \rangle}{4\pi D^2}, \tag{7.30}$$

where

$$\langle L \rangle = \frac{\sum \overline{N}_i L_i^2}{\sum \overline{N}_i L_i} \tag{7.31}$$

is a weighted mean stellar luminosity (Tonry & Schneider 1988). Thus, the RMS fluctuations are the same as those which we would observe from a galaxy populated entirely by stars of luminosity $\langle L \rangle$. If two galaxies contain the same fractional mix of stellar species, then $\langle L \rangle$ is the same for both of them, and so we can calculate their relative distances from surface brightness fluctuations using equation (7.30).

In order to use this method to measure the absolute distances to galaxies, we need the value of $\langle L \rangle$. If we know the stellar luminosity function of a galaxy, we can calculate $\langle L \rangle$ directly from equation (7.31).[8] Alternatively, measurements of the surface brightness fluctuations in a galaxy whose distance has been measured using a different method can be used to calibrate

---

[8] For a continuous luminosity function, we replace the summations over discrete species in equation (7.31) by integrals over luminosity.

its value. This calibration can then be applied to determining the distances to other galaxies if it is assumed that they have the same stellar luminosity functions. Note, however, that the numerator in equation (7.31) depends on $L_i^2$, so the average is heavily weighted toward the brightest stars in the galaxy: for any reasonable luminosity function, $\langle L \rangle$ will be similar to the luminosity of the brightest red giant stars, and does not change significantly if the luminosity function of fainter stars varies from galaxy to galaxy.

Since the dominant giant stars are brightest in the red, the fluctuations in surface brightness will be most readily detectable when observed in the $I$-band. This choice of filter also has the advantage that the absolute magnitudes of giant stars in this band do not vary much with the stars' metallicity and age. The value of $\langle L \rangle$ in the $I$-band should thus depend little on the properties of the galaxy's stellar population.

The method clearly works best when looking at the outer parts of galaxies where the number of stars is low, and the Poisson fluctuations are relatively large. However, the finite brightness of the night sky places a limit as to how far out it is possible to look. In the $I$-band, the sky at a dark site has an intrinsic brightness of $\sim 19\,\mathrm{mag\,arcsec^{-2}}$, and it is difficult to observe regions of galaxies very much fainter than this limit. A square arcsecond is typical of the resolution possible with ground-based optical telescopes; with this image quality, the sky brightness restricts accurate $I$-band photometry to a limit of $\sim 20\,\mathrm{mag\,arcsec^{-2}}$.

For a galaxy at a distance of 20 Mpc, this surface brightness corresponds to the $I$-band light from $\overline{N} \sim 10^4$ giant stars per square arcsecond. In images limited by seeing to a resolution element of $1\,\mathrm{arcsec^2}$, the Poisson variations in this number cause the smooth light profile of the galaxy to fluctuate at a level of $\sim 1\%$. Thus, fluctuations are detectable in galaxies at a distance of $\sim 20$ Mpc if we can measure the photometry of these systems to better than 1%.

Various sources of noise combine to make it difficult to do photometry to this level of accuracy. The simplest to deal with are the purely observational errors: Poisson statistics from the finite number of photons detected, and the intrinsic noise of the detector (such as the read noise in a CCD) both contribute fluctuations to the observed photometry. These sources of noise can be effectively eliminated by obtaining long-exposure images. The signal arising from real features in a galaxy's photometry, including surface brightness fluctuations, will grow linearly with the integration time, $t_{\mathrm{int}}$. However, the Poisson noise grows only as $t_{\mathrm{int}}^{1/2}$, and the intrinsic noise of the detector will increase little, if at all, with integration time. In a long integration, these sources of noise can thus be made negligible relative to the signal.

A rather more troublesome source of "noise" arises from other objects that give rise to fluctuations in the observed galaxy photometry. In particular, unresolved faint background galaxies and globular clusters associated with the target galaxy will produce fluctuations of their own in the smooth brightness profile of the galaxy. These fluctuations must be characterized by measuring individual luminosities for those contaminating objects which are bright enough to be identified. The luminosity function for these contaminants can be modeled using the observed brighter objects; extrapolating the luminosity function to fainter luminosities then allows us to predict how the faint undetected contaminants will contribute to the observed brightness fluctuations. After these effects have been allowed for, any residual fluctuations can be attributed to the intrinsic surface brightness fluctuations of the galaxy.

This process is described in detail by Tonry & Schneider (1988), who present the first detection of intrinsic surface-brightness fluctuations using CCD photometry of two nearby galaxies. Tonry (1991) has since calibrated the value of $\langle L \rangle$ using galaxies with Cepheid distances. By looking at a number of galaxies in the same cluster, he also established that $\langle L \rangle$ varies somewhat with the color of the observed galaxy, and derived the empirical relation

$$\langle M_I \rangle = 3.0(V - I) - 4.8, \tag{7.32}$$

where $\langle M_I \rangle$ is the $I$-band absolute magnitude corresponding to $\langle L \rangle$. His comparisons between the distances to galaxies obtained using this effective standard candle and those derived from planetary-nebulae luminosity functions (§7.3.2) revealed satisfyingly identical results.

## 7.4 Results

As illustrated in Figure 7.7, there is now a wide array of different techniques that can be used to measure distances. As we discussed at the beginning of this chapter, the most secure absolute distance estimates have traditionally been obtained for nearby objects. In order to bridge the gap to objects at large distances, we have to work our way outwards using a whole series of techniques for estimating relative distances, in order to construct the complete cosmic distance ladder. The ultimate goal of this process is to reach out to distances where redshifts are dominated by the expansion of the Universe, so that we can derive an unambiguous value for the Hubble constant. It is clear from Figure 7.7 that there are many ways of combining the different overlapping distance estimators in order to connect the known local distances to the largest cosmic scales. Unfortunately, different combinations of distance estimators have produced radically different answers: the Hubble constant derived from these analyses takes values from $\sim 40 \, \mathrm{km \, s^{-1} \, Mpc^{-1}}$ (e.g. Sandage 1993b) up to $\sim 80 \, \mathrm{km \, s^{-1} \, Mpc^{-1}}$ (e.g. van den Bergh 1994). As the distance to an object of given redshift scales as $H_0^{-1}$ [see equation (7.2)], these extremes are known as the **long distance scale** and **short distance scale**, respectively. All of the methods for deriving distances are prone to sizeable systematic errors, and so it is not surprising that such a wide range of values for $H_0$ can be derived by employing different techniques.

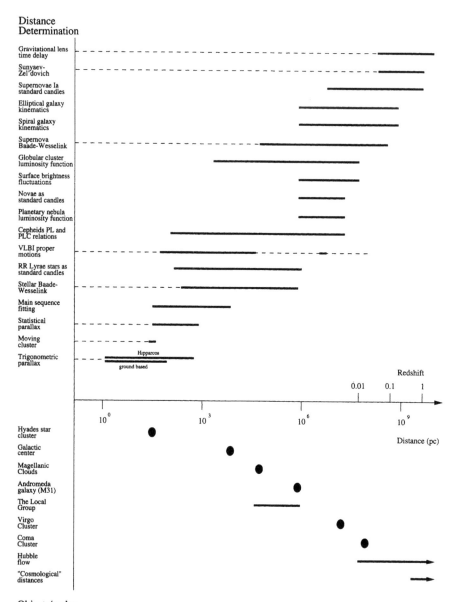

**Figure 7.7** Summary of the cosmic distance scale. The upper half of the plot shows absolute and relative distance measures: solid bars indicate the range of distances over which the measure has been applied; dashed bars indicate that the method is "anchored" by an absolute calibration. The lower half of the figures lists objects whose distances we seek. Bars indicate the approximate range of distances over which these objects can be found.

What is rather more disturbing is the fact that different authors using approximately the same set of steps also fail to agree on distances by significant factors. For example, de Vaucouleurs & Peters (1986) have analyzed a sample of spiral galaxies using Tully-Fisher distances (§7.3.4) to obtain $H_0 = 110 \pm 10\,\mathrm{km\,s^{-1}\,Mpc^{-1}}$, whereas Sandage (1988) has argued that the true Hubble constant derived from these data should be $56 \pm 13\,\mathrm{km\,s^{-1}\,Mpc^{-1}}$. Clearly, the errors quoted on these two estimates are not mutually consistent, and at least one of the analyses must be wrong.

In fact, historically, the derived distance scale has been much more closely correlated with the author of the work than it is with the particular choice of distance estimator. The advocates for the long and short distance scales are firmly set in their respective views, and have found support for their preferred answer using a wide range of techniques. It is only the complexity of the cosmic distance ladder that allows this dichotomy in scale to persist. Each step on the ladder is accompanied by a degree of uncertainty. Part of this uncertainty is made up of the usual random errors due to finite number statistics, etc, but part arises from ill-constrained systematic effects. As we have seen, different techniques for calibrating a standard candle can easily give results that differ systematically by a few tenths of a magnitude. Since the origin of these differences are not well understood, it is impossible to decide objectively which calibration to adopt, and so a subjective choice must be made. By the time that three or four such choices have been made in the construction of the cosmic distance ladder, it is easy to build up distance estimates that differ by a factor of two (see Problem 7.6). Individuals naturally tend to give more weight to the particular distance calibrations that fit in with their own previous work, causing the observed strong correlation between author and result.

As we discuss below, this conflict is finally beginning to be resolved. The improved quality of data available from instruments such as HST means that distance-measuring techniques can be pushed to fainter magnitudes, and correspondingly greater distances. It is thus possible to reduce the number of steps that one has to make in order to define the cosmic distance ladder. By reducing the number of points in the analysis at which small systematic errors can creep in, the freedom to combine these uncertainties to produce a pre-ordained result is removed. Taking this approach to its logical conclusion, the best determination of the distance scale would be one based on a technique that measures the absolute distance to remote extragalactic objects in a single step. Although the systematic uncertainties in techniques for measuring large absolute distances are still too large to provide an unequivocal answer, advances are also being made in beating these errors down to manageable levels.

The growing profusion of techniques for measuring relative distances is also helping to clarify the situation, since it is possible to detect systematic errors by comparing the distances predicted by different methods in the range where they overlap. For example, if we derive distances to a sample of galaxies using two different methods such as the Tully-Fisher relation and the surface brightness fluctuation technique, then a plot of one distance estimate versus the other should give a straight line of unit slope. Any significant departure from this linear relation implies that at least one of the methods is compromised. By carrying out such comparisons for a number of distance estimators, we can isolate those methods that fail.

As we shall see below, these advances mean that we are now close to being in the remarkable position of being able to measure the distances to objects right across the Universe to within an accuracy of $\sim 10\%$.

### 7.4.1 Distances within the Local Group

Historically, distances to objects within our own Local Group have played a major role in determining the extragalactic distance scale. As we have seen, it was the analysis of Cepheids in the Magellanic Clouds which determined their usefulness as distance indicators, and observations of these stars in M31 which established the existence of extragalactic scales. Reviews of distance determinations to other galaxies within the Local Group can be found in van den Bergh (1989) and van den Bergh (1994). In this section, we discuss the measurement of distances both to the center of our own galaxy and to two local group members. As some of the more reliable distance determinations, these measurements still provide a crucial anchor point for establishing the extragalactic distance scale.

**Distance to the Galactic center**    Measurements of the distance to the Galactic center, $R_0$, have recently been reviewed by Reid (1993). This relatively short distance is disproportionately difficult to measure, mainly because our view of the Galactic center is so heavily obscured. One must therefore either observe in one of the few relatively unobscured lines of sight that passes near the Galactic Center (such as "Baade's Window"), or use distance estimates that do not require optical observations.

The only useful absolute distance estimator for $R_0$ is provided by $H_2O$ masers in Sgr B2 and other star-forming regions [see §7.2.4; $H_2O$ maser emission in the Milky Way has been reviewed by Reid & Moran (1988)]. Statistical parallax applied to the line-of-sight velocities and proper motions of these sources yield Galactocentric distances that range from $R_0 = 6.5 \pm 1.5\,\text{kpc}$ (Reid *et al.* 1988) to $R_0 = 8.1 \pm 1.1\,\text{kpc}$ (Gwinn, Moran & Reid 1992).

A variety of calibrated relative distance estimators have also been employed in the estimation of $R_0$. Classically, the best candidates are provided by populations of halo objects – we shall see in §10.5, the density of halo objects peaks at the Galactic center, so $R_0$ can be estimated by determining the distance to this density peak. As we have seen in Chapter 1, the vast scale of the Milky Way was first established by Shapley (1921), who applied this approach using globular clusters as standard candles and rulers. Reid (1993) reviews the numerous attempts to refine his analysis. One major

**Table 7.1**   Galactic center distance estimates

| Method | Distance/kpc |
|---|---|
| $H_2O$ masers | $7.2 \pm 0.7$ |
| RR Lyrae stars | $7.8 \pm 0.4$ |
| Globular clusters | $8.0 \pm 0.8$ |
| Cepheids | $8.0 \pm 0.5$ |
| Red clump stars | $8.4 \pm 0.4$ |

source of systematic error in these studies is uncertainty in the distances to individual clusters. These distances are usually determined from the apparent magnitudes of horizontal branch stars, including RR Lyrae variables (§6.1.2). Hence any uncertainty in the absolute magnitudes of RR Lyraes will inevitably show up in an error in the value of $R_0$ that one obtains from globular clusters. In these circumstances, the best procedure is to determine $R_0$ directly from field RR Lyraes, of which enormous samples are now available (Figure 5.14).

The two biggest problems to be overcome when determining $R_0$ from RR Lyraes or any other halo population are: (i) the effects of extinction; and (ii) uncertainties in the absolute magnitudes of the stars observed. Both of these problems are substantially mitigated if one observes in the near-IR (Fernley et al. 1987). In the fields studied near the Galactic center, extinction in the visual band amounts to $A_V \sim 1.6$ magnitudes, while in the near infrared, it is only $A_K \sim 0.17$ magnitudes. Moreover, the absolute magnitudes of RR Lyrae stars seem to vary less with [Fe/H] in the near-IR than they do in the visual band. Hence, the uncertain metallicities of RR Lyraes near the Galactic center have less impact on distance determinations made in the infrared than those based on the optical distance modulus.

By fitting the observed apparent-magnitude distributions of RR Lyraes in two fields to those predicted if the density of stars is given by

$$\nu(\mathbf{r}) \propto \frac{1}{(x^2 + y^2 + z/q^2)^{\lambda/2}}, \tag{7.33}$$

Carney et al. (1995) concluded that $q \simeq 0.8$, $\lambda \simeq 2.3$ and $R_0 = 7.8 \pm 0.4 \,\mathrm{kpc}$.

A problem with this measurement of $R_0$ is that it is based on an absolute-magnitude calibration of RR-Lyraes that appears to conflict with the generally accepted absolute-magnitude calibration of the Cepheids in that it yields a distance modulus to the LMC that is $0.2 - 0.3$ magnitudes smaller than that obtained from Cepheids. When Carney et al. take the distance to the LMC to be that given by the Cepheids and then determine $M_K$(RR Lyrae) from stars in the LMC, $R_0$ increases to $8.9 \pm 0.5 \,\mathrm{kpc}$. This simple example illustrates the difficulty in determining the distance scale discussed above. One is often confronted by mutually-inconsistent data sets, and the derived distances depend critically on which data one chooses to trust, which calibration of a relative distance estimator one adopts, etc. The sample estimates for $R_0$ listed in Table 7.1 should therefore be viewed with a healthy degree of skepticism.

When comparing different estimates for $R_0$, one must take care to consider the "band-wagon effect," whereby new published values show a tendency to agree with existing estimates, even if a completely different method has been employed. In fact, such a collective phenomenon is difficult to avoid: if a new method returns a value for $R_0$ in line with conventional wisdom, one is likely to write up and publish the result with little further thought; if, on the other hand, the method returns a discrepant value, one will search carefully for possible explanations, and will probably not produce a paper until the discrepancy can be explained. Thus, results in the literature will always tend to cluster around a current "accepted" value for $R_0$. Reid (1993) found clear evidence for this effect by plotting estimates for $R_0$ as a function of their publication date; he showed that there was a clear trend in this plot, with estimates for $R_0$ decreasing systematically with time from $R_0 \sim 9\,\mathrm{kpc}$ in 1975 to $R_0 \sim 7.5\,\mathrm{kpc}$ in 1985. More recently, the published values for $R_0$ have recovered slightly to cluster around a best current estimate of $R_0 \sim 8\,\mathrm{kpc}$.

**Distance to the Large Magellanic Cloud**    The Large Magellanic Cloud (LMC) is sufficiently nearby that a range of stellar calibrators can be used to determine its distance. Since we can resolve individual stars in the globular clusters surrounding the LMC, we can estimate the distance to the system by main-sequence fitting (§3.5.2). For example, Walker (1985) observed the young globular cluster NGC 1866, and used the position of its main sequence in the CM diagram to determine a distance modulus for the cluster of $m - M = 18.5 \pm 0.2$, which corresponds to a distance of $50 \pm 5\,\mathrm{kpc}$.

Analyses of variable stars (§7.3.1) yield consistent results with somewhat smaller uncertainties. In their review of the use of Cepheids as distance indicators, Feast & Walker (1987) find that the period-luminosity relation of equation (5.19) places the LMC Cepheids at a mean distance of $50.3\,\mathrm{kpc}$. Interestingly, the derived distance depends on which part of the LMC is observed: Cepheids at the eastern edge of the LMC are 5% closer than those at the western edge (Gascoigne & Shobbrook 1978). This differential arises because the Cepheids populate the plane of the highly-flattened LMC, which is tilted with respect to the line of sight. Our ability to measure this small difference illustrates once again the power of relative distance indicators when absolute calibration is not required.

The properties of fainter variable stars provide a further check on the distance to the LMC. A compilation of the RR Lyrae stars detected in 6 LMC globular clusters by Walker (1992a) reveals that on average they have mean apparent magnitudes of 18.9, with a dispersion from cluster to cluster of $\sim 0.1\,\mathrm{mag}$. The average metallicities of these RR Lyrae stars were measured to be $[\mathrm{Fe/H}] \sim -1.9$, and so adopting relation (5.24) we find that their mean

**Table 7.2**  LMC distance estimates

| Method | Distance/kpc |
|---|---|
| Main Sequence Fitting | $50 \pm 5$ |
| Cepheids | $50 \pm 2$ |
| RR Lyrae | $44 \pm 2$ |
| SN1987a time delay | $52 \pm 3$ |
| SN1987a Baade–Wesselink method | $55 \pm 5$ |

absolute magnitudes are $\langle M_V \rangle = 0.7$. The clusters must thus lie at distances in the range $44 \pm 2$ kpc. Since the LMC subtends an angle of some 10 degrees on the sky, it must have a radius of $\sim 5$ kpc, and so the scatter in distance estimates to globular clusters can all be explained if they are distributed in a halo around the LMC. The difference between the average RR Lyrae distance to the LMC and that derived from the Cepheid variables and main-sequence fitting led Walker (1992a) to suggest that the absolute calibration of equation (5.24) should be adjusted by $\sim 0.2$ magnitudes.

The detonation of Supernova 1987A provided independent methods for determining the distance to the LMC. The application of the Baade–Wesselink method to the supernova's expanding photosphere returned a distance of $55 \pm 5$ kpc (see §7.2.1), while the circum-supernova ring time delay gave a value of $52 \pm 3$ kpc (see §7.2.3). These absolute distance determinations are not dependent on any intermediate calibration, but, as we have discussed, they are subject to various systematic errors. Since the sources of uncertainty in the calibrated and absolute distance determinations are so different, their close agreement as to the distance to the LMC suggests that neither approach has been unduly compromised by errors.

A summary of the results from the various methods for determining the distance to the LMC can be found in Table 7.2. There is a reasonable consensus from these disparate measurements that the center of the LMC lies approximately 50 kpc away from us. Averaging together the estimates in Table 7.2, we obtain a best estimate for the distance to the LMC of $50 \pm 2$ kpc, or a distance modulus of $m - M = 18.5 \pm 0.1$. We can thus determine this distance to better than $\sim 5\%$.

**Distance to M31**   The Andromeda Galaxy, M31, provided the first proof of the existence of an extragalactic distance scale through the detection by Hubble (1922) of its Cepheids. As the nearest large spiral galaxy (besides the Milky Way), it still provides an important calibration point on the extra-galactic scale. Since it lies significantly further than the LMC, several of the fainter classes of object used as distance estimators are no longer applicable. There is, however, still a respectable range of applicable distance estimators.

Recent observations of Cepheids in M31 have tied down the relation between their periods and apparent magnitudes quite closely. The data are now so good, in fact, that the error in the derived distance is dominated by

**Table 7.3**  Andromeda Galaxy (M31) distance estimates

| Method | Distance/kpc |
|--------|--------------|
| Cepheids | $760 \pm 50$ |
| RR Lyrae | $750 \pm 50$ |
| Planetary Nebula Luminosity Function | $750 \pm 50$ |
| Globular Cluster Luminosity Function | $700 \pm 60$ |
| Novae | $710 \pm 80$ |

the uncertainty in the zero-point calibration. Distances are therefore often quoted with respect to the LMC, so that there is no contribution to the error in distance from the zero-point; a recalibration of the scale only affects the distance to the LMC, and all other distances scale up or down accordingly. For example, recent comparisons between Cepheids in the LMC and those in M31 imply that M31 is $15.3 \pm 0.8$ times as far away (Freedman & Madore 1990). If we adopt a distance of $50 \pm 2$ kpc for the LMC, this places M31 at $760 \pm 50$ kpc. Similarly, observations of the apparent magnitudes of RR Lyrae stars in M31 compared with those in the LMC implies that the ratio of distances is $15 \pm 1$.

The distance estimates that are based on the properties of luminosity functions (§7.3.2) have also been applied to M31. Jacoby, Walker & Ciardullo (1990) have compared the planetary-nebula luminosity functions of the LMC to that of M31, and find that the distributions of apparent brightness for planetary nebulae are consistent with a single intrinsic luminosity function if the distance to M31 is $15 \pm 1$ times the distance to the LMC, in excellent agreement with the other estimators. Similar results have also been obtained from globular clusters. The peak of the globular cluster luminosity function lies at an apparent $V$-band magnitude (corrected for reddening) of $16.8 \pm 0.2$ (Racine & Harris 1992). Assuming that this luminosity function is intrinsically the same as the Milky Way's which peaks at an absolute magnitude of $-7.4 \pm 0.1$ (see §7.3.2), we obtain a distance modulus for M31 of $24.2 \pm 0.2$, which corresponds to a distance of $700 \pm 60$ kpc. This last value, which is consistent with the other cited estimates, is calibrated directly with respect to the Milky Way rather than by way of the LMC. The good agreement between the values adds confidence not only to the determination of the distance to M31, but also to the adopted distance to the LMC and the assumption that the globular cluster luminosity function is universal.

A summary of recent determinations of the distance to M31 is presented in Table 7.3. As these results show, there is excellent agreement between the different methods. Combining the results weighted appropriately by their errors, we obtain a best estimate of $740 \pm 40$ kpc, with a corresponding distance modulus of $m - M = 24.3 \pm 0.1$. Where appropriate, the errors also include the contribution from the uncertainty in the distance to the LMC.

Due to its proximity, M31 has the best-determined distance of any large galaxy. It therefore plays a key role in calibrating distance estimators that are based on such galaxies. For example, for the surface-brightness fluctuation method (§7.3.5) we can calibrate the variation with $V - I$ color in equation (7.32) using galaxies in clusters at unknown distances, but the absolute calibration of the relation – the constant on the right-hand side of this formula – requires surface brightness fluctuation analysis for a galaxy at known distance. The bulge of M31 was found by Tonry (1991) to have surface brightness fluctuations [as defined by equation (7.30)] with an amplitude in magnitudes of $\langle m_I \rangle = 23.16$. Assuming that M31 lies at a distance of 740 kpc, this value corresponds to an absolute magnitude of $\langle M_I \rangle = -1.2$. Since the bulge of M31 has a color of $V - I = 1.19$, the constant in equation (7.32) follows.

### 7.4.2 Distances beyond the Local Group

As can be seen from Figure 7.7, a good range of techniques is available for measuring distances to galaxies immediately beyond the local group. Some of these methods are based on the properties of the galaxies themselves, while others use their brightest inhabitants. When more than one method can be used to calculate the distances to galaxies, the results for the different techniques give respectable agreement out to distances beyond 20 Mpc (Jacoby *et al.* 1992). Figure 7.8 shows the comparison between distances to clusters of galaxies obtained by using the surface brightness fluctuation method on their elliptical members (§7.3.5) and those obtained by a number of other methods. The agreement with distances obtained by analyzing planetary nebula and globular cluster luminosity functions (§7.3.2) is very good. The distances derived using the Tully-Fisher relation (§7.3.4) for spiral galaxies in these clusters show significantly poorer agreement. Since there are also systematic differences in redshift between ellipticals and spirals in the clusters for which the agreement is poorest (Jacoby *et al.* 1992), it seems likely that some of the disagreement arises because the ellipticals from which surface-brightness distances are derived and the spirals are not intermixed in the same physical association. This disagreement suggests that the Tully-Fisher relation is not suitable for obtaining the distances to clusters of galaxies. The distances to elliptical galaxies obtained via the $D_n$–$\sigma$ relation also show a degree of scatter relative to the surface-brightness distances, although the larger errors on individual $D_n$–$\sigma$ relation distances is at least partially responsible.

**Distance to the Virgo Cluster**    The Virgo Cluster is the nearest moderately large cluster of galaxies. Since the cluster contains both spiral and elliptical galaxies in large numbers, and has also been the site of a number of well-observed supernovae, it provides an ideal calibration point for comparing the results obtained using different distance estimators. Moreover, once its distance has been determined, we can also infer the absolute properties of what is presumably a fairly average cluster of galaxies; such properties

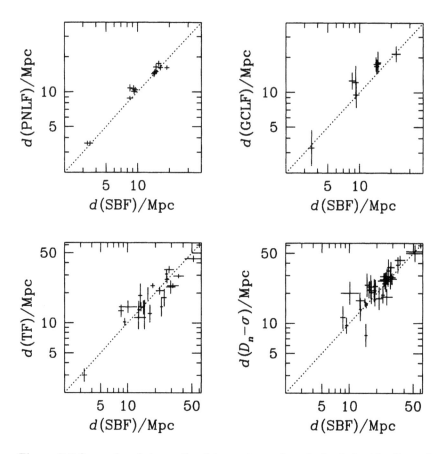

**Figure 7.8** Comparison between the distances to nearby galaxies derived by the surface brightness fluctuation method, and those obtained using the planetary nebula and globular cluster luminosity functions, the Tully-Fisher relation for spiral galaxies, and the $D_n$-$\sigma$ relation for ellipticals. [From data kindly provided by J. Tonry]

might then form the calibration point for techniques to derive the distances to other, more distant clusters.

A wide range of techniques have been applied to measuring the distance to the Virgo Cluster. In their extensive review of the subject, Jacoby *et al.* (1992) showed that the three methods with the smallest uncertainties (surface-brightness fluctuations, planetary-nebula luminosity function, and the Tully-Fisher relation) all provided consistent distance estimates of $\sim$ $16 \pm 1\,\mathrm{Mpc}$ ($m - M = 31.0 \pm 0.1$).

The one seriously conflicting measurement comes from the analysis of type Ia supernovae. The Cepheid calibration of the brightnesses of nearby supernovae implies that their absolute magnitudes are $M_B = -19.7 \pm 0.1$ at maximum light (§7.3.3), which implies that at a Virgo Cluster distance of 16 Mpc, their brightest apparent magnitudes should be $m_B = 11.3 \pm 0.1$. In fact, they should appear somewhat fainter than this value due to absorption both in the Milky Way and in their host galaxies. Capaccioli *et al.* (1990) have compiled a sample of 10 type Ia supernovae in Virgo Cluster galaxies. Using the colors of the supernovae to measure the amount of reddening, they find that the extinction-corrected apparent magnitudes of these objects have a mean of $m_B = 12.1 \pm 0.1$. This value is significantly fainter than the prediction. In order to reconcile the observed apparent magnitudes with the calibrated absolute magnitudes of the supernovae, the Virgo Cluster would have to lie at a distance of $23 \pm 2$ Mpc ($m - M = 31.8 \pm 0.2$). As we have seen in §7.3.3, some doubt has now been cast on the role of type Ia supernovae as standard candles. Non-standard luminosities for the few supernovae used to calibrate their absolute magnitudes could well explain the discrepancy, but this suggestion has yet to be explored in detail.

In an effort to resolve this conflict once and for all, enormous efforts have been expended to detect Cepheid variable stars in the Virgo Cluster. Meticulous observations of the Virgo galaxy NGC 4571 using the Canada-France-Hawaii Telescope by Pierce *et al.* (1994) revealed three candidate Cepheids at an apparent magnitude of $m_R \approx 24$. The periods derived for these very faint stars consistently follow the expected Cepheid period-luminosity relation if the galaxy is placed at a distance of $14.9 \pm 1.2$ Mpc. Subsequently, the unparalleled imaging quality of the Hubble Space Telescope enabled Freedman *et al.* (1994) to identify 20 Cepheids in the Virgo galaxy M100, and the period-luminosity relation for these stars implies that this galaxy is at a distance of $17.1 \pm 1.8$ Mpc. Given the errors on these estimates and the intrinsic size of the Virgo Cluster (a few megaparsecs in depth), these measurements are entirely consistent with a Virgo-Cluster distance of $16 \pm 1$ Mpc.

A summary of the recent determinations of the distance to the Virgo Cluster is presented in Table 7.4. Note that the Cepheid determination involves fewer steps to establish the distance scale, since it goes directly from Galactic measurements of Cepheid variables to observations of these stars in the Virgo Cluster; all of the other methods involve calibrations using objects at intermediate distances. Since systematic errors at each calibration point generate the worst uncertainties in the cosmic distance scale, correspondingly greater weight should be given to the Cepheid measurement.

**Peculiar velocity field**    We can use the various distance indicators to search for local departures from the Hubble law. If the Hubble law is obeyed exactly, then a plot of the derived distance versus redshift for a sample of galaxies should obey the straight-line relation of equation (7.2). Any departures from this straight line can be attributed to the contribution to the redshift from local **peculiar velocities**. Happily, this procedure is not critically dependent on the uncertain absolute calibration of the distance indicators: a change in the calibration will change the slope of the distance-redshift relation, but any departures from the straight-line relation will still

**Table 7.4**  Virgo Cluster distance estimates

| Method | Distance/Mpc |
|---|---|
| Surface Brightness Fluctuations | $16 \pm 1$ |
| Planetary Nebula Luminosity Function | $15 \pm 1$ |
| Cepheids | $16 \pm 1$ |
| Tully-Fisher Relation | $16 \pm 2$ |
| $D_n-\sigma$ Relation | $17 \pm 2$ |
| Type Ia Supernovae | $23 \pm 2$ |
| Globular Cluster Luminosity Function | $19 \pm 4$ |
| Novae | $21 \pm 4$ |

SOURCE: adapted from Jacoby *et al.* (1992)

be apparent. The best available distance estimators have uncertainties of $\sim$ 10%, so only relatively large disturbances to the Hubble flow are measurable. The large errors also mean that Malmquist effects can significantly distort the apparent departures from the Hubble law, and steps must be taken to minimize these biases (see §3.6.1).

The best-established large-scale departure from the Hubble flow is the **Virgocentric infall**, which has been investigated by many authors. By comparing distances to 300 spiral galaxies derived from the infrared Tully-Fisher relation (§7.3.4) to those predicted from the galaxies' redshifts, Aaronson *et al.* (1982) found systematic residuals, which can be understood if there is a large-scale motion of galaxies towards the Virgo Cluster. Since this cluster is the closest large concentration of visible material, it should not be surprising if it also represents a local maximum in mass density which gravitationally attracts surrounding material, causing the observed departures from the Hubble flow.

Using the $D_n-\sigma$ relation (§7.3.4) for a sample of 400 elliptical galaxies, Lynden-Bell *et al.* (1988) looked at the distribution of peculiar velocities out to higher redshifts. In addition to the Virgocentric infall, they found strong evidence for a bulk flow of galaxies towards a point in the constellation of Centaurus, which has been dubbed **the Great Attractor**. This region is heavily obscured since it lies close to the Galactic plane, so it is difficult to observe the attractor itself in any detail. However, there is certainly a large concentration of clusters of galaxies in this part of the sky, and so it seems likely that the Great Attractor represents the center of this concentration, the **Local Supercluster**. To give some idea of the gravitational importance of this supercluster on the Hubble flow, the influence of the Virgo cluster has induced an infall of $\sim 250\,\mathrm{km\,s^{-1}}$, whereas the motion of the Local Group caused by the much more distant Great Attractor is close to $600\,\mathrm{km\,s^{-1}}$.

### 7.4.3 The asymptotic Hubble constant

The ultimate aim in establishing the cosmic distance scale was to reach to sufficient distances to measure the Hubble flow directly. Not only does the Hubble law set the ultimate scale of the Universe via equation (7.2), but it also provides us with an estimate of the age of the Universe via equation (7.4).

Unfortunately, as we have seen in §7.4.2, even the Virgo Cluster is not sufficiently distant for its motion to be dominated by the Hubble flow, and so we cannot estimate the Hubble constant simply by substituting this cluster's distance and redshift into equation (7.2). We could use a model of the peculiar velocity field in order to try to remove this contaminant from the Virgo Cluster's cosmological redshift. However, the size of the correction required is rather uncertain: estimates of the size of the Local Group's peculiar infall velocity towards Virgo have ranged from $-90\,\mathrm{km\,s^{-1}}$ to $520\,\mathrm{km\,s^{-1}}$ (Davis & Peebles 1983), and there is still an uncertainty of several hundred kilometers per second in the size of this motion. Since the total redshift of the Virgo Cluster is only $1035\,\mathrm{km\,s^{-1}}$ (Huchra 1984), the uncertainty in the infall correction produces a considerable uncertainty in the cluster's cosmic redshift, so it is not possible to obtain an unambiguous determination of the Hubble constant from Virgo Cluster data.

We must therefore look to more distant objects. One obvious target is the Coma Cluster. This rich cluster of galaxies is the closest such system and has therefore been studied in great detail. With a heliocentric velocity of $v_C = 6925\,\mathrm{km\,s^{-1}}$ (Fukugita $et\ al.$ 1991), the uncertain correction for the Local Group's infall towards Virgo is not a significant factor when deriving the Hubble constant. The Coma Cluster is sufficiently distant that many of the techniques which have been used to measure distances beyond the Local Group can no longer be applied (see Figure 7.7). For example, although globular clusters have now been detected around Coma Cluster galaxies (see §7.3.2), the observations do not yet reach faint enough magnitudes to measure the peak of their luminosity function, so this standard candle is not available to us.

Since rich galaxy clusters like Coma are mostly populated by elliptical galaxies, the various elliptical kinematic estimators such as the $D_n$–$\sigma$ relation (see §7.3.4) provide the best constraints on the Coma Cluster's distance. The poor absolute calibration of these relations means that they are best suited to measuring the Coma Cluster's distance relative to that of the Virgo Cluster. This comparison, based on the $D_n$–$\sigma$ relation, yields the result that the Coma Cluster is $\sim 5.5$ times more distant than the Virgo Cluster (Dressler $et\ al.$ 1987). In fact, because relative distance measurements such as this one are not dependent on uncertain zero-point calibrations, there is generally good agreement between all the different methods for obtaining this distance ratio. A compilation by van den Bergh (1992) showed that 8 out of 12 distance determinations based on cluster galaxy properties returned relative distances to these two clusters in the range $5.3 - 5.8$. A standard candle analysis of the type Ia supernovae that have exploded in both clusters yielded a consistent value for the ratio of 5.6 (Capaccioli *et al.* 1990). Combining these various analyses, we obtain a best estimate for the Coma–Virgo distance ratio of $5.5 \pm 0.2$.

Taking this relative distance measure in conjunction with the observed Hubble-flow velocity of the Coma Cluster, we can parameterize the Hubble constant in terms of the distance to the Virgo cluster, $D_V$, as follows:

$$H_0 = (63 \pm 2)(D_V/20\,\mathrm{Mpc})^{-1}\,\mathrm{km\,s^{-1}\,Mpc^{-1}} \qquad (7.34)$$

Thus, if we adopt $D_V = 16 \pm 1$ Mpc as the distance to the Virgo Cluster then $H_0 = 79 \pm 6\,\mathrm{km\,s^{-1}\,Mpc^{-1}}$. If, on the other hand, the Virgo distance determination based on type Ia supernovae, $D_V = 23 \pm 2$ Mpc, is correct, then $H_0 = 55 \pm 5\,\mathrm{km\,s^{-1}\,Mpc^{-1}}$.

Type Ia supernovae are sufficiently bright that they can be observed directly out to distances where their redshifts are dominated by the Hubble flow (see Figure 7.7). Since all methods agree fairly closely on relative distances beyond the Virgo Cluster, the calibration of type Ia supernovae which returns a large distance to the Virgo Cluster also produces larger estimates than other methods for more distant objects. The Hubble constant derived directly from type Ia supernovae thus also lies at the low end of the range, with a typical estimate of $H_0 = 47 \pm 5\,\mathrm{km\,s^{-1}\,Mpc^{-1}}$ (Sandage & Tammann 1993).

Here, then, is where most of the residual discrepancy between the long and short distance scales rests. The advocates for the short distance scale point out the close agreement between a wide range of distance estimators which all imply a small distance to the Virgo Cluster, and a correspondingly large Hubble constant. Those who favor the long distance scale use the supernova type Ia distance scale as the primary evidence for a low value of the Hubble constant, and emphasize the systematic errors in less direct standard candles to explain their discrepant distance estimates to the Virgo Cluster and beyond.

There are, however, distinct signs of convergence between these two deeply-entrenched camps. Recalibration of the absolute magnitudes of seven type Ia supernovae using HST observations of Cepheids in their host galaxies led Saha *et al.* (1995) – staunch advocates for the long distance scale – to estimate that $H_0 = 58^{+7}_{-8}\,\mathrm{km\,s^{-1}}/\mathrm{Mpc}$. One can, therefore, now quote a value of

$$H_0 = (65 \pm 10)\,\mathrm{km\,s^{-1}\,Mpc^{-1}} \qquad (7.35)$$

without upsetting anyone unduly.

It is also notable that the absolute distance measures are starting to converge toward similar values. In the case of the SZ effect (see §7.2.2), better observations of both the X-ray properties of clusters and the distortions in the microwave background are leading to larger inferred values of the Hubble constant. Thus, instead of obtaining estimates like $H_0 = 24^{+13}_{-10}\,\mathrm{km\,s^{-1}\,Mpc^{-1}}$

(McHardy *et al.* 1990), more recent SZ measurements result in values of between 60 and $80 \, \mathrm{km \, s^{-1} \, Mpc^{-1}}$ (Holzapfel *et al.* 1997). Similarly, Kundić *et al.* (1997) have found that the gravitational lens time delay method (§7.2.3) applied to QSO 0957+561 yields $H_0 = (64 \pm 13) \, \mathrm{km \, s^{-1} \, Mpc^{-1}}$. The skeptical reader might invoke the band-wagon effect discussed above in the context of determining $R_0$ (see §7.4.1) to explain this convergence. Nevertheless, it is heartening that such a broad variety of methods can return values consistent with equation (7.35).

However, there remains one factor which strongly suggests that we should favor values for $H_0$ toward the lower end of the range in equation (7.35). For any adopted value of the Hubble constant, one can estimate the age of the Universe using equation (7.4). If we adopt $H_0 = 65 \, \mathrm{km \, s^{-1} \, Mpc^{-1}}$ then $t_0 = H_0^{-1} = 1.5 \times 10^{10}$ years. This value is alarmingly close to the ages of the oldest globular clusters, which stellar evolution calculations find to be also $\sim 1.5 \times 10^{10}$ years (§6.1.4). If both these numbers are correct, then globular clusters must have formed surprisingly early in the evolution of the Universe. In fact, as we discuss below, the Hubble time $t_0$ provides only an upper limit to the age of the Universe, so even this marginal consistency between cosmological and stellar ages breaks down. As we have seen in Chapter 6, future progress in reconciling these numbers requires that we look carefully at the globular cluster age determinations as well as refining the more direct determinations of $H_0$.

One remarkably simple error, which would go a long way toward explaining the conflict between globular cluster age estimates and the age of the Universe, is that we may have adopted too faint an absolute magnitude calibration for the main sequence. Such an error would have two effects. First, it would mean that we have underestimated the distances to clusters obtained by MS fitting. Since the cosmic distance scale has traditionally been anchored by such MS fits, this error will propagate out to large scales, so that all distances will be similarly underestimated. Thus, the current measurements for $H_0$ will be overestimated, and the calculated age of the Universe, $t_0 \propto H_0^{-1}$ [equation (7.4)] will be too small. Some of the age estimates for globular clusters will also be affected by adopting too faint an absolute magnitude scale for the MS. Specifically, the adopted absolute magnitudes of the turn-off points from the MS will be too faint, and thus the ages calculated from equation (6.2) will be systematically too small. Hence, if the absolute magnitude calibration for the MS is made brighter, then the estimates for the age of the Universe become larger and the ages of globular clusters become smaller. A change of only $-0.2$ magnitudes in this calibration would be enough to reconcile the difference (see Problem 7.7), and there are already indications that an alteration of this order may be appropriate (Feast & Catchpole 1997).

### 7.4.4 The deceleration parameter and cosmic density

As we have seen in §7.1, the value of the deceleration parameter, $q_0$, is as intriguing a cosmological quantity as the Hubble constant. Just as $H_0$ can be used to estimate the age of the Universe, so the value of $q_0$ predicts its ultimate fate: if $q_0 < 0.5$ then the Universe will expand forever, whereas if $q_0 > 0.5$ then the Universe will ultimately stop expanding and collapse back on itself. The value of $q_0$ also has implications for the exact age of the Universe. Gravitational deceleration means that the Universe is currently expanding at a lower velocity than was the case in the past. In using the Hubble time, $t_0$, as an estimate for the age of the Universe, we neglect this effect and overestimate the true value. For example, if $q_0 = 0.5$ then the actual age of the Universe is only $\frac{2}{3}t_0$ (Problem 7.1).

We already know that the deceleration parameter cannot differ from the critical value of $q_0 = 0.5$ by orders of magnitude. If $q_0$ were very large, then the expansion of the Universe would already have been halted and the Universe would have collapsed back to on itself. Conversely, a very small value of $q_0$ would imply that there was almost no gravitating matter in the Universe, and this implication conflicts with the large amount of material in galaxies, clusters of galaxies, etc, that we observe. Is it just a coincidence that $q_0$ lies reasonably close to the critical value? We can appeal to anthropocentric arguments to explain the coincidence: if $q_0$ had a more extreme value, then there would either be no Universe or no galaxies at the current time, and hence no human life could have evolved to ponder the question. However, there may also be fundamental physical reasons why the density of the Universe should be close to the critical value, $\rho_c$, and hence why $q_0 \approx 0.5$. In particular, inflationary theories of cosmology imply that the early Universe underwent a period of enormously rapid expansion which should have driven its average density very close to $\rho_c$ [see Guth (1989) for a non-mathematical description of this theory]. The observational determination of $q_0$ would enable us to test these theories directly and see whether the Universe really is critically balanced between eternal expansion and ultimate collapse.

**Standard candles and rulers**    Given the current uncertainties in the determination of the Hubble constant, it seems unlikely that the subtler cosmological effects of the deceleration parameter will be detectable in observations of standard candles and rulers. In one sense, however, $q_0$ is easier to measure than $H_0$. The Hubble constant can only be calculated from the apparent brightness of a standard candle via equation (7.7) if the source's intrinsic luminosity is known. However, $q_0$ can in principle be derived by observing a sample of standard candles of unknown luminosity. If we produce a plot of distance modulus versus the logarithm of redshift (known as a **Hubble diagram**) for a set of standard candles, then at low redshifts the Hubble law implies that we should see a straight line. However, at high redshifts the departure from the linear Hubble law in equation (7.7) means that the relation will curve in a way which depends solely on $q_0$. If the absolute magnitude of the standard candles is unknown, then there will be an indeterminate offset in the vertical position of the locus occupied by the standard candles in the Hubble diagram, but the characteristic curvature in the locus due to $q_0$ will be preserved. We should thus be able to solve directly for $q_0$ from the shape of this relation without knowing the intrinsic luminosity of the standard candles.

Unfortunately, such analyses are plagued by a deep-rooted problem. Specifically, the light travel time from putative standard candles at high redshift is a significant fraction of the age of the Universe, and almost all objects will have evolved significantly over this timescale. The unknown variation in intrinsic luminosity during this evolutionary process is likely to dominate the departure from a straight line in the Hubble diagram, swamping the subtler cosmological effects. For example, Hoessel, Gunn & Thuan (1980) have found that if the brightest galaxies in rich clusters are assumed to be standard candles, then the cosmological model which provides formally the best fit to the observed Hubble diagram of these objects would imply that $q_0 \sim -0.5$ – the Universe appears to be expanding at an accelerating rate. This unphysical result can be explained straightforwardly if brightest cluster galaxies are not standard candles, but are intrinsically fainter at high redshifts. This luminosity evolution is exactly what we would expect if such galaxies grow over time by "cannibalizing" some of their neighbors (Hausman & Ostriker 1978). Since we do not understand the details of this evolutionary process, it is not possible to disentangle its effects from those of the cosmological model, and so $q_0$ cannot be estimated from these data.

Recently, some progress has been made toward measuring $q_0$ by focusing on objects for which we expect the evolutionary effects to be small. One appealing class of object for such a study is the type Ia supernova. The simplicity of the explosion producing such objects suggests that evolutionary effects should be small, and they are bright enough to be seen right across the Universe. Garnavich et al. (1998) and Perlmutter et al. (1998) have made very high quality observations with HST and ground-based instruments, which have enabled them to follow the light curves of several type Ia supernovae at high redshift. Allowing for the possible variations in supernova luminosity with color (see §7.3.3), these collaborations were able to calculate the distance moduli for these supernovae. As Figure 7.9 illustrates, the remarkably tight linear relation at low redshift in the resulting Hubble diagram shows what good standard candles type Ia supernovae can provide. Although the data are still rather sparse, there are strong indications that the departures from the linear Hubble law at large redshifts follow what one would expect if $q_0$ is small; the data in Figure 7.9 rule out a value of $q_0 = 0.5$ at the 95% confidence level.

A conflicting result, which favors a high value of $q_0$, has been obtained by Kellerman (1993). He has examined the angular extents of compact active galaxy radio sources using very high resolution imaging techniques. These sources have typical angular extents of $\sim 0.01$ arcseconds, and it has been

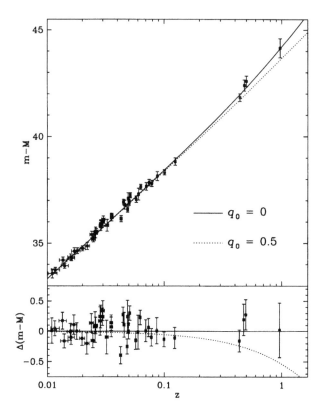

**Figure 7.9** Hubble diagram, showing the distance moduli of type Ia supernovae as a function of their redshifts. The solid line shows the predicted relation for $q_0 = 0$ ($\Omega = 0$), while the dotted line shows the relation for $q_0 = 0.5$ ($\Omega = 1$). The bottom panel shows the difference between data and models. [After Garnavich *et al.* (1998) courtesy of P. Garnavich]

suggested that their small size means that their properties do not depend on the largescale structure of their surroundings. These sources might, therefore, make useful standard rulers which are relatively unaffected by large-scale evolution in the surrounding environment. Figure 7.10 shows the angular diameters for a sample of 82 such radio sources as a function of their redshifts. Although the angular extents of these sources initially decreases with redshift, they seem to stop getting smaller at a redshift of around unity. In fact, within the uncertainties the plot seems to follow the relation predicted by equation (7.8) for $q_0 = 0.5$. Either unexpected evolutionary effects have somehow conspired to produce a variation in apparent angular size with redshift which mimics the effects of a critically-bound Universe, or these results imply that $q_0 \sim 0.5$.

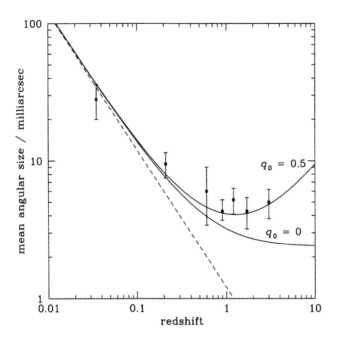

**Figure 7.10** Angular size versus redshift for a sample of compact radio sources. Solid lines indicate the expected variation in angular size of a standard ruler with redshift for different values of $q_0$. The dashed line shows the "Euclidean" result for which angular size varies as $z^{-1}$. [After Kellerman (1993)]

Clearly, we are only just beginning to make measurements of the quality required to tie down $q_0$. If we are to determine whether we live in a critically-bound Universe with $q_0 = 0.5$, as many theorists and aesthetes would like, we really require other measures of the mean density of the Universe.

**Peculiar velocity field**    In the above analysis, we have described how $q_0$ might be measured using properties of the global geometry of the Universe. We can also approach this problem using more local observations of the peculiar velocity field (§7.4.2). Since these departures from the Hubble flow are believed to be induced gravitationally by large-scale variations in density, galaxies' peculiar motions provide a direct probe of the cosmic density and hence the deceleration parameter.

As we have seen in §7.4.2, major components of the peculiar velocity field involve large-scale streaming towards the Virgo Cluster and the Great Attractor, and these motions can be explained by large mass concentrations at these locations. However, attempts to model the peculiar velocity field using isolated gravitationally-attractive mass concentrations are clearly over-simplifications. In reality, the cosmic density distribution will be a continuously fluctuating function containing regions of over- and under-density on many different scales. Recent analyses have therefore concentrated on using non-parametric techniques for recovering the complete three-dimensional distribution of density from the observed three-dimensional distribution of peculiar line-of-sight velocities. A description of the practical implementation of such techniques and their results can be found in the review by Dekel (1994). These methods confirm the presence of large mass concentrations in the direction of the Virgo Cluster and the Great Attractor, but they also detect the influence of other gravitational perturbations such as the Perseus-Pisces supercluster.

These analyses are very effective at measuring the fluctuations in the density distribution. Unfortunately, they are insensitive to the contribution to the cosmic density from any uniformly-distributed component, as such a mass component does not produce local peculiar velocities. Thus, we cannot measure the total density directly using this approach. However, we can infer the cosmic density if we assume that the density distribution is related to the distribution of galaxies. If we compare the variations in the number density of galaxies to the fluctuations in the density distribution derived from the peculiar velocities, we find that the over- or under-density of galaxies is approximately proportional to the density fluctuations (Dekel $et$ $al.$ 1993). It therefore seems that the distribution of galaxies traces the mass in the Universe quite closely, so we might legitimately calculate a cosmic mass-per-galaxy ratio by comparing these two distributions. As long as the same ratio applies to the fraction of galaxies that are more uniformly distributed, we can calculate the total cosmic mass density at any point by multiplying the total number density of galaxies by this mass-per-galaxy ratio. Calculations along these lines have been made by a number of authors; see Dekel (1994) for a compilation of the results. The consensus from these analyses is that the cosmic density lies quite close to $\rho_c$, consistent with a value of $q_0$ equal to, or perhaps somewhat below, a half.

If the value of $q_0$ is really as high as a 0.5, then the conflict between the age of the oldest globular clusters ($\sim 1.5 \times 10^{10}$ years) and the age implied by $H_0$ (§7.4.3) is exacerbated. For $q_0 = 0.5$, the age of the Universe is only $\frac{2}{3}H_0^{-1}$; if $H_0 = 65 \pm 10\,\mathrm{km\,s^{-1}\,Mpc^{-1}}$, the implied age of the Universe is a mere $(1.0 \pm 0.2) \times 10^{10}$ years. In order to make the Universe older than the globular clusters, we require a long distance scale with $H_0 \lesssim 45\,\mathrm{km\,s^{-1}/Mpc}$. We must therefore either adopt this long distance scale (and be prepared to explain why so many observations favor a shorter distance scale), or we must look critically at the arguments for a high value of $q_0$ and the stellar evolution calculations which give the ages of globular clusters.[9]

---

[9] The Friedmann cosmological models that we have adopted in this chapter neglect the possibility of a non-zero cosmological constant, $\Lambda$. If this extra term is added into the equations governing the expansion of the Universe, then the deceleration parameter is replaced by $q_0 = 0.5\rho_0/\rho_c - \Lambda/(3H_0^2)$. Setting the arbitrary constant $\Lambda$ to a positive

# Problems

**7.1** Many of the properties of an isotropic cosmological model can be derived using a simple Newtonian formalism. In particular, consider a spherical region in the Universe of radius $r$ and mass $M$. An object of mass $m$ lying at the edge of this sphere only feels the gravitational influence of the material within the sphere [**Birkhoff's theorem**, the relativistic analog of Newton's first theorem (BT§2.1.1)]. If the Universe is critically-bound, the total energy of the object will be zero, and we can write

$$\frac{1}{2}m\dot{r}^2 - \frac{GMm}{r} = 0. \tag{7.36}$$

By integrating this equation forward in time from the Big Bang ($r = 0$, $t = 0$) to the present day when $t = t_{now}$ and $v = H_0 r$, show that $t_{now} = (2/3)H_0^{-1} = (2/3)t_0$.

**7.2** A star expands in a spherically-symmetric manner with radial velocity $v_r$. Defining a spherical coordinate system on the surface of a star with the polar axis aligned along the line of sight, show that the measurable flux-weighted mean line-of-sight velocity will be

$$v_{los} = v_r \frac{\int_0^{\frac{\pi}{2}} I(\theta) \cos^2\theta \sin\theta d\theta}{\int_0^{\frac{\pi}{2}} I(\theta) \cos\theta \sin\theta d\theta}. \tag{7.37}$$

Hence show that, for a star of uniform brightness, $p = v_r/v_{los} = 1.5$. In reality, a star will not appear uniformly bright: its opacity means that near the edge of the star (its "limb") one cannot peer so far into its atmosphere, so one sees the less bright outer layers. A reasonable analytic approximation to this **limb darkening** is given by $I(\theta) = I(0)(0.4 + 0.6\cos\theta)$. In this approximation, show that $p = 24/17$.

**7.3** If one were to make X-ray and S-Z decrement observations along the long axis of a prolate cluster of galaxies with axis ratio $a : b : b$, but were to assume erroneously that the cluster is a sphere of radius $a$, show that the derived distance to the cluster would be wrong by a factor of $b/a$.

**7.4** Consider a disk of potentially-masing material following circular Keplerian orbits about its center. Assume that the disk is viewed edge-on, and consider a line of sight that passes through the disk a distance $R$ from its center. Measuring the distance along this line of sight, $S$, such that center

---

value can decrease the rate of deceleration in the cosmic expansion, and perhaps reconcile the age of the Universe with a high value for $H_0$. As yet, there is no compelling reason to complicate the geometry of the Universe by setting $\Lambda$ to a non-zero value [see Carroll, Press & Turner (1992) for a review of the effects of this extra term and the current observational constraints on its amplitude.]

of the disk lies in the plane $S = 0$, show that the line-of-sight velocity of material at position $S$ along the line of sight is given by

$$v_{\text{los}} \propto R(R^2 + S^2)^{-3/4}. \tag{7.38}$$

The production of maser emission requires a long path length of material all travelling with the same line-of-sight velocity (so that the emission occurs at a single frequency). Hence, show that emission will preferentially be seen from the region $R \simeq 0$ and from the region $S \simeq 0$. If the disk is only populated with material at radii $r > r_{\text{in}}$, at what values of $v_{\text{los}}$ and $R$ will we detect maser emission?

**7.5** From the statistical parallaxes of 162 RR Lyrae stars of mean metallicity $[\text{Fe/H}] = -1.61$, Layden *et al.* (1996) find $\langle M_V \rangle = 0.71 \pm 0.12$. From this calibration they derive $R_0 = 7.6 \pm 0.4\,\text{kpc}$ and a distance modulus to the LMC of $18.28 \pm 0.13$. By comparing these values with ones given in §7.4, determine by how much the value of $\langle M_V \rangle$ for RR Lyraes given above differs from the more conventional value.

**7.6** Suppose that a series of four different standard candles are used to step out along the cosmic distance ladder as far as the Hubble flow, and that the calibration of each standard candle carries an uncertainty of 0.2 magnitudes. Show that, by changing the calibration for each step within the range allowed by its uncertainty, it is possible to derive values for the Hubble constant that lie anywhere between $\sim 0.7$ and $\sim 1.4$ of the nominal value.

**7.7** Current estimates suggest that the Universe has an age of $t_0 \approx 12\,\text{Gyr}$, while globular clusters have ages of $t_g \approx 15\,\text{Gyr}$. Show that changing the absolute magnitude calibration of the MS by $\Delta M$ magnitudes has the effects of: (a) changing the age estimate for the Universe (derived using MS-fitting as one of the steps on the distance ladder) by $\Delta t_0 = -0.46\Delta M t_0$; (b) changing the age estimate for globular clusters (derived from the MS turnoff point, equation (6.2)) by $\Delta t_g = 0.85\Delta M t_g$. Hence show that the age of the Universe and the ages of globular clusters can be reconciled by a change of $\Delta M \approx -0.2$.

# 8

# The Interstellar Media of Galaxies

The space between the stars is by no means empty. It contains rarefied gas, dust particles, a magnetic field, and relativistically moving electrons, protons and other atomic nuclei. These various forms of interstellar energy density prove to be quite tightly coupled together, with the result that for many purposes they together form a single dynamical entity. The terms **interstellar medium** and **interstellar matter**, or **ISM** for short, are useful short-hands for this composite dynamical entity.

Although the ISM accounts for only a small fraction of the mass in galaxies, one of the most fundamental differences between galaxies of different Hubble types concerns the distribution of their interstellar media. Moreover, differences in the stellar contents and light distributions of early- and late-type galaxies can in large measure be traced to differences in their interstellar media. Hence a study of the ISM is a prerequisite for understanding why galaxies are as we observe them.

Until the early 1980s it was thought that only late-type galaxies had significant interstellar media. Advances in X-ray and microwave astronomy have now demonstrated that many early-type galaxies also have rich interstellar media. Hence the difference between early- and late-type galaxies lies not so much in the *quantity* of interstellar matter, as in how the ISM is structured.

Spitzer (1982) provides an exceptionally lucid and non-technical account of the theory and observations of the Milky Way's ISM, while Spitzer (1978)

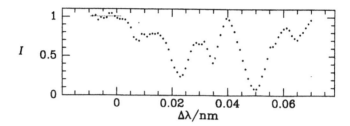

**Figure 8.1** Absorption by interstellar Na atoms in spectrum of $\epsilon$ Ori. [After Hobbs (1968)]

provides a wealth of technical detail on the physics of the ISM. An outstanding collection of reviews of every aspect of the ISM will be found in Thronson & Shull (1992).

## 8.1 How interstellar matter is detected

We start with a discussion of how interstellar matter is detected and its temperature, density and so on are measured. Besides its obvious observational significance, this review will explain how our understanding of interstellar matter has developed gradually as technological advances have enabled us to probe more of the many different forms which interstellar matter can take. Judging by the number of poorly understood observations of interstellar matter in both the Milky Way and external galaxies, our understanding of what is going on between the stars is still far from complete and remains liable to significant revision.

### 8.1.1 Absorption of starlight

In §3.7 we discussed the absorption and scattering of starlight by dust grains. We saw that dust absorbs and scatters blue light more than red, but it generally does not discriminate strongly between photons of similar wavelength. Interstellar dust is always associated with interstellar gas, which betrays its presence by giving rise to sharp absorption lines in stellar spectra.

Figure 8.1 shows the effect of absorption by interstellar sodium atoms on the optical spectrum of $\epsilon$ Ori. About six different V-shaped features, each about 0.01 nm wide, are apparent. Each such feature is thought to arise from a different interstellar cloud: neutral sodium atoms absorb light of wavelength 589.0 nm in their rest frame, and the differences in the central wavelengths of the various features are due to differences of $\sim 10\,\mathrm{km\,s^{-1}}$ in the line-of-sight velocities of the clouds. This interpretation is confirmed by the fact that a very similar pattern of absorption occurs near 589.6 nm, which is the wavelength of the other line of the sodium D line-doublet.

A line-doublet at 393.3 nm of once-ionized calcium atoms ($Ca^+$) gives rise to a similar phenomenon, and our understanding of interstellar space took a great stride forward during World War II, when W.S. Adams studied $Ca^+$ absorption in the spectra of about 300 stars. This study revealed for the first time the cloudy nature of the ISM.

Unfortunately, calcium and sodium are the only elements that give rise to readily measurable interstellar absorption features in optical spectra and their lines do not provide reliable diagnostics of the density and temperature of the interstellar medium. The problem is two-fold. Most obviously, one expects hydrogen and helium to dominate the ISM, and any extrapolation from the mass of trace elements such as Na and Ca to the total mass of the ISM is liable to be hazardous. Worse still, only a small fraction of all Na and Ca atoms are expected to be in the ionization states that give rise to the line-doublets described above; most free sodium atoms will be at least once ionized, and others will be locked up in dust grains. Similarly, only a small fraction of Ca atoms will be exactly once ionized. Little can be inferred about the ISM as a whole from measured column densities of neutral Na atoms and $Ca^+$ ions unless the fractions of all Na and Ca atoms that these measured species represent can be estimated. A reliable technique for doing this has not been found.

As Spitzer (1946) early appreciated, ultraviolet spectra are required to resolve this problem. The essential point is that, for most atoms and ions, the energy required to promote an electron from the ground state to the lowest-lying excited state is characteristic of ultraviolet rather than optical photons.[1] Observations sufficiently far into the ultraviolet ($\lambda < 300$ nm) cannot be made from the ground, and it was only in the 1970s that they could be made from space. In the eight years from its launch in August 1972, the 32-inch *Copernicus* orbiting telescope obtained ultraviolet spectra for over 100 stars, mostly in the spectral region 90 nm $< \lambda <$ 115 nm. In January 1978 the *International Ultraviolet Explorer* (IUE) was launched. It had lower spectral resolution than *Copernicus* and was confined to longer wavelengths ($\lambda > 120$ nm), but it had high sensitivity and could therefore study fainter objects.

The strength of an absorption feature is quantified by the feature's equivalent width $W$ – see Figure 3.3. Obviously, as the column density $N$ of absorbing atoms along a line of sight to a star increases, the equivalent width of any absorption feature to which they give rise increases. A plot of $\log(W)$ against $\log(N)$ is called the **curve of growth** for the line – see Figure 8.2. Near the origin, the slope of the curve of growth is always unity; at small column densities the number of absorbed photons is simply proportional to

---

[1] We are thinking here of the gross structure of atoms. When fine and hyperfine splittings are taken into account, much lower-lying excited states often exist. Not only do the relevant energies then lie far below the energies of optical electrons, but the transitions are invariably forbidden. The 21-cm line of hydrogen is a case in point.

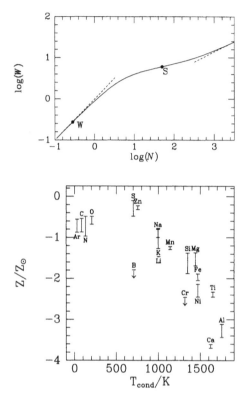

**Figure 8.2** A curve of growth. The point W shows the location of a weak line on the linear portion of the curve of growth, while the point S shows the location of a saturated line on the flat part of the curve of growth. The lower and upper dashed lines have slopes of unity and a half, respectively.

**Figure 8.3** The depletion factor (abundance relative to solar) of various elements along the line of sight to the star $\zeta$ Ophiuchi plotted against an estimate of the Temperature at which each element would condense onto dust grains. [After Morton (1974)]

the number of atoms available to absorb them. At higher column densities, atoms begin to 'shadow' one another. Thus once most photons of the right energy have been absorbed, adding further atoms to the column does not appreciably increase the number of photons absorbed because atoms which are added near the observer are hardly exposed to absorbable photons, while atoms added near the source star cast the rest of the column into the shade. When most absorbable photons are absorbed, one says that the line is **saturated**. In a well-resolved spectrum, a saturated line reaches almost to zero intensity, while in the $(\log(N), \log(W))$ plane it lies on the almost horizontal or 'flat' portion of the curve of growth, as shown in Figure 8.2. Since the measured quantity $W$ is very insensitive to $N$ on the flat part of the curve of growth, column densities cannot normally be reliably estimated from saturated absorption lines. Fortunately, a given species is usually capable of absorbing photons at a series of different energies, and the cross-section for absorption generally decrease strongly with increasing energy. Hence if one spectral line is saturated, another at shorter wavelength will not be. In this way the abundances of many elements have been determined along the lines of sight to several bright stars.

Figure 8.3 displays the results for one particular star as a plot of the ratio of the abundance of each element to the cosmic abundance[2] versus the temperature at which that element is expected to condense onto dust grains. Nearly all elements are less abundant in interstellar gas than in the solar system; one says that the elements are **depleted** in interstellar gas. Moreover, the degree of depletion is larger for elements that can condense onto grains at high temperature than for more volatile elements.

Although the curve of growth cannot be used to determine $N$ when a line is strongly saturated, in favorable cases accurate values of $N$ can nevertheless be extracted from saturated lines. The general idea is to focus attention on the 'wings' of the line, that is, wavelengths far enough from line-center that only a small fraction of the available photons are absorbed. In effect, the wings are always unsaturated, and therefore exhibit absorption that *is* proportional to the column density.

Clearly, to exploit this idea one has to be in a position to determine theoretically the relationship between $N$ and the strength of the absorption far from line-center. This is possible in the case of the Ly$\alpha$ transition of neutral hydrogen at $\lambda_0 = 121.6$ nm. Since hydrogen is so abundant and the cross-section per atom for absorption of Ly$\alpha$ photons is large ($\sim 4\,000$ nm$^2$), the feature at 121.6 nm is generally extremely strongly saturated. In fact, the saturation is such that there can be appreciable absorption up to one nm from line-center.

Two factors determine how far from line-center a line will extend. The first is Doppler broadening; atoms moving at different speeds absorb photons of different energies. The velocity spread of hydrogen atoms in the interstellar medium is of order $10\,\mathrm{km\,s^{-1}}$, and this dispersion gives rise to a wavelength spread $\Delta_v \simeq (v/c)\lambda \approx 0.004$ nm.[3] Since the distribution of atomic velocities should be Gaussian, whether because they are thermal velocities, or because they arise from the random motions of clouds, Doppler broadening yields a net absorption cross-section for photons with wavelengths in the interval $(\lambda + \mathrm{d}\lambda, \lambda)$ proportional to

$$\phi_\mathrm{D}(\lambda)\,\mathrm{d}\lambda \equiv \exp\left[-\frac{(\lambda - \lambda_0)^2}{2\Delta_v^2}\right]. \tag{8.1}$$

Unfortunately, one cannot accurately predict $\Delta_v$ because it depends both on the unknown temperature of the absorbing gas and on the random velocities of the clouds that lie along the given line of sight.

The second factor that determines how far from line-center absorption extends, is the line's 'natural width'. The excited state of the transition is not a stationary state of a real atom but only of an idealized atom which does not interact with the ambient electromagnetic field. When this interaction is

---

[2] The cosmic abundance is defined to be the abundance in the early solar system. For most elements this is determined from meteorites.

[3] Throughout this chapter we approximate the relativistic Doppler-shift formula by its value to first order in $v/c$

taken into account, the excited state is not a stationary state and therefore not a state of well-defined energy. By Heisenberg's uncertainty principle, the excited state's energy is uncertain by an amount $\delta E = h/\tau$, where $\tau$ is of order of the lifetime of the excited state. Since the energy of the excited state is ill-determined, the wavelengths of the photons that excite the transition are uncertain too, by an amount $\Delta_n$, the **natural width** of the line. By the uncertainty principle, $\Delta_n/\lambda_0 = \lambda_0/c\tau$. The fraction of photons absorbed by an atom at rest that have wavelength in the interval $(\lambda + \mathrm{d}\lambda, \lambda)$ is given by the **Lorentzian profile**

$$\phi_{\mathrm{L}}(\lambda)\,\mathrm{d}\lambda \equiv \frac{\Delta_n}{\pi}\frac{\mathrm{d}\lambda}{\Delta_n^2 + (\lambda - \lambda_0)^2}. \tag{8.2}$$

The natural width of the Ly$\alpha$ line, $\Delta_n = 6 \times 10^{-4}$ nm, is significantly smaller than the width $\Delta_v \approx 4 \times 10^{-3}$ nm typical of Doppler broadening. Consequently, for $|\lambda - \lambda_0| \lesssim \Delta_v$ the frequency dependence of the net absorption cross-section along a typical line of sight is essentially determined by Doppler broadening. Sufficiently far from line-center the sharp decline of equation (8.1) for $|\lambda - \lambda_0| > \Delta_v$ will give way to the more gradual asymptotic form $\propto |\lambda - \lambda_0|^2$ of equation (8.2). Consequently, sufficiently far from line center the absorption cross section will always follow the readily-calculated Lorentzian profile. Since $\Delta_n \ll \Delta_v$, $\phi_{\mathrm{L}}$ will be small in this region and the absorption can be readily measured only if the line is strongly saturated. The Ly$\alpha$ line generally satisfies this criterion and has been used to determine the column density of hydrogen atoms required, for example, for the construction of Figure 8.3.

Much of the mass of the ISM turns out to be in molecular rather than atomic form. Simple molecules, like atoms, rarely have strong absorption lines in the optical wave-band, so that most of what we know about interstellar molecules derives from spectroscopy at ultraviolet and radio frequencies. Therefore we now discuss the ultraviolet spectra.

Molecules can store energy in three ways: its electrons can be promoted to an excited state; the distance between the nuclei can oscillate as if the electron cloud that unites them were a spring and they were vibrating masses; the nuclei can orbit around each other as if the molecule were a rigid rotating body. The complete set of energies permitted by these three quantized possibilities is complex. Fortunately only a small fraction of all possible energy levels plays a significant role astrophysically.

Not surprisingly, $H_2$ is the most abundant molecule. All the transitions of $H_2$ that give rise to measurable ultraviolet absorption lines, start from either the ground state, or a state reached by adding a small amount of spin to the ground state. The molecule's spin is measured by the quantum number $J = 0, 1, \ldots$ which can only change by $\pm 1$ when a photon is absorbed or emitted. Moreover, all observed transitions involve the promotion of the

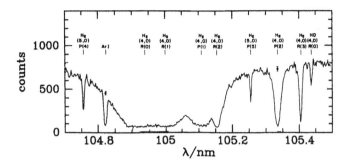

**Figure 8.4** Absorption by $H_2$ molecules on the line of sight to $\zeta$ Ophiuchi. [After Spitzer & Jenkins (1975)]

electrons to their first excited state.[4] In addition, on absorbing a photon the molecule may be left vibrating. The amplitude of vibration is characterized by the quantum number $v = 0, 1, \ldots$, which can change by any amount on absorption or emission of a photon.

Promoting the electrons to their first excited state requires 11.2 eV of energy. Increasing $v$ by one requires about 0.15 eV, while changing the angular-momentum quantum number from $J - 1$ to $J$ requires about $0.015J$ eV. Hence the relevant photons are ones with energies $\sim 11.5$ eV and wavelengths $\sim 100$ nm.[5] Changing $v$ by 1 will shift the wavelength by $\sim 1.3$ nm, while changing $J$ will induce a further wavelength shift of order $0.13J$ nm. Hence the absorption spectrum of $H_2$ consists of clusters, or **bands**, of a handful of lines each, each band being separated by $\sim 1.3$ nm from the next.

Figure 8.4 shows an example of absorption by $H_2$ molecules. The two most prominent features are clearly saturated and the column density of molecules giving rise to them can be determined from their wings, in the same way that the column density of hydrogen atoms is found from the wings of the Ly$\alpha$ line. One of these saturated features is caused by molecules in the ground level, and the other by molecules in the first spin-excited level, $J = 1$. Knowing the relative populations $N(J)$ of these two levels, enables one to determine the temperature of the absorbing gas from Boltzmann's formula

$$\frac{N(J)}{N(0)} = (2J + 1) \exp\left[- J(J + 1)85\,\mathrm{K}/T\right]. \tag{8.3}$$

(The factor of $2J + 1$ arises from the degeneracy of the $J^{\mathrm{th}}$ level.) The temperatures determined in this way range from 45 K to 130 K with a typical value of 80 K. From equation (8.3) it is clear that at such temperatures few molecules will be in levels with $J > 1$. A complication that should be

---

[4] Rotation transitions can be made radiatively only if the molecule has a non-zero electric dipole moment. A homonuclear molecule such as $H_2$ has a non-zero electric dipole moment only in excited electronic states.

[5] The energies and wavelengths of photons are related by $E/\,\mathrm{eV} = 1240\,\mathrm{nm}/\lambda$.

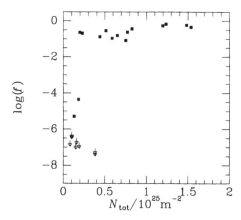

**Figure 8.5** The fraction $f$ of hydrogen nuclei in molecular form is very small in clouds with total column densities $N(\mathrm{H_{tot}}) \lesssim 2 \times 10^{24}\,\mathrm{m^{-2}}$ and approaches unity in clouds with $N(\mathrm{H_{tot}}) \gtrsim 10^{25}\,\mathrm{m^{-2}}$. [From data published in Spitzer *et al.* (1973)]

mentioned is that in lower-density clouds the ratio $N(1)/N(0)$ may deviate significantly from the thermal value of equation (8.3) because molecules tend to be left in states of high $J$ when they return to their lowest electronic and vibrational level after being excited by a photon. Such non-thermal photo-excitation is overwhelmed by thermal collisional excitation only when the particle density is higher than $\sim 10^{24}\,\mathrm{m^{-3}}$.

Figure 8.5 plots values obtained from the ultraviolet spectra of 23 stars of the quantities

$$N(\mathrm{H_{tot}}) \equiv N(\mathrm{HI}) + 2N(\mathrm{H_2}) \quad \text{and} \quad f_{\mathrm{H_2}} \equiv \frac{2N(\mathrm{H_2})}{N(\mathrm{H_{tot}})}, \tag{8.4}$$

where $N(\mathrm{H_2}) \equiv N(0) + N(1)$. Notice that the horizontal axis is linear, while the vertical axis is logarithmic. For $N(\mathrm{H_{tot}}) < 2 \times 10^{24}\,\mathrm{m^{-2}}$ the fraction of molecular hydrogen is very small and extremely sensitive to $N(\mathrm{H_{tot}})$. For $N(\mathrm{H_{tot}})$ larger than $\sim 10^{25}\,\mathrm{m^{-2}}$, a large fraction of the matter detected is in molecular form.

A relationship of the type shown in Figure 8.5 arises naturally if the matter observed on each line of sight is concentrated into clouds. $\mathrm{H_2}$ molecules form most rapidly in dense gas, where hydrogen atoms frequently collide with dust grains, which catalyze the conversion of hydrogen from atomic into molecular form. This fact alone leads one to expect that $\mathrm{H_2}$ will be most abundant in the dense cores of clouds. Moreover, gas near a cloud's surface is exposed to ultraviolet radiation from early-type stars that lie outside the cloud. This radiation increases the fraction of hydrogen in atomic form by dissociating $\mathrm{H_2}$ molecules. The deeper one goes into the cloud, the smaller the ultraviolet flux becomes, and in a large cloud the flux may eventually dry up because all the ultraviolet photons incident on the cloud's surface have been absorbed. In such shaded regions the abundance of $\mathrm{H_2}$ shoots up, since $\mathrm{H_2}$ molecules form rapidly and have long lifetimes.

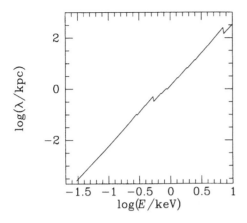

**Figure 8.6** The mean-free path of a photon of energy $E$ that moves through an ISM of particle density $10^6 \, \mathrm{m}^{-3}$. [From data published in Morrison & McCammon (1983)]

### 8.1.2 Extreme UV and X-ray observations

In all galaxies a significant fraction of interstellar space is occupied by gas that is so hot ($T > 10^5 \, \mathrm{K}$) that hydrogen and helium are fully ionized. Such hot gas is commonly called a **plasma** and emits predominantly in the extreme ultraviolet (EUV) and X-ray bands. The energies of EUV photons lie in the range $13.6 - 100 \, \mathrm{eV}$, while X-ray photons are generally classified as soft, medium or hard depending on whether their energies lie in the ranges $0.1 - 1 \, \mathrm{keV}$, $1 - 10 \, \mathrm{keV}$ or $> 10 \, \mathrm{keV}$.

$13.6 \, \mathrm{eV}$ is the binding energy of a hydrogen atom. Interstellar hydrogen atoms offer a large cross-section for the absorption of photons with energies somewhat in excess of this value, an absorbed photon's energy being used to eject the atom's electron. The energy required to eject the first and second electrons from a helium atom are $24.6 \, \mathrm{eV}$ and $54.4 \, \mathrm{eV}$, respectively. Consequently, He atoms and $\mathrm{He}^+$ ions also offer large cross-sections to the absorption of EUV photons. Since significant column densities of HI, He and $\mathrm{He}^+$ accumulate over short distances along any line of sight from the Sun, the local ISM is highly opaque to EUV photons and it is extremely difficult to probe the larger Milky Way or external galaxies with observations in the EUV band.

The cross section for the absorption of a photon of energy $E$ by an ion with ionization energy $E_i$ falls off as $(E - E_i)^{-3}$. Hence the opacity of the local ISM due to hydrogen and helium is falling steeply as one enters the soft X-ray band. In fact, the opacity of the local ISM soon becomes dominated by ions such as $\mathrm{O}^{5+}$, which has ionization energy $138 \, \mathrm{eV}$ and is relatively abundant. Although the local ISM is by no means transparent at $\sim 0.6 \, \mathrm{keV}$, we do receive significant numbers of soft X-rays from extragalactic objects that lie at high galactic latitudes $b$, since the line of sight to such objects does not run far through the disk of the Milky Way. Figure 8.6 indicates the distances that photons of various energies can travel through an ISM of particle density $10^6 \, \mathrm{m}^{-3}$.

No reasonably abundant species in the local ISM offers a large cross-section for absorption of photons with energy $E \gtrsim 2\,\mathrm{keV}$, with the consequence that at such high energies we can study sources throughout the Milky Way and in external galaxies almost irrespective of their latitudes $b$. Moreover, at these energies we may usually treat the interstellar media of external galaxies as optically thin,[6] which greatly simplifies the extraction of astrophysically interesting parameters from observed X-ray spectra.

The X-rays emitted by a plasma are generated by two predominant processes: (i) the **bremsstrahlung**[7] process in which a free electron scatters off an ion, and (ii) the bound-bound transitions of ions, in which an excited state of an ion decays to a less energetic state.

The rate per unit volume of collisions between the free electrons and the ions of a fully ionized plasma is proportional to the square of the plasma's density $\rho$ because the electron density is proportional to $\rho$ and the rate at which each electron suffers collisions is proportional to the ion density, which is itself proportional to $\rho$. Consequently, the bremsstrahlung emissivity of a hot plasma is proportional to $\rho^2$. Quantitatively, for a plasma at temperature $T$ the emissivity is

$$j_\nu = 5.44 \times 10^{-52} \overline{Z^2} n_e n_i T^{-1/2} g e^{-h\nu/kT} \, \mathrm{W \, m^{-3} \, Hz^{-1}}, \qquad (8.5)$$

where $\overline{Z^2}$ is the mean-square atomic charge on the ions. The **Gaunt factor**, $g$, is a dimensionless quantity of order unity, which depends on the range of distances between electrons and ions at which the scattering process occurs. Since the energy of radiation produced depends on the closeness of the encounter between electron and ion, $g$ depends on the photon energy $E = h\nu$: Figure 8.7 shows $g(E)$ for bremsstrahlung produced by plasmas at various temperatures. It is apparent that $g$ depends fairly weakly on $\nu$, and so equation (8.5) implies that $j_\nu$ is almost constant at low frequencies and exponentially cut-off above $\nu \simeq kT/h$.

In an optically-thin medium, ions are excited by collisions with energetic electrons and de-excited either by collision with a second electron, or radiatively through the emission of a photon. At the densities characteristic of the hot component of the ISM, collisional de-excitation is much less important than radiative decay, with the result that essentially every collisional excitation gives rise to the emission of a photon. From this it follows that the luminosity per unit volume in line radiation is also proportional to $\rho^2$.

In a hot astrophysical plasma, bremsstrahlung is not the only X-ray emission process. At high temperatures low-mass atoms are fully ionized, but heavier, highly-charged nuclei are able to hold on to some of their most tightly bound electrons. These remaining bound electrons are excited to higher energy levels by collisions with energetic free electrons in the hot

---

[6] See §8.1.4 for a discussion of optical depth in emission lines.

[7] Bremsstrahlung is German for 'braking' (bremsen) 'radiation' (Strahlung).

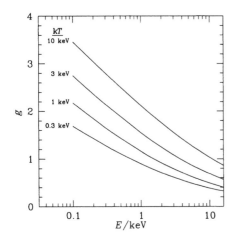

**Figure 8.7** The Gaunt factor as a function of photon energy $E = h\nu$ for bremsstrahlung produced by plasmas (which contain a helium mass fraction $Y = 0.27$) at various temperatures $T$. The Gaunt factors were calculated using the prescription given by Kellogg, Baldwin & Koch (1975).

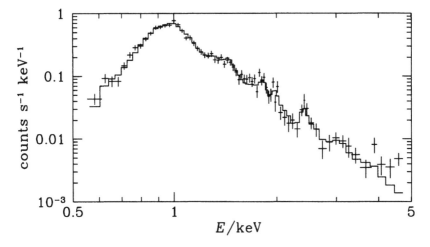

**Figure 8.8** X-ray spectrum of the elliptical galaxy NGC 4472. The horizontal extent of each cross shows the energy binning of the data, while the vertical extent indicates the RMS error in the signal. The histogram shows a model spectrum for a two-temperature plasma. [From data taken by the ASCA satellite and kindly provided by D. Buote]

gas. The subsequent radiative decays of such excited atoms produce X-ray emission at specific energies. Thus, the X-ray spectra generally contain discrete emission lines from various highly-ionized heavy elements on top of the bremsstrahlung continuum.

In the past, X-ray detectors have generally had very limited energy resolution. Emission lines in spectra were therefore smeared out to a point where they were difficult to detect for any but the strongest features. In recent years, this situation has improved dramatically. Figure 8.8 shows an X-ray spectrum of the elliptical galaxy NGC 4472 obtained using the Japanese ASCA satellite. The CCDs that form the detectors in this satellite have better energy resolution than previous instruments, and it is apparent that this improved resolution makes it possible to detect a number of discrete emission lines. For example, the strong line at $\sim 2.5\,\mathrm{keV}$ arises from K-series[8] line emission from highly-ionized sulfur. Even with these improved data, many lines are still blended: the broad peak at $\sim 1\,\mathrm{keV}$ in Figure 8.8 is attributed to a mixture of L-series emission from Fe in a variety of ionization states and K-series emission from Ne. Nevertheless, there is clearly a great deal of information to be gleaned from such high quality spectra.

In principle, it is possible to determine the temperature, density and metallicity of a plasma from its X-ray spectrum. A further parameter that alters the observed spectrum at low energies is the amount of absorption by cold gas along the line of sight, either in the local ISM or in any material surrounding the emitting object. All these quantities can be estimated by comparing the observed spectrum to ones simulated using the full range of plausible parameter values. Simulating the X-ray spectrum of a plasma requires a large and complex computer program, which draws on a mass of atomic data. Unfortunately, many of the required data are not accurately known, and it is not practicable to include all possible atomic processes in a single computer program; one merely tries to include the most important processes. Uncertainties in the available atomic data and incompleteness of the codes in which they are used combine to produce significant uncertainties in the values of physical variables inferred from X-ray spectra. As the quality of X-ray data has improved, the shortcomings of widely-used spectral interpretation codes have become apparent, so parallel efforts have been made to improve the models [see, for example, Liedahl, Osterheld & Goldstein (1995)].

As well as forcing us to improve the spectral fitting codes, higher quality X-ray data have also made it necessary to consider more sophisticated astrophysical models. While the data were poor, it was possible to explain the observations using simple-minded physical interpretations, but such models are frequently ruled out when the data quality improves. In the case of NGC 4472, for example, observed spectra cannot be reproduced by modeling the emission as arising from a single temperature plasma. As shown in Figure 8.8, a plasma with two distinct temperature phases could explain the observations of this galaxy, but even this model is probably an oversimplification; the plasma might contain gas at a whole range of temperatures, or there might be non-negligible contributions to the spectrum from other sources, such as X-ray binaries in the galaxy. Testing these yet more complex

---

[8] X-ray spectroscopists have their own notation for defining spectral lines: transitions in which an electron moves down into an $n = 1$ state (see Box 8.2) produce "K-series lines," those from electrons ending in an $n = 2$ state produce "L-series lines," and so on. These series are analogous to the Lyman, Balmer, etc., series in hydrogen.

models must await still higher-quality X-ray data from the next generation of instruments.

### 8.1.3 Optical emission lines

The optical spectra of many galaxies contain conspicuous emission lines. These lines are generally formed in excited interstellar gas. The gas has usually been excited by ionizing radiation from either hot stars or an active galactic nucleus, but it has sometimes been excited by the passage of a powerful shock wave. We shall concentrate on the more important case of photo-excited regions. These regions are characterized by the presence of atoms in higher states of ionization than would occur in thermodynamic equilibrium at the kinetic temperature of the gas ($T \simeq 8000\,\mathrm{K}$).

The spectra of many hot stars are such that radiation from these stars ionizes essentially all hydrogen atoms within some limiting distance from the stars, while material beyond this distance is largely neutral. The region of ionized hydrogen is called an **HII region** and the volume within which it is contained is called a **Strömgren sphere**. A great deal of valuable information about both the ISM and hot stars has been gleaned from the spectra of such photo-excited HII regions.

**Hydrogen lines**     Lines of the Balmer series such as Hα (656 nm), Hβ (486 nm) and Hγ (434 nm) (see Figure 8.9) are prominent in the optical spectra of HII regions. In thermodynamic equilibrium at the kinetic temperature of a typical HII region ($\sim 8000\,\mathrm{K}$), there would be negligible flux in these lines since they arise from states that lie more than 12 eV above hydrogen's ground state. It follows that the observed lines are produced by H atoms cascading to their ground state following capture of an electron by a proton. The relative intensities of the different lines formed by this recombination process can be accurately calculated, and depend only weakly on $T$ and (for interstellar densities) negligibly on the density. The relevant theory is simple because observed nebulae are invariably optically thin in these lines.

The observed relative intensities of the Balmer lines do not agree with the predictions of theory. In particular $I(\mathrm{H}\alpha)/I(\mathrm{H}\beta)$, which is called the **Balmer decrement**, is larger than the predicted value $I(\mathrm{H}\alpha)/I(\mathrm{H}\beta) \simeq 3$. This discrepancy between simple theory and observation is attributed to absorption of line radiation by dust: since dust absorbs blue light more strongly than red (see §3.7.1), the fraction of Hβ photons that are absorbed by dust is larger than the fraction of the redder Hα photons that are absorbed. This conjecture can be tested by comparing the observed value of $I(\mathrm{H}\beta)/I(\mathrm{H}\gamma)$ to that which it predicts; the agreement is generally found to be good.

The number of ionizing photons absorbed by an HII region per unit time can be deduced from the observed intensity of any of the Balmer lines. The argument, which was first given by Zanstra (1927), is as follows. In a steady state, each photo-ionization of a hydrogen atom may be associated with

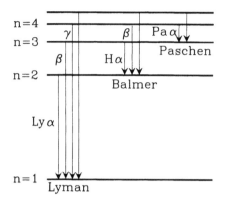

**Figure 8.9** Schematic diagram of the Lyman, Balmer and Paschen series of spectral lines in the spectrum of hydrogen.

a hydrogen atom reaching its ground state by emission of a Lyman-series photon. Now a Lyman-series photon emitted within a HII region is very unlikely to escape from the HII region because the ground-state hydrogen atoms, which are so abundant in and around the HII region, offer a large cross-section for its absorption. Suppose the photon is not a Ly$\alpha$ photon but one of the higher members of the series. Then the excited hydrogen atom produced by its absorption has a finite chance of emitting two or more photons before returning to its ground state. For example, when a Ly$\gamma$ photon is absorbed, there is a non-trivial probability for the emission of (a) an H$\beta$ photon followed by a Ly$\alpha$ photon, or (b) a Paschen $\alpha$ photon ($n = 4 \rightarrow n = 3$) followed by either (i) a Ly$\beta$ photon or (ii) an H$\alpha$ photon plus a L$\alpha$ photon. Thus each time a Lyman-series photon is absorbed, there is non-negligible probability for the emission of a less energetic Lyman-series photon plus one or more photons of the Balmer, Paschen ..., series. It follows that after a sufficient number of absorptions, the energy of *any* Lyman-series photon will have been divided between a Ly$\alpha$ photon and some less energetic photons. Moreover, the creation of a Ly$\alpha$ photon by any process other than the destruction of a Ly$\alpha$ photon is associated with the creation of a Balmer-series photon. Thus the rate of creation of all Balmer-series photons should equal the rate of ionization of hydrogen atoms. Finally, we recall that the relative strengths of the Balmer lines depend only weakly on temperature and density. Hence the rate of creation of all Balmer-series photons, and thus the ionization rate, can be estimated from the strength of any single Balmer line.

The ionization rate within an HII region depends upon the luminosity of the exciting source in the EUV band. When knowledge of this rate is combined with the assumption that the source radiates as a black body and with an estimate of the flux at much longer wavelengths, for example in the optical or infrared bands, the temperature of the ionizing source can be determined.

**Metal lines**    When HII regions were first studied spectroscopically, astronomers were puzzled by the two most prominent lines in the spectra. These green 'nebular' lines, $N_1$ at 500.7 nm and $N_2$ at 495.9 nm, defied interpretation until it was demonstrated by Bowen (1927) that they are 'forbidden' lines of $O^{2+}$. (See Box 8.1 for the meaning of 'forbidden'.)

In practice the spectra of interstellar nebulae are dominated by forbidden and semi-forbidden lines of the ions of heavy elements such as N, O and S. At first sight this may seem a very strange phenomenon: not only are these ions over 1000 times rarer than those of H and He, but the transitions that are dominating the observed radiation are precisely the least probable transitions of these ions! Actually it is not hard to understand the dominance of forbidden transitions.

The only excited states of ions that will be significantly populated in an HII region are those whose energies lie within a few times $kT \simeq 1$ eV of the relevant ground-state energy. All such low-lying states have the same electronic 'configuration' as the ground state,[9] with the result that all transitions between them and the ground state are forbidden.

The forbidden nature of these transitions effectively prevents their giving rise to interstellar absorption – as we saw in §8.1.1, significant interstellar absorption from ground states only occurs in the UV. But in a low-density medium a forbidden transition can cause significant emission simply because once an ion is knocked into an excited state, it will hang there until it decays radiatively. Thus the forbiddenness of the transition delays rather than prevents the emission of a photon. Notice, however, that this is true only if the half-life of the excited state is short compared to the collision time in the plasma; at high enough density the excited state is liable to be de-excited by a second collision before it has had time to decay radiatively. Consequently, if the emissivity per atom $j$ of a plasma is plotted against density $\rho$, $j$ will be proportional to $\rho$ for $\rho$ less than some critical density $\rho_{\text{crit}}$ at which the collision time becomes comparable to the radiative decay time.[10] At higher densities $j$ will be constant because the number of ions in the excited state will have leveled off at the value appropriate to thermodynamic equilibrium, and each excited ion has some probability per unit time of emitting a photon. Thus the emissivity of a dense plasma, such as that in a fluorescent light, is liable to be dominated by allowed transitions, which have high critical densities and peak emissivities.

The temperature and density of an HII region can be determined from optical lines of O, N and S. The relevant ions have energy-level diagrams of two types. Figure 8.10 is a schematic energy-level diagram for $N^+$, $O^{2+}$, and $S^{2+}$, which have equivalent electronic configurations. For our purposes it suffices to distinguish three levels: that associated with the $^3P$ ground term and those associated with the $^1D$ and $^1S$ terms, which lie 2.47 eV and 5.32 eV

---

[9] See Box 8.2 for the meaning of 'configuration'.

[10] Astronomers sometimes refer to $\rho_{\text{crit}}$ as the "density required for excitation." It is more accurately called the "density required for *full* excitation."

## Box 8.1: Allowed and Forbidden Spectral Lines

Standard time-dependent perturbation theory yields an expression for the Einstein $B$ coefficients [see equation (8.6)] that is proportional to the squared absolute value of an infinite sum. The first term in this sum is proportional to the quantum-mechanical matrix element $\langle 1|\mathbf{d}|2\rangle$, where $|1\rangle$ and $|2\rangle$ are the quantum states of the ion before and after emission, and $\mathbf{d}$ is the operator for the ion's electric dipole moment. The two next terms in the series are proportional to $\langle 1|\mathbf{q}|2\rangle$ and $\langle 1|\mathbf{m}|2\rangle$, where $\mathbf{q}$ and $\mathbf{m}$ are the ion's electric quadrupole and magnetic dipole moments. On dimensional grounds the relative magnitudes of these quantities are of order $|\langle 1|\mathbf{d}|2\rangle| : |\langle 1|\mathbf{q}|2\rangle| : |\langle 1|\mathbf{m}|2\rangle| \approx \lambda/a_0{:}1{:}1$, where $\lambda$ is the wavelength of the photon concerned and $a_0$ is the Bohr radius. Since $\lambda/a_0 \approx 5000$ in the optical band, the first (dipole) contribution to $B$ dominates unless it vanishes by virtue of some exact or near symmetry of the ion. **Forbidden** transitions are those for which $\langle 1|\mathbf{d}|2\rangle$ vanishes. Their forbidden nature is indicated by surrounding the symbols of the ions in which they occur by square brackets thus: $[O^{2+}]$. If $\langle 1|\mathbf{d}|2\rangle$ is much smaller than expected on dimensional grounds, the transition is called **semi-forbidden** and a square bracket is put to the right of the ion involved: $N^+]$.

   A transition is allowed (i.e., not forbidden) if the differences in the quantum numbers of the states $|1\rangle$ and $|2\rangle$ obey the electric-dipole selection rules. The most important quantum numbers characterizing these states are the atom's total angular momentum $F$, its total electronic angular momentum $J$, its total electronic spin $S$ and its total orbital angular momentum $L$. The selection rules for these quantities are

$$\delta F = 0, 1 \text{ but } F = 0 \not\to F = 0,$$
$$\delta J = 0, 1 \text{ but } J = 0 \not\to J = 0,$$
$$\delta L = 0, 1 \text{ but } L = 0 \not\to L = 0, \tag{1}$$
$$\delta S = 0.$$

$F$ and $J$ are always good quantum numbers but $L$ and $S$ are good quantum numbers only to the extent that the Russell-Saunders or $L - S$ coupling scheme is applicable. The quantum states of atoms are generally classified under the assumption that $L - S$ coupling applies, although this coupling scheme is never more than an approximation, and one that deteriorates as one proceeds down the periodic table. In reality a state $|1\rangle$ that is classified as having, say, $S = 1$, may be a linear combination of a predominant state that has $S = 1$ and a state that has $S = 0$. This mixed state can decay by emission of electric dipole radiation to a state classified as having $S = 0$ without violation of the electric-dipole selection rule. Such **intercombination transitions** are semi-forbidden.

## Box 8.2: Spectroscopic Notation

Standard spectroscopic notation presumes that the $L-S$ coupling scheme is valid. The **electronic configuration** is a specification of the principal $n$ and orbital angular momentum $l$ quantum numbers of the individual electrons of the outermost shell. Within a configuration a **spectroscopic term** specifies definite values for the total orbital $L$ and spin $S$ angular momenta of the outer electrons. Within each term a **fine-structure level** specifies a definite value for the total electronic angular momentum $J$. Within a fine-structure level may be distinguished different **hyperfine levels** that differ in total angular momentum $F$.

A typical configuration is denoted $2s2p^3$ meaning one electron has $n = 2$, $l = 0$, and three electrons have $n = 2$, $l = 1$.

Terms are denoted by $^{(2S+1)}L_J$; for example $^4P_{1/2}$ means $S = \frac{3}{2}$, $L = 1$, $J = \frac{1}{2}$.

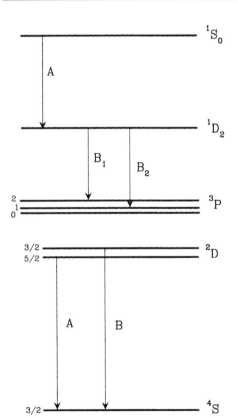

**Figure 8.10** Schematic energy-level diagram of $O^{2+}$ and $N^+$. In the case of $O^{2+}$ transition A has wavelength 436.3 nm while the wavelengths of transitions $B_1$ and $B_2$ are 500.7 nm and 495.9 nm, respectively. In the case of $N^+$ the wavelengths are A: 575.5 nm, $B_1$: 658.3 nm, $B_2$: 654.8 nm.

**Figure 8.11** Schematic energy-level diagram of $O^+$ and $S^+$. In the case of $O^+$ transition A has wavelength 372.9 nm while transition B has wavelength 372.6 nm. In the case of $S^+$ the wavelengths are A: 673.1 nm, B: 671.6 nm.

above the ground term. The nebular lines that puzzled astronomers before 1927 are produced by transitions in $O^{2+}$ between the top and middle levels (at 436.3 nm) and between the middle and bottom levels (at $\sim 500$ nm). Clearly the intensities of these lines will be proportional to the occupation probabilities of their respective upper levels. In thermodynamic equilibrium these would be in the ratio top:middle = $1:e^{2.84\,eV/kT}$, and thus be sensitive to the plasma's temperature $T$. As we have seen, the densities of HII regions are so low that even states with long radiative decay times are out of thermodynamic equilibrium, and we must calculate the required occupation probabilities by considering the rates at which collisions and radiative decays populate each state rather than simply invoking the Boltzmann factor. Detail theory shows that the relative intensities of the lines is nevertheless approximately proportional to $e^{2.84\,eV/kT}$.

Figure 8.11 shows the second type of energy-level diagram that plays an important rôle in studies of HII regions. The energy-level diagrams of N, $O^+$ and $S^+$ are of this type. The density of an HII region can be estimated from the relative intensities of the lines that connect the two middle levels, $^2D_{3/2}$ and $^2D_{5/2}$ to the ground level. Again we argue that the intensities of these lines is proportional to the occupation probabilities of their upper levels. Since the energies of these differ by much less than $kT$, in thermodynamic equilibrium the occupation probabilities will be in the ratio 4:6 of the degeneracies of the $J = \frac{3}{2}$ and $J = \frac{5}{2}$ levels. At lower densities the occupation probabilities will be determined by the relevant rate equations. One finds that in the limit of low plasma density it tends to 1:0.32.

### 8.1.4 Radio observations

Radio telescopes provide some of the most powerful diagnostics of interstellar gas. To interpret radio-frequency data correctly, it is necessary to delve rather more deeply than we have so far into the way in which a beam of radiation interacts with material along its path. Our analysis of this problem is entirely general and not restricted in its validity to radio frequencies; we present it here merely because we wish to apply it to radio-frequency observations.

The **specific intensity** at frequency $\nu$, $I_\nu$, of a beam of radiation is the energy flux in the beam per unit solid angle and per unit frequency.[11] In distance d$s$ along the beam, decays of atoms from excited states cause $I_\nu$ to increase, while excitations of atoms to excited states cause $I_\nu$ to decrease.

---

[11] The units of $I_\nu$ are usually $W\,Hz^{-1}\,m^{-2}sterad^{-1}$.

Mathematically, we write[12]

$$\frac{\mathrm{d}I_\nu}{\mathrm{d}s} = \left(\frac{h\nu}{4\pi}\right)\left[n_2^{(\nu)}A_{21} - (n_1^{(\nu)}B_{12} - n_2^{(\nu)}B_{21})\frac{4\pi I_\nu}{c}\right]. \tag{8.6}$$

Here $n_i^{(\nu)}$ is the number density of atoms capable of emitting ($i = 2$) or absorbing ($i = 1$) a photon of frequency $\nu$, $A_{21}$ is the Einstein coefficient that gives the probability for spontaneous emission of a photon, and $B_{21}$ is the corresponding coefficient for stimulated emission. $B_{12}$ is the Einstein coefficient for absorption of a photon of frequency $\nu$. Defining

$$\chi_\nu \equiv \frac{h\nu}{c}(n_1^{(\nu)}B_{12} - n_2^{(\nu)}B_{21}),$$

$$S_\nu \equiv \frac{c}{4\pi}\frac{n_2^{(\nu)}A_{21}}{n_1^{(\nu)}B_{12} - n_2^{(\nu)}B_{21}}, \tag{8.7}$$

$$\tau_\nu \equiv \int \chi_\nu\,\mathrm{d}s,$$

equation (8.6) becomes[13]

$$\frac{\mathrm{d}I_\nu}{\mathrm{d}\tau_\nu} + I_\nu = S_\nu. \tag{8.8}$$

$\tau_\nu$ is called the **optical depth** and $S_\nu$ the **source function** at frequency $\nu$. Equation (8.8) is a first-order linear equation such as may be integrated with an integrating factor $\mathrm{e}^{\tau_\nu}$. In the especially simple case in which $S_\nu$ is constant along the line of sight, the solution may be written

$$I_\nu(\tau_\nu) = I_\nu(0)\mathrm{e}^{-\tau_\nu} + S_\nu(1 - \mathrm{e}^{-\tau_\nu}). \tag{8.9}$$

Equation (8.9) states that the intensity at the observer is equal to the intensity at some distance decreased by an exponential factor plus a fraction of the source function.

We now assume that the material through which the beam passes is in thermodynamic equilibrium at temperature $T$. Then the value of $I_\nu(\tau_\nu)$

---

[12] Einstein's equation was $\dot{N}_2 = -A_{21}N_2 + \left(N_1B_{12} - N_2B_{21}\right)\rho$, where $N_i$ is the total number of atoms in the $i^{\mathrm{th}}$ energy level and $\rho$ is the radiation's energy density per unit frequency at the frequency of the transition. We obtain equation (8.6) by (a) focusing on the atoms in unit volume; (b) recognizing that only $n_i^{(i)}\mathrm{d}\nu$ atoms per unit volume could emit/absorb a photon with frequency in $(\nu + \mathrm{d}\nu, \nu)$; (c) figuring that each such transition will contribute $\pm h\nu$ to the energy density $\rho\mathrm{d}\nu$; (d) recalling that for the isotropic radiation field considered by Einstein $\rho = 4\pi I_\nu/c$; (e) setting $\mathrm{d}s = c\mathrm{d}t$.

[13] In the theory of stellar atmospheres, $\tau$ is generally defined such that it increases away from rather than towards the observer, as here. Defining $\tau$ in that way changes signs in equation (8.9).

given by (8.9) must tend to the Planck intensity[14] $B_\nu$ in the limit of long path length $s$, that is the limit $\tau_\nu \to \infty$. Clearly, this implies that $S_\nu = B_\nu$ in any material that is in thermodynamic equilibrium, and equation (8.7) becomes

$$\frac{dI_\nu}{d\tau_\nu} + I_\nu = B_\nu(T). \tag{8.10}$$

Finally, when we replace $S_\nu$ in equation (8.7) by $B_\nu$ and exploit the fact that in thermodynamic equilibrium

$$\frac{n_2^{(\nu)}}{n_1^{(\nu)}} = \frac{g_2}{g_1} \exp\left(-\frac{h\nu}{kT}\right), \tag{8.11}$$

where $g_i$ is the number of quantum states associated with energy level $i$, we obtain Einstein's relations

$$A_{21} = \frac{8\pi h\nu^3}{c^3} B_{21} \quad , \quad g_1 B_{12} = g_2 B_{21}. \tag{8.12}$$

At radio frequencies, we are generally working in the Rayleigh–Jeans regime, $h\nu \ll kT$, in which we have the approximation

$$B_\nu(T) \simeq \frac{2kT\nu^2}{c^2}. \tag{8.13}$$

Equation (8.9) becomes

$$I_\nu(\tau_\nu) = I_\nu(0)e^{-\tau_\nu} + \frac{2kT\nu^2}{c^2}\left(1 - e^{-\tau_\nu}\right). \tag{8.14}$$

Equation (8.14) has led radio astronomers to report surface brightness measurements in terms of the **brightness temperature**, $T_B$, defined by

$$T_B \equiv \frac{c^2}{2k\nu^2} I_\nu. \tag{8.15}$$

With this definition equation (8.14) becomes

$$T_B(\tau_\nu) = T_B(0)e^{-\tau_\nu} + \left(1 - e^{-\tau_\nu}\right)T. \tag{8.16}$$

In the **optically-thin** limit of small $\tau_\nu$, we have for the case of negligible $T_B(0)$

$$T_B(\tau_\nu) \simeq \tau_\nu T. \tag{8.17}$$

---

[14] $B_\nu = \frac{2h\nu^3}{c^2}(e^{h\nu/kT} - 1)^{-1}$ is the specific intensity radiated by a black body. When multiplied by $\cos\theta$ and integrated over a hemisphere and all frequencies we obtain the Stefan–Boltzmann law for the energy radiated per unit area per unit time from a black body: $f = \int d\Omega d\nu \cos\theta B_\nu = \sigma T^4$.

Under our assumption of constant $S_\nu$, $\tau_\nu$ is, by its definition (8.7), proportional to the column density of material along the line of sight. Thus when the medium is optically thin and $T_B(0)$ is negligible, $T_B$ is likewise proportional to the column density along the line of sight. When the medium is optically thick, $T_B$ measures the medium's temperature rather than its column density. We shall see that the ISM is often optically thin in the 21-cm line of atomic hydrogen and optically thick in the widely-observed 2.6 mm line of carbon monoxide.

Observers frequently report results in terms of the measured **antenna temperature** $T_A$. This is the temperature required of a black-body that fills the telescope beam in order to generate the same signal as the source being measured. If the source fills the beam (including any side-lobes), then $T_A = T_B$. But if the source fills only a fraction $\alpha$ of the beam, $T_A = \alpha T_B$ and $T_A$ is said to be depressed below $T_B$ by **beam dilution**. Observers tend to report $T_A$ because it is the directly measurable quantity and the degree of beam dilution is frequently difficult to determine. Since the angular resolution of many radio telescopes is comparatively poor, in many studies the object does not fill the beam, with the result that $T_A$ is significantly smaller than $T_B$.

**The 21-cm line of atomic hydrogen**     The ground state of the gross structure of atomic hydrogen is split into two hyperfine levels by the interaction between the spins of the electron and proton; the atom's energy is $6 \times 10^{-6}$ eV lower when the two spins are antiparallel than when they are parallel. Thus the true ground state has total angular momentum $F = 0$, above which lie three degenerate states with angular momentum $F = 1$. Photons with frequency $\nu = 1.4204$ GHz or wavelength 21.105 cm mediate transitions between the two levels. Since the temperature of the ISM cannot fall below that of the cosmic background radiation, 2.7 K, which corresponds to $kT = 2.3 \times 10^{-4}$ eV, the approximation $h\nu \simeq 5.9 \times 10^{-6}$ eV $\ll kT$ is invariably valid in connection with the 21-cm line.

Since neither the $F = 0$ nor any of the $F = 1$ states of atomic hydrogen has an electric dipole moment, the absorption or emission of 21-cm photons is 'forbidden'. Thus the lifetime of the excited level is long $(1.1 \times 10^7$ yr) and the values of the associated Einstein constants $A_{21}$ and $B_{ij}$ are extremely small. At densities of interest $(n \sim 10^6$ m$^{-3})$, the rate of collisional excitation and de-excitation is very fast $(\sim 400$ yr$^{-1})$ compared to the spontaneous decay rate, so equation (8.11) holds accurately with $T$ equal to the kinetic temperature of the gas.

Let us now calculate the optical depth in the 21-cm line. From equations (8.7), (8.11) and (8.12) we obtain in the Rayleigh–Jeans regime

$$\tau_\nu = \left(\frac{B_{12}h^2\nu^2}{ck}\right)\frac{N_1^{(\nu)}}{T}, \tag{8.18}$$

where $N_1^{(\nu)} \equiv \int ds\, n_1^{(\nu)}$ is the column density of hydrogen atoms capable of

---

## Box 8.3:  Spin Temperature

The populations $n(F = 1)$ and $n(F = 0)$ of the $F = 0, 1$ levels of hydrogen are characterized by the **spin temperature**, $T_{\text{spin}}$, which is defined by

$$\frac{n(F = 1)}{n(F = 0)} = 3 \exp\left(\frac{-h\nu_0}{kT_{\text{spin}}}\right), \tag{1}$$

where $\nu_0$ is the frequency of the 21 cm transition. At the low densities that occur astrophysically $T_{\text{spin}}$ usually differs from the kinetic temperature. Naively, one would expect $T_{\text{spin}} < T_{\text{kinetic}}$ at densities, $n \lesssim 5 \times 10^3 \, \text{m}^{-3}$, so low that the collisional excitation time exceeds $10^7$ yr. However, Watson & Deguchi (1984) show that at very low densities, $T_{\text{spin}} \gtrsim T_{\text{kinetic}}$ as a result of the tendency of resonant scattering of Ly$\alpha$ photons to equalize the occupation probabilities of the hyperfine states. In the presence of a powerful 21-cm continuum radiation field, such as that emitted by an AGN, one may even have $T_{\text{spin}} \gg T_{\text{kinetic}}$.

---

absorbing at frequency $\nu$. An individual hydrogen atom has negligible probability of absorbing photons whose frequencies lie outside an extremely narrow frequency range ($\delta\nu \lesssim 5 \times 10^{-13}$ Hz) in the atom's rest frame. Consequently, $N_1^{(\nu)}$ is determined by the distribution of atoms over radial-velocity:

$$N_1^{(\nu)} d\nu = N_1 f\left[(1 - \nu/\nu_0)c\right] dv, \tag{8.19}$$

where $N_1 \equiv \int d\nu \, N_1^{(\nu)}$, $d\nu = c \, dv/\nu$ and $f(v) \, dv$ is the fraction of all atoms on the line of sight with radial velocity $v$ in the range $(v + dv, v)$. Since $kT \gg h\nu_0$, we have from equation (8.11) that $N_2/N_1 \simeq g_2/g_1 = 3$, and thus that $N_1 \simeq \frac{1}{4}N_H$. Inserting this approximation and the appropriate numerical values into equation (8.19), we have finally

$$\tau_\nu = Cf\left[(1 - \nu/\nu_0)c\right]\frac{N_H}{T}, \quad \text{where} \quad \begin{aligned} C &\equiv \frac{B_{12}h^2\nu}{4k} \\ &= 5.489 \times 10^{-20} \, \text{m}^3 \, \text{s}^{-1} \, \text{K}. \end{aligned} \tag{8.20}$$

Three factors determine $f(v)$: (i) the random thermal motions associated with the gas temperature $T$; (ii) the random velocities that different macroscopic volumes of gas possess because the ISM is turbulent; (iii) the velocities that different macroscopic volumes of gas have because there are large-scale ordered velocity gradients in the ISM, for example on account of differential Galactic rotation. Thermal motions contribute a characteristic width $\sim 1 \, \text{km s}^{-1}$ to $f(v)$ (see Problem 8.1), while studies of 21-cm emission from face-on galaxies suggest that turbulence contributes a width of order

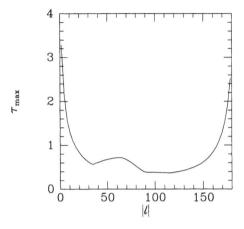

**Figure 8.12** Peak optical depth in the 21-cm line as a function of longitude for a naive model of the galactic disk. The number density of hydrogen atoms has been assumed to equal $2 \times 10^5 \, \mathrm{m}^{-3}$ at $R_0$ and to vary with $R$ as $\exp(-R/2R_0)$. The circular speed is given by $v_c(R) = (R/R_0)^{0.1} \times 210 \, \mathrm{km \, s}^{-1}$ and the emitting atoms have a velocity dispersion $\sigma = 0.04v_c(R_0) \simeq 8.4 \, \mathrm{km \, s}^{-1}$.

$7 \, \mathrm{km \, s}^{-1}$ (van der Kruit & Shostak 1984, Hanson, Dickey & Helou 1989). The width contributed to $f(v)$ by Galactic rotation varies enormously with the longitude $l$ of the line of sight. In the center and anticenter directions it is negligible, and it can exceed $100 \, \mathrm{km \, s}^{-1}$ at intermediate directions (see Problem 9.2). Figure 8.12 shows $\tau_{\nu_0}(l)$ for a simple model of the galactic disk. In the center and anticenter directions the optical depth exceeds unity with the result that the brightness temperature lies close to the gas temperature $T$, while at $l \simeq 120°$ the optical depth falls to $\tau_{\nu_0} \simeq 0.5$.

Since the radial-velocity distribution $f(v)$ is normalized to ensure that $\int \mathrm{d}v \, f = 1$, the value of the total neutral hydrogen column density $N_H$ can be obtained by integrating equation (8.20) over velocity. With velocities measured in $\mathrm{km \, s}^{-1}$, this yields

$$N_H = 1.82 \times 10^{22} \int_{-\infty}^{\infty} \mathrm{d}v \, T\tau(v) \, \mathrm{atoms \, m}^{-2}, \qquad (8.21)$$

If the material is optically thin, we may replace $T\tau(v)$ by the brightness temperature $T_B$ [see equation (8.17)]:

$$N_H(l, b) = 1.82 \times 10^{22} \int_{-\infty}^{\infty} \mathrm{d}v \, T_B(l, b, v) \, \mathrm{atoms \, m}^{-2}. \qquad (8.22)$$

When applicable, this equation is to be preferred to equation (8.21) because $T_B$ is directly measured while $T$ and $\tau(v)$ are not easily determined.

When we observe an external galaxy, we can determine its total atomic hydrogen content $\mathcal{N}_H$ by integrating the HI column density over the surface area of the system. If we write an element of area as $\mathrm{d}S = D^2 \mathrm{d}\Omega$, where $D$ is the distance to the galaxy and $\mathrm{d}\Omega$ is an element of solid angle on the sky, and if we write the angle- and velocity-dependent brightness temperature as $T_B(l, b, v)$, then

$$\mathcal{N}_H = 1.82 \times 10^{22} D^2 \int_{-\infty}^{\infty} \mathrm{d}v \int_{\Omega} \mathrm{d}\Omega \, T_B(l, b, v) \, \mathrm{atoms}, \qquad (8.23)$$

where $\Omega$ is the solid angle subtended by the whole galaxy. The corresponding mass of hydrogen is

$$\frac{\mathcal{M}_H}{\mathcal{M}_\odot} = 2.36 \times 10^5 \left(\frac{D}{\text{Mpc}}\right)^2 \frac{\int_{-\infty}^{\infty} dv\, S(v)}{\text{Jy km s}^{-1}}. \tag{8.24}$$

Notice that both $\mathcal{M}_H$ and the total luminosity $L$ of a galaxy are proportional to $D^2$, so that the ratio $(\mathcal{M}_H/L)$ is independent of the often uncertain distance to an external galaxy.

**Rotation transitions of heteronuclear molecules**    The spectra of several fairly abundant molecules such as CO, CS and HCN have readily observed mm-band lines. These lines are formed by molecules changing their angular momentum by $\hbar$ while retaining the same electronic and vibrational quantum numbers. On account of the constancy of these quantum numbers, the energies of these transitions are much smaller than those of the compound transitions of $H_2$ studied in §3.7, and the lines fall in the mm wavebands rather than in the ultraviolet. These mm-band lines constitute powerful probes of the denser and colder components of the ISM. Much the most important lines are those of CO at wavelengths 2.6 mm and 1.3 mm. We shall concentrate on these lines, but most of our analysis applies with minor modification to lines of other molecules.

Unlike homonuclear molecules such as $H_2$ and $O_2$, a heteronuclear molecule such as CO has a net dipole moment, and consequently should radiate when it spins. Quantum-mechanically, this fact manifests itself in the Einstein constants, $A_{12}$ etc., for the mm-lines of CO being much larger than the Einstein constants associated with the 21-cm line.

The differences between the relevant Einstein constants for CO and for HI have two important consequences. First the lifetimes of rotationally excited levels of CO are relatively short, and unless the gas density and collision frequency are high, equation (8.11) may be violated in a way that it is not in the case of HI. Second, a much smaller column density of CO molecules than of HI atoms is required to establish a given optical depth in the relevant lines [see equation (8.7)]. In fact, the intrinsic strength of the rotation transitions of CO effectively compensates for the low abundance of CO molecules relative to hydrogen, so that the strongest CO lines saturate at similar column densities of all atoms as does the 21-cm line ($\sim 2 \times 10^{23}$ atoms m$^{-2}$/ km s$^{-1}$). Table 8.1 gives the column densities of CO and $H_2$ molecules, distributed over a velocity range of 1 km s$^{-1}$ that will yield unit optical depth at 20 K, a typical temperature for molecular material.[15] These values apply only if the gas is at a sufficiently high spatial density for equation (8.11) to hold; at lower spatial densities, unit optical depth will be achieved at higher column densities than those given the table. Notice that the 1.3-mm line arising from

---

[15] The column density required for a given line to saturate scales as $T^2$ (Goldreich & Kwan 1973).

**Table 8.1** CO rotation lines

| Species | Relative abundance | Transition | $N_{\tau=1}(^{12}\text{CO})$ | $N_{\tau=1}(\text{H}_2)$ |
|---------|--------------------|-----------|------------------------------|--------------------------|
| $^{12}\text{C}^{16}\text{O}$ | 1 | $J=2 \rightarrow J=1$ | $5.5 \times 10^{19}$ | $3.9 \times 10^{23}$ |
|  |  | $1 \rightarrow 0$ | $2.2 \times 10^{20}$ | $1.7 \times 10^{24}$ |
| $^{13}\text{C}^{16}\text{O}$ | 1/65 | $2 \rightarrow 1$ | $3.6 \times 10^{21}$ | $2.8 \times 10^{25}$ |
|  |  | $1 \rightarrow 0$ | $1.5 \times 10^{22}$ | $1.1 \times 10^{26}$ |
| $^{12}\text{C}^{18}\text{O}$ | 1/500 | $2 \rightarrow 1$ | $2.8 \times 10^{22}$ | $2.2 \times 10^{26}$ |
|  |  | $1 \rightarrow 0$ | $1.1 \times 10^{23}$ | $8.5 \times 10^{26}$ |

NOTES: For each species, $N_{\tau=1}(^{12}\text{CO})$ is the density in atoms $\text{m}^{-2}/\text{km s}^{-1}$ of $^{12}\text{C}^{16}\text{O}$ molecules at $T = 20\,\text{K}$ that yields optical depth unity in the given line; $N_{\tau=1}(\text{H}_2)$ is the corresponding column density of molecular hydrogen for an assumed ratio $N(^{13}\text{CO})/N(\text{H}_2) = 2 \times 10^{-6}$ (Dickman 1978).

the $J = 2 \rightarrow J = 1$ transition saturates at a column density that is a factor 4 smaller than the column density at which the 2.6-mm $J = 1 \rightarrow J = 0$ line saturates.

Besides the common isotope $^{12}\text{C}^{16}\text{O}$, (or $^{12}\text{CO}$ in astronomical shorthand) there are the rarer forms $^{13}\text{C}^{16}\text{O}$ ($^{13}\text{CO}$) and $^{12}\text{C}^{18}\text{O}$ ($\text{C}^{18}\text{O}$), whose abundances are locally approximately 65 and 500 times smaller than that of the common isotope (Langer 1997). Each of these species forms lines of slightly different frequency, and since each line saturates at a different total column density of all material (Table 8.1), valuable information about the structure of clouds can be deduced by comparing the lines associated with each species. Phillips *et al.* (1979) made such comparisons for lines of sight through several dark clouds in the Milky Way. They find that the antenna temperature in the 1.3-mm $^{12}\text{CO}$ line is invariably very similar to that in the 2.6-mm $^{12}\text{CO}$ line. From this observation, it follows that the cloud must be optically thick in both these lines because, if it were thin in the less opaque line (the 2.6-mm line), the latter would yield a lower antenna temperature. Figure 8.13 shows the spatial profiles of the dark cloud in NGC 2024 when observed in the 1.3-mm lines of $^{12}\text{CO}$, $^{13}\text{CO}$, and $\text{C}^{18}\text{O}$. The peak intensities of these radiations are in the ratio $1:\frac{1}{2}:\frac{1}{5}$, from which it follows that the cloud must be optically thick in the 1.3-mm line of $^{13}\text{CO}$ because, if it were thin in this line, it would be thin also in the corresponding line of $\text{C}^{18}\text{O}$, which would then be fainter than the $^{13}\text{CO}$ line by a factor of $500/65 \simeq 8$, compared to the observed value of 2.5. Comparison of the 1.3-mm and 2.6-mm $^{13}\text{CO}$ antenna temperatures of several clouds confirms that $^{13}\text{CO}$ is usually optically thick at line center. If either of these lines were optically thin, the 1.3-mm temperature would be nearly a factor of four higher than the 2.6-mm temperature, whereas they are, in fact, comparable. Nevertheless, the $^{13}\text{CO}$ temperatures are invariably at least a factor two lower than the $^{12}\text{CO}$ temperatures. This behavior suggests that the unit-optical-depth surface of $^{13}\text{CO}$ has either a smaller radius or a lower brightness temperature than that

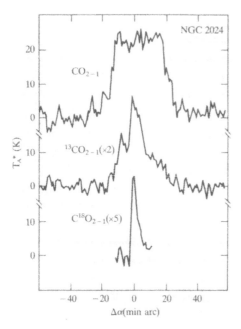

**Figure 8.13** The dark cloud NGC 2024 traced in the 1.3-mm line of three species of CO. The top profile suggests that the region that is optically thick in the $^{12}$CO line is spatially resolved and the gas has kinetic temperature near 20 K. Only a smaller portion of the cloud is optically thick in the $^{13}$CO line, and the C$^{18}$O line may be optically thin everywhere. Notice that the peak antenna temperatures differ by much less than the abundances of the various species. [After Phillips *et al.* (1979)]

of $^{12}$CO.

From these results it follows that no molecular line is ideally suited to a determination of $N(H_2)$: a line that is weak enough to be optically thin through a cloud's core will be too weak to measure accurately towards the cloud's periphery, where much of the mass resides. Moreover, the abundances of the different isotopes studied are uncertain. One cause of uncertainty is that the abundance ratios $n(H_2)/n(^{12}C)$ and $n(^{12}C)/n(^{13}C)$ are affected by the passage of interstellar material through stars (**astration**) with the consequence that $n(^{12}C)/n(^{13}C)$ is believed to decrease from a value as low as 20 at the Galactic center to perhaps 90 at large Galactic radii (Langer 1997). Another source of uncertainty is the possibility that one isotope may have a greater propensity than another to be locked up in CO molecules. Finally, variations in temperature and density within a cloud both affect the emissivity of CO molecules in a way that contrasts with the near temperature- and density-independence of the 21-cm emissivity of interstellar H atoms. To solve these problems properly, one would have to model the radiative transfer of each line studied through a detailed model of each observed cloud, adjusting the temperatures and densities within the clouds until the predicted maps of $T_B$ agreed with those observed. Not only would this be a formidable undertaking, but it presupposes that no brightness temperature varies on angular scales too small to be resolved. In practice even the nearest sources seem to contain unresolved structure – see §9.6.1. Consequently, detailed modeling of the type described is not feasible. Nonetheless, we shall see in §9.6.1 that when clouds in the Milky Way are observed at relatively low angular resolution, the antenna temperatures of weak lines are roughly proportional to the antenna temperatures of strong lines. This suggests that the average column density CO within a large beam *is* roughly proportional to the integral over velocities of the antenna temperature in a strong line such as the 2.6-mm line of $^{12}$CO.

Let us define

$$I_{CO} = \int dv \, T_A (2.6 \, \text{mm line of } ^{12}\text{CO}). \qquad (8.25)$$

The units of $I_{CO}$ are $\text{K km s}^{-1}$. The following naive theoretical argument suggests that the mass of a molecular cloud will be proportional to its value of $I_{CO}$. When large molecular clouds are observed at high resolution, they are found to be rich in internal structure; knots and filaments are apparent and there is every reason to believe that within these there is even smaller-scale, unresolved structure – see §9.6.1. A crude model of this complexity is to imagine that the cloud is made up of a large number of cloudlets. Irrespective of whether it is optically thick in a given CO line, each cloudlet would in isolation contribute a definite amount to $I_{CO}$. Now consider the value of $I_{CO}$ that we will measure for a big cloud. The contributions to $I_{CO}$ from individual cloudlets will add unless they frequently shadow each other in the sense that behind the optically thick portion of one cloudlet there is another with a velocity which differs from the velocity of the first cloudlet by less than twice the line-width $\sigma$ of a cloudlet. Clearly, if shadowing is unimportant, the value of $I_{CO}$ measured for the entire cloud will be proportional to the number of cloudlets and therefore to the mass $\mathcal{M}$ of the cloud. In fact, $I_{CO}$ will be proportional to $\mathcal{M}$ even if shadowing is a significant effect, providing that the fraction of all clouds that are shadowed at a given time does not vary with $\mathcal{M}$.

Suppose the cloud has radius $R$ and each cloudlet has mass $m$. Then there are $\mathcal{M}/m$ cloudlets in the cloud, and along a line of sight through the center of the cloud there are $\sim \mathcal{M}/(mR^2)$ cloudlets per unit area. Suppose each cloudlet is optically thick in the relevant line over area $A$, and let $\Delta$ be the velocity dispersion associated with the random motion of cloudlets through the cloud. Then shadowing will be important if and only if the 'shadowing factor'

$$s \equiv \left( \frac{\mathcal{M}A}{mR^2} \right) \left( \frac{\sigma}{\Delta} \right) \qquad (8.26)$$

is greater than unity: the first factor gives the mean number of cloudlets whose centers project within the opaque area of a randomly chosen cloudlet, and the second factor gives the probability that the velocity ranges of the two cloudlets significantly overlap.

Now suppose that the overall dynamics of the cloud resemble those of a star cluster, with the mutual gravitational attraction of the cloudlets balanced by the random motions of the cloudlets. Then by the virial theorem (e.g., BT §4.3) $\Delta$ will satisfy $\Delta \sim (GM/R)^{1/2}$. Using this to eliminate $\Delta$ from (8.26) and rearranging slightly, we obtain

$$s \simeq \left(\frac{A\sigma}{mG^{1/2}}\right)\rho^{1/2} \qquad (8.27)$$

where $\rho \equiv M/R^3$ is a characteristic mean density of the cloud. In equation (8.27) for the shadowing factor, the bracketed term depends only on the characteristics of individual cloudlets, which it is reasonable to assume do not vary from cloud to cloud. Hence we expect crowding to become more or less important with increasing $M$ according to whether $\rho$ increases or decreases with $M$. Observations show that $\rho$ *decreases* with $M$ – see §9.6.1. Moreover, crowding is clearly not significant if $M$ is so small that the cloud is made up of only one, two or a handful of cloudlets. So we conclude that crowding is *never* important, and that $I_{CO}$ is *always* proportional to $M$. All that remains is to calibrate this proportionality.

Studies of molecular clouds in the Milky Way that are discussed in §9.6.1 yield the calibration (Strong *et al.* 1988)

$$X \equiv \frac{N(H_2)/\,m^{-2}}{I_{CO}/\,K\,km\,s^{-1}} \simeq 2.3 \times 10^{24}. \qquad (8.28)$$

It must be emphasized that this is a rough empirical relation rather than one such as equation (8.22) that is based on good physics. Moreover, it is based on studies of the Milky Way (see §9.6.1) and it cannot be applied with confidence to galaxies of very different luminosities or morphological type. In particular, one would expect $X$ to increase with metallicity, and observation confirms this expectation. Wilson (1995) finds that

$$X \simeq 3 \times 10^{24}\{(6 \pm 0.9) - (0.7 \pm 0.1)[12 + \log_{10}(O/H)]\}. \qquad (8.29)$$

This fit to the data corresponds to a five-fold decrease in $X$ for a tenfold decrease in the metallicity.

**Synchrotron radiation**    A charged particle that moves in a magnetic field spirals around field lines and radiates because the Lorentz force is accelerating it towards the field line along which it is spiralling. If the particle's velocity is sub-relativistic, the radiation it emits is nearly monochromatic; it is called **cyclotron radiation** and its frequency is the **gyro-frequency** $\nu_g \equiv qB/2\pi m$, where $q$ and $m$ are the particle's charge and mass, respectively. If the particle is at relativistic velocity $\mathbf{v}$, it radiates broad-band radiation which peaks at the **critical frequency** $\nu_c \equiv \frac{3}{2}\gamma^2\nu_g\sin\theta$, where $\gamma = (1 - v^2/c^2)^{-1/2}$ is the particle's Lorentz factor and $\theta = \cos^{-1}(\mathbf{v}\cdot\mathbf{B}/vB)$ is the **pitch angle** of the helix it describes while traveling through the magnetic field. The power radiated is

$$-m_0 c^2 \frac{d\gamma}{dt} = 2\left(\frac{q^4}{6\pi\epsilon_0^2 c^4 m^2}\right)c\rho_B\gamma^2(v/c)^2\sin^2\theta, \qquad (8.30)$$

where $\rho_B = B^2/2\mu_0$ is the energy density of the magnetic field. Since this power is proportional to $(q/m)^2$, electrons are more than three million times more effective at generating synchrotron radiation than protons, and nearly 13 millions times as effective as most other nuclei. Consequently, observed synchrotron radiation is generally assumed to have been generated by electrons. In the case of electrons, the term in the large round bracket in equation (8.30) evaluates to the Thomson cross-section, $\sigma_T = 6.7 \times 10^{-29}\,\mathrm{m}^2$.

The solar neighborhood proves to contain relativistic particles ('cosmic rays') gyrating around a pervasive magnetic field of strength $B \simeq 0.6\,\mathrm{nT}$ (Rand & Kulkarni 1989). Electrons make a comparatively small contribution to energy density of the cosmic rays that reach the Earth, but in interstellar space, far from the disturbing influences of the solar wind and the Earth's magnetic field, electrons are probably a major constituent of cosmic rays. At the Earth the overwhelming majority of cosmic-ray particles are protons. Helium nuclei are 7 times less common than protons and most other nuclei more than a thousand times less common.[16]

The numbers of particles $N(\gamma)\mathrm{d}\gamma$ with Lorentz factors in the range $(\gamma + \mathrm{d}\gamma, \gamma)$ is approximately given by

$$N(\gamma) \propto \begin{cases} \gamma^{-2.5} & \text{for } \gamma \lesssim 10^6, \\ \gamma^{-3.7} & \text{for } 10^6 \lesssim \gamma \lesssim 10^{11}. \end{cases} \tag{8.31}$$

The total energy density due to these particles is $\rho_E \simeq 3.9 \times 10^{-13}\,\mathrm{J\,m}^{-3}$ (Holzer 1989), which is comparable to the energy density $\rho_B \simeq 1.4 \times 10^{-13}\,\mathrm{J\,m}^{-3}$ of the interstellar magnetic field. This energy is concentrated in the lowest-energy particles, which are at most mildly relativistic. Hence **cosmic rays** should be thought of as comprising **suprathermal** particles; only some cosmic-ray particles are truly relativistic, but all have kinetic energies much greater than the kinetic energies of the thermal particles past which they stream.

A population of gyrating particles with Lorentz factors that are distributed as a power law $N(\gamma) \sim \gamma^{-x}$ generates synchrotron radiation with an emissivity $j_\nu \sim \nu^{-(x-1)/2}$ that is a power law in frequency. At high frequencies the spectra of many synchrotron sources, for example young supernova remnants and the lobes of radio galaxies, are observed to fall off as $f_\nu \sim \nu^{-0.75}$. If these sources are optically thin to synchrotron radiation, it follows that the underlying electron spectrum has $x \simeq 2.5$, consistent with equation (8.31). At low frequencies the spectra of synchrotron sources tend to $f_\nu \sim \nu^{5/2}$ as a result of self-absorption:[17] a synchrotron photon from one electron is liable to be absorbed by another. It turns out that the fraction

---

[16] The exceptions are C and O nuclei, which are less common by factors of $\sim 250$ and $\sim 300$, respectively.

[17] Note that the self-absorbed synchrotron spectrum does not follow the Rayleigh–Jeans form $\sim \nu^2$ because the source is far from thermodynamic equilibrium.

of the photons that are emitted into the line of sight that do not reach us because they are absorbed en route increases with diminishing frequency $\nu$. Consequently, an optically thick source, in which the absorption of photons is important, becomes faint at low frequencies.

**Radio-frequency bremsstrahlung and recombination lines**   A valuable diagnostic of comparatively dense ionized gas such as that of HII regions is provided by observations of radio-frequency bremsstrahlung. By equation (8.5) the emissivity of hot plasma is approximately independent of frequency for $\nu \ll kT/h$. Consequently, the observed radio-frequency spectrum will be flat if the plasma is optically thin. By contrast, we have seen that the black-body spectrum $B_\nu(T)$ rises as $\nu^2$ in the Rayleigh–Jeans regime – see equation (8.13). Since the specific intensity that emerges from any body that is in thermal equilibrium tends to the Planck function at the body's temperature as the optical depth through the body becomes large, it is clear that at sufficiently low frequency, every thermal plasma must become optically thick to bremsstrahlung. The value of the frequency $\nu_1$ at which the spectrum passes from the form $f_\nu \sim$ constant characteristic of bremsstrahlung, to the Rayleigh-Jeans form $f_\nu \sim \nu^2$ that holds below $\nu_1$, allows one to determine the **emission measure** of the plasma $\int ds\,\overline{Z^2}n_e n_1$. The plasma's temperature is the observed brightness temperature at $\nu \ll \nu_1$.

In a plasma with temperature $T \approx 10^4\,$K, significant numbers of free electrons will be captured by protons into states that have values of the principal quantum number $n \gtrsim 50$. The highly excited hydrogen atoms thus formed are liable subsequently to decay by cascading down through states of ever smaller $n$, each decay being accompanied by the emission of an appropriate photon. Quantitatively, on moving from the level $n + \delta n$ to the level $n$, an atom emits a photon of frequency

$$\nu = 3.29 \times 10^{15}\left(\frac{1}{n^2} - \frac{1}{(n+\delta n)^2}\right) \simeq 6.58\left(\frac{100}{n}\right)^3 \delta n\,\text{GHz}. \qquad (8.32)$$

Analysis of the **radio recombination lines** formed in this way yields a number of astrophysically interesting quantities.

An estimate of the plasma's temperature $T$ can be obtained from the ratio

$$q \equiv \frac{I_l}{I_c}\frac{b}{c} \qquad (8.33)$$

where $I_l$ is the peak intensity of the line, $I_c$ is the intensity of the bremsstrahlung continuum at the line's central frequency and $b/c$ is the ratio of the line's velocity-width to the speed of light. The precise temperature dependence of $q$ is not easy to calculate because it is sensitive to departures from thermodynamic equilibrium in the plasma, but crudely $q \propto T^{-1}$.

A quantity that is easier to estimate from radio recombination lines is the helium abundance of the plasma. When a $\text{He}^+$ ion captures an electron into a level $n \gtrsim 50$, the highly excited He atom that results is scarcely distinguishable from a similarly excited H atom; the cloud formed by the inner electron of the He atom is of negligible size compared to an orbital with $n \gtrsim 50$ and effectively screens one of the nucleus's two units of charge. The frequencies at which it emits photons are different, however, by virtue of the difference in the reduced masses $\mu_H = m_e/(1 + m_e/m_p)$ and $\mu_{He} = m_e/(1 + \frac{1}{4}m_e/m_p)$ of the two atoms. Hence in the optically-thin limit, the ratio of the intensities of H and He recombination lines for a common value of $n$, is accurately proportional to ratio of the number of $H^+$ and $He^+$ ions in the telescope beam. Moreover, it turns out that in HII regions excited by stars of spectral type earlier than O7, the volume within which nearly all hydrogen is in the form of $H^+$ coincides with the volume within which all helium is in the form $He^+$. Consequently, for such nebulae the abundance by number of helium is directly given by the observed ratio of recombination line strengths. Independent estimates are provided by each measurable pair of corresponding H and He lines.

**Dispersion and Rotation Measures**    When plane polarized radiation of wavelength $\lambda$ propagates through a plasma in which there is a component $B_{\parallel}$ of magnetic field parallel to the direction of propagation, the radiation's plane of polarization slowly rotates. This phenomenon is called **Faraday rotation**. The angle of rotation $\psi$ is wavelength-dependent, being given by

$$\psi = R_M \lambda^2, \tag{8.34}$$

where

$$R_M = \frac{e^3}{2\pi m_e^2 c^4} \int ds\, n_e B_{\parallel}$$
$$= 2.64 \times 10^{-13} \int ds\, n_e B_{\parallel}\ \text{m}^{-2}. \tag{8.35}$$

The coefficient $R_M$ is called the **rotation measure** of the path.

Radiation from pulsars and all sorts of synchrotron sources is significantly polarized. When measurements of the direction of the electric vector $\mathbf{E}$ are available at several wavelengths, it is found to be wavelength-dependent. When the position angle $\phi$ of $\mathbf{E}$ is plotted against $\lambda^2$ an approximate straight line is obtained, and by equation (8.34) the slope of this line is $R_M$.

If the distance to the source is known, $R_M$ yields an estimate of the mean value of the product $n_e B_{\parallel}$ between the observer and the source. If the mean value of $n_e$ is known from some other measurement, for example from the timing of pulsars (see below), then estimates of $n_e$ and $B_{\parallel}$ can be obtained. For example, Rand & Kulkarni (1989) obtained the estimate $B \simeq 0.6\,\text{nT}$ for the local magnetic field strength cited above by measuring the rotation measures of 200 pulsars.

The refractive index of plasma differs from unity; at frequency $\nu$ it is

$$n(\nu) = \sqrt{1 - \nu_p^2/\nu^2}, \tag{8.36}$$

where $\nu_p$ is the **plasma frequency**

$$\nu_p \equiv \sqrt{\frac{n_e e^2}{4\pi^2\epsilon_0 m_e}} = 8.97\left(\frac{n_e}{\mathrm{m}^{-3}}\right)^{1/2} \mathrm{Hz}. \tag{8.37}$$

On account of the frequency dependence of $n$, the group velocity $c_g = \mathrm{d}\omega/\mathrm{d}k$ of radiation is frequency-dependent. One finds

$$c_g = \frac{c}{n}\left(1 + \frac{\mathrm{d}\ln n}{\mathrm{d}\ln\nu}\right)^{-1} = c\sqrt{1 - \nu_p^2/\nu^2}. \tag{8.38}$$

The time for a pulse of central frequency $\nu$ to arrive from a source at distance $D$ is

$$t(\nu) = \int_0^D \mathrm{d}s/c_g(\nu) = \frac{D}{c}\left[1 + \tfrac{1}{2}\overline{\nu_p^2/\nu^2} + \cdots\right]$$

$$\simeq \frac{D}{c}\left(1 + \frac{e^2}{8\pi^2\epsilon_0 m_e D}\frac{D_\mathrm{M}}{\nu^2}\right), \tag{8.39}$$

where $D_\mathrm{M} \equiv \int \mathrm{d}s\, n_e$ is the **dispersion measure** of the source and $\nu \gg \nu_p$ has been assumed. $D_\mathrm{M}$ can be determined for the line of sight to a pulsar by fitting to equation (8.39) the times of the pulses observed at different central frequencies. Numerically, if $D_\mathrm{M}$ is expressed in units of $\mathrm{m}^{-3}\,\mathrm{pc}$, the difference $\Delta t$ between the arrival times of a pulse at two frequencies is given in ms and the frequencies are given in MHz, we have

$$D_\mathrm{M} = 0.241\Delta t(\nu_1^{-2} - \nu_2^{-2})^{-1}. \tag{8.40}$$

### 8.1.5 $\gamma$-ray emission

As they speed through interstellar space, cosmic rays are liable to smash into other nuclei at relativistic speed. Such traffic accidents have two observable consequences: (i) the emission of $\gamma$-rays from the colliding particles; (ii) the **spallation**, that is fragmentation, of one of the nuclei. The first effect causes regions in which the ISM is densest to glow when observed by a $\gamma$-ray telescope. Since the intensity $Q_\gamma$ with which $\gamma$-rays are emitted must be proportional to the product $q_\gamma n$ of the local cosmic-ray flux $q_\gamma$ and the density $n$ of the ISM, $n$ can be estimated by measuring $Q_\gamma$ if one knows $q_\gamma$. It is frequently assumed that $q_\gamma$ varies only slowly from point to point within the ISM because fast-moving cosmic-ray particles diffuse fairly fast.

This idea is the basis for the calibration of equation (8.28) to be discussed in §9.6.1. Cosmic-ray induced spallation of nuclei increases the abundance of light nuclei in cosmic rays relative to the abundance of brittle heavy nuclei. By estimating what fraction of heavy nuclei have been broken up by collisions, one can determine how long a typical cosmic ray has been streaming through space.

Below 50 MeV $\gamma$-rays from the ISM are largely bremsstrahlung emission from encounters between nuclei and electrons with $\gamma$-factors in the range $1 - 200$. Photons which have gained energy by scattering off even more energetic electrons (**Inverse-Compton scattering**) also make a contribution to the observed $\gamma$-ray fluxes. The mutual annihilation of positrons and electrons gives rise to a prominent line at 0.511 MeV. Line-radiation from decaying nuclei contribute in the range $(1\,\mathrm{MeV} < E_\gamma < 50\,\mathrm{MeV})$. Above 50 MeV decaying pions make an important contribution to the observed fluxes.

### 8.1.6 Radiation by dust

In §3.7 we studied the extinction and reddening of starlight by dust. We saw that absorption is characterized by the extinction curve, which is a plot of the ratio $A_\nu/A_J$ of the extinction at arbitrary frequency $\nu$ to the extinction in some fiducial band, for example the $J$ band. Dust does not merely absorb light; it re-radiates much of the energy it absorbs. Here we discuss radiation by dust and what can be learned from it about the ISM.

The key to understanding emission by dust is Kirchhoff's law, which may be loosely phrased "good absorbers are good emitters." Quantitatively, Kirchhoff's law states that if a body of temperature $T$ absorbs a fraction $Q_\nu$ of the radiation of frequency $\nu$ that hits it, then the outgoing flux at its surface is $Q_\nu B_\nu(T)$, where $B_\nu(T)$ is the Planck function. We apply this law to a box containing a sample of the ISM. For simplicity we assume that the grains in the box are all at the same temperature $T$, and that they have a frequency-independent albedo $\gamma$ – see §3.7.1. Then the fraction of the radiation of frequency $\nu$ that enters the box on a given line of sight, which is absorbed in the box is [cf. equation (3.68)] $(1 - \gamma)(1 - 10^{-0.4A_\nu})$, where $A_\nu$ is the extinction along that line of sight. So by Kirchhoff's law the flux out of the box along the given line of sight is

$$(1 - \gamma)(1 - 10^{-0.4A_\nu})B_\nu(T). \tag{8.41}$$

It is useful to reformulate this physical argument in the language of radiative transfer introduced in §8.1.4. Suppose the grains have a number density $n(x)$, where $x$ measures position along a line of sight from the observer, and an average absorption cross-section per particle $K_a(\nu)$ at frequency $\nu$. Then, on traversing a slab of thickness $dx$ a beam of light with wavevector $\mathbf{k}$ and intensity $I_\mathbf{k}$ suffers a change in intensity

$$dI_\mathbf{k} = \left[ S_\mathbf{k} - n(x)K_a(\nu)I_\mathbf{k} \right] dx, \tag{8.42}$$

where $\nu \equiv k/2\pi c$ and $S_{\mathbf{k}} dx$ is the intensity added to the beam by scattering and by emission. Now the contribution from scattering is

$$S_s(\mathbf{k}) = -n(x)K_s(\nu)\Big[I_{\mathbf{k}} - \int d\mu\, I_{\mathbf{k}'}\rho(\mu)\Big], \qquad (8.43)$$

where $K_s$ is a coefficient that characterizes the reflectivity of the grains. The first term in the square brackets of equation (8.43) describes scattering out of the direction of $\mathbf{k}$ while the integral describes scattering into the direction of $\mathbf{k}$; the variable $\mu \equiv \mathbf{k}' \cdot \mathbf{k}/k^2$ is the cosine of the angle between the wavevectors $\mathbf{k}'$ and $\mathbf{k}$ of light before and after scattering. The function $\rho$ describes whether the scattering is isotropic ($\rho = $ constant), or predominantly forward, etc. Let $S_e(\mathbf{k})$ be the contribution to $S_{\mathbf{k}}$ in equation (8.42) from emission and define the optical depth $\tau$ by

$$d\tau_\nu \equiv n(x)K_a(\nu)\, dx. \qquad (8.44)$$

Then equation (8.42) may be written

$$\frac{dI_{\mathbf{k}}}{d\tau_\nu} + I_{\mathbf{k}} + \frac{K_s}{K_a}\Big[I_{\mathbf{k}} - \int d\mu\, I_{\mathbf{k}'}\rho(\mu)\Big] = \frac{S_e(\mathbf{k})}{K_a n(x)}. \qquad (8.45)$$

In the limit of large optical depth, we know that $I_{\mathbf{k}}$ tends to the Planck function $B_\nu(T)$. Moreover, the contents of the square bracket in (8.45) then vanish since as much radiation is scattered into the beam as is scattered out of it. Clearly if $I_{\mathbf{k}} = B_\nu(T)$ is to satisfy (8.45), we must have

$$S_e(\mathbf{k}) = n(x)K_a(\nu)B_\nu(T). \qquad (8.46)$$

To determine the relation between $\tau_\nu$ and $A_\nu$ we assume that $I_{\mathbf{k}}$ is non-zero only for $\mathbf{k}$ parallel to the direction from some star and integrate equation (8.45) in the limit $B_\nu \to 0$; this is the appropriate limit if the dust is too cold to emit significantly at frequency $\nu$, as will be the case if $\nu$ lies in the visual band and $T \sim 10\,\mathrm{K}$. Then we find

$$I_\nu(\tau_\nu) = I_\nu(0)e^{-(1+K_s/K_a)\tau_\nu} \qquad (8.47)$$

Hence

$$\begin{aligned} A_\nu &\equiv -2.5\log_{10}\big[I_\nu(\tau_\nu)/I_\nu(0)\big] = 2.5\log_{10} e^{(1+K_s/K_a)\tau_\nu}\\ &= 1.086(1 + K_s/K_a)\tau_\nu. \end{aligned} \qquad (8.48)$$

This shows that $A_\nu$ is proportional to $\tau_\nu$, which is itself [see equation (8.44)] proportional to the column density of grains along the line of sight. Moreover, the ratio $A_\nu/A_{\nu'}$ of the extinctions at two frequencies is just the ratio of the values taken by $(1 + K_s/K_a)$ at those frequencies. Consequently,

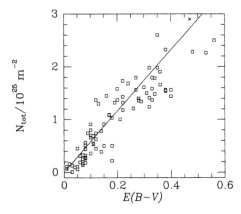

**Figure 8.14** The reddening $E(B - V)$ down various lines of sight is approximately proportional to the column density of hydrogen, $N(H_{tot})$, along that line of sight. The straight line is given by equation (8.49). [From data published in Bohlin *et al.* (1978)]

if the distribution of grain properties were universal, the extinction curve (Figure 3.17) would be the same down all lines of sight.

Figure 8.14 plots for various lines of sight the reddening $E(B - V)$ (see §3.7.1) against $N(H_{tot})$, the column density of hydrogen in all its forms. Although there is appreciable scatter of the points, the linear relationship

$$N(H_{tot}) = 5.8 \times 10^{25} E(B - V) \, \text{m}^{-2} \text{mag}^{-1}, \qquad (8.49)$$

provides a fair summary of the data. With equation (3.63), this implies that

$$N(H_{tot}) = 1.9 \times 10^{25} A_V \, \text{m}^{-2} \text{mag}^{-1}. \qquad (8.50)$$

Thus, since we have shown that $A_V$ is proportional to the column density of dust, to a first approximation, a given mass of gas contains a characteristic mass of dust, independent of whether the gas is in molecular or atomic form. Closer examination of data for individual interstellar clouds shows that fast-moving clouds tend to contain less dust than clouds that move more slowly relative to the surrounding material (Spitzer 1982). This difference is thought to arise because fast-moving clouds have been recently shocked, and dust grains tend to fall to pieces in the hot gas downstream from a shock wave.

Equations (8.46) and (8.48) establish an intimate connection between emission by dust and extinction. In fact, measurements of $A_\nu$, the albedo $\gamma_\nu$, and the emissivity enable one to predict the temperature of the emitting dust.

Several lines of argument indicate that there are significant numbers of grains comparable in size to the wavelength of blue light. For example, in §3.7.1 we mentioned that in the $B$ band the albedo is high $\sim 0.6$ and that there is evidence for forward rather than isotropic scattering. Grains much smaller than 500 nm would scatter little light, and do so isotropically.

Grains that are a fraction of a micron in size radiate fairly efficiently at temperatures much in excess of tens of Kelvin, with the result that they are expected to be cool $(T \lesssim 20\,\mathrm{K})$. Now the Planck function $B_\nu(T)$ peaks at the Wien frequency $\nu_W = 2.82kT/h$, which corresponds to wavelength $\lambda_W = 0.25(20\,\mathrm{K}/T)\,\mathrm{mm}$. So by equation (8.46) a grain at 20 K will radiate most strongly near 200 $\mu$m, i.e., in the sub-mm band.

Infrared astronomy was revolutionized by the launch in January 1983 of the Infrared Astronomy Satellite (IRAS). IRAS carried a 0.57 m diameter mirror, and thermal radiation from the optics was suppressed by cooling them to $\lesssim 10\,\mathrm{K}$ with liquid helium. IRAS surveyed almost the entire sky with an angular resolution of $\sim 0.5$ arcmin and measured fluxes in bands centred on 12 $\mu$m, 25 $\mu$m, 60 $\mu$m and 100 $\mu$m. The spectral region longward of 100 $\mu$m, which, as we have just seen, is where many grains are expected to radiate most strongly, was surveyed during 1989–1995 by the Diffuse Infrared Background Experiment (DIRBE) aboard the Cosmic Background Explorer satellite (COBE). Broad summaries of the findings of IRAS can be found in Soifer, Neugebauer & Houck (1987) and Beichman (1987), while Boulanger & Pérault (1987) and Boulanger et al. (1996) give details of results from IRAS and COBE that relate to dust in the Milky Way.

IRAS detected diffuse infrared radiation in all four of its wavebands. This demonstration that diffuse interstellar dust radiates significantly at wavelengths as short as 12 $\mu$m and 25 $\mu$m precipitated a radical restructuring of models of interstellar dust. It is now believed that the ISM contains extremely small grains, containing $\lesssim 100$ atoms, in addition to the 'classical' dust grains, 0.1 $\mu$m in size and containing $\gtrsim 10000$ atoms, that must be responsible for the high albedo and forward scattering of blue light discussed above. These small grains dominate the signal in the shorter IRAS wavebands for the following reason. When a UV photon is absorbed by a grain, its $\sim 10\,\mathrm{eV}$ of energy heats the grain. A grain with $N$ atoms has $\sim 3N$ vibrational degrees of freedom and therefore maximum specific heat $\sim 3Nk$.[18] Consequently, the temperature to which the absorption of a single UV photon would heat a grain that is initially at $T = 0$ is

$$T \simeq \frac{10\,\mathrm{eV}}{3Nk} = \frac{38\,670}{N}\,\mathrm{K}. \tag{8.51}$$

For example, if $N = 40$ a 10 eV photon will heat the grain from absolute zero to nearly 1000 K! A grain as warm as this will soon cool to a very much lower temperature, but in doing so it will radiate most of its 10 eV windfall at a few microns. This process of randomly heating dust grains to high temperatures for short periods is called **stochastic heating**.

Strong evidence for the importance of the stochastic heating of very small grains is provided by observations of reflection nebulae. Sellgren (1984) showed that the near-infrared spectra of three reflection nebulae, NGC 7023,

---

[18] Modes with frequencies $\nu$ so high that $h\nu > kT$ will contribute less than the classical $3k$ to the specific heat. We assume that $T$ is large enough for this phenomenon to be negligible.

NGC 2023 and NGC 2068, are independent of distance from the illuminating stars, and can be fitted by a model spectrum in which emission features are superposed on the Planck function for $T \simeq 1000\,\text{K}$. Moreover, the integrated $2.2\,\mu\text{m}$ and $3.8\,\mu\text{m}$ fluxes from the nebulae significantly exceed the fluxes at these wavelengths from the illuminating stars. This last observation proves that the observed fluxes are emitted rather than reflected by grains, and the large temperature characteristic of the spectrum proves that the grains are very small. The stochastic nature of grain heating follows from the failure of the spectrum of the emitted radiation to redden with increasing distance from the illuminating star: *any* body whose temperature is determined by a balance between thermal radiation from its surface and radiation received from a star distance $r$ away, will become colder as $r$ is increased. On the other hand, the mean temperature of radiation from a stochastically heated grain will be independent of $r$; removing the grain from the illuminating star merely decreases the frequency with which it will be hit by a photon and therefore the fraction of the time during which it will be effectively radiating.

A grain containing only $N \sim 50$ atoms is as much a large molecule as a grain. Léger & Puget (1984) have suggested that many of these objects may be **polycyclic aromatic hydrocarbons** (**PAHs**), which are fragments of graphite sheets onto which hydrogen and perhaps some other atoms have adhered here or there.

In view of the discoveries just described, the derivation of equation (8.46) above is over-simplified in that it assumes that all grains are at the same temperature. Observations discussed below show that at a given point in the ISM grains of different sizes have different temperatures. Fortunately, the derivation given can be generalized to include a range of grain sizes and temperatures simply by treating the ISM as a superposition of idealized media, in each of which all grains have the same size and temperature. Let $n^{(i)}$ and $k_\nu^{(i)}$ be the density of the $i^{\text{th}}$ such medium and the extinction coefficient of its grains. Then equation (8.46) generalizes to

$$S_e(\mathbf{k}) = \sum_i n^{(i)}(x) K_a^{(i)}(\nu) B_\nu(T_i), \qquad (8.52)$$

where, as always $\nu \equiv k/2\pi c$.

## 8.2 The ISM in Disk Galaxies

Observations in the Hα and 21-cm lines of HI and in the mm-lines of CO have yielded quite detailed maps of the distribution and kinematics of the cooler components of the ISM in over a hundred nearby galaxies – see Young & Scoville (1991) and Huchtmeier & Richter (1989) for references to the literature. Our knowledge of the hot components of the ISM of nearby galaxies is much more rudimentary, although useful data are available from radio-continuum observations, and, to a lesser degree, X-ray observations.

Some understanding of how surface densities and velocities of gas are extracted from line observations is necessary for the correct interpretation of the final data. The raw data are gathered by instruments of several types. Traditionally Hα measurements are made with a long-slit spectrograph but scanning Fabry–Perot interferometers are now increasingly used because they yield full two-dimensional information. The first radio-line observations were made with single dish instruments. In the early 1970s aperture-synthesis telescopes started taking 21-cm data, while the first aperture-synthesis observations in the mm band date from the late 1980s. Consequently, the fraction of the available data that has the poor angular resolution characteristic of a single-dish instrument is much larger in the case of mm lines than in the case of the 21-cm line.

With the Very Large Array (VLA), a giant aperture-synthesis telescope in New Mexico, it is possible to observe galaxies with a beam as small as ∼ 1 arcsec. However, for most galaxies an acceptable signal-to-noise ratio can be achieved even by the VLA at such high resolution only after inconveniently long integration times. Consequently, most aperture-synthesis observations have been made with beams of diameter ∼ 20 arcsec. Many of the best 21-cm data derive from the Westerbork telescope in the Netherlands. Hα measurements can be made at angular resolutions of ∼ 1 arcsec but suffer from the severe limitation that only a very small fraction of the gas in galaxies emits Hα radiation, and the Hα-emitting gas is usually sparsely distributed.

A significant weakness of every aperture-synthesis telescope should be noted: it is insensitive to large-scale smooth distributions of material, since these produce no fringes even on the telescope's smallest base-lines. Moreover, the sensitivity of the largest single-dish instruments is significantly greater than that obtainable with even the largest aperture-synthesis telescopes. Consequently, single-dish observations are still useful for determining how much cool gas a galaxy contains and how the gas is distributed over velocity.

Observations made with either a scanning Fabry–Perot interferometer or an aperture-synthesis telescope yield a **data cube**: the intensity of radiation received at each frequency from each position in the field of view. These data are often displayed as a series of channel maps; each channel map shows the intensity received from every point in the field in a certain narrow range of frequencies. By the Doppler effect, each such range of frequencies corresponds to a range of velocities. A typical velocity range might

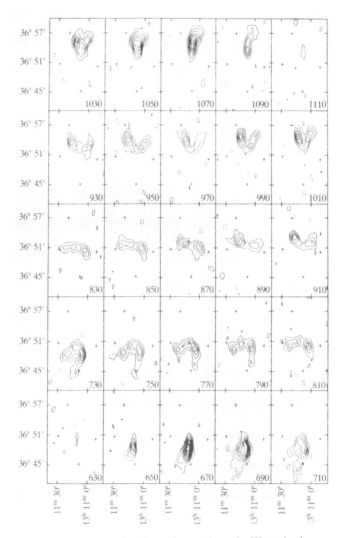

**Figure 8.15** Twenty-five channel maps from the Westerbork aperture synthesis telescope for the Sc galaxy NGC 5033. Each channel map gives the brightness temperature measured through a 129 kHz-wide filter centered on the frequency corresponding to the heliocentric velocity quoted in $\mathrm{km\,s^{-1}}$ at bottom right. The sequence of maps should be read from top right to bottom left. Notice the asymmetry of many of the profiles; this is evidence that the disk of NGC 5033 is warped. [After Bosma (1978)]

be $20\,\mathrm{km\,s^{-1}}$ wide, and $\sim 30$ channels might be employed to ensure that all gas in the galaxy, no matter what its heliocentric velocity, will register in at least one channel. Figure 8.15 shows 25 channel maps obtained in the 21-cm

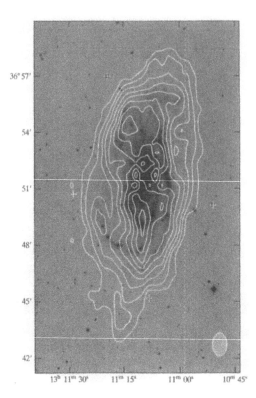

**Figure 8.16** Map of the HI column density in the Sc galaxy NGC 5033 superposed on a photograph of the system. Notice that the gas appears to cover a much larger area than the optical galaxy. [After Bosma (1978)]

line by the Westerbork aperture-synthesis telescope for the Sc galaxy NGC 5033. At the bottom right-hand corner of each map is given the heliocentric velocity corresponding to the middle of the channel associated with that particular map. Consider now how one might derive for this galaxy the HI surface density shown in Figure 8.16 and the contour map of the lines of constant velocity shown in Figure 8.17. Essentially identical procedures would be involved in the interpretation of channel maps obtained in any spectral line.

The first step is to isolate the line radiation from the continuum radiation. In this step, one examines each spectrum to decide which channels are line-free and then subtracts the average intensity of these continuum channels from each of the channels that contain line radiation. The next step involves assigning a typical velocity to each point in the galaxy by identifying the frequency at which the line radiation has peak intensity, or some other suitably chosen central frequency. Finally, one integrates over frequency the total line emission from each point to determine, via equation (8.22) for example, the column density at that point on the sky. A glance at Figure 8.18 shows that each of these steps is subject to an uncertainty that increases as the signal-to-noise ratio of the spectrum declines. When line centers and column densities have been assigned in this way to a grid of points on the sky, it is

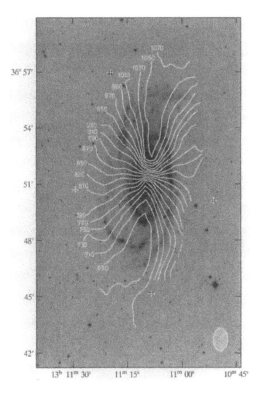

**Figure 8.17** Contours of constant HI velocity in NGC 5033. Notice the curvature of the kinematic principal axes. [After Bosma (1978)]

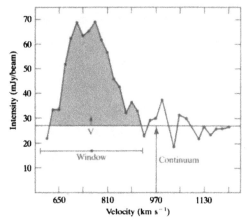

**Figure 8.18** A typical velocity profile constructed from the channel maps shown in Figure 8.15. One possible estimate of the line center is marked V, and an estimate of the area of the line profile associated with the emission is shaded. [After Bosma (1978)]

straightforward to construct HI density and velocity contour maps such as those shown for NGC 5033 and M81 in Figures 8.17 and 8.19.

Notice that a high-quality line profile contains much more information than the values of the total column density and central velocity just discussed. If the profile is approximately Gaussian, it is natural to measure

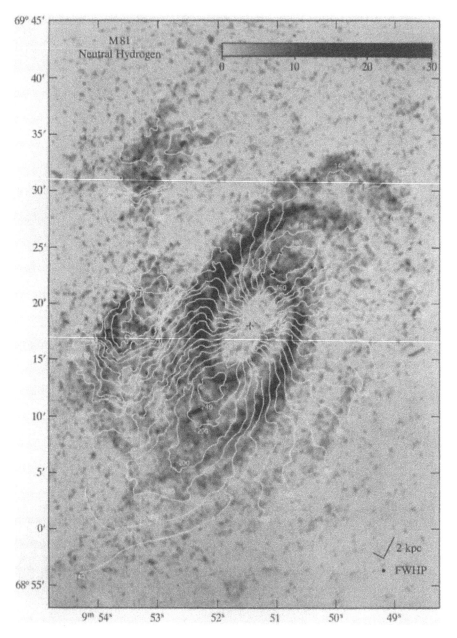

**Figure 8.19** Contours of constant HI velocity in M81 overlaid on a gray-scale representation of the surface density of HI. The scale at top gives column densities in units of $10^{24}$ atoms m$^{-2}$. Notice that in this high-resolution map, irregular motions associated with spiral arms are evident (cf. Figure 8.17). A central hole in the HI distribution is evident and the kinematic axes are curved. [After Rots (1975)]

its dispersion. If it is double-peaked, two characteristic velocities should be extracted, and so on. In general the number of parameters that should be extracted from a line profile increases with the signal-to-noise ratio of the data, and the complexity of the simplest galaxy model that fully exploits the extracted numbers will likewise increase with the quality of the data. (See §11.1.2 for a discussion of this type of problem.)

Data cubes typically lack the spatial resolution of photographs, so that interpreting line data can be likened to trying to make out the structure of a galaxy from a plate which is badly out of focus. The natural measure of the resolution of observations of some galaxy is the ratio $R_d/B$ of the scale-length $R_d$ of the disk [see equation (4.20)] to the telescope's half-power beam width $B$. Radial density and velocity profiles will not be significantly affected by beam-smearing provided $R_d/B \gtrsim 2$ (Bosma 1978). The sample of 16 nearby disk galaxies studied optically and in the 21-cm line by Wevers, van der Kruit & Allen (1986) will give an idea of the restrictiveness of this condition; Wevers *et al.*'s values of $R_d$ ranged from 25 arcsec to 140 arcsec, while their beams had 27 arcsec $< B <$ 60 arcsec. Thus a study such as that of Wevers *et al.* will not always yield accurate radial profiles. Moreover, any structure smaller than $R_d$, for example spiral arms, will be heavily smoothed by the beam in all but the nearest galaxies.

Low-resolution or noisy line data are best interpreted by a model-fitting procedure in which one attempts to optimize the fit between the observations and pseudo-data derived from a suitably parameterized model galaxy. The 'tilted-ring' model to be described in §8.2.4 below is frequently employed for this purpose.

### 8.2.1 Global measures

**HI and H$_2$ in disk galaxies**    Since the total masses $\mathcal{M}_{HI}$ and $\mathcal{M}_{H_2}$ of HI and H$_2$ in a galaxy can be derived from data of low spatial resolution, these masses have been determined for several hundred galaxies. The optical luminosities and values of $\mathcal{M}_{HI}$ and $\mathcal{M}_{H_2}$ of these galaxies all cover enormous ranges. Consequently, the absolute values of the gas masses are of less interest than their relative magnitudes and the ratio of $\mathcal{M}_{gas} \equiv \mathcal{M}_{HI} + \mathcal{M}_{H_2}$ to a dynamical estimate $\mathcal{M}_{dyn}$ of the galaxy's total mass.

Figures 8.20 and 8.21 show the distributions of these ratios as functions of Hubble type. Although the spreads in the ratios at a given Hubble type are large, very clear trends of the ratios with Hubble type are apparent.

Figure 8.20 shows that the ISM of a later-type spiral tends to account for more of the galaxy's total mass than does the ISM of an early-type spiral. In fact, the median value of $\mathcal{M}_{gas}/\mathcal{M}_{dyn}$ increases by a factor of 5 from Sa to Sc. Figure 8.21 indicates that the increase in fractional gas content as one goes to later types is mainly a reflection of an increase in $\mathcal{M}_{HI}$: from Sa to Sc the median value $\mathcal{M}_{HI}/\mathcal{M}_{H_2}$ increases by just over a factor 5. Hence while

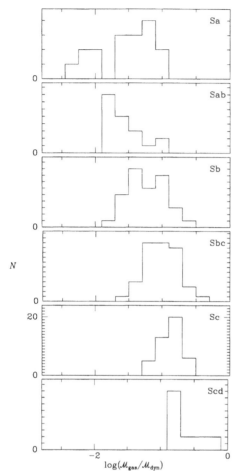

$N$

$\log(\mathcal{M}_{\rm gas}/\mathcal{M}_{\rm dyn})$

**Figure 8.20** Early-type spirals are less gas-rich than late-type spirals. Here we show for galaxies of various Hubble types the ratio of gas mass to dynamical mass. The latter is estimated by multiplying the luminosity by the typical mass-to-light ratio of the type from Rubin *et al.* (1985). [After Young & Scoville (1991) with additional data from the FCRAO Extragalactic CO Survey (Young *et al.* 1995) using data kindly provided by J. Young]

$\mathcal{M}_{\rm HI}/\mathcal{M}_{\rm dyn}$ increases quite strongly down the Hubble sequence, $\mathcal{M}_{\rm H_2}/\mathcal{M}_{\rm dyn}$ is approximately independent of Hubble type. This extremely interesting conclusion should be accompanied by a word of warning: $\mathcal{M}_{\rm H_2}$ has been obtained under the assumption that the CO-to-$\mathcal{M}_{\rm H_2}$ ratio is the same for all galaxies. Later-type galaxies tend to have lower luminosities and metallicities than earlier-type galaxies, and this might lead to the median CO-to-$\mathcal{M}_{\rm H_2}$ ratio decreasing down the Hubble sequence. Another source of anxiety is that observations of external galaxies yield systematically larger values of the ratio $T_A(^{12}{\rm CO})/T_A(^{13}{\rm CO})$ than do observations of molecular clouds in the Galaxy, upon which the standard CO-to-$\mathcal{M}_{\rm H_2}$ ratio, equation (8.28), rests. From §8.1.4 it will be recalled that larger values of $T_A(^{12}{\rm CO})/T_A(^{13}{\rm CO})$ imply either that the radiation is emitted by more optically thin material, or that beam dilution is relatively important for the $^{13}{\rm CO}$ measurements. It is natural that when a large telescope beam is turned on an external galaxy, a larger fraction of the CO radiation received derives from the optically

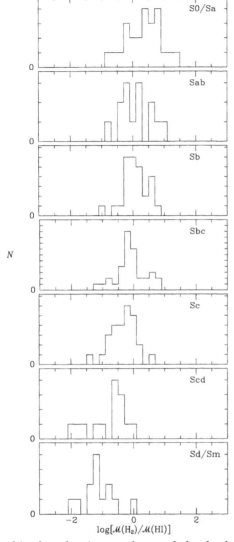

**Figure 8.21** The ISM of an early-type spiral tends to be predominantly molecular, while that of a late-type spiral tends to be atomic. Here we show for galaxies of various Hubble types the ratio of masses of interstellar gas in molecular and atomic form. [After Young & Scoville (1991) using data kindly provided by J. Young]

thin, low-density envelopes of clouds than is the case when the same beam is turned towards a molecular cloud complex in the Milky Way. One must worry, however, that any difference between the way in which molecular material is sampled in the Milky Way and in external galaxies should be reflected in a difference in the way $\mathcal{M}_{H_2}$ is derived from $T_A(CO)$.

Galaxies in the Virgo cluster have less HI, for any given luminosity, than non-cluster galaxies of the same Hubble type (Giovanelli & Haynes 1983, van Gorkom & Kotanyi 1985, Warmels 1986). On the other hand, their $H_2$ contents are approximately normal for their Hubble type (Kenney & Young 1989, Stark *et al.* 1986). We shall give a possible explanation of

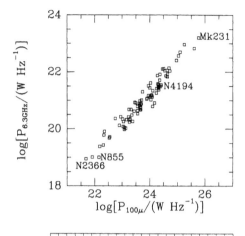

Figure 8.22 Radio continuum luminosity is tightly correlated with IR luminosity for a surprisingly large range of galaxies. Here $P_{6.3GHz}$, the luminosity at 6.3 GHz, is plotted against luminosity at 100 μm for 98 galaxies. [After Knapp (1990) using data kindly supplied by G. Knapp]

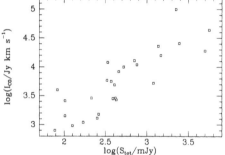

Figure 8.23 For a sample of 31 spiral galaxies the strength of line emission by $^{12}$CO is proportional to radio-continuum luminosity. $I_{CO}$ is the integral over all velocities of the flux in the 2.6-mm $^{12}$CO line. $S_{tot}$ is the flux at 1.5 GHz. [After Adler, Allen & Lo (1991) using data kindly supplied by D. Adler]

these phenomena in §8.4 below.

**Radio-continuum and IR luminosities**    Figure 8.22 shows a tight correlation between the radio-continuum and infrared luminosities of galaxies. The luminosities shown here are at 6.3 GHz and 100 μm, but a similar correlation could be shown for, say, the 1.4 GHz and 12 μm luminosities. The correlation shown in Figure 8.22 is the more striking because of the wide range of galaxies to which it applies: at the low end NGC 2366 is an SBb galaxy while NGC 855 is a dwarf elliptical; at the high end NGC 4194 is a Magellanic irregular and Mk 231 is a starburst galaxy with a strong UV continuum. The fact that these very different galaxies lie along the same correlation suggests that their radio continuum and IR luminosities are generated by a single process. That process is thought to be the formation of massive stars. UV photons from these stars are converted into IR radiation by dust grains, while the supernovae that mark the death throes of the most massive stars generate the cosmic rays that are responsible for the observed radio continuum radiation.

Figure 8.23 shows another interesting correlation between radio continuum luminosity and a measure of the cold component of the ISM by plotting the 1.5 GHz radio-continuum flux, $S_{tot}$, against the flux $I_{CO}$ of 2.6 mm ra-

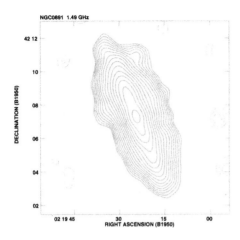

**Figure 8.24** NGC 891 at 1.49 GHz [After Condon (1987) courtesy of J. Condon]

diation by $^{12}$CO molecules. These two variables prove to be approximately proportional to one another: $I_{CO}/S_{tot} = 1.3 \pm 0.6\,km\,s^{-1}$.

The correlations of Figures 8.22 and 8.23 suggest a close physical connection between cool gas, star formation and supernovae. The unusually tight correlation of Figure 8.22 probably arises because both the plotted variables directly reflect the rate of formation of massive stars by measuring energy released by these stars: UV photons to heat the dust in the case of the IR luminosity and bulk kinetic energy to accelerate cosmic rays in the case of the radio luminosity. Is the correlation of Figure 8.23 of the same nature? That is, does increased CO luminosity reflect increased heating of molecular gas by young stars, perhaps through heating by cosmic rays? In this case, some galaxies might possess clouds that are rich in CO but inconspicuous in the spectral lines of CO because they are deficient in luminous stars to heat the CO (Allen 1992). Alternatively, the correlation of Figure 8.23 may merely indicate that galaxies with lots of cool gas form lots of stars and thus generate plenty of cosmic rays.

Although radio-continuum luminosity is tightly correlated with diagnostics of cold gas, it does not come from the thin disk to which cold gas is confined. Figure 8.24 illustrates this phenomenon by showing the radio-continuum surface brightness of the edge-on disk galaxy NGC 891: the optical image of the same galaxy that is shown in Figure 1.4 makes it clear that the emission of Figure 8.24 is not confined to the disk. The actual distribution of radio-continuum luminosity is naturally explained if cosmic rays diffuse more than a kiloparsec from their places of birth before they cease to radiate at a few GHz; this diffusion is most apparent in making the disk of synchrotron emission a few kiloparsecs thick, but it also causes the scale-length of the synchrotron disk to be larger than that of the parent distribution of luminous stars.

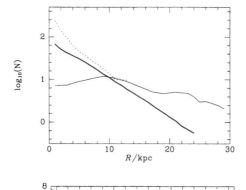

**Figure 8.25** The heavy line shows the $B$-band surface brightness of the Sc galaxy NGC 6946 expressed as a number of solar luminosities per $pc^2$. The dashed and light curves show, respectively, the surface densities of $H_2$ and HI in units of $10^{24}$ atoms per $m^2$. [From data published in Tacconi & Young (1986)]

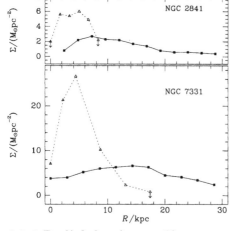

**Figure 8.26** Surface densities of $N(H_2)$ (triangles) and $N(HI)$ (squares) in the Sb galaxies NGC 2841 and NGC 7331. [After Young & Scoville (1983) using data kindly supplied by J. Young]

## 8.2.2 Radial density profiles

Figure 8.25 shows the surface densities $N(H_2)$ and $N(HI)$ as a function of radius in the Sc galaxy NGC 6946. Also shown is the galaxy's $B$-band optical surface-brightness profile. These data illustrate several general features of the distribution of gas in late-type galaxies. First, $H_2$ is very much more strongly concentrated towards the center of the galaxy than is HI. Second, the distribution of HI shows a slight depression at the center. Third the HI distribution is more extended than the distribution of star-light.

Figure 8.26 shows $H_2$ and HI profiles for two Sb galaxies, NGC 7331 and NGC 2841. Again $H_2$ is very much more centrally concentrated than HI. However in these two galaxies there are central depressions in both the $H_2$ and the HI distributions rather than in the latter only. Young & Scoville (1991) report that not quite half of all Sb galaxies have such central depressions in their $H_2$ distributions. The Andromeda galaxy, M31, provides the classic example of this phenomenon. Early-type disk galaxies generally have central depressions in their radial HI profiles. The few radial $H_2$ profiles that are available for early-type disk galaxies resemble the $H_2$ profiles of Sc galaxies such as that shown Figure 8.25. In particular, the $H_2$ profiles of early-type galaxies peak at the center and at values of $N(H_2)$ similar to the peak values $(10^{26} \, m^{-2})$ attained at the centers of Sc galaxies.

The great extent of the HI distributions of some galaxies has been of crucial importance for studies of dark matter since HI provides the perfect probe of a galaxy's circular speed and thus of its mass as a function of radius. With this application in mind, Broeils & van Woerden (1994) searched for objects with unusually extended HI distributions. Their data were obtained with the Westerbork telescope, but they used it in a quick-look mode pioneered by Warmels (1986), rather than as an aperture synthesis instrument. In this mode the Westerbork array can produce a radial surface-density profile in two hours rather than the twelve hours required for an aperture-synthesis map. The HI surface-density profiles obtained by Broeils & van Woerden (1994) show a wide variety of shapes, and are frequently quite different on the two sides of the galaxy. All galaxies in the sample have more than $1 \, \mathcal{M}_\odot \, pc^{-2}$ of HI at $R_{25}$ and NGC 6674, for example, reaches this level only at $R/R_{25} \simeq 1.8$.[19] In fact, the mean relation between the radius $R_{HI}$ at which $1 \, \mathcal{M}_\odot \, pc^{-2}$ of HI is reached and $R_{25}$ is[20]

$$\log_{10} R_{HI} = (0.9 \pm 0.1) \log_{10} R_{25} + (0.26 \pm 0.07). \tag{8.53}$$

Few galaxies in the Broeils & van Woerden sample show extended emission at surface densities below $10^{24} \, m^{-2}$. In fact, there is now considerable evidence that HI disks cut off very abruptly at a characteristic surface density $N_{HI} = 10^{23} \, m^{-2}$ (van Gorkom 1993). Maloney (1993) shows that the likely cause of this cutoff is a phenomenon predicted by Sunyaev (1969): HI exists only where it is shielded from the cosmic background flux of ionizing photons by a layer of nearly fully ionized material. For plausible parameters Maloney finds that this shielding layer has column density $\sim 3 \times 10^{23} \, m^{-2}$. Once the column density of a galactic disk approaches this value, the HI layer at its center becomes very thin, and altogether vanishes at lower column densities. If this model is right, HI disks are embedded in ionized gas that must glow in the H$\alpha$ line. Moreover, these H$\alpha$ disks will extend to considerably greater radii than the HI disks. We shall see in §9.3.3 that the disk of the Milky Way is indeed enveloped in H$\alpha$ emission.

We have seen (i) that late-type spirals are on the average more gas-rich than early-type spirals because they are richer in HI rather than in $H_2$, and (ii) that HI tends to be stored at large radii, often outside a galaxy's optical image. These facts suggest that a galaxy's Hubble type may be at least in part determined by its HI content and that the latter may be susceptible to environmental influences. This conjecture is supported by the morphology-density relation (§4.1.2) which states that late-type galaxies are

---

[19] See §2.3.6 for the definition of $R_{25}$.
[20] The values of $R_{25}$ for which equation (8.53) was established were corrected for inclination and internal extinction.

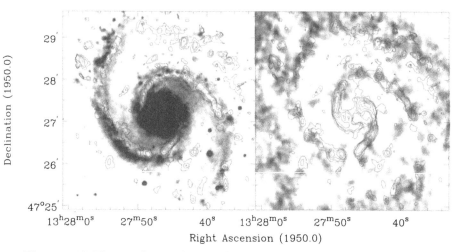

**Figure 8.27** The correlation of CO and HI emission with spiral arms in the tidally disturbed galaxy M51. The left panel shows contours of CO emission overlaid on a gray-scale representation of the red-continuum light. The right panel shows the same contours overlaid on a gray-scale representation of the 21-cm emission. [After Rand & Kulkarni (1990) courtesy of R. Rand]

comparatively rare in crowded regions. Moreover, we have seen that spiral galaxies in the Virgo cluster tend to be deficient in HI, while Szomoru, van Gorkom & Gregg (1996) find that galaxies in the Boötes void are unusually gas-rich. In §8.4 we shall discuss several processes that would tend to deprive cluster galaxies of outlying HI.

### 8.2.3 Azimuthal distributions

**Bars and oval distortions**   Gas (mostly $H_2$) near the centers of galaxies is frequently found to form a bar, even in systems whose optical luminosity does not appear to be barred. When an optical bar is present, the $H_2$ distribution is usually strongly peaked along the bar.

It has been speculated that the outer disks of galaxies might be mildly elliptical rather than axisymmetric (e.g., Binney 1995). The difficulty that tests of this conjecture have to overcome is resolving the ambiguity between an intrinsically elliptical disk and a circular disk seen at a slightly different inclination angle. We shall see below that this is best done by modeling the disk's velocity field.

**Spiral structure**   Spiral structure may be traced in both $H_2$ and HI distributions. Figure 8.27 illustrates this phenomenon in the case of the 'whirlpool' galaxy, M51. M51 has an unusually beautiful two-armed spiral structure because it is being tidally disturbed by a companion – several dynamical studies of this system can be found in Wielen (1990). It is likely that M51 displays the basic physics of spiral structure in exaggerated but nonetheless general form. The left panel of Figure 8.27 shows contours of CO emission superimposed on the distribution of red light, which should be a reliable tracer of the underlying mass distribution in the galaxy's disk. Careful inspection of this panel reveals that the CO emission peaks just on the concave side of the ridge-line of the spiral arms as delineated by red light. Dynamical studies indicate that gas flows through the arms from concave side to convex side. Hence the CO emission peaks just upstream of the arm's crest in the mass distribution.

The right panel of Figure 8.27 shows contours of CO emission superimposed on the HI distribution. The HI can be seen to peak slightly towards the arm's convex side relative to the peak in the CO emission. Thus the HI density appears to peak slightly downstream of the peak in CO. $H\alpha$ emission similarly peaks downstream of the peak in CO. The near coincidence of the $H\alpha$ and HI emission suggests that both arise through heating and dissociation of molecular gas by hot short-lived stars that form where the molecular density and CO emission are greatest.

It must be emphasized that Figure 8.27 considerably exaggerates the arm/inter-arm contrast in the CO emission; the data were gathered with an interferometer which was insensitive to smoothly distributed emission. It turns out that $\sim 75\%$ of the CO emission is smoothly distributed and only $\sim 25\%$ is concentrated into the arms.

The HI surface density distribution of M81 depicted in Figure 8.19 shows one of the most distinct and regular spiral patterns seen in 21-cm line radiation (Rots 1975, Rots & Shane 1975). Out to $r = 10$ kpc, the HI distribution is dominated by a clear two-arm spiral pattern, which is well approximated by a logarithmic spiral[21] with a pitch angle of about $15°$. The ratio of the peak HI density in the arms to the average HI density near them is about a factor of 2. The HI arms show an excellent correlation with optical arms; the HI ridge line tends to lie on the inner (concave) edge of the bluest part of the optical arm (which presumably is the region containing the highest density of young blue stars). A narrow dust lane, seen optically, lies along the inner edge of the eastern HI arm. The phase lag from the HI peak to the blue light peak is about $10°$ to $15°$. Using an estimated pattern speed (§4.4.7) for the spiral structure (chosen to place the corotation radius at $R_{CR} \simeq 10.5$ kpc), one deduces that the time required for an element of gas to pass from the maximum HI surface density to the highest density of young stars is $10^7$ yr. The ratio of hydrogen density to stellar mass density increases progressively along the arms; the HI arms can be traced far beyond the end of the optical arms. Outside $R \simeq 10$ kpc, two outer-arm features develop. The northern arm is crisply defined and can be followed easily over several kiloparsecs. The southern arm is fragmented, and it joins into the two large HI concentrations to the east of the galaxy.

---

[21] If $(R, \phi)$ are polar coordinates, then along a logarithmic spiral, $\ln(R) = \tan(\psi)\phi +$ constant, where $\psi$ is the pitch angle.

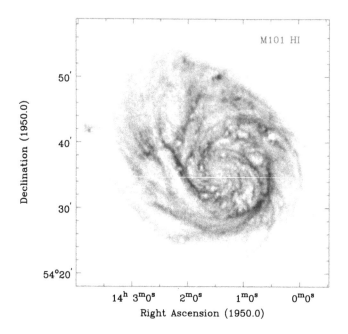

Declination (1950.0)

M101 HI

50'

40'

30'

54°20'

14^h 3^m0^s    2^m0^s    1^m0^s    0^m0^s

Right Ascension (1950.0)

**Figure 8.28** A map of M101 in the 21-cm line made with the Westerbork Synthesis Radio Telescope. [After Kamphuis, Sancisi & van der Hulst (1991) using data kindly provided by T. van der Hulst]

In addition to the spiral structure just discussed, Figure 8.19 shows that on a small scale, the HI distribution of M81 appears to break up into a large number of small blobs The patchiness of the HI distribution of Figure 8.19 is commonly observed in all types of late-type galaxy. Indeed, one finds that even in low-mass systems like the Magellanic Clouds (McGee & Milton 1966, Hindman 1967) and other low-luminosity systems (Tully *et al.* 1978), the neutral hydrogen is concentrated into complexes that each contain around $5 \times 10^6 \, \mathcal{M}_\odot$ of HI and are a few hundred parsecs in diameter. Probably the fact that low-mass systems can contain only a few of these complexes ($\lesssim 50$ for an Im galaxy), whereas giant galaxies like M81 contain many hundred such complexes, contributes to the irregular appearance characteristic of low-mass, late-type systems.

**Lop-sidedness** Gas that lies outside a galaxy's optical image frequently displays features that would not arise if the gas disk were axisymmetric and planar. The HI distribution of M101 shown in Figure 8.28 illustrates the phenomenon of **lop-sidedness**: the HI appears to extend much further towards the north-east than in the opposite direction. In this particular example, the contours of constant HI density are approximately circular at all radii, but at large radii they are not concentric with the galactic nucleus.

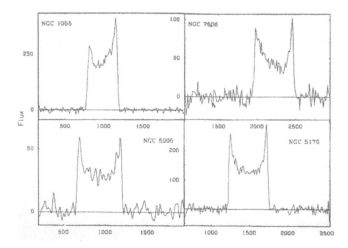

**Figure 8.29** Each panel shows the integrated 21-cm line profile of a disk galaxy. The fairly symmetrical profiles in the lower two panels are for galaxies that are not lop-sided, while the asymmetric profiles shown in the upper two panels are for significantly lop-sided galaxies. [After Richter & Sancisi (1994) courtesy of R. Sancisi]

This observation suggests that gas at large radii might be orbiting around a point offset from the nucleus. Actually, maps of the velocity field of the gas (see below) indicate that even at large radii, gas is moving on circular orbits around the nucleus. Thus the lop-sidedness of the HI distribution must arise because gas is non-uniformly distributed around each circular orbit, rather than because the orbits are strangely centered. Since the period of rotation of a gas cloud is roughly proportional to its galactocentric distance, in a few rotation periods ~ 1 Gyr, differential rotation should smear an initially lop-sided distribution of gas into an axisymmetric distribution. Hence it surprising that lop-sided distributions of HI are common. Galaxies such as the LMC of type Sdm show this phenomenon to a particularly marked extent (de Vaucouleurs & Freeman 1970) and in §9.2.6 we shall see that the Milky Way is itself probably lop-sided.

Richter & Sancisi (1994) pointed out that lop-sidedness can be detected from velocity profiles such as those shown in Figure 8.29: typically, lop-sidedness causes there to be more gas on one side of the kinematic minor axis than on the other, with the result that one horn of the integrated velocity profile is larger than the other. Richter & Sancisi (1994) exploited this fact to determine the frequency of occurrence of significant lop-sidedness in 1371 galaxies for which good-quality integrated velocity profiles were available. They found that at least 50% of the galaxies are lop-sided. There is no generally accepted explanation of why lop-sidedness is so common given the comparatively short lifetime of an initially lop-sided distribution mentioned above.

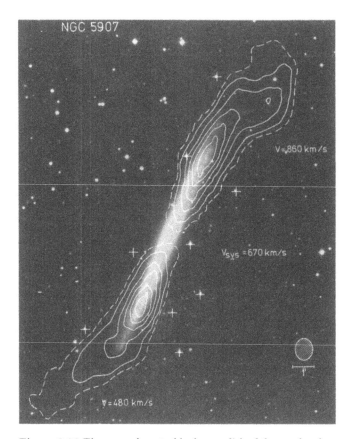

**Figure 8.30** The warped neutral hydrogen disk of the nearly edge-on galaxy NGC 5907. Emission by gas that is moving near the systemic velocity of the galaxy has been suppressed for clarity. NGC 5907 has no nearby neighbors that could have recently disturbed it tidally. [After Sancisi (1976) courtesy of R. Sancisi]

One point that may be significant in this context, is that the surface density of *all* gas, neutral and ionized, could be nearly axisymmetric even in lop-sided galaxies. Indeed, at the large radii and low column densities $\sim 10^{24} \, \mathrm{m}^{-2}$ of HI at which lop-sidedness is commonly observed, HI may constitute only a small fraction of the total column density, and large fluctuations in its density may reflect comparatively small fractional variations in either the total column density or the ionizing flux incident on the disk – see Problem 8.9.

In Figure 8.30 two channel-maps of the edge-on spiral galaxy NGC 5907 are superimposed on a photograph of the system. The velocities of the channel-maps shown are centered on $190 \, \mathrm{km \, s^{-1}}$ larger and $190 \, \mathrm{km \, s^{-1}}$

smaller than the systemic velocity of NGC 5907.[22] Channel-maps at velocities that differ by more than $190 \, \mathrm{km \, s^{-1}}$ from the systemic velocity show
little emission. Hence rotation velocity of gas in the disk of NGC 5907 must
lie near $190 \, \mathrm{km \, s^{-1}}$. But if the disk were planar and its material rotating
on circles, the material displayed in Figure 8.30 would be symmetrically dis-
posed along the galaxy's apparent major axis, rather than to the left of the
major axis in the south and to the right of it in the north. The obvious
explanation of this anomalous disposition of material in Figure 8.30 is that
the disk is not planar, but warped.

Only a small fraction of galaxies are viewed as nearly edge-on as NGC
5907 and any warp in a less edge-on galaxy would not be immediately apparent in the angular distribution of HI. In the next subsection we shall learn
how to diagnose warps from the observed velocity fields of galaxies, and see
that they are as common as lop-sided galaxies.

### 8.2.4 Velocity fields of disks

We have so far concentrated on the total *intensity* of radio-line radiation
coming from each point in a disk, regardless of frequency. We now turn our
attention to the central *frequency* of the received radiation. As usual, we
interpret the frequency of the radiation detected via the Doppler formula
as a velocity. In many cases a central velocity and a dispersion about this
velocity provide an adequate description of the observed line profile. Occasionally we shall find that the line profile requires a more sophisticated
velocity distribution (cf. §11.1.2).

Galactic disks are frequently modeled by a collection of concentric, circular rings – see Figure 8.36. Let $\mathbf{r}$ be a vector from the galactic nucleus to
an arbitrary point in the disk, and let $\hat{\mathbf{n}}(r)$ be the unit vector normal to the
ring of radius $r$. Then if $\Omega(r)$ is the angular velocity of this ring, the velocity
of material at position $\mathbf{r}$ is $\mathbf{v} = \Omega(r) \, \hat{\mathbf{n}} \times \mathbf{r}$ and the line-of-sight velocity is

$$v_{\mathrm{los}} \equiv \hat{\mathbf{R}} \cdot \mathbf{v} = \Omega(r)\mathbf{r} \cdot (\hat{\mathbf{R}} \times \hat{\mathbf{n}}), \qquad (8.54)$$

where $\hat{\mathbf{R}}$ is the unit vector from the observer towards the galaxy. The general
modeling procedure involves varying for each ring $\Omega$ and the polar angles
$(\theta, \phi)$ of $\hat{\mathbf{n}}$ so as to optimize the fit between the values of $v_{\mathrm{los}}$ predicted by
equation (8.54) and those extracted from the line profiles for a grid of points
on the sky.

It is instructive to consider the predictions of equation (8.54) in the case
of a flat disk. The normals $\hat{\mathbf{n}}$ of all rings then coincide, and we may simplify
equation (8.54) to

$$v_{\mathrm{los}} = \Omega(r) \sin i \, \mathbf{r} \cdot \hat{\mathbf{k}}, \qquad (8.55)$$

---

[22] See page 507 for the definition of systemic velocity.

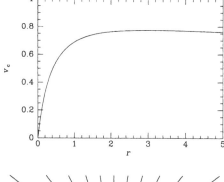

**Figure 8.31** A typical galactic circular-speed curve.

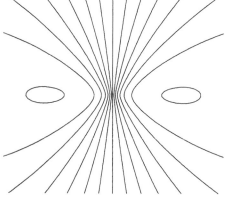

**Figure 8.32** The spider diagram generated by the circular-speed curve of Figure 8.31 when the system is viewed at inclination $i = 30°$ with the apparent major axis horizontal. The area contoured is a square 10 distance units on a side.

where $i$ is the galaxy's inclination and

$$\hat{\mathbf{k}} \equiv \hat{\mathbf{R}} \times \hat{\mathbf{n}}/\sin i \qquad (8.56)$$

is the unit vector perpendicular to both $\hat{\mathbf{R}}$ and $\hat{\mathbf{n}}$; $\hat{\mathbf{k}}$ runs parallel to the disk's the line of nodes, which is also the galaxy's apparent major axis.

Figure 8.31 shows a plot of the rotation velocity $v_c(r) \equiv r\Omega(r)$ of a typical galaxy. The central part of the galaxy, in which $\Omega(r) \simeq$ constant and $v_c \propto r$, is called the **region of solid-body rotation** because here the galaxy rotates almost as if it were rigid. Figure 8.32 shows contours of constant $v_{\text{los}}$ in the sky for the rotation curve of Figure 8.31; such a velocity-contour plot for a disk is called a **spider diagram**. Notice that in Figure 8.32, the velocity contours run straight and parallel to the apparent minor axis of the galaxy in the region of solid-body rotation. The decline in the rotation curve of Figure 8.31 beyond $r \simeq 3$, gives rise in the spider diagram to a closed contour that is elongated along the apparent major axis of the galaxy. If the rotation curve of Figure 8.31 had not peaked but had gone on rising as some rotation curves do, or if it had risen to some constant value, there would be no closed contours in the spider diagram. The **kinematic minor axis**

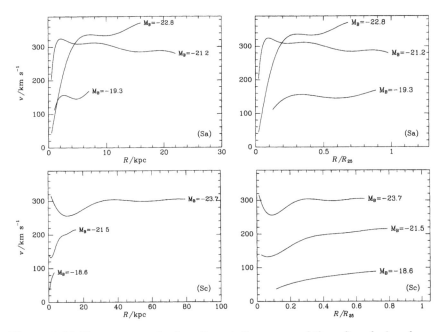

**Figure 8.33** The upper panels show the rotation curves of three Sa galaxies of very different luminosities from the sample of Rubin *et al.* (1985) plotted both on the same linear scale (left) and rescaled by their optical radii, $R_{25}$ (right). The lower panels show similar plots for three Sc galaxies from the sample of Burstein *et al.* (1982).

is the locus of points having the same velocity as the nucleus of the system (the **systemic velocity**), whereas the **kinematic major axis** is the curve which runs through nucleus and is everywhere perpendicular to the local constant-velocity contours. In Figure 8.32, the kinematic major and minor axes coincide with the apparent optical axes. In the case of a warped or elliptical disk, the two pairs of axes are distinct.

**Circular-speed curves**     The circular-speed curve $v_c(r)$ of a galaxy is of fundamental importance because from it one can deduce the mass $\mathcal{M}(r)$ interior to radius $r$ – see BT §2.6. Burstein *et al.* (1982) and Rubin *et al.* (1985) have extracted circular-speed curves from long-slit optical spectra of about 60 spiral galaxies of all types. Their circular-speed curves rarely extend to $R_{25}$,[23] but usually extend beyond $\frac{1}{2}R_{25}$. Figure 8.33 shows the curves of their most and least luminous Sa and Sc galaxies, and of randomly chosen Sa and Sc galaxies of intermediate luminosity. Each curve is plotted twice, once as a function of radius measured in kiloparsecs (left panels) and once as a function of radius scaled by $R_{25}$ (right panels). The only clear generalization that is suggested by Figure 8.33 is that the speed achieved by a rotation curve at large $R/R_{25}$ is positively correlated with luminosity – equation (7.26) puts

---

[23] See §2.3.6 for the definition of $R_{25}$.

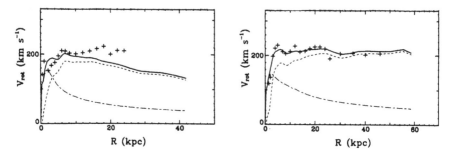

**Figure 8.34** In NGC 753 (left panel) the shape of the optically observed circular-speed curve (crosses) cannot be reproduced by assigning a mass-to-light ratio $\Upsilon$ to each of the disk and the bulge and the calculating the resulting circular speed (full curve). In the case of NGC 801 (right panel), by contrast, the two curves are in excellent agreement. The dashed and dot-dashed curves show the calculated contributions to $v_c$ from the disk and the bulge, respectively. [After Kent (1986)]

this important correlation onto a quantitative basis. Amongst the Sc rotation curves there is a suggestion that luminous Sc galaxies attain their maximum rotation speeds at smaller values of $R/R_{25}$ than do faint Sc galaxies, with the result that at large values of $R/R_{25}$ the rotation curve of a luminous Sc galaxy is more likely to be flat or falling than is the rotation curve of a faint Sc galaxy. This type of luminosity dependence in the form of $v_c(R/R_{25})$ is not evident in the Sa rotation curves of Figure 8.33.[24] Indeed, upper right-hand panel in Figure 8.33 shows that the general shapes of the rotation curves of the faintest and most luminous Sa galaxies in the Rubin *et al.* sample are rather similar.

The correlation between a galaxy's luminosity and the amplitude of its circular-speed curve that is so apparent from Figure 8.33, is quantified by the Tully–Fisher relation that was discussed in §7.3.4.

Kent (1986) has presented $r$-band CCD photometry for most of the galaxies in the Rubin *et al.* (1985) sample and has shown that the overwhelming majority of the observed circular-speed curves are in reasonable agreement with the circular-speed curves predicted by assigning plausible mass-to-light ratios $\Upsilon(d)$ and $\Upsilon(b)$ to the galaxies' disks and bulges. In many cases the last point of the observed curve is higher than theory predicts, and in two cases theory and observation are in significant disagreement. Figure 8.34 shows one of these cases, that of NGC 753, and a case in which the agreement is good. Kent finds that $\Upsilon_r(d)$ lies in the range $1.2\Upsilon_r(\odot)$ to $7.8\Upsilon_r(\odot)$ and in all but two cases $\Upsilon_r(b)$ lies in the range $1.2\Upsilon_r(\odot)$ to $9\Upsilon_r(\odot)$. The last point on the circular-speed curves usually lies inside $0.7R_{25} \simeq 2.3R_d$, where $R_d$ is the scale length defined by equation (4.20).

21-cm observations enable one to determine circular-speed curves outside the optically conspicuous parts of galaxies. At radii sufficiently large

---

[24] But see Rubin *et al.* (1985) for a contrary point of view regarding Sa rotation curves.

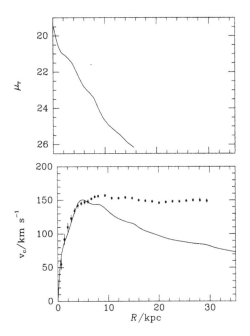

**Figure 8.35** The 21-cm circular-speed curve of the Sc galaxy NGC 3198 implies that most of the galaxy's mass lies beyond $R_{25}$. The upper panel shows the $r$-band surface brightness profile from Kent's CCD photometry. The curve in the lower panel shows the circular-speed curve derived from this and the observed HI mass under the assumption that $\Upsilon_r = 3.8\Upsilon_r(\odot)$. The dots in the lower panel show the circular-speed curve derived from the 21-cm velocity field. [After Begeman (1987) using data kindly provided by K. Begeman]

to contain all but a negligible fraction of the galaxy's luminosity, any model which assigns fixed mass-to-light ratios to disk and bulge will predict that the circular speed declines as $v_c \sim r^{-1/2}$. Actually, the observed circular-speed curves very rarely show a significant decline at the largest observed radii. Moreover, the mass required to keep the circular-speed curve flat or rising at large radii is in many cases comparable to or larger than the masses of the disk and the bulge that one infers from the structure of the circular-speed curve inside $R_{25}$.

The most carefully documented case of this phenomenon is that of NGC 3198 studied by van Albada *et al.* (1985). NGC 3198 is an Sc galaxy at a distance of $\sim 9.4\,\mathrm{Mpc}$. Its velocity field can be accurately fitted by a tilted-ring model in which the apparent major axes of the rings are aligned to within $1°$ and the inclinations of the rings vary systematically with radius by about $8°$. The circular-speed curves independently derived from the northern and southern halves of the galaxy agree to within $2\,\mathrm{km\,s^{-1}}$. Thus there can be little doubt that the observed HI is on circular orbits about the galactic center. The lower panel in Figure 8.35 shows the observed circular-speed curve (points) as well as that predicted by adopting a mass-to-light ratio $\Upsilon_r = 3.8\Upsilon_r(\odot)$ for the stellar component and taking into account the HI mass revealed by the 21-cm observations. The upper panel in Figure 8.35 gives the $r$-band brightness profile from Kent (1987). From Kent's CCD photometry the disk scale-length is $R_d = 55\,\mathrm{arcsec} = 2.5\,\mathrm{kpc}$ and last point on the measured circular-speed curve is at $11.2\,\mathrm{arcmin} = 30\,\mathrm{kpc}$, so the circular-speed curve extends to $12R_d$. Van Albada *et al.* show that there must be at least three times as much mass within $12R_d$ as within $R_{25}$.

Kent (1987) and Begeman (1987) showed that circular-speed curves that extend to more than $\sim 5R_d$ generally require significant masses of material to reside beyond $R_{25}$. Several authors have endeavored to map the distribution of this 'dark matter.' The procedure adopted is to assume that the galaxy is a superposition of a bulge, a disk and a dark halo. The luminosity distributions of bulge and disk are inferred from photometry, so their mass distributions are determined by their assumed mass-to-light ratios $\Upsilon(i)$. The dark halo emits no light and thus has a completely unknown spatial distribution. Most authors assume that it is spherical with density

$$\rho(r) = \frac{\rho_0}{1 + (r/a)^2}, \tag{8.57}$$

where $\rho_0$ and $a$ are parameters to be determined. Unfortunately, this functional form is physically unmotivated. N-body simulations of gravitational clustering suggest that dark halos have density profiles that can be approximated by the **NFW law** (Navarro, Frenk & White 1997)

$$\rho(r) = \frac{\mathcal{M}_0}{r(a + r)^2}, \tag{8.58}$$

where $\mathcal{M}_0$ and $a$ are free parameters. Consequently, several studies have employed this law. Unfortunately, whichever profile one adopts, one finds that most circular-speed curves can be adequately fitted by models that have wildly differing values of the free parameters, depending on what assumptions are made regarding the values of $\Upsilon(i)$ for disk and bulge.

In general there is no reason to assume that dark matter is distinct from luminous matter. All that we know for certain is that $\Upsilon$ increases sharply as the surface brightness of galaxies falls below a fraction of a percent of the brightness of the night sky. Whether this increase reflects a change in the mass-to-light ratio of stellar populations or the emergence of a tenuous distribution of utterly dark material cannot be ascertained by studying circular-speed curves and photometric profiles alone.

**Kinematic warps**   The observed spider diagrams of most galaxies show significant deviations from the classical form shown in Figure 8.32. So consider how a galaxy's spider diagram will be affected by a warp such as that evident in Figure 8.30. We suppose that this warp can be well represented by the tilted-ring model (right-hand panel of Figure 8.36). Each ring projects into an ellipse on the sky, and within this ellipse the velocity contours will be as in Figure 8.32. In particular, within each ellipse, the kinematic axes will be mutually perpendicular. But on account of the warp, the directions of the kinematic axes will vary from ellipse to ellipse. Hence, the vector $\hat{\mathbf{k}}$ defined by equation (8.56) becomes a function of $r$ through $\hat{\mathbf{n}}(r)$, and the kinematic major axis will twist so as to follow the changes in the direction of $\hat{\mathbf{k}}$. Figure 8.37 illustrates this phenomenon in the case of M83.

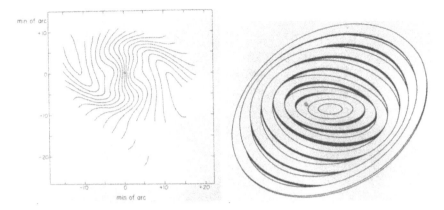

**Figure 8.36** A tilted ring model of M83 (right) and the spider diagram predicted by this model (left). [After Rogstad, Lockhart & Wright (1974)]

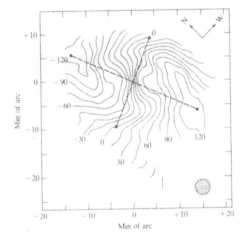

**Figure 8.37** The observed spider diagram of M83. [After Rogstad, Lockhart & Wright (1974)]

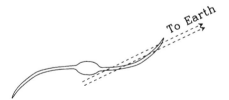

**Figure 8.38** When a strongly warped disk is viewed nearly edge-on, some lines of sight cut the warped disk twice, and the line-profile is liable to be multi-peaked.

If the amplitude of a warp is sufficiently large and the galaxy is sufficiently highly inclined, the line of sight may in places pass through the disk twice – see Figure 8.38. Along such lines of sight the line profile will normally have more than one peak, and it will be necessary to associate more than one velocity with a given line of sight. Both our close neighbors, M31 and

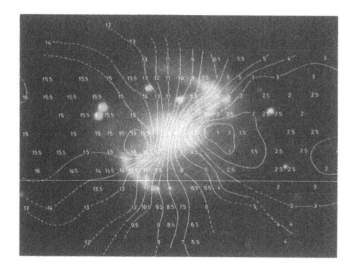

**Figure 8.39** The HI constant-velocity contours of NGC 5383 superimposed on an optical image of the galaxy. The spider diagram is approximately twofold symmetric, but the kinematic major and minor axes are by no means perpendicular. This is the signature of an elliptical disk. [After Sancisi, Allen & Sullivan (1979) courtesy of R. Sancisi]

M33, exhibit this phenomenon (Rogstad, Wright & Lockhart 1976, Brinks 1984).

Evidence of warps of the type just described led Bosma (1978) to conclude that 8 of a sample of 25 galaxies are warped. Briggs (1990) studied a sample of 12 galaxies with high-quality 21-cm data and concluded that warps tend to develop between $R_{25}$ and the Holmberg radius (§2.3.6). Our understanding of the dynamics of warps, which is still primitive, was reviewed by Binney (1992).

**Oval distortions** In observed spider diagrams the kinematic axes are sometimes clearly not mutually perpendicular – see Figure 8.39. This phenomenon is the signature of gas moving on elliptical rather than circular orbits. Franx, van Gorkom & de Zeeuw (1994) analyze the appearance of elliptical disks. The details of this analysis are unfortunately complex and model-dependent; complex because cumbersome rotation matrices are involved, and model-dependent because the relationship between a disk's shape and velocity field depends on the nature of the driving non-axisymmetric gravitational potential. By way of illustration, Figure 8.40 displays two spider diagrams of an elliptical disk that is seen nearly face on ($i = 1°$).[25] Gas in this disk moves on oval orbits, one of which is shown. Notice that in both panels the systemic-velocity contour is straight and cuts the oval where it

---

[25] Figure 9.35 shows the spider diagram of a highly inclined elliptical disk.

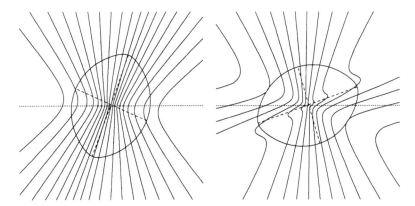

**Figure 8.40** The effect of an oval distortion on the spider diagram. In both panels the normal to the plane of the disk is inclined at 1° to the line of sight. In the left panel the major axis of the elliptical disk makes an angle of 20° with the projection of the line of sight into the disk plane, while this angle is 70° in the right panel. The dotted line shows the line of nodes while the ovals show the projections of a single orbit. The dashed lines are the projected principal axes of the orbit.

runs parallel to the line of nodes (shown dotted). Contours of equal projected surface density within the disk will not necessarily coincide with the projections of individual orbits, but they will have the same principal-axis directions as the orbits (dashed lines in Figure 8.40).

The key points about the velocity fields of elliptical disks that are illustrated by Figure 8.40 are: from most viewing angles (i) the kinematic axes are not perpendicular, although the photometric axes are for face-on orientation, (ii) the kinematic and photometric minor axes do not coincide, and (iii) the kinematic major axis lies close to the line of nodes, especially at high inclinations. It follows from point (ii) that there will usually be a velocity gradient along the photometric minor axis. Measurements of velocity gradients along photometric minor-axes have been used to constrain the ellipticities of disks; if the outer disks of many spiral galaxies are elliptical, their average ellipticity cannot exceed $\epsilon \simeq 0.05$ (Franx & de Zeeuw 1992).

### 8.2.5 S0 galaxies

21-cm observations of S0 galaxies confirm that these systems do indeed occupy the intermediate position between ellipticals and spirals. Wardle & Knapp (1986) studied statistically the results of several studies of HI in S0 galaxies with sensitive single-dish radio telescopes. Their sample consisted of 305 galaxies, of which 85 had detectable quantities of HI. This detection rate is about twice that encountered in studies of HI in ellipticals – see below. The detection rate of galaxies whose classification is ambiguous between S0 and Sa, designated S0/a, is roughly twice that for unambiguous S0 galaxies.

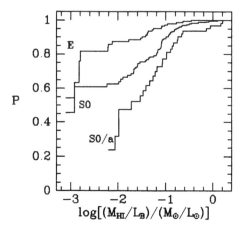

**Figure 8.41** The three curves give for each of the galaxy types E, S0, S0/a, the probability that a randomly chosen galaxy of the given type will have $M_{\rm HI}/L_B$ smaller than any given value. [After Wardle & Knapp (1986) from data kindly provided by M. Wardle]

Thus the detection rate increases steadily as one proceeds down the Hubble sequence.

The lower two curves in Figure 8.41 give the probability that a given galaxy of type S0 or S0/a will be found to have $M_{\rm HI}/L_B$ smaller than any given value. From this it can be seen that half of S0 galaxies have $M_{\rm HI}/L_B < 10^{-3}\,\mathcal{M}_\odot\,L_\odot^{-1}$, while a quarter have $M_{\rm HI}/L_B > 0.03\,\mathcal{M}_\odot\,L_\odot^{-1}$. The corresponding numbers for S0/a galaxies are that half have $M_{\rm HI}/L_B < 0.02\,\mathcal{M}_\odot\,L_\odot^{-1}$ and a quarter have $M_{\rm HI}/L_B > 0.08\,\mathcal{M}_\odot\,L_\odot^{-1}$.

van Driel & van Woerden (1991) and their collaborators have mapped the distribution of HI in 24 early-type disk galaxies. They found that in these galaxies HI is frequently concentrated into rings rather than forming a filled disk, as it usually does in unbarred spiral galaxies. The HI-velocity fields are consistent with the gas clouds that make up the rings being on circular orbits. The disks of SB0 galaxies are generally virtually free of HI inside a circle whose radius is of order three times the length of the bar's semi-major axis. In a few cases tails of HI suggest that gas is being accreted by a lenticular from a neighboring object.

The azimuthally-averaged surface density of HI in the gaseous rings is significantly smaller than in later-type disks: in S0 galaxies peak values are of order $1\,\mathcal{M}_\odot\,{\rm pc}^{-2}$, while in S0a galaxies values up to $3\,\mathcal{M}_\odot\,{\rm pc}^{-2}$ are achieved. This may be compared with the value $\sim 10\,\mathcal{M}_\odot\,{\rm pc}^{-2}$ which is characteristic of the peak in the HI profile of an Sc galaxy such that shown in Figure 8.25. van Driel & de Jong (1990) have confirmed that the dust responsible for the weak dust lanes observed in S0$_3$ galaxies gives rise to detectable IR emission. Sofue & Wakamatsu (1993) present evidence that dust in early-type galaxies is also associated with emission by CO, at roughly the same level as in spirals. We shall see below that some S0 galaxies contain significant quantities of very hot, X-ray emitting gas.

Van Driel & van Woerden divide HI rings in lenticulars into inner and outer rings depending on whether their radii are less than or greater than

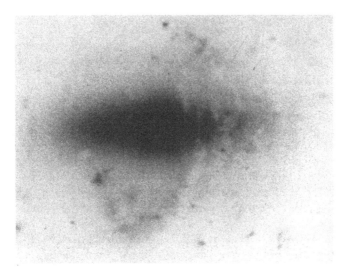

**Figure 8.42** The spindle galaxy, NGC 2685. [Courtesy A. Sandage]

$R_{25}$. Inner rings have typical radius $0.7R_{25}$, while outer rings have typical radius $1.8R_{25}$. About half of both inner and outer rings are significantly tilted with respect to the plane of the stellar disk; the outer rings are frequently tilted through large angles ($\gtrsim 30°$). Figure 8.42 shows an extreme example of this phenomenon, NGC 2685. Sandage (1961–1995) dubbed this the **spindle galaxy** because on photographic plates the object looks like a ring around a spindle. Optical and 21-cm spectroscopy subsequently showed that the 'spindle' is, in fact, an edge-on S0 galaxy whose rotation axis is almost perpendicular to the axis about which a gas-rich ring rotates. Galaxies like this, in which a ring runs nearly over the poles of the galaxy, are called **polar-ring galaxies**. Whitmore *et al.* (1990) have catalogued $\sim 70$ galaxies that are, or may well be, of this class. They conclude that $\sim 5\%$ of S0 galaxies either have, or have had, a polar ring.

The angular-momentum vector of a polar ring is perpendicular to the angular-momentum vector of the bulk of the stars. A remarkably large fraction of S0 galaxies display such misalignment of gaseous and stellar angular-momentum vectors in an even more extreme form: Bertola, Buson & Zeilinger (1992) find that in 9 of 33 S0 galaxies for which adequate data are available, stars and gas rotate in opposite senses, so that their angular-momentum vectors are (approximately) *anti-parallel*. Thus counterrotation of stars and gas is not uncommon in S0 galaxies. Since gas shed by a galaxy's stars will inevitably co-rotate with the stars, while accreted gas will sometimes co-rotate and sometime counter-rotate, the high incidence of counter-rotation in S0s implies that the gas in a significant fraction ($\gtrsim 40\%$) of S0 galaxies has an external origin. In §11.3.2 we shall see that a few galaxies contain two stellar disks, which counter-rotate with respect to each other.

Such objects presumably form through star formation in a counter-rotating gaseous disk.

### 8.2.6 Metallicities of disk galaxies

The metallicity of the ISM of a disk galaxy can be estimated from the strengths of emission lines in the spectra of HII regions. The idea underlying this approach is that the strengths of lines due to metals relative to those due to hydrogen, increase with metallicity. The most commonly measured lines are $H\beta$ and forbidden lines of $O^+$, $O^{2+}$, $N^+$, and $S^+$ at 372.7 nm, 500.7 nm, 658.3 nm, and 672.4 nm, respectively – see §8.1.3. The measured intensities of these lines have to be corrected for the effects of reddening and contamination of the nebular spectrum by starlight, which usually shows $H\beta$ in *absorption*.

Aller (1942) noted that in M33 the forbidden lines of $O^{2+}$ tend to be strong in the HII regions more distant than 20' from the nucleus, and they tend to be weak or absent in regions close to the nucleus. Searle (1971) discussed the strengths of emission lines from HII regions in six Sc galaxies, including M33, M81, and M101. He showed that different lines of sight through the same HII region produce similar spectra, enabling one to characterize an entire HII region with a single spectral type, and HII regions situated at similar distances from the nuclei tend to have similar spectra, but the spectra of regions located at different galactocentric radii are generally quite unlike one another. In other words, the spectrum coming from an HII region depends systematically on the region's galactocentric radius and on little else. The sense of this dependence is that, in the spectra of HII regions which are located far from the center, the lines emitted by $O^{2+}$ are strong, whereas the spectra of regions at smaller galactocentric distances have relatively strong lines of $O^+$ and $N^+$.

These trends are illustrated by Figure 8.43 from the work of Henry *et al.* (1992): the top two panels show that in the giant Sbc galaxy NGC 4303 the strengths of the lines of both $O^+$ and $N^+$ fade with increasing galactocentric radius $R$, but the lines of N fade fastest. The bottom panel of Figure 8.43 plots the radial variation of the quantity

$$R_{23} \equiv \frac{I(O^+) + I(O^{2+})}{I(H\beta)}, \tag{8.59}$$

where $I(O^+)$ is the intensity of the line doublet at 673 nm and $I(O^{2+})$ is the intensity of the doublet at 500 nm. Since most O atoms in the nebular gas will be either once or twice ionized and the intensities of both $H\beta$ and the O lines are proportional to electron density, $R_{23}$ is an indicator of the O abundance.[26] Hence the bottom panel of Figure 8.43 suggests that the

---

[26] By combining calibrations in the literature of $R_{23}$ versus [O/H], Zaritsky, Kennicutt & Huchra (1994) obtain $-[O/H] \simeq 2.73 + 0.33x + 0.202x^2 + 0.207x^3 + 0.333x^4$ where $x \equiv \log R_{23}$.

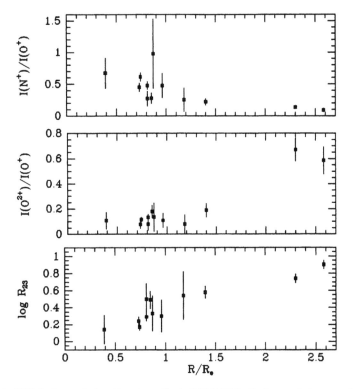

**Figure 8.43** Nebular emission-line strengths in NGC 4303. Each panel shows as functions of radius a ratio of line strengths. The top panel shows that lines of N become relatively weaker with increasing galactocentric radius $R$, while the middle panel shows that the ionization level of O increases with $R$. The bottom panel shows that the O–abundance indicator that is defined by equation (8.59) rises with $R$. [From data published in Henry *et al.* (1992)]

abundance of O falls from the center of NGC 4303 outwards. Similar results have been obtained for about three dozen disk galaxies – see Vila-Costas & Edmunds (1992) for references.

The radial variation of the spectral type of HII regions just described is evidence that the interstellar abundances of metals in a disk galaxy tend to fall with increasing radius. Unfortunately quantifying this decline is not easy. A full account of the relevant theory is beyond the scope of this book – the interested reader is referred to the book by Osterbrock (1989) – but the basic principles are as follows.

The temperature of interstellar gas depends on its metallicity because heavy-element ions and the dust grains emit most of the radiation that cools the ISM at temperatures below a million degrees. The temperature of the ISM in turn controls the strengths of the emitted spectral lines by regulating the frequency with which free electrons knock ions into the excited states from which the different spectral lines are emitted. The other factor that controls the observed nebular spectrum is the fraction of the atoms of a given element which are in each ionization state. This clearly depends on what fraction of the atoms is locked up in dust grains – see the discussion of depletion in §8.1.1. It also depends on the hardness of the stellar radiation field that ionizes the HII region because a harder radiation field will push atoms into higher states of ionization. The hardness of the radiation field is determined by (a) the nature of the nebula's exciting stars, and (b) the dust-to-gas ratio within the nebula. In particular, if the dust-to-gas ratio is high, the hardest exciting photons are likely to be absorbed by dust rather than by gas-phase atoms, and ionization levels will tend to be low. Hence low ionization levels are a sign of metal-richness. In particular, the rise with radius $R$ in the ionization level of O suggested by the middle panel of Figure 8.43 is consistent with the outwards decline in metal abundance that is implied by the fall with $R$ in the abundance parameter $R_{23}$ (bottom panel of Figure 8.43).

Given the complexity of the physical processes that govern nebular spectra, it is clear that extensive and high-quality observational data are required for the determination of accurate abundances. In particular, it is highly desirable to measure the relative intensities of a pair of lines such as those of $S^+$ at 407 nm, and 672 nm or those of $O^{2+}$ at 436.3 nm and 500.7 nm – as we saw in §8.1.3, from the ratio of the strengths of the members of such pairs, one can accurately estimate the electron temperature in the gas. Once the temperature is known, reliable values for the abundances of ions whose line intensities have been observed follow straightforwardly by equating the number of collisional excitations of each excited state to the observed number of radiative decays. Unfortunately, both these temperature-sensitive line pairs mentioned above involves a very weak line, whose intensity is often unmeasurably small, especially from metal-rich, and therefore cool, nebulae. Hence most analyses of nebular spectra determine the temperature from a model of the nebula's energy balance. The parameters of the model, which include the effective temperature of the exciting stars and the extinction by dust, are adjusted until the predicted spectrum agrees with the observed one.

In his seminal paper, Searle (1971) used model nebulae to conclude that his observations of normal Sc galaxies indicated that a modest increase in the O/H abundance ratio toward the galactic centers is accompanied by a substantial rise in the N/O abundance ratio; that is, he concluded that the nitrogen abundance relative to hydrogen rises more steeply toward galactic centers than does the oxygen abundance. Subsequent analyses, using either more refined model-fitting techniques or direct electron temperature measurements, have broadly confirmed Searle's results. For NGC 4303 Henry *et al.* (1992) estimate

$$[O/H] = -0.42R/R_e + 0.64,$$
$$[N/H] = -0.6R/R_e + 0.55. \tag{8.60}$$

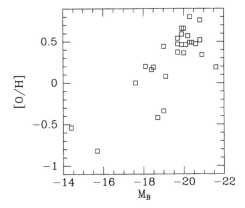

**Figure 8.44** The central O abundances of disk galaxies are correlated with the galaxies' absolute magnitudes $M_B$. [From data published in Vila-Costas & Edmunds (1992)]

In particular, the O abundance in NGC 4303 falls by roughly an order of magnitude over the measured range in $R$, while the N abundance falls by a factor of order of twenty in this range.

N is largely a secondary element, so the theory of §5.2.1 implies that its abundance should be proportional to the square of the abundance of a primary element such as O, and thus that $[N/H] \propto 2[O/H]$. By contrast, equation (8.60) implies that $[N/H] \propto 1.5[O/H]$. In §10.4 we shall see that abundances in stars near the Sun confirm that the N abundances rises significantly more slowly than as the square of the O abundance.

Vila-Costas & Edmunds (1992) summarize the literature of abundance gradients within galactic disks and explore the degree to which the available data can be understood in terms of models of chemical evolution such as those discussed in §5.2. One clear result is that barred galaxies show much smaller abundance gradients than do non-barred galaxies (Martin & Roy 1994). Figure 8.44 shows that another clear correlation is between the absolute magnitude of a galaxy and the metallicity that one obtains by extrapolating $[O/H]$ within its disk to the galactic center: low-luminosity galaxies tend to be metal-poorer than luminous galaxies (e.g., Miller & Hodge 1996). Since the asymptotic circular speed $v_c$ of a disk galaxy is strongly correlated with its absolute magnitude by the Tully–Fisher relation (7.26), it follows that $[O/H]$ must be correlated with $v_c$. Indeed, Zaritsky, Kennicutt & Huchra (1994) find that this correlation is tighter than that between $[O/H]$ and $M_B$.

In §5.3.1 we explored the predictions of the simplest model of chemical evolution within an annulus of a disk galaxy – the closed-box model. This model predicts that the metallicity is related to the fraction $f_{gas}$ of the local mass density that is in gaseous form by

$$Z = -p \ln f_{gas}, \tag{8.61}$$

where $p$ is the nucleosynthetic yield [cf. equation (5.37)]. We have seen above that in Sa and Sb galaxies, the gas fraction tends to be much lower near the center than at large radii. Hence (8.61) suggests that the metallicity of the interstellar gas in these galaxies should decrease from the center outward just as the observations suggest. However, Vila-Costas & Edmunds (1992) show that the closed-box model cannot explain the data quantitatively, although it predicts the observed trend qualitatively. Specifically, they estimated the **effective yield**

$$p_{\text{eff}} \equiv -\frac{Z}{\ln f_{\text{gas}}} \tag{8.62}$$

as a function of radius for each of the galaxies in their sample – if the closed-box model applies, $p_{\text{eff}}$ should be everywhere the same. The derived values of $p_{\text{eff}}$ are very uncertain because $f_{\text{gas}}$ is hard to estimate for two reasons. First, the masses of stellar disks are extremely ill determined because of uncertainty regarding the contribution of possible massive halos to observed circular-speed curves (§§8.2.4 and 10.6). Second, a significant fraction of the gas mass will typically be in molecular form and be traced through observations of emission by CO. The factor $X$ [equation (8.28)] that connects the intensity of this emission to gas mass almost certainly varies with metallicity, and this variation is not well understood.

Vila Costas & Edmunds found that even allowing for the uncertainties just mentioned, $p_{\text{eff}}$ clearly varies systematically both within a single galaxy and from galaxy to galaxy in the sense that it tends to be larger at small radii and at high metallicity. Specifically, dwarf galaxies such as the SMC and LMC, which are metal poor [$Z(\text{LMC}) \simeq 0.4Z_{\odot}$, $Z(\text{SMC}) \simeq 0.1Z_{\odot}$], imply $p_{\text{eff}} \sim 0.005$, while the inner disks of luminous galaxies require $p_{\text{eff}} \sim 0.012$.

It is possible that the yield $p$ varies with metallicity $Z$, for example because the IMF (§5.1.9) depends upon $Z$. However, in Chapter 10 we shall see that the observed metallicity distribution of stars near the Sun does not favor this possibility. A more plausible explanation for the inward increase of $p_{\text{eff}}$ within galaxies is that gas drifts inwards through galactic disks and at large radii is replaced by the accretion of metal-poor gas from intergalactic space. By carrying metals inwards, this drift of gas through the disk increases $p_{\text{eff}}$ at small radii and reduces it at large radii. In dwarf galaxies the inward drift of cold material within the disk may be counteracted by a galactic wind or fountain which carries metal-rich material high above the disk and then out to large radii.

### 8.2.7 Magnetic fields

The magnetic-field structures of a few nearby disk galaxies have been probed in several ways. The classical approach involves determining the polarization of starlight that has suffered extinction by dust – in §3.7.1 we saw that the Davis–Greenstein effect causes dust grains to align perpendicular to a magnetic field **B**, and that this causes the electric vector of radiation from

**Figure 8.45** Contours of constant $j_{\rm sync}$ and the direction of the magnetic field in M51 superimposed on an optical image of the galaxy. The tendency of **B** to run parallel to spiral arms is evident. [After Neininger & Herellou (1996) courtesy of N. Neininger]

obscured sources to be aligned with **B**. This approach suffers from the drawback that radiation *reflected* by dust is polarized even in the absence of a field **B**. Fortunately, the Davis–Greenstein effect causes the far-IR emission from dust to be polarized perpendicular to **B** and reflected starlight gives rise to no complications at far-IR wavelengths.[27]

At radio frequencies magnetic fields have been probed by measuring the polarization of synchrotron emission from galactic disks and/or measuring the rotation measures of radio sources that lie behind the disks. As we saw in §8.1.4, synchrotron emission is polarized perpendicular to the direction of the local magnetic field, so from measurements of synchrotron radiation one may readily determine the direction of the projection of **B** onto the plane of the sky. Moreover, Figure 3.19 indicates that in the Milky Way the mean direction of **B** lies within the disk. If we assume that the same is true in external galaxies, we can infer the three-dimensional direction of **B** from synchrotron observations.

Figure 8.45 shows contours of constant $j_{\rm sync}$ and the direction of the magnetic field in M51 superimposed on an optical image of the galaxy. There is a clear tendency of **B** to run parallel to the spiral arms. Studies of other spiral galaxies show that this is the normal state of affairs (Beck *et al.* 1996). In the Andromeda galaxy, M31, which differs from M51 in lacking grand-design spiral structure, **B** aligns with many small dust lanes and cloud complexes, and at $R \simeq 20\,{\rm kpc}$ the galaxy is surrounded by a torus of field lines (e.g., Ruzmaikin *et al.* 1990).

It is much harder to determine the magnitude of **B** than its direction. A widely used method exploits the fact that the intensity $j_{\rm sync}$ of synchrotron emission is proportional to the product $\mathcal{E}B^2$ of the energy density $\mathcal{E}$ of synchrotron-emitting electrons and the square of the strength the magnetic

---

[27] Hough (1996) reviews determinations of **B** from the polarizing effects of dust.

field, including both ordered and random components. $\mathcal{E}$ is estimated by assuming that the energy density of all cosmic-ray particles (most of which are of too low energy to contribute to the measured synchrotron radiation) is equal to the energy density $\frac{1}{2}B^2/\mu_0$ of the magnetic field. This hypothesis of **equipartition of energy** between magnetic field and cosmic-ray particles is not self-evidently valid, but the derived field strength $B$ is proportional to only the fourth root of the measured quantity $j_{\text{sync}}$, so significant errors in the input parameters lead to relatively small errors in the recovered value of $B$. For a sample of 146 late-type galaxies Fitt & Alexander (1993) find that $B \simeq (1 \pm 0.4)\,\text{nT}$. In spiral arms **B** tends to be larger in magnitude but more disordered in direction than between spiral arms. In §9.3.2 we shall see these results entirely consistent with our understanding of the Galactic magnetic field.

Measurements of the rotation measures, $R_{\text{M}}$, of background sources provide a valuable check on the method of estimating $B$ described above: from $R_{\text{M}}$ one can determine the integral of $n_e B_{\parallel}$ through the disk [equation(8.35)], where $B_{\parallel}$ is the component of **B** parallel to the line of sight. Since $R_{\text{M}}$ is proportional to $B_{\parallel}$ rather than proportional to $B^2$ as $j_{\text{sync}}$ is, contributions to $R_{\text{M}}$ from different points along a line of sight will have a tendency to cancel unless **B** is highly ordered. Hence estimates of $B$ that are derived from measurements of $R_{\text{M}}$ will be lower than those derived from $j_{\text{sync}}$ unless fields are highly ordered. If, by analogy with observations of the Milky Way [equation (9.21)], one adopts $n_e \simeq 3 \times 10^4\,\text{m}^{-3}$, then measurements of $R_{\text{M}}$ imply values of $B$ that are smaller than the values of $B$ that are inferred from $j_{\text{sync}}$, just as we expect if **B** wiggles to the degree that Figure 3.19 suggests. This satisfyingly consistent result supports the hypothesis of field–particle equipartition from which it is has been derived.

It is interesting to compare the sign of $B_{\parallel}$ at points that lie at equal galactocentric distances on opposite sides of a galaxy's major axis. In most galaxies these signs are opposite, as would be the case if field lines simply encircled the galaxy. In a few cases the signs are found to be the same. This result may indicate a fundamentally more complex field geometry, or it may arise because (i) field lines move in and out as they run round a galaxy, and (ii) the direction in which **B** encircles the galaxy is not the same at all radii – in §9.3.2 we shall find that this is the case in the Milky Way. If these two conditions are fulfilled, the sign of $B_{\parallel}$ will sometimes be the same on opposite sides of the galaxy even though individual field lines run around the disk in a fixed sense.

### 8.2.8 Star formation in disk galaxies

We saw in §8.1.3 that the H$\alpha$ luminosity of a galaxy provides a direct measure of the number of ionizing photons that impact on its ISM. At distances in excess of some tens of parsecs from the galactic nucleus the vast majority

**Table 8.2**  Median values of the birthrate parameter for galactic disks

| Type | Sa | Sab | Sb | Sbc | Sc | Scd/Sd | Sm/Im |
|------|-----|------|------|------|------|--------|-------|
| $b$  | < 0.07 | 0.17 | 0.33 | 0.84 | 0.99 | 0.69 | 1.67 |

SOURCE: From data published in Kennicutt, Tamblyn & Congdon (1994)

of ionizing photons come from massive, young stars. If we assume some distribution of these stars over spectral type, we can determine the total mass of gas in these stars from the rate at which they produce ionizing photons. If we assume, moreover, that the rate of formation of massive stars has been approximately constant over the relatively short lifetimes of these objects ($\lesssim 10\,\mathrm{Myr}$ – see Table 5.2), we can infer the rate at which massive stars are forming from the numbers currently present. Thus, observations of H$\alpha$ luminosity of a galactic disk lead rather directly to the rate at which massive stars are forming within it. In §5.4.2 we defined the stellar birthrate parameter $b$ as the ratio of the present star-formation rate to its value averaged over the galactic lifetime, and saw that in models $b$ is tightly correlated with both the equivalent width EW(H$\alpha$) of the H$\alpha$ line and $B - V$ color (Figure 5.22).

Kennicutt, Tamblyn & Congdon (1994) estimated $b$ for the disks of 210 galaxies from these relations. They found that in Sa and Sab disks, $b$ is always small ($b \lesssim 0.4$). The spread of $b$ values at a given Hubble type is generally large, but there is a clear tendency for larger values of $b$ to occur at later Hubble types. This trend can be clearly seen in Table 8.2, which gives median values of $b$ for each Hubble type. The large scatter in measured $b$ values presumably arises because in any given disk, the star-formation rate goes up and down in an irregular way. The median values listed in Table 8.2 suggest that, averaged over a sufficient period, the disks of galaxies later than Sb have been forming stars at a fairly constant rate throughout their lifetimes. By contrast, star formation has effectively ceased in the disks of Sa and Sab galaxies.

From the position of their galaxies in the $[B - V, \mathrm{EW}(\mathrm{H}\alpha)]$ plane, Kennicutt, Tamblyn & Congdon (1994) argued that the IMF in these systems has to be about as rich in massive stars as the Salpeter IMF predicts. In particular, they suggest that if star formation followed the Scalo IMF, EW(H$\alpha$) would be too small at a given $B - V$ color. Unfortunately, this conclusion is sensitive to assumptions that must be made about the history of star formation in disks.

If one knows both the IMF and the rate at which massive stars form, one can predict the rate at which gas is being locked up in long-lived stars, and compare this to a disk's present stock of interstellar gas. When proper account is taken of the rate at which dying stars return gas to the ISM, Kennicutt, Tamblyn & Congdon (1994) find that spirals can sustain their present star-formation rates for $5 - 15\,\mathrm{Gyr}$ without accreting additional material.

The same techniques that are used to determine the global star-formation rates in galaxies of different Hubble types can be used to estimate the ra-

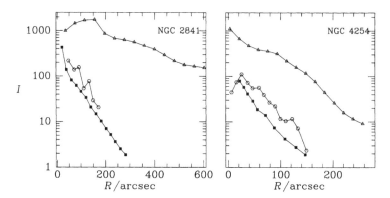

**Figure 8.46** The total gas density (triangles), the blue continuum surface brightness (squares) and the Hα intensity (hexagons) in NGC 4254 and NGC 2841. The zero-points of the vertical scales are arbitrary. [From data published in Kennicutt (1989)]

dial distribution of star formation within a given galaxy. Figure 8.46 shows the results of such an investigation for an Sb galaxy, NGC 2841 (cf. Figure 8.26), and an Sc galaxy, NGC 4254. Outside the core of each galaxy, the surface density of Hα emission (open hexagons) is roughly proportional to the surface-brightness in blue continuum light (filled squares), just as we would expect if the number of stars forming at each radius is proportional to the number of stars that have already formed there. In both galaxies this proportionality breaks down at small radii: here the star-formation rate is currently peculiarly small.

Since stars form through the gravitational instability of the ISM, Schmidt (1959) early speculated that the star-formation rate is proportional to a power of the surface density of the ISM. Where the combined surface density of atomic and molecular hydrogen is sufficiently high, Kennicutt (1989) finds that the Hα intensities of every galaxy in a sample of seven can be fitted by the **Schmidt law**

$$I(H\alpha) \propto \Sigma_{\text{gas}}^{1.3}. \tag{8.63}$$

The Hα intensity invariably falls below the value predicted by this law at large radii and the largest radius at which the law holds coincides, to within the substantial uncertainties, with the radius at which the gas density falls below $\sim 0.63$ times the critical value

$$\Sigma_{\text{crit}} \equiv \frac{\kappa v_{\text{s}}}{\pi G}, \tag{8.64}$$

where $\kappa$ is the epicycle frequency (§3.2.3 of BT) and $v_{\text{s}} \sim 6\,\text{km s}^{-1}$ is the sound speed in the gas. In BT §6.2.3 it is demonstrated that a gas disk is unstable to axisymmetric disturbances only when its surface density exceeds $\Sigma_{\text{crit}}$. Thus, the observations suggest that the ISM is unstable even at slightly

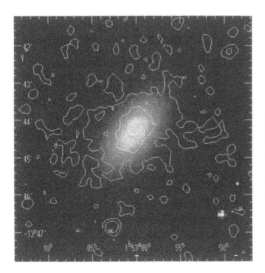

**Figure 8.47** Contours of constant X-ray surface brightness in NGC 720 overlaid on the optical image of the galaxy from the Digital Sky Survey. Adjacent contours are separated by a factor of 4 in X-ray surface brightness. [The X-ray contours are from the analysis of ROSAT data by Buote & Canizares (1996) courtesy of D. Buote]

lower surface densities, presumably because the gas represents only part of the mass of the disk.[28]

Although the gas density rises above $0.63\Sigma_{\rm crit}$ in the star-forming regions of disks, it does not do so by a big factor, even when $\Sigma_{\rm gas}$ varies by more than an order-of-magnitude. This finding suggests that if $\Sigma_{\rm gas}$ ever significantly exceeds $\Sigma_{\rm crit}$, star formation depletes the ISM rapidly until $\Sigma_{\rm gas} \sim \Sigma_{\rm crit}$.

In a few galaxies, especially early-type ones, $\Sigma_{\rm crit}$ rises more rapidly near the center than does $\Sigma_{\rm gas}$, with the result that $\Sigma_{\rm gas}$ falls below the threshold for star formation at both small and large radii. In these cases there is usually a central depression in the H$\alpha$ profile, just as theory predicts.

# 8.3 The ISM in elliptical galaxies

### 8.3.1 X-ray emitting plasma

Whereas the bulk of the ISM in spiral galaxies is in the form of HI and $H_2$, the ISM in ellipticals consists primarily of hot ($T \gtrsim 10^6$ K) plasma. This plasma produces X-rays by bremsstrahlung, bound–free, and line emission processes. Figure 8.47 is an X-ray image of NGC 720 obtained with the ROSAT satellite. From the X-ray spectrum of this object one infers that the temperature of the plasma, $T \simeq 7 \times 10^6$ K (Buote & Canizares 1997). This temperature is rather higher than the characteristic kinetic temperature that one associates with a population of stars with line-of-sight velocity dispersion

---

[28] For a quantitative discussion of how a stellar disk affects the stability of an embedded gaseous one, see Jog & Solomon (1984).

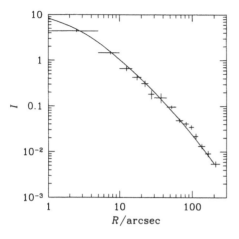

**Figure 8.48** X-ray and optical brightness profiles of NGC 720. The X-ray data are shown as crosses, which indicate the radial averaging and uncertainty in the data. These data came from ROSAT HRI and PSPC observations which have been normalized to agree at $R = 30$ arcsec. The optical profile is a combined fit to the HST data from Lauer *et al.* (1995) and the ground-based photometry of Peletier *et al.* (1990). The profiles have been normalized to unity at $R = 10$ arcsec. [From data published in Buote & Canizares (1996)]

$\sigma$, $T_* = 2.9 \times 10^6 (\sigma/200 \,\mathrm{km\,s^{-1}})^2$ K, since for NGC 720 $\sigma \simeq 200 \,\mathrm{km\,s^{-1}}$ (Fried & Illingworth 1994). On account of the high interstellar pressure to which this high temperature gives rise, the plasma is approximately spherically distributed in the galaxy's potential well, rather than organized into a thin, centrifugally supported disk as in a spiral galaxy.

Figure 8.48 shows that in NGC 720 the X-ray surface brightness is almost exactly proportional to the optical surface brightness. This coincidence is characteristic of elliptical galaxies. At first glance it is rather surprising, since the optical emissivity of a galaxy, $j_{\mathrm{opt}}$, is directly proportional to the stellar density, $n_*$, whereas its X-ray emissivity, $j_X$, is proportional to the square of the plasma density, $n_e$, (§8.1.2). Thus, if $j_X \propto j_{\mathrm{opt}}$, then $n_e \propto n_*^{1/2}$; the gas density falls off with radius more slowly than the stellar density in such a way that it conspires to make the brightness profiles similar. This difference occurs because the plasma temperature is higher than the kinetic temperature of the stellar distribution. In fact, it is straightforward to show that if $n_* \propto r^{-\alpha}$ then $n_e \propto r^{-\alpha\beta}$, where $\beta \equiv T_*/T$ is the ratio of the stellar and plasma temperatures – see Problem 8.11. Thus, since generically $j_X \propto j_{\mathrm{opt}}$ in elliptical galaxies, we know that $\beta \sim 0.5$ in these systems.

Figure 8.49 shows the degree to which the X-ray luminosities of the elliptical galaxies that were detected by the Einstein X-ray observatory are correlated with (i) their optical, and (ii) radio-continuum luminosities. At low luminosities $L_X$ and $L_B$ are fairly tightly correlated and roughly proportional to one another. The X-ray luminosity at $L_B \lesssim 10^{37}$ W is probably dominated by the emission from mass-transfer binaries rather than the ISM[29]; the spectra of low-luminosity galaxies resemble the rather hard spectra of low-luminosity X-ray binaries in the Milky Way, while the spectra of luminous galaxies are characteristic of cooler, more rarefied plasma. Above

---

[29] X-ray binaries probably dominate the X-ray emissions of all disk galaxies (Long & van Speybroek 1983)

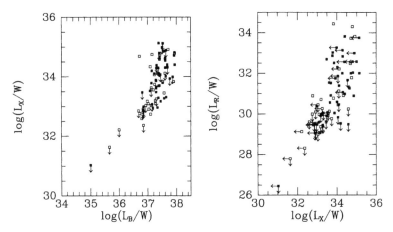

**Figure 8.49** Correlations between the luminosities of E and S0 galaxies in the optical (*B*), X-ray (Einstein) and radio (5 GHz) bands. [From data published in Fabbiano, Gioia & Trinchieri (1989)]

$L_B \simeq 10^{37}$ W, where the ISM is dominant, the mean X-ray luminosity increases roughly as $L_B^2$ but with considerable scatter.

The right panel of Figure 8.49 shows that the radio-continuum and X-ray luminosities of elliptical galaxies are correlated. As in the case of the radio–IR correlation of Figure 8.22, the natural inference to draw from this correlation is that there is a causal connection between the X-ray emitting plasma and the radio-continuum emission. This point is highly controversial, however (e.g., Binney 1996).

By modeling the radial X-ray brightness and temperature profiles one can infer the mass of the X-ray emitting plasma in an elliptical galaxy. The inferred masses range up to $10^{11}\,\mathcal{M}_\odot$, so the most X-ray luminous ellipticals have as large a fraction of their mass in gas as a typical Sc galaxy. Other ellipticals, especially low-luminosity ones, have very small gas-fractions. We saw in §5.4 that the rate at which an elliptical's dying stars pump gas into the ISM can be calculated quite reliably. One can easily show that if all this gas were to accumulate in the galaxy at the temperature $T \simeq 2T_*$ similar to that observed, the galaxy would lie close to the top of the $L_X - L_B$ correlation of Figure 8.49 – see Problem 8.12.

### 8.3.2 Cool gas in ellipticals

The great majority of elliptical galaxies are much poorer in cool gas than a spiral galaxy of comparable luminosity. This poverty, combined with the fact that comparatively few giant ellipticals lie conveniently close to the Milky Way, makes them hard to observe at 21 cm, 2.6 mm or the far infrared and hinders quantification of their cool ISM content.

The uppermost curve in Figure 8.41 summarizes the incidence of HI in elliptical galaxies. It shows that three quarters of elliptical galaxies

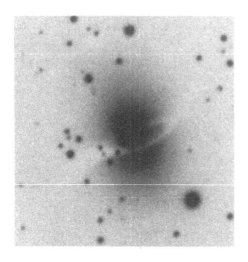

**Figure 8.50** The dust lane around NGC 5266. This negative image is based on a plate taken with the Anglo-Australian Telescope and has been processed to heighten the dust lane. [Courtesy of D. Malin]

have $\mathcal{M}_{HI}/L_B \lesssim 1.5 \times 10^{-3} \mathcal{M}_\odot L_\odot^{-1}$, while only 5% have $\mathcal{M}_{HI}/L_B > 0.05 \mathcal{M}_\odot L_\odot^{-1}$.

CO has been detected in several ellipticals (Knapp 1990). Roughly 40% of a sample of bright ellipticals were detected by IRAS Knapp *et al.* (1989) but mostly at very low values of $L_{60\,\mu m}/L_B$. About 40% of bright ellipticals have dust lanes. These consist of cool, dusty gas and the gas frequently shows H$\alpha$ or similar optical emission lines (Bertola 1987, Bertola 1990).

The origin of cool gas in ellipticals is unclear. One would expect gas ejected by dying stars rapidly to join the pervasive hot plasma, although measurable IR or CO radiation might be emitted by shells of gas between their ejection by stars and their absorption into the general plasma. Some cool gas may be material that has cooled from the hot plasma. One would expect this gas to be very centrally concentrated.

Observations of dust lanes in ellipticals such as that shown in Figure 8.50 suggest, however, that they do not comprise material which has condensed out of the galaxy's ISM. Notice that the dust lane of Figure 8.50 lies far out in the optical image and runs around the apparent minor-axis of the galaxy. Observations of emission lines from the lane's cool gas show that the gas is rotating around the galaxy's major axis, as one might expect from Figure 8.50. By contrast, stellar absorption-line spectra show that the stars in this galaxy rotate around the galaxy's *minor* axis. Thus in NGC 5266 the angular-momentum vectors of the dust and the stars are *perpendicular*. This proves to be the arrangement in about half of dust-lane ellipticals.[30] In other cases the angular momentum vectors appear to be inclined at an arbitrary angle. Sometimes they are anti-parallel, so that the stars circulate

---

[30] A word of caution is in order here regarding the statistics of dust-lane orientations: there is a danger that ellipticals that have dust lanes along their major axes will be classified as S0 galaxies.

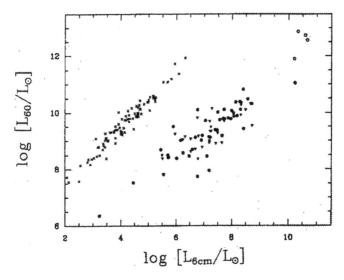

**Figure 8.51** Galaxies with active nuclei are on average 1000 times more luminous at radio wavelengths than normal galaxies of the same IR luminosity. [After Knapp (1990)]

one way and the gas another! Only in a small minority of cases are the two vectors parallel. This morphology is hard to reconcile with the gas having been ejected by the galaxy's stars. Moreover, the presence of dust in the gas and the large radius of the dust lane are both hard to reconcile with the gas having condensed out of the hot plasma; dust grains would be rather quickly sputtered out of existence in the X-ray emitting plasma. Neither finding is strange if the gas has fallen in from outside.

If dust lanes are formed from infalling material, there are two reasons why it is natural for strangely oriented dust lanes to be common in ellipticals but unknown amongst spirals. First, the conventional gaseous disk of a spiral will absorb any infalling material that comes in with an anomalous angular momentum vector – the orbital plane of such gas will intersect the disk's plane, and at the points of intersection infalling material will collide and merge with more massive clouds in the disk. Second, it is widely believed that many elliptical galaxies have been formed in a merger. If one of the merging galaxies is a spiral, part of its disk is liable to be flung out in a tidal tail. Much of this tail will fall into the elliptical which evolves out of the merger remnant over a period of several Gyr. Thus of all galaxies ellipticals may be especially well placed to possess a steady stream of infalling dusty gas (Schweizer 1990).

Figure 8.51 shows that although the range of galaxies that conform to the correlation shown in Figure 8.22 is broad, it is not universal, by showing that quasars (open circles) and radio galaxies (filled circles) lie on a different correlation line from that shown in Figure 8.22. These galaxies with active nuclei have roughly three orders of magnitude more radio luminosity per unit IR luminosity than the normal and star-bursting galaxies. It is natural that galaxies with active nuclei should lie to the right of normal galaxies in Figure 8.51: in their nuclei they have an additional powerful source of cosmic rays and thus radio luminosity. The interesting and poorly understood fact is that this luminosity is still correlated with infrared luminosity.

## 8.4 Intergalactic gas

The space between galaxies is not empty. The existence of intergalactic gas is important because cool gas is liable to fall into a neighbouring galaxy, while very hot gas can sweep the ISM out of a galaxy that moves through it. In either case intergalactic gas has the potential profoundly to modify the nature of nearby galaxies.

### 8.4.1 The Magellanic Stream

The Magellanic Stream consists of a series of six elongated HI clouds that are aligned with each other to form a swath of 21-cm emission that stretches over more than $60°$ in the southern sky. One end of this swath merges with the diffuse HI envelope around the Magellanic Clouds (Figure 8.52). This structure has been successfully modeled as gas that has escaped from the Small and Large Magellanic Clouds, and is now trailing the Clouds in their orbit about the Milky Way. This model, which was introduced by Murai & Fujimoto (1980), predicts that the Clouds are currently near the pericenter of their orbit and are moving eastwards. The proper motion of the Clouds was subsequently found to be $\mu_\alpha = (1.20 \pm 0.28)\,\mathrm{mas\,yr^{-1}}$ and $\mu_\delta = (0.26 \pm 0.27)\,\mathrm{mas\,yr^{-1}}$ per century (Jones, Klemola & Lin 1994) in agreement with Murai & Fujimoto's model.

Early 21-cm surveys of the Milky Way revealed HI emission at high latitudes and, at low latitudes, at velocities that differed by more than $50\,\mathrm{km\,s^{-1}}$ from those associated with simple differential rotation (Verschuur 1975). The gas responsible for this emission does not form part of the Galaxy's differentially rotating disk. There is a preponderance of these **high-velocity clouds** with negative heliocentric velocities, and less than $\sim 3\%$ of them are found at heliocentric velocities of more than $300\,\mathrm{km\,s^{-1}}$ (Wakker 1991). Both of these results indicate that high-velocity clouds are either associated with the Milky Way or with the Local Group.

The Local Group is not the only galaxy group that is known to contain intergalactic HI. A spectacular example of intergalactic gas is provided by the M81/M82/NGC 3077 group, in which neutral hydrogen of surface density $N_{\mathrm{HI}} > 10^{22}\,\mathrm{m^{-2}}$ envelops several adjacent galaxies.

We have seen that 21-cm measurements do not provide a good way of detecting gas whose total surface density $(N_{\mathrm{HI}} + N_{\mathrm{H_2}})$ is less than $\sim 10^{24}\,\mathrm{m^{-2}}$.

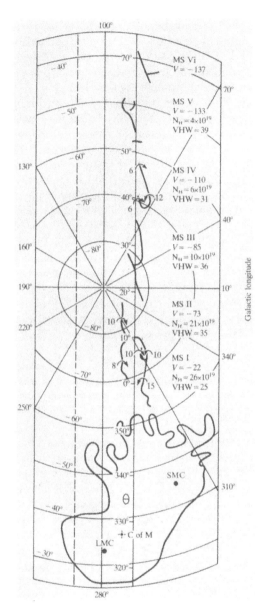

**Figure 8.52** The Magellanic Stream. The contour surrounding the Clouds is that on which the HI surface density equals $5 \times 10^{23}\,\mathrm{m}^{-2}$. [After Mathewson, Schwarz & Murray (1977)]

Studies of absorption lines in the spectra of extragalactic sources enable us to detect much thinner layers of gas in the same way that studies of the ultraviolet spectra of bright Galactic stars enable us to probe the ISM (see §8.1.1). Unfortunately, no extragalactic source is bright enough to have been studied by the *Copernicus* satellite, so until the advent of the HST it was impossible to study nearby intergalactic gas as thoroughly as the ISM has been studied by this technique. In fact, astronomers had to choose between studying nearby intergalactic gas through lines of trace ions in particular ionization stages – as we saw in §8.1.1, the physical interpretation of such lines is problematic – and studying intergalactic gas at redshifts $z > 2.5$ large enough to allow the Ly$\alpha$ line to be redshifted into the optical bands.

One very important result which did emerge from studies of absorption by ions such as $C^{3+}$ was that the gaseous disks of typical bright galaxies have to extend to $R \simeq 30h^{-1}$ kpc. The simplest argument for this great extent involves determining the number of such lines per unit velocity interval in a typical quasar spectrum and calculating the number of galactic disks our line of sight would intersect over the corresponding distance in space if each bright galaxy had a disk of radius $p$ kpc. The number of lines observed and disks intersected agrees for $p \simeq 100$ kpc (Spitzer & Bahcall 1969). This basic argument is confirmed by an experiment carried out by Bergeron & Boissé (1991): for each low-redshift absorption line in a quasar spectrum, try to find the responsible galaxy. From the redshift of the absorption line one can estimate the apparent magnitude of the galaxy. Then candidate galaxies are sought in the field around the quasar and finally their redshifts are measured. This work has been confirmed and extended by Steidel, Dickinson & Persson (1994), who concluded that QSO absorption lines are formed in disks that extend to $\sim 30h^{-1}(L/L^*)^{0.2}$ kpc. These disks belong to luminous ($L \simeq L^*$) galaxies.[31]

UV observations with HST make it possible to study the nearby intergalactic gas through Ly$\alpha$ absorption, which enables us to detect gas with column densities $N_{\rm HI} \gtrsim 10^{18}$ m$^{-2}$.[32] Spectra of nearby quasars reveal a surprisingly large number of clouds with column densities $10^{18}$ m$^{-2} < N_{\rm HI} < 10^{20}$ m$^{-2}$ (Bahcall *et al.* 1993).

X-ray observations reveal that most rich clusters of galaxies are filled with plasma with temperatures $T > 10^7$ K. Such plasma within a cluster is called the **intracluster medium**. The mass of the intracluster medium is enormous – it exceeds the total mass in the luminous parts of the cluster galaxies by a factor of several. Since its mass is so great, at most a minority of this material can ever have been in a galaxy. The intracluster medium is not a simple primordial hydrogen–helium plasma, however: observations of X-ray spectral lines reveal that it contains significant numbers of heavy-element ions. It seems likely that these heavy elements were formed in galaxies and then transferred to the intracluster medium either by being blown out of the galaxy by the cumulative effect of many supernovae, or by being stripped from galactic disks as their parent galaxy sped through the intracluster medium.

What impact do these various detections of intergalactic gas have on the study of galaxies themselves? This question raises the further question: is

---

[31] See §4.1.3 for the definition of $L^*$.

[32] Column densities $N_{\rm HI} \gtrsim 10^{20}$ m$^{-2}$ are required to generate detectable absorption lines of ions such as Mg$^+$.

the observed gas entering or leaving galaxies? There is significant evidence that much of the gas observed in groups of galaxies has been stripped from galaxies. In the case of the Magellanic stream, the model of Murai & Fujimoto (1980) indicates that dynamical interaction between the two Clouds and the Milky Way led to gas detaching itself from the Clouds to form the stream. Similarly, a great tail of HI near NGC 3628 in the Leo group of galaxies has been successfully linked by dynamical modeling to a stellar tail that is observed in this system (Haynes, Giovanelli & Roberts 1979). We have also seen that some part of the hot gas in clusters of galaxies has probably been stripped from cluster galaxies. Yet one galaxy's loss is likely to be another galaxy's gain. In clusters of galaxies the hot plasma may be condensing into stars in the giant galaxies that are often found at the centers of clusters. The gas of the Magellanic stream may eventually stream into the disk of the Milky Way. Similarly gas that currently forms the polar rings around S0 galaxies is likely eventually to sink into the underlying galaxy. In all these cases the orientation and structure the existing stellar disks will be profoundly affected by the acquisition of polar gas (see Problem 8.13). Finally, prolonged infall of streamers of gas into elliptical galaxies may sustain a low level of star formation within these galaxies, with significant effect upon the galaxy's colors.

## Problems

**8.1** Calculate the width contributed to the quantity $f(v)$ [defined by equation (8.19)] by thermal motions of hydrogen atoms at temperature 80 K.

**8.2** Show that the slope of a curve of growth (§8.1.1) tends to 0.5 when the equivalent width of the line is much greater than its Doppler width.

**8.3** The Arecibo telescope has a collecting area of $7 \times 10^4 \, \mathrm{m}^2$, and is often used to observe the 21-cm line at a resolution of $5 \, \mathrm{km \, s^{-1}}$ per channel. Calculate the power received in a single channel when the optical depth along the line of sight is $\tau_\nu = 0.2$ and the material observed has spin temperature $T = 80 \, \mathrm{K}$.

**8.4** Estimate the power received in a single channel of the type specified in the previous question by simply calculating the rate of spontaneous decays of excited atoms in the beam. Explain why this estimate is accurate even though the vast majority of photons emitted are absorbed before they reach the observer. (Note that this last statement is true even for the moderate value of $\tau_\nu$ specified.)

**8.5** Radio astronomers frequently report specific intensities, $I_\nu$, in units of mJy arcsec$^{-2}$. Show that if $I_\nu$ is measured in these units and the wavelength, $\lambda$, is in meters, then the antenna temperature is given by $T = 1.54 \times 10^4 \lambda^2 I_\nu \, \mathrm{K}$.

**8.6** A galaxy at a distance $D$ Mpc is observed to have 21-cm flux $S(v)$ Jy at velocity $v$ km s$^{-1}$. Given that the 21-cm line is optically thick, show that the mass of HI in this galaxy is $M_{\rm HI}/\mathcal{M}_\odot = 2.36 \times 10^5 D^2 \int dv\, S$.

**8.7** Assuming the average absorption cross-section per dust grain, $K_a$, [equation (8.42)] varies as $\lambda^{-\alpha}$, and that all grains have a common temperature $T$, show that the total emissivity is proportional to $T^{4+\alpha}$. For $\alpha = 1.5$, find the value of $hc/(k_{\rm B}T\lambda)$ at which the dust's emissivity, $K_a B_\nu(T)$, peaks.

**8.8** Calculate the rate per unit surface area at which a grain is heated by impacts of H atoms; assume that impacts are totally inelastic but do not cause H atoms to stick. Find the grain temperature at which radiative cooling would be in equilibrium with such collisional heating. Explain why this temperature is much lower than the observed temperatures of interstellar dust.

**8.9** The total gaseous column density in a galactic disk is

$$N(R, \phi) = N_0[1 + 0.2\cos(kR + 2\phi)]e^{-R/R_{\rm d}}, \qquad (8.65)$$

where $k = 3/R_{\rm d}$. Plot the HI column density as a function of $R$ for $\phi = 0$ and $\phi = 90°$ under the assumption that the ionized layers above and below the disk have uniform column density $N_{\rm HII} = 0.1N_0$. In the light of these plots, comment on the finding that at large galactocentric radii 21-cm observations yield much more regular velocity contours than contours of $N_{\rm HI}$.

**8.10** Consider gas that is ejected from stars in an elliptical galaxy in which the stars move isotropically with one-dimensional velocity dispersion $\sigma$. Given that radiative losses are unimportant and that the speeds of individual stellar winds are negligible compared with $\sigma$, show that the equilibrium temperature of interstellar gas in the galaxy is $T_* = 2.8 \times 10^6 (\sigma/200\,{\rm km\,s^{-1}})^2$ K. (Assume that the mean molecular weight in a fully ionized plasma is $\mu = 0.6$ – see Problem 5.2.)

**8.11** Suppose that the stellar density in the elliptical galaxy of Problem 8.10 falls with radius as $n_* \propto r^{-\alpha}$. Show that the density of the galaxy's isothermal ISM will vary as $n \propto r^{-\alpha\beta}$, where $\beta \equiv T_*/T$.

**8.12** Assume that gas accumulates for a Hubble time in the galaxy of Problem 8.10 at a temperature $T \simeq 2T_*$, similar to that observed. Show that the galaxy should lie close to the top of the $L_X - L_B$ correlation of Figure 8.49.

**8.13** The mass of HI in the Magellanic Stream is probably $\sim 10^8\,\mathcal{M}_\odot$. Show that this gas probably has more angular momentum about the Galactic center than all the gas in the Milky Way.

# 9

# The Milky Way's ISM

In §8.1 we discussed the techniques used to track interstellar material and to determine its physical characteristics. We drew upon many results of studies of the Milky Way' ISM because we can study nearby interstellar material in very much more detail than the interstellar media of external galaxies. In fact many of the techniques mentioned in §8.1 can currently only be used to study the local ISM. One might have hoped that from this great wealth of information on the properties of the local ISM would have flowed a clear and definitive picture. Sadly this is not the case; studies of the Milky Way's ISM suffer badly from the syndrome of not being able to see the wood for the trees.

The problem is two-fold. First, the wealth of the available information gets in the way of sweeping generalizations of the sort that may be plausibly made about external galaxies, about which we know so much less. Second, we have the misfortune to be located in the plane of the Milky way, and suffer dreadfully from the particular bugbear of astronomy: ignorance of the distance to which we are peering. In this chapter our aim is to synthesize our knowledge of the local ISM into a more-or-less coherent, three-dimensional picture. We start with an analysis of the kinematics of Galactic rotation, since this provides us with much of our knowledge of the distances to gas clouds.

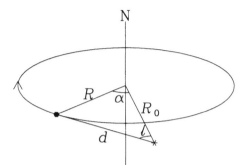

**Figure 9.1** Geometry of a disk of
material in circular rotation.

## 9.1 The kinematics of differential rotation

We saw in §8.2 that line-observations give rise to a 'data cube,' and that,
for an optically thin line, this cube specifies the amount of material moving
at each velocity down each line of sight. When using line observations of
material that is confined to a thin layer near the plane to study the large-
scale structure of the Milky Way, it is natural to average the data for a small
range of values of galactic latitude $b$ around $b = 0$. The slice through the data
cube that is isolated by this averaging procedure is conveniently displayed as
a longitude–velocity plot [$(l, v)$ **plot**] like that shown in Figure 9.13 below.
Our first task is to learn to interpret such plots.

Our discussion is significantly simplified if we work with velocities relative to the **local standard of rest** (**LSR**). This is the current velocity of a
fictional particle that moves around the plane of the Milky Way on the closed
orbit in the plane that passes through the present location of the Sun.[1] If
the Milky Way is axisymmetric, this orbit is circular, otherwise it is oval.
In §10.6 we shall see how the Sun's velocity relative to the LSR may be
determined. Clearly, once it is known, it is trivial to convert any measured
heliocentric velocity into a velocity relative to the LSR. In this chapter we
shall work exclusively with velocities relative to the LSR.

### 9.1.1 The naive $(l, v)$ plot

Consider the $(l, v)$ plot that we would observe if all the gas were in circular
rotation. Let $\mathbf{R}$ be the position vector of material with respect to the Galactic
center, and let the angular velocity of rotation of this material be $\mathbf{\Omega}(R)$.
The velocity of this material is then $\mathbf{v}_c = \mathbf{\Omega} \times \mathbf{R}$. When observed from the
particle which defined the LSR, whose rotation velocity is $\mathbf{v}_0 \equiv \mathbf{\Omega}(R_0) \times \mathbf{R}_0$,
the line-of-sight velocity, $v_{\text{los}}$, of this material is the projection of the velocity
difference $\mathbf{v}_c - \mathbf{v}_0$ on to the vector $(\mathbf{R} - \mathbf{R}_0)$ that runs from the Sun to the

---

[1] It is conceivable that more than one orbit could fulfill our defining condition for the
LSR. In this case it would be natural to choose the most nearly circular orbit.

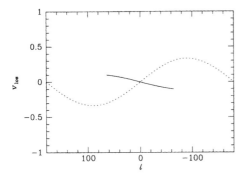

**Figure 9.2** The traces in the $(l, v)$ plot of three rings. The radii of the rings are $r = 0.9R_0$ (full curve), $r = 0.1R_0$ (dotted curve) and $r = 1.5R_0$ (dashed curve). The circular speed in the disk is unity at all $R$.

material:

$$v_{\mathrm{los}} = \frac{\mathbf{R} - \mathbf{R}_0}{|\mathbf{R} - \mathbf{R}_0|} \cdot \big[\boldsymbol{\Omega}(R) \times \mathbf{R} - \boldsymbol{\Omega}(R_0) \times \mathbf{R}_0\big]. \tag{9.1}$$

Using the vector identities $\mathbf{a} \cdot (\mathbf{a} \times \mathbf{b}) = 0$ and $\mathbf{a} \cdot (\mathbf{b} \times \mathbf{c}) = \mathbf{b} \cdot (\mathbf{c} \times \mathbf{a})$, we can rewrite equation (9.1) as

$$v_{\mathrm{los}} = \frac{\big[\boldsymbol{\Omega}(R) - \boldsymbol{\Omega}(R_0)\big] \cdot (\mathbf{R}_0 \times \mathbf{R})}{|\mathbf{R} - \mathbf{R}_0|}. \tag{9.2}$$

In the notation of Figure 9.1, $\mathbf{R}_0 \times \mathbf{R} = -R_0 R \sin\alpha\,\hat{\mathbf{n}}$, where $\hat{\mathbf{n}}$ is the unit vector that points perpendicular to the disk. Also, by the law of sines, $\sin\alpha/|\mathbf{R} - \mathbf{R}_0| = \sin l/R$. Finally, since the Milky Way rotates in the sense shown in Figure 9.1 when viewed from the northern hemisphere, we have $\boldsymbol{\Omega} = -|\boldsymbol{\Omega}|\hat{\mathbf{n}}$. Inserting these results into equation (9.2), we have finally

$$v_{\mathrm{los}}(l) = \big[\Omega(R) - \Omega(R_0)\big] R_0 \sin l. \tag{9.3}$$

Now imagine that all the gas in the Milky Way were concentrated into a ring of radius $R$. Then equation (9.3) states that, along any line of sight that intersects the ring, we will detect material at just one velocity, and, moreover, that this velocity will be proportional to $\sin l$. Figure 9.2 shows the **traces** in the $(l, v)$ plot of three such rings. Each is a section of a sine curve. If $R/R_0$ is significantly smaller than unity, the section is small and approximately straight; quantitatively, it runs from $l = -l_{\max}$ to $l = l_{\max}$, where $l_{\max} \equiv \arcsin(R/R_0)$ is half the angular extent of the ring. If $R/R_0$ is less than but of order unity, the section is distinctly curved. If the ring lies outside the solar circle, then it traces a full sine curve in the $(l, v)$ plot. For any value of $R/R_0$ the slope of the trace at the origin is proportional to the difference $\Omega(R) - \Omega(R_0)$. For any plausible circular-speed curve, $\Omega(R)$ increases inwards, and therefore this slope is largest for the smallest rings. As $R/R_0$ increases from small values towards unity, this slope continuously declines, and is zero for $R/R_0 = 1$; points on the solar circle do not move relative to the LSR. For $R/R_0 > 1$ the slope of the sine curve near the

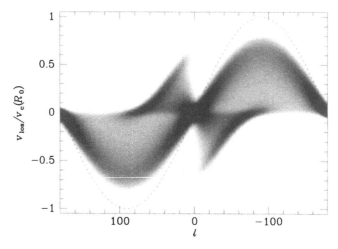

**Figure 9.3** Grayscale of optical depth in the $(l, v)$ plane. As
for Figure 8.12, the number density of hydrogen atoms has been
assumed to equal $2 \times 10^5 \, \text{m}^{-3}$ at $R_0$ and to vary with $R$ as
$\exp(-R/2R_0)$. The circular speed is given by $v_c(R) = (R/R_0)^{0.1} \times 210 \, \text{km s}^{-1}$. The sine curve $v_{\text{los}} = -v_c(R_0) \sin l$ (shown dashed) lies
just outside the boundary of the occupied band. Figure 9.4 shows a
section of the surface $\tau(l, v)$ along the vertical dotted line $l = 20°$.

origin is opposite to that for $R/R_0 < 1$, and the absolute value of the slope
increases steadily as $R/R_0$ increases. Since, for any plausible circular-speed
curve, $\Omega(R) \to 0$ as $R \to \infty$, it is clear from equation (9.3) that the amplitude
of the sine curve tends in the limit $R \to \infty$ to minus the circular speed at
the solar circle, $R_0\Omega_0(R_0)$.

The Milky Way's disk may be considered to be made up of infinitely
many rings, so much of any observed $(l, v)$ plot is occupied. Material interior to the solar circle should occupy the quarters $0 \le l \le 90°, v_{\text{los}} \ge 0$ and
$-90° \le l \le 0°, v_{\text{los}} \le 0$ at top left and bottom right of Figure 9.2. Gas outside the solar circle should occupy the remaining two quarters of the figure.
We shall see that this theoretical division of the $(l, v)$ plane is only partially
confirmed by observation, and we speak of material that lies outside the
quarters specified above as being at **forbidden velocities**. We mentioned
above that the central slope of the curve associated with a ring of radius
$r$ can be large when $r$ is small, but is never greater than $R_0\Omega(R_0)$ when
$r > R_0$. Hence for $l > 0$ velocities less than $-R_0\Omega(R_0) \sin l$ are strongly
forbidden in the sense that material should not be found at such velocities
regardless of whether it lies inside or outside the solar circle. On the other
side of the Galactic center there is an equivalent range of strongly forbidden
positive velocities.

Figure 9.3 shows the magnitude of the optical depth $\tau$ of the 21-cm line
in the $(l, v)$ plane for a model disk in which all material moves on perfectly

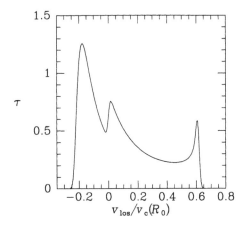

**Figure 9.4** A section of the surface $\tau(l, v)$ that is shown in Figure 9.3. This section is at longitude $l = 20°$ and corresponds to the vertical dotted line in Figure 9.3.

circular orbits. As we anticipated in the discussion above, $\tau$ is non-zero only in two broad arcs that are bounded on one side by sections of a sine curve. When we point a radio telescope to the Milky Way at some longitude $l_1$, the resulting spectrum will show emission at velocities corresponding to the intersection in Figure 9.3 of the vertical line $l = l_1$ with the populated band. The vertical dotted line in Figure 9.3 and the plot of optical depth versus $v_{los}$ shown in Figure 9.4 illustrates the case $l_1 = 20°$. Since this line lies in the first quadrant $(0 < l_1 < 90°)$, it runs through the occupied band from $v_{los} = -v_c(R_0) \sin l_1$ below to

$$v_{los}^{(t)}(l_1) = [\Omega(r_t) - \Omega_0] R_0 \sin l_1 \qquad (9.4)$$

above, where $r_t$ is the radius of the smallest ring (with the largest value of $\Omega$) that can be seen along that line of sight. Clearly the line of sight is tangent to this ring, and simple geometry demonstrates that

$$r_t = R_0 \sin l_1. \qquad (9.5)$$

The point at which the line of sight touches the ring of radius $r_t$ is called the **tangent point** and the quantity $v_{los}^{(t)}$ is called the **terminal velocity**. Empirically, the terminal velocity can be determined by examining an observational analog of Figure 9.4: it is the velocity of the abrupt decrease in emission at $v > 0$ for $l > 0$ and $v < 0$ for $l < 0$. In Figure 9.4 $v_{los}^{(t)}$ lies near $0.6v_c(R_0)$. Combining equations (9.4) and (9.5) we have

$$v_c(r_t) = v_{los}^{(t)}(l_1) + v_c(R_0) \sin l_1. \qquad (9.6)$$

Since $v_{los}^{(t)}$ is in principle readily determined for any value of $l_1$ between 0 and 90°, this relation enables us to determine the rotation curve $v_c(R)$ for $R < R_0$ provided some means can be found to determine $v_c(R_0)$. The data for $l < 0$

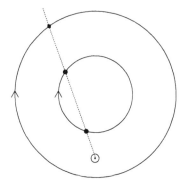

**Figure 9.5** Lines of sight at $|l| < 90°$ cut each ring of the inner disk twice but rings of the outer disk only once. At $l > 0$ the inner rings contribute emission at $v > 0$, while outer rings contribute emission at $v < 0$.

and $l > 0$ yield independent determinations of $v_c(R)$, and comparison of these two determinations provides a valuable check on the validity of the derived circular-speed curve.

Determining $v_c(R_0)$ is extremely difficult and we shall defer full discussion of this problem to §10.6. However, we have already seen that $v_c(R_0)$ is in principle simply the amplitude of the sine curve that forms one boundary of the occupied band in Figure 9.3. We shall discuss this idea further in §9.2.1.

**Radii and distances from the $(l, v)$ plot**     Equation (9.6) provides a means of determining the circular-speed curve $v_c(R)$ without any prior knowledge of the distances to the material that fills the $(l, v)$ plot from which the circular-speed curve is determined. *After* we have determined the circular-speed curve and therefore the function $\Omega(R)$, we can hope to infer the radius of material observed at any given value of $v_{\rm los}$ by solving equation (9.3) for $r$. Since $\Omega(R)$ is almost certain to be a monotonic function, $r$ is in principle uniquely determined by $(l, v_{\rm los})$. Of course, deviations of the measured values of $v_{\rm los}$ from the true values will lead to errors in the values of $r$ inferred from these measurements. The resulting errors in $r$ are liable to be large for material that lies outside the solar circle, because in this region an infinite range in $r$ corresponds to a finite, and, at many values of $l$, narrow range in $v_{\rm los}$. Consequently a small error in the measured value of $v_{\rm los}$ will lead to a large error in the recovered value of $r$. Inside the solar circle the ranges in $r$ and $v_{\rm los}$ are both finite, so the fractional error in $r$ will not be large if $v_{\rm los}$ is accurately known. In this connection, it should be noted that the error in $v_{\rm los}$ is likely to be dominated by the effects of non-circular velocities to be discussed below, rather than by inaccuracies in determining the central frequency of the observed line profile.

Once we have determined $r$ as described above, we often wish to go on to determine the distance $d$ of the observed material from the Sun. Any given line of sight intersects rings that lie outside the solar circle just once. So $d$ follows immediately from $r$ when the observed material lies in bottom left or top right quarters of Figure 9.3. By contrast, any given line of sight

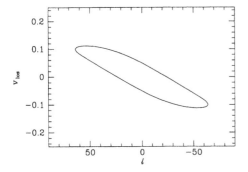

**Figure 9.6** The $(l, v)$ trace of a ring interior to the solar circle that is expanding radially as well as rotating. The expansion velocity is just 1/20 of the circular speed in the disk. the full curve in Figure 9.2 shows the trace of the ring in the absence of expansion.

intersects rings that lie inside the solar circle either twice or not at all – see Figure 9.5. Consequently, when material lies in the top left or bottom right quarters of Figure 9.3, *two* values of $d$ are possible. It is sometimes possible to resolve the ambiguity between these values by seeing how far the emission in question extends in $|b|$: if the nearer value of $d$ is the correct one, the material will subtend a bigger solid angle at the Sun, and therefore extend further in $|b|$ than if the larger value of $d$ is correct (Schmidt 1957).

### 9.1.2 Non-circular motion and the $(l, v)$ plot

We shall see that observed $(l, v)$ plots deviate significantly from the naive $(l, v)$ plot described above, so we must now look at the implications for the $(l, v)$ plot of more complex motions in the disk than simple circular rotation. There are four topics to discuss: (i) axisymmetric radial expansion; (ii) oval distortions; (iii) spiral structure; (iv) random motions. Since we know that many disk galaxies are barred and/or display spiral structure, the motivation for discussing topics (ii) and (iii) is self-evident. Similarly, it is obvious that gas clouds must have at least small random motions, so it is clear that we must understand how these affect the $(l, v)$ plot. The motivation for discussing topic (i), axisymmetric expansion, is rather different: this is the simplest possible deviation from perfect circular motion and serves as a useful introduction to the other topics. Moreover, on account of its simplicity, it has often been suggested as an explanation of structures in observed $(l, v)$ plots (e.g., Uchida, Morris & Bally 1994).

**Axisymmetric expansion**    We have seen that a circularly-rotating ring interior to the solar circle traces a section of a sine curve in the $(l, v)$ plot. Now let this ring be expanding radially as well as rotating. The added expansion velocity runs perpendicular to our line of sight at the tangent points, and hence the observed values of $v_{\text{los}}$ at these points are unaffected by expansion. Thus the end points of the ring's trace in the $(l, v)$ plot are unchanged by the addition of expansion. By contrast, the line of sight through the Galactic center cuts the ring at points at which the added expansion velocity runs parallel to the line of sight. At the nearer intersection of the line $l = 0$ with

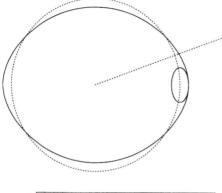

**Figure 9.7** In the epicycle approximation [equations (9.7)] an elliptical orbit (large full curve) is formed by adding anti-clockwise motion on an elliptical epicycle to clockwise motion around the dotted circle of radius $R_g$. The small full curve shows the location of the epicycle at $t = 0$; the center of this ellipse moves anti-clockwise around the circle as the object moves clockwise around the epicycle.

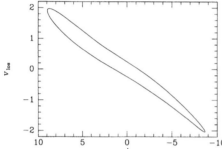

**Figure 9.8** The trace in the $(l, v)$ plot of the elliptical orbit shown in Figure 9.7 when viewed down the sloping dotted line in that figure. The distance from the viewing point to the center of the orbit is six times the radius of the dotted circle in Figure 9.7.

the ring, the expansion velocity will cause $v_{\text{los}}$ to be negative, while at the further intersection of this line with the ring, expansion causes $v_{\text{los}}$ to become positive. Along lines of sight either side of $l = 0$ the single value of $v_{\text{los}}$ expected for pure circular motion is similarly split into two values, the larger one deriving from the further intersection of the ring and the line of sight. Arguing from continuity, it is now clear that expansion causes the section of a sine curve predicted by circular rotation to open up into a curved loop – see Figure 9.6.

If an expanding ring is larger than the solar circle, its expansion is immediately apparent by making the average value of $v_{\text{los}}$ around the ring positive – see Problem 9.1.

**Oval distortions**     We now suppose that the ring under observation is oval. The ellipticity of the ring is obviously liable to change the angle it subtends on the sky, but more important than this change in angle are the changes in velocity that are associated with the ellipticity. Clearly material circulating on the ring has to move in and out as it goes around, but the speed with which it circulates around the ring also varies from one point to another, and we have to draw on dynamical theory to get these changes right.

We treat the problem in the weak-bar approximation discussed in §3.3 of BT. In this approximation, the gravitational potential, $\Phi(R, \phi)$, in which objects orbit is the sum of two terms, a dominant axisymmetric term, $\Phi(R)$, and a weaker barred term of the form $\Phi_b(R)\cos[2(\phi - \Omega_b t)]$. Here $\Phi_b$ determines the strength of the bar and $\Omega_b$ is the **pattern speed** of the bar; that is, the angular velocity with which the barred potential rotates like a rigid body. The coordinates in position and velocity of an orbiting object are, like the potential, written as sums of two terms, one due to $\Phi_0$ and one due to $\Phi_b$. In the frame that rotates with the figure of the potential, equations (3-120a) and (3-117) of BT yield for motion of the type of interest here

$$
\begin{aligned}
R &= R_g + C\cos[2(\Omega - \Omega_b)t], \\
\phi &= -(\Omega - \Omega_b)t + \frac{D}{R_g}\sin[2(\Omega - \Omega_b)t], \\
v_R &= -2(\Omega - \Omega_b)C\sin[2(\Omega - \Omega_b)t], \\
v_\phi &= R\dot\phi \\
&\simeq -(\Omega - \Omega_b)\big\{ R_g + (C - 2D)\cos[2(\Omega - \Omega_b)t] \big\}.
\end{aligned}
\tag{9.7}
$$

where $C$ and $D$ are positive constants determined by the potential. The terms in equations (9.7) that are independent of $C$ and $D$ describe clockwise motion around a circle of radius $R_g$ at angular velocity $\Omega$. The point $R = R_g, \phi = \Omega t$ is called the **guiding center** of the orbit. The terms in equations (9.7) that depend on $C$ or $D$ describe anti-clockwise motion around an elliptical **epicycle**. The orbit is a compound of clockwise motion around a circle and anti-clockwise motion about the epicycle, which is centered on the guiding center and has principal axes aligned with the local radial and tangential directions – see Figure 9.7. The orbit is at **pericenter** (closest to the Galactic center) when the epicycle displaces the object inwards, and at **apocenter** (furthest from the Galactic center) when the epicycle displaces the object outwards. Since the epicycle is traversed in the opposite sense to the circle, the contributions to $v_\phi$ from circle and epicycle add at pericenter and subtract at apocenter.

The axis ratio of the epicycle is determined by the constants $C$ and $D$ and thus by the potential. The details need not bother us here; we need only note that usually $D > C$, with the result that the epicycle is elongated in the tangential rather than in the radial direction – see Figure 9.7. In consequence, the ellipticity of the orbit affects $v_\phi$ more than it does $v_R$.

The effect of the epicycle upon the $(l, v)$ plot will clearly depend on the angle $\phi_\odot$ between the Sun-center line and the major axis of the orbit. If this is 0 or $90°$, the epicycle merely changes the extent of the orbit's trace in the $(l, v)$ plot; if $\phi_\odot = 0$, the trace extends to smaller values of $l$ and larger values of $v_{\rm los}$ than if the orbit were circular, and vice versa if $\phi_\odot = 90°$. For general viewing angle $\phi_\odot$, the trace becomes a closed curve that is reminiscent of a cross-section through an aerofoil – see Figure 9.8. For the viewing direction shown by the dotted line in Figure 9.7 ($\phi_\odot = 20°$), the lower tangent point generates the blunt end of the aerofoil, while the upper

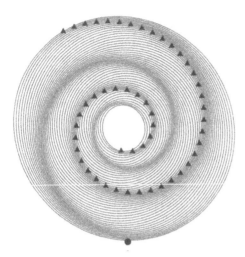

**Figure 9.9** Spiral arms are formed when the directions of the long axes of oval orbits described by equations (9.7) change systematically with radius. Here the angle between the horizontal and an oval's long axis is proportional to the logarithm of the length of the oval's long axis. The circular speed is assumed independent radius and the peak value of $v_R$ is for any star 10% of $v_\phi$ in the rotating frame of the spiral. The blob at the bottom marks the location of 'sun' at which the $(l, v)$ plot of Figure 9.10 would be obtained. One of the spiral arms is decorated with a series of triangles.

tangent point generates the curve's sharp end. For $\phi_\odot \simeq 70°$ the positions of the sharp and blunt ends of the curve in the $(l, v)$ plot are reversed.

**Spiral structure**   The central idea of the theory of spiral structure is that spiral arms will form if stars and gas clouds move on elliptical orbits whose major-axis positions vary smoothly as a function of radius. Figure 9.9 illustrates this idea, which is discussed in detail in §6.2 of BT. Figure 9.9 shows a number of oval orbits similar to that of Figure 9.7. Each oval should be imagined to be uniformly populated with gas clouds. Because the directions of the long axes of the ovals rotate by more than 540° from the smallest to the largest ovals, the ovals are in some places much closer to one another than in other places. If each oval were uniformly populated by gas clouds, the surface density of clouds would be higher where the ovals approach one another than elsewhere. In Figure 9.9 it is apparent that the crowded parts of the disk form two spiral arms.

When we use the epicycle approximation to model the kinematics of a disk of gas that has a well-defined spiral pattern, we obtain an $(l, v)$ plot like that of Figure 9.10. Each oval in Figure 9.9 yields a long, thin, diagonally oriented trace in Figure 9.10. As one progresses from small ovals to large ones, the ovals' traces in the $(l, v)$ plot vary from nearly vertical to nearly horizontal as their extents in $l$ increase and their extents in $v_{\text{los}}$ decrease. Simultaneously, the elongation of the trace varies, in the same way that the shape of the trace in Figure 9.8 varies with $\phi_\odot$. In Figure 9.10 variation in the shapes of the $(l, v)$ traces leads to crowding along curves that fan out from the origin. Along some of these crowded curves triangles are plotted. These mark the images in the $(l, v)$ plot of points in Figure 9.9 that lie on one of the spiral arms. Thus in the real world, material that is seen on a ridge in the $(l, v)$ plot, lies on a spiral arm. Notice that in Figure 9.10 there is a bump in the terminal-velocity curve wherever the latter is touched by one of these ridges.

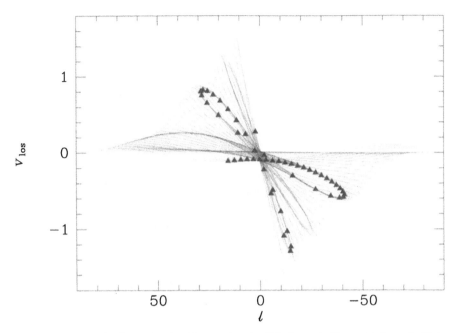

**Figure 9.10** The $(l, v)$ plot obtained by viewing the disk shown in Figure 9.9 from the blob at the bottom of that figure. The spiral arms in Figure 9.9 give rise to strong crowding in the $(l, v)$ plot along curves which fan out from the origin. Triangles mark the images in this $(l, v)$ plot of the points in Figure 9.9 that are also marked by triangles.

When observational $(l, v)$ plots first became available, considerable effort went into tracing the Milky Way's spiral arms as follows (e.g. Kerr & Westerhout 1965). Crowded regions in the $(l, v)$ plot similar to those seen in Figure 9.10 were identified as spiral arms. These arms were mapped back into real space by finding the real-space points that would map to them *if the disk were in circular rotation*. However, spiral structure affects the velocities of clouds just as much as their real-space coordinates, so the existence of spiral structure in the disk is incompatible with material in the disk being on circular orbits. Burton (1971) pointed out that the assumption of circular orbits can lead to serious errors in the real-space positions assigned to the clouds that are responsible for emission at $(l, v)$. These errors are most serious when the cloud lies near the tangent point, because then the value of $v_{\rm los}$ in circular rotation varies very slowly with distance down the line of sight, with the consequence that a large displacement in real space will be required to compensate for a small contribution to $v_{\rm los}$ from spiral structure.

This argument indicates that spiral arms detected in $(l, v)$ plots need to be modeled dynamically as in Figure 9.10, rather than with the simple kinematic approach described above. In particular, the density contrast between arm and inter-arm regions is greatly over-estimated by naive kinematic modeling, as is hinted at by Figures 9.9 and 9.10; the amplitude of the arms is larger in the $(l, v)$ plot than in real space, which suggests that the velocity shifts associated with spiral structure tend to enhance the strengths of arms in $(l, v)$ plots.

In the light of this discussion, it is surprising to discover that naive kinematic modeling leads to accurate maps of the ridge-lines of arms in real space. Figure 9.11 illustrates this result by showing the contribution to $v_{los}$ from spiral structure for some of the points in Figure 9.10: each triangle indicates the location in the $(l, v)$ plot of some point in the disk when spiral structure is present, while the corresponding locations in the absence of spiral structure lie at the further ends of the lines that are attached to the triangles. Many (but not all) of the points that lie along spiral arms, appear to have no attached lines because the true value of $v_{los}$ differs negligibly from the value for circular rotation. Hence the real-space positions of these points can be accurately recovered under the assumption of circular rotation, and, when thus recovered, will faithfully trace the spiral arms in real space.

In Figure 9.11 there are in the neighborhood of each spiral arm many points that have long lines attached, indicating that for the associated clouds the true value of $v_{los}$ differs significantly from that associated with circular orbits. Consequently, the real-space locations of these clouds cannot be accurately recovered under the assumption of circular rotation. It turns out that the errors in their positions made by assuming they are in on circular orbits nudges the recovered real-space points towards the ridges of the nearby arms, thus enhancing the recovered arm/inter-arm contrast as mentioned above.

**Random motions**     From §8.2 we know that clouds do not move on perfectly circular orbits because the clouds at any given location have velocities that spread over a few $km\,s^{-1}$. Moreover, within a given cloud there is a spread in the velocities of individual H atoms. Qualitatively it is easy to see that this velocity spread will modify $(l, v)$ plots by smoothing them in velocity. Such smoothing is potentially important wherever the brightness temperature is varying rapidly with velocity, as it does, for example, near the terminal velocity.

For simplicity we assume that in each small volume $d^3\mathbf{x}$ of the disk the distribution of atomic velocities is Maxwellian – the volume should be large enough to encompass several clouds but small enough to be unresolved by the telescope. The number of atoms in the volume element with velocities in $d^3\mathbf{v}$ is then

$$dn = \frac{n(\mathbf{x})\,d^3\mathbf{x}\,d^3\mathbf{v}}{(2\pi\sigma^2)^{3/2}} \exp\left\{ - \frac{[\mathbf{v} - \overline{\mathbf{v}}(\mathbf{x})]^2}{2\sigma^2}\right\}. \qquad (9.8)$$

Here $\sigma$ is the atoms' velocity dispersion (which may in principle be a function of $\mathbf{x}$) and $\overline{\mathbf{v}}(\mathbf{x})$ is the mean velocity of the atoms in $d^3\mathbf{x}$. Now if $\overline{\mathbf{v}}(\mathbf{x})$ is the circular velocity at $\mathbf{x}$, we know by equation (9.3) that an atom that moves with this velocity has line-of-sight velocity

$$v_{los}^{(0)}(\mathbf{x}) \equiv \{\Omega[R(\mathbf{x})] - \Omega(R_0)\}R_0 \sin l. \qquad (9.9)$$

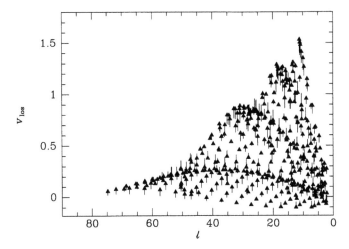

**Figure 9.11** The triangles show the locations of a selection of points on the curves of Figure 9.10. If the disk were in pure circular rotation, material at the corresponding real-space points would have the velocities that lie at the ends of the lines that are attached to each triangle. Thus the length of each line gives an indication of the magnitude of the error in the real-space position that one would obtain for that cloud by assuming it was on a circular orbit.

It follows that an atom with a random velocity $\delta\mathbf{v} \equiv \mathbf{v} - \bar{\mathbf{v}}$ has line-of-sight velocity

$$v_{\text{los}} = v_{\text{los}}^{(0)}(z) + \delta v_z, \tag{9.10}$$

where we have oriented our coordinate system such that the $x$ and $y$ axes run perpendicular to the line of sight and the $z$ axis runs parallel to it. We obtain the surface density of atoms in the distance interval $\mathrm{d}z$ and with line-of-sight velocities within $\mathrm{d}\delta v_z$ of the value of $v_{\text{los}}$ given by equation (9.10), by dividing equation (9.8) by $\mathrm{d}x\mathrm{d}y$ and integrating the result over $\mathrm{d}\delta v_x \mathrm{d}\delta v_y$:

$$
\begin{aligned}
\mathrm{d}^2 N[v_{\text{los}}^{(0)}(z) + \delta v_z] &= n(z)\mathrm{d}z\,\mathrm{d}\delta v_z \int \mathrm{d}\delta v_x \int \mathrm{d}\delta v_y \frac{e^{-|\delta\mathbf{v}|^2/2\sigma^2}}{(2\pi\sigma^2)^{3/2}} \\
&= \mathrm{d}z\,n(z)\mathrm{d}\delta v_z \frac{e^{-\delta v_z^2/2\sigma^2}}{\sqrt{2\pi\sigma^2}}.
\end{aligned}
\tag{9.11}
$$

Finally, we find the total column density of atoms with any given line-of-sight velocity $u_{\text{los}}$ by summing the contributions $\mathrm{d}N$ over increments in the path length $\mathrm{d}z$ with $\delta v_z$ chosen such that $v_{\text{los}}^{(0)}(z) + \delta v_z = u_{\text{los}}$:

$$\frac{\mathrm{d}N(u_{\text{los}})}{\mathrm{d}u_{\text{los}}} = \int \mathrm{d}z\,n(z)\frac{\exp\{-[u_{\text{los}} - v_{\text{los}}^{(0)}(z)]^2/2\sigma^2\}}{\sqrt{2\pi\sigma^2}}, \tag{9.12}$$

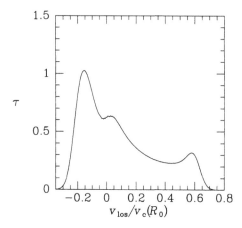

**Figure 9.12** Optical depth as a function of velocity along the line of sight $l = 20°$ for a simple model hydrogen disk in which the velocity dispersion is $\sigma = 0.04v_c(R_0) \simeq 8.4\,\mathrm{km\,s^{-1}}$. Comparison with Figure 9.4 shows that non-zero $\sigma$ significantly modifies $\tau(v)$.

where we have used $\mathrm{d}\delta v_z = \mathrm{d}u_{\mathrm{los}}$. Comparison of Figures 9.4 and 9.12 shows for one longitude, $l = 20°$, the effect of random velocities on optical depth $\tau$, which, by equation (8.16), is proportional to $\mathrm{d}N/\mathrm{d}u_{\mathrm{los}}$: both figures are for the same distribution of hydrogen within the disk, but Figure 9.4 shows $\tau$ for the case of negligible random velocity, while Figure 9.12 shows $\tau$ for $\sigma = 0.04v_c(R_0) \simeq 8.4\,\mathrm{km\,s^{-1}}$. It can be seen that random velocities significantly reduce the peaks in $\tau$.

In the limit $\sigma \to 0$, the fraction in equation (9.12) tends to a Dirac $\delta$-function. We can then use a standard formula[2] for a $\delta$-function of a function to evaluate the integral over $z$:

$$\frac{\mathrm{d}N(u_{\mathrm{los}})}{\mathrm{d}u_{\mathrm{los}}} = \sum_{z_i \text{ s.t. } v_{\mathrm{los}}^{(0)}(z_i)=u_{\mathrm{los}}} \frac{n(z_i)}{|\mathrm{d}v_{\mathrm{los}}^{(0)}/\mathrm{d}z_i|}. \tag{9.13}$$

Figure 9.3 was plotted by substituting values of $N$ from this equation into equation (8.20). The presence of $|\mathrm{d}v_{\mathrm{los}}^{(0)}/\mathrm{d}z_i|$ in the denominator of equation (9.13) has the important implication that points at which the velocity gradient is small can make very large contributions to $N$. At the tangent point, $|\mathrm{d}v_{\mathrm{los}}^{(0)}/\mathrm{d}z_i|$ actually vanishes and equation (9.13) fails. Hence, the optical depth near the tangent velocity is sensitive to the magnitude of the random velocities in the gas, no matter how small these are.

---

[2] For any function $f(x)$ which never touches the $x$ axis we have

$$\delta[f(x)] = \sum_{x_i \text{ s.t. } f(x_i)=0} \frac{\delta(x - x_i)}{|\mathrm{d}f/\mathrm{d}x|}.$$

# 9.2 The large-scale distribution of HI and CO

The disk of the Milky Way falls naturally into three parts: the outer disk at $R > R_0$; the middle disk, $3\,\text{kpc} < R < R_0$; and the central disk, $R < 3\,\text{kpc}$. The central disk, which is complex and only partially understood, is discussed in §9.4. In this section we focus on the middle and outer parts of the disk.

The main distinction between the middle and the outer disk relates to the techniques used to study them: at $R < R_0$ we concentrate on the neighborhood of the terminal-velocity curve in the $(l, v)$ plane, and thus study the disk primarily at the tangent points. At $R > R_0$ there are no analogs of tangent points, and we are heavily reliant on the rather uncertain determination of the Milky Way's circular-speed curve.

However, besides this operational distinction, there prove to be objective differences between the middle and the outer disk. The Sun happens to lie near the outer edge of the part of the disk that contains most of the Milky Way's molecular gas and luminous star formation. Also, we shall see that inside the solar circle the disk is extremely flat, and by $1.5R_0$ it is significantly warped.

### 9.2.1 The 21-cm line in emission

The HI in the Milky Way has been surveyed many times in the last half-century. Table 9.1 lists the characteristics of the surveys that are of more than historical interest. Since aperture-synthesis telescopes are insensitive to faint, diffusely-distributed emission, all these surveys have been done with single-dish instruments. It would, moreover, be extremely time-consuming and of limited scientific interest to map the entire Milky Way with the angular resolution characteristic of the largest single-dish instruments. Consequently, these surveys have employed relatively small dishes of diameter $\sim 25\,\text{m}$. The angular resolution of the surveys is thus of order a degree, with the consequence that they each contain in excess of ten thousand resolution elements and require of order one year of observing time for their construction.

An ideal radio telescope would have a simple and compact 'beam,' or pattern of sensitivity on the sky. That is, when pointed in a given direction, it would respond sensitively to a source that lay in a narrow cone about that direction, and be highly insensitive to sources elsewhere. Unfortunately this ideal is hard to realize in practice, and most real telescopes have complex beams that vary significantly with the orientation of the telescope with respect to the ground, surrounding buildings, and so on. Radio astronomers frequently adopt a simple model of a telescope's beam and then consider the telescope's response to sources that lie outside this beam to arise from 'leakage' of radiation into the beam. At low flux levels the uncertainties in most of the surveys listed in Table 9.1 are dominated by the leakage of radiation into the adopted telescope beam rather than by thermal noise in the telescope and receiver. That is, the quality of the published data is determined

**Table 9.1**   HI surveys of the Milky Way

| Survey | $\Delta\theta$ | $l$ | $b$ | $v$ ($km\,s^{-1}$) | $T_{min}$ |
|---|---|---|---|---|---|
| Weaver & Williams | 36′ | 10° to 250° | −10° to 10° | −100 to 100 | 1.7 K |
| (1973,1974) | | $\Delta l = 0.5°$ | $\Delta b = 0.25°$ | $\Delta v = 2.1$ (6.3) | |
| Heiles & Habing | 36′ | all $l$ at $\delta > -30°$ | all $b$ at $|b| > 10°$ | −92 to 75 | 1.2 K |
| (1974) | | $\Delta l = 0.3°/\cos b$ | $\Delta b = 0.6°$ | $\Delta v = 2.1$ (6.3) | |
| Bania & Lockman | 4′ | 32° to 64° | −3° to 3° | −250 to 250 | 0.6 K |
| (1984) | | $\Delta l = 1.3°$ | $\Delta b = 2′$ | $\Delta v = 1.03$ | |
| Burton (1985) | 21′ | all $l$ at $\delta > -46°$ | all $b$ at $|b| > 20°$ | −250 to 250 | 0.2 K |
| | | $\Delta l = 1°$ | $\Delta b = 1°$ | $\Delta v = 1$ (2.1) | |
| Kerr et al. (1986) | 48′ | 240° to 350° | −90° to 10° | −150 to 150 | 0.8 K |
| | | $\Delta l = 0.5°$ | $\Delta b = 0.25°$ | $\Delta v = 2$ | |
| Stark et al. (1992) | 150′ | all $l$ at $\delta > -30°$ | all $b$ | −300 to 300 | 0.1 K |
| | | $\Delta l = 0.5°/\cos b$ | $\Delta b = 0.5°$ | $\Delta v = 5.3$ (2.1) | |
| Leiden–Dwingeloo | 36′ | all $l$ at $\delta > -30°$ | all $b$ | −460 to 460 | 0.07 K |
| | | $\Delta l = 0.5°/\cos b$ | $\Delta b = 0.5°$ | $\Delta v = 1$ (2.1) | |

NOTES: $\Delta\theta$ gives the beam diameter and $T_{min}$ the sensitivity of each survey.

SOURCE: Adapted from Burton (1992)

by the accuracy with which it has been possible to determine the telescope's beam and thus to determine where the radiation that was detected was actually coming from. This problem of beam determination is most acute in directions in which the HI column density is small, but which are close to directions in which the column density is large. The survey of Stark *et al.* (1992), which was done with the horn antenna with which Penzias and Wilson discovered the cosmic background radiation in 1965, is least affected by radiation leakage because the antenna was originally designed to have the simplest possible, and therefore most easily determined, beam.

Figure 9.13 is the $(l, v)$ plot for $b = 0$ from the Leiden–Dwingeloo–IAR survey of the Galaxy in the 21-cm line. We recognize several features from the theoretical discussion of §9.1.1. In particular, the curves which bound the band of significant emission in Figure 9.13 are similar in shape to those which bound the occupied zone of Figure 9.3. Both figures show strongly discontinuous behavior near $l = 0$, with the bounding curves showing marked horns either side of the origin: in Figure 9.13 there are horns at $v \simeq 140\,km\,s^{-1}$ near $l = 15°$ and $v \simeq -150\,km\,s^{-1}$ near $l = -15°$. As one moves away from the Galactic center, the observed terminal velocities fall steeply from the values associated with the horns to near zero at $l = \pm 90°$, just as expected from Figure 9.3. Moreover, in Figure 9.13 the boundary at $(l > 0, v < 0)$ and at $(l < 0, v > 0)$ can be approximated by a sine curve, just as in Figure 9.3.

Quantitatively, there are important differences between Figures 9.13 and 9.3. For example, in Figure 9.13 the horns of emission at small values of $|l|$ are much fainter, relative to the peak intensity at, say, 30° than are the corresponding horns in Figure 9.3. The obvious interpretation of

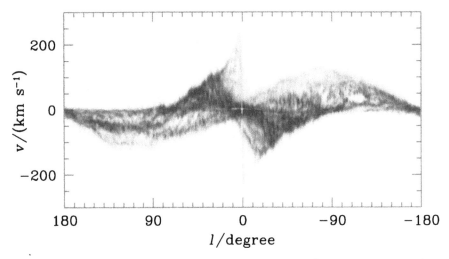

**Figure 9.13** Brightness temperature arising from the 21 cm line in the Galactic plane ($b = 0$), plotted as a function of Galactic longitude, $l$, and line-of-sight velocity, $v$. The data come from the Leiden–Dwingeloo Survey (Hartmann & Burton 1997) and the complementary southern hemisphere IAR survey by Bajaja *et al.*. They were kindly provided by D. Hartmann.

the faintness of the horns in Figure 9.13 is that there is very little HI at radii $R \lesssim 3\,$kpc, which correspond to $|l| < 20°$. Figure 9.3, by contrast, was calculated for a disk in which HI continues to the very center.

Another significant quantitative difference between Figures 9.13 and 9.3 concerns the amplitudes of the sine curves that best fit the boundaries of the band of emission at ($l > 0$, $v < 0$) and at ($l < 0$, $v > 0$). In Figure 9.3 this amplitude is $\sim 85\%$ of the circular speed at the solar radius, whereas in Figure 9.13 it is $\sim 130\,$km s$^{-1}$, which is only $\sim 60\%$ of the local circular speed given in Table 10.7. Does this unexpectedly small value indicate that the observations have not been pushed to sufficiently high sensitivity to reveal faint emission beyond the apparent boundary in Figure 9.13, or is there really negligible emission beyond a sine curve of amplitude $\sim 130\,$km s$^{-1}$? Knapp, Tremaine & Gunn (1978) have investigated this question and concluded that there really is no significant emission beyond the fitted sine curve, presumably because the Milky Way's HI disk does not extend beyond $R \simeq 21\,$kpc (see Problem 9.4).

A significant feature in Figure 9.13 that has a counterpart in Figure 9.3 is the ridge along $v \sim 0$ at $-90° < l < 90°$. In Figure 9.3, this ridge is associated with a jump in the optical depth by a factor of 2 – see Figure 9.4. Figure 9.5 explains the origin of this jump: for $0 < l < 90°$, two identical points in the disk contribute to emission at $v > 0$, whereas only one point contributes to emission at $v < 0$. The appearance of this jump in a brightness-temperature map such as Figure 9.13 demonstrates that $T_B$ de-

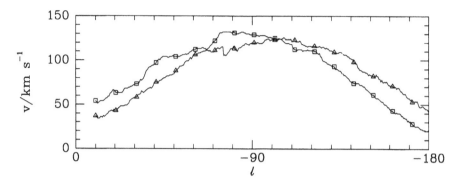

**Figure 9.14** The curve studded with squares shows the 2 K contour that bounds the zone of emission in the upper-right quadrant of the $(l, v)$ plot for 21-cm emission averaged over $|b| < 10°$. The other curve shows the 2 K contour that bounds the emission in the lower-left quadrant of the $(l, v)$ plane after inversion through $(l, v) = (0, 0)$. The systematic displacement of one curve with respect to the other is evidence that the Milky Way's outer HI disk is not axisymmetric. [After Blitz & Spergel (1991a) from data kindly provided by L. Blitz]

pends strongly on the optical depth $\tau$ even near $v = 0$, where the emission is strongest. From this sensitivity of $T_B$ to $\tau$ it follows that in most directions the disk is not very optically thick in the 21-cm line.

The maximum brightness temperature $T_B^{(\mathrm{max})}(l)$ observed in the plane at longitude $l$ peaks at $\sim 135$ K for some lines of sight near $l = 0$, and rarely falls below $\sim 80$ K. While there is considerable scatter in $T_B^{(\mathrm{max})}(l)$ between neighbouring lines of sight, large values typically occur near $l = 0$, $l = 180°$ and $l = \pm 75°$, where emission tends to be concentrated into small velocity ranges and therefore the disk is most nearly optically thick (Burton 1992). Since $T_B$ should approach the spin temperature $T_{\mathrm{spin}}$ defined in Box 8.3 at high optical depths, these observations suggest that $T_{\mathrm{spin}} \simeq 135$ K.

If the disk were axisymmetric and in simple circular rotation, any observed $(l, v)$ plot would be symmetric under inversion through the origin, that is under $l \to -l \equiv 360° - l$ and $v \to -v$. Figure 9.14 reveals an intriguing breakdown of this symmetry. Each curve in the figure shows the location of a 2 K contour in the $(l, v)$ plot for 21-cm emission averaged over $|b| < 10°$. The curve studded with squares shows the contour that bounds the zone of emission in the upper right part of the $(l, v)$ plane, while that studded with triangles shows the result of inverting through $(l, v) = (0, 0)$ the contour that bounds the zone of emission in the lower left part of the $(l, v)$ plane. If the zone of emission in the $(l, v)$ plane were symmetrical as in Figure 9.3, the two curves would lie on top of each other. In reality the curve for $l < 0$ is displaced to the left with respect to the curve for $l > 0$. Blitz & Spergel (1991a) have argued from this phenomenon that the outer Milky Way is oval, and that the LSR is currently receding from the Galactic center on an oval orbit. Kuijken & Tremaine (1994) argue that a more plausible

**Table 9.2**  CO surveys of the Milky Way

| Survey | Beam | $l$ | $b$ | $v$ ( km s$^{-1}$) | Sensitivity |
|---|---|---|---|---|---|
| Clemens et al. | 0.8′ | 8° to 90° | −1.05° to 1.05° | −50 to 150 | 1.2 K |
| (1986) | | $\Delta l = 0.125°$ | $\Delta b = 0.05°$ | $\Delta v = 1$ | |
| Dame et al. | 8.7′ | −180° to 180° | −3.3° to 3.3° | −80 to 80 | 0.2 K |
| (1987) | | $\Delta l = 0.5$ | $\Delta b = 0.5°$ | $\Delta v = 1.3$ | |
| Stark et al. | 0.6′ | −5° to 122° | −1° to 1° | −80 to 80 | 0.3 K |
| (1988) | | $\Delta l = 3'$ | $\Delta b = 3'$ | $\Delta v = 0.7$ | |

SOURCE: Adapted from Burton (1992)

interpretation of the data is that the outer parts of the Milky Way, like those of so many external galaxies, are lop-sided (cf. §8.2.3).

**Measuring the spin temperature**    While the estimate $T_{\rm spin} \simeq 135\,{\rm K}$ obtained above is interesting, studies of 21-cm absorption in the spectra of extragalactic radio sources reveals that it is rather crude. From equations (8.18) and (8.16) we may show that the intensity of 21-cm absorption in the spectrum of a continuum source that is observed through a foreground hydrogen cloud always depends sensitively on the value of $T_{\rm spin}$ in the cloud. In equation (8.16) we equate $T_B(0)$ with the brightness temperature of the background source. Typically this temperature will be very much larger than the spin temperature in the ISM, with the consequence that, when the beam includes the background source, we may neglect the second term on the right-hand side of equation (8.16) and approximate the measured brightness temperature by

$$T_{\rm on} = T_{\rm source} e^{-\tau_\nu}. \tag{9.14}$$

This equation, taken with the assumption that $T_{\rm source}$ is a weak function of $\nu$ in the neighborhood of $\nu = 1.42\,{\rm GHz}$, enables us to determine $\tau_\nu$ as a function of $\nu$. Applied to a line of sight slightly away from the background source along which $T_B(0) = 0$, equation (8.16) yields

$$T_{\rm spin} = \frac{T_B}{1 - e^{-\tau_\nu}}. \tag{9.15}$$

If the angular size of the cloud of foreground hydrogen is much greater than the beam, $\tau_\nu$ will not vary significantly as we slew the beam to a position that just excludes the source. Hence we may use our previously determined values of $\tau_\nu$ in equation (9.15) to derive $T_{\rm spin}$ from the measured value of $T_B$ for an off-source line of sight.

The values of $T_{\rm spin}$ that are obtained by the procedure just described are found to vary systematically with $\tau$: the more optically thick a line of sight is in the 21-cm line, the lower is the recovered value of $T_{\rm spin}$. Specifically, clouds with peak optical depth $\tau > 0.2$ yield $20\,{\rm K} \lesssim T_{\rm spin} \lesssim 110\,{\rm K}$ with a mean value $\sim 80\,{\rm K}$. As the peak optical depth decreases towards 0.01, the mean value of $T_{\rm spin}$ rises to $\sim 300\,{\rm K}$ (Spitzer 1978, Kulkarni & Heiles 1988).

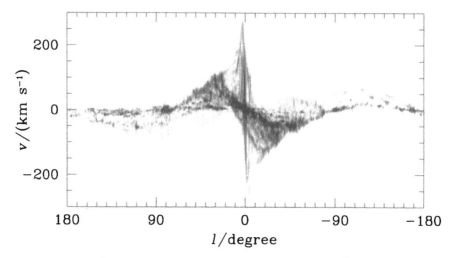

**Figure 9.15** An $(l, v)$ plot of the Milky Way in the 2.6-mm line of $^{12}$CO. The grayscale shows the intensity of the line averaged over a $\sim 4°$ strip centered on the Galactic plane. [After Dame *et al.* (1987), but incorporating more recent, higher sensitivity data kindly provided by T. Dame]

### 9.2.2 CO lines in emission

In §8.1 we saw that the most important probes of molecular gas in galaxies are the rotation lines of various isotopes of CO, and especially the 2.6-mm line of $^{12}$CO. Table 9.2 lists some important surveys of the Milky Way in this line. The velocity resolution and sensitivity of these surveys are comparable to those of the 21-cm surveys listed in Table 9.1. By contrast, the angular resolution of the CO surveys is significantly higher than that of the 21-cm surveys. This difference is a consequence of the fact that a dish of given size yields nearly two orders of magnitude greater angular resolution at 2.6 mm than at 21 cm, and good sensitivity cannot be obtained with a very small dish. At the high angular resolution of the CO surveys, the Milky Way contains so many resolution elements that observing them all would require an unobtainable allocation of telescope time. Hence none of the surveys listed in Table 9.2 covers more than a small fraction of the sky. The best coverage is provided by the survey of Dame *et al.* (1987), who went to some trouble to use unusually small dishes. Fortunately, CO is more strongly confined to the plane than HI, and it unlikely that the Dame *et al.* survey missed significant CO emission.

Figure 9.15 shows the $(l, v)$ plot of the Milky Way in the 2.6-mm line of $^{12}$CO from Dame *et al.* (1987). The emission is more patchy than the 21-cm emission shown in Figure 9.13, but certain prominent features are apparent in both plots. In particular, in both plots the envelopes at $(l > 0, v > 0)$ and at $(l < 0, v < 0)$ decline to the origin from prominent horns, although the horns of the CO emission are at larger longitudes $|l| \simeq 25°$, and slightly

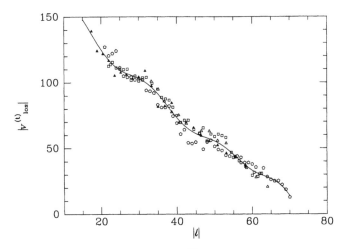

**Figure 9.16** Terminal velocities $v_{\text{los}}^{(\text{t})}$ as a function of $|l|$ for HI (open symbols) and CO (filled symbols). Data for $l > 0$ from Bania & Lockman (1984) (open triangles), Weaver & Williams (1973,1974) (open squares) and Malhotra (1995) (filled triangles); for $l < 0$ from Kerr *et al.* (1986) (open pentagons). The curve is a fit to the data by the sum of a linear function of $|l|$ and a cosine series in $|l|$. [From data kindly supplied by S. Malhotra]

smaller velocities $|v| \simeq 130\,\text{km s}^{-1}$. These horns and the diagonal swath of emission that joins them are often interpreted as arising from a **molecular ring** of radius $R_0 \sin 25° \simeq 3.6\,\text{kpc}$, which we discuss further below.

At positive longitudes the terminal-velocity curve in Figure 9.15 seems to fall to zero before $l = 90°$ is reached, although at $l < 0$ the terminal-velocity curve reaches $v = 0$ near $l = -90°$ as expected. There is very little emission in the regions $90° < l < 180°$, $v < 0$ and $-180° < l < -90°$, $v > 0$ associated with gas that lies beyond the solar circle, although there is emission near $v = 0$ at all longitudes. This emission presumably comes from the solar circle and probably from the solar neighborhood. What emission there is at $R > R_0$ seems to be associated with the ridges of intense 21-cm emission in Figure 9.13 that were identified with spiral arms. In a detailed study of CO emission from just outside the solar circle, Digel *et al.* (1996) found that the intensity of CO emission is at least 25 times smaller between spiral arms than in the arms.

### 9.2.3 The Milky Way's circular-speed curve

The 21-cm and CO-line data reviewed above provide the key to determining the circular speed $v_c(R)$, and hence the radial distribution of mass, in the Milky Way. In §9.1.1 we saw that for $R < R_0$ we determine $v_c(R)$ by estimating the terminal velocities $v_{\text{los}}^{(\text{t})}(l)$ from the $(l, v)$ plot and inserting them

**Table 9.3**  Mean terminal velocities from 21-cm data

| $l$ | 15° | 20° | 30° | 40° | 50° | 60° | 70° |
|---|---|---|---|---|---|---|---|
| $v/\,\mathrm{km\,s^{-1}}$ | 148 | 125 | 103 | 73 | 56 | 33 | 14 |

SOURCE: Data plotted in Figure 9.16

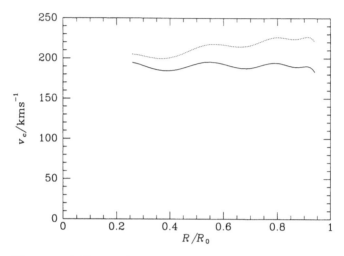

**Figure 9.17** The circular-speed curves that follow from the tangent velocities given by the curve in Figure 9.16 and equation (9.6) for $v_c(R_0) = 180\,\mathrm{km\,s^{-1}}$ (full curve) and $v_c(R_0) = 220\,\mathrm{km\,s^{-1}}$ (dashed curve).

into equation (9.6).[3]  The terminal velocities determined from 21-cm and CO line data are in good general agreement with one another. Figure 9.16 shows the quality of this agreement: both the general trend and small-scale variations in $v_{\mathrm{los}}^{(t)}$ with $l$ are reproduced in both the HI and CO data. The small-scale fluctuations of $v_{\mathrm{los}}^{(t)}$ about its the smooth underlying trend with $l$ have amplitude of order $7\,\mathrm{km\,s^{-1}}$. This value may be considered to be an estimate of the influence of the Milky Way's spiral structure on orbital velocities.

The differences in Figure 9.16 between the values of $v_{\mathrm{los}}^{(t)}(l)$ for HI and CO at fixed $l$ are distributed as a Gaussian centred on zero and with a dispersion of $4.1\,\mathrm{km\,s^{-1}}$ (Burton 1992). This value provides an estimate in the dispersion in the random tangential velocities of interstellar clouds.

The values of $v_{\mathrm{los}}^{(t)}$ determined at $\pm l$ generally agree to better than $10\,\mathrm{km\,s^{-1}}$ (Kerr & Rogers 1964, Blitz & Spergel 1991a). While more accurate agreement is not expected if each side is independently perturbed by spiral structure of amplitude $7\,\mathrm{km\,s^{-1}}$, it is perhaps surprising that there is

---

[3] See Celnik, Rholfs & Braunsfurth (1979) and Malhotra (1995) for details of how $v_{\mathrm{los}}^{(t)}(l)$ can be estimated from the data.

a tendency for $v_{\text{los}}^{(t)}$ to be systematically lower for $l$ in the range $(-30°, -50°)$ than in the corresponding range of positive longitudes. Table 9.3 gives mean values of $v_{\text{los}}^{(t)}$ for $0 < |l| < 90°$, and Figure 9.17 plots the resulting circular-speed curves for two values of $v_c(R_0)$. Notice that the larger the assumed value of $v_c(R_0)$, the larger is the predicted gradient $dv_c/dR$ at the Sun. This behavior reflects the fact that equation (9.6) expresses $v_c$ as the sum of a declining function of $l$ and a rising function of $l$ whose coefficient is $v_c(R_0)$.

Beyond the solar circle we cannot determine $R$ in equation (9.3) kinematically as we do when using equation (9.6), and must devise an alternative way of estimating the Galactocentric radius $R$ of each object for which we measure $v_{\text{los}}$. The usual procedure is to determine the distance $d$ of the object from the Sun, and to obtain $R$ from the law of cosines,

$$R^2 = R_0^2 + d^2 - 2R_0 d \cos l. \tag{9.16}$$

The standard technique is to determine the distance to a young cluster by main-sequence fitting (§6.1.2) and then to measure the cluster's velocity from radio-line observations of associated molecular gas – see Brand & Blitz (1993) for references. An alternative approach involves observations of Cepheid variables (§5.1.10), which simultaneously yield both distances and velocities (e.g., Pont *et al.* 1997). Planetary nebulae and Carbon stars have also been as tracers of the outer circular-speed curve – see the review by Fich & Tremaine (1991) for references to the literature.

Interstellar absorption gives rise to significant errors in the distances $d$ to tracers of Galactic rotation at $R > R_0$ and it is important to understand how distance errors bias the circular-speed curve that one derives from such tracers. To this end, we define

$$W \equiv \frac{v_{\text{los}}}{\sin l} = R_0[\Omega(R) - \Omega(R_0)]. \tag{9.17}$$

Here $v_{\text{los}}$ is the line-of-sight component of the relative velocity between the LSR at the Sun and at the location of a tracer that lies distance $d$ from the Sun at longitude $l$, and we have used equation (9.3). Since $l$ is measured precisely, errors in $W$ simply reflect errors in $v_{\text{los}}$, which are usually dominated by the peculiar velocities of tracers rather than by measurement errors.

Equation (9.17) shows that what we measure is a simple multiple of the difference between the circular frequency at the estimated radius $R$ of the tracer and at the Sun. Hence, an error in the distance $d$ of a tracer, which translates through equation (9.16) into an error in $R$, leads us to assign the measured frequency difference $W/R_0$ to the wrong radius $R$. In particular, tracers that in reality all lie on a ring of radius $R_1$, and therefore lead to similar measured values of $W$, will appear to be distributed throughout an annulus whose width depends on the magnitude of the errors in their distances; the measurements of the velocities of such tracers will falsely imply that $\Omega(R)$ is constant through this annulus and therefore that $v_c(R) \propto R$.

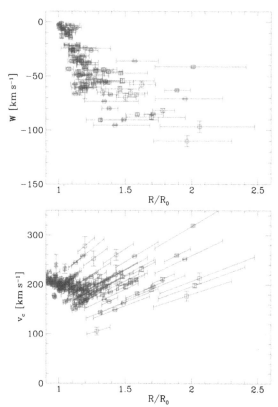

**Figure 9.18** The top panel shows 98 estimates of $W(R)$ [equation (9.17)] from Brand & Blitz (1993). The lower panel shows the values of $v_c(R)$ that one derives from these data. [After Binney & Dehnen (1997)]

The top panel of Figure 9.18 shows measurements of $W$ for 99 tracers from Brand & Blitz (1993). Most errors in $R$ are uncomfortably large, especially at $R \gtrsim 1.2R_0$, while the errors in $W$ are rather small even though they include both the measurement errors and a contribution of $7\,\mathrm{km\,s^{-1}}$ from the estimated peculiar velocities of the tracers. The top panel of Figure 9.18 suggests that $W$ falls steeply at $R_0 < R \lesssim 1.2R_0$ and then falls gradually or not at all. The lower panel of Figure 9.18 shows the values of $v_c$ that correspond through equation (9.17) to the data shown in the top panel. For the reason explained above, the errors in $d$ now give rise to diagonal error bars along which $v_c \propto R$. Partly on account of the slope of these error bars, this panel conveys the impression that the circular-speed curve rises steadily at $R \gtrsim 1.2R_0$ after falling at $R \lesssim 1.2R_0$. This behavior of $v_c(R)$ is dynamically implausible, however, and Binney & Dehnen (1997) argue that in a more plausible interpretation of the data, most of the points at $R > 1.25R_0$ are for objects that lie around a ring of radius $R \simeq 1.6R_0$ whose circular speed is lower than that at $R_0$ by $\sim 8\,\mathrm{km\,s^{-1}}$.

If the circular speed of the Galaxy were falling off in Keplerian fashion, $v_c(R) \propto R^{-1/2}$ at $R > R_0$, it would fall by $\gtrsim 50\,\mathrm{km\,s^{-1}}$ between $R_0$ and $1.6R_0$. Such a rapid decline is ruled out even by the very uncertain

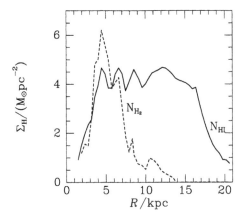

**Figure 9.19** Full curve: the average density of neutral hydrogen atoms in an annulus around the Galactic center, versus radius of the annulus. Dashed curve: the corresponding surface density of molecular gas for the assumption that $X$ is given by equation (8.28). $R_0$ has been assumed to be 8.5 kpc and emission from the innermost 1.5 kpc is not shown. [After Dame (1993) from data kindly provided by T. Dame]

data shown in Figure 9.18. In §8.2.4 we saw that the circular-speed curves of external galaxies are generally flat or rising to the last observed point. Therefore it would be surprising if Figure 9.18 did not tell a similar story. Historically, however, the discovery by Jackson, Fitzgerald & Moffat (1979) that the rotation curve to $R \simeq 1.6R_0$ does not decline beyond $R_0$ as the classic galaxy model of Schmidt (1965) requires, played a significant role in undermining the traditional belief that light would be a good tracer of mass.

### 9.2.4 Radial distributions of HI and CO

As we explained in §9.1.1, once the Milky Way's circular-speed curve has been determined, we can immediately deduce the radius $R$ that is responsible for emission at any point in the $(l, v)$ plane. Then if the disk is optically thin at the given $l$ and $v$, equation (8.20) relates the measured brightness temperature $T_B(l, v)$ to the column density $N$, and this is in turn related to the atomic density in the disk, $n(R)$, by equation (9.13). The relation between $T_B(l, v)$ and $n(R)$ is significantly more complex at points in the $(l, v)$ plane that are optically thick, but at least a lower limit on $n(R)$ may be set by proceeding as if the disk were optically thin. If an estimate of the optical depth can be obtained, equation (8.21) can be used to obtain the column density $N$, from which $n(R)$ follows as before.

In §8.1.4 we argued that it is probable that the optically-thick cores of molecular clouds do not strongly shadow one another. In this case $N(l, v)$ is proportional to the measured brightness temperature $T_B(l, v)$ and equation (9.13) can again be used to derive $n(R)$.

The full curve in Figure 9.19 shows the azimuthally-averaged distribution of HI in the Milky Way that one obtains as described above. Interior to $R \sim 4$ kpc the density falls from a wide plateau to near zero inside 1.5 kpc. Beyond $R_0$ the density remains large to $R \simeq 16$ kpc, where it turns sharply downwards. Notice that the *mass* of HI is proportional to the integral $\int n_H R \, dR$, in which values of $n_H$ are weighted by $R$. Consequently, the

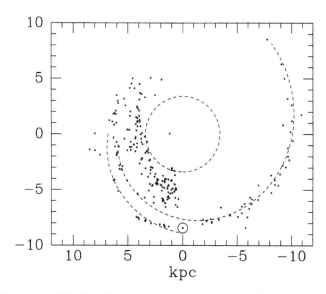

**Figure 9.20** The dots show the estimated positions within the Galaxy of giant molecular clouds; the Galactic center lies at the coordinate origin and the Sun's location is marked by ⊙. A logarithmic spiral with a pitch angle of 10° delineates the Sagittarius–Carina arm. A short section of a similar logarithmic spiral suggests the location of the Orion–Cygnus arm. A dotted circle of radius $0.4R_0$ emphasizes the elongated distribution of clouds in the molecular ring. [From data kindly provided by P. Solomon and data published in Grabelsky *et al.* (1988)]

proportion of the Milky Way's HI that lies at $R > R_0$ is larger than casual inspection of Figure 9.19 would suggest: Figure 9.19 implies that 80% of the $\sim 4.3 \times 10^9 \, \mathcal{M}_\odot$ of Galactic HI lies beyond $R_0$.

The dashed curve in Figure 9.19 is the corresponding plot for molecular hydrogen. The most striking feature of this curve is the implication that $H_2$ is almost entirely confined to $R < R_0$: 77% of the $\sim 10^9 \, \mathcal{M}_\odot$ of molecular material in the Galaxy appears to lie inside $R_0$. This narrower confinement of $H_2$ relative to HI is in line with the results for external galaxies reviewed in §8.2.2.

The Milky Way seems to be one of the minority of galaxies that have a central depression in their $H_2$-disks. In fact, at $l > 0$ a large fraction of the $H_2$ appears to be concentrated into a ring of mean radius $\sim 4.5$ kpc and full width at half maximum $\sim 2$ kpc. The data for $l < 0$ suggest that the $H_2$ distribution is bimodal, peaking around $R \simeq 2.5$ kpc and $R \simeq 6$ kpc. It has been suggested that the $H_2$ may not be concentrated into a ring as much as into two giant spiral arms, and that these arms may be connected with the central bar to be discussed below (Dame 1993).

### 9.2.5 Evidence for spiral structure

Some of the clearest evidence of spiral structure in the Milky Way comes from $(l, v)$ plots of emission by CO. In §9.6.1 we shall see that catalogs of giant molecular clouds can be constructed by analyzing such plots. Figure 9.20 shows the result of plotting the positions in the Galaxy of the most massive clouds in two such catalogs for a constant circular speed, $v_c = 220\,\mathrm{km\,s^{-1}}$ and $R_0 = 8.5\,\mathrm{kpc}$ in the way described in §9.1.2. Two factors complicate the interpretation of this figure. First, it is based on the questionable assumption that gas moves on circular orbits. Second, the catalogs from which it is constructed by no means cover the entire Galaxy. In particular, at positive longitudes coverage is confined to $R \lesssim R_0$, while at negative longitudes only $R > R_0$ is covered. Consequently, the figure surely omits some features that are important in the Galaxy. Nevertheless, the figure clearly shows that giant molecular clouds have a tendency to concentrate into elongated structures.

At Galactocentric radii $R \sim \frac{1}{2} R_0$, a broad roughly elliptical ridge in the density of molecular clouds is apparent towards the left of Figure 9.20. This feature, which accounts for roughly half of the molecular gas in the Galaxy, is identical with the molecular ring discussed above. While the molecular ring may really be elliptical as the figure suggests, one should not feel much confidence in Figure 9.20 at small radii because in §§9.4 and 10.1 we shall see that the inner few kiloparsecs of the Milky Way are dominated by a bar. Consequently, clouds at radii as small as $\frac{1}{2} R_0$ are probably on significantly non-circular orbits; Figure 9.20, by contrast, is based on the assumption of perfectly circular orbits. In particular, it is perfectly possible that the molecular ring is in reality a pair of massive spiral arms as we suggested at the end of the last subsection.

Figure 9.20 suggests that the Sun lies near the outer edge of a spiral arm that can be traced through nearly one turn about the Galaxy. The ridge-line of this spiral arm is quite accurately represented by the dotted curve in Figure 9.20, which is a logarithmic spiral with a $10°$ pitch angle. For historical reasons, the part of this arm that lies inside $R_0$ and at positive longitudes is called the **Sagittarius arm**, while the part that lies at negative longitudes is called the **Carina arm**.

Another, less conspicuous, spiral arm is just evident at the bottom left of Figure 9.20; a short section of a logarithmic spiral very approximately follows a chain a clouds. These clouds form part of the Orion–Cygnus arm, which probably passes just outside the Sun. Unfortunately, this arm lies near the edge of the region covered by the catalogs from which the figure is constructed, so no physical significance attaches to the complete absence of clouds further from the center than this arm. In §10.4.5 we shall see that young star clusters are concentrated along three ridges: the Sagittarius–Carina arm, the Orion–Cygnus arm, and a third arm, the Perseus arm, that passes the Sun significantly further out.

To search for spiral structure well outside $R_0$ we must turn from the distribution of molecular clouds, which are scarce at large radii, to $(l, v)$

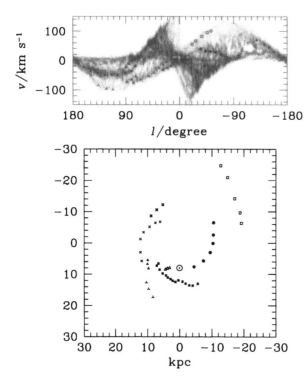

**Figure 9.21** Mapping of spiral arms from the $(l, v)$ plot to real space. The top panel shows a section of Figure 9.13, with points marking the most prominent linear features. The lower panel shows how these points map into real space. This mapping assumes that the hydrogen follows circular orbits at $v_c = 220 \, \text{km s}^{-1}$, and that $R_0 = 8.5 \, \text{kpc}$. The Galactic center lies at the origin and the Sun's location is marked by $\odot$.

plots for HI emission. Careful examination of Figure 9.13 reveals that the terminal-velocity curve at $(0 < l < 90°, v > 0)$ and at $(0 > l > -90°, v < 0)$ is bumpy, and points of unusually high intensity seem to delineate loops reminiscent of those generated by spiral structure in Figure 9.10. The upper panel of Figure 9.21 shows ridge-lines of spiral arms in the HI $(l, v)$ plot that have been formed by connecting local maxima in $T_B$. The lower panel shows the result of mapping these arms from $(l, v)$ space to real space under the assumption of circular orbits discussed in §9.1.2. A series of trailing spiral arms is apparent. One of these is clearly coincident with the Carina arm. Another arm passes about 2 kpc from the Sun in the direction of the anticenter. This is the **Perseus arm**.

### 9.2.6 Vertical distributions of HI and CO

Hitherto we have concentrated entirely on that part of the $(l, b, v)$ data cube that lies near the plane $b = 0$. Now we ask what can be learnt from the variation of observed brightness temperature with $b$. Let $(R, \phi, z)$ be a system of Galactocentric cylindrical coordinates oriented such that the plane $z = 0$ is identical with the plane $b = 0$. We wish to determine the value $z_c$ of $z$ at which the density of some tracer of the ISM such as CO emission peaks for given values of $(R, \phi)$. We call $z_c$ the central value of $z$ at $(R, \phi)$. In addition, we want to know how far in $z$ one must move from $(R, \phi, z_c)$ for the density of the given tracer to fall to a half of its value at $z = z_c$. We call this distance the half-thickness at half-intensity of the tracer, $z_{1/2}$.

Since the HI and CO layers turn out to be very thin, it is not surprising to find that the mean velocity of the ISM at $(R, \phi, z)$ lies near the circular speed at the corresponding point $(R, \phi, 0)$ in the plane (Lockman 1984). That is, the ISM approximately rotates as if it were fixed on a nested sequence of rigid cylinders, which are all rotating about the same axis through the Galactic center. In this approximation, every feature in the $(l, v)$ plane generalizes to a feature in the full $(l, b, v)$ data cube. In particular, the terminal-velocity curve generalizes to a terminal-velocity surface.

Since we wish to determine the extent of gas in $z$ from measurements of its extent in $b$, any ambiguity in the distance from which observed line emission comes is highly undesirable. In §9.1.1 we saw that, for a disk in simple circular rotation, a given point in the $(l, v)$ plane corresponds to a unique location in the disk at $R < R_0$ only if $(l, v)$ lies on the terminal-velocity curve. Consequently, the vertical structure of the ISM at $R < R_0$ is best determined by focusing on points in the $(l, b, v)$ data cube that lie near the terminal velocity surfaces. Malhotra (1995) presents such an analysis.

Beyond the solar circle, each $(l, b, v)$ point corresponds to at most one location in the Milky Way at $R > R_0$ and the determination of vertical structure from the observed distribution in $b$ is relatively straightforward once the circular-speed curve has been determined.

At the start of this section we remarked that the disk of the Milky Way falls into three parts: the inner $(R < 3\,\text{kpc})$, middle $(3\,\text{kpc} < R < R_0)$, and outer $(R > R_0)$ disk. We postpone discussion of the vertical structure of the disk at $R < 3\,\text{kpc}$ to §9.4.

**The middle disk**   The middle disk, which contains most of the molecular gas and star formation, is remarkably flat. More precisely, when one determines the central value of $z$ for some tracer of the ISM such as CO, one finds that $z_c(R, \phi)$ defines a surface that is very nearly a plane for $3\,\text{kpc} < R < R_0$. In 1958, the Galactic coordinate system $(l, b)$ was reoriented such that this plane was well approximated by the plane $z = 0$ (§2.1.2).

Figure 9.22 shows the degree to which the HI and CO layers deviate from the plane $z = 0$ by plotting, for HI and CO at $3\,\text{kpc} < R < R_0$, values of $z_c$ that have been estimated from the $l > 0$ tangent points. These oscillate around zero with rms amplitude $\sim 30\,\text{pc}$. Since $30\,\text{pc}$ is less than 1% of $R_0$, the middle disk is *extremely* flat. The oscillations apparent in Figure 9.22 appear to be coherent, however, in the sense that the values of $z_c$ for adjacent data points are highly correlated. Moreover, the two layers oscillate roughly in phase. Hence, it would seem that some dynamical process **corrugates** the disk. It is unclear what this process is. A full description of corrugations in the disk can be found in Spicker & Feitzinger (1986).

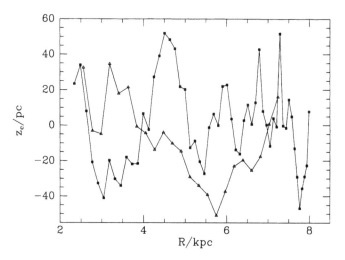

**Figure 9.22** The middle disk is very nearly planar. At a given radius $R$, $z_c$ is the value of $z$ at which the emission peaks for HI (squares) and CO (triangles). [After Malhotra (1994) and Malhotra (1995) from data kindly provided by S. Malhotra.]

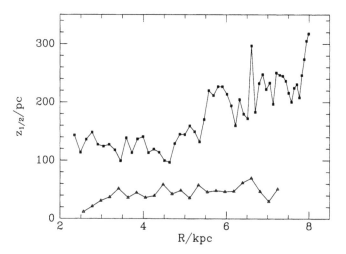

**Figure 9.23** The half-height at half-intensity of the HI (squares) and CO (triangles) layers. Open symbols are for $l > 0$ and filled symbols are for $l < 0$. [After Malhotra (1994) and Malhotra (1995) from data kindly provided by S. Malhotra]

Figure 9.23 shows the half-thicknesses at half-intensity $z_{1/2}$ of the HI and CO layers. The thickness of the CO layer is seen to increase slowly with $R$ from $\sim 35\,\text{pc}$ at $R = 3\,\text{kpc}$ to $\sim 70\,\text{pc}$ at $R = R_0$. The HI layer is about three times as thick at a given radius. Thus in the middle disk, the ISM forms an extremely thin layer.

The thickness of the HI and CO layers reflects a balance between the tendency of gravity to pull clouds into the equatorial plane and the tendency of the energy densities associated with the interstellar magnetic field, cosmic rays and the random motions of clouds to push clouds up out of the plane. By modeling the effect of random motions on the structure of the $(l, v)$ plot near the terminal-velocity curve (see §9.1.2), Malhotra (1995) has shown that the line-of-sight motions of HI clouds may be characterized by a velocity dispersion $\sigma_{HI} = 9 \pm 1\,\mathrm{km\,s^{-1}}$ independent of $R$ for $0.3 \lesssim R/R_0 < 1$. This dispersion is in good agreement with the line-of-sight velocity dispersions of HI clouds observed at high latitudes $|b|$ (Merrifield 1993) and also with the measurements of the 21-cm line widths of face-on galaxies that were cited in §8.1.4.[4] Of course, in the Milky Way the line of sight usually lies within the disk, while for a face-on galaxy it lies perpendicular to the disk. Thus it seems likely that the velocity dispersion of clouds in all galaxies is nearly isotropic.

If the pull of gravity on HI clouds in the Milky Way towards the plane were balanced only by a velocity dispersion of $9\,\mathrm{km\,s^{-1}}$ perpendicular to the plane, it is easy to show that the HI layer would be roughly one third as thick as it actually is – see Problem 9.7. This calculation demonstrates the importance of the magnetic and cosmic-ray pressures for the dynamics of the HI layer.

In §9.2.3 we saw that the random velocities of CO clouds are of order $7\,\mathrm{km\,s^{-1}}$. Since the random velocities of clouds contribute to the thickness of a gas layer as the square of their dispersion, from the dispersions of the HI and CO clouds alone we would predict that the HI layer would be nearly twice as thick as the CO layer, just as the observations indicate. Since magnetic and cosmic-ray pressures play important roles, this calculation is clearly too naive. The correct interpretation of the harmony between the relative dispersions and layer-thicknesses of the HI and CO clouds is probably that the velocity dispersions of the two types of cloud are in dynamical equilibrium with the magnetic and cosmic-ray forces on the clouds, with the result that the total pressure from all three sources is proportional to the square of the velocity dispersion.

**The outer disk**   In §8.2.4 we saw that the majority of galactic disks are warped at large radii. The disk of the Milky Way is no exception, and was, in fact, the first warped disk to be discovered: its warp was apparent as soon as 21-cm data became available for the southern hemisphere (Burke 1957, Kerr 1957, Westerhout 1957). Our location in the Milky Way proves to be ideally suited for the study of the warp, and it has been reanalyzed several times

---

[4] Figure 5.18 shows that the cooling rate climbs steeply near $T = 10^4$ K. Consequently, it is very hard to heat the ISM above this temperature and much of the diffuse ISM is at temperatures near $10^4$ K. At this temperature the sound speed in the ISM lies close to $10\,\mathrm{km\,s^{-1}}$. Velocities in excess of the sound speed are strongly damped by shocks.

since the pioneering investigations of Oort, Kerr & Westerhout (1958) and
Gum, Kerr & Westerhout (1960), most recently by Diplas & Savage (1991),
who employed the HI survey of Stark *et al.* (1992).

The structure of the warp is most easily understood by first transforming
the data cube of a 21-cm survey from $(l, b, v)$ coordinates, which are natural
for the observations, into the usual Galactocentric polar coordinate system
$(R, \phi, z)$ (see Problem 9.8). In this transformation velocities are converted
to radii using equation (9.3) in conjunction with an assumed circular-speed
curve $v_c(R) = R\Omega(R)$. This step involves the implicit assumption that the
ISM rotates as if it were made up of a series of concentric rigid cylinders. This
assumption is very unlikely to be precisely valid where the disk is strongly
warped, but the errors in the derived values of $(R, \phi, z)$ to which it gives rise
are probably insignificant: the motions that are generated by the warp should
lie nearly perpendicular to the plane, and therefore make no contribution to
$v_{\rm los}$.

The classical analyses of the warp assumed that beyond $R_0$ the circular speed was slowly falling, while modern analyses assume that $v_c$ is flat
or slowly rising to the largest observed radii. Changing the shape of the
assumed rotation curve changes the scale associated with the observed warp
as the following argument shows. Equation (9.17) is used to infer $R$ from a
measured value of $W$, and the slower $\Omega$ decreases with $R$, the larger will be
the inferred value of $R$ for given $W$. Consequently, the radial scale of the
warp was smaller in earlier work than in more recent analyses.

Figure 9.24 shows the intersections of the HI layer with four cylinders
that have $R$ equal to 12, 16, 20 and 24 kpc. The figure, in which the azimuthal
coordinate $\phi$ is plotted horizontally, and $z$ vertically, is for $R_0 = 8.5$ kpc and
a circular-speed curve that is perfectly flat at $v_c = 220 \, {\rm km \, s^{-1}}$ for $R > R_0$.
The warp is just discernible at $R = 12$ kpc, and becomes pronounced by
$R = 16$ kpc. For $R \lesssim 17$ kpc the warp is well approximated by a simple
sine curve with an amplitude that rises linearly in $R$. Beyond 17 kpc, a
pronounced asymmetry develops between $\phi > 0$ and $\phi < 0$. For $R > 11$ kpc
a reasonable approximation to the figure of the warp is given by

$$z_c = \frac{R/\,{\rm kpc} - 11}{6} \sin \phi + 0.3 \left( \frac{R/\,{\rm kpc} - 11}{6} \right)^2 (1 - \cos 2\phi). \qquad (9.18)$$

In particular, at $\phi > 0$, $R = 24$ kpc the HI layer rises more than 4 kpc out
of the plane $z = 0$ defined by the middle disk, while at $\phi < 0$ the HI layer
approximately lies within the plane $z = 0$. The Sun happens to lie near
the warp's line of nodes, that is, the line in which the warped outer disk
intersects the plane of the inner Galaxy.

Mapping the warp in CO is a painful process for two reasons: (i) in
the outer disk CO clouds are sparse and have small CO luminosities; (ii)
the warp is expected to carry clouds well out of the great circle $b = 0$, with
the result that a thorough survey must cover a large area of the sky. This

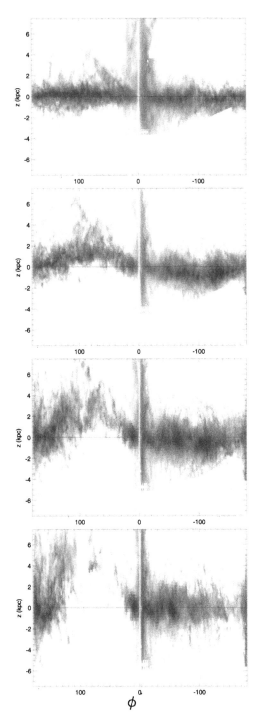

**Figure 9.24** In each panel the grayscale shows the HI density on a cylinder of radius $R$ with the Galactic center on the cylinder's axis. The horizontal and vertical coordinates are cylindrical polar coordinates $\phi$ and $z$, respectively, with the Sun at $(R, \phi, z) = (R_0, 0, 0)$ and $\phi$ increasing in the direction of Galactic rotation. Hence, the northern data are at $\phi > 0$. From top to bottom $R$ takes the values 12, 16, 20 and 24 kpc. [From data published in Burton (1985), Hartmann & Burton (1997) and Kerr *et al.* (1986) courtesy of T. Voskes and B. Burton]

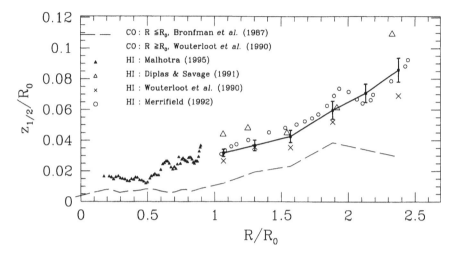

**Figure 9.25** The half-thickness at half-maximum intensity of the HI and CO layers, showing how the increase in thickness with radius continues beyond $R = R_0$. Results from various data sets are shown for the HI layer thickness; the solid line and error bars indicate the average and standard error of these results. [Figure kindly provided by R. Olling]

requirement is especially difficult to satisfy with a dish of the size needed to achieve high sensitivity, because this will measure emission from a small region with each pointing. On account of these difficulties, the warp has been mapped much less extensively in CO than in HI, but the available data suggest that the CO layer has the same shape as the HI layer (Grabelsky *et al.* 1987, Digel *et al.* 1991). Wouterloot *et al.* (1990) circumvented the need for a systematic survey in CO by searching in the IRAS catalogue for IR sources that are embedded in molecular clouds. They obtained CO spectra in the directions of these objects, and thus deduced their velocities, from which kinematic distances followed in the usual way. With this technique they were able to trace the warp to $R \simeq 20\,\mathrm{kpc}$, and measure the thickness of the CO layer to this distance.

The warp has also been traced in relatively nearby OB stars (Miyamoto, Yoshizawa & Suzuki 1988, Smart & Lattanzi 1996), red-giant stars (Carney & Seitzer 1993), discrete sources in the IRAS catalogue (Djorgovski & Sosin 1989) and diffuse IR radiation from stars (§10.1). There is still no generally accepted explanation for the cause of the warp (Binney 1992).

The increase with $R$ in the thickness of the HI layer that we encountered above for $R < R_0$ continues, and indeed gathers pace, at $R > R_0$. Figure 9.25 illustrates this phenomenon: $z_{1/2}$ for HI rises from $\sim 250\,\mathrm{pc}$ at $R_0$ to $\sim 600\,\mathrm{pc}$ at $2R_0$ and to even larger values further out. The variation in $z_{1/2}$ with $R$ is of considerable dynamical interest because $z_{1/2}$ is dictated by the hydrostatic balance between the pull of gravity toward the Galactic plane and the countervailing pressure within the ISM. Thus, $z_{1/2}(R)$ provides a direct measure of the total mass near the plane of the Galaxy as a function of radius

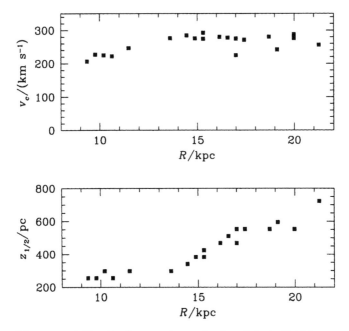

**Figure 9.26** The circular-speed curve (top) and the half-thickness at half intensity (below) of the HI layer as determined by Merrifield (1992) using only the 21-cm data of Weaver & Williams (1973,1974) and Kerr *et al.* (1986).

(Olling 1995). As we saw in §9.2.3, the Milky Way's rotation curve yields the total mass as a function of radius in the Galaxy; combining this information with the measurement of how much mass lies close to the Galactic plane then allows us to measure the *shape* of the mass distribution. For $R > R_0$, where most of the mass cannot be explained by the visible components, this analysis gives a direct measure of the shape of the massive dark halo. The procedure is somewhat complicated by the uncertain contribution to pressure support from cosmic rays and magnetic fields, and the warp in the gas layer, but the analysis suggests that the dark halo of the Milky Way is slightly oblate, with a shortest-to-longest axis ratio of $\sim 0.8$ (Olling & Merrifield 1998).

Before we leave the topic of the thickness of the HI layer, we should mention a method of determining $z_{1/2}(R)$ and $v_c(R)$ outside the solar circle without knowledge of the distances to individual objects. Merrifield (1992) shows that if the disk is in circular rotation and $z_{1/2}$ depends only on $R$ and not on azimuth $\phi$, one can determine both $z_{1/2}(R)$ and $v_c(R)$ from 21-cm data alone. The essence of this method is as follows: we pick a particular value of angular velocity, $\Omega_i$ and at each point $(l,b)$ on the sky we determine the intensity of HI emission at velocity $v_{\rm los} = (\Omega_i - \Omega_0)R_0 \sin l$. By equation (9.3), this slice through the $(l, b, v_{\rm los})$ data cube originates from material with angular velocity $\Omega_i$. So long as $\Omega(R)$ is a monotonic function, $\Omega_i$ corresponds to a single radius $R_i$, with the result that all emission in this slice through the data cube originates from a cylindrical shell with an (as-yet unknown) radius $R_i$.

If $R_i < R_0$, we lie outside the shell, so the emission is confined to a finite range in $l$. We can determine $R_i$ from the extent of this range, and the analysis is essentially identical to the tangent point method of §9.1.1.

If $R_i > R_0$, we lie inside the cylindrical shell of emitting gas, so the emission extends to all $l$. From these data we may determine for each $l$ the distance in $|b|$ at which the emission falls to half its peak intensity. This quantity is a function $b_{1/2}(\Omega_i, l)$ of the chosen angular velocity and $l$. The reason $b_{1/2}$ varies systematically with $l$ is that the distance $d$ to the ring of emission varies with Galactic latitude, and so the angle subtended by $z_{1/2}(R_i)$ varies with $l$. Quantitatively,

$$b_{1/2}(R_i, l) = \arctan\left(z_{1/2}/d\right)$$
$$= \arctan\left(\frac{z_{1/2}(R_i)}{R_0 \cos l + \sqrt{R_i^2 - R_0^2 \sin^2 l}}\right), \qquad (9.19)$$

where we have used equation (9.16). The quantities $z_{1/2}(R_i)$ and $R_i(\Omega_i)$ can be estimated by varying them until equation (9.19) most nearly predicts the observationally determined function $b_{1/2}(\Omega_i, l)$. Of course, once $R(\Omega)$ has been determined by calculating $R_i$ for a range of values of $\Omega_i$, $v_c(R)$ follows trivially.

Figure 9.26 shows the circular-speed curve and disk-thickness profile obtained by this method. The method breaks down for $R \gtrsim 2R_0$ because the HI layer is too patchy at these large distances for $b_{1/2}$ to be well defined in the HI data. Out to these radii the values of $v_c(R)$ and $z_{1/2}(R)$ compare reasonably well with those shown in Figures 9.18 and 9.25. There is, however, a noticeable discrepancy between the circular-speed curves derived by these different methods. Kuijken & Tremaine (1994) pointed out that this discrepancy is expected if the outer disk of the Milky Way is slightly elliptical in shape rather than circular, as we have assumed.

## 9.3 Other tracers of the ISM

### 9.3.1 Diffuse infrared emission

As we mention in §8.1.6, our knowledge of the infrared appearance of the Universe in general, and the Milky Way in particular, was revolutionized in the 1980s by the IRAS satellite. In addition to cataloging $\sim 250\,000$ point sources, IRAS detected diffuse emission over almost the entire sky. In the 1990s these measurements were confirmed and extended to longer wavelengths by the DIRBE experiment aboard the COBE satellite. Boulanger &

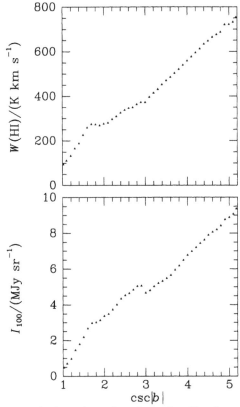

**Figure 9.27** Lower panel: 100 μm flux density, averaged over longitude $l$, measured by COBE as a function of galactic latitude $b$. The $x$-coordinate is linear in $\csc |b|$, so that emission from an optically-thin, uniform layer of dust would yield a straight line. Upper panel: the corresponding plot for the 21-cm line as measured in the Leiden–Dwingeloo survey. [After Boulanger *et al.* (1996) from data kindly supplied by F. Boulanger]

Pérault (1987) analyzed the implications of the IRAS data for the Galaxy's diffuse infrared emission, while Boulanger *et al.* (1996) analyzed the COBE data.

In Box 3.2 we saw that if we model the distribution of dust in the disk by a layer of uniform thickness, the column density of dust along the line of sight is proportional to $\csc |b|$. Hence if the diffuse IR radiation detected by IRAS and COBE is emitted by dust, we might expect its intensity, when averaged over latitude $l$, to be proportional to $\csc |b|$. The lower panel of Figure 9.27 tests this hypothesis by plotting against $\csc |b|$ the $l$-averaged 100 μm flux density measured by COBE. In preparing this plot, Boulanger *et al.* (1996) omitted data for directions to the Magellanic Clouds and to a few prominent molecular clouds, and they have subtracted the contribution of dust in the solar system ('zodiacal emission' – see §4.2.1).

For $\csc |b| \lesssim 3$ ($|b| < 20°$) all curves fall nearly linearly in Figure 9.27 in accordance with the prediction of the slab model of Box 3.2. For $\csc |b| \lesssim 2.2$ ($|b| \gtrsim 30°$), the flux densities in the two longer wavebands lie slightly above straight lines fitted to the data for $\csc |b| \gtrsim 3$. Thus at all latitudes the data for these longer wavebands is in reasonable agreement with the prediction of the model in which the detected IR radiation is emitted by a thin, uniform

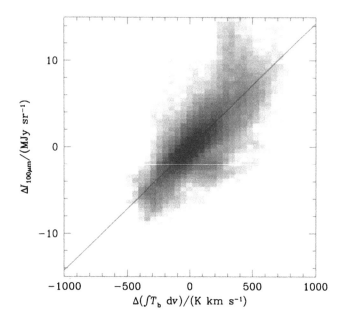

**Figure 9.28** Plot showing how the small-scale fluctuations in
$100\,\mu$m emission (as measured by the COBE satellite) are correlated with those in the integrated flux from the 21-cm line (as
measured in the Leiden–Dwingeloo survey). The small-scale fluctuations have been calculated by subtracting a smooth model from
the infrared and HI data. The grayscale shows the number of 0.5°-
square pixels on the sky where the residual fluctuations have the
indicated values. The line shows the mean relation between $I_{100\,\mu m}$
and 21-cm flux determined from Figure 9.27. [After Boulanger &
Pérault (1987) using data kindly provided by F. Boulanger and
G. Lagache]

dust layer such as that assumed in Box 3.2.

The distributions of the diffuse IR radiation shown in Figure 9.27
strongly suggests that this radiation comes from the Galactic plane, but
do unresolved stellar sources contribute significantly to it? Boulanger &
Pérault (1987) argue that the answer to this question is 'no' because even in
the $12\,\mu$m waveband, to which stars (being hot) would contribute most, the
integrated radiation from all resolved stars amounts to only 8% of the observed diffuse flux density, and the emission of unresolved stars is estimated
to account for $\lesssim 3\%$ of the diffuse flux density. Thus the ISM accounts for
$\sim 90\%$ of the IR luminosity of the Milky Way.

This conclusion is reinforced by comparison of the $100\,\mu$m flux-density
profile shown in the lower panel of Figure 9.27 with the analogous profile

for the 21-cm line that is shown in the top panel. The shapes of these two profiles are similar. Figure 9.28 strongly reinforces this conclusion by plotting for each 0.5°-square pixel on the sky, the difference between the observed 100 $\mu$m flux density and the mean flux density at the pixel's latitude $b$, versus the analogous quantity for the 21-cm line. The diagonal elongation of the densely populated part of the figure demonstrates that a pixel that is brighter at 100 $\mu$m than normal for its latitude, is overwhelmingly likely to yield a stronger than average 21-cm signal. The diagonal straight line in Figure 9.28 shows the relationship between 100 $\mu$m flux and 21-cm flux that one derives from the $l$-averaged data plotted in Figure 9.27. In so far as this line lies close to the ridge-line of the densely-populated part of Figure 9.28, there is a universal relationship between 21-cm line strength and 100 $\mu$m flux density. If we use equation (8.22) to convert 21-cm line strengths into column densities of hydrogen atoms, we may express this relationship as

$$j_{100\,\mu\mathrm{m}} = 6.66 \times 10^{-44}\,\mathrm{W\,Hz^{-1}\,atom^{-1}}. \tag{9.20}$$

That is, a box of ISM that contains $n_H$ hydrogen atoms on the average radiates $1.07 \times 10^{-43} n_H\,\mathrm{W\,Hz^{-1}}$ at 100 $\mu$m. In particular, there is no room for doubt that the ISM is as responsible for the diffuse 100 $\mu$m emission as it is for the 21-cm line.

We saw in §8.1.6 that the brightness of the ISM in the shorter IRAS wavebands came as a considerable surprise because the importance of small grains had not been appreciated prior to the launch of the IRAS satellite. It is still true, however, that most of the luminosity of the ISM emerges at $\lambda > 60\,\mu$m: roughly half of the luminosity observed by IRAS falls in the 100 $\mu$m band, with a sixth only in each of the other three bands. COBE was sensitive at the longer wavelengths, where most of the flux from the ISM is expected to lie. Sodroski *et al.* (1994) presented a preliminary analysis of the data for $|b| < 10°$. From a comparison of the 140 $\mu$m and 240 $\mu$m fluxes for different fields with the corresponding fluxes in the 21-cm and CO lines and in 5 GHz radio-continuum radiation, they concluded that the larger grains that dominate the far-IR fluxes, have temperatures $\sim 20\,\mathrm{K}$ and together radiate about $1.1 \times 10^{10}\,L_\odot$ in the far-IR. For comparison, the Milky Way's $V$-band luminosity is about $L_V = 1.4 \times 10^{10}\,L_\odot$. Since the dust only emits because it is heated by stars, we conclude that *interstellar dust reprocesses nearly a half of the energy produced by stars.*

Figure 9.29 shows the variation with $l$ of the far-IR luminosity per unit gas mass determined by Sodroski *et al.* (1994). Near the solar circle ($l \lesssim 90°$) this is $\sim 1.5-3\,L_\odot/\mathcal{M}_\odot$, but it is significantly greater further in.[5] With data for only two wavebands, Sodroski *et al.* (1994) could not map the temperature of the dust with precision, but they do conclude that its temperature rises

---

[5] Boulanger & Pérault (1987) estimated a similar total IR luminosity from the shorter-wavelength IRAS data.

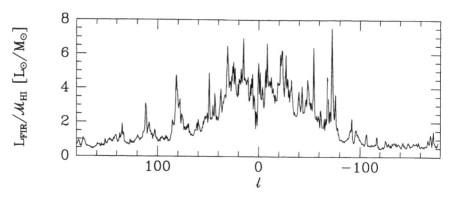

**Figure 9.29** Far-IR luminosity per unit gas mass as a function of longitude $l$. [After Sodroski *et al.* (1994) from data kindly provided by T. Sodroski]

from $\sim 18\,\mathrm{K}$ in the anticenter direction to $\sim 21\,\mathrm{K}$ near the center. From more extensive COBE data Boulanger *et al.* (1996) measured the dust temperature to be $17.5\,\mathrm{K}$. Since the emissivity of dust depends on temperature roughly as $T^6$ (see Problem 8.7), these temperatures imply that the emissivities of dust near the center and beyond the solar circle should be in the ratio $\sim 2.5 : 1$, which is crudely consistent with Figure 9.29.

On small scales the Galaxy's IR emission is irregular, and much of it is concentrated into wisps or streaks on the sky. Low *et al.* (1984) dubbed this wispy component of the emission **infrared cirrus**. Equation (9.20) gives the *mean* $100\,\mu\mathrm{m}$ emissivity per interstellar hydrogen atom. On small scales this emissivity varies by up to a factor of 3. It tends to be highest in the vicinity of OB associations. A natural explanation of this phenomenon is that the emissivity at any point depends on the local flux of UV photons, which dominate the supply of energy to grains – see Problem 8.8.

### 9.3.2 Pulsars and the Galactic magnetic field

Measurements of the dispersion measure $D_\mathrm{M}$ (§8.1.4) along lines of sight to pulsars that have known distances have been used to determine the mean value of the free-electron density, $n_\mathrm{e}$, in the disk. In most cases the distance to the pulsar is estimated from the structure of 21-cm absorption in its spectrum. In a few cases the pulsar is associated with a nebula whose distance can be determined by some other method.[6] A model of the variation of $n_\mathrm{e}$ with Galactocentric radius $R$ and vertical position $z$ within the Galaxy that accounts for the available data is

$$n_\mathrm{e}(R, z) = \left(2.5 + 1.5\mathrm{e}^{-|z|/70\,\mathrm{pc}}\right)\left(\frac{2}{1 + R/R_0}\right) \times 10^4\,\mathrm{m}^{-3}. \qquad (9.21)$$

---

[6] The distances to most pulsars are determined from their dispersion measures by *assuming* the validity of a model such as that given by equation (9.21).

The Galactic magnetic field **B** has been probed by all the methods that have been used to probe magnetic fields in external galaxies (§8.2.7). In particular, the tendency of the Davis–Greenstein effect (§3.7.1) to align the electric vector of radiation from obscured stars parallel to **B**, has been used to obtain Figure 3.19, which shows the direction of the Galactic magnetic field near the Sun. Averaged over kiloparsec scales, **B** usually points within the plane. On smaller scales, plumes of magnetic field rise tens of parsecs above the plane in a way reminiscent of magnetic loops observed in the chromosphere of the Sun.

In §8.1.4 we saw that when both the dispersion measure and the rotation measure of a pulsar are known, one can determine the mean value along the line of sight (weighted by $n_e$) of the component of **B** that is parallel to the line of sight. Such measurements confirm that the Galactic magnetic field is highly organized on kiloparsec scales and also enable us to determine the magnitude of the field. The mean field runs azimuthally around the Galaxy. At the solar radius and beyond, the field points in the direction of Galactic rotation. Roughly 500 pc inside $R_0$, the direction of the field reverses, so that it points in the opposite direction to Galactic rotation in the Sagittarius spiral arm (§9.2.1). At a Galactocentric radius of $\sim 5.5$ kpc, **B** seems again to reverse its direction (Rand & Lyne 1994).

As Figure 3.19 makes clear, on small scales **B** wiggles. The importance of the wiggles is usually quantified in terms of a picture in which the field is a superposition of an ordered large-scale component and a random small-scale component. The ratio of the RMS values of the ordered and small-scale fields is a measure of the importance of wiggles. Measurements of $D_M$ and $R_M$ for pulsars are sensitive only to the ordered field. They indicate that this has magnitude $\sim 0.2$ nT near the Sun, rising to $\sim 0.6$ nT at $R = 4.5$ kpc (Rand & Lyne 1994).

The magnitude of the disordered component of $B$ is much harder to estimate. Probably the most reliable estimates are obtained from the typical intensity $j_{\text{sync}}$ of Galactic synchrotron emission by the argument given in §8.2.7. Berkhuijsen (1998) finds that $B$ decreases from $\sim 1$ nT at $r = \frac{1}{2}R_0$ through $\sim 0.6$ nT at $R_0$ to $\sim 0.3$ nT at $R = 2R_0$. Taken with the estimates of the magnitude of the ordered component of $B$ cited above, these values for $B$ imply that the random component of $B$ is at least two and a half times larger than the ordered field. As in external galaxies, it seems that $B$ is most highly ordered between spiral arms and becomes relatively disordered in spiral arms.

In a few cases there is convincing evidence that $B$ is larger in molecular clouds than in lower-density interstellar space. For example, from a study of the polarization of 21-cm radiation from the Orion B cloud, van der Werf *et al.* (1993) concluded that, as one probes deeper into the cloud and $n_H$ rises from $10^9$ m$^{-3}$ to $10^{11}$ m$^{-3}$, $B$ increases from $\sim 3$ nT to 6 nT.

The strongest interstellar magnetic field known occurs in the filaments of synchrotron emission that are pictured in Figure 9.30. These filaments are

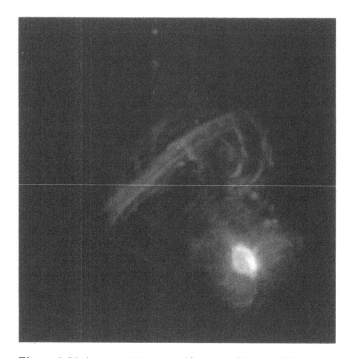

**Figure 9.30** A square 85 pc on a side mapped in 1.65 GHz continuum emission. The plane runs from top left to bottom right and the radio source Sgr A, which contains the Galactic center, visible as a big blob of emission just below and to the right of the center of the panel. About 30 pc from Sgr A several filaments of synchrotron emission extend perpendicular to the disk for a few parsecs. [After Yusef-Zadeh (1989) using data at http://imagelib.ncsa.uiuc-.edu/project/preview/95.FY.01.01]

located about 30 pc from the Galactic center and extend perpendicular to the Galactic plane for a few parsecs. From the brightness of the synchrotron emission in them one infers $B \sim 100$ nT. Since magnetic stresses scale as $B^2$, such an intense field would exert forces that are 10 000 times larger than those exerted by $\mathbf{B}$ at a typical point in the ISM, where $B \lesssim 1$ nT. These objects are not fully understood, but their straightness probably reflects the ability of a dynamically dominant magnetic field to resist the tendency of interstellar turbulence to tangle field lines. A few cases of filaments of synchrotron emission that run perpendicular to the plane are known nearer to the Sun (Wieringa *et al.* 1993).

### 9.3.3 Diffuse Hα radiation

Fabry-Perot spectrometers detect diffuse Hα radiation at all Galactic latitudes (Reynolds 1993). The intensity of this radiation varies irregularly over the sky, but the average intensity within a window that is several degrees in size is roughly proportional to csc $|b|$, where $b$ is the latitude of the window's center. This scaling with csc $|b|$ suggests that the radiation originates in a layer centred on the plane – see Box 3.2. The production of the observed H$\alpha$ emission within this layer requires a large power input to the gas of the layer: when the data are analyzed along the lines of §8.1.3, one finds that the minimum power requirement is $\sim 10^{-7}\,\mathrm{W\,m^{-2}}$ of the disk, or $\sim 4 \times 10^{34}\,\mathrm{W}$ for the disk out to a Galactocentric distance of 12 kpc. It is interesting to compare this power requirement with the average power input by supernovae, which are thought to be one of the major energy sources for the ISM. If there is one supernova per 30 years in the Milky Way, then their average power input to the ISM is $\sim 5 \times 10^{34}\,\mathrm{W}$. Thus the diffuse Galactic H$\alpha$ emission is associated with a major item in the energy budget of the ISM.

The observed H$\alpha$ emission is not associated with detectable emission in the 630 nm and 520 nm lines of neutral oxygen and nitrogen, respectively. This absence of N and O lines suggests that the diffuse H$\alpha$ emission comes from gas that is nearly fully ionized. Consequently, in this gas the free-electron density is a reliable guide to the total mass density. We saw in §8.1.5 that the column density of free electrons along the line of sight to a pulsar can be determined from the pulsar's dispersion measure. Column densities determined in this way for lines of sight that reach far out of the plane are $N \sim 7 \times 10^{23}$ csc $|b|\,\mathrm{m^{-2}}$. When the contribution to the mass density of helium atoms is taken into account, one concludes that there is $1.6\,\mathcal{M}_\odot\,\mathrm{pc^{-2}}$ of ionized gas in the disk near the Sun.

If we assume that the particle-density $n$ of the ionized gas varies with height $z$ above the plane as

$$n(z) = n(0)e^{-z/z_0}, \tag{9.22}$$

then we can estimate the scale-height $z_0$ of the gas from the value of $n(0)$ that is obtained by observing nearby pulsars, and the total column density $N_\perp$ along latitude $b = 90°$. Reynolds (1993) concludes that $n(0) \simeq 2.5 \times 10^{-8}\,\mathrm{m^{-3}}$, and thus that $z_0 \simeq 1\,\mathrm{kpc}$.

From the irregularity of the observed H$\alpha$ flux on the sky it follows that the ionized gas from which it comes is inhomogeneously distributed. It is likely that it fills only a small fraction $\lesssim 10\%$ of space near the plane, but $\gtrsim 50\%$ of space at $z \sim 1\,\mathrm{kpc}$.

### 9.3.4 Diffuse synchrotron and $\gamma$-radiation

Until recently, the angular resolution with which $\gamma$-rays have been detected was extremely poor, and even now it remains poor by comparison with the angular resolutions routinely employed in other branches of astronomy. A consequence of this lack of resolution is that it is not clear how much of the observed $\gamma$-radiation is contributed by point sources. That point sources do

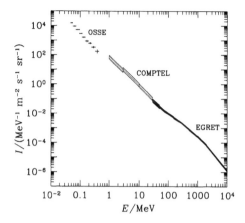

**Figure 9.31** Energy spectrum of diffuse $\gamma$-rays from the Milky Way, as measured by various instruments (OSSE, COMPTEL, EGRET) on the Compton Gamma-Ray Observatory. [From data presented in Strong *et al.* (1996), and references therein]

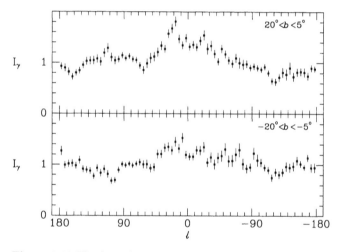

**Figure 9.32** The dependence upon longitude $l$ of the $70-150\,\mathrm{MeV}$ $\gamma$-ray flux measured by the COS-B satellite in the northern and southern Galactic hemispheres. [After Bloemen (1993) from data kindly provided by H. Bloemen]

contribute significantly follows from the fact that observed $\gamma$-ray fluxes are time-variable (Skibo, Ramaty & Leventhal 1992).

Figure 9.31 shows the spectrum of the Milky Way's diffuse $\gamma$-radiation. It extends from $0.1 - 10^4\,\mathrm{MeV}$ and can be crudely characterized as a power law $q_\gamma(E) \propto E^{-2}$. The shape of this spectrum is reasonably well understood in terms of the processes discussed in §8.1.4.

Figure 9.32 shows the flux of $\gamma$-rays as a function of longitude $l$ for both northern and southern Galactic hemispheres. The flux is weakly peaked around $l = 0$. Theoretical models of the $\gamma$-ray intensity tend to predict a more centrally peaked $\gamma$-ray flux than that shown in Figure 9.32; as we saw in §8.1.4, the $\gamma$-ray emissivity should be proportional to the product $q_\gamma n$ of the cosmic-ray flux $q_\gamma$ and the density $n$ of the ISM, and both of these are expected to peak near the Galactic center.

Cosmic rays not only give rise to $\gamma$-rays, but also to synchrotron radiation. Consequently, by observing diffuse synchrotron radiation, we can obtain additional information regarding the Galactic distribution of cosmic rays. In §8.2.1 we saw that in disk galaxies the 1.4 GHz radio-continuum brightness temperature $T_{1.4}$, which measures the strength of synchrotron radiation, is accurately proportional to the intensity $I_{CO}$ of 2.6 mm radiation by $^{12}CO$ molecules. Allen (1992) shows that the same proportionality holds within the disk of the Milky Way, point by point, for Galactocentric radii in the range $2 - 12$ kpc. This result is striking because over this radius range $T_{1.4}$ and $I_{CO}$ vary by two orders of magnitude. The most natural interpretation of the correlation is that the cosmic rays responsible for the synchrotron emission also heat the molecular clouds, increasing the intensity of their CO emission (Suchkov, Allen & Heckman 1993). Thus, the value of $I_{CO}$ from any point depends on the cosmic-ray environment as well as the total amount of CO at that point. The common use of $I_{CO}$ as a direct measure of the molecular content of the Milky Way and other galaxies (§§9.6.1 and 8.1.4) should therefore be viewed with some caution.

### 9.3.5 Diffuse X-rays

The distribution of diffuse X-radiation in a band centred on 0.2 keV is strongly concentrated towards the Galactic poles. More generally, in this band the X-ray flux in each direction is strongly anti-correlated with the strength of the 21-cm line in that direction. A natural interpretation of this finding is that the X-rays come from outside the relatively thin layer to which neutral hydrogen is confined, and at low latitudes are absorbed by cool gas on their way to us. If this interpretation were correct, X-rays of lower energy, which are more strongly absorbed by cool gas, would be even more strongly concentrated to high latitudes. Actually the distribution of lower-energy X-rays is remarkably similar to that observed for the band centred on 0.2 keV.

These findings led Snowden *et al.* (1990) to hypothesize that the Sun sits in a 'local bubble' of hot X-ray emitting gas that is embedded in the layer of cool gas. In this model, the X-ray and 21-cm fluxes are anticorrelated because the X-ray flux is small and the 21-cm flux is large in directions in which we are near the bubble's edge, while in directions in which we are far from the bubble's edge the X-ray flux is large and the 21-cm flux is small. Subsequently, the ROSAT satellite demonstrated that at least part of the correlation between X-ray and 21-cm fluxes *is* generated by absorption, so the situation is now rather confused.

A major problem that dogs attempts to study the ISM with X-rays is that of determining how much of the observed diffuse flux is of extragalactic origin. It is known that a large part of the high-latitude soft X-ray flux derives from unresolved quasars and Seyfert galaxies (Hasinger *et al.* 1993).


## 9.4 The central disk

We now turn to the study of the ISM inside $R \simeq 3\,\mathrm{kpc}$, which corresponds to $|l| \lesssim 20°$. The structure of this region proves to be highly complex, and after forty years of study there is barely a consensus as to even its basic form. Here we shall argue the case for what seems to us the most plausible interpretation of the observations, while trying to deal fairly with some alternative interpretations. Moreover, "most plausible" should not be read as "plausible." If the picture we shall present is at all close to the truth, the center of our galaxy is a very strange place indeed.

Before we plunge into the detail of the observations, it is worth taking an inventory of the relevant data. The most useful data are from line observations of various species, especially HI, CO, CS, OH, and $H_2CO$ (formaldehyde), which we now discuss.


### 9.4.1 21-cm observations

The 21-cm line of HI is important because it can be detected over almost the entire region ($|l| < 20°$, $|b| < 10°$). However, the interpretation of the 21-cm spectra for some crucial lines of sight is delicate because the observed spectra are significantly affected by absorption. Two phenomena contribute to absorption in the observed spectra. First and most obviously, some of the 21-cm line radiation emitted in the center region is absorbed in the great column of gas that lies in the middle disk between us and the center. Since this material is moving on nearly circular orbits, it contributes strong absorption only at the velocities that lie in the interval $(0, 300\sin l)\,\mathrm{km\,s^{-1}}$ characteristic of circular orbits in the middle disk (see Problem 9.10).

A second contribution to absorption in spectra observed at small $|l|$ and $|b|$ arises from diffuse 21-cm continuum emission from the region ($|l| \lesssim 1.2°$, $|b| \lesssim 0.2°$) (Reich *et al.* 1984). As we saw in §8.2, when a background continuum source is observed through a screen of HI atoms, it contributes to the spectrum more strongly far from line-center, where there is negligible absorption, than it does at line-center. Consequently, the line appears weaker in emission than it would if the continuum source were not present. If the brightness temperature of the weakened line is blindly inserted into equation (8.22), a spuriously low value of the column density $N_H$ will be derived.

There are two possible strategies for dealing with this problem of pervasive absorption. One is to calibrate the absorption with the aid of lines of molecules such as OH and $H_2CO$ that absorb radiation without themselves

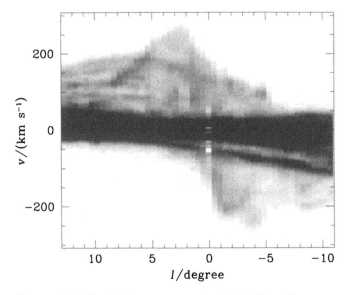

**Figure 9.33** The brightness temperature in the 21-cm line averaged over $|b| < 0.5°$ the central region of the Galaxy. [After Liszt & Burton (1980) and Burton & Liszt (1983) from data kindly provided by H. Liszt]

emitting significantly. The other strategy is to abandon the 21-cm data and observe the region of interest in lines of molecules such as CO that are less affected by absorption; at the high frequencies of these lines, the brightness temperature of continuum emission is negligible. A major problem with both strategies is that tracers other than HI have very low densities in many areas of interest, so some dynamical information can only be gleaned from the 21-cm line, with all its attendant problems.

In Figure 9.13 we saw that the envelope of the 21-cm emission at $(l > 0, v > 0)$ has a pronounced horn at $(l = 15°, v = 140 \,\mathrm{km\,s^{-1}})$ and falls steeply from there towards the center, and we interpreted this phenomenon as evidence for a hole of radius $\sim 3\,\mathrm{kpc}$ in the centre of the HI disk. This hole is not *entirely* devoid of HI, however. Figure 9.33 is an $(l, v)$ plot for the region $(|l| < 12°, |b| < 0.5°)$, which is said to be occupied by the **central disk**. (The grayscale of this plot corresponds to much lower brightness temperatures than the faintest levels shown in Figure 9.13 with the result that matching Figures 9.13 and 9.33 is difficult.) At $(l > 0, v > 0)$ the envelope of the emission in Figure 9.33 sweeps up from $v \sim 180\,\mathrm{km\,s^{-1}}$ at $l = 12°$ to reach $v \sim 270\,\mathrm{km\,s^{-1}}$ at $l \sim 3°$. The envelope then rounds a horn and plunges steeply *but crosses* $l = 0$ at $v \sim 200\,\mathrm{km\,s^{-1}}$. Recall from §9.1.1 that when $l$ is small and negative, positive velocities are strictly forbidden if the disk is in circular rotation. In reality, the upper envelope of the emission in Figure 9.33 reaches $l \sim -5°$ before dropping below $v = 100\,\mathrm{km\,s^{-1}}$. This constitutes unmistakable evidence that within the central disk the orbits of HI clouds are not even approximately circular.

Since so many external disk galaxies have bars, the natural reaction to this conclusion is to ask whether the central region of our own galaxy does not contain a bar, and to interpret Figure 9.33 in the light of our discussion of the $(l, v)$ traces of oval orbits in §9.1.1. In particular, we have seen that the $(l, v)$ traces of oval orbits cut the line $l = 0$ at non-zero $v$ (see Figure 9.8), so an ensemble of oval orbits may be able to account of the fact that the upper envelope in Figure 9.33 cuts $l = 0$ at $v \sim 200\,\mathrm{km\,s^{-1}}$. Recall from Figure 9.8 that the $(l, v)$ trace of an oval orbit is roughly symmetrical under inversion through the origin. From this it follows that the envelope in the $(l, v)$ plot of emission from a disk made up of oval orbits should also be symmetrical on inversion through the origin.

Now consider the lower envelope in Figure 9.33. At $l = -11°$ this lies at $v \sim -200\,\mathrm{km\,s^{-1}}$ and drops under a marked bump to a horn at $l \sim 4°$, $v \sim -270°$, very approximately as if it were the inversion through the origin of the envelope at $(l > 0,\ v > 0)$. But from $(l, v) = (0, -200)$ the lower contour moves almost vertically towards the origin *exactly as expected if the central disk were in circular rotation* (cf. Figure 9.3). Moreover, Figure 9.33 is significantly unsymmetrical on inversion through the origin in that there is no counterpart at $(l > 0,\ v < 0)$ of the forbidden emission at $(l \sim -5°,\ v > 100\,\mathrm{km\,s^{-1}})$. In fact, we are confronted with a paradox: the upper envelope in Figure 9.33 suggests that the HI moves on oval orbits, and the lower envelope seems to imply circular orbits.

This paradox has perplexed astronomers since it first emerged when the northern-hemisphere data taken in the Netherlands were compared with the southern-hemisphere data taken in Australia. Burton & Liszt (1978) and Burton (1992) argue powerfully that the paradox can be solved by taking into account two factors: (i) 21-cm emission from the part of the central disk that lies in front of the Galactic center is strongly masked by absorption of continuum radiation, and (ii) the central disk, amazingly, does not lie in the plane $b = 0$.

Since the HI of the middle disk contributes significantly to absorption only at velocities in the interval $(0, 300 \sin l)\,\mathrm{km\,s^{-1}}$, the effects of absorption by such gas can be largely eliminated by focusing on emission at velocities outside this interval. Figure 9.34 shows a map of the 21-cm brightness temperature integrated over $-300\,\mathrm{km\,s^{-1}} < v < -100\,\mathrm{km\,s^{-1}}$ and $100\,\mathrm{km\,s^{-1}} < v < 300\,\mathrm{km\,s^{-1}}$. Absorption against the Galactic-centre continuum emission is clearly apparent around the origin of the panel for $v < 0$, while the other panel shows no trace of such absorption.[7] Evidently the near side of the central disk is approaching us at $(l, b) = (0, 0)$. At $l > 0$

---

[7] $(l, v)$ plots of absorption by OH and $H_2CO$ molecules (Cohen & Few 1976, Cohen & Few 1981) confirm that in front of the nucleus there is more material at $v < -100\,\mathrm{km\,s^{-1}}$ than at $v > 100\,\mathrm{km\,s^{-1}}$.

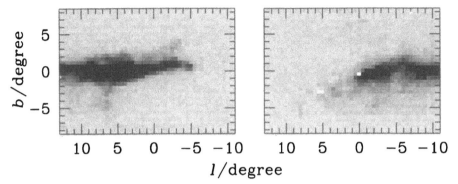

Figure 9.34 The 21-cm brightness temperature integrated over $-300\,\mathrm{km\,s^{-1}} < v < -100\,\mathrm{km\,s^{-1}}$ (right) and $100\,\mathrm{km\,s^{-1}} < v < 300\,\mathrm{km\,s^{-1}}$ (left) [After Burton (1992) from data kindly supplied by H. Liszt]

Figure 9.35 The spider diagram of a nearly edge-on elliptical disk.

the ridge-line of emission at allowed velocities $(v > 0)$ runs almost straight along $b = 0$, while at $l < 0$ the ridge-line of emission at allowed velocities has a slight tendency to run $0.5°$ above $b = 0$. Almost all the material with forbidden velocities at $l > 0$ lies at $b < 0$, while all the material at $l < 0$ with forbidden velocities lies at $b > 0$.

This last antisymmetry between $b > 0$ and $b < 0$ suggests that the central disk may not lie exactly in the plane $b = 0$. In §8.2.4 we investigated the projected velocity fields or spider diagrams of inclined disks. What would the spider diagram of the central disk look like if the disk were arbitrarily thin and could be studied at arbitrary angular resolution? Figure 9.35 shows the spider diagram of an elliptical disk that is viewed at inclination $i = 82°$. The long axis of the disk makes an angle of $20°$ with the projection of the line of sight into the disk's plane, and the orbits of clouds in the disk have axis ratio 0.74. The systemic-velocity contour is the straight contour that slopes up and to the right. If the disk were circular, this contour would be perpendicular to the (horizontal) line of nodes. Forbidden velocities occur in the wedge between the systemic-velocity contour and a vertical line through the center. Thus if the central disk were elliptical and oriented like the disk that underlies Figure 9.35, the forbidden velocities would be confined to $(l > 0, b < 0)$ and $(l < 0, b > 0)$, just as observed. In particular, the nearer end of the disk's long axis lies at $l > 0, b < 0$.

Three angles are required to specify the orientation of an elliptical disk: two angles specify the orientation of the normal to the disk, and the third

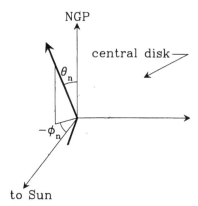

**Figure 9.36** The geometry of the central disk and the Galactic plane.

specifies the direction of the disk's long axis within its plane. Two observations indicate likely values of this last angle:

- The values of $|v|$ at which the envelope of emission in Figure 9.33 crosses $l = 0$ are large, $\sim 150 \, \mathrm{km \, s^{-1}}$. To achieve such substantial radial velocities $v_r$ anywhere around an elliptical orbit, the latter has to be pretty elongated (see Problem 9.11). In view of this, it is likely that the orbits happen to be orientated such that the line of sight to the center cuts the orbits near where $v_r$ peaks. Given that the orbits are quite elongated, it follows that the angle between the line of sight and their long axes is $\sim 20°$ (Binney *et al.* 1991).

- The conclusion we have just reached, that the line of sight lies near to the long axis of the bar, is confirmed by the shape of the terminal-velocity curves at $2° < |l| < 12°$. In fact, it naturally reconciles the strong rise in $\left| v_{\mathrm{los}}^{(\mathrm{t})} \right|$ as $|l|$ decreases in this interval with the *outwards*-increasing circular-speed predicted by the near-IR brightness profile of the Milky Way – see §10.1.

What about the orientation of the normal to the central disk? Let $(\theta_n, \phi_n)$ be the usual polar angles of the normal in a system of spherical polar coordinates oriented such that $b = 0$ and $\theta = \pi/2$ are the same plane and the Sun–center line coincides with $(\theta = \pi/2, \phi = 0)$ – see Figure 9.36. Then it is clear that $\theta_n$ is small because we see the disk nearly edge-on. Now consider two contrasting cases as regards $\phi_n$. If $|\phi_n| \simeq \pi/2$, we would see the central disk nearly edge-on, and emission at the largest values of $|l|$ and large allowed velocities would come from well above or well below $b = 0$, unless $\theta_n$ were negligible. Since Figure 9.34 shows that emission at allowed velocities peaks near $b = 0$ we can rule out this case. If $\phi_n \simeq 0$, the further end of the disk will be at $b > 0$ and the nearer at $b < 0$, just as we deduced above. Moreover, in this case, emission at large $|l|$ and large allowed velocities will come from points near $b = 0$, as observed. Finally, in order to ensure that forbidden velocities occur at $b > 0$ for $l < 0$ we must have $\phi_n < 0$. Thus both $\theta_n$ and $|\phi_n|$ are small, and $\phi_n$ is negative.

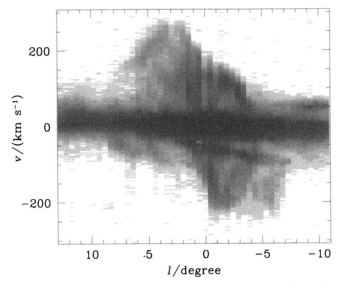

**Figure 9.37** A slice of the 21-cm $(l, b, v)$ data cube parallel to the plane $b = -l \tan 22°$. Data for the pixel around $l = b = 0$ are taken to be the average of the data for $l = 0, b = \pm 0.5°$ because at $l = b = 0$ the spectrum is strongly affected by absorption. Notice that this slice, in contrast to that parallel to $b = 0$ shown in Figure 9.33, is nearly symmetric under inversion through the origin. [After Burton & Liszt (1978) from data kindly supplied by H. Liszt]

To quantify $\theta_n$ we argue from Figure 9.35 that the extent $b_{\max}$ in $b$ of the forbidden emission is $\sim a_x \sin \theta_n / R_0$, where $a_x$ is the length of the disk's semi-major axis and we have approximated by unity the cosine of the angle between the Sun–center line and the bar's long axis. In this approximation, the extent $l_{\max}$ in $|l|$ of the emission at permitted velocities is $a_y / R_0$, where $a_y$ is the length of disk's semi-minor axis. Figure 9.34 suggests $b_{\max} \sim 3°$ and $l_{\max} \gtrsim 12°$. Thus

$$\frac{a_x}{a_y} \sin \theta_n \simeq \frac{b_{\max}}{l_{\max}} < \tfrac{1}{4}. \tag{9.23}$$

From Figure 9.43 below we conclude that $a_x/a_y \sim 2$ and hence that $\theta_n < 7°$.

This analysis is not the only possible interpretation of the data. In a series of papers summarized in Burton (1992), Burton & Liszt have argued for a slightly different orientation of the central disk than that derived above: they favor $(\theta_n = 24°, \phi_n = -33°)$ and envisage a rather large angle (51°) between the bar's major axis and the sun–center line (see Problem 9.12).

Conventional $(l, v)$ plots display slices through the $(l, b, v)$ data cube that are parallel to a plane $b = $ constant, but we can make an $(l, v)$ plot by slicing through the data cube in almost any direction. Figure 9.37 shows the result of slicing the data cube parallel to the plane $b = -0.4l$, close to the

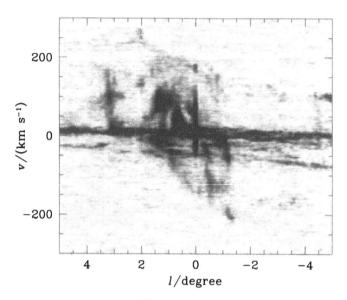

**Figure 9.38** $(l, v)$ plot of $^{13}$CO emission at $|l| \lesssim 5°$ averaged over $|b| < 0.6°$. [After Bally *et al.* (1988) from data kindly supplied by A. Stark]

line of nodes inferred by Burton & Liszt (1978). The resulting plot shows a high degree of symmetry on inversion through the origin, just as we would expect if the central disk were made up of oval orbits. Switching from the plane $b = 0$ to the plane $b = -0.4l$ modifies the $(l, v)$ plot in two ways. First, it diminishes the importance of absorption by concentrating on lines of sight that lie further from the plane, and therefore pass through less foreground HI. Second, it enhances the strength of the forbidden emission because, as we have seen, this is concentrated below and above $b = 0$ at $l > 0$ and $l < 0$, respectively.

### 9.4.2 Observations in lines of CO and CS

CO emission from the central disk is clearly visible in Figure 9.15 in the form of great spikes at $|l| \lesssim 4°$ that reach to $|v| > 180 \, \text{km s}^{-1}$. At $v < 0$ these spikes are confined to smaller values of $|l|$ than at $v > 0$. Figure 9.38 shows at higher angular resolution the $^{13}$CO emission in the region $|l| \lesssim 5°$, which corresponds to $R \lesssim 700 \, \text{pc}$ at the Galactic center. A band of strong emission runs horizontally through the figure. This emission is associated with gas that lies well outside the inner 700 pc and serves only to confuse study of the central region. Emission from central material can be most easily traced at $|v| \gtrsim 30 \, \text{km s}^{-1}$. On the extreme left in Figure 9.38 a feature centred on $(l, v) = (3°, 100 \, \text{km s}^{-1})$ is conspicuous. This tall thin feature consists of emission from gas that is confined to $\sim 0.6°$ (80 pc) in $l$ yet distributed over $\sim 100 \, \text{km s}^{-1}$ in velocity – this gas is called **Bania's Clump 2**. Several

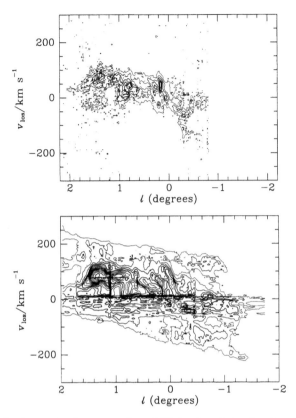

**Figure 9.39** $(l, v)$ plot for $J = 2 \to 1$ line of CS (top) and the $J = 1 \to 0$ line of $^{12}$CO (bottom). In both panels emission has been averaged over $|b| < 0.1°$. No CS spectra are available at $l < -1°$. Contours are spaced by 0.15 K and 1 K in the upper and lower panels, respectively. [After Binney *et al.* (1991)]

similar clumps of gas are seen in the inner few degrees of the Galaxy, while nothing of comparable scale is seen further out.

The intensity of the $^{13}$CO emission in Figure 9.38 also peaks strongly at $(l, v) = (0.66°, 55 \,\mathrm{km\,s^{-1}})$. This peak is due to the **Sgr B2 complex**. It has an angular diameter of $\sim 0.25°$ (35 pc) and several lines of argument suggest that its mass lies near $3 \times 10^6 \, \mathcal{M}_\odot$ (Bally *et al.* 1988).

Sgr B2 lies along a ridge of intense molecular emission that is most clearly seen in the upper panel of Figure 9.39. This diagram shows emission in the $J = 2 \to 1$ line of CS. Since CS is a highly dipolar molecule, its emission traces the densest gas (§8.1.4): whereas CO molecules are collisionally de-excited at densities $n_H \gtrsim 3 \times 10^8 \,\mathrm{m^{-3}}$, CS molecules are only collisionally de-excited for $n_H \gtrsim 10^{12} \,\mathrm{m^{-3}}$. Hence Figure 9.39 traces the distribution of the densest gas and shows that this gas is confined in the $(l, v)$ plane to a ridge, which slopes gently downward through the location of Sgr B2 towards a hot spot at $(l, v) \simeq (-0.1, 55\,\mathrm{km\,s^{-1}})$. This hot spot marks the location of the Sgr A molecular complex. In Figure 9.39 data are not available for $l < -1°$, but there is a striking absence of significant emission at $-1° < l < -0.2°$. Other $(l, v)$ plots confirm that three-quarters of the molecular gas at the Galactic center lies at positive longitudes. We do not know why.

Much of the emission in Figure 9.38 falls within a parallelogram that extends from $(l, v) \simeq (1.8°, 200\,\mathrm{km\,s^{-1}})$ at top left to $\simeq (-1.4°, -200\,\mathrm{km\,s^{-1}})$ at bottom right. The lower panel in Figure 9.39 shows this parallelogram more clearly by plotting the intensity of $^{12}\mathrm{CO}$ emission averaged over $|b| < 0.1°$. By emphasizing emission from more tenuous gas this plot reveals that the parallelogram extends as far as $l = 2$ at positive longitudes and to $l \simeq -1.4$ at negative longitudes, and to more than $\pm 200\,\mathrm{km\,s^{-1}}$ in $v$.

Bally et al. (1988) find that the line on the sky of peak molecular emission is aligned with the line $b = 0$ to within $\sim 1.5°$, but lies about $0.03°$ below $b = 0$. In §9.5 below we shall see that Sgr A*, which is thought to mark the dynamical center of the Galaxy, lies $0.046°$ below $b = 0$. The width of the emitting layer varies with $l$, but between $l = 0.2°$ and $l = 0.6°$ is typically $0.13°$ (18 pc at the Galactic center). If this width reflects the random vertical velocities of emitting gas clouds, the velocity dispersion of the clouds would have to be of order $\sigma_z = 54\,\mathrm{km\,s^{-1}}$. This dispersion is considerably larger than that of giant molecular clouds near $R_0$, but comparable to the line-of-sight velocity dispersion within features such as Bania's Clump 2.

### 9.4.3 A dynamical model of the central disk

We have seen that 21-cm data suggest that gas observed at longitudes of a few degrees is moving on highly eccentric orbits. Binney et al. (1991) proposed a dynamical model of the inner few kiloparsecs of the galaxy that connects this phenomenon both with the morphology of molecular emission at $|l| < 2°$ and with the near-infrared photometry of the Galaxy that is the subject of §10.1. Figure 9.40 illustrates the idea behind this model by showing a series of closed orbits in a simple rotating barred potential. Two distinct types of orbit are evident: (i) at the largest radii shown the orbits are distinctly elliptical and become progressively more elongated as one goes further in – eventually an orbit is reached that has sharp cusps at its apocenters; (ii) inside this **cusped orbit** the orbits are again elongated, but form ovals that are oriented perpendicular to the cusped orbit. The family of orbits that culminates in the cusped orbit is called the family of $x_1$ **orbits**, while the family of orbits that lie inside the cusped orbit is called the family of $x_2$ **orbits**. The transition from the $x_1$ family to the $x_2$ family is associated with the **inner Lindblad resonance** of the given bar potential – see §3.3 of BT. The $x_2$ orbits are typically less elongated than the $x_1$ orbits, and their elongation increases rather than decreases with increasing radius.

If cold gas is injected into in a barred potential, it will settle quickly to approximately closed orbits and then drift slowly inwards by moving continuously through a sequence of ever smaller closed orbits. This pattern of

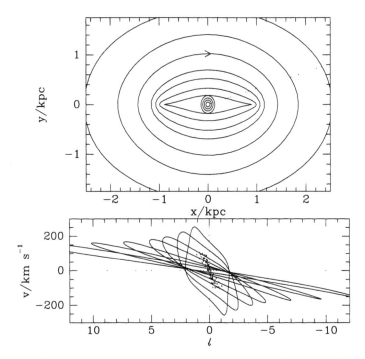

**Figure 9.40** Closed orbits in a simple rotating bar potential (upper panel) and their traces in the $(l, v)$ plane (lower panel) when viewed from a point that lies 8 kpc down the dotted line in the upper panel. [After Blitz *et al.* (1993).]

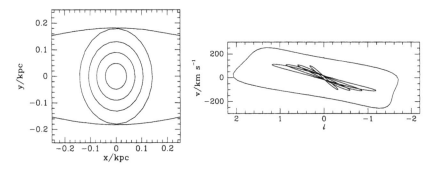

**Figure 9.41** The smallest of the orbits of Figure 9.40 shown on an enlarged scale. The biggest orbit shown is the cusped orbit and the orbits inside it belong to the $x_2$ orbit family.

gradual orbital evolution breaks down when the cusped orbit is reached, because smaller orbits of the $x_1$ sequence have small loops at their apocenters – see Figure 3-16 of BT. Hence they are self-intersecting orbits, and loss of kinetic energy as parcels of gas at different points around such an orbit crash into each other will quickly drive the gas to substantially smaller radii.

Thus once gas reaches the cusped orbit, it quickly transfers to much smaller orbits of the $x_2$ family, and resumes its slow inward drift. Once gas is on $x_2$ orbits, its inward drift slows because energy is lost more slowly on rounder orbits, and the $x_2$ orbits are not only rounder than the majority of $x_1$ orbits, but become even rounder as one goes to smaller radii. If the disk is in an approximately steady state, the surface density of gas will be highest where gas drifts inwards most slowly, and there should be a significant build-up of gas on $x_2$ orbits.

In the model of Binney *et al.*, the Sun-center line makes an angle of about $20°$ with the long axis of the bar, which coincides with the long axis of the $x_1$ orbits. Most 21-cm emission is interpreted as arising from gas on $x_1$ orbits, while the dense molecular gas that is traced by CS emission is on $x_2$ orbits. Consequently, the distribution of emission in Figure 9.37 is interpreted in terms of the $(l, v)$ traces of $x_1$ orbits, while that in the upper panel of Figure 9.39 is interpreted in terms of the $(l, v)$ traces of $x_2$ orbits. These traces are shown in the lower panel of Figure 9.40 and the right-hand panel of Figure 9.41.

As we expect from the discussion of §9.1, the $(l, v)$ traces of a series of $x_1$ orbits of decreasing size and therefore increasing elongation, become steadily more open, and the $(l, v)$ trace of the cusped orbit approximates a parallelogram. Binney *et al.* suggested that the parallelogram apparent in the lower panel of Figure 9.39 is a modified form of the parallelogram associated with the cusped orbit – the observational and theoretical parallelograms differ significantly because gas-dynamical effects such as shocks cause gas to move on significantly non-closed orbits near the transition from $x_1$ to $x_2$ orbits. Since the parallelogram in Figure 9.39 extends to $l \simeq 2°$, the potential is probably such that the $(l, v)$ trace of the cusped orbit extends to $l \simeq 2°$. This model does not explain why the CO emission is so very asymmetrical in longitude.

If the 21-cm emission does come from material on $x_1$ orbits, the upper-left and lower-right envelopes of the emission in Figure 9.37 should correspond to envelopes of the $(l, v)$ traces of $x_1$ orbits in the lower panel of Figure 9.40. A good family likeness between these envelopes is apparent – see Figure 9.42. In particular, the downward sweep away from $l \simeq 2°$ of the upper-left envelope in the observational figure is reproduced by the theoretical figure. This downward sweep does not reflect a strong decrease with $R$ in the strength of the inward gravitational pull of the Galaxy, but arises because as $l$ increases we are seeing less and less elongated orbits rather nearly end-on; at $l \simeq 2°$ emission near the envelope in the observational $(l, v)$ plot comes from material that is zipping past the center on its way from the near cusp to the far one, while further out emission near the envelope comes from material on more nearly circular orbits.

The emission from dense gas in Figure 9.39 nearly all arises at velocities $v \lesssim 100 \, \text{km s}^{-1}$ that are markedly smaller than those, $v \gtrsim 200 \, \text{km s}^{-1}$, associated with the corresponding envelope for HI in Figure 9.37. Figure 9.42

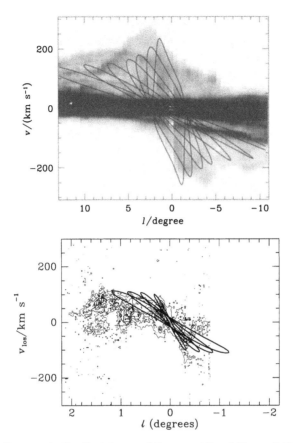

**Figure 9.42** Top panel: the $(l, v)$ traces of the $x_1$ orbits of Figure 9.40 superimposed on the HI distribution of Figure 9.33. Lower panel: the $(l, v)$ traces of the $x_2$ orbits of Figure 9.41 superimposed on the CS distribution of Figure 9.39.

shows that the Binney *et al.* model provides a natural explanation for this contrast: on account of the orientation of the $x_2$ orbits, the highest-velocity CS emission is dominated by material that is near apocenter and therefore going most slowly around its orbit, rather than being near pericenter as is the HI that is seen on $x_1$ orbits further out. Moreover, this consequence of the different orientations the two families of orbits is amplified by another circumstance. The $x_2$ orbits that are associated with CS emission are physically much smaller than the $x_1$ orbits that are associated with 21-cm emission, even though the largest $x_2$ orbit does not differ greatly in projected size from the smallest $x_1$ orbit. Consequently, the CS velocities probe the Galactic force-field at substantially smaller radii than do the 21-cm velocities. In the Binney *et al.* model, the circular speed associated with the axisymmetric component of the Galaxy's potential increases slowly with $R$ (quantitatively, $v_c \propto R^{0.12}$), so the small $x_2$ orbits have smaller characteristic velocities than do the much bigger $x_1$ orbits.

If the Galaxy were axisymmetric, the downward sweep of the envelope of 21-cm emission at $l > 2°$ in Figure 9.37 would imply that the circular speed was a decreasing function of $R$ at $0.2\,\mathrm{kpc} \lesssim R \lesssim 1\,\mathrm{kpc}$. Older mass models of the Galaxy usually include a compact central mass component to ensure that the circular speed behaves in this way. In the Binney *et al.* model, this downward sweep is entirely due to the decreasing elongation of closed $x_1$ orbits as $R$ increases, and occurs despite $v_c$ being an increasing function of $R$.

As we saw in §9.4.1, the concentration of forbidden velocities at $b > 0$ for $l < 0$ and at $b < 0$ for $l > 0$ arises naturally if the plane of the $x_1$ orbits is tipped by a few degrees about a line that lies nearly in the plane of the sky so that the far ends of these orbits lie at $b > 0$. It is not clear whether the plane of the $x_2$ orbits is similarly tipped, but the relatively small extent of these orbits along the line of sight ensures that a small tip of their plane would not be conspicuous.

As we noted above, pressure forces will ensure that gas clouds do not move on exactly closed orbits, especially near the transition from the $x_1$ to $x_2$ orbits. Therefore it is important to test a simple picture such as that of Binney *et al.* (1991), which is based on the characteristics of closed orbits, with hydrodynamical simulations of gas flow in barred potentials. Many such simulations has been performed, using a variety of numerical techniques to simulate the hydrodynamics of the ISM (which is in detail an extremely complex fluid). By and large, there is reasonable agreement between the results of these simulations.[8] Figure 9.43 shows a simulation of gas flow in a model of the Galactic potential that is based on near-infrared photometry of the Milky Way – see §10.1. Comparison of the $(l, v)$ plot of this simulation, which is shown in the lower panel of Figure 9.43, with the observational $(l, v)$ plots of Figures 9.13 and 9.15 shows that the most prominent features in the observational plots have theoretical counterparts. Unfortunately, the observational features tend to occur at larger longitudes and velocities than their theoretical counterparts. Hence, while it is clear that the general picture of a rotating bar at the Galactic center is correct, we do not yet have a satisfactory understanding of its shape and kinematics.

---

[8] See Englmaier & Gerhard (1997) for a discussion of critical parameters of gas-dynamical simulations.

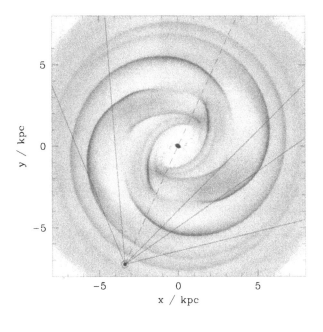

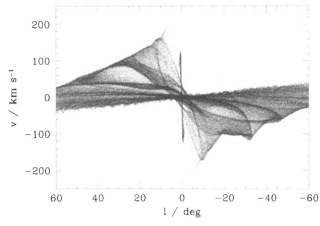

**Figure 9.43** A hydrodynamical simulation of gas flow in a model of the Galactic potential that is based on near-infrared photometry of the Milky Way (§10.1). Upper panel: the density of the ISM in real space. The bar's long axis lies along the line $x = y$ and the location of the Sun is marked along the line that is inclined by $20°$ to bar's axis. The straight lines from the position of the Sun indicate the approximate directions of tangents to spiral arms in the Galaxy (Vallée 1995); that immediately to the right of the center is tangent to the 3 kpc arm. Lower panel: the $(l, v)$ plot obtained when the model is viewed from the Sun's location. The Sun's rotation speed is $v_c = 208\,\mathrm{km\,s^{-1}}$ and the bar's pattern speed is $57.3\,\mathrm{km\,s^{-1}\,kpc^{-1}}$, which places corotation at 3.4 kpc. [After Englmaier & Gerhard (1997) from data kindly supplied by O. Gerhard]

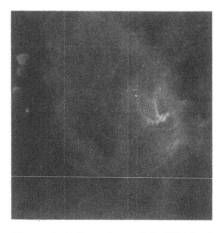 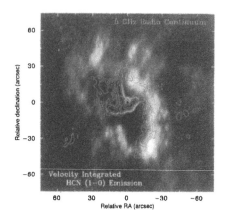

**Figure 9.44** Two views of the ISM in the vicinity of the Galactic nucleus. Left panel: 6-cm radio-continuum emission from Sgr A. The mapped region, 10 pc on a side, shows Sgr A East as an extended region of non-thermal emission centered on the middle of the frame and the characteristic spiral pattern of Sgr A West to the right of the frame. Right panel: An expanded view of Sgr A West. A radio-continuum contour map is superposed on contours of the intensity of emission from HCN molecules. [Left panel: VLA data of Killeen & Lo courtesy of K.Y. Lo; right panel after Güsten *et al.* (1987)]

## 9.5 The nucleus

Probably the most important gap in our understanding of the Universe relates to what goes on inside galactic nuclei. Observations of quasars and other active galactic nuclei (AGN; §4.6) indicate that more energy has been released by AGN since the Big Bang than by all the stars that have ever lived (see Problem 9.13), yet we have only vague theories of how this energy is released and know little of how it affects the surrounding Universe. Two major reasons for this sad state of ignorance are that whatever machinery powers an AGN is very small, and that only one galactic nucleus lies within the Galaxy. The study of this one readily accessible nucleus is clearly of the utmost importance. Our discussion closely follows that of Blitz *et al.* (1993), who give many references to the literature.

The Galactic nucleus is far from active. In fact, understanding how it contrives to be so quiescent is a major puzzle. However, we have reason to believe that all nuclear activity in galaxies is episodic, so understanding the low luminosity of our own nucleus may enable us to understand the hyperactivity of other, more distant, nuclei.

Figures 9.30 and 9.44 show that interesting things go on in the inner $\sim 30\,$pc of the Galaxy. Figure 9.30, which covers a region 85 pc on a side, gives an overview of radio-continuum emission from this region. The Galactic plane runs from top left to bottom right of the panel, whose coordinates are right ascension and declination. As we saw in §9.3.2, the filaments of synchrotron emission seen up and to the left of the center of this figure are unique in the strength of the magnetic field in them and their orientation perpendicular to the plane.

At the lower right of Figure 9.30 we see the radio source Sgr A, which surrounds the nucleus of the Galaxy. The bright dot at its center is the unusual non-thermal radio source **Sgr A\***, which is thought to mark the very center of the Galaxy. This dot lies at $(l, b) = (-0.054°, -0.046°)$ to the right (West) of a crescent of enhanced emission. The left-hand panel of Figure 9.44 reveals the structure of this crescent, which extends about 8 pc from top to bottom. Sgr A\* lies near the center of a spiral of emission that is evident below and to the right of the picture – we refer to this as the **mini-spiral**. This emission is predominantly Bremsstrahlung radiation from ionized gas and its source is called **Sgr A West** to distinguish it from the more extensive region of non-thermal emission that dominates the rest of the picture; this non-thermal source is called **Sgr A East**. It is probably a shell of shocked gas, which is blasting outwards through the ISM that surrounds the nucleus. This blast wave may have been powered by supernovae and fast stellar winds, or by something more exotic.

The right-hand panel of Figure 9.44 reveals some of the structure of Sgr A West by superposing a contour map of 5 GHz continuum emission on a gray-scale representation of the intensity of emission by HCN molecules. One can trace the mini-spiral in the 5 GHz emission. It just about fits inside an elliptical ring of HCN emission, whose semi-major axis is $\sim 2\,\mathrm{pc}$ long. From observations in other molecular lines we know that the HCN emission comes from the inner edge of a ring of molecular material that extends to $\sim 7\,\mathrm{pc}$ from Sgr A\* – this is called the **nuclear molecular ring**.[9] This ring clearly does not lie within the plane, for we do not see it edge-on. Its velocity field has been successfully modeled by circular rotation at $v_c = 110\,\mathrm{km\,s^{-1}}$ in a plane on inclination $i = 65°$ and position angle $\sim 20°$. Since $v_c$ is inferred to be approximately independent of radius, in the range $2\,\mathrm{pc} < r < 7\,\mathrm{pc}$ the mass of the Galaxy must increase as $\mathcal{M}(r) \simeq 2.8 \times 10^6 (r/\mathrm{pc})\,\mathcal{M}_\odot$.

Within the hole at the center of the nuclear ring, filaments of ionized gas form the mini-spiral. Observations of infrared and radio-frequency lines show that velocity varies systematically with position along the spiral arms. Lacy, Achtermann & Serabyn (1991) found that they could obtain an acceptable fit to this velocity field if each parcel of gas was assumed to move within the plane of the nuclear ring on a circular orbit about Sgr A\*: in this model gas does not flow along the spiral arms; rather the arms are filaments that are being wound into ever tighter spirals by differential rotation. The model requires a central mass $2 - 3 \times 10^6\,\mathcal{M}_\odot$ with up to $3 \times 10^6\,\mathcal{M}_\odot$ of further material distributed through the inner 2 pc.

Gas within the nuclear ring is ionized by radiation from a remarkable star cluster whose center lies within 0.04 pc of Sgr A\* and whose half-mass radius is $\sim 0.5\,\mathrm{pc}$. This cluster can only be studied at infrared wavelengths,

---

[9] See Lacy (1994) for a review and references.

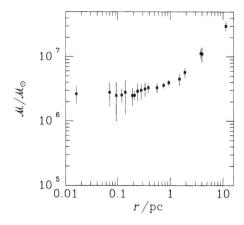

**Figure 9.45** The mass $\mathcal{M}(r)$ within distance $r$ of Sgr A* seems to tend to a limit $\sim 2.6 \times 10^6 \mathcal{M}_\odot$ as $r$ approaches zero. All but two of the mass estimates plotted here were obtained by modeling the kinematics of stars. [From data published in Genzel *et al.* (1997)]

in which it appears as a tight clustering of hot-spots. About a dozen of these hot-spots have been shown to be sources of He I emission at 2.06 $\mu$m, which implies that the responsible stars are hot ($T_{\text{eff}} \simeq 20\,000\,$K). If these stars are supergiants or Wolf-Rayet stars, they are young, which implies that about $10^7$ yr ago there was a burst of massive star formation at the center. Alternatively, they may be stars that have been profoundly modified by mass exchanges in close collisions with other stars. Indeed, the density towards the center of the nuclear cluster is estimated to be $\gtrsim 10^6$ pc$^{-3}$, which may be compared with $\lesssim 0.1$ pc$^{-3}$ in the solar neighborhood. In these circumstances serious traffic accidents are only to be expected. Allen (1994) discusses these possibilities and gives references to the literature.

Sgr A* is a variable non-thermal radio source of size $0.3 \pm 0.1$ mas (2.4 AU). Its luminosity is hard to determine accurately because it may be dominated by optical and UV radiation that is completely obscured. Mezger (1994) places an upper limit on it, $L \lesssim 7 \times 10^5\,L_\odot$. In principle, Sgr A* could be a neutron star, but it is more likely to be a massive black hole. First and foremost, observations of other galaxies make it likely that there is a massive black hole in the nucleus (§11.2.2).

There are two ways to estimate the mass of this putative black hole. One is to determine the velocities of stars and gas in its vicinity and to infer from them the magnitude of the gravitational field that confines them. We have seen above that gas velocities at $r \simeq 1$ pc imply that about $10^6\,\mathcal{M}_\odot$ lies within the inner parsec. Measurements of stellar velocities enable us to probe the Galactic potential much closer to Sgr A*. Genzel *et al.* (1997) find that the proper motions of stars that are located within $r \sim 0.016$ pc of Sgr A* have dispersion $\sigma_{\text{cl}} \sim 1150\,$km s$^{-1}$. From such data one can estimate the mass $\mathcal{M}(r)$ contained within a sphere of radius $r$ about Sgr A*. Figure 9.45 shows that $\mathcal{M}$ seems to converge at small $r$ to $\mathcal{M}(0) \sim 2.6 \times 10^6\,\mathcal{M}_\odot$.

Does this mass reside *in* Sgr A* or simply around it? If Sgr A* is a black hole of mass $2 \times 10^6 \mathcal{M}_\odot$, its Schwarzschild radius[10] would be a mere $2 \times 10^{-7}$ pc, much smaller than the radius of the last point in Figure 9.45. Hence, in principle the data plotted in Figure 9.45 are consistent with $\mathcal{M}(r)$ turning down towards zero inside the last observed point, but the mass density implied by such a turn-down would exceed $\rho \sim 2 \times 10^{12} \mathcal{M}_\odot \, \text{pc}^{-3}$. Moreover, a simple dynamical argument enables us to constrain directly the mass of Sgr A* itself. Because the stellar density around the nucleus is very high, the local dynamical relaxation time (§8.3.4 of BT) is short and Sgr A* should be in dynamical equilibrium with the nuclear cluster. Consequently, its random velocity should be drawn from a Maxwellian whose dispersion is smaller than that of the nuclear cluster, $\sigma_{cl}$ by a factor $(\mathcal{M}_*/\mathcal{M}_{A*})^{1/2}$, where $\mathcal{M}_* \simeq 10 \mathcal{M}_\odot$ is the mass of a typical star in the cluster (Genzel *et al.* 1997) and $\mathcal{M}_{A*}$ is the mass of Sgr A*. We cannot determine the radial velocity of Sgr A* because its spectrum contains no lines, but we can measure its proper motion – see §10.3.3 below. We may resolve this into components parallel and perpendicular to the plane. Only the perpendicular component is useful for our present purpose: it implies that the perpendicular velocity of Sgr A* is $v_\perp = -11.6 \pm 8.9 \, \text{km s}^{-1}$. Thus the data are essentially consistent with $v_\perp$ being vanishingly small, and $\mathcal{M}_{A*}$ being much larger than $\mathcal{M}_\odot$. Suppose on the other hand that further measurements demonstrated that $v_\perp = 10 \, \text{km s}^{-1}$, say. Then we could argue that the probability of $v_\perp$ being so low is

$$P(< v_\perp) = \text{erf}\left(\left(\frac{\mathcal{M}_{A*}}{\mathcal{M}_*}\right)^{1/2} \frac{v_\perp}{\sigma_{cl}}\right). \tag{9.24}$$

For $\sigma_{cl} = 1150 \, \text{km s}^{-1}$ this yields probabilities $P(< 10 \, \text{km s}^{-1}) = 0.003, 0.03$ and $0.3$ for $\mathcal{M}_{A*} = \mathcal{M}_\odot, 100 \mathcal{M}_\odot$ and $10^4 \mathcal{M}_\odot$, respectively. Thus it is highly implausible that Sgr A* is a neutron star, and extremely unlikely that it is the remnant of a supergiant star. Moreover, the errors in the determination of $v_\perp$ are rapidly decreasing, so these constraints on $\mathcal{M}_{A*}$ should become even stronger within a few years.

## 9.6 Small-scale structure of the ISM

The nature of a galaxy's ISM profoundly affects all the observable properties of the galaxy because it is from the ISM that stars form. In this section we examine the small-scale structure of the ISM, and investigate the connection between clouds of dense, cool gas and the formation of stars. Most of our knowledge of the small-scale structure of the ISM comes from studies of the Milky Way's ISM because we can study this at very much higher spatial resolution than the ISM of any other galaxy.

---

[10] The **Schwarzschild radius** of a body of mass $\mathcal{M}$ is defined to be $r_s = 2G\mathcal{M}/c^2$. Even photons do not move fast enough to escape a body's gravitational field at radii smaller than $r_s$.

The ISM is extremely inhomogeneous. Studies of interstellar absorption along lines of sight to nearby stars indicate that the Sun is located in an irregular region of low density. In most of this region, which extends to over $100\,\mathrm{pc}$ from the Sun in some directions, but only $\sim 10\,\mathrm{pc}$ in others, the number density of H nuclei is $n_{\mathrm{H}} \lesssim 10^5\,\mathrm{m}^{-3}$ and the ionization level is high (e.g. Pounds *et al.* 1993). Scattered throughout this low-density region there are many clouds of denser material. The clouds have radii of a few parsecs and column densities $N_{\mathrm{H}}$ in the range $10^{24} - 10^{25}\,\mathrm{m}^{-2}$. From these numbers one infers particle densities $n_{\mathrm{H}} \gtrsim 10^7\,\mathrm{m}^{-3}$ and masses $\mathcal{M} \sim 100\,\mathcal{M}_\odot$. The Sun is located in one of these small clouds, which occupy only a small fraction ($\lesssim 5\%$) of the volume of the Galactic disk, but contain perhaps 90% of the mass of the ISM near the Sun. In the clouds the temperature is $\sim 100\,\mathrm{K}$, while the temperature in the low-density material is $\sim 10\,000\,\mathrm{K}$.

At distances of order $100\,\mathrm{pc}$ from the Sun there are large volumes of much hotter plasma. This plasma must have a temperature $\sim 10^6\,\mathrm{K}$ because it is rich in highly ionized species such as $\mathrm{O}^{5+}$, and emits soft X-rays. The plasma must be located more than $100\,\mathrm{pc}$ from the Sun because in X-ray maps made with the ROSAT observatory, nearby, cool interstellar clouds show up as dark filaments on a bright background; these clouds lie in front of the X-ray emitting plasma.

Regions filled by million-degree plasma are thought to form when gas that has been accelerated to velocities in excess of $1000\,\mathrm{km\,s}^{-1}$ by supernovae and fast winds from hot, young stars, crashes into essentially stationary gas. This collision gives rise to a shock front that thermalizes the kinetic energy of the high-velocity gas. Outflows that are driven by many young stars and supernovae often give rise in this way to long thin regions of million-degree plasma, which are called **worms**. The largest worms extend from the plane right up into the Galactic halo, and are conspicuous features of 21-cm maps of the nearby Galaxy (Heiles 1984).

### 9.6.1 Molecular gas in the Galaxy

Stars do not form from in clouds of ionized gas, or even in clouds of HI, but in molecular clouds. Hence we now examine these enigmatic objects in some detail. Solomon *et al.* (1987) and Solomon & Rivolo (1989) studied the properties of 439 molecular clouds that they identified in a survey of the region ($8 \leq l \leq 90, |b| \leq 1$) in the $2.6\,\mathrm{mm}$ line of $^{12}\mathrm{CO}$. Their beam had a half-power full width of $47\,\mathrm{arcsec}$ and their spectra were taken $3\,\mathrm{arcmin}$ apart, so their survey was slightly undersampled, but could resolve a cloud bigger than $20\,\mathrm{pc}$ in diameter anywhere inside the solar circle. Clouds were identified by searching for closed surfaces of constant antenna temperature $T_A$ in three-dimensional $(l, b, v)$-space. Such a closed surface was catalogued as a cloud providing: (i) it contained at least 25 points of the $(l, b, v)$ grid; and (ii) the total intensity contained within it exceeded a given threshold.

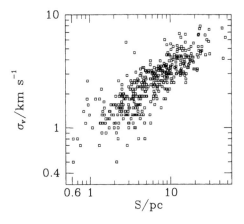

**Figure 9.46** Physically bigger molecular clouds cover a wider range in velocity. The size parameter $S$ is defined by equation (9.25). [From data kindly provided by P. Solomon]

For example, surfaces on which $T_A = 4\,\text{K}$ were required to enclose at least $60\,\text{K km s}^{-1}$ of emission. The value of $T_A$ at which closed surfaces were identified varied with $(l, b)$ so as to ensure that it was everywhere above the background intensity. Inevitably, this procedure enables smaller clouds to be cataloged in regions of low background intensity (such as the anticenter direction) than in regions of high background intensity.

Each catalogued cloud was characterized by its dispersions, weighted by $T_A$, in $l$, $b$ and $v$. A model circular-speed curve was used to associate each cloud with a location $(R, \phi)$ in the Galaxy as described in §9.1.1. Once this location had been determined, the distance $d$ to each cloud could be calculated and the cloud's physical size $S$ estimated as

$$S = d \tan(\sqrt{\sigma_l \sigma_b}), \tag{9.25}$$

where $\sigma_l$ and $\sigma_b$ are the cloud's dispersions in longitude and latitude. Figure 9.46 shows that a cloud's dispersion in $v$, $\sigma_v$, is quite tightly correlated with $S$; when a straight line is fitted to the points in Figure 9.46, one obtains the mean relation

$$\frac{\sigma_v}{\text{km s}^{-1}} = (1 \pm 0.1)\left(\frac{S}{\text{pc}}\right)^{0.5 \pm 0.05} \tag{9.26}$$

The median value of $\sigma_v$ for the clouds is $\sim 3\,\text{km s}^{-1}$, which is much larger than the sound speed in gas as cold ($T \lesssim 30\,\text{K}$) as that of the clouds. Consequently, $\sigma_v$ must be dominated by turbulent velocities within the clouds. The turbulent pressure within a cloud is thus $P \sim n_H m_p \sigma_v^2$. In molecular clouds $n_H \gtrsim 10^8\,\text{m}^{-3}$ and this turbulent pressure exceeds by more than an order of magnitude the pressure in the ISM between molecular clouds, which is of order $\sim n_H m_p c_s^2$, with $n_H \sim 10^6\,\text{m}^{-3}$ and $c_s \sim 10\,\text{km s}^{-1}$. If molecular clouds are not confined by pressure from the surrounding ISM, they must be either freely expanding, or be confined by gravity. The second possibility

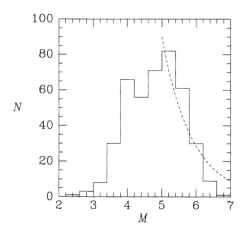

**Figure 9.47** The mass spectrum of molecular clouds derived from a sample of 439 clouds. The dashed line shows a power-law fit to the data for high-mass clouds. [From data kindly provided by P. Solomon]

seems the most plausible. By the virial theorem, it implies that the mass of a cloud of size $S$ and velocity dispersion $\sigma_v$ is of order

$$\mathcal{M} = K \frac{\sigma_v^2 S}{G}, \tag{9.27}$$

where $K$ is a dimensionless numerical factor that depends on the geometry of the clouds – Solomon *et al.* (1987) suggest that $K \sim 8.7$, and we adopt this value in the following.

Figure 9.47 shows the resulting distribution over mass of the clouds in the samples of Solomon *et al.* sample. At the high-mass end, this distribution is consistent with a power-law relation

$$\frac{dN}{d\mathcal{M}} \propto \mathcal{M}^{-\alpha} \quad \text{with} \quad \alpha = 1.5 . \tag{9.28}$$

At low masses the distribution falls well below this relationship because low-mass clouds cannot be detected throughout the inner Galaxy and no attempt has been made to correct the data for this bias. In a study of clouds in the nearby Perseus arm (see below) Digel *et al.* (1996) found that the best power-law fit to the mass spectrum in the range $2000 < \mathcal{M}/\mathcal{M}_\odot < 40\,000$ has exponent $\alpha = 1.7$, which is close to the value of equation (9.28). If the mass distribution of the clouds follows equation (9.28) at high masses, the mass contained in clouds less massive than $\mathcal{M}$ diverges as $\mathcal{M}^{1/2}$. This result indicates that *most of the Galaxy's molecular gas is concentrated in the most massive clouds.*

Equation (9.27), together with the result $\sigma \propto S^{0.5}$, implies that all clouds have the same mean surface density, $\sim 170\,\mathcal{M}_\odot\,\mathrm{pc}^{-2}$, and that the characteristic density of a cloud decreases with increasing mass.

We conclude this chapter by returning to the crucial task of understanding the connection between the amount of matter in a given cloud and the measured intensity of the CO line radiation from it. The latter is conventionally measured in terms of the quantity $I_{CO}$ that is defined by equation (8.25), and its relation to the total column density of H is expressed in terms of the quantity $X$ that is defined by equation (8.28). $X$ has been determined in three ways. We discuss each in turn.

**$X$ from virial masses** Solomon *et al.* (1987) show that the mass of a cloud is tightly correlated with the integral, $L_{CO}$, of $I_{CO}$ over its surface. The mean relation is

$$\frac{\mathcal{M}}{\mathcal{M}_\odot} = 39 \left( \frac{L_{CO}}{\mathrm{K\,km\,s^{-1}\,pc^2}} \right)^{0.81 \pm 0.03} \quad \text{where} \quad L_{CO} \equiv \int \mathrm{d^2x}\, I_{CO}. \quad (9.29)$$

In an individual cloud, both $N(H_2)$ and $I_{CO}$ will decline from maxima near the center to small values at the edge. If $X$ were constant, $L_{CO}$, which is the area-integral of $I_{CO}$, would be proportional to $\mathcal{M}$, which is proportional to the area integral of $N(H_2)$. Since the exponent in equation (9.29) is not unity, $I_{CO}$ cannot be precisely proportional to $N(H_2)$. However, since 0.8 is close to unity and, as we have seen, most molecular gas is in the largest clouds, we overlook this problem and evaluate $X$ as the ratio of the mean value of $N(H_2)$ to the mean value of $I_{CO}$ in a typical large cloud of mass $\mathcal{M} = 10^6\, \mathcal{M}_\odot$. We then find $X = 2.5 \times 10^{24}$.

**$X$ from $\gamma$-rays** In §8.1.5 we saw that $\gamma$-rays are produced when cosmic rays collide with protons and other atomic nuclei. Hence, at each point in the Galaxy, the $\gamma$-ray emissivity, $j_\gamma$, should be proportional to the product $q_\gamma n_H$, where $q_\gamma$ is the cosmic-ray flux. Cosmic rays are thought to diffuse quite rapidly through space, so $q_\gamma$ should vary much less rapidly with position than $n_H$. If $q_\gamma$ were constant throughout the disk, $j_\gamma$ would be proportional to $n_H$, and the observed $\gamma$-ray intensity down any line of sight, $I_\gamma$, would be proportional to $N(H_{tot})$. In reality, $q_\gamma$ does vary through the disk, but, to a reasonable approximation, it can be assumed to be a function only of Galactocentric radius. With this assumption, Strong *et al.* (1988) estimated $X$ by decomposing $I_\gamma(l,b)$ for $l \geq 12°$ into a sum of contributions $I_{\gamma i}(l,b)$ from each of a series of annuli in the disk, and then writing

$$I_{\gamma i}(l,b) = q_{\gamma i}\left( N_{HIi}(l,b) + 2X_i I_{COi}(l,b)\right) + I_{0i}. \quad (9.30)$$

Here $N_{HIi}(l,b)$ is the column density of HI through the annulus in the direction $(l,b)$, $X_i$ is the value of $X$ for the annulus, $I_{COi}$ is the 2.6 mm brightness temperature at $(l,b)$ integrated over the velocities of the annulus, and $I_{0i}$ is the (small) contribution to $I_{\gamma i}$ from inverse-Compton scattering and point $\gamma$-ray sources such as pulsars. Finally, Strong *et al.* optimized the fit between the observed and predicted values of $I_\gamma(l,b)$ by varying the parameters $q_{\gamma i}$ and $X_i$. This procedure was be carried out for each range of $\gamma$-ray energies for which data were available from the COS-B satellite.

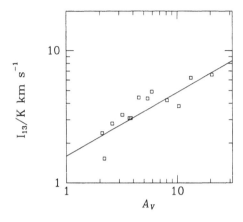

**Figure 9.48** The dependence upon visual extinction $A_V$ of $I_{13} \equiv \int dv\, T_A(^{13}CO)$ for lines of sight through the Taurus molecular cloud. The straight line shows equation (9.31). [After Sanders, Solomon & Scoville (1984) from data published in Frerking, Langer & Wilson (1982)]

Strong *et al.* first assumed that $X$ was the same for every annulus and showed that the best fit to the data is obtained when the $q_{\gamma i}$ are taken to be independent of energy, while $X$ decreases slightly with increasing energy. Then they asked whether significantly better fits to the data could be obtained by allowing $X$ to be smaller by a factor $\alpha$ at $R < 0.8R_0$ than at $R > 0.8R_0$. They concluded that the low-energy data favor large values of $\alpha$, while the high-energy data, which they consider most reliable, are consistent with $\alpha = 1$. They conclude that $q_\gamma$ decreases by about a factor 2 between $0.2R_0$ and $1.5R_0$ and that the best value of $X$ is $(2.3 \pm 0.3) \times 10^{24}$, in remarkably good agreement with the result of Solomon *et al.* (1987).

**$X$ from $A_V$**    In §8.1.4 we saw that the 2.6 mm line of $^{12}CO$ is optically thick in many directions, with the result that $I_{CO}$ is not proportional to $N(H_2)$ when data of high spatial resolution are examined in detail, line of sight by line of sight. Notwithstanding this conclusion, we argued that when $I_{CO}$ is averaged over many lines of sight, the average value of $I_{CO}$ may be proportional to the average value of $N(H_2)$. In the last two subsections we have seen how this proportionality between averaged quantities has been established and calibrated. Now we establish a relationship for individual lines of sight between the intensities of CO lines and values of $N(H_{tot})$ that are estimated from the extinction, $A_V$, and equation (8.50).

Since a spectral line that has completely saturated conveys no information about the emitting column density, we focus on the intensity of the 2.6 mm line from the relatively rare species $^{13}CO$. In Figure 9.48 the velocity-integrated antenna temperature of this line, $I_{13}$, is plotted against the visual extinction, $A_V$, for lines of sight through the Taurus molecular cloud. The two variables are evidently fairly tightly correlated, with a mean relation

$$I_{13} = 1.6 A_V^{0.48}\, \mathrm{K\, km\, s^{-1}}. \tag{9.31}$$

Hence, the optical thickness of the spectral line causes the line's intensity to increase roughly as the square root of $A_V$, and thus also as the square root

of $N(\mathrm{H_{tot}})$.[11] If we could identify a particular point on the mean relation
(9.31) as the one responsible for the bulk of the CO emission in the Galaxy, we
could derive the most appropriate value of $X$. Sanders, Solomon & Scoville
(1984) argue that the relevant point is at $I_{13} = 7\,\mathrm{K\,km\,s^{-1}}$ because this value
of $I_{13}$ is characteristic of giant molecular clouds in large-scale surveys of the
Galaxy. For this value of $I_{13}$, equation (9.31) yields $A_V = 21.6$, and, when we
insert this extinction into equation (8.50), we obtain $N(\mathrm{H_2}) \simeq \frac{1}{2}N(\mathrm{H_{tot}}) =
2.05 \times 10^{26}\,\mathrm{m^{-2}}$. Hence, we tentatively write

$$\frac{N(\mathrm{H_2})}{\mathrm{m^{-2}}} = 2.0 \times 10^{26}\frac{I_{13}}{7\,\mathrm{K\,km\,s^{-1}}}.$$

When giant molecular clouds are identified in large-scale surveys, the velocity-
integrated antenna temperatures of the $2.6\,\mathrm{mm}$ lines of $^{12}\mathrm{CO}$ and $^{13}\mathrm{CO}$ are
typically in the ratio 5.5:1. Hence, in equation (9.31) we replace $I_{13}$ by
$I_{\mathrm{CO}}/5.5$ and conclude that $X = 5.2 \times 10^{24}\,\mathrm{K\,km\,s^{-1}\,m^2}$. This value is twice
as large as our other estimates. Clearly, a lower value would have been ob-
tained if we had started from a lower point on the line in Figure 9.48 than
that at $I_{13} = 7\,\mathrm{K}$. This method of estimating $X$ has the virtue of forcing us
to recognize that *no* value of $X$ can yield more than a crude approximation
to the truth.

## Problems

**9.1** The velocity at point in an expanding ring is $\boldsymbol{\Omega} \times \mathbf{R} + (v_e/R)\mathbf{R}$, where
$\mathbf{R}$ is the Galactocentric position vector of the point. For such a ring, show
that equation (9.3) may be written

$$v_{\mathrm{los}}(l) = \left[\Omega(R) - \Omega(R_0)\right]R_0\sin l + \frac{d^2 + R^2 - R_0^2}{2Rd}v_e, \qquad (9.32)$$

where $d$ is the heliocentric distance to the point in the ring that lies along
the line of sight at Galactic latitude $l$. Show that, in the limit $R \to \infty$, the
last term tends to $v_e$.

**9.2** Estimate the path-length one must go at (a) $l = 30°$ and (b) $l = 90°$ for
the spread in the radial velocities of hydrogen atoms due to Galactic rotation
to exceed the $\simeq 9\,\mathrm{km\,s^{-1}}$ that arises from interstellar turbulence. [Assume
that the rotation curve is perfectly flat and of amplitude $v_c = 220\,\mathrm{km\,s^{-1}}$.]

**9.3** Determine which way around the aerofoil-shaped curve of Figure 9.8 a
cloud moves as it orbits in Figure 9.7 clockwise around the large full curve.

**9.4** If the circular speed of the Milky Way is constant at $v_c = 220\,\mathrm{km\,s^{-1}}$
from $R_0$ outwards, and no 21-cm emission is seen beyond $v_{\mathrm{los}} = 130\,\mathrm{km\,s^{-1}}$
at $l = 225°$ (Knapp, Tremaine & Gunn 1978), derive an upper limit to the
radius of the HI disk under the assumption that the disk is perfectly flat.

---

[11] There is an interesting parallel between this result and the result of Problem 8.2.

**9.5** Estimate the telscope time that would be required for a fully sampled survey of the Milky Way at $|b| < 5°$ in the 2.6 mm CO line with a 10 m single-dish telescope. Assume that each pointing takes one minute.

**9.6** The HI disk is seriously warped at large $R$. In which direction has one the best chance of detecting 21-cm emission from the largest possible Galactocentric radii?

**9.7** We will see in §10.4.3 that the thin component of the stellar disk in the solar neighborhood has a characteristic thickness of $z_{1/2} \simeq 250$ pc, and a velocity dispersion perpendicular to the Galactic plane of $\sigma_z \simeq 18 \, \text{km s}^{-1}$. Use these values, together with the hydrostatic equilibrium relation of equation (11.15), to predict what the local thickness of the HI disk should be if it were also in hydrostatic equilibrium. Hence, show that the observed thickness of the HI disk is approximately a factor of three too large, and that additional forces must be responsible for its support.

**9.8** Express Galactic latitude, $l$, as a function of the Galactocentric polar coordinates, $(R, \phi)$.

**9.9** Assuming that one third of the vertical support of the HI layer comes from the random motions of clouds, and neglecting the complication introduced by the Milky Way's warp, use the data in Figure 9.25 to estimate how the surface density of material near the Galactic plane, $\Sigma(R)$, varies with radius.

**9.10** Let $R_{\text{in}}$ be the inner radius of the middle disk and assume that the circular speed is constant at $v_c$ at radii in the interval $R_{\text{in}} \leq R \leq R_0$. Show that the maximum line-of-sight velocity at which 21-cm emission from the middle disk will be observed at longitude $l$ is

$$v_{\text{max}} = \left( \frac{R_0}{R_{\text{in}}} - 1 \right) v_c \sin l. \tag{9.33}$$

**9.11** Assume that the Galactocentric radius, $R$, of an HI cloud is given as a function of time by $R = R_g + X \cos \kappa t$, where $R_g$, $X$ and $\kappa$ are constants (cf. BT §3.2.3). Assume further that $\kappa \simeq \sqrt{2}\Omega$, where $\Omega$ is the circular frequency at $R_g$. Show that

$$\frac{X}{R_g} = \frac{v_r(\text{max})}{\sqrt{2}v_c}, \tag{9.34}$$

where $v_r(\text{max})$ is the largest line-of-sight velocity at which the cloud could possibly be detected against the Galactic center and $v_c = R_g \Omega$. Hence, estimate the elongation of the orbits that would be required if clouds are to be observed at 150 km s$^{-1}$ along $l = 0$.

**9.12** Burton & Liszt (1978) argue that the Milky Way's nuclear disk is tilted by $22°$ with respect to the plane $b = 0$ and $78°$ with respect to the plane of the sky. Using the notation of §9.4, show that this orientation corresponds to $(\theta_n = 22°, \phi_n = 56°)$. Show further that the line of nodes is the line $\tan b = -0.335 \tan l$.

**9.13** The observed luminosity density of quasars as a function of redshift implies that AGN have radiated $E_{\mathrm{AGN}} \gtrsim 2 \times 10^5 \, \mathcal{M}_\odot c^2 \, \mathrm{Mpc}^{-3}$ (Chokshi & Turner 1992). Compare this energy density with that yielded by the nucleosynthesis of the metals required to raise all luminous galaxies to solar metallicity. Assume that the galaxy luminosity function is given by equation (4.3), that galaxies have mass-to-light ratio $\Upsilon_B = 3$, and that conversion of a mass $\mathcal{M}$ of primordial gas to solar-metallicity gas releases energy $E_{\mathrm{NS}} = 1.4 \times 10^{-4} \mathcal{M} c^2$.

# 10
# Components of the Milky Way

We are able to study the workings of one particular galaxy, the Milky Way, in extraordinary detail. In this chapter we shall examine observations of stars within the Milky Way with two goals in mind. First, we wish to determine what kind of galaxy the Milky Way is. Where does it lie on the Hubble sequence? How large is it? Second, we want to learn as much as we can about galaxies in general by closely examining our particular specimen. What kinds of stars are disks made of? Are thick disks made of the same kinds of stars as thin disks? How are the different morphologies of thin disks, thick disks and bulges reflected in the patterns of motion of their constituent stars? Do the sort of high-resolution measurements that we can make of the Milky Way provide any clues as to how galaxies formed and evolve?

Ideally, the existence of the Milky Way's components would emerge naturally from the observational data rather than being assumed at the outset. It may be that the data are now so extensive that the components can be made to emerge in this way, but to approach the data in a totally unbiased way would be a large, and, we fear, confusing enterprise. It is didactically much easier to say 'here we have stars belonging to the thick disk, over there you see characteristic bulge objects,' and so forth. At the end of the day the reader must judge whether the picture that has been painted is a coherent and convincing one, or a modernist daub that only a pseud or an art gallery director would buy. In particular, controversy rages as to whether the components into which we shall divide the Galaxy are really distinct objects or just the beginning the middle and the end of an inherently continuous structure. By constantly referring to these components, we do not wish to prejudice the outcome of this controversy.

What makes the task in hand exceptionally confusing is the great variety of the available data. The crudest data are surface-brightness measurements analogous to those we have of external galaxies. Although these data are crude in the sense that they distinguish very little between different kinds of stars, they are hard to gather: the problem of sky subtraction, which is difficult in the case of external galaxies (see §4.2), is horrendous in the case of the Milky Way, which covers the whole sky rather than just a fraction of a square degree. Fortunately, in the last few years advances in the surface photometry of the Milky Way have been made which have allowed us to probe for the first time the surface brightnesses of the disk and bulge. The key has been to observe in the infrared from space: by working in the infrared one can largely overcome the problem posed by the huge quantities of dust that lie between us and the inner Galaxy, and by working from space one can make the problem of sky-subtraction manageable.

The earliest work on Galactic structure was exclusively concerned with **star counts**; that is, the numbers of stars with given colors per unit area of sky per magnitude interval in each of several directions on the sky. In the 1980s this field was revived, after decades of torpor, by two advances: (i) the advent of automatic plate-measuring machines and friendly computers to analyze the data they produce; (ii) models of the Milky Way that are inspired by observations of external galaxies, and allow one to predict the star counts for any field given assumed disk and bulge parameters.

The spectra of stars provide crucial insights into Galactic structure, because they betray not only the star's radial velocity, but also its chemical composition. Unfortunately, spectroscopy is time consuming even with modern techniques. Therefore one has to be selective in deciding which stars should be observed. Two very different strategies are commonly employed. The selected stars are those in a given field and magnitude range that either have specified colors, or have specified proper motions. These selection criteria are invaluable because they allow one to focus on one component of the Milky Way in preference to the others. However, they greatly complicate the statistical interpretation of spectroscopic surveys because one knows exactly which portion of the Milky Way a given survey has brought into focus only when one has a complete understanding of the structure of the Milky Way as a whole – and, of course, it precisely because one lacks that knowledge that the survey was undertaken in the first place! In these circumstances, models play an essential rôle in the interpretation of data; the data are interpreted in terms of the model that was used to design the selection criteria, and any discrepancies lead to a revision of the model and perhaps the definition of a new survey. This process is, unfortunately, both laborious and potentially confusing.

The confusion that is introduced by the interpretation of data in terms of inaccurate models is exacerbated by a fundamental weakness in the available models, namely, their near total disregard for Newton's laws of motion. In reality the positions and velocities of stars are tightly connected by dynamics, but the techniques of dynamical modeling have developed so slowly that most current models reflect this coupling only to an extremely limited degree. Not only does this make the models internally inconsistent, but it endows them with more free parameters than are strictly necessary, and this circumstance obviously makes it harder to search for a model that can account satisfactorily for any given body of data. Fortunately, the stage does now seem set for the advent of dynamically consistent models – see Dehnen & Binney (1996) for details.

The layout of this chapter is the following. We start by examining the Milky Way using the kinds of data that are available for external galaxies – surface photometry and line-of-sight velocity distributions. Specifically in §10.1 we examine the structure of the Milky Way from surface photometry, and in §10.2 we use velocity data to examine the bulge. Then we turn to an analysis of the solar neighborhood, which we can observe by means of tools that will probably never be available for the examination of most external galaxies. In §10.3 we analyze the velocities of common stars near the Sun. These measurements enable us to determine the peculiar velocity of the Sun itself, and to probe the differential rotation of the Galactic disk. In §10.4 we describe correlations that exist between the ages, chemical compositions and kinematics of stars that are confined to the disk. Combining insights obtained in this way with results obtained from star-counts, we shall conclude that the disk is a composite object, and determine its surface density. Next, in §10.5, we examine the halo. This component can be traced to large distances through globular clusters, field HB stars and metal-weak giants, and can also be studied through its rare representatives in the immediate solar neighborhood. §10.6 describes computer models of the Milky Way. These models combine all the Galaxy's components in such a way that the results of various types of surveys can be predicted before observing begins, and interpreted physically after the observations are complete. Thus, in a sense, they summarize our understanding of the Milky Way. Finally, in §10.7, we discuss the implications of all that we have learned for theories of how the Galaxy came to be structured the way it is. This is a question on which there is still no consensus within the scientific community, so this section strays further from the bedrock of secure knowledge than does most of the rest of this book.

## 10.1 Gross Structure from Surface Photometry

Most of the Milky Way is heavily obscured by dust – its center is obscured by $\sim 30$ magnitudes in $V$. Hence, surface photometry that contains useful information on the large-scale structure of the Galaxy can only be obtained at infrared wavelengths, at which dust is much less opaque than in the visible. While obscuration by dust decreases monotonically with increasing wavelength, the optimum waveband for observations of the Galaxy lies in the near infrared. Typical stars radiate a substantial fraction of their luminosity in the near infrared, while contributing very little luminosity to the mid and far infrared. Interstellar dust, by contrast, emits most strongly in the mid and far infrared (§8.1.6). Hence the stellar system stands out most prominently against the background from dust at $2 - 4\,\mu\mathrm{m}$.

The Galaxy has been observed in the near-infrared with an angular resolution of order a degree by several balloon-borne telescopes, from the Space Shuttle and by the COBE satellite – for references see Kent, Dame & Fazio (1991) and Freudenreich (1998). Structure on scales smaller than a degree, which is important close to the center (where one degree corresponds to 140 pc), has only been studied from the ground (Becklin & Neugebauer 1968, Allen *et al.* 1983, Eckart *et al.* 1994).

Even at $2 - 4\,\mu\mathrm{m}$ there is non-negligible extinction over much of the bulge, and this absorption must be taken into account when interpreting observations. This has been done in three entirely different ways.

In the approach of Kent, Dame & Fazio (1991), one starts by deriving a model of the distribution of interstellar gas in the Galaxy from radioline observations of HI and CO. The optical depth $\tau(R, \phi, z)$ to absorption between the point $(R, \phi, z)$ in the Galaxy and the Sun then follows from the assumption that the standard gas-to-dust ratio, equation (3.65), holds everywhere. Once $\tau$ has been determined, the surface brightness predicted at $(l, b)$ is simply

$$I(l, b) = \int_0^\infty \mathrm{d}s\, \mathrm{e}^{-\tau} j_*(R, \phi, z), \tag{10.1}$$

where $(R, \phi, z)$ is the point at a distance $s$ from the Sun in the direction $(l, b)$, and $j_*$ is the stellar luminosity density. Kent *et al.* found that they could accurately reproduce small-scale structure in the $2.4\,\mu\mathrm{m}$ brightness distribution that was observed from the Space Shuttle with a smooth distribution of luminosity density. That is, small-scale irregularities in the observed $2.4\,\mu\mathrm{m}$ surface brightness arise naturally from patchiness in the predicted obscuration rather than from irregularities in the stellar distribution.

An objection to the analysis of Kent, Dame & Fazio (1991) is that it employs a model of the interstellar hydrogen density that is uncertain for two reasons: first, it depends upon both the assumption of circular rotation, which we have seen is significantly in error for $R \lesssim 4\,\mathrm{kpc}$ (§9.4); second, it involves the conversion of CO linewidths to hydrogen column densities $N_\mathrm{H}$ and, as we saw in §9.6.1, this conversion is problematical, especially

near the Galactic center, where obscuration is heaviest. Spergel, Malhotra
& Blitz (1996) estimated the obscuration by a different technique that is
not subject to this objection. They first modeled the emission observed by
COBE at 240 $\mu$m as thermal radiation from dust – see §9.3.1. As a first
approximation, the dust was taken to be distributed in an axisymmetric
double-exponential disk – see equation (4.45) – in which the scale height
$z_0$ of the dust disk increased linearly with radius. The free parameters in
this model were adjusted to optimize the fit between the observed 240 $\mu$m
emission and that predicted by the model. Let $\rho_{\rm axi}(R, z)$ be best-fit model's
dust density and $I_{\rm axi}(l, b)$ its 240 $\mu$m surface-brightness distribution. Then
a non-axisymmetric model of the dust density that *exactly* reproduces the
observed brightness distribution $I_{\rm obs}(l, b)$ is

$$\rho_{\rm dust}(R, \phi, z) \equiv \rho_{\rm axi}(R, z) \frac{I_{\rm obs}(l, b)}{I_{\rm axi}(l, b)}, \tag{10.2}$$

where $(l, b)$ is the direction from the Sun to the point $(R, \phi, z)$ in the Galaxy. Physically, this equation distributes any additional dust that may be
required along a given line of sight in proportion to the dust density of the
axisymmetric model. Although this assignment must be wrong in detail, it
represents as reasonable a guess as can be made without using velocity information of the type employed by Kent *et al.*.[1] Spergel *et al.* used this simple
dust model to calculate for each COBE waveband optical depths, $\tau(R, \phi, z)$,
to absorption from each point in the Galaxy. A valuable check on the accuracy of the model is available, since it predicts, from the 240 $\mu$m data alone,
the amount by which light will be reddened in passing from any point in the
Galaxy to the Sun. This prediction may be compared with the reddening
inferred from the near-infrared data, which are available in the $J$, $K$, $L$ and
$M$ bands.

Whereas Kent *et al.* and Spergel *et al.* modeled the distribution of obscuring dust *before* fitting a model to the stellar luminosity density, Freudenreich (1998) modeled the dust and stellar distributions *simultaneously*. However, Freudenreich's modeling procedure employs neither radio-frequency line
data nor far-IR data, but depends only on the near-IR data. The procedure's
key assumption is that neither the bulge nor the disk has an intrinsic near-IR
color gradient.

Freudenreich started by defining parameterized models of the dust density and the luminosity densities of the bulge and disk. From these densities
and equation (10.1), he then predicted for each COBE waveband $X$ the observed intensity $I_X(l, b)$ for many directions $(l, b)$ on the sky. Finally, the
parameters of all three density distributions were simultaneously adjusted to

---

[1] If a good model of the Galaxy's non-axisymmetric velocity field were available, a
better dust model could be obtained by distributing excess dust at $(l, b)$ along the line of
sight so as to optimize the fit between the predicted HI and CO line profiles at $(l, b)$ and
the observed ones.

**Table 10.1**   Parameters of photometric models of the Milky Way

| Kent, Dame & Fazio (1991) | | Freudenreich (1998) | |
|---|---|---|---|
| $L_d$ | $4.9 \times 10^{10} L_\odot$ | $L_d$ | $2.2 \times 10^{10} L_\odot$ |
| $I_d$ | $(1000 \pm 200) L_\odot \, pc^{-2}$ | $I_d$ | $867 \, L_\odot \, pc^{-2}$ |
| $R_d$ | $(3 \pm 0.5) \, kpc$ | $R_d$ | $2.5 \, kpc$ |
| $z_0$ | $[0.165 + 0.21(R/R_0 - \frac{5}{8})] \, kpc$ | $z_0$ | $0.167 \, kpc$ |
| | | $R_h$ | $2.94 \, kpc$ |
| | | $(q_h, \beta)$ | $(0.87, 1.72)$ |
| $L_b$ | $1.1 \times 10^{10} L_\odot$ | $L_b$ | $0.62 \times 10^{10} L_\odot$ |
| $I_b$ | $(7400 \pm 1500) L_\odot \, pc^{-2}$ | $I_b$ | $1300 \, L_\odot \, pc^{-2}$ |
| $s_b$ | $(0.7 \pm 0.2) \, kpc$ | $(a_x, a_y, a_z)$ | $(1.69, 0.63, 0.42) \, kpc$ |
| $q_b$ | $0.61 \pm 0.18$ | $(c_\perp, c_\parallel)$ | $(1.55, 3.43)$ |
| | | $(R_{bar}, h_{bar})$ | $(2.97, 0.44) \, kpc$ |
| | | $\phi_0$ | $14°$ |

NOTES: Both models assume $R_0 = 8 \, kpc$. The Kent *et al.* model is for the $K$ band while Freudenreich's is for the $L$ band.

optimize the fit for each waveband between $I_X$ and the intensity measured by COBE.

Kent, Dame & Fazio (1991) assumed that both bulge and disk are axisymmetric bodies, with luminosity densities of the form

$$j_d(R, z) = \frac{I_d}{2z_0} e^{-R/R_d - |z|/z_0},$$

$$j_b(R, z) = \frac{I_b}{\pi s_b} K_0(s/s_b) \quad \text{where} \quad s^4 \equiv R^4 + (z/q_b)^4.$$

$(10.3)$

Here $I_d$ is the central face-on surface brightness of the disk and $I_b$ is the central surface brightness of the bulge. $K_0$ is a modified Bessel function, which ensures that the bulge has an exponential surface-brightness profile – see Problem 10.1. The variable $s$ is defined so that the bulge has boxy isodensity surfaces of axis ratio $q_b$. The left-hand column of Table 10.1 gives the best-fitting values of the various parameters.

The parametric model of Freudenreich (1998) differs from that of Kent *et al.* in several important respects. First, the disk is assumed to have an elliptical hole at its center,[2] and to have a vertical density profile that follows the $sech^2$ law [equation (4.44)] rather than a simple exponential. Second, in view of the kinematic evidence that the inner Galaxy is triaxial that was reviewed in §9.4, Freudenreich assumed that the bulge is triaxial – the Sun–center line is assumed to make an angle $\phi_0 \simeq 15°$ with the bulge's longest

---

[2] If the disk really has a central hole, an external observer would consider the Milky Way to be a type II galaxy (§4.4.3).

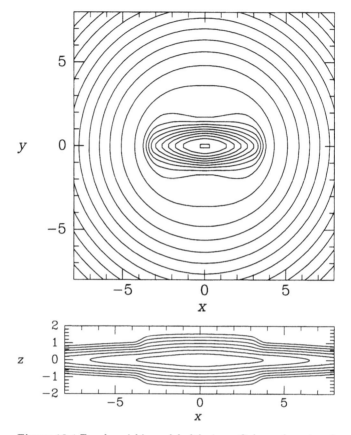

**Figure 10.1** Freudenreich's model of the inner Galaxy when viewed along the shortest axis of the bar (top) and along to the bar's intermediate axis (bottom). There are ten contours per decade in surface brightness in the upper panel and three per decade in the lower panel. The axes are marked in kpc.

principal axis. In detail, for $R < R_0$ Freudenreich's luminosity densities are given by

$$
\begin{aligned}
j_{\rm d}(x,y,z) &= \frac{I_{\rm d}}{2z_0} \, {\rm sech}^2(z/2z_0) \, {\rm e}^{-R/R_{\rm d}} \left(1 - {\rm e}^{-m_{\rm h}^\beta}\right), \\
j_{\rm b}(x,y,z) &= \frac{I_{\rm b}}{a_z} \, {\rm sech}^2(m_{\rm b}) \, T(R),
\end{aligned}
\tag{10.4}
$$

where $I_{\rm b}$ is the central surface brightness of the bulge when it is viewed down its shortest axis, $I_{\rm d}$ is the corresponding surface brightness for the disk in the absence of the central hole, and

$$m_h \equiv \sqrt{(x^2 + q_h^2 y^2)/R_h^2}$$

$$m_b \equiv \left\{ \left[ \left( \frac{|x|}{a_x} \right)^{c_\perp} + \left( \frac{|y|}{a_y} \right)^{c_\perp} \right]^{c_\parallel/c_\perp} + \left( \frac{|z|}{a_z} \right)^{c_\parallel} \right\}^{1/c_\parallel} \tag{10.5}$$

$$T(R) \equiv \begin{cases} 1 & \text{for } R \le R_{bar} \\ e^{-(R-R_{bar})^2/h_{bar}^2} & \text{otherwise} \end{cases}$$

The hole in the disk, which has radius $R \sim R_h$, is created by the factor $(1 - e^{-m_h^\beta})$ in the first of equations (10.4). The long axis of this hole lies along the $y$ axis for $q_h < 1$. At Galactocentric radii $R < R_{bar}$, the luminosity density of the bulge is constant on surfaces of constant $m_b$, which have principal axes with lengths in the ratios $a_x:a_y:a_z$. The shape of these surfaces is controlled by the parameters $c_\perp$ and $c_\parallel$. For the best-fitting values of these parameters ($c_\perp < 2$ and $c_\parallel > 2$ – see Table 10.1), the bulge appears boxy when viewed perpendicular to its shortest ($z$) axis, and pointy when viewed from along this axis. The Sun–center line makes an angle of $\sim 14°$ with the long axis of the bar, which is taken to be the $x$ axis. The function $T(R)$ sharply truncates the bulge's density near $R_{bar}$. Figure 10.1 illustrates these points by showing face-on and edge-on views of Freudenreich's model.

A significantly better fit of the model to the data can be obtained if the Sun is assumed to lie $\sim 16$ pc above the Galactic plane. At large values of $|l|$ the fit to the data can be further improved if the model described by equations (10.4) is modified in two ways: (i) the exponential disk is truncated near $R = 1.5R_0$; (ii) the disk is warped at $R \gtrsim R_0$. The sense of the warp in the stellar disk is the same as that of the warp in the HI disk that was discussed in §9.2.6. In particular, the line of nodes nearly coincides with the Sun–center line, and the disk bends towards the north at $l > 0$. Since the stellar disk does not extend as far as the HI disk, the maximum deviation of the stellar disk from the plane of the inner Galaxy is not large ($\sim 300$ pc).

It is instructive to compare the parameters listed in Table 10.1 for the models of Freudenreich (1998) and Kent, Dame & Fazio (1991). In part the differences reflect the fact that the Kent $et\ al.$ model is of the Galaxy's $K$-band luminosity distribution, while Freudenreich's model is for the $L$ band. In view of this difference, there is reasonable agreement between the values given for the disk's extrapolated central surface brightness $I_d$, scale length, $R_d$ and scale height, $z_0$. Comparison of the parameters for the bulge is not straightforward because the functional forms used in the two studies are very different. It is noteworthy, however, that Freudenreich's bulge has a central, face-on surface brightness that is smaller than that of the bulge of Kent $et\ al.$ by a factor $\sim 5.7$. This difference reflects the fact that Freudenreich's bulge/bar is long and diffuse. It is disconcerting that the numbers given in Table 10.1 for the total luminosity of the disk differ by a factor two. Actually, the numbers quoted in the published papers differ by an even larger factor;

our values are based on the values of the $K$- and $L$-band luminosities of the Sun that are given in Table 2.1. Several factors account for the lower luminosity of Freudenreich's disk: its scale length is shorter; its central surface brightness is lower; and it has a hole of radius $\sim 3\,\mathrm{kpc}$ at its center. Freudenreich's bulge is about 50% less luminous than that of Kent *et al.* because its greater radial extent does not fully compensate for its lower characteristic surface brightness.

Given that it is so hard to determine the luminosity of the Galaxy photometrically, it is interesting to estimate it from the Tully–Fisher relation (§7.3.4) and the local circular speed (§10.6). Equations (7.26) and (10.71) with $W_R^i = 2v_c$ together imply that the $H$-band absolute magnitude of the Galaxy is $M_H = -23.0 \pm 0.2$. Adopting the value of the $H$-band absolute magnitude of the Sun given in Table 2.1, this absolute magnitude corresponds to a luminosity of $(3.3 \pm 0.6) \times 10^{10}\,L_\odot$. This value lies between the values give in Table 10.1 and is probably our best estimate of the Galaxy's luminosity.

### 10.1.1 The Galaxy at optical wavelengths

To compare the Milky Way with external galaxies, it is desirable to know what the Galaxy would look like when viewed from outside at optical wavelengths. From our location it is extremely hard to construct such a model directly. The best approach is to assume that the $B - K$ and $V - K$ colors of the Milky Way are similar to those of galaxies of similar Hubble type, and to use this assumption to derive the $B$- and $V$-band luminosity densities from the near-IR models discussed above. From a study of 86 nearly face-on spiral galaxies, de Jong (1996) concludes that late-type galaxies like the Milky Way have disk scale lengths that are larger in the $B$ band than in the $K$ band by a factor $(1.25 \pm 0.25)$. He finds, moreover, that the disks of these galaxies have extrapolated central surface brightnesses that are fainter in the $B$ band than in the $K$ band by $(3.6 \pm 0.6)\,\mathrm{mag}$, while the bulges have $B - K \sim 4.5$. From these numbers and those presented in Table 10.1 one can infer the $B$-band luminosity density at any point in the Milky Way. The appearance of the Galaxy to an external observer could then be calculated by integrating equation (10.1), which includes the effects of obscuration by dust, along the observer's lines of sight.

There is one place in which we can directly measure the $B - K$ color of the Galaxy, namely the solar neighborhood. From the IR modeling described above, we know that the face-on integrated $K$-band surface brightness of the Galaxy at $R_0$ is $\mu_K = 20.62$ – see Problem 10.3. To calculate the corresponding $B$-band surface brightness, we proceed as follows. We count stars in the direction of a Galactic pole down to some limiting apparent magnitude $B_{\mathrm{lim}}$ and sum their fluxes. Figure 10.2 is the resulting plot of integrated surface brightness, $\mu_B$, versus $B_{\mathrm{lim}}$. It shows that $\mu_B$ converges to a value $\sim 24.7$ by $B_{\mathrm{lim}} \sim 16$. De Vaucouleurs & Pence (1978) argue that,

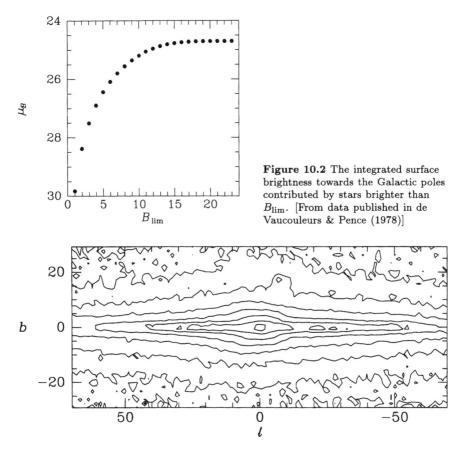

**Figure 10.2** The integrated surface brightness towards the Galactic poles contributed by stars brighter than $B_{\text{lim}}$. [From data published in de Vaucouleurs & Pence (1978)]

**Figure 10.3** A reconstruction of how the Milky Way would appear in the absence of dust. The $L$-band COBE data have been corrected for absorption using equation (10.6) and the dust and stellar luminosity models of Spergel, Malhotra & Blitz (1996). Contours are spaced by 1 mag. [After Binney, Gerhard & Spergel (1997)]

in the absence of extinction, the surface brightness would be brighter by $\sim 0.2\,\text{mag}$. If the local disk were viewed from outside, its surface brightness, after correction for extinction, would be brighter by $2.5\log 2\,\text{mag}$ because one would be looking through all the disk rather than half of it. Hence, the face-on, extinction-corrected surface brightness of the disk is $\mu_B = 23.75$. It follows that the solar neighborhood has color $B - K = 3.13$. This value is in excellent agreement with the $B - K$ colors that de Jong (1996) measures in galaxies similar to the Milky Way at radii comparable to that of the Sun.

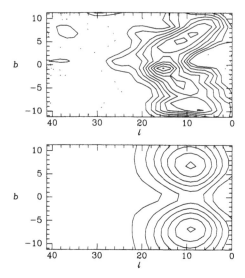

**Figure 10.4** The COBE surface photometry displays marked asymmetries in $l$ of the type that would occur if the Galaxy were barred, with the nearer end of the bar at positive $l$. The top panel shows the ratio $I(l, b)/I(-l, b)$ for the data shown in Figure 10.3. The lower panel shows the prediction of this difference for the bar model that is defined by equation (10.4). The COBE data have been smoothed. Contours are spaced by 0.05 mag. Dotted contours indicate that $I(l, b) < I(-l, b)$.

## 10.2 The bulge

### 10.2.1 Integrated surface photometry

Freudenreich's model of the Galaxy includes a triaxial bulge because the kinematic evidence reviewed in §9.4 indicates that we inhabit a barred galaxy. In the last section we saw, however, that the axisymmetric model of Kent, Dame & Fazio (1991) also provides a comparable fit to the near-infrared photometry of the Galaxy. This situation prompts us to ask whether the near-IR photometry, taken alone, provides compelling evidence that the Galaxy is barred. In this enquiry it is important not to restrict oneself to a particular parametric form of the Galaxy's luminosity density. Indeed, unless the Galaxy precisely follows the model's particular functional form, which is entirely improbable, the fit to high-quality data provided by the model will inevitably be formally poor. This fact deprives us of an objective criterion for distinguishing between the fits furnished by an axisymmetric model and a barred model. Hence, we now examine the COBE photometry for direct indications that the Galaxy is barred.

The possession of models for the optical depth to absorption $\tau(R, \phi, z)$ and the emissivity $j_*(R, \phi, z)$ associated with each point in the Galaxy enables us to infer what the Milky Way would look like in the absence of dust. Specifically we write

$$I_{\text{dustfree}}(l, b) = \frac{I_{\text{dustless model}}(l, b)}{I_{\text{dusty model}}(l, b)} I_{\text{obs}}(l, b). \tag{10.6}$$

Unless our models of $\tau$ and $j_*$ are seriously in error, $I_{\text{dustfree}}$ should yield quite an accurate estimate of what a dust-free Milky Way would look like. Figure 10.3 shows such a dust-free view of the Milky Way.

**Figure 10.5** The nearer end of a bar generally has higher surface brightness than the further end because for given $|l|$ the left-hand line of sight crosses the long axis of the bar closer to the center than does the right-hand line of sight.

Although the brightness distribution shown in Figure 10.3 is fairly symmetrical in $b$, significant asymmetries are apparent in $l$. The upper panel of Figure 10.4 quantifies this phenomenon by plotting on a magnitude scale the ratio $I(l, b)/I(-l, b)$ of the brightnesses at $\pm l$ in Figure 10.3. In an elongated region around the center that extends to $|l| \simeq 10°$, both sides have about the same brightness.[3] At larger $|l|$ or $|b|$, the side at $l > 0$ is brighter than the other side. Beyond $|l| \sim 20°$, this trend reverses, and it is the side at $l < 0$ that is brighter. Figure 10.5 shows schematically why the nearer end of the bar is brighter than the further end: at a given value of $|l|$ the left-hand line of sight crosses the long axis of the bar closer to the center than does the right-hand line of sight.

The lower panel in Figure 10.4 shows the degree to which Freudenreich's triaxial bulge is able to reproduce the observed asymmetry distribution that is shown in the top panel. The model correctly predicts that the brighter side is at $l > 0$ for $|l| \lesssim 20°$, but it incorrectly predicts that the Galaxy is symmetrical further out. This failure of the model reflects the fact that at $|l| \gtrsim 20°$ lines of sight do not intersect the model bulge and the model disk is very nearly axisymmetric.

This discussion makes it clear that the triaxiality of any Galaxy model derives entirely from the differences between the Galactic luminosity density at $l > 0$ and at $l < 0$ that are contoured in Figure 10.4. Unfortunately, these differences are much smaller than the corrections made to the observed surface-brightness distribution to allow for obscuration by dust. Hence, a subtle change in the dust model is liable to change significantly the assumed left-right asymmetry in appearance of the Galaxy from which the bar's shape is derived. We can estimate the importance of this sensitivity by comparing Freudenreich's model with a triaxial luminosity distribution that has been derived using a different model for the dust. The data shown in Figure 10.3 constitute an appropriate data set, and the luminosity distribution can be constructed by applying a Richardson–Lucy deconvolution algorithm (Appendix C) to these data. In order to focus on the differences implied by the alternative dust model, we use Freudenreich's model as the starting point of the Richardson–Lucy iterations.

---

[3] As Blitz & Spergel (1991b) pointed out, if the Galaxy had an approximately homogeneous core, its right-hand side would be brighter than its left close to the Center, because lines of sight on the right make smaller angles with the bar's long axis, with the result that they remain within a region of high luminosity density for longer than do lines of sight on the left.

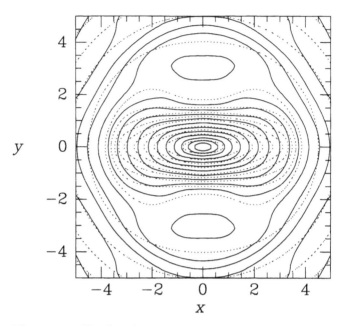

**Figure 10.6** The dotted contours show the central part of Figure 10.1. The full contours show the model that is produced from this model and the data of Figure 10.3 by seven iterations of the Richardson–Lucy algorithm of Binney & Gerhard (1997).

Comparison of the full and dotted contours in Figure 10.6 show how seven iterations of the deconvolution technique modify the model of Figure 10.1 to produce a better fit to the data of Figure 10.3. Each set of contours shows the surface brightness of a model when the latter is viewed perpendicular to the disk; the dotted contours are for Freudenreich's model, while the full contours are for the iterated model, which fits the data significantly better. One striking difference between the two models is that, at small radii, the full contours are boxy while the dotted contours are pointy.

Another important, but less obvious, difference between the models shown in Figure 10.6 is that the full contours are more crowded towards the center than the dotted contours. This phenomenon arises because the iterations have modified the bulge from an exponential profile (which yields uniformly spaced logarithmic contours) to a profile that is better represented by a power law. In §4.3 we saw that the luminosity densities of faint elliptical galaxies behave as power laws, $j_* \sim r^{-1.9}$, at very small radii (see Figure 4.32). Interestingly, ground-based near-infrared observations, which have much higher spatial resolution than the space-based data, show that the luminosity density of the Milky Way conforms to this pattern in as much as, in the angular range $2'' < R < 4°$, the azimuthally-averaged $K$-band surface brightness fits a power law $I(R) \sim R^{-0.8}$ (Eckart et al. 1994), which naively implies $j_* \propto r^{-1.8}$ (§4.2.3). For $R_0 = 8\,\mathrm{kpc}$, $2'' < R < 4°$ corresponds to $0.8\,\mathrm{pc} \lesssim r \lesssim 560\,\mathrm{pc}$. Beyond a few degrees, the surface brightness falls much more steeply: in the range $-4° < b < -12°$, Frogel *et al.* (1990) find that power laws $I \propto |b|^{-\alpha}$ with index $\alpha$ in the range $2.5 < \alpha < 3.4$ fit the projected density of various tracers, with larger values of $\alpha$ associated with more metal-rich tracers. Thus it seems that, to within the substantial errors, the luminosity density of the inner Galaxy can be fitted by equation (4.15) with $\gamma \simeq 1.8$ and $a \simeq 0.56\,\mathrm{kpc}$. The dependence of $\alpha$ upon metallicity probably arises because at $4° < |b| < 12°$ both the bulge and the halo contribute significantly to the observed tracers, and the surface density of the halo is falling less rapidly with $|b|$ than is the surface density of the bulge. As we shall see below, the bulge has significantly higher metallicity than the halo, so metal-rich tracers should provide the truest picture of the surface density of the bulge.

### 10.2.2 Evidence for a bar from individual stars

Further evidence that the Galaxy is barred comes from studies of individual stars rather than integrated photometry. Unfortunately, most of the relevant observations have been made at optical wavelengths. Consequently, observations of bulge stars in fields that lie sufficiently far from the plane to be observable at optical wavelengths are important. Since the distribution of dust is highly non-uniform, we can see much further in some directions that lie near the center than in others. One such direction, at $(l, b) = (1°, -3.9°)$ is especially important: it is known as **Baade's Window** in honor of W. Baade, who discovered it. Along lines of sight that lie in Baade's Window, the extinction to the point of closest approach to the center varies in the range $1.26 < A_V < 2.79$ (Stanek 1996). Several important observational programs have involved studies of stars in Baade's Window and similar windows on to the bulge.

One class of studies of stars in Baade's window involves monitoring the brightnesses of several million stars in order to catch the rare occasions when a star is gravitationally lensed by a foreground star – see §2.4. Such searches for microlensing events provide an important constraint on several characteristics of bulge stars. First and foremost, the optical depth to microlensing, $\tau$, that a model predicts for a given line of sight increases with both the surface density, $\Sigma_b$, of bulge stars and the effective depth, $d_b$, of the bulge along that line of sight. It is clear that $\tau$ must increase with the surface density since lensing occurs when one star passes behind another star. The sensitivity of $\tau$ to the depth of the bulge arises because the cross-section for lensing, $\pi r_E^2$, contributed by any given foreground star is greatest when it lies half way between us and the lensed star – see §2.4. Quantitatively, $\tau$ is approximately proportional to $\Sigma_b d_b$.

Now suppose that we looked exactly down the long axis of a prolate spheroidal bulge and consider the variation in $\tau$ as we elongate the bulge whilst leaving its mass and mean density unchanged. Let $q < 1$ be the bulge's axis ratio. Mass conservation implies that $\Sigma_b q^2 d_b^2 = $ constant while volume conservation implies $q^2 d_b^3 = $ constant. Combining these relations, we find that $\tau \propto d_b^2 \propto q^{-4/3}$. Figure 10.6 suggests that, if the angle $\phi_0$ between the Sun–center line and the long axis of the bar is 15°, then $q \simeq 0.3 - 0.4$. The calculation above suggests that in this case $\tau$ will be $\gtrsim 3.5$ times larger than it would be if the bulge were spherical and equally massive. Moreover, the smaller $\phi_0$ is, the smaller the COBE photometry requires $q$ to be and therefore the larger $\tau$ will be. In Baade's Window Udalski et al. (1994) find $\tau = (3.3 \pm 1.2) \times 10^{-6}$ for $8.6\,d < t < 62\,d$ while at $(l, b) = (2.3°, -2.65°)$ Alcock et al. (1995b) find $\tau \simeq 4 \times 10^{-6}$ for a similar range of values of $t$. Axisymmetric models of the Milky Way predict optical depths that are smaller by roughly a factor of 3 (Paczynski et al. 1994). Thus the fact that the measured value of $\tau$ is of order 3 times larger than axisymmetric models predict, supports the proposition that $\phi_0 \sim 15°$. Zhao, Spergel & Rich (1995) find that a barred bulge of the type that is suggested by the COBE photometry yields $\tau \simeq 2 \times 10^{-6}$ in Baade's Window, which is just compatible with the data of Udalski et al..

Photometry of the millions of stars in these studies that are not lensed yields an additional important constraint. When these stars are plotted in the $(V - I, V)$ plane, there is a pronounced clustering of stars at the location expected for a red clump star (§5.1.2) that is $\sim 8\,\mathrm{kpc}$ distant and obscured by $A_V \simeq 2\,\mathrm{mag}$ of extinction. However, for fields at $l \simeq +5°$ and $b \simeq -3.5°$ the center of this clustering is brighter by 0.4 mag than for fields at the same latitude but $l \simeq -5°$ (Stanek et al. 1994). Bar models predict just such a difference because they imply that stars observed at $l > 0$ are, on average, closer than those observed at $l < 0$ – see Figure 10.5. Quantitatively, the two models shown in Figure 10.6 predict differences between the red-clump magnitudes for Stanek's fields that amount to 0.29 mag for Freudenreich's model, and 0.21 mag for the model shown by full contours. Thus, the COBE-based bar models predict values for Stanek's magnitude difference that are smaller than the observed value. This discrepancy will probably be resolved by improved estimates of the extinction towards the Galactic center.[4]

Further evidence that objects in the bulge that are seen at $l > 0$ tend to be closer than corresponding objects seen at $l < 0$ comes from studies of several classes of objects of known absolute magnitude in addition to red clump stars. These objects include AGB stars (Nakada et al. 1991, Weinberg 1992), OH/IR stars (Sevenster 1996), and Mira variables (Whitelock & Catchpole 1992)

As we saw in §2.4, microlensing events have a characteristic timescale $t$ which depends upon the mass of the lensing star and the transverse velocities of the lensing relative to the lensed star. Consequently, microlensing surveys

---

[4] Wozniak & Stanek (1996) discuss the effects of a systematic variation in the extinction across the Galactic center.

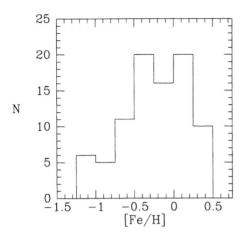

**Figure 10.7** The distribution of 88 bulge K giants in [Fe/H]. [After McWilliam & Rich (1994)]

provide information on both the kinematics and the mass function of bulge stars. Unfortunately, the study of microlensing is still in its infancy and neither the models nor the data analysis has developed to the point at which the full potential of this technique can be exploited.

### 10.2.3 Age and metallicity of the bulge

Studies of the distribution in age and metallicity of bulge stars are fundamental for our understanding of the structure and formation of all galaxies, including the Milky Way. The reason that such studies enjoy this fundamental rôle is simply that the bulge is the only stellar system of its type that is close enough to permit the detailed study of individual stars.

Rich (1988) analyzed the spectra of 88 bulge K giants in Baade's Window and compared them with similar spectra for K giants in the disk and globular clusters. Subsequently, McWilliam & Rich (1994) obtained spectra of twelve of these stars at 14-fold greater resolution and concluded that Rich (1988) had overestimated [Fe/H] by $\sim 0.3$ for the most metal-rich stars. Figure 10.7 shows a histogram of [Fe/H] for these stars on the revised scale of McWilliam & Rich (1994).[5] The distribution is wide, reaching from [Fe/H] $= -1.25$ to [Fe/H] $= 0.5$ with mean value $\langle$[Fe/H]$\rangle = -0.25$. We shall see below (§10.4) that this mean value is similar to that of solar-neighborhood K giants, and is considerably higher than the mean values of [Fe/H] for either the halo or the disk globular clusters (§10.5.1).

On account of the width of the bulge's metallicity distribution, the CM diagram for the bulge differs from that of any globular cluster. In particular, although it contains RR Lyrae stars and K giants similar to those in globular clusters, it also contains M giants such as we find in the solar neighborhood. Tiede, Frogel & Terndrup (1995) show that the morphology of the bulge's

---

[5] Medium resolution spectra of a large sample of K giants confirm the metallicity distribution of Figure 10.7 (Minniti *et al.* 1995).

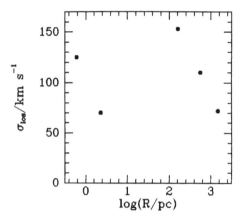

**Figure 10.8** Five measurements of the velocity dispersion of stars along lines of sight at $l < 10°$ indicate that the velocity dispersion rises towards the center both within the bulge (outer three points) and within the nuclear star cluster (inner two points), but drops to a minimum between these two dynamically distinct entities.

CM diagram is consistent with the value of $\langle[Fe/H]\rangle$ that McWilliam & Rich derived spectroscopically.

The K giants analyzed by McWilliam & Rich (1994) have abundance ratios that differ from those in the Sun. The $\alpha$ element Mg and the iron-peak element Ti are enhanced relative to Fe; $[Mg/Fe] \simeq [Ti/Fe] \simeq 0.3$ at all values of $[Fe/H]$.[6] On the other hand, $[Si/Fe] \simeq [Ca/Fe] \simeq 0$ even though both Si and Ca are $\alpha$ elements. Thus the abundance ratios in bulge giants are not naturally explained by the theory of §5.2. The reason for this failure is unknown.

### 10.2.4 Bulge kinematics

To probe the very center of the bulge one has to work at infrared or radio wavelengths because the inner few degrees are extremely highly obscured by dust at optical wavelengths. The advent of sensitive imaging infrared detectors has made infrared studies possible. A convenient spectral feature for velocity determinations is caused by absorption at 2.3 μm by CO molecules: this absorption trough has a very sharp edge – the **CO bandhead** – at its blue end, which can be accurately located in typical spectra. Sellgren *et al.* (1990) used this feature to study the dynamics of stars in the innermost 1' (2.3 pc) of the Galaxy, which is dominated by the nuclear star cluster (§9.5). They concluded that any systematic rotation within 2.3 pc has amplitude $\lesssim 15\,km\,s^{-1}$. The measured velocity dispersion rises from $\sigma_{los} \simeq 70\,km\,s^{-1}$ at $R = 1'$ to $\sigma_{los} \simeq 125\,km\,s^{-1}$ at $R < 16''$ (0.6 pc). They found $\sigma_{los}$ to be constant inside $R = 16''$, but concluded that this constancy probably reflects an absence from the innermost 0.6 pc of stars with CO lines in their spectra: the strength of the CO bandhead decreases at $R < 16''$ as if all the light

---

[6] Sadler, Rich & Terndrup (1996) derived a slightly smaller Mg enhancement in bulge giants from low-resolution spectra.

emitted within 0.6 pc of Sgr A* were free of CO lines. This conclusion is consistent with the idea that the central star cluster is dominated by early-type stars – see §9.5.

Blum et al. (1995) used the CO bandhead to determine $\sigma_{los}$ for the bulge in four fields, the most central of which lies at $(l, b) = (-0.59°, 0.98°)$, which corresponds to a distance $r \sim 160$ pc from the center. They found evidence that $\sigma_{los}$ rises as one approaches the center: in their innermost field, $\sigma_{los} = 153 \pm 17$ km s$^{-1}$. This value is twice as large as the dispersion reported by Sellgren et al. (1990) at $r \simeq 2.3$ pc. Evidently Blum et al. measured the velocity dispersion of the bulge, while even the outermost point of Sellgren et al. relates to the central star cluster, which is a dynamically distinct entity.

Radial velocities are available for several hundred bulge stars that lie in Baade's Window (Mould 1983, Rich 1988, Terndrup, Sadler & Rich 1995). If $\sigma_{los}$ is to be measured accurately, care has to be taken to eliminate from the sample foreground disk stars, which tend to bias the measurement towards low values. When this correction is made, one finds for the bulge $\sigma_{los} = 110 \pm 10$ km s$^{-1}$.

Minniti (1996) studied the kinematics of giants in a field at $(l, b) = (8°, 7°)$, to which halo stars make a significant contribution. A sample of 194 K giants with [Fe/H] $> -1$ had $\langle v_{los} \rangle = 66 \pm 5$ km s$^{-1}$ and $\sigma_{los} = 72 \pm 4$ km s$^{-1}$, while the 31 most metal-weak K giants ([Fe/H] $< -1.5$) had $\langle v_{los} \rangle = -6 \pm 20$ km s$^{-1}$ and $\sigma_{los} = 113 \pm 14$ km s$^{-1}$. If we identify the stars with [Fe/H] $> -1$ as bulge stars and the metal-weak stars as halo objects (see below), then these data indicate that: (i) the outward decrease in the bulge's velocity dispersion that we encountered above continues to at least $R \simeq 1.5$ kpc, and (ii) the bulge has significant rotational velocity. Figure 10.8 summarizes measurements of the line-of-sight velocity within the bulge and the nuclear star cluster.

OH/IR stars (see Table 3.5) provide another important probe of the bulge because: (i) their detectability at infrared and radio wavelengths allows them to be studied throughout the Galaxy; and (ii) their velocities can be measured accurately. te Lintel-Hekkert et al. (1991) and Nakada et al. (1993) searched for maser emission from selected IRAS sources, while Lindqvist, Habing & Winnberg (1992) and Sevenster et al. (1997) made flux-limited surveys for 18 cm maser emission. Unlike the radio-selected surveys, the IRAS-based surveys cover the whole inner Galaxy but suffer from a bias against sources near the plane. The velocity dispersions of the IRAS-based samples both lie near $\sigma_{los} = 110$ km s$^{-1}$. Both IRAS samples show clear signs of Galactic rotation; when a straight line is fitted to the values of $v_{los}$ versus longitude $l$, one obtains $\langle v_{los} \rangle \simeq 10l$ km s$^{-1}$, with $l$ measured in degrees. The significance of this result is unclear, however, since the data are not well fitted by a straight line, and we do not expect the bulge to rotate like a solid body.

The line profile of an OH/IR star is double peaked because the masing gas is contained in a shell that is expanding away from the central star.

The expansion velocity of this shell, which can be measured from the line profile, is thought to be a diagnostic of the age of the exciting star: initially more massive and therefore shorter-lived stars give rise to larger expansion velocities. When the radio-selected sample of Lindqvist, Habing & Winnberg (1992) is divided into two according to whether the expansion velocity is greater than or less than $18 \, \mathrm{km \, s^{-1}}$, the velocity dispersions of the subsamples are $\sigma_{\mathrm{los}} = 82 \pm 7 \, \mathrm{km \, s^{-1}}$ for the slowly expanding (old) stars and $\sigma_{\mathrm{los}} = 65 \pm 6 \, \mathrm{km \, s^{-1}}$ for the rapidly expanding (young) stars. The rapidly expanding stars are significantly more strongly concentrated towards the plane than are the slowly expanding stars. Both samples show evidence for Galactic rotation. In fact, the mean values of $\langle v_{\mathrm{los}} \times \mathrm{sign}(l) \rangle$ are very similar for this sample and for the samples of IRAS-selected OH/IR stars studied by Nakada *et al.* and te Lintel-Hekkert *et al.*, even though the Lindqvist *et al.* sample covers a much smaller range in $|l|$ ($|l| < 1°$) than do the IRAS-selected surveys.

If the transverse velocities of bulge stars exceed $100 \, \mathrm{km \, s^{-1}}$, their proper motions should be measurable: $120 \, \mathrm{km \, s^{-1}}$ at a distance of $R_0 = 8 \, \mathrm{kpc}$ yields $3.1 \, \mathrm{mas \, yr^{-1}}$. Spaenhauer, Jones & Whitford (1992) have measured proper motions for 429 stars in Baade's Window. These stars were selected to have $B - V > 1.4$, with the intention of including within the sample all K and M giants in the field.[7] The proper motions measured by Spaenhauer *et al.* are *relative*, that is, the mean proper motion of the sample is unknown. The dispersions of $\mu_l$ and $\mu_b$ within the sample are $(\sigma_l, \sigma_b) = (3.2 \pm 0.1, 2.8 \pm 0.1) \, \mathrm{mas \, yr^{-1}}$. The subsample formed by stars fainter than $B = 18$ has only very slightly larger dispersions. Since any disk stars in the sample should be relatively nearby and therefore mostly brighter than $B = 18$, this result suggests that disk contamination is not a major problem. The rough agreement between the line-of-sight and transverse velocity dispersions shows that any velocity anisotropy within the bulge is minor. It does not follow that the velocity ellipsoids (see below) within the bulge are isotropic, however, because these ellipsoids will in most places not be aligned with the direction to Baade's Window and the $l$ and $b$ directions.

## 10.3 Kinematics of stars near the Sun

### 10.3.1 The solar motion

We now fill a gap in our knowledge which we papered over in Chapters 2 and 9: determining the velocity of the Sun relative to the LSR. Our strategy for this project is the following. We identify a series of types of spectroscopically similar stars – gK stars, dM stars etc. – and then determine the velocity

---

[7] The first epoch plates were exposed by W. Baade in 1950.

of the Sun relative to the mean velocity of each type. Each such velocity is called the **solar motion** relative to that type of stars. We shall find that the solar motion varies systematically with the type studied, and that the velocity of the LSR can be inferred from this systematic variation.

The solar motion can be determined from either radial velocities or proper motions – see Jaschek & Valbousquet (1991) and Dehnen & Binney (1998a) for recent determinations by each method. To see this, let $\mathbf{v}$ denote the velocity of an object in the frame of rest of the given type; in this frame the Sun has velocity $\mathbf{v}_\odot$ and the average of any component of velocity, $v_i$, over stars of the type is zero. Now the heliocentric velocity of the $k^{\text{th}}$ star is $\mathbf{u}_k = \mathbf{v}_k - \mathbf{v}_\odot$, so, by equation (2.23), its radial velocity is

$$v_{\text{los }k} = \hat{\mathbf{x}}_k \cdot \mathbf{v}_k - v_\odot \cos \psi_k, \qquad (10.7)$$

where $\hat{\mathbf{x}}_k$ is the unit vector from the Sun to the star and $\psi_k$ is the angle between this vector and $\mathbf{v}_\odot$. When equation (10.7) is averaged over a large number of stars that are seen in almost the same direction $\hat{\mathbf{x}}$, we have that $\hat{\mathbf{x}}_k \cdot \mathbf{v}_k$ approximately averages to zero because the mean of $\mathbf{v}_k$ is, by construction, zero and $\hat{\mathbf{x}}_k \simeq \hat{\mathbf{x}}$ for all stars. Hence

$$\langle v_{\text{los}} \rangle \simeq -v_\odot \cos \psi. \qquad (10.8)$$

This equation states that the mean radial velocity will be largest when $\hat{\mathbf{x}}$ points in the opposite direction to $\mathbf{v}_\odot$ ($\psi = 180°$), and smallest when $\hat{\mathbf{x}}$ and $\mathbf{v}_\odot$ are parallel ($\psi = 0$) – in this direction the Sun is on the average approaching the stars, so they tend to have negative radial velocities. The direction of $\mathbf{v}_\odot$ is called the **apex of the solar motion**, and the opposite direction is called the **antapex of the solar motion**.

Suppose now that we wish to determine $\mathbf{v}_\odot$ from proper motions. We average equation (2.19) for a star's proper motion over a large number of stars of the type that are all seen near the direction of $\hat{\mathbf{x}}$ and all lie approximately at distance $d$, so that $|\mathbf{x}_k| \simeq d$:

$$\begin{aligned}
\langle \boldsymbol{\mu} \rangle &= \left\langle \frac{((\mathbf{v}_k - \mathbf{v}_\odot) \times \hat{\mathbf{x}}_k) \times \hat{\mathbf{x}}_k}{|\mathbf{x}_k|} \right\rangle \\
&\simeq \frac{(\langle (\mathbf{v}_k - \mathbf{v}_\odot) \rangle \times \hat{\mathbf{x}}) \times \hat{\mathbf{x}}}{d} \\
&= -\frac{1}{d} (\mathbf{v}_\odot \times \hat{\mathbf{x}}) \times \hat{\mathbf{x}} \\
&= \frac{1}{d} (\mathbf{v}_\odot - v_\odot \cos \psi \, \hat{\mathbf{x}}).
\end{aligned} \qquad (10.9)$$

From this equation we may deduce the physically obvious fact that the mean proper motion of members of the type vanishes in the direction of either the apex or the antapex of the solar motion ($\psi = 0, 180°$), and is largest in the

---

## Box 10.1: Determining $\mathbf{v}_\odot$ by Least-Squares

It is not immediately obvious from equations (10.8) and (10.9) how $\mathbf{v}_\odot$ should be determined from a given body of data. We address this problem for the case of radial velocities – a closely analogous procedure works for proper motion data (Dehnen & Binney 1998a). We start by rewriting (10.7) in the form

$$\hat{\mathbf{x}}_k \cdot \mathbf{v}_k = \hat{\mathbf{x}}_k \cdot \mathbf{v}_\odot + v_{\mathrm{los}\,k} \qquad (k = 1, \ldots, N),$$

where we assume that $N$ stars have been observed. These $N$ equations have infinitely many solutions, since if $\mathbf{v}_k$ and $\mathbf{v}_\odot$ solve them, so do $\mathbf{v}'_k \equiv \mathbf{v}_k + \overline{\mathbf{v}}$ and $\mathbf{v}_\odot + \overline{\mathbf{v}}$; this change corresponds to a simple boost in our reference frame. To pick out the solution of interest, we have to impose the condition that the $\mathbf{v}_k$ have zero mean. Suppose that this condition is satisfied in the unprimed frame. Then it is easy to see that

$$S \equiv \sum_k^N (\hat{\mathbf{x}}_k \cdot \mathbf{v}'_k)^2 = \sum_k^N (\hat{\mathbf{x}}_k \cdot \mathbf{v}_k)^2 + \sum_k^N (\hat{\mathbf{x}}_k \cdot \overline{\mathbf{v}})^2 > \sum_k^N (\hat{\mathbf{x}}_k \cdot \mathbf{v}_k)^2.$$

It follows that the desired frame is that which minimizes $S$. Therefore we determine $\mathbf{v}_\odot$ by minimizing $S = \sum_k^N (\hat{\mathbf{x}}_k \cdot \mathbf{v}_\odot + v_{\mathrm{los}\,k})^2$ with respect to the three components of $\mathbf{v}_\odot$. That is, we solve the three simultaneous linear equations

$$0 = \frac{\partial S}{\partial v_{\odot i}} = 2 \sum_k^N (\hat{\mathbf{x}}_k \cdot \mathbf{v}_\odot + v_{\mathrm{los}\,k}) \hat{\mathbf{x}}_{k\,i} \qquad (i = 1, 2, 3)$$

for the three components $v_{\odot i}$.

**Figure 10.9** Definition of a coordinate system centered on the Sun.

perpendicular directions ($\psi = 90°$). Moreover, if we further average equation (10.9) over stars that lie in the same direction but at different distances $d$, it remains true that $\langle \mu \rangle$ vanishes in the apex and antapex directions. Hence, the direction of the solar motion can be deduced from proper motions even if one does not know any stellar distances. To obtain the magnitude of $\mathbf{v}_\odot$, however, it is essential to have a distance estimate for each star in one's sample.

Box 10.1 and Problem 10.5 describe the mechanics of extracting the

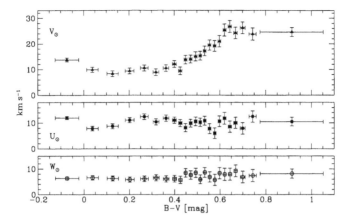

**Figure 10.10** The three components of the solar motion as functions of $B - V$. [After Dehnen & Binney (1998a)]

solar motion from a large sample of stellar velocities. Figure 10.9 defines the coordinate system that is generally used in kinematic studies of the solar neighborhood. The $x$ direction points towards the Galactic center, the direction of Galactic rotation is the $y$ direction, and the $z$ direction points towards the NGP. It is conventional to use the letters $U, V$ and $W$ for $v_x, v_y$ and $v_z$, respectively. Figure 10.10 plots the $U, V$ and $W$ components of the solar motion relative to MS stars for which the Hipparcos satellite has determined accurate parallaxes. Different colors yield fairly similar values of $U_\odot$ and $V_\odot$, so we may estimate the radial and vertical components of the solar motion with respect to the Galactic center by averaging individual values of $U_\odot$ and $V_\odot$ in Figure 10.10. For reasons that will emerge below, we exclude the bluest data-point from this average and find the radial and vertical components of this motion to be

$$U_\odot \equiv v_{\odot x} = 10.0 \pm 0.4 \, \text{km s}^{-1},$$
$$W_\odot \equiv v_{\odot z} = 7.2 \pm 0.4 \, \text{km s}^{-1}. \tag{10.10}$$

In Figure 10.10 it is striking that for $B - V < 0.61$, $V_\odot$ is increases steadily with increasing $B - V$, while for $B - V > 0.61$, $V_\odot$ is independent of $B - V$. Figure 10.11 shows that these variations in $V_\odot$ are linearly related to the squared random velocity $S^2$ for each stellar group – see Problem 10.5 for the definition of $S^2$. In fact, theory predicts this dependence of $V_\odot$ on $S^2$ – see equation (4-35) of BT. The straight line in Figure 10.11 shows a suitable linear fit to the data, with the data point for the bluest stars again excluded from the fit. From the $y$ intercept of this line we infer the value, $5.2 \pm 0.6 \, \text{km s}^{-1}$, of $V_\odot$ for a hypothetical stellar type that had $S^2 = 0$. In §9.1 we defined the local standard of rest (LSR) as the velocity of the closed orbit within the plane that passes through the present location of the Sun.

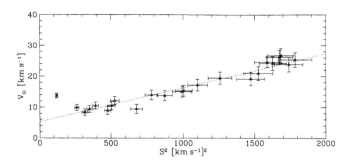

**Figure 10.11** The $V$ component of the solar motion relative to different stellar types is a linear function of the random velocity $S^2$ of each type. [After Dehnen & Binney (1998a)]

Since a class of stars that moved on such closed orbits would have $S^2 = 0$, we conclude that the solar motion relative to the LSR has $V_\odot = 5.2 \pm 0.6 \, \mathrm{km \, s^{-1}}$. In summary, the Sun's motion relative to the LSR is

$$\left.\begin{array}{rl} U_\odot = & 10.0 \pm 0.4 \, \mathrm{km \, s^{-1}} \\ V_\odot = & 5.2 \pm 0.6 \, \mathrm{km \, s^{-1}} \\ W_\odot = & 7.2 \pm 0.4 \, \mathrm{km \, s^{-1}} \end{array}\right\} \Rightarrow |\mathbf{v}_\odot| = 13.4 \, \mathrm{km \, s^{-1}}. \qquad (10.11)$$

Thus, the Sun is moving in toward the Galactic center and up toward the north Galactic pole and away from the plane. It is also moving around the Galactic center faster than it would if it were on a circular orbit. It follows that the Sun is currently inside its guiding-center radius $R_g$ and is approaching, but has not yet quite reached, the pericenter of its orbit – see equation (9.7) and §3.3.3 of BT for a discussion of orbits like the Sun's.

From this point on we will whenever possible refer the velocities of stars to the velocity of the LSR rather than that of the Sun. In particular, we shall assume that the means $\langle U \rangle$ and $\langle W \rangle$ vanish for stars near the Sun.

The systematic trend of $V_\odot$ with $S^2$ shown in Figure 10.11 is a reflection of a phenomenon called **asymmetric drift**, which is the tendency of the mean rotation velocity of a stellar population to lag behind that of the LSR more and more with increasing random motion within the population – see §10.3.2 below and §4.2.1 of BT. When the Sun's velocity is referred to such a lagging reference frame, it acquires a value of $V$ which grows as the lag increases. It is easy to see that the magnitude of the asymmetric drift, $V_\mathrm{a}$, of any stellar type is given by the difference between that type's $y$-coordinate in Figure 10.11 and the $y$-intercept of the straight line through the points. We shall see that the squared random velocity, $S^2$, is linearly related to $\langle U^2 \rangle$, so a linear relation holds between $V_\mathrm{a}$ and $\langle U^2 \rangle$. This relation takes the form

$$V_\mathrm{a} = \frac{\langle U^2 \rangle}{80 \pm 5 \, \mathrm{km \, s^{-1}}} \qquad (10.12)$$

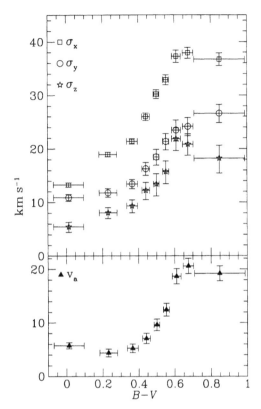

**Figure 10.12** Top panel: the principal velocity dispersions $\sigma_x$, $\sigma_y$, $\sigma_z$ of MS stars as a function of $B - V$. Bottom panel: the asymmetric drift, $V_\mathrm{a}$, of MS stars as a function of $B - V$. [After Dehnen & Binney (1998a)]

(Dehnen & Binney 1998a). The bottom curve in Figure 10.12 shows the variation of $V_\mathrm{a}$ with $B - V$ for MS stars.

### 10.3.2 Random velocities of stars

As we have already seen, important dynamical information is contained in the dispersion of velocities,

$$\sigma_i \equiv \left\langle (v_i - \langle v_i \rangle)^2 \right\rangle^{1/2}, \tag{10.13}$$

of each stellar type about the mean velocity of that type. The upper three curves in Figure 10.12 show, as a function of $B - V$, values for $\sigma_x$, $\sigma_y$ and $\sigma_z$ for MS stars. For every color we have $\sigma_x > \sigma_y > \sigma_z$. Roughly speaking, $\sigma_z \simeq 0.5\sigma_x$ while the ratio $\sigma_y/\sigma_x$ lies in the range from 0.55 to 0.7. The fact that $\sigma_x \neq \sigma_z$ has important dynamical implications – see Chapters 3 and 4 of BT. The value of the ratio $\sigma_y/\sigma_x$ has often been used to constrain the shape of the Milky Way's circular-speed curve – see below.

In Figure 10.12 all three velocity dispersions increase with $B - V$ color up to $B - V \simeq 0.6$ and are approximately constant redward of $B - V \simeq 0.6$,

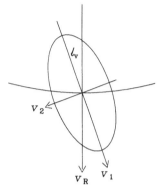

**Figure 10.13** Orientation of the velocity ellipsoid. The direction to the Galactic center is upwards and the Sun lies at the center of the ellipse and moves towards the left. A portion of its orbit is shown.

just as the asymmetric drift $V_a$ increases with $B - V$ to $B - V = 0.61$ and then remains constant. From the theory of stellar evolution (§5.1), we know that MS stars bluer than $B - V \simeq 0.6$ are all younger than $\sim 10\,\mathrm{Gyr}$. In contrast, redder stars are a mixture of a few young stars and mostly old stars, and the fact that they have systematically larger velocity dispersions suggests the operation of a mechanism that leads to a progressive increase of the dispersion with time. We shall return to this idea in §10.7 below.

**Vertex deviation**    In addition to squares of velocity components such as $v_x^2$, one can average products of velocity components such as $v_x(v_y - \langle v_y \rangle)$. For all stellar types, the averages of the two products of this type that involve $v_z$ are smaller than the errors. However, for many stellar types the average of the third product, $v_x(v_y - \langle v_y \rangle)$, is significantly different from zero. Evidently for stars of these types $v_x$ and $v_y$ are not statistically independent: if $\langle v_x(v_y - \langle v_y \rangle) \rangle > 0$, then if $v_y - \langle v_y \rangle$ is measured for some star and found to be positive, that star is more likely have a positive than a negative value of $v_x$. When confronted with such correlated observables, it can be helpful to find linear combinations of the observables which are statistically independent. Therefore we define

$$v_1 \equiv v_x \cos l_v - (v_y - \langle v_y \rangle) \sin l_v,$$
$$v_2 \equiv v_x \sin l_v + (v_y - \langle v_y \rangle) \cos l_v, \tag{10.14}$$

where the angle $l_v$ is called the **vertex deviation**. Now on multiplying these equations together we find that

$$\langle v_1 v_2 \rangle = \tfrac{1}{2} \left( \sigma_x^2 - \sigma_y^2 \right) \sin 2l_v + \langle v_x(v_y - \langle v_y \rangle) \rangle \cos 2l_v. \tag{10.15}$$

Hence $v_1$ and $v_2$ will be statistically independent in the sense that $\langle v_1 v_2 \rangle = 0$ if we set

$$l_v = \tfrac{1}{2} \arctan \left( \frac{2 \langle v_x(v_y - \langle v_y \rangle) \rangle}{\sigma_x^2 - \sigma_y^2} \right). \tag{10.16}$$

## Box 10.2: The Velocity Ellipsoid

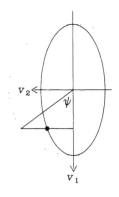

Suppose we wish to determine the mean-square speed in a direction $\hat{n}$ that lies in the plane but makes an angle $\psi$ with the $v_1$-direction. (The direction to the Galactic center is upwards.) Then since the velocity in the direction of $\hat{n}$ is $v_n = v_1 \cos\psi + v_2 \sin\psi$ and $\langle v_1 v_2 \rangle = 0$, we have

$$\langle v_n^2 \rangle = \langle v_1^2 \rangle \cos^2\psi + \langle v_2^2 \rangle \sin^2\psi$$
$$= \langle v_1^2 \rangle (\cos^2\psi + q^2 \sin^2\psi),$$

where $q$ is the axis ratio of the ellipse shown in the figure. It is easy to see that the coordinates of the point marked with a blob are $(v_1, v_2) = \langle v_1^2 \rangle^{1/2} \times (q \sin\psi, \cos\psi)$. Hence $\langle v_n^2 \rangle^{1/2}$ is the distance from the origin to the blob.

The arctan function introduces some ambiguity in the definition of $l_v$. By convention this ambiguity is resolved such that $\sigma_1 \equiv \langle v_1^2 \rangle^{1/2} > \sigma_2 \equiv \langle v_2^2 \rangle^{1/2}$.

The linear transformation (10.14) is a rotation of coordinates, so $v_1$ and $v_2$ are simply components of velocity in directions that are inclined by $l_v$ to the center-anticenter direction. Figure 10.13 shows a geometrical interpretation of what is going on. If we draw the ellipse marked with semi-axes $\sigma_1$ and $\sigma_2$, then the mean-square of the component of velocity in any direction can be obtained by the construction described in Box 10.2. In fact, we can generalize this construction by replacing the circle in Box 10.2 with a sphere and the ellipse with the ellipsoidal surface that has semi-axes $\sigma_1$, $\sigma_2$ and $\sigma_z$. This surface is called the **velocity ellipsoid**. For all types of stars in the solar neighborhood, one of its principal planes coincides with the plane of the Milky Way, but its longest axis deviates from the center-anticenter direction by the vertex deviation $l_v$. Table 10.2 gives, as a function of $B - V$, values of the $\sigma_i$ and values of $l_v$. The latter decreases with increasing $B - V$ from $\sim 30°$ for the bluest stars to $\sim 10°$ for stars redder than $B - V \sim 0.45$. We discuss the causes of the vertex deviation below.

Our discussion of solar-neighborhood kinematics has focused on MS stars because the largest body of accurate and homogeneous data is available for these objects. Table 10.3 shows that other stellar groups display exactly the same general kinematic trends as MS stars, however. In particular, older stellar groups are characterized by larger velocity dispersions and asymmetric drift velocities than are younger stellar groups, and vertex deviation tends

**Table 10.2**   Velocity dispersions and vertex deviations for MS stars

| $(B-V)_{\mathrm{min,max}}$ | | $\sigma_1/\mathrm{km\,s}^{-1}$ | $\sigma_2/\sigma_1$ | $\sigma_z/\sigma_1$ | $l_{\mathrm{v}}/\mathrm{deg}$ |
|---|---|---|---|---|---|
| $-0.238$ | $0.139$ | $14.35^{+0.49}_{-0.40}$ | $0.65^{+0.07}_{-0.06}$ | $0.38^{+0.05}_{-0.10}$ | $30.2^{+4.7}_{-5.3}$ |
| $0.139$ | $0.309$ | $20.17^{+0.50}_{-0.43}$ | $0.47^{+0.07}_{-0.03}$ | $0.40^{+0.02}_{-0.10}$ | $22.8^{+2.8}_{-3.0}$ |
| $0.309$ | $0.412$ | $22.32^{+0.56}_{-0.47}$ | $0.53^{+0.06}_{-0.04}$ | $0.42^{+0.03}_{-0.09}$ | $19.8^{+3.2}_{-3.4}$ |
| $0.412$ | $0.472$ | $26.26^{+0.80}_{-0.59}$ | $0.60^{+0.06}_{-0.04}$ | $0.46^{+0.03}_{-0.09}$ | $10.2^{+5.0}_{-5.4}$ |
| $0.472$ | $0.525$ | $30.37^{+0.96}_{-0.70}$ | $0.60^{+0.06}_{-0.04}$ | $0.44^{+0.04}_{-0.11}$ | $6.9^{+5.1}_{-5.3}$ |
| $0.525$ | $0.582$ | $32.93^{+1.09}_{-0.75}$ | $0.66^{+0.06}_{-0.05}$ | $0.46^{+0.04}_{-0.10}$ | $1.9^{+6.0}_{-6.1}$ |
| $0.582$ | $0.641$ | $37.64^{+1.37}_{-0.94}$ | $0.62^{+0.08}_{-0.03}$ | $0.56^{+0.01}_{-0.12}$ | $10.2^{+5.6}_{-6.0}$ |
| $0.641$ | $0.719$ | $38.13^{+0.71}_{-0.31}$ | $0.62^{+0.06}_{-0.04}$ | $0.54^{+0.02}_{-0.10}$ | $7.6^{+5.9}_{-6.0}$ |
| $0.719$ | $1.543$ | $37.20^{+1.41}_{-0.93}$ | $0.69^{+0.06}_{-0.05}$ | $0.49^{+0.04}_{-0.11}$ | $13.1^{+6.7}_{-7.6}$ |
| $0.610$ | $1.543$ | $37.91^{+0.79}_{-0.63}$ | $0.65^{+0.04}_{-0.03}$ | $0.54^{+0.02}_{-0.05}$ | $10.3^{+3.9}_{-3.9}$ |

SOURCE: Data published in Dehnen & Binney (1998a)

to decline with increasing age.

**The Schwarzschild distribution**      Each component of the velocity distribution of, say, a population of oxygen molecules at room temperature has a Gaussian probability distribution. Schwarzschild (1907) pointed out that a similar probability distribution can account for many aspects of the probability distribution of stellar velocities. The main difference between the case of molecules in air and stars is that in the former case the velocity dispersion is independent of direction, whereas, as we have seen, in the stellar case the dispersion of a component of velocity depends strongly on direction. Schwarzschild postulated that the probability that the components of velocity $(v_1, v_2, v_z)$ defined above lie in the element of velocity space $d^3\mathbf{v} \equiv dv_1 dv_2 dv_z$ is

$$P(\mathbf{v})d^3\mathbf{v} = \frac{d^3\mathbf{v}}{(2\pi)^{3/2}\sigma_1\sigma_2\sigma_z} \exp\left[ -\left( \frac{v_1^2}{2\sigma_1^2} + \frac{v_2^2}{2\sigma_2^2} + \frac{v_z^2}{2\sigma_z^2} \right) \right]. \qquad (10.17)$$

This probability distribution is know as the **Schwarzschild distribution**. In §7.5 of BT it is shown to have a natural dynamical interpretation. Notice that $P$ is constant on ellipsoids in velocity space.

Figure 10.14 shows the distributions in $U, V$ and $W$ of two samples of nearby stars for which accurate radial velocities and Hipparcos parallaxes are available.[8] The upper panels are for MS stars bluer than $B - V = 0.34$ – these are mostly A stars. The lower panels are for K and M dwarfs from Vyssotsky (1963). From the shapes of the upper histograms it is plausible that the Schwarzschild distribution gives a reasonable model of the velocity

---

[8] See Jahreiss & Wielen (1997) for details of these samples.

**Table 10.3** Kinematics of non-MS stars

| Stellar type | Asymm. drift, $V_a$ | Dispersions $\sigma_R$ | $\sigma_\phi$ | $\sigma_z$ | Vertex dev., $l_v(°)$ |
|---|---|---|---|---|---|
| *Giants* | | | | | |
| A | 6.4 | 22 | 13 | 9 | 27 |
| F | 13.3 | 28 | 15 | 9 | 14 |
| G | 5.9 | 26 | 18 | 15 | 12 |
| K0 | 13.4 | 31 | 21 | 16 | 14 |
| K3 | 11.5 | 31 | 21 | 17 | 4 |
| M | 13.1 | 31 | 23 | 16 | 7 |
| *Supergiants* | | | | | |
| Classical Cepheids | 6.8 | 13 | 9 | 5 | – |
| O–B5 | 8.2 | 12 | 11 | 9 | 36 |
| F–M | 6.5 | 13 | 9 | 7 | 18 |
| *Other* | | | | | |
| Carbon stars | 27 | 48 | 23 | 16 | – |
| Subgiants | 23 | 43 | 27 | 24 | – |
| Planetary Nebulae | 24 | 45 | 35 | 20 | – |
| White Dwarfs | 32 | 50 | 30 | 25 | – |
| Variables, $P > 300^d$ | 22 | 50 | 40 | 30 | – |
| Variables, $P < 300^d$ | 37 | 80 | 60 | 60 | – |
| RR Lyrae, $P < 0.45^d$ | 26 | 45 | 40 | 25 | – |
| RR Lyrae, $P > 0.45^d$ | 220 | 160 | 100 | 120 | – |
| Subdwarfs | 145 | 100 | 75 | 50 | – |

SOURCE: From data published in Delhaye (1965)

distributions of the A stars. However, from the shape of the lower histogram of $V$ components it is clear that Schwarzschild's model cannot provide an adequate representation of the data for stellar types that have higher velocity dispersions, such as M dwarfs. Specifically, it is inherently unable to reproduce the characteristic asymmetry of the measured distribution of $V$ components: whereas the distribution cuts off sharply at positive $V$, it has a long tail towards negative $V$.

The origin of this asymmetry in $V$ is easy to understand. Solar-neighborhood stars with negative $V$ have less tangential motion than is required to be on a circular orbit at $R_0$. Hence they are closer to apocenter than pericenter, and we may think of them as being at home interior to the solar circle. Moreover, the smaller a star's value of $V$ is, the further inside $R_0$ its home lies. Now from §4.4 we know that the surface densities of galactic disks increase exponentially towards the center. Moreover, in §11.3.2 we shall find that the velocity dispersions within a disk also increase exponentially as we move inwards. For both these reasons we expect many more stars to visit us

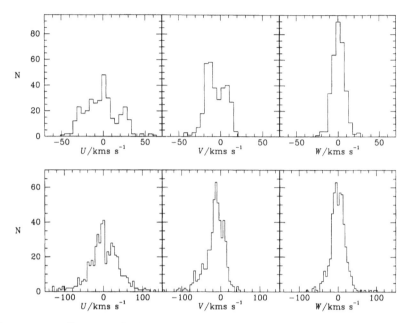

**Figure 10.14** Upper panels: histograms of $U, V$ and $W$ for a sample of 323 nearby MS stars of MK type F1 and earlier. Lower panels: similar data for 510 K and M dwarfs. Velocities are with respect to the LSR that is defined by equations (10.11). [From data kindly supplied by H. Jahreiss]

in the solar neighbourhood from small radii than from large radii: not only does the density of tourists increase inwards, but so does the size of their travel budgets and therefore the distance from which they can visit us. The skewness of the $V$ distributions in Figure 10.14 is simply a reflection of these basic facts.[9] The asymmetric drift discussed above is another consequence of these facts, for the skewness of the $V$ distribution drags the mean value of $V$ for a stellar type to more and more negative values the larger the type's velocity dispersions and therefore the skewer its $V$ distribution.

**Star streams**     Figure 10.15 shows the densities of MS stars of various colors in the $(U, V)$ plane. Each panel is for a different range in $B - V$, with the bluest and youngest stars at top left and the reddest and oldest stars at bottom right. We can immediately recognize in these plots phenomena with which we are already familiar.

1. As one proceeds to redder groups, the distribution of stars spreads. The increase in the $\sigma_i$ with $B - V$ quantifies this phenomenon.
2. Three of the maps show a clear tendency for the region of highest stellar density to be elongated along a line that slopes from bottom left to top right. The vertex deviation quantifies this effect.

---

[9] For mathematical models of the $V$ distributions, see Cuddeford & Binney (1994).

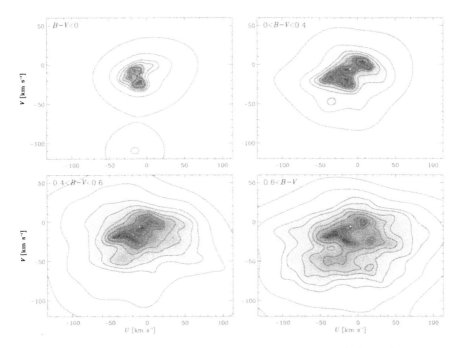

**Figure 10.15** The density of stars near the Sun in velocity space. Each panel shows the density of MS stars projected onto the $(U, V)$ plane for a different range of $B - V$ color, with the bluest stars at top left and the reddest stars at bottom right. The Sun's velocity is at $U = V = 0$ and the velocity of the LSR that is defined by equations (10.11) is marked by a triangle. [After Dehnen (1998) from data kindly supplied by W. Dehnen]

3. The lower two distributions clearly peak at $V < 0$ and extend further to negative $V$ than to positive $V$. The asymmetric drift and the skewness of the $V$ distributions in Figure 10.14 reflects this asymmetry.

In addition to these familiar phenomena, the maps of Figure 10.15 reveal the presence of tight clumps of stars in velocity space over and above the large-scale structure that the Schwarzschild distribution reproduces. In this connection, it is important that in Figure 10.15 each map is constructed from a separate sample of stars. Consequently, structure that is present in more than one map is almost certainly real rather than an artifact introduced by sampling noise.

The top right-hand panel of Figure 10.15 shows four strong peaks in the stellar density. The peak at $(U, V) = (-10, -5)\,\mathrm{km\,s^{-1}}$ is associated with the LSR, which is marked with a triangle. The peak at $(U, V) = (9, 3)\,\mathrm{km\,s^{-1}}$ corresponds to the **Sirius stream**, that at $(-12, -22)\,\mathrm{km\,s^{-1}}$ to the **Pleiades stream** and that at $(-35, -18)\,\mathrm{km\,s^{-1}}$ corresponds to the **Hyades stream**. Careful examination of the lower two panels of Figure 10.15 reveals that several wiggles and lumps that might be dismissed as noise if they occurred in only one plot, are evident in both plots and therefore probably reflect real concentrations of stars – examples are evident near $(U, V) = (-40, -45), (15, -60)$ and $(0, -100)\,\mathrm{km\,s^{-1}}$.

These streams, or **moving groups** as they are often called, are thought to be vestiges of the clusters and associations in which most stars form (§6.2.2). Long after their stars have drifted far apart and they are no longer easily distinguishable by their space concentration, the $U$ and $V$ velocities of their stars remain very similar and betray the stars' common origin. Conclusive proof of this common origin comes from the fact that when all the stars of a moving group are plotted in a CM diagram, one sees the characteristic structure of the CM diagram of a coeval cluster (§6.2.2 and Eggen 1965).

Given these facts, one would expect moving groups to be most important for young stars and therefore most apparent in the distribution of early-type stars. This expectation is borne out by Figure 10.15, the top left panel of which, for stars bluer than $B - V = 0$, is dominated by just two moving groups. Moreover, the brightest early-type stars in the night sky do not lie in the Galactic plane, but rather are found in **Gould's belt** in a plane inclined by about $16°$. A detailed kinematical analysis of these stars (Lesh 1968) suggests that they form an expanding group.

Interestingly, Dehnen (1998) shows that star streams do not show up clearly in plots that involve the vertical component of velocity, $W$. This fact presumably arises because the Galactic potential changes most rapidly in the vertical direction with the consequence that as the stars of an association drift apart, they experience significantly different vertical forces when they are still subject to very similar radial and azimuthal forces. Clearly, once two stars experience very different vertical forces, their vertical velocities rapidly move apart.

**Causes of vertex deviation**    If we imagine our Galaxy to be an axisymmetric system in a steady state in which stars are distributed at random in their orbits, then, from considerations of symmetry alone, we would expect to find one axis of the velocity ellipsoid of stars in the Galactic plane pointing exactly at the Galactic center. This expectation is borne out by dynamical theory (see BT §7.5), but, as was mentioned earlier, the vertex deviation, $l_v$, often differs significantly from zero.

The existence of moving groups is significant for our understanding of the vertex deviation for the following reason. If a significant fraction of the objects upon which the numbers in Table 10.2 are based are, in fact, members of a few distinct moving groups, then the number of truly independent velocities employed is strongly reduced, and the statistical errors in the results will be seriously underestimated. In these circumstances, estimates of the orientation of the velocity ellipsoid axes and the velocity dispersion along these axes will be noisy. For example, if in Figure 10.15 the Pleiades stream lay at $(+12, -20)\,\mathrm{km\,s^{-1}}$ rather than at $(-12, -20)\,\mathrm{km\,s^{-1}}$, the vertex deviation would be drastically reduced, especially for stars bluer than $B - V = 0.6$.

It is not clear, however, that moving groups are entirely responsible for the non-zero values of $l_v$ in Table 10.2. A significant additional cause of vertex deviation could be the non-axisymmetric component of the Galactic potential. In §4.4.6 we encountered evidence that spiral structure affects the distribution of old as well as young stars, and in §8.2.3 we saw how the resulting spiral perturbation in a galaxy's gravitational potential imposes a pattern on the velocity field of the ISM. The velocities of stars must also be changed by the same spiral gravitational field. One can show that stellar types with small random velocities will suffer larger velocity changes than types with large velocity dispersions – see §6.2 of BT. One may also show that the spiral field will produce a vertex deviation (Kuijken & Tremaine 1994). Thus observations of spiral structure and dynamics together predict that all stellar types will show a vertex deviation, but that this will be largest for stellar types with low velocity dispersions, just as is observed. It is not yet clear whether spiral structure can account satisfactorily for all the observed vertex deviations, or whether moving groups also play an important rôle.

### 10.3.3 The Oort constants

Before we go on to discuss correlations between the kinematic properties of stars and their ages and chemical compositions, we should plug a gap in our discussion. This gap concerns differential rotation within the solar neighborhood. We have proceeded above as though the solar neighborhood is of negligible size compared to the Milky Way, so that one standard of rest applies throughout. Actually, the 'solar neighborhood' must be considered to be a sphere large enough to contain an adequate sample of whatever stars are under investigation. If these stars are intrinsically rare, for example most kinds of giants, the sphere may have to have a non-negligible radius and the effects within it of differential rotation may be significant.

We now show how these effects may be quantified and thus allowed for in kinematic studies. One valuable spin-off of the analysis will be evidence that the mean motion at each point of the disk is circular rather than significantly elliptical, for example. Another will be important information regarding the circular-speed curve of the Milky Way.

By analogy with the definition of the LSR (see §9.1), we associate a **standard of rest (SR)** with any point in the disk: $\mathbf{v}_{SR}(\mathbf{x})$, is the velocity at $\mathbf{x}$ of the closed orbit that passes through $\mathbf{x}$. With this definition, $\mathbf{v}_{LSR} = \mathbf{v}_{SR}(\mathbf{x}_\odot)$. If the Milky Way were axisymmetric, $\mathbf{v}_{SR}$ would be everywhere perpendicular to the local direction to the Galactic center and equal in magnitude to the local circular speed $v_c(R)$.

By a direct extension of the logic employed in §10.3.1, we argue that at any point in the disk, $\mathbf{v}_{SR}(\mathbf{x})$ should lie close to the mean velocity at $\mathbf{x}$ of any stellar type that has small random velocities. Hence key information regarding $\mathbf{v}_{SR}$ can be gleaned from studies of the proper motions and radial velocities of stars near the Sun. With this idea in mind, we now ask how the variation of $\mathbf{v}_{SR}$ from point to point will be reflected in the pattern of radial velocities of stars near the Sun.

Let $\delta\mathbf{v}(\mathbf{x}) \equiv \mathbf{v}_{SR}(\mathbf{x}) - \mathbf{v}_{LSR}$ and let the $(x, y, z)$ coordinate system be defined as in Figure 10.9. Each of the two components of $\delta\mathbf{v}$ is a function of the two variables $(x, y)$ and may be Taylor expanded about the origin of the $(x, y)$ system. Since $\delta\mathbf{v}$ vanishes at the origin by construction, we may write this series in the form

$$
\begin{pmatrix} \delta v_x \\ \delta v_y \end{pmatrix} = \begin{pmatrix} \dfrac{\partial \delta v_x}{\partial x} & \dfrac{\partial \delta v_x}{\partial y} \\ \dfrac{\partial \delta v_y}{\partial x} & \dfrac{\partial \delta v_y}{\partial y} \end{pmatrix} \begin{pmatrix} x \\ y \end{pmatrix} + O(x^2 + y^2)
$$

$$
= \begin{pmatrix} k + c & a - b \\ a + b & k - c \end{pmatrix} \begin{pmatrix} x \\ y \end{pmatrix} + O(x^2 + y^2). \tag{10.18}
$$

Here the partial derivatives are evaluated at the origin and $a, b, c, k$ are linear combinations of the values of these derivatives – the reason for writing the partial-derivative matrix in this way will emerge shortly. We expect the mean radial velocity of the stars at $\mathbf{x} = (x, y)$ to be the projection of $\delta\mathbf{v}$ on to $\mathbf{x}$:

$$
v_{los} = \frac{1}{d}\mathbf{x} \cdot \delta\mathbf{v}
$$

$$
\simeq \frac{1}{d}\left[(k + c)x^2 + (k - c)y^2 + 2axy\right] \tag{10.19}
$$

where $d \equiv (x^2 + y^2)^{1/2}$ is heliocentric distance as usual. In terms of Galactic longitude $l$, we have $x = d\cos l$ and $y = d\sin l$, so with the aid of two trigonometric identities equation (10.19) becomes

$$
v_{los} = d(k + c\cos 2l + a\sin 2l) + O(d^2). \tag{10.20}
$$

This equation suggests that we could determine the constants $a$, $c$ and $k$ by measuring the radial velocities of stars of similar distances and plotting the results as a function of longitude $l$. The stars could be selected to be stars of given spectral type and apparent magnitude, thus guaranteeing that they all have similar distances.

The constant $b$ can be determined from the proper motions of stars. To see this, let $\delta v_t$ be the component of velocity perpendicular to the line of sight. Then

$$
\delta v_t = \frac{1}{d}(\mathbf{x} \times \delta\mathbf{v})_z = \frac{1}{d}(x\delta v_y - y\delta v_x)
$$

$$
= d(b + a\cos 2l - c\sin 2l) + O(d^2). \tag{10.21}
$$

Hence from a plot of the proper motion $\mu_l = v_t/d$ as a function of $l$ we can determine $a$, $b$ and $c$.

The relations (10.20) and (10.21) above are completely general – they assume only that $\mathbf{v}_{SR}$ varies smoothly in the solar neighborhood. What do they reduce to if we assume that the disk is in circular rotation with angular speed $\Omega(R)$? For this case, equation (9.3) gives an exact relation for $v_{los}(l, R)$. To obtain a relation equivalent to equation (10.20), we need to Taylor expand the right-hand side of equation (9.3) in powers of $d$. On Taylor expanding $\Omega(R)$ to first order in $R - R_0$, equation (9.3) yields

$$
\begin{aligned}
v_{los}(l, R) &\simeq \left.\frac{d\Omega}{dR}\right|_{R_0} (R - R_0)R_0 \sin l \\
&= -2A(R - R_0)\frac{R_0}{R} \sin l,
\end{aligned}
\tag{10.22}
$$

where **Oort's constant** $A$ is defined by

$$
A \equiv -\tfrac{1}{2}\left(R\frac{d\Omega}{dR}\right)_{R_0} = \tfrac{1}{2}\left(\frac{v_c}{R} - \frac{dv_c}{dR}\right)_{R_0}.
\tag{10.23}
$$

On neglecting $d^2$ in equation (9.16), we have

$$
(R - R_0)(R + R_0) = R^2 - R_0^2 \simeq -2R_0 d \cos l,
\tag{10.24}
$$

so approximating $(R + R_0)$ by $2R_0$, we may write to first-order in $d$

$$
(R - R_0) = -d \cos l.
\tag{10.25}
$$

Combining equations (10.22) and (10.25) we find

$$
v_{los} \simeq Ad \sin 2l.
\tag{10.26}
$$

This is equation (10.20) for the special case of circular rotation. Evidently in this case $c = k = 0$, and $a = A$ is given by equation (10.23).

To derive equation (10.21) for the case of circular rotation we have to go back to equation (9.1). We write

$$
\begin{aligned}
v_t &= \left(\frac{\mathbf{R} - \mathbf{R}_0}{|\mathbf{R} - \mathbf{R}_0|} \times [\mathbf{\Omega}(R) \times \mathbf{R} - \mathbf{\Omega}(R_0) \times \mathbf{R}_0]\right)_z \\
&= \left(\frac{\mathbf{R} - \mathbf{R}_0}{d} \times \{\mathbf{\Omega}(R) \times (\mathbf{R} - \mathbf{R}_0) + [\mathbf{\Omega}(R) - \mathbf{\Omega}(R_0)] \times \mathbf{R}_0\}\right)_z \\
&\simeq \Omega_z(R)d + \left.\frac{d\Omega_z}{dR}\right|_{R_0} \frac{(\mathbf{R} - \mathbf{R}_0) \cdot \mathbf{R}_0}{d}.
\end{aligned}
\tag{10.27}
$$

---

### Box 10.3: The $AR_0$ Formula

We can derive a useful formula from equation (9.4) for the terminal velocity (see §9.2.3) in the direction $l$ by writing

$$\Omega(R_0 \sin l) - \Omega(R_0) \simeq \left(\frac{d\Omega}{dR}\right)_{R_0} R_0(\sin l - 1) \quad \text{for} \quad l \simeq 90°. \quad (1)$$

Substituting for $d\Omega/dR$ from (10.23), equation (9.4) becomes, to first order in $(1 - \sin l)$,

$$v_{\text{los}}^{(t)}(l) = 2AR_0(1 - \sin l) \quad \text{for} \quad l \simeq 90°. \quad (2)$$

This formula is useful because it links two quantities that are key but hard to measure accurately, $A$ and $R_0$, to the terminal velocity curve, which can be measured quite accurately – see §9.2.1. Kerr & Lynden-Bell (1986) list several determinations of $AR_0$, and find that recent observations give $AR_0 = 108 \pm 3\,\text{km s}^{-1}$. The data plotted in Figure 9.16 yield a rather larger value, $AR_0 \simeq 115\,\text{km s}^{-1}$, but with considerable uncertainty.

---

By drawing the appropriate triangle it is easy to see that $(\mathbf{R} - \mathbf{R}_0) \cdot \mathbf{R}_0 = -dR_0 \cos l$. When we substitute this relation and equation (10.25) into equation (10.27), we find

$$
\begin{aligned}
v_t &\simeq \Omega_z(R_0)d + dR_0 \left.\frac{d\Omega_z}{dR}\right|_{R_0} \cos^2 l \\
&= d\left(\Omega_z(R_0) + \tfrac{1}{2}R_0 \left.\frac{d\Omega_z}{dR}\right|_{R_0}(1 + \cos 2l)\right).
\end{aligned}
\quad (10.28)
$$

We now recall that, because the Milky Way rotates clockwise, $\Omega_z = -\Omega$, and define **Oort's constant** $B$ by

$$
\begin{aligned}
B &\equiv -\left(\Omega + \tfrac{1}{2}R\frac{d\Omega_z}{dR}\right)_{R_0} \\
&= -\tfrac{1}{2}\left(\frac{v_c}{R} + \frac{dv_c}{dR}\right)_{R_0}.
\end{aligned}
\quad (10.29)
$$

Equation (10.28) then becomes

$$\mu_l = \frac{v_t}{d} = B + A \cos 2l. \quad (10.30)$$

This is equation (10.21) for the case of circular rotation. Evidently in the case of circular rotation $b = B$ is given by equation (10.29).

$A$ measures the **shear** in the disk at the Sun. It would be zero if the disk rotated like a solid body, such as a compact disk, for then $\Omega$ would be independent of radius and, by equation (10.23), $A$ would vanish. $B$ measures the **vorticity** of the material of the disk, that is, its tendency to circulate about any given point – see Problem 10.8. From equations (10.23) and (10.29) it immediately follows that

$$v_c = R_0(A - B) \quad \text{and} \quad \left.\frac{dv_c}{dR}\right|_{R_0} = -(A + B). \tag{10.31}$$

In §8.2.4 we saw that the circular-speed curves of spirals are flat or gently rising at large radii. If the Milky Way's circular speed curve followed this pattern at the Sun, we would have $A + B \lesssim 0$.

**Estimating the Oort constants** Kuijken & Tremaine (1991) reviewed the observational constraints on $c$ and $k$, and concluded that

$$c = 0.6 \pm 1.1 \,\text{km s}^{-1}\,\text{kpc}^{-1}, \qquad k = -0.35 \pm 0.5 \,\text{km s}^{-1}\,\text{kpc}^{-1}. \tag{10.32}$$

Thus these two constants are zero to within the errors, just as they should be if the Milky Way is axisymmetric. In view of this result, we shall assume that $a = A$ and $b = B$, and henceforth make no distinction between these logically distinct entities.

$A$ can be determined either from radial velocities [equation (10.20)], or from proper motions [equation (10.21)]. Since the determination of $v_c(R_0)$ is one of the harder problems in Galactic structure, equation (10.23) is usually used to determine $v_c(R_0)$ from $A$ rather than vice versa. Notice that the value of $A$ derived from radial velocities is inversely proportional to the assumed distances $d$ of the observed stars, and must be updated whenever there is an improvement in the accuracy of the standard distance scale.

Fundamentally, $B$ can only be obtained from proper motions. Traditionally, equation (10.21) has been used. The great drawback of this technique is that any rotation of one's astrometric system (§2.1.5) will directly contribute to the measured value of $B$. To show this, we suppose that our astrometric system rotates at angular velocity $\omega$, and calculate the contribution $\delta\mu_l$ that this rotation makes to the left-hand side of equation (10.21). The rotation of the system adds $\hat{\mathbf{x}} \times \omega$ to the measured proper motion of a star that lies in the direction $\hat{\mathbf{x}}$. We wish to calculate the component of this proper motion that is perpendicular to the normal to the Galactic plane, $\hat{\mathbf{z}}$. This component is

$$\hat{\mathbf{z}} \times [(\hat{\mathbf{x}} \times \omega) \times \hat{\mathbf{z}}] = (\hat{\mathbf{z}} \cdot \hat{\mathbf{x}})\hat{\mathbf{z}} \times \omega - (\hat{\mathbf{z}} \cdot \omega)\hat{\mathbf{z}} \times \hat{\mathbf{x}}. \tag{10.33}$$

For stars that lie within the plane, $\hat{\mathbf{z}} \cdot \hat{\mathbf{x}} \simeq 0$ and $\hat{\mathbf{z}} \times \hat{\mathbf{x}}$ is just the unit vector in direction of increasing longitude $l$. Hence

$$\delta\mu_l = -\hat{\mathbf{z}} \cdot \omega, \tag{10.34}$$

which is obviously independent of $l$. However, in equation (10.21), $B$ is precisely the term that is independent of $l$, so rotation of one's astrometric system directly affects the measured value of $B$. Moreover $B$ is small; expressed in sensible units (like $A$ and the Hubble constant it is a frequency) its value is $\simeq -4 \times 10^{-16}$ Hz. So one has to be very sure that one's frame is not rotating before reporting a measurement of $B$! It is a tribute to the diligence of astronomers that the errors in astrometric systems such as the FK5 that are based on the dynamics of the solar system (§2.1.5), are only of this order. Even so, values of $B$ that are based upon an extragalactic system such as the ICRS (§2.1.5) are clearly to be preferred.

Kerr & Lynden-Bell (1986) reviewed the observational constraints on $A$ and $B$ and concluded that

$$A = 14.4 \pm 1.2 \, \text{km s}^{-1} \, \text{kpc}^{-1}, \qquad B = -12.0 \pm 2.8 \, \text{km s}^{-1} \, \text{kpc}^{-1}. \quad (10.35)$$

From the proper motions of Cepheids that were measured by the Hipparcos satellite (§2.1.3), Feast & Whitelock (1997) found

$$A = 14.8 \pm 0.8 \, \text{km s}^{-1} \, \text{kpc}^{-1}, \qquad B = -12.4 \pm 0.6 \, \text{km s}^{-1} \, \text{kpc}^{-1}. \quad (10.36)$$

Since Hipparcos measured proper motions on the ICRS, these estimates are the most reliable values of the Oort constants, and should supersede all earlier values. Since $A > |B|$ they imply a gently falling circular-speed curve at $R_0$. Numerically, $v_c(R_0) = 218(R_0/8\,\text{kpc}) \, \text{km s}^{-1}$.

Recently an intriguing alternative to equation (10.21) has arisen, namely measurement by the VLA of the proper motion of the compact radio source Sgr A*, which is generally thought to mark the Galactic center (§9.5). Radio observations at seven epochs between 1981 and 1994 of the position of Sgr A* relative to extragalactic sources show that it has proper motion (Backer 1996)

$$(\mu_l, \mu_b) = (-6.55 \pm 0.17, -0.48 \pm 0.12) \, \text{mas yr}^{-1}. \quad (10.37)$$

This proper motion is the sum of the proper motion of Sgr A* with respect to the Galactic center and the reflex of the Sun's velocity, which is itself the sum of $v_c(R_0)$ and the solar motion with respect to the LSR. When we subtract the contribution of the latter for $R_0 = 8\,\text{kpc}$, we are left with $(\mu_l', \mu_b') = (-6.24 \pm 0.17, -0.30 \pm 0.12) \, \text{mas yr}^{-1}$. Clearly $\mu_b'$ must reflect the intrinsic motion of Sgr A*, and the fact that it does not differ significantly from zero encourages us to believe that Sgr A* is essentially stationary with respect to the Galactic center. In this case, from the first of equations (10.31) it follows that

$$\mu_l' = A - B = 30.1 \pm 0.8 \, \text{km s}^{-1} \, \text{kpc}^{-1}, \quad (10.38)$$

which is marginally inconsistent with equation (10.36).

# 10.4 The structure of the stellar disk

In the last section we saw that different spectral types have characteristically different velocity dispersions. The discovery of this and related correlations in the 1950s constituted one of the major advances in our understanding of the structure and history of the Milky Way. In this section we investigate these correlations in detail and explore their implications for Galactic structure and evolution.

We speculated above that the correlations between spectral type and kinematics might be connected with the different mean ages of stars of different spectral types. We now pursue this hint by looking at samples of stars to which ages can be assigned.

### 10.4.1 Ages and metallicities of nearby stars

The stars which can most readily and reliably be assigned ages are dwarfs of MK type G and earlier: such a star begins to turn off the ZAMS within 10 Gyr (§5.1), and as it moves away from the ZAMS it can be dated from its position in the CM diagram. Moreover, an interesting upper limit can be placed on the age of any such star that is still on the ZAMS. For our present purposes, F and G stars are most interesting because their lifetimes are comparable to the long time-scales characteristic of Galactic evolution.

Edvardsson *et al.* (1993) report the results of a major study of such objects. They obtained high-quality spectra and space velocities for 189 nearby F and G dwarfs. By fitting sophisticated model atmospheres to their spectra they obtained for each star an age and the abundances of the elements O, Na, Mg, Al, Si, Ca, Ti, Fe, Ni, Y, Zr, Ba and Nd. From §5.2.1 we recognize Na and Al as odd-light elements, Mg, Si and Ca as $\alpha$ elements, Ti, Fe and Ni as iron-peak elements and Y, Zr, Ba and Nd as s-process elements. Their sample of stars was selected by searching a large catalog of stars with Strömgren $uvby\beta$ photometry (§2.3.2) for stars of appropriate spectral type and age. Consequently, no inferences can be drawn from their results concerning the *fractions* of stars that have given ages or abundances, but currently they provide some of the best information regarding the mutual dependencies between abundances, velocity dispersion and age.

**Correlations between abundances**   The left-hand panel of Figure 10.16 demonstrates that the abundances of the light and heavy s-process elements Y and Ba are tightly correlated, as theory predicts. The right-hand panel of Figure 10.16 shows that the abundances of secondary elements such as Ba increase faster than those of primary elements such as Mg – the slope of the mean relation in this panel implies that the abundance of Ba varies as the abundance of Mg to the power 1.7. In §5.2.1 we argued that, in the absence of neutron poisons, the slope of this relation would be 2, and will be smaller if poisons are present. Therefore a slope of 1.7 is entirely consistent with theory.

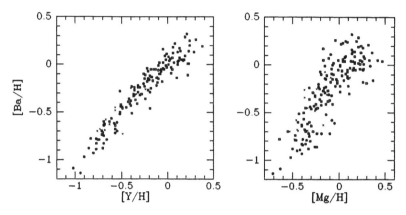

**Figure 10.16** Left panel: the abundances of yttrium and barium in disk stars are very tightly correlated. This is interesting in itself and also, crucially, demonstrates that the abundances derived by Edvardsson *et al.* have small experimental errors. Right panel: the secondary element Ba increases faster than the primary element Mg. Symbols in each panel indicate range of $R_m$ as in Figure 10.18. [After Edvardsson *et al.* (1993) from data kindly supplied by B. Edvardsson]

We saw in §5.2 that plots of relative abundances such as [O/Fe] versus [Fe/H] provide clues as to the sources of heavy-element enrichment. In particular, the abundances of O and the $\alpha$-elements such as Mg are expected to decrease relative to Fe with increasing [Fe/H] because they are quickly synthesized by core-collapse supernovae, whereas the iron-group elements are predominantly produced by type Ia supernovae, whose progenitors can be long-lived. Figure 10.17 shows that this expectation is amply borne out in the Edvardsson *et al.* data: in the most metal-poor stars, [O/Fe] and [Mg/Fe] are both larger by $\sim 0.4$ than in the most metal-rich stars. By contrast, the abundance of the iron-peak element Ni varies in step with [Fe/H], exactly as theory predicts.

**Correlations between age and abundance**     Figure 10.18 shows that younger stars tend to be more metal-rich. However, there is a considerable spread in metallicity at a given age. This spread cannot be attributed to observational error: the tight correlation in the left-hand panel of Figure 10.16 would not have been found if there were substantial uncertainties in the metallicity estimates.[10] Notice in the upper panel in Figure 10.18 that stars that have a higher iron abundance than the Sun have been forming for at least 9 Gyr. Nonetheless, $\tau$ and [Fe/H] are correlated for two reasons: (i) stars older that 10 Gyr almost all have [Fe/H] $\lesssim -0.5$;[11] and (ii) stars less than 5 Gyr old almost all have [Fe/H] $\gtrsim -0.5$. The pattern of barium abundances shown in the lower panel of Figure 10.18 is similar.

---

[10] Much of the small scatter in Figure 10.16 is also cosmic.

[11] Edvardsson *et al.* caution that the paucity of old, metal-rich stars may in part reflect their selection criteria, which discriminate against them.

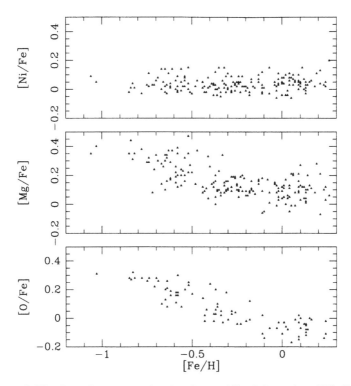

**Figure 10.17** The dependence upon the abundance of Fe of the ratios of Ni, O and Mg to Fe. The ratio of Ni to Fe is flat because these elements are produced alongside one another. The ratios of O and Mg to Fe decline with increasing Fe abundance because O and Mg, which are produced by short-lived stars, were formed before the disk became heavily polluted by iron from type Ia supernovae, which have long-lived progenitors. [After Edvardsson *et al.* (1993) from data kindly supplied by B. Edvardsson]

From their space velocities and a model of the Milky Way's potential, Edvardsson *et al.* calculated the apo- and peri-centric radii of each star and defined its characteristic Galactocentric radius, $R_m$, to be the average of these quantities. If, as is probable, stars form on nearly circular orbits and only later acquire random velocities, $R_m$ may be a reasonable estimate of the radius of a star's original orbit. In Figures 10.16 to 10.18 different symbols indicate the range within which each star's $R_m$ lies. From the concentration of filled dots at large $\tau$ it is evident that the oldest stars in this sample have that smallest characteristic Galactocentric radii. This correlation suggests that the disk of the Milky Way did not extend as far as the Sun 10 Gyr ago, and that we only see very old stars in the solar neighborhood because old stars that formed at small radii sometimes pay us a fleeting visit.

The most metal-poor stars in Figure 10.18 have the smallest values of $R_m$ and it has been suggested (Grenon 1989) that this could be because at any given radius the metallicity of newly formed stars is always the same, but

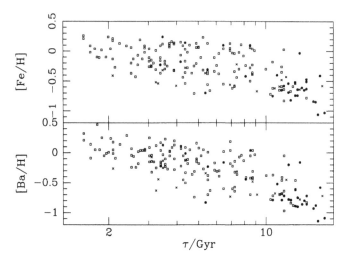

**Figure 10.18** The dependence upon age of the abundances of iron and barium. The characteristic Galactocentric radius of each star, $R_m$, is indicated by different symbols: • ⇒ $R_m < 7\,\mathrm{kpc}$; □ ⇒ $7\,\mathrm{kpc} < R_m < 9\,\mathrm{kpc}$; × ⇒ $R_m > 9\,\mathrm{kpc}$. [After Edvardsson *et al.* (1993) from data kindly supplied by B. Edvardsson]

is smaller at smaller radii. In this theory, the apparent correlation between metallicity and age arises because only old stars have acquired sufficient random velocity to visit the solar neighborhood from small radii. The a priori probability of this theory is low, given that it requires metallicity to increase outwards, whereas all the indications are that metallicity decreases outwards in external galaxies (§§4.4.5 and 8.2.6), but it does nicely illustrate the difficulty of reaching firm conclusions from studies of the solar neighborhood alone. All too frequently several physically distinct processes are active simultaneously, and it is not clear what is cause and what is side-effect.

To distinguish between temporal and metallicity gradients, Edvardsson *et al.* examine means and dispersions of various quantities in each of 20 cells in the $(R_m, \tau)$ plane. Since their entire sample consists of only 189 stars, the statistical fluctuations within some cells are uncomfortably large. However, they conclude that at a given radius and age there is a significant dispersion in [Fe/H]. The standard deviation of this scatter increases from $\sim 0.15$ at 2 Gyr to $\sim 0.25$ for the oldest stars. This conclusion is strongly confirmed by the observation that the Orion nebula, a major site of star formation at $R_0$ in our own epoch, is significantly metal-weaker than the Sun, which formed about 5 Gyr ago. Moreover, we shall see in §10.4.2 below that studies of star clusters in the disk indicate that clusters of similar ages and Galactocentric radii differ significantly in [Fe/H].

To investigate further the relative importance of different enrichment

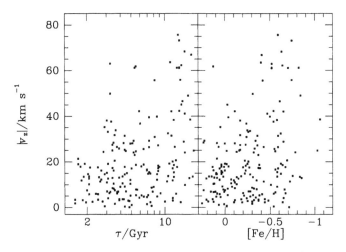

**Figure 10.19** $|v_z|$ versus $\tau$ (left-hand panel) and $|v_z|$ versus [Fe/H] (right-hand panel). [After Edvardsson *et al.* (1993) from data kindly supplied by B. Edvardsson]

mechanisms at various times and places, Edvardsson *et al.* define a quantity

$$[\alpha/\text{Fe}] \equiv \tfrac{1}{4}\big([\text{Mg/Fe}] + [\text{Si/Fe}] + [\text{Ca/Fe}] + [\text{Ti/Fe}]\big). \tag{10.39}$$

They find that there is negligible scatter in $[\alpha/\text{Fe}]$ in a given cell of the $(R_m, \tau)$ plane. Given the finding above that [Fe/H] *does* show scatter, this conclusion is important. On the one hand, the scatter in [Fe/H] indicates that the ISM at a given radius is imperfectly mixed, simultaneously containing regions that differ in [Fe/H] by a factor 2. On the other hand, the absence of scatter in $[\alpha/\text{Fe}]$ implies that the products of a massive supernova are thoroughly stirred into the ISM before significant star formation can occur in the region enriched by it.

The mean values of $[\alpha/\text{Fe}]$ tend to increase with age at a fixed value of $R_m$. For the youngest stars, $[\alpha/\text{Fe}] \simeq 0$, and for the oldest, $[\alpha/\text{Fe}] \simeq 0.25$. Since essentially all the oldest stars have small $R_m$, it is not possible to be sure whether the main dependence of $[\alpha/\text{Fe}]$ is upon radius or age, but the data suggest the latter. In fact, the natural interpretation of these data is that early in the life of the Milky Way the star formation rate was high, and large numbers of stars formed after the ISM had been heavily polluted by massive supernovae but before type Ia supernovae had produced much iron.

Let us now return to random velocities. The variation of random velocity with age is harder to study than the corresponding metallicity variations because one cannot measure the velocity dispersion of a single star; many stars are required for a single determination of $\sigma_z$. Since the Edvardsson sample contains only 189 stars, subdividing the sample to the required extent is scarcely practicable. Therefore in Figure 10.19 we plot for each star $|v_z|$

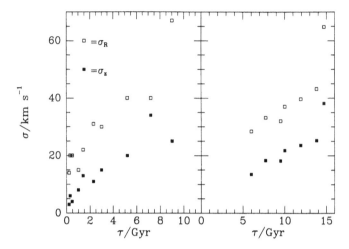

**Figure 10.20** Velocity dispersion versus age as determined by Jahreiss & Wielen (1983) (left panel) and by the Danish group (right panel). Aside from a difference in the age scales used in these studies, the two determinations are broadly consistent with one another. [From data published in Jahreiss & Wielen (1983) and in Strömgren (1987)]

against [Fe/H] and $\tau$. By estimating the vertical extent of the distribution in this figure at each value of $\tau$, the eye sees at a glance that the dispersion $\sigma_z$ increases with age. The right-hand panel suggests that the dispersion in $\sigma_z$ is fairly constant for [Fe/H] $\gtrsim -0.3$ and increases rapidly with falling [Fe/H] at [Fe/H] $\lesssim -0.3$.

Additional light is cast on this problem by samples of stars that have accurate space velocities and estimates of age and/or metallicity that are based on low-dispersion spectra or intermediate-band photometry rather than on high-dispersion spectra. Such data are available for several thousand stars. Jahreiss & Wielen (1983) estimated the ages of stars in the Gliese Catalog by one of two methods. For early-type stars they estimated ages from the stars' positions in the CM diagram (all these stars have reliable distances). The ages of those K and M stars which were in the McCormack Catalog (Vyssotsky 1963) were estimated from the strength of their Ca H and K lines. The left-hand panel of Figure 10.20 shows their result: both $\sigma_R$ and $\sigma_z$ increase steadily with age to quite large velocities ($\sigma_R \simeq 60\,\mathrm{km\,s^{-1}}$, $\sigma_z \simeq 30\,\mathrm{km\,s^{-1}}$).

Data from a large Danish program of $uvby\beta$ photometry provides an entirely independent check on the trend of velocity dispersion with age shown in Figure 10.20 as well as a check on the variation of velocity dispersion with metallicity that is suggested by Figure 10.19. The right-hand panel of Figure 10.20 shows the relation between age and velocity dispersion that emerges from this study. The adopted age-scale differs from that used by Jahreiss & Wielen by a factor of almost 1.5, but when allowance is made for

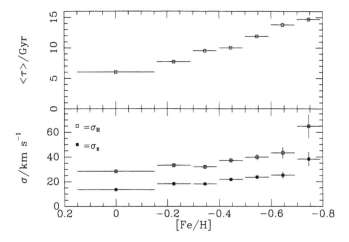

**Figure 10.21** The dispersions in $v_R$ and $v_z$ and the mean age $\langle\tau\rangle$ of 1294 stars binned by metallicity. [From data published in Strömgren (1987)]

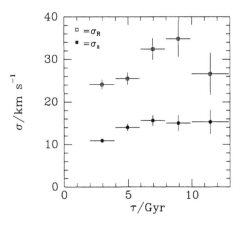

**Figure 10.22** The variation with age $\tau$ of the radial and vertical velocity dispersions of 556 stars with $-0.15 < [\text{Fe/H}] < 0.15$. For this sample $\sigma_\phi/\sigma_R = 0.61 \pm 0.02$ with no statistically significant dependence of the ratio upon $\tau$. [From data published in Strömgren (1987)]

this difference, the age–velocity-dispersion relations shown in the left- and right-hand panels of Figure 10.20 are broadly consistent.

The lower panel in Figure 10.21 shows the variation of $\sigma_R$ and $\sigma_z$ with [Fe/H] in data gathered by the Danish group. Both components of velocity dispersion increase as [Fe/H] decreases, very much as Figure 10.19 suggests. The upper panel in Figure 10.21 shows the mean age of the stars in each metallicity group. We see that metal-poorer stars tend to be older, and, as Figure 10.20 implies, older stellar populations have larger velocity dispersions. Hence, Figures 10.20 and 10.21 tell a consistent story.

From Figures 10.19 , 10.20 and Figure 10.21, it is clear that the velocity dispersion of a group of stars increases with its age and decreases with its metallicity. What is the correct physical interpretation of this result? Are

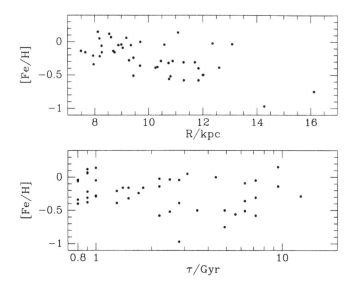

**Figure 10.23** The lower panel plots values of [Fe/H] for 44 old disk clusters against estimates of the clusters' ages. The upper panel plots [Fe/H] against the Galactocentric radius of each cluster. [From data published in Friel (1995)]

metal-poor stars born with larger random velocities and at earlier times than metal-rich stars? Or are both metal-rich and metal-poor stars born with small random velocities and then gradually increase their random velocities as a result of some dynamical process that is at work within the disk? In this second case, the observed correlation between random velocities and metallicity might be simply a consequence of the tendency of metal-rich stars to be, on the average, younger than metal-poor ones.

Figure 10.22 suggests that neither of these pictures represents the whole truth by showing the variation with $\tau$ of $\sigma_R$ and $\sigma_z$ for a sample of relatively metal-rich stars ($-0.15 <$ [Fe/H] $< 0.15$). Both components of velocity dispersion increase with age for $\tau \lesssim 7\,\mathrm{Gyr}$, but there is a suggestion that they level off at $\sigma_R \simeq 35\,\mathrm{km\,s^{-1}}$ and $\sigma_z \simeq 16\,\mathrm{km\,s^{-1}}$ for $\tau > 7\,\mathrm{Gyr}$. These asymptotic values of the velocity dispersions are almost a factor of 2 lower than the dispersions of the oldest stars in Figure 10.20. Hence it seems that by merely hanging around in the disk, stars do not acquire velocity dispersions as large as those observed in the more metal-weak stellar groups. This conclusion is very much in line with the predictions of the theory of the stochastic acceleration of disk stars – see §7.5 of BT. Indeed Jenkins (1992), in a very complete and careful study of this problem, concluded that a combination of excitation by spiral structure and scattering by giant molecular clouds can nicely explain the trends seen in Figure 10.22, but cannot plausibly account for the existence of groups of stars with $\sigma_R \simeq 60\,\mathrm{km\,s^{-1}}$.

## 10.4.2 The old disk clusters

As we saw in Chapter 6, a star cluster can be dated relative straightforwardly from the structure of its CM diagram, and of course we can obtain a reasonable estimate of the mean metallicity of its stars. Moreover, many, perhaps most, field stars in the disk were formed in a cluster or association, which may by now have completely dissolved. Hence studies of old disk clusters provide valuable information about the history of the disk.

In the lower panel of Figure 10.23 we plot [Fe/H] for 44 old disk clusters against estimates of the clusters' ages $\tau$. It is clear that clusters of very different metallicities have formed at similar times. The upper panel suggests that the further a cluster is from the Galactic center, the lower its metallicity is likely to be, and it is found that metallicity also tends to decrease with increasing distance from the plane. In fact, from a study of red giants in 63 open clusters, Piatti, Claria & Abadi (1995) estimate that

$$\langle [\mathrm{Fe/H}] \rangle = 0.6 \pm 0.1 - (0.07 \pm 0.01)\frac{R}{\mathrm{kpc}}$$
$$- (0.34 \pm 0.03)\frac{|z|}{\mathrm{kpc}} - (0.002 \pm 0.005)\frac{\tau}{\mathrm{Gyr}}. \tag{10.40}$$

Notwithstanding these trends, clusters with very different values of [Fe/H] can be found at any given distance from the center and/or height above the plane (Nissen 1988, Boesgaard 1989, Garcia Lopez, Rebolo & Beckman 1988). Thus old disk clusters display the same chemical inhomogeneity and mild metallicity gradients that we encountered above in the Edvardsson *et al.* (1993) sample of field stars. This finding adds weight to the idea that field stars were formed in clusters and subsequently either escaped their parent cluster, or became free when the cluster dissolved.

## 10.4.3 Star counts and the thick disk

The variation with distance $z$ from the Galactic plane of the density of various types of stars can be determined by counting faint stars in fields centered on a Galactic pole. The classic study of this problem is that of Gilmore & Reid (1983), who determined the photometric distances of more than 12 000 stars over 18 square degrees of the sky from photographic plates of the south Galactic pole that had been taken with a Schmidt telescope. Their key plates were in the $V$ and $I$ bands since they wished to obtain accurate results for lower-MS stars, which are red in color. On these plates, an automatic measuring machine determined the $V$ and $I$ magnitudes of all objects brighter than $V = 19$ and $I = 18$. The measuring machine was able to determine

rather reliably which of these relatively bright images were fuzzy and therefore galaxy images – in this way it produced a list of $\sim 12\,500$ stars with
known $V$ and $I$ magnitudes. Reasonably accurate absolute trigonometric
parallaxes and therefore $V$-band magnitudes, $M_V$, were available for about
200 of these stars. These values of $M_V$ confirmed that most of the stars in
the sample are MS stars, and defined the relationship between $M_V$ and $V - I$
in the photometric system employed. Most of the stars that are giants rather
than MS stars will have color $V - I \simeq 2.3$, which on the MS is associated
with $M_V \simeq 6.5$. Since in the absence of spectra the analysis must proceed
under the assumption that all stars are dwarfs, the inferred stellar density
at $M_V \simeq 6.5$ is most suspect, and should be treated with caution.

    Clearly, all stars with trigonometric parallaxes are located close to the
Sun, and it is not clear that the $M_V/(V - I)$ relation defined by such stars
applies well away from the plane. This doubt arises because the mean metallicity of stars might be expected to decrease with distance $z$ from the plane,
and we know that a metal-poor MS star is fainter than a metal-rich one of
the same color (see §5.1.5). Since Gilmore & Reid did not know the magnitude of the vertical metallicity gradient in the disk, they analyzed their
data twice, once under the assumption that the gradient was negligible, and
once under the assumption that $d[\text{Fe/H}]/dz = -0.3\,\text{kpc}^{-1}$ for $z < 5\,\text{kpc}$ and
vanishes at larger $z$. Fortunately, most of their conclusions stand for either
assumed metallicity gradient. For simplicity the results presented below are
for the case of no metallicity gradient.

    Under the assumptions that every star is a MS star and that there is
negligible vertical metallicity gradient, the distance to each star can be trivially derived from equation (2.36). When a metallicity gradient is assumed,
distances are obtained iteratively. $M_V$ is first obtained from $V - I$ under
the assumption of solar metallicity. Then $z$ is inferred from equation (2.36),
and $M_V$ is changed to the value appropriate to the metallicity of the stars
at the derived value of $z$. Now $z$ is updated, and the cycle repeated until the
derived values of $M_V$ and $z$ have converged.

    Once a value of $z$ has been associated with each star, a table can be
constructed in which each entry shows the number of stars observed in given
small ranges of $M_V$ and $z$. Down a column of fixed $z$, this table gives the
luminosity function $\Phi(M_V)$ that applies at a given distance from the plane.
Along a row of fixed $M_V$, the table gives the number of stars of a given kind
that are observed in thin slices of constant $z$ across the cone of observation
– clearly, this cone has the Sun at its apex.

    Figure 10.24 shows the luminosity functions derived by Gilmore & Reid
(1983) for different distances $z$ from the plane – each luminosity function
is labeled by the logarithm of $(z/1\,\text{pc})$. At large values of $z$, the luminosity
function is determined only at bright magnitudes because less luminous stars
fall below the apparent magnitude limit of the survey. Although the different
lengths of the curves in Figure 10.24 make it hard to compare the shapes
of the underlying luminosity functions, it is apparent that at $\log(z/1\,\text{pc}) \lesssim$

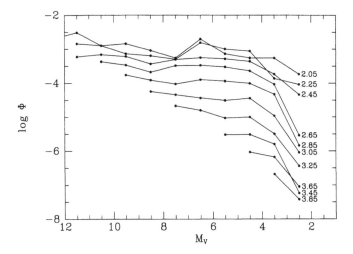

**Figure 10.24** Each curve shows the luminosity function that was determined by Gilmore & Reid (1983) for the distance in parsecs from the plane whose logarithm is given at the right-hand end of the curve. Notice that the three uppermost curves fall less steeply to the right of $M_V \simeq 3.5$ than do the lower curves.

2.45 the luminosity function falls less steeply at bright magnitudes than it does further from the plane. Physically, stars brighter than $M_V \simeq 3.5$ are more strongly confined to the plane than are fainter stars. Since stars with $M_V \lesssim 3.5$ have lifetimes that are significantly shorter than the Hubble time, they constitute a relatively young population. From Figure 10.22 we know young stellar populations have smaller velocity dispersions. Thus these stars are strongly confined to the plane because when they pass through the plane they have insufficient kinetic energy to rise high above the plane.

Figure 10.25 shows the density profile $\nu(z)$ derived by Gilmore & Reid for stars with absolute magnitudes in the range $4 - 5$. This profile clearly differs from an exponential $\nu \propto e^{-z/z_0}$, which in this log–linear plot would be a straight line. The dashed curve in Figure 10.25 shows that the vertical density profile can, however, be accurately represented as the sum of two exponentials, one of which has scale height $z_0 = 300\,\text{pc}$ and dominates the overall density distribution at $z \lesssim 1200\,\text{pc}$, while the other exponential has scale height $z_0 = 1350\,\text{pc}$. The vertical density profiles of less luminous stars can also be successfully modeled as a sum of these two exponentials, although the evidence for the exponential with scale height $\sim 1350\,\text{pc}$ is weak or absent for stars much less luminous than $M_V = 5$ because the density profile of these stars cannot be followed far from the plane. The vertical density profiles of stars that are more luminous than $M_V \simeq 3$ require a significantly smaller scale height than $z_0 = 300\,\text{pc}$ for the exponential that dominates the model density near to the plane. Moreover, for these luminous, short-lived stars the requirement for a second exponential with a large scale height is not

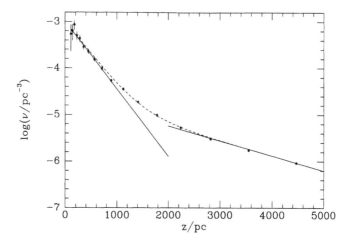

**Figure 10.25** The space density as a function of distance $z$ from
the plane of MS stars with absolute magnitudes $4 \leq M_V \leq 5$. The
full lines are exponentials with scale heights $z_0 = 300$ pc (at left)
and $z_0 = 1350$ pc (at right). The dashed curve shows the sum of
these two exponentials. [From data published in Gilmore & Reid
(1983)]

clear-cut. Ojha *et al.* (1996) found that the vertical distribution of stars
fainter than $M_V = 3.5$ is well represented by the sum of two exponentials,
one having scale height $z_0 = 260 \pm 50$ pc and the other having scale height
$z_0 = 760 \pm 50$ pc. In the plane the luminosity density of the second component
is smaller than that of the first by a factor $13.5^{+3.4}_{-3.3}$, which implies that $\sim 20\%$
of the luminosity of the disk resides in the thicker component.

**The thick disk**    The fact that the vertical density profile $\nu(z)$ cannot be
fitted by a single exponential has two possible physical interpretations. In
the simplest picture, the Galaxy possesses a single disk, and it just happens
that the vertical density profile of this disk is non-exponential. A more elaborate interpretation is that the Galaxy possess both thin and thick disks. In
this picture both disks have approximately exponential profiles $\nu_i(z)$, and
the observed profile is non-exponential because it is dominated by the thin
disk at small $z$ and the thick disk at large $z$. Since we know of no reason
why the vertical profile of a disk should be exponential rather than following some other functional form, the non-exponential nature of the observed
density profile by itself hardly calls for decomposing the observed disk into
two component disks. What *would* justify such a decomposition is evidence
that the two disks are made of different kinds of stars. Hence we now ask
whether it is possible to assign stars to either one disk or the other according
their *intrinsic* properties, such as age or metallicity.

    Notice that to be meaningful, the division of stars between the thin
and thick disks must be independent of either position or velocity. To see

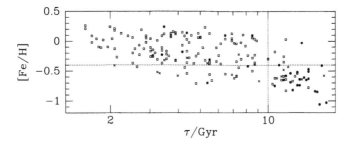

**Figure 10.26** A potential division of stars between the thin and thick disks. Stars above and to the left of the dotted lines are assigned to the thin disk, while those below and to the right of the lines are assigned to the thick disk.

this, consider the possibility of decomposing a spherical stellar system whose density profile $\nu(r)$ follows the modified Hubble law [equation (4.13)] into a **Plummer law** $\nu(r) = \nu_0(1 + r^2/r_0^2)^{-5/2}$ and the difference between this profile and the Hubble law. This decomposition could be made, and then the observed system would be dominated by the Plummer component at small radii and the difference component at large radii. Nonetheless, if all the system's stars were identical, it would be a senseless exercise that could only be inspired by some theoretical prejudice in favor of Plummer models. On the other hand, if the stars that comprised the Plummer model were all older than the remaining stars, the division into two components would be natural and convey important information about the manner of the system's formation.

A key point about the first, artificial, division of the modified-Hubble system is that it is not unique: if all stars are identical, a given star could be assigned a *probability* of belonging to one disk rather than the other on the basis of its position and velocity. For example, a star of low energy, which is confined to small radii, would have a high probability of belonging to the Plummer component, while a high-energy star would have a high probability of belonging to the other component. However, two stars of identical energies but different component assignments could always be swapped from one component to the other; it would be *in principle* impossible to know which was the 'correct' assignment. By contrast, if the stars of one component are older than those of the other, any uncertainty in the assignments will merely reflect imperfections in the data.

With this example in mind, let us now return to the data plotted in Figures 10.17 to 10.19 to see whether we can distinguish stars of the thin disk from those of the thick disk. Figure 10.26 shows one of these data sets with a potential division into thin- and thick-disk stars. According to this division, thick disk stars are older than 10 Gyr and more metal-poor than [Fe/H] = −0.4. The great majority of the remaining stars, which are by definition all younger than 10 Gyr, are also more metal-rich than

Table 10.4   Characteristic velocities of the disks
(km s$^{-1}$)

|            | $\langle v_\phi \rangle$ | $\sigma_R$ | $\sigma_\phi$ | $\sigma_z$ |
|------------|------|------|------|------|
| Thin disk  | $-6$ | 34   | 21   | 18   |
| Thick disk | $-36$ | 61  | 58   | 39   |

SOURCE: Data published in Edvardsson *et al.* (1993)

[Fe/H] $= -0.4$. We identify stars younger than 10 Gyr and more metal-rich than [Fe/H] $= -0.4$ with the thin disk and reserve judgment on the small number of stars that fall outside these two areas of Figure 10.26. Interestingly, the middle panel of Figure 10.17 shows that nearly all stars with [Fe/H] $> -0.4$ have [Mg/Fe] $< 0.2$, while most of the more metal-poor stars have significantly larger over-abundances of the $\alpha$-element Mg. This division suggests that type Ia supernovae contributed significantly fewer iron-peak elements to thick-disk stars than they did to thin-disk stars.

Now that we have divided stars between the disks, it is interesting to turn to Figure 10.19 to see what the kinematic *consequences* of this division are. Although the case is not compelling, a widening of the velocity distribution to the right of $\tau = 10$ Gyr is discernible. This impression is confirmed by calculating the velocity dispersions of stars to the right and left of the line $\tau = 10$ Gyr: the 44 stars to the right of this line have a dispersion of 38 km s$^{-1}$, while those to the left have a dispersion of 21 km s$^{-1}$. Indeed, even if we exclude from the left-hand group stars with $\tau < 3$ Gyr, where there appears to be an abrupt increase in the velocity dispersion, the velocity dispersion of the left-hand group is still only 22 km s$^{-1}$, little over half that of the right-hand group. Finally, if we refer all velocities to the LSR as defined by equation (10.11) and calculate the velocity dispersions of just the stars that can be definitely assigned to one disk or the other according to the definition of Figure 10.26, we find the numbers given in Table 10.4, which may be considered to be characteristic of the two disks.

Is the thick disk a real physical entity or merely a useful short-hand for the tail of the stellar distribution of the disk as a whole? We doubt that this question can yet be answered conclusively. Figure 10.26 makes the strongest case for the thick disk of which we are aware. While it certainly *suggests* that the thick disk is real, data from just a handful of extra stars could substantially weaken the case. A man could not be safely hanged on such evidence.

### 10.4.4 The local mass density of the disk

An obvious question is 'how much of the Galaxy's mass resides in the disks, thick and thin?' The *only* secure way to determine the mass of an astronomical body is to measure the strength of the gravitational field **F** to which it gives rise. In this section we explain how studies of the kinematics of stars enable us to estimate $\mathbf{F}$ in the solar neighborhood.

Once $\mathbf{F}$ has been determined, we can (in principle) determine the local mass density, $\rho$, from Poisson's equation,

$$\nabla \cdot \mathbf{F} = -4\pi G \rho. \tag{10.41}$$

If we assume that the Galaxy is approximately axisymmetric at $R_0$, the natural coordinate system in which to write (10.41) is a cylindrical system $(R, \phi, z)$. In this system, Poisson's equation for an axisymmetric system reads

$$\frac{1}{R}\frac{\partial}{\partial R}(RF_R) + \frac{\partial F_z}{\partial z} = -4\pi G \rho. \tag{10.42}$$

Now the circular speed is given by $v_c^2/R = -F_R$, so we may rewrite equation (10.42) in the form

$$\rho = -\frac{1}{4\pi G}\left(\frac{\partial F_z}{\partial z} - \frac{1}{R}\frac{\partial v_c^2}{\partial R}\right). \tag{10.43}$$

Observations tell us that $v_c$ is approximately independent of $R$ (see §9.2.3), so the second term on the right-hand side of (10.43) is small compared to the first (see Problem 10.10). Moreover, simple galactic mass-models suggest that this term does not depend much on $z$ for $z \ll R$. Therefore it is an excellent approximation to assume that it is independent of $z$, and to find on integrating (10.43) with respect to $z$ that the mass per unit area within distance $z$ of the plane is

$$\Sigma(R, z) \equiv 2\int_0^z \mathrm{d}z'\, \rho(R, z') \simeq -\frac{1}{2\pi G}\left(F_z(R, z) - \frac{z}{R}\frac{\partial v_c^2}{\partial R}\right). \tag{10.44}$$

Here we have assumed that the Galaxy is symmetrical in $z$, with the consequence that $F_z(R, 0) = 0$. Near the Plane we expect the mass density to be dominated by the thin disk, because it is so flattened. Therefore we may crudely identify, say, $\Sigma(R_0, 3z_0)$, where $z_0$ is the scale height of the thin disk, with the local surface density of this disk. In a more sophisticated analysis we can subtract the contribution of any massive halo from $\Sigma(R_0, 3z_0)$ to obtain a more accurate estimate of the density of the disk – see Problem 10.11 and the discussion below.

From equation (10.44), it is clear that the key to estimating the surface density of the layer that lies at $|z'| \leq z$ is the determination of $F_z(R, z)$.[12] The physical idea that underlies the determination of $F_z$ is the following. We identify some population of stars lying $\sim z$ from the plane and determine both their typical random $z$-velocities $\sigma_z$, and how fast the number density $\nu$ of this population declines with increasing $|z|$. We then argue that this

---

[12] $F_z$ is often denoted $K_z$, where $K$ stands for *Kraft*, the German for force.

decline in $\nu$ arises because the slower stars observed at $z$ have insufficient kinetic energy to reach $z + dz$, where the potential energy of stars is greater than at $z$ by $|F_z|dz$. Hence the larger $F_z$ is, the more steeply $\nu$ declines with $|z|$.

Unfortunately, the accurate implementation of this simple physical idea is fraught with difficulty. The essential problem is that stars have significant kinetic energy associated with their radial motions, and this energy should not be completely neglected when deciding to what altitude a given star can rise. Most studies have neglected the coupling of the radial and vertical motions, however.

The classical study of this problem was made by Oort (1960), who worked from the Jeans equation (§4.2 of BT)

$$\nu F_z = \frac{\partial \nu \sigma_z^2}{\partial z} + \frac{1}{R}\frac{\partial}{\partial R}(R\nu\sigma_{Rz}^2). \tag{10.45}$$

Here $\sigma_{Rz}^2$ is the average over the population of the product of velocity components $v_R v_z$, and the term containing it arises from the coupling between radial and vertical motions. Oort dropped this term, as did Bahcall (1984) in an extensive re-examination of the problem that will be described below.

Kuijken & Gilmore (1989) obtained a first estimate of $F_z$ by neglecting the $v_R - v_z$ coupling, and then made a slightly obscure correction for it. We shall describe their discussion in some detail, as it is the most recent major study of the problem. Significant commentaries on it can be found in Gould (1990b) and Kuijken & Gilmore (1991).

For the $z$ motions of stars to decouple from their radial motions, the gravitational potential $\Phi(R, z)$ must be of the form

$$\Phi(R, z) = \Phi_R(R) + \Phi_z(z), \tag{10.46}$$

where $\Phi_R$ and $\Phi_z$ are suitable functions of one variable. If $\Phi$ is assumed to be of this form, we can focus on the distribution of stars in the $(z, v_z)$ plane – as we shall see below, it is in principle possible to determine the location within this plane of every star in a suitable sample. Since our population should by now have achieved a steady state, Jeans' theorem (§4.2.1 of BT) assures us that the density $f(z, v_z)$ of stars within this plane must be a function of $z$ and $v_z$ only through the energy of $z$-motion,

$$E_z \equiv \tfrac{1}{2}v_z^2 + \Phi_z(z). \tag{10.47}$$

That is,

$$f(z, v_z) = f_z(E_z). \tag{10.48}$$

The stellar number density is

$$\nu(z) = \int dv_z\, f(z, v_z)$$
$$= 2\int_{\Phi_z(z)}^{\infty} dE_z\, \frac{f_z(E_z)}{\sqrt{2[E_z - \Phi_z(z)]}}, \tag{10.49}$$

where the second line has been obtained from the first by a simple change of integration variable.

Since we may measure $\nu(z)$ fairly accurately, Kuijken & Gilmore argue that we should regard equation (10.49) as an integral equation for $f(E_z)$ given $\nu(z)$. If $\Phi_z(z)$ were known, then $\nu$ could be considered to be a function $\nu(\Phi_z)$. Equation (10.49) would then become an Abel equation (see §1B.4 of BT) with solution

$$f_z(E_z) = -\frac{1}{\pi} \int_{E_z}^{\infty} d\Phi_z \, \frac{d\nu/d\Phi_z}{\sqrt{2(\Phi_z - E_z)}}. \tag{10.50}$$

If we were to use our known form $\Phi_z(z)$ and equation (10.47) to eliminate $E_z$ from equation (10.50) in favour of $z$ and $v_z$, we would obtain a prediction for the entire distribution of the stars in the $(z, v_z)$ plane. This line of argument motivated Kuijken & Gilmore to proceed as follows. They assumed that $\Phi_z$ is given by

$$\frac{\Phi_z(z)}{2\pi G} = \left(\sqrt{z^2 + D^2} - D\right)K + Fz^2, \tag{10.51}$$

where $D$, $F$ and $K$ are constants whose physical motivation will emerge shortly. When this form of $\Phi_z$ is adopted, and the observed form of $\nu(z)$ and the definition (10.47) are used in equation (10.50), we obtain a prediction for $f(z, v_z)$ that depends on the parameters $(D, F, K)$. Kuijken & Gilmore calculated the likelihood[13] $L$ of their observed distribution of stars in $(z, v_z)$ for each set of values of the parameters. The optimum parameters are those that maximize $L$. Once $\Phi_z(z)$ has been determined in this way, $F_z(z)$ follows by differentiation of equation (10.51), and then $\Sigma(z)$ can be found by substitution of $F_z$ into equation (10.44).

The physical motivation of equation (10.51) is as follows. Differentiating it we find

$$\frac{|F_z|}{2\pi G} = \frac{Kz}{\sqrt{z^2 + D^2}} + 2Fz. \tag{10.52}$$

The first term on the right-hand side rises linearly as $Kz/D$ for $z \ll D$ and tends to $K$ at $z \gg D$. This is the characteristic $z$-dependence of the gravitational force from a disk whose scale height is of order $D$ and equatorial density is $K/D$ – $K$ is the surface density of this disk. The second term on the right of equation (10.52) represents the contribution to $F_z$ of the Galaxy's spheroidal components. Indeed, the potential $\Phi_S$ of such a component may be taken to be constant on spheroidal surfaces with axis ratio $a : c = 1 : q$, where $q$ is of order 1. By the chain rule, $\Phi_S(R^2 + z^2/q^2)$ contributes a vertical force

$$F_S = -\frac{\partial \Phi_S}{\partial z} = \left(\frac{\partial \Phi_S}{\partial m}\right)\frac{2z}{q^2} \qquad \text{where} \qquad m \equiv R^2 + z^2/q^2. \tag{10.53}$$

---

[13] The likelihood of given data is simply the unnormalized probability of the data given that they are drawn from some model probability distribution.

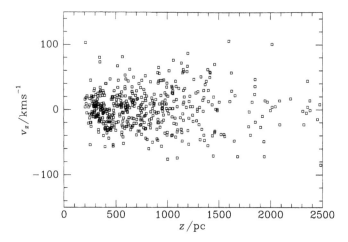

**Figure 10.27** The distribution in the $(z, v_z)$ plane of 472 K dwarfs seen towards the SGP. [From data published in Kuijken & Gilmore (1989)]

If we evaluate $(\partial \Phi_S / \partial m)$ at $(R_0, 0)$, equation (10.53) for $F_S$ becomes identical with the last term in equation (10.52) with $F = (\partial \Phi_S / \partial m)/q^2$. The dimensions of $F$ are those of mass density, and we may loosely consider it to be the local density contributed by the spheroidal components provided we bear in mind that $F$ does not actually vanish in the complete absence of a spheroidal component.

After determining $F_z$ in the way we have described, Kuijken & Gilmore (1989) corrected their raw value for the effects of the neglected $v_R - v_z$ coupling. The idea underlying this correction is that uncoupled motion in the $(z, v_z)$ plane of the type they assumed is associated with a Jeans equation that is identical with the correct Jeans equation (10.45) with the second term on the right dropped. Therefore the raw value of $F_z$ determined by Kuijken & Gilmore may be identified with $\nu^{-1}[\partial(\nu \sigma_z^2)/\partial z]$, and the true value of $F_z$ is the raw value plus the contribution from the second term on the right of equation (10.45). The magnitude of this term (which turns out to be negative) depends on the shape of the velocity ellipsoid (see §10.3.2) and the way in which its orientation changes as one moves away from the plane – see §4.2.1(c) of BT. Although the shape of the velocity ellipsoid is known, the dependence of its orientation upon $z$ is unknown. It may, however, be guessed with reasonable plausibility. In this way Kuijken & Gilmore arrive at a range of plausible values for the required correction.

Kuijken & Gilmore applied their technique to a sample of 512 K dwarfs. These stars were selected from 11.5 square degrees at the south Galactic pole as follows. Spectra were obtained of all stars with $V < 19$ and $B < 20.06$ and $B - V$ in the range $(0.8, 1.2)$. From these spectra $v_z$ could be estimated to better than $10 \, \mathrm{km \, s^{-1}}$. Giants were eliminated from this sample as apparently bright stars that did not show significant MgH absorption; the quality of the spectra made it difficult to identify MgH absorption in apparently fainter stars, but it is inherently unlikely that a faint star is a giant. The color–magnitude relation for solar-metallicity dwarfs was then used to determine the distance $z$ to each star, and then the presence of a metallicity gradient was allowed for by an iterative procedure very similar to that described in §10.4.1. Finally, the standard correction for Malmquist bias [equation (3.34)] was applied to the derived $M_V$ of each star.

Figure 10.27 shows the final distribution of stars in the $(z, v_z)$ plane. The distribution in $z$ is determined by competition between the increase $\propto z^2$ in the volume that lies within the cone of observation within $dz$ of $z$, and the declining density of stars as one moves away from the plane. For $z \lesssim 200\,$pc there is little decline in the density, and the number of stars in the sample increases rapidly. There follow a broad maximum and a long steady decline. When the distribution of Figure 10.27 was analyzed as described above, Kuijken & Gilmore (1991) concluded that within $1.1\,$kpc of the plane the surface density of material is

$$\Sigma_{1.1}(R_0) = 71 \pm 6\,\mathcal{M}_\odot\,\mathrm{pc}^{-2}. \tag{10.54}$$

Uncertainty regarding the variation of the velocity ellipsoid near the Sun contributes an uncertainty of about 15% to this number. Bahcall *et al.* (1992) obtained very similar estimates of $\Sigma(R_0)$ from a sample of 125 K giants.

Significant controversy surrounds the question of how much of the surface density (10.54) is attributable to the disks rather than to spheroidal components (chiefly the 'dark halo'). Mathematically, the problem is that the data constrain the values of the right-hand side of equation (10.52) at $z \simeq 1.1\,$kpc, but permit variations in the parameters $D$, $F$ and $K$ that are consistent with this value. Since, as we have seen, the physical interpretation of $K$ is the surface density of the disk, uncertainty in $K$ implies uncertainty in the surface density of the disk.

Kuijken & Gilmore (1989) resolved this question by remarking that, since the circular-speed curve is known, mass taken from the disk must be, by and large, added to the spheroidal components. Hence acceptable solutions with small $K$ should have large $F$, and vice versa. A study of specific Galactic mass-models led them to the constraint

$$(15 \pm 3)\,\mathcal{M}_\odot\,\mathrm{pc}^{-2} = 0.094K + F \times 1\,\mathrm{kpc}. \tag{10.55}$$

When $F$ and $K$ are constrained by equation (10.55) as well as by the observationally determined value of equation (10.52) at $z \simeq 1.1\,$kpc, Kuijken & Gilmore (1991) find that the surface density of the local disk is

$$\Sigma_d(R_0) = 48 \pm 9\,\mathcal{M}_\odot\,\mathrm{pc}^{-2}. \tag{10.56}$$

Unfortunately, the uncertain behavior of the velocity ellipsoid near the Sun, which contributes an uncertainty of only $\sim 15\%$ to equation (10.54), renders equation (10.56) extremely uncertain. It appears that at present we cannot strongly constrain the mass density near the Sun.

When our estimate (10.56) of the local mass-density of the disk is combined with estimates of the disk's local surface brightness (§10.1.1), we conclude that the mass-to-light ratios of the disk are

$$\Upsilon_B(R_0) \simeq 2.3 \, \mathcal{M}_\odot \, L_\odot^{-1}$$
$$\Upsilon_K(R_0) \simeq 1.0 \, \mathcal{M}_\odot \, L_\odot^{-1}. \tag{10.57}$$

The errors in these mass-to-light ratios are large because we are forming ratios of numbers that are themselves subject to large errors.

How much mass do we know to reside in the local disk in the form of known stars, interstellar gas and so forth? Figure 9.19 implies that the ISM contributes $\Sigma_{\rm ISM}(R_0) \simeq 5.5 \, \mathcal{M}_\odot \, {\rm pc}^{-2}$. The mass contributed by stars is extremely difficult to assess. From the Gliese catalog, we have a fair idea of what stars lie within $10 \, {\rm pc}$ of the Sun. In particular, we saw in §3.6.4 that stars contribute about $\rho_* = 0.1 \, \mathcal{M}_\odot \, {\rm pc}^{-3}$ at the Sun. Star counts such as those analyzed by Kuijken & Gilmore (1989) enable us to determine the ratio of the number density of stars at altitude $z$ to their number density near the Sun. Hence determining the contribution of stars to the local surface density would be simple if the mix of stars were the same at all altitudes. Unfortunately, the mix of stars must vary systematically with $z$, since, as we saw in §10.3.2, different stellar types have different values of $\langle v_z^2 \rangle$ in the solar neighborhood. Thus each stellar population must have its own vertical density profile. We cannot determine all these profiles by direct observation, but must resort to dynamical models in which the vertical density profiles of most populations are determined by requiring that each population is in dynamical equilibrium in a common potential. The structure of the potential, on the other hand, is determined by requiring the density profile of a selected population, for example K dwarfs, agrees with an observationally determined profile. That is, the contribution of stars to the local surface density can in practice not be determined independently of the problem of determining $F_z$.

Since Kuijken & Gilmore (1989) consider only a single tracer population, their analysis cannot be used to predict the density of all types of stars throughout the disk. The methodology of Bahcall (1984) is, by contrast, well suited to this problem. In the standard way, Bahcall reduced the problem to a one-dimensional one by neglecting the $v_R - v_z$ coupling. Then he decomposed the matter near the Sun into a series of isothermal populations – each such population has a distribution function of the form[14]

$$f_z(E_z) = \frac{\rho_{i0}}{\sqrt{2\pi\sigma_i^2}} \exp(-E_z/\sigma_i^2), \tag{10.58}$$

---

[14] It is now convenient to define the distribution function $f$ to be the mass density in the $(z, v_z)$ plane.

**Table 10.5** Estimates of the mass density of the disk at the Sun

| Model | $\dfrac{\rho_{\rm tot}(0)}{\mathcal{M}_\odot\,{\rm pc}^{-3}}$ | $\dfrac{\rho_{\rm dark}(0)}{\mathcal{M}_\odot\,{\rm pc}^{-3}}$ | $\dfrac{\Sigma_{\rm tot}(\infty)}{\mathcal{M}_\odot\,{\rm pc}^{-2}}$ | $\dfrac{\Sigma_{\rm dark}(\infty)}{\mathcal{M}_\odot\,{\rm pc}^{-2}}$ |
|---|---|---|---|---|
| Dark matter has $\langle v_z^2\rangle^{1/2}=4\,{\rm km\,s}^{-1}$ | 0.89 | 0.79 | 70 | 20 |
| Dark matter has $\langle v_z^2\rangle^{1/2}=20.3\,{\rm km\,s}^{-1}$ | 0.21 | 0.10 | 88 | 38 |
| Dark and luminous matter have identical profiles | 0.26 | 0.16 | 84 | 34 |

SOURCE: Data published in Bahcall *et al.* (1992)

where $\rho_{i0}$ and $\sigma_i$ are constants characteristic of the $i^{\rm th}$ population. It is easy to show: (i) that $\rho_{i0}$ is the mass density of stars where $\Phi_z=0$, which we may place at $z=0$; and (ii) that the velocity dispersion $\langle v_z^2\rangle^{1/2}$ of each isothermal population is independent of $z$, and has value $\sigma_i$. With this result, equation (10.45) simplifies to

$$F_z=\sigma_i^2\frac{\partial\ln\rho_i}{\partial z},\qquad(10.59)$$

where the density $\rho_i(z)$ is given by

$$\rho_i(z)=\rho_{i0}\exp(-\Phi_z(z)/\sigma_i^2).\qquad(10.60)$$

Now the total density of matter near the Sun is just the sum over $i$ of $\rho_i$, and $F_z=-\partial\Phi_z/\partial z$, so equation (10.43) becomes

$$4\pi G\sum_i\rho_{i0}\exp(-\Phi_z(z)/\sigma_i^2)=\frac{\partial^2\Phi_z}{\partial z^2}+\frac{1}{R}\frac{\partial v_c^2}{\partial R}.\qquad(10.61)$$

Once the constants $\rho_{i0}$ and $\sigma_i$ have been chosen, equation (10.61) becomes an ordinary differential equation for $\Phi_z(z)$. The density $\rho_i(z)$ of each component follows when our solution, $\Phi_z(z)$, is substituted into equation (10.60).

The density profiles $\rho_i(z)/\rho_{i0}$ of certain tracer populations can be determined from star counts. If the observationally determined profiles agree with the theoretical ones from equation (10.61), all is well. However, Bahcall found that the theoretical profiles decrease with increasing $|z|$ less rapidly than do the observational profiles. This discrepancy is what would occur if some components had been missed out, for the more mass there is, the faster $\Phi_z$ increases away from the plane, and the more steeply the densities (10.60) decrease with increasing $|z|$. Bahcall tried to determine how much missing matter there was by adding a single isothermal 'dark-matter' component and adjusting its mid-plane density $\rho_{\rm DM0}$ until an acceptable fit was obtained to the observed profile of a tracer population. The amount of dark matter required varies with the assumed dark-matter velocity dispersion $\sigma_{\rm DM}$: larger values of $\sigma_{\rm DM}$ yield lower densities $\rho(0)$ in the plane but larger total column densities $\Sigma_\infty(R_0)$ through the disk. The first two rows of Table 10.5 gives results from Bahcall *et al.* (1992) for the case that the dark matter has a low velocity dispersion and a velocity dispersion characteristic of K giants. The third row gives densities for the case that the density of dark matter is a fixed multiple of the density of luminous matter. The volume densities are uncertain by at least a factor of 2, and the surface densities by at least 0.2 dex.

The difficult step in the practical application of equation (10.61) is the identification of populations, for they must be *isothermal* populations rather than mere collections of physically similar stars. Consider, for example, white dwarfs as a candidate population. Do they form an isothermal population? We have no reason to believe so a *priori*, so this question should be settled observationally by determining the velocity distribution of white dwarfs near the Sun, and deciding whether it is Maxwellian. If it proves to be significantly non-Maxwellian, we will have to treat the white dwarfs as a superposition of more than one isothermal component. For example, if there are more high-speed white-dwarfs than a Maxwellian predicts, we shall have to add to the basic isothermal population a subsidiary population that has a larger value of $\sigma_i$. Near the plane, the white dwarf density may be dominated by the basic population, but at sufficiently large $|z|$ the subsidiary population will be dominant. Therefore serious errors in $\rho_i(z)$ can arise from small deviations of the local velocity distribution from a Maxwellian. Gould (1990a) discusses whether a given population is sufficiently Maxwellian.

Kuijken (1991) describes a technique in which $\Phi(z)$ is estimated from the distributions of tracer populations in the $(z, v_z)$ plane without making any assumption of isothermality. By applying this technique to a sample of K dwarfs, he concluded that $\rho_{\rm tot}(0) < 0.22\,\mathcal{M}_\odot\,{\rm pc}^{-3}$ and that there is no requirement for dark matter in the Galactic disk. At present the amount of stellar matter in the disk is significantly uncertain.

### 10.4.5 Distribution of the youngest stars

The appearance at optical wavelengths of external spiral galaxies is profoundly affected by the large-scale distribution of young stars because these objects and their associated HII regions trace out spiral arms. What can we say about the large-scale distribution of young stars in the Milky Way?

Much of the disk, within which stars form, is obscured from the Sun by several magnitudes of optical extinction. This extinction both confines our field of study to within a radius of $\sim 4\,{\rm kpc}$ around Sun, and distorts the apparent distribution of young stars. Indeed, extinction is generally patchy, so we can see to greater distances along some lines of sight than along other nearby ones, and the apparent distribution of young stars will inevitably be influenced by the distribution of extinction. To compound these already

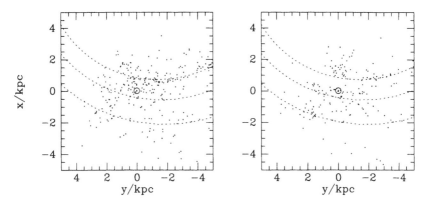

**Figure 10.28** The distribution of Cepheid variables within 5 kpc of the Sun (left) and of open clusters in Lynga (1987) with ages < 30 Myr (right). The upper two dotted curves in each panel show the ridge-lines of the Sagittarius–Carina arm (top) and the Orion–Cygnus arm (middle) from Figure 9.20.

formidable difficulties, the literature is deficient in systematic surveys for objects associated with star formation, so we have to rely on catalogs that have been put together from a variety of sources that contain all manner of unquantified selection effects. For all these reasons, and the usual difficulty of determining distances to astronomical objects, we shall outline a picture of the distribution of young stars that is woefully out of focus so that a measure of imagination will be required to make out features that may in reality be well defined.

Cepheid variables are useful tracers of star formation since their ages are $\lesssim 0.1\,$Gyr (half a rotation period of the Sun about the Galaxy), and their positions within the disk can be determined with reasonably accuracy from their period–luminosity relation (§5.1.10). The left-hand panel of Figure 10.28 shows the distribution of 278 Cepheid variables with reliable distances in a coordinate system in which the Sun is at the origin and the direction to the Galactic center is upwards. Two dotted curves from Figure 9.20 are superimposed on this distribution: the top curve shows the ridge-line of the Sagittarius–Carina arm while the middle curve shows the ridge-line of the Orion–Cygnus arm. The distribution of Cepheids is clearly enhanced along the Sagittarius–Carina arm.

The right-hand panel of Figure 10.28 shows the distribution of open clusters younger than 30 Myr. In addition to being enhanced near the ridge-line of the Sagittarius–Carina arm, the density of young clusters shows signs of being enhanced also along both the Orion–Cygnus arm and a third arm, the Perseus arm, that passes the Sun in the anticenter direction at a distance of $\sim 2\,$kpc.

Additional evidence that star formation is enhanced along the Perseus arm is furnished by Figure 10.29, which shows the distribution of HII regions with photometric distances: there is a distinct enhancement in the density of

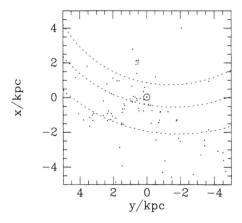

**Figure 10.29** Dots mark the locations of 113 HII regions. The Sun, marked by ☉, is at the origin and the direction to the Galactic center is upwards. From top to bottom, dotted curves indicate the ridge-lines of the Sagittarius–Carina, Orion–Cygnus and Perseus arms. [From data published in Blitz, Fich & Stark (1982) and Gillespie *et al.* (1977)]

HII regions in an elongated region around $(y, x) = (3, -1)$. The Sagittarius-Carina arm, by contrast, is not evident in Figure 10.29.

Alfaro *et al.* (1991) have shown that the Galaxy's layer of young clusters is not perfectly flat but corrugated in much the same way that the Galactic gas layer is corrugated (§9.2.6). Thus, the region of highest cluster density oscillates above and below the plane by ∼ 50 pc with a characteristic wavelength of ∼ 2 kpc. Since clusters form by the fragmentation of dense molecular clouds, the corrugation of the cluster layer is presumably a simple consequence of the poorly understood corrugation of the gas layer.

## 10.5 The halo

We saw in §4.5 that most galaxies have a globular cluster system that extends to beyond $R_e$. It is obviously of interest to see how the Milky Way's globular cluster system compares with those of external galaxies. The Milky Way's globular cluster is especially important because it is associated with a halo of field stars. We shall find that field-star halo is nearly two orders of magnitude more luminous than the globular cluster system, but its surface brightness at the solar radius is so small (∼ 30 mag arcsec$^{-2}$ – see Problem 10.13) that similar halos could not be studied at comparable radii in external galaxies. Hence, if external galaxies do possess field-star halos, we can only trace them through the association between the local globular-cluster system and field-star halo, which we now investigate.

### 10.5.1 The globular cluster system

As we saw in §6.1, there are about 150 globular clusters in the Milky Way. Kinman (1959) pointed out that the metal-richer globular clusters seem to form a more highly flattened and more rapidly rotating system than the least

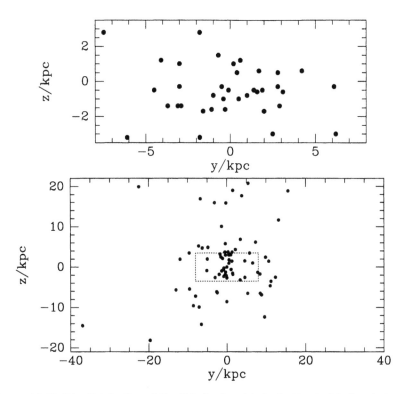

**Figure 10.30** The distribution of the disk (top) and halo (bottom) globular clusters as viewed by an observer located at a large distance in the anticenter direction. The dotted rectangle in the lower panel indicates the region shown in the upper panel. In the upper panel we have superimposed the distribution that would be seen by an observer located at a large distance in the direction $l = 270°$. In each panel the Galactic center lies at the origin. [From data published in Armandroff (1989)]

metal-rich globular clusters. Zinn (1985) established beyond doubt that such a division of the globular clusters into sub-systems with $[Fe/H] \geq -0.8$ and $[Fe/H] < -0.8$ is real – clusters of the former, metal rich sub-system, are called **disk globular clusters** or **G clusters**, while clusters of the metal-weak sub-system are called **halo globular clusters** or **F clusters**. Since Zinn's analysis, significant improvements in the quality of the data available for the metal-rich clusters – which tend to occur in crowded and reddened low-latitude fields – prompted Armandroff (1989) to re-evaluated the parameters of the two sub-systems, and we shall largely follow his analysis.

Figure 10.30 shows the spatial distribution of well-studied disk and halo globular clusters. In the upper panel each cluster is shown twice, once as it would be observed by an observer who was located far from the Milky Way in the anticenter direction, and once as viewed by an observer located far down the line $l = 270°$. Since the halo clusters are more numerous, the lower panel shows them only as seen from the anticenter direction. From this figure one sees that the halo clusters form an essentially spherical distribution that extends to at least 25 kpc, whereas the disk clusters form a flattened distribution that does not extend to very large radii.

The disk clusters are too few in number to define adequately a density distribution in the $(R, z)$ plane. However, the idea that the disk sub-system might be the analog for globular clusters of the thick stellar disk introduced in §10.4.3, suggests that we assume that the density of disk clusters falls off with $z$ as $e^{-z/z_0}$ and try to determine the scale height $z_0$. Two facts make the value of $z_0$ very uncertain. The first is the awkward presence of six points along the upper and lower edges of the top panel of Figure 10.30. These points, which are associated with the clusters, NGC 6356, 47 Tuc and Pal 11, lie at some remove from the main clump of disk clusters, and fit poorly with the idea that the cluster density declines exponentially with $z$. The second source of uncertainty in $z_0$ is the expectation that clusters are preferentially missing from the data set at small $z$, because it is precisely near the plane that obscuration and confusion make it hard to study clusters. Subject to these large uncertainties, Armandroff (1989) concludes that $z_0 = 1.1 \pm 0.35\,\mathrm{kpc}$. To within the errors this value is the same as the scale height $z_0 \simeq 1.3\,\mathrm{kpc}$ of the thick disk that is inferred from Figure 10.25.

There are too few globular clusters in each sub-system for the kinematics of these systems to be well defined. Nevertheless it is possible to estimate a mean rate of rotation and a characteristic velocity dispersion for each sub-system. In early work it was generally assumed that each sub-system rotates like a solid body at an angular frequency $\Omega_i$, but, following the work of Frenk & White (1980), it is now usually assumed that each sub-system rotates differentially, in such a way that the rotation speed takes the same value, $v_{\rm rot}$, everywhere. In addition to this systematic rotation, each globular cluster has a random velocity $\mathbf{v}_{\rm rand}$. Compounding these velocities, the total Galactocentric velocity of the $i^{\rm th}$ cluster is

$$\mathbf{v}^{(i)} = v_{\rm rot}\mathbf{e}_\phi^{(i)} + \mathbf{v}_{\rm rand}^{(i)}, \tag{10.62}$$

where the expectation value of $\mathbf{v}_{\rm rand}^{(i)}$ is zero, and $\mathbf{e}_\phi^{(i)}$ is the unit vector in the azimuthal direction at the cluster's location. While undoubtedly an over-simplification, this kinematic model is a reasonable approximation to the observed rotation fields of elliptical galaxies (see §11.2.1), which are the archetypical hot stellar systems.

Frenk & White (1980) fitted $v_{\rm rot}$ to the line-of-sight velocities of clusters because only a few clusters have useful proper motions. Let $\hat{\mathbf{x}}^{(i)}$ be the direction from the Sun to the $i^{\rm th}$ cluster. The line-of-sight velocity of this cluster is $v_{\rm los}^{(i)} = \hat{\mathbf{x}}^{(i)} \cdot (\mathbf{v}^{(i)} - \mathbf{v}_\odot)$. Taking the scalar product of equation (10.62) with $\mathbf{x}^{(i)}$ and rearranging, we obtain

$$\mathbf{x}^{(i)} \cdot \mathbf{v}_{\rm rand}^{(i)} = v_{\rm los}^{(i)} - v_{\rm rot}\mathbf{e}_\phi^{(i)} \cdot \hat{\mathbf{x}}^{(i)} + \mathbf{v}_\odot \cdot \hat{\mathbf{x}}^{(i)}. \tag{10.63}$$

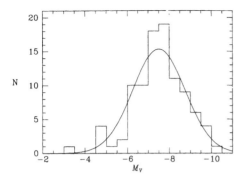

**Figure 10.31** The distribution over luminosity of the Milky Way's globular clusters. The curve is the Gaussian distribution given by equation (10.65). [From data published in Armandroff (1989)]

By the argument explained in Box 10.1, we now choose the value of $v_{\rm rot}$ that minimizes the sum over $i$ of the square of the left of equation (10.63). One easily finds that the required value is

$$v_{\rm rot} = \frac{\sum_i v^{(i)}_{\rm los} + \mathbf{v}_\odot \cdot \hat{\mathbf{x}}^{(i)}}{\sum_i \left(\mathbf{e}^{(i)}_\phi \cdot \hat{\mathbf{x}}^{(i)}\right)^2}. \tag{10.64}$$

Substituting the derived value of $v_{\rm rot}$ into equation (10.63) one obtains the line-of-sight component of the random velocity of each cluster, and can hence derive $\sigma_{\rm los}$ for the clusters.

The greatest source of uncertainty in equation (10.64) for $v_{\rm rot}$ is the cosines $\mathbf{e}^{(i)}_\phi \cdot \hat{\mathbf{x}}^{(i)}$ – especially for clusters that lie near the Galactic center, where the direction of $\mathbf{e}_\phi$ changes rapidly, so these cosines depend sensitively on the rather uncertain cluster distances. The derived value of $v_{\rm rot}$ also depends upon the assumed value of $\mathbf{v}_\odot$.

For the halo sub-system, Armandroff (1989) finds $v_{\rm rot} = 43 \pm 29\,{\rm km\,s}^{-1}$ and $\sigma_{\rm los} = 116 \pm 11\,{\rm km\,s}^{-1}$, while for the disk sub-system he finds $v_{\rm rot} = 193 \pm 29\,{\rm km\,s}^{-1}$ and $\sigma_{\rm los} = 59 \pm 14\,{\rm km\,s}^{-1}$. Thus the kinematics of the globular-cluster sub-systems bear out the impression one gains from Figure 10.30 that the halo sub-system is dominated by random motions while the disk sub-system is flattened because it rotates rapidly. For Armandroff's assumed value of the local circular velocity, $v_c(R_0) = 220\,{\rm km\,s}^{-1}$, the rotation rates of the halo and disk globular-cluster sub-systems relative to the LSR correspond to $\langle v_\phi \rangle = -177\,{\rm km\,s}^{-1}$ and $\langle v_\phi \rangle = -27\,{\rm km\,s}^{-1}$, respectively.[15] Comparing this last number with the value $\langle v_\phi \rangle = -36\,{\rm km\,s}^{-1}$ given in Table 10.4 for the thick disk, one sees that the disk globular-cluster sub-system rotates marginally even faster than the local thick disk.

The data are consistent with both sub-systems having a common luminosity function. Figure 10.31 shows the histogram of luminosities for the two

---

[15] Armandroff adopted a larger solar motion [$20\,{\rm km\,s}^{-1}$ towards $(l, b) = (57°, 22°)$] than that given by equation (10.11).

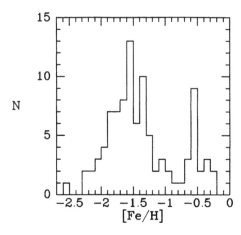

Figure 10.32 The metallicity distribution of the globular clusters. [From data published in Armandroff (1989)]

sub-systems combined. It is adequately fitted by [cf. equation (4.49)]

$$\frac{dN}{dM_V} = A \exp\left(-\frac{(M_V - \overline{M})^2}{2\sigma^2}\right) \text{ with } \overline{M} = -7.5, \ \sigma = 1.25. \quad (10.65)$$

There is no statistically significant evidence for a correlation of luminosity with either Galactocentric distance or distance from the plane.

Although there are accurate data for only 19 disk clusters, so that statistical uncertainties are large, the mean metallicity of the disk clusters seems to decrease outwards. There is no evidence of a corresponding metallicity gradient in the halo sub-system, even though it is more numerous and so gives rise to smaller statistical uncertainties.

Figure 10.32 shows the metallicity distribution of 96 globular clusters for which reliable data are available. The division between the disk and the halo sub-systems at $[Fe/H] = -0.8$ is evident. The metallicities of the disk clusters are strongly peaked around $[Fe/H] = -0.5$, while the halo clusters are distributed over a much wider range in $[Fe/H]$ that reaches down to $[Fe/H] = -2.6$. In §10.7.2 below we shall see that the leaky-box model of §5.3.2 provides a reasonable fit to the metallicity distribution of the halo clusters.

### 10.5.2 Field halo stars

Most globular clusters are more metal-poor than any star in the Edvardsson et al. (1993) sample that was discussed in §10.4. In the 1940s it was discovered that some field stars closely resemble globular-cluster stars, in particular by being very metal-poor. These stars are of considerable interest for two reasons. First, it is useful to identify very metal-poor stars in the solar neighborhood, which can be studied in greater detail than stars located in even the nearest globular cluster. Second, very metal-poor stars must have either formed outside the Milky Way or before the Milky Way's ISM became significantly polluted with metals. In either case they must carry significant information about the Milky Way's origins. In particular, we would like to know the connection between these stars and the globular-cluster system – for example, are they debris from shattered globular clusters? To answer such questions we need to know how they are distributed in real space and in velocity space, and their distribution in metallicity.

In the solar neighborhood, the space-density of very metal-poor stars is at least two orders of magnitude smaller than the space-density of relatively metal-rich stars (Dahn *et al.* 1995). In view of this scarcity, effective strategies for locating these objects have had to be developed. A classical strategy is to look for RR Lyrae stars whose periods lie in the range defined by the RR Lyrae stars of halo globular clusters. These stars have the same metallicity distribution as the halo clusters (Suntzeff, Kinman & Kraft 1991) and can be identified from their characteristic variability – see §5.1.10. The surfaces on which the density of these RR Lyrae stars is constant are approximately spherical at $r > R_0$ but become flattened (axis ratio $q \simeq 0.6$) at $r < R_0$ (Kinman, Mahaffey & Wirtanen 1982, Saha 1985, Hartwick 1987). Interior to $r \simeq 25\,\mathrm{kpc}$, the space-density of these stars falls off with Galactocentric radius $r$ as $r^{-\alpha}$ with exponent $\alpha$ in the range $3 \leq \alpha \leq 3.5$. Beyond $25\,\mathrm{kpc}$, the density may fall more steeply with increasing radius. The local space-density of RR Lyraes is $\sim 10 \pm 5\,\mathrm{kpc}^{-3}$ (Layden 1995).

An objection to using RR Lyrae stars as tracers of the halo is that they make up only a fraction of the HB stars, and we know from §6.1.2 that small changes in the abundances and/or ages of these stars give rise to significant and ill-understood changes in the disposition of stars on the HB. Consequently, one cannot safely infer the total number of halo stars at some location from the local density of RR Lyrae stars. Fortunately, from the ratio of the numbers of blue HB stars and RR Lyrae stars in a cluster one can predict with reasonable accuracy the numbers of red HB stars – see Lee, Demarque & Zinn (1994) and Chapter 6. Therefore, by combining a search for RR Lyrae stars with one for blue HB stars, one can estimate the total number of HB stars, and this, as we saw in §5.1, is tightly coupled to the density of all types of stars in an old population.

Several large-scale searches for blue HB stars have been conducted in recent years (Sommer-Larsen, Christiansen & Carter 1989, Preston, Shectman & Beers 1991, Arnold & Gilmore 1992, Kinman, Suntzeff & Kraft 1994). These studies identify blue HB stars from a combination of objective-prism images and the broad-band colors of large numbers of stars. Kinman, Suntzeff & Kraft (1994) conclude that at distances $z$ above the plane in excess of $5\,\mathrm{kpc}$ the space-density of HB stars falls off as $r^{-3.5}$. The density $\nu_0$ of HB stars in the solar neighborhood has been determined by Green & Morrison (1993) from $uvby\beta$ photometry followed by high-resolution spectroscopy – they find $\nu_0 \simeq 51 \pm 17\,\mathrm{kpc}^3$. This value is considerably larger than is expected by extrapolating the trend at large $z$ to the plane under the

## Box 10.4:   Velocity dispersions from line-of-sight velocities

Let the $k^{\text{th}}$ star in a survey lie in the direction $\hat{\mathbf{x}}^{(k)}$, and let the principal axes of the velocity ellipsoid at its location have directions $\mathbf{e}_1^{(k)}$, $\mathbf{e}_2^{(k)}$ and $\mathbf{e}_3^{(k)}$. The line-of-sight velocity of this star is

$$v_{\text{los}}^{(k)} = \hat{\mathbf{x}}^{(k)} \cdot \mathbf{v}^{(k)} = \hat{\mathbf{x}}^{(k)} \cdot (v_1^{(k)}\mathbf{e}_1^{(k)} + v_2^{(k)}\mathbf{e}_2^{(k)} + v_3^{(k)}\mathbf{e}_3^{(k)}).$$

Squaring this equation we find

$$v_{\text{los}}^2 = \sum_{i,j=1,}^{3} v_i^{(k)} v_j^{(k)} \hat{x}_i^{(k)} \hat{x}_j^{(k)}, \tag{1}$$

where $\hat{x}_j^{(k)} \equiv \hat{\mathbf{x}}^{(k)} \cdot \mathbf{e}_j^{(k)}$ etc. If we have many stars at a given location, we can average (1) over these stars. Then $v_i^{(k)} v_j^{(k)}$ averages to zero for $i \neq j$, since the components of velocity parallel to the velocity ellipsoid's principal axes are statistically independent, and we find that

$$\langle v_{\text{los}}^2 \rangle_{\hat{\mathbf{x}}} = \sum_i \sigma_i^2 \hat{x}_i^2.$$

Usually we have insufficient stars to average them point by point. In this case we multiply (1) by $(\hat{x}_1^{(k)})^2$ and average over the whole sample. We argue that the resulting average will be dominated by intrinsically non-negative terms. Hence

$$\langle v_{\text{los}}^2 \hat{x}_1^2 \rangle_{\text{sample}} \simeq \sigma_1^2 \hat{x}_1^4 + \sigma_2^2 \hat{x}_2^2 \hat{x}_1^2 + \sigma_3^2 \hat{x}_3^2 \hat{x}_1^2.$$

Two similar equations are obtained by multiplying (1) by $(\hat{x}_2^{(k)})^2$ and $(\hat{x}_3^{(k)})^2$ before averaging, and then the resulting set of three equations can be solved simultaneously for the velocity dispersions $\sigma_i$.

assumption that the system is spherical or mildly elliptical. Hence it is likely that a large contribution to the local space-density of HB stars comes from a highly flattened component, such as the thick disk. Kinman (1994) estimates that half to two-thirds of the Milky Way's $\sim 150\,000$ blue HB stars lie in this thick-disk component. Since the visual luminosity of globular clusters is related to the number of their HB stars by $L_V/N_{\text{HB}} \simeq 540\,L_\odot$ (Preston, Shectman & Beers 1991), $\sim 75\,000$ HB stars should correspond to a total halo luminosity $L_V \simeq 4 \times 10^7\,L_\odot$. If the mass-to-light ratio of the halo resembles that of a globular cluster, $\Upsilon_V \simeq 2.5\Upsilon_\odot$ (§6.1.11), the mass of the halo is $\sim 1.1 \times 10^8\,\mathcal{M}_\odot$.

Since the local density of halo stars is rather small, many studies of the

**Table 10.6** Characteristic parameters of the halo

| $\rho(r)$ | $L_V / L_\odot$ | $\langle v_\phi \rangle$ | $\sigma_R$ | $\sigma_\phi$ | $\sigma_z$ |
|---|---|---|---|---|---|
| | | | km s$^{-1}$ | | |
| $\propto r^{-3.5}$ ($r \lesssim 25\,\mathrm{kpc}$) | $4 \times 10^7$ | $-185$ | $135$ | $105$ | $90$ |

halo are based on relatively distant stars. Reliable proper motions will often not be available for such stars, so the structure of the halo's velocity ellipsoid must frequently be derived from radial velocities alone. Box 10.4 explains how this is done. Unfortunately, the technique described in Box 10.4 requires that one know a priori the orientation of the principal axes of the velocity ellipsoid at the location of each star. In practice, one usually simply assumes that the velocity ellipsoid is oriented parallel to the coordinate directions of either cylindrical or spherical polar coordinates, since these are the most extreme of all plausible geometries.

From the radial velocities of a kinematically unbiased sample of 174 red giants with [Fe/H] $\leq -1.5$, Carney & Latham (1986) concluded that $(\sigma_r, \sigma_\theta, \sigma_\phi) = (154 \pm 18, 102 \pm 27, 107 \pm 15)\,\mathrm{km\,s^{-1}}$, and $\langle v_\phi \rangle = -206 \pm 23\,\mathrm{km\,s^{-1}}$, where all velocities are with respect to the LSR. Norris (1986) finds $(\sigma_R, \sigma_\phi, \sigma_z) = (131 \pm 6, 106 \pm 6, 85 \pm 4)\,\mathrm{km\,s^{-1}}$, and $\langle v_\phi \rangle = -183 \pm 10\,\mathrm{km\,s^{-1}}$ for objects more metal-poor than [Fe/H] $= -1.2$, with no indication that any sub-population of such metal-poor objects (such as HB stars or globular clusters) has different kinematics from the halo as a whole. Norris finds that samples of stars more metal-rich than [Fe/H] $= -1.2$ have values of $\langle v_\phi \rangle$ that rise steadily with [Fe/H] to $\langle v_\phi \rangle \simeq -60\,\mathrm{km\,s^{-1}}$ at [Fe/H] $= -0.6$. The line-of-sight velocity dispersion of stars with $-0.9 <$ [Fe/H] $< -0.6$ is $\sigma_{\mathrm{los}} = 70 \pm 10\,\mathrm{km\,s^{-1}}$. Zinn (1985) derives similar numbers for the more metal-rich RR Lyrae stars: stars with $\Delta S \leq 2$ ([Fe/H] $\lesssim -0.8$; §5.1.10) yield $\langle v_\phi \rangle = -32 \pm 20\,\mathrm{km\,s^{-1}}$, $\sigma_{\mathrm{los}} = 64 \pm 16\,\mathrm{km\,s^{-1}}$. Layden et al. (1996) find $(\sigma_r, \sigma_\theta, \sigma_\phi) = (168 \pm 13, 102 \pm 8, 97 \pm 7)\,\mathrm{km\,s^{-1}}$, and $\langle v_\phi \rangle = -210 \pm 12\,\mathrm{km\,s^{-1}}$ for 162 RR Lyrae stars of mean metallicity [Fe/H] $= -1.61$ and $(\sigma_r, \sigma_\theta, \sigma_\phi) = (56 \pm 8, 51 \pm 8, 31 \pm 5)\,\mathrm{km\,s^{-1}}$, and $\langle v_\phi \rangle = -48 \pm 9\,\mathrm{km\,s^{-1}}$ for 51 RR Lyrae stars of mean metallicity [Fe/H] $= -0.76$. Table 10.6 summarizes the characteristic properties of the halo.

**Kinematically selected samples**     A widely employed strategy for identifying field halo stars is to select as candidates for spectroscopic investigation stars with large proper motions. The reasoning behind this method is that halo stars tend to have large heliocentric velocities because they are on orbits that differ greatly from the Sun's. Consequently, they have large proper motions out to a greater distance than do disk stars. This strategy is especially valuable for identifying halo stars that are still on the MS – the subdwarfs – because, unlike HB stars, such stars are too faint to be seen to great distances. Hence we are obliged to study them when they pass close to the Sun, which is precisely when they have large proper motions – for a recent

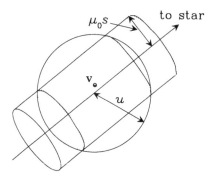

**Figure 10.33** The velocity-space geometry of selecting stars with large proper motions.

determination of the solar-neighborhood density of subdwarfs, see Dahn *et al.* (1995). Unfortunately, great care has to be taken in the interpretation of kinematic data obtained from proper-motion samples. An example will best explain the problem.

Let us imagine that we have a population of stars that all have the same absolute magnitude $M$. From this population we select all stars brighter than some apparent magnitude $m_0$ with proper motions larger than $\mu_0$. At each $m < m_0$ candidate stars lie on the spherical shell S that is centered on the Sun and whose radius $s$ satisfies equation (2.36). Let $\mathbf{u}$ be the heliocentric velocity of a star and $\mathbf{u}_t$ be the component of this velocity perpendicular to the direction $\hat{\mathbf{x}}$ to the star. Then from the stars on S we select those with $|\mathbf{u}_t| > \mu_0 d$. It is helpful to interpret this constraint in terms of the geometry of velocity space: we select only stars that lie outside a cylinder of radius $\mu_0 s$ whose axis is parallel to $\hat{\mathbf{x}}$ and passes through the Sun's velocity $\mathbf{v}_\odot$ – see Figure 10.33. Now let us assume that $s \ll R_0$ so that we may assume that stars are uniformly distributed over S. Then we may calculate the probability of selecting a star at a given point in velocity space, irrespective of its direction $\hat{\mathbf{x}}$, by averaging over directions $\hat{\mathbf{x}}$ the function which is unity outside the cylinder and zero inside it. The result of this averaging is constant on spherical shells in velocity space, and is therefore a function $P_s(u)$ of $u = |\mathbf{u}|$. It is not hard to see that $P_s(u)$ is equal to the fraction of a spherical shell of radius $u$ that lies outside the cylinder of radius $\mu_0 s$. A simple integral shows that this fraction is

$$P_s(u) = \begin{cases} \sqrt{1 - \dfrac{\mu_0^2 s^2}{u^2}} & \text{for } u > \mu_0 s \\ 0 & \text{otherwise.} \end{cases} \qquad (10.66)$$

Thus, our probability of selecting a star is zero if its heliocentric speed $u$ is smaller that $\mu_0 s$ and rises gradually to unity as $u$ becomes large compared to $\mu_0 s$.

If the space-density of stars of the type we are studying is $n_0$, and of these stars a fraction $f(u)\mathrm{d}u$ have heliocentric speeds in the range $(u+\mathrm{d}u, u)$,

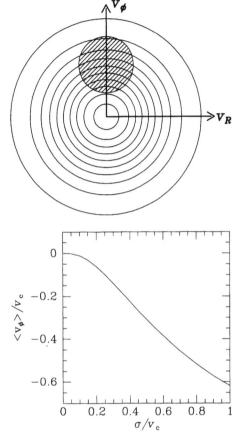

**Figure 10.34** Contours of equal stellar density in velocity space are assumed to center on the velocity of the Galactic center. The Sun's velocity is located at the center of the shaded region. Stars with velocities within that region are excluded from a proper-motion selected sample. The mean rotation velocity of stars outside the shaded region is negative.

**Figure 10.35** Stars selected from proper-motion surveys give biased kinematic results. Here we plot the mean azimuthal velocity of a sample selected by proper motion from a Maxwellian population of dispersion $\sigma$. The population-mean of zero is attained only when the Sun's rotation velocity $v_c$ is much larger than $\sigma$.

then there are $4\pi n_0 s^2 \mathrm{d}s\, f(u)\, \mathrm{d}u$ stars in this speed-range in the shell of radius $s$ around the Sun. Multiplying by $P_s(u)$, dividing by $\mathrm{d}u$ and integrating over $s$, we obtain the number of stars we expect to pick up in our proper-motion survey:

$$\frac{\mathrm{d}N_{\mathrm{cat}}}{\mathrm{d}u} = 4\pi n_0 f(u) \int_0^{s_{\max}} \mathrm{d}s\, s^2 P_s(u), \tag{10.67}$$

where $s_{\max} = 10^{1+0.2(m_0-M)}$ pc is the maximum distance at which stars are bright enough to enter the catalog. For simplicity we now assume that our magnitude limit $m_0$ is faint so that $P_{s_{\max}}(u) = 0$ for even the fastest stars – in practice this assumption is realistic. Then the upper limit of the integral in (10.67) can be set to $u/\mu_0$, and it is easy to see that the integral evaluates to $\frac{1}{3}u^3/\mu_0^3$. We then have

$$\frac{\mathrm{d}N_{\mathrm{cat}}}{\mathrm{d}u} = \frac{4\pi n_0}{3\mu_0^3} u^3 f(u). \tag{10.68}$$

Now consider the application of this formula in a simple case. Let the stars have a Maxwellian distribution of velocities in the rest frame of the Galactic center – in particular we are imagining that the population has no net rotation around the Galaxy – and let us neglect the solar motion with respect to the LSR. With these assumptions, we calculate the mean value $\langle v_\phi \rangle$ of the azimuthal velocity $v_\phi$ for stars in our sample. Figure 10.34 makes it intuitively clear that the result of this calculation is going to be negative, because the mean of $v_\phi$ over the entire population vanishes, and we will be calculating the mean of a sample from which stars with positive values of $v_\phi$ have been preferentially excluded. Quantitatively, $f(u)$ is the integral over a spherical shell of radius $u$ of the probability density $p(\mathbf{v})$ of finding a star at Galactocentric velocity $\mathbf{v}$. Thus

$$f(u) = u^2 \int d^2\Omega \, \frac{e^{-v^2/2\sigma^2}}{(2\pi\sigma^2)^{3/2}} \tag{10.69}$$

Substituting this function into equation (10.68) and bearing in mind that $v_\phi$ is a function $v_\phi(u, \Omega)$ of both $u$ and position $\Omega$ on the sphere of radius $u$ about the Sun's velocity (see Figure 10.34), we conclude that the required mean is

$$\begin{aligned} \langle v_\phi \rangle &= \frac{\int du\, v_\phi (dN_{\text{cat}}/du)}{\int du\, (dN_{\text{cat}}/du)} \\ &= \frac{\int du\, v_\phi u^5 \int d^2\Omega\, e^{-v^2/2\sigma^2}}{\int du\, u^5 \int d^2\Omega\, e^{-v^2/2\sigma^2}} \\ &= \frac{\int d^3\mathbf{v}\, v_\phi u^3 e^{-v^2/2\sigma^2}}{\int d^3\mathbf{v}\, u^3 e^{-v^2/2\sigma^2}}, \end{aligned} \tag{10.70}$$

where the third line follows from the second because $d^3\mathbf{v} = u^2 du d^2\Omega$. Figure 10.35 plots $\langle v_\phi \rangle$ as a function of $v_c/\sigma$ (see Problem 10.14). From this figure and the fact that $\sigma/v_c \simeq 0.5$ for a typical halo population, it follows that the rotation velocity of a proper-motion selected sample is liable to be in error by $\sim 0.31 v_c \simeq 60\,\text{km s}^{-1}$. Thus, raw means and dispersions of samples selected from proper-motion catalogs are in themselves meaningless; it is essential to correct such quantities for biases introduced by the selection criteria (Ryan & Norris 1993).

Since the selection criteria have blanked off a part of velocity space, and the desired means and dispersions are over all velocity space, the necessary corrections can, in principle, not be determined from the data. In practice, one uses a reasonable model to interpolate from the studied portions of velocity space into the portion that was blanked off by the selection criteria.

Samples of nearby high-velocity stars are particularly suited to the investigation of the metallicities of high-velocity halo stars. Figure 10.36 shows plots of $v_\phi$ versus [Fe/H] for the samples of Nissen & Schuster (1989) and Carney et al. (1996). At the top right corner of each panel there is a dense

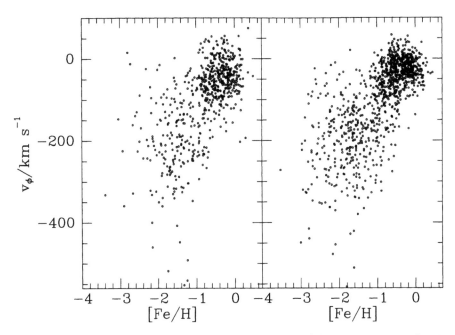

**Figure 10.36** [Fe/H] versus $v_\phi$ from Nissen & Schuster (1989) (left panel, 611 stars) and Carney *et al.* (1996) (right panel 1022 stars). [From data kindly supplied by B. Carney and P. Nissen]

clump of points that is centered on $([\text{Fe/H}], \langle v_\phi \rangle) \simeq (-0.4, -40\,\text{km}\,\text{s}^{-1})$. Stars whose points fall in this clump are considered to be disk stars. In both plots this clump is bounded at lower left by a lane of reduced stellar density that slopes downwards from left to right and passes through the points $([\text{Fe/H}], \langle v_\phi \rangle) \simeq (-0.3, -225\,\text{km}\,\text{s}^{-1})$ and $(-1.5, -50\,\text{km}\,\text{s}^{-1})$ – it is significant that this lane is apparent in both plots because the two samples plotted in Figure 10.36 have only 130 stars in common. Stars that lie below and to the left of the lane are considered to be halo stars. Since the lane does not run vertically, the halo and the disk overlap in metallicity, roughly between $[\text{Fe/H}] = -0.9$ and $[\text{Fe/H}] = -1.3$.

The full curve in Figure 10.37 shows the metallicity distribution of 372 kinematically-selected halo MS stars. The distribution extends to lower values of [Fe/H] than does the distribution of globular clusters plotted in Figure 10.32. The dashed curve in Figure 10.37 shows that, on account of this fact, the leaky-box model of §5.3.2 can fit the data better than it can the metallicity distribution of the halo globular clusters. The prediction of the leaky-box model lies below the observed metallicity distribution at $[\text{Fe/H}] \gtrsim -1$. This discrepancy indicates that the formation of the thick disk involves processes that lie beyond the scope of the leaky-box model. Presumably, the missing ingredients are the accumulation within the disk of

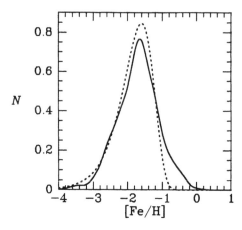

**Figure 10.37** Full curve: a normalized, generalized histogram of the metallicity distribution of 372 kinematically-selected halo MS stars. Dashed curve: the distribution predicted by the leaky-box model with $y_{eff} = 0.025$. [After Ryan & Norris (1991) from data kindly supplied by S. Ryan]

metal-rich debris from the formation of the halo, and the tendency for the disk to form a closed rather than a leaky box.

Since halo stars are much rarer in the solar neighborhood than disk stars, the detailed pattern of abundances in halo stars is harder to study than it is in disk stars. Notwithstanding the difficulties, Gilroy et al. (1988) have obtained high-resolution spectra of 20 halo stars and studied their elemental abundances in detail. They conclude that there are real variations from star to star in the relative abundances of the elements. In the most metal-weak stars, r-process elements are relatively more abundant than s-process elements. This result suggests that very massive and short-lived stars were relatively important contributors of heavy elements to the most metal-poor stars.

## 10.6 Galaxy models

Now that we have completed our survey of the structure of the Milky Way, we are ready to discuss models of the Galaxy. Such models play an important rôle in astronomical research by facilitating the interpretation of observational data. For example, one might use a model to estimate the location in the Galaxy of an HII region from the wavelength of it emission lines, or one might use a model to estimate how many stars of a given color and apparent magnitude one would detect in a field at some given values of $l$ and $b$.

The construction of Galaxy models is also an important discipline in itself because it obliges one to assemble a coherent set of values of the various parameters that describe the Galaxy – the sizes, masses, shapes, and so forth, of each of its components. Many of these numbers must be inferred by rather indirect arguments, and there are frequently several different ways of estimating a given number. Building a model Galaxy forces one to decide on

**Table 10.7**  Some local parameters of the Milky Way

| $(\mathrm{km\,s^{-1}})$ | $(\mathrm{km\,s^{-1}\,kpc^{-1}})$ | | $(\mathcal{M}_\odot\,\mathrm{pc^{-2}})$ | | |
|---|---|---|---|---|---|
| $v_c(R_0)$ | $A$ | $B$ | $\Sigma_{\mathrm{tot}}(R_0)$ | $\Sigma_{\mathrm{gas}}(R_0)$ | $\Upsilon_B(R_0)$ |
| 220 | 14.8 | −12.4 | 48 | 5.5 | 2.3 |

**Table 10.8**  Location of Milky-Way parameters

| Type of parameter | Location |
|---|---|
| Distance to the Galactic center, $R_0$ | Table 7.1 |
| Photometric parameters, $R_d$ etc. | Table 10.1 |
| Disk velocity dispersions, $\sigma_R$ etc. | Table 10.4 |
| Tangent velocity as a function of $l$ | Table 9.3 |
| Halo velocity dispersions, $\sigma_R$ etc. | Table 10.6 |

a single value for each number, and to check that any relations that should hold between these numbers are, in fact, valid.

Table 10.7 gives values for six of the Galactic parameters that relate to the solar neighborhood, while Table 10.8 indicates where the values of many other parameters are tabulated elsewhere in this book. Methods for determining all but one of these parameters have already been discussed elsewhere, and here we need only discuss $v_c(R_0)$, the local circular speed.

**The local circular speed**  The obvious strategy for determining $v_c(R_0)$ is to measure the Oort constants and then to note from equations (10.23) and (10.29) that $v_c(R_0) = R_0(A - B)$ [cf. equation (10.31)]. Using the values of $A$ and $B$ determined by Feast & Whitelock (1997) from Hipparcos proper motions [equation (10.36)], we find

$$v_c(R_0) = (218 \pm 8)(R_0/8\,\mathrm{kpc}). \tag{10.71}$$

The Oort constants describe the streaming of stars in a small patch of the disk around the Sun, so it is surprising that we can determine from them the rate at which the entire solar circle rotates. What allows us to make this leap from the small- to the large-scale is our *hypothesis* that stars stream precisely on circles. In §10.3.3 we saw that this hypothesis is supported observationally by the fact that the quantities $c$ and $k$ defined by equation (10.18) are zero to within the errors. Nonetheless, we know that the Galaxy is not exactly axisymmetric, if only because it has a bar at the center and spiral structure further out. So we should feel uncomfortable about estimating a parameter as fundamental as $v_c(R_0)$ from equation (10.31) alone.

The measurement of the proper motion of Sgr A* mentioned at the end of §10.3.3 provides a direct route to $v_c(R_0)$. From equation (10.38) we have

$$v_c(R_0) = (241 \pm 7)(R_0/8\,\mathrm{kpc}). \tag{10.72}$$

In connection with this estimate, it should be stressed that the quoted error is a formal one: it takes no account of the possibility that Sgr A$^*$ has a non-zero peculiar velocity.

Prior to the advent of the Hipparcos catalog, $B$ was so uncertain as to make equation (10.31) of little use in the determination of $v_c(R_0)$. Moreover, it is only recently that accurate VLBI measurements of the proper motion of Sgr A$^*$ have been available. Consequently, astronomers have traditionally had to estimate $v_c(R_0)$ from a variety of more-or-less unsatisfactory arguments. These include: an argument based on the assumption that the system of halo globular clusters and RR-Lyraes has no net rotation; an argument based on the speeds of the fastest stars seen near the Sun (Problem 10.16); an argument that requires an estimate of the systemic velocity of the Milky Way with respect to the rest of the Local Group; and the use of the $AR_0$ formula (Box 10.3). Since none of these methods is competitive with the two methods described above (both of which ultimately exploit radio-frequency VLBI), we shall say no more about them here.

**Mass models**     Several distinct types of model can be identified. Schmidt (1956) and Caldwell & Ostriker (1981) pioneered the most basic type of model, **mass models**. A mass model consists of specified density distributions for one or more components of the Galaxy. The combined gravitational force-field of these components is calculated and the parameters of the density distributions are adjusted to optimize the agreement between the kinematic predictions of the model and observables such as the values of the Oort constants, the run of terminal velocities with radius, the total surface density of matter near the Sun, $\Sigma_{tot}(R_0)$, and so on.

In the mass models of Dehnen & Binney (1998b), the Galaxy is represented by a bulge, a halo and three disks: thin and thick stellar disks and the disk of the ISM. The density distributions of the two stellar disks are given by the first of equations (10.3) with $R_d$ the same for both disks and $z_0 = 180\,\mathrm{pc}$ and $z_0 = 1\,\mathrm{kpc}$ for the thin and thick disks, respectively. The gas disk is assumed to contribute 25% of the local surface density and to have a value of $R_d$ twice as great as that for the stellar disk, while its scale height is small: $z_0 = 40\,\mathrm{pc}$. The gas disk is assumed to have a hole at its center. The density distribution of the bulge is assumed to be given by the second of equations $r_{max} = 1.9\,\mathrm{kpc}$ and $c_b = 0$.

The density distribution of the halo is taken to be of the form

$$\rho(R, z) = \rho_0 \left(\frac{m}{r_0}\right)^{-\gamma} \left(1 + \frac{m}{r_0}\right)^{\gamma-\beta} \mathrm{e}^{-m^2/r_t^2}, \text{ where } m \equiv (R^2 + z^2/q^2)^{1/2}.$$

(10.73)

Hence, at small radii the density is proportional to $r^{-\gamma}$ for $r \ll r_0$ and proportional to $r^{-\beta}$ for $r \gg r_0$. Results are given for two values of the axis ratio $q$: 0.8 and 0.3. The exponents are constrained to satisfy $-2 \le \gamma \le 1.8$ and $\beta \ge 1$ to limit the sharpness of the inner and outer edges of the halo, respectively.

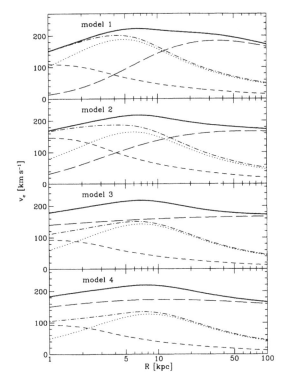

**Figure 10.38** Circular-speed curves of four mass models. In each panel the full curve shows $v_c(R)$ for a complete model, while the other curves show contributions: short-dashed, bulge; dotted, disks; dot-dashed, disks plus bulge, long dashed, halo. All models assume $R_0 = 8\,\text{kpc}$ and halo axis-ratio $q = 0.8$. From top to bottom the values of $R_d/R_0$ are 0.25, 0.3, 0.35 and 0.4. [After Dehnen & Binney (1998b)]

The free parameters of the model (the values of the scale lengths and surface densities of the disks, the normalization of the bulge, and the normalization, scale length and exponents of the halo) are constrained by the following observational data:

1. the observed terminal velocities (Figure 9.16);
2. the run of $W(R)$ at $R > R_0$ (Figure 9.18);
3. the values of the Oort constants [equation (10.36)];
4. the total surface density $\Sigma_{1.1}(R_0)$ [equation (10.54)];
5. the contribution of the disks to $\Sigma_{1.1}$, which was required to exceed $40\,\mathcal{M}_\odot\,\text{pc}^{-2}$;
6. the velocity dispersion of bulge stars in Baade's window (§10.2.4);
7. the total mass inside $100\,\text{kpc}$, which was taken to be $(7\pm2.5)\times10^{11}\,\mathcal{M}_\odot$.

Figure 10.38 illustrates the fact that very different mass models prove compatible with these constraints. Each panel shows the circular-speed of a model (full curve) and the contributions to it from the bulge (short-dashed curve), the disks (dotted curve) and the halo (long-dashed curve). The top model has a small disk scale length ($R_d = 0.2\,\text{kpc}$) and its circular speed is dominated by the disks and bulge out to $R \gtrsim 10\,\text{kpc}$. (The dot-dashed curves show the contribution to $v_c$ from bulge and disks together.) By contrast, the bottom model has $R_d = 3.2\,\text{kpc}$ and a circular-speed curve that is dominated by the halo at all radii. In fact, as one proceeds from top to bottom in Figure 10.38, $R_d$ is steadily increasing, and this causes mass to move from the disks, first to the bulge and later to the halo. What drives this transfer of mass is the fact that the shape of the terminal-velocity curve in Figure 9.16 is appropriate for disks with $R_d/R_0 \simeq 0.25$ rather than any larger value. Since the terminal velocities have been measured with high precision, they carry great weight in the determination of the model's parameters, with the result that a component that can fit all terminal velocities at once receives plenty of mass. If one specifies a value of $R_d/R_0$ that does not suit the terminal velocities, the fitting procedure brings the bulge and the halo into play so that it can make up the tangent velocities from a suitable combination of all three components.

Notice that there are significant differences between the four circular-speed curves in Figure 10.38 at $R > R_0$. These differences reflect the fact that $v_c$ has not been accurately measured at $R > R_0$ – in fact, from Figure 9.18 it is clear that the data are essentially non-existent for $R > 2R_0$.

By showing that the inner Galaxy can be dominated by either the disks or a nearly spherical halo, Figure 10.38 indicates that we know next to nothing about the Galaxy's vertical distribution of mass. This ignorance arises because the only constraints that bear on this distribution are items (4) and (5) above, and these constraints apply too close to the plane to provide much information about what lies kiloparsecs above the plane. In the coming years, this crucial lack of information may be filled by dynamical models of the Galaxy (see below), and studies of gravitational lensing by disk galaxies.

**Star-count models**    In §10.1 we discussed models in which each component of the Galaxy consists of a smooth distribution of luminosity. Models of this type are suitable for analyzing data from satellites that use small telescopes to measure the Galaxy's surface-brightness distribution with low angular resolution, but are of limited value for studies that employ large-aperture telescopes. In connection with such studies, one needs a model that predicts star counts, such as those described in §10.4.3. A model with this capability specifies not merely how much luminosity is produced at each point in the Galaxy, but the precise distribution over absolute magnitude of the stars that produce the light. That is, a star-count model specifies for all $M$ and $\mathbf{x}$ the luminosity function, $\Phi(M, \mathbf{x})$, defined by equation (3.20).

Star-count models, which are reviewed by Bahcall (1986), were pioneered by Bahcall & Soneira (1980). They specified $\Phi(M, \mathbf{x})$ by decomposing the Galaxy into two components and assigning a position-independent luminosity function $\Phi(M)$ to each component. Since the proportion of the total luminosity at $\mathbf{x}$ that comes from each component varies with $\mathbf{x}$, this procedure makes the overall luminosity function position-dependent. Bahcall & Soneira (1984) concluded that star-count data for five widely dispersed fields in several wavebands are consistent with standard disk and globular-cluster luminosity functions provided:

1. the luminosity density of the disk is given by the first of equations (10.3) with $R_d \geq 2.5\,\mathrm{kpc}$, and $z_0 = 350 \pm 50\,\mathrm{pc}$ for MS stars fainter than $M_V = 5$, and $z_0 = 250 \pm 100\,\mathrm{pc}$ for more luminous stars;
2. the luminosity density of the halo obeys the $R^{1/4}$ law [equation (4.18)] with effective radius $R_e \simeq \frac{1}{3}R_0$, axis ratio $q = 0.8^{+0.2}_{-0.05}$, and normalized such that near the Sun one star in $\sim 500$ is a halo star.

Bahcall & Soneira (1984) and Bahcall (1986) argue that star counts in different colors are compatible with the existence of a thick disk only if the luminosity function of this component is similar to that of the halo.

**Kinematic models**    Many studies involve either proper motions or radial velocities in addition to positions and apparent magnitudes. Consequently, there is a need for models that predict the velocity distribution of stars at each point in the Galaxy, in addition to the number density of stars as a function of absolute magnitude.

The simplest models with this capability assume that at each point $\mathbf{x}$ the velocity distribution of any given component can be represented by the Schwarzschild distribution [equation (10.17)]. In general, this Schwarzschild distribution must be assumed to be about a mean-streaming velocity $\bar{\mathbf{v}}$ that depends on $\mathbf{x}$, and the principal velocity dispersions $\sigma_i$ will also depend on $\mathbf{x}$. We call such a model a **kinematic model** to distinguish it from models that explicitly take into account dynamical constraints – see below. Ratnatunga, Bahcall & Casertano (1989) describe a kinematic model and use it to predict for several directions the distribution of proper motions of stars listed in the Bright Star Catalog (Appendix B).

**Dynamical models**    An accurate kinematic model is exceedingly hard to construct because $\bar{\mathbf{v}}$, the $\sigma_i$ and the orientation of the axes of the velocity ellipsoid all have to be specified as functions of $\mathbf{x}$. Moreover, Schwarzschild's ellipsoidal hypothesis is only an approximation to the truth, so after an immense amount of work specifying the model, one is liable to be left with a inadequate representation of the Galaxy.

If we assume that the Galaxy is in an approximate steady-state, the value of $\mathbf{v}$ at one point, $\mathbf{x}$, cannot be varied independently of the value of $\mathbf{v}$ at neighboring points, $\mathbf{x}'$, because stars that are at $\mathbf{x}$ at one instant are at $\mathbf{x}'$ somewhat later. This argument suggests that, in a steady-state Galaxy, $\mathbf{v}$ and the $\sigma_i$ satisfy differential equations. §4.2 of BT describes the **Jeans equations** for $\mathbf{v}$ and $\sigma_i$ that would hold in a satisfactory model. If there were a clean procedure for finding solutions of the Jeans equations, this procedure would eliminate most of the freedom that we have when specifying a kinematic model. We might then, at acceptable cost, be able to find a model that provided a reasonable representation of the Galaxy. Unfortunately, the Jeans equations do not pose a tractable boundary-value problem, so they have proved of limited value in the construction of Galaxy models.

A more promising, but still undemonstrated, way to incorporate dynamical constraints in Galaxy models is based on the 'Jeans theorem' that

is discussed in §4-4 of BT. Dehnen & Binney (1996) describe this approach
and present some preliminary results. An alternative, conceptually simpler,
approach is to construct an $N$-body model that looks like the Galaxy – see
Sellwood (1993) or Fux (1998) for promising examples of this procedure.
The central difficulty with this approach is the problem of choosing initial
conditions from which the $N$-body model will relax to a configuration like
that of the Milky Way.

# 10.7 Formation and evolution of the Milky Way

What can we infer from the observations of disk, halo and bulge that we have
studied in this chapter about the history of the Milky Way? Astronomers
have now debated this question for several decades without reaching a consensus. In view of this unsatisfactory situation, we cannot present an authoritative discussion of this fascinating topic. We start by presenting some
of the more frequently cited lines of argument. We then outline what seems
to us the most likely route by which the Milky Way arrived at its present
state, and discuss some respects in which this picture pulls together disparate
lines of evidence from observations of both the Milky Way and of external
galaxies.

### 10.7.1 Formation scenarios

Discussion of the formation of the Milky Way has been profoundly influenced
by a paper by Eggen, Lynden-Bell & Sandage (1962). By studying model
orbits of solar-neighborhood stars for which space velocities had been determined, these authors reformulated the classical conclusion of Roman (1955)
that high-velocity stars near the Sun tend to be metal-poor.[16] They showed
that the eccentricities, $e$, energies of vertical oscillation, $E_z$, and angular
momenta, $L_z$, of their model orbits are correlated with stellar metallicity in
the sense that as metallicity decreases from solar, $e$ and $E_z$ increase, while
$L_z$ decreases.[17] From these correlations Eggen *et al.* inferred that either the
Milky Way had a violent history, or most low-metallicity stars cannot have
formed in a centrifugally supported disk. The argument from which they
drew this conclusion is as follows.

Both $e$ and $L_z$ are 'adiabatic invariants' in the sense that their values for
a given star change negligibly if the potential in which the star moves changes
slowly – see §3.6 of BT for more detail. Hence, either low-metallicity stars
formed on orbits that have large values of $e$, or the potential of the Milky

---

[16] The metal-poverty of a star was inferred from its ultra-violet excess – see §5.1.5.

[17] Since the possession of a large velocity with respect to the Sun is synonymous for a
solar-neighborhood star with moving on either an eccentric or a highly inclined orbit, the
correlations of Eggen *et al.* did not represent an important advance on the work of Roman
(1955).

Way has changed rapidly at some time since they formed. In the former case these stars cannot have formed in a centrifugally supported disk, and in the latter case the Milky Way has had a violent history.

On the strength of these observations, Eggen *et al.* presented a picture of the formation of the Galaxy. In this picture, the Milky Way formed from an approximately spherical, spinning protogalactic cloud. Initially this cloud was metal-poor and in near free-fall. As it collapsed, its rate of spin increased to conserve its angular momentum, and from it condensed the most metal-poor stars and halo globular clusters. The present eccentric orbits of these objects were a direct consequence of the free-fall of the protogalactic cloud. The approximately coeval nature of globular clusters (§6.1.4) was believed to be a consequence of the rapidity of this collapse. Supernovae increased the metallicity of the cloud as it collapsed. After shrinking in radius by a factor of order 10, the cloud became metal-rich and flattened into a centrifugally supported disk. At this point disk formation commenced and the Milky Way settled down to something like its present configuration.

A very different picture of the formation of the Milky Way's halo was suggested by Searle (1977). This picture was inspired by the discovery that the distribution of metallicities among halo globular clusters is wide and approximately independent of Galactocentric radius (see §10.5.1). This phenomenon would be expected if the halo formed from a large number of independent fragments of mass $\sim 10^8 \, \mathcal{M}_\odot$, which evolved independently of one another. The metallicity of a given fragment would be determined by the number of 'enrichment events' that took place within it before supernovae swept it clean of gas. Searle envisaged an enrichment event to consist of the formation of an entire globular cluster. This assumption ensured that each fragment experienced only a handful of enrichment events, and Poisson statistics in the number of enrichment events that occurred in different fragments gave rise to significant scatter in the fragments' final metallicities.

Since the papers of Eggen, Lynden-Bell & Sandage (1962) and Searle (1977), $N$-body simulations of gravitational clustering in an expanding Universe have heavily influenced the conventional picture of galaxy formation. In these simulations, stellar systems are constantly merging to form larger ones. It is very tempting to identify Searle's fragments with the stellar systems that early on merged to form the Milky Way. In simulations, as fragments merge, the potential of the forming Galaxy is very far from axisymmetric and the angular momenta of individual stars are by no means conserved, contrary to the assumption of Eggen *et al.*. Hence, in its original form, the picture of Eggen *et al.* is largely obsolete.

It would, however, be a mistake to conclude that galaxies such as the Milky Way formed simply by the merging of pre-existing stellar systems. In particular, whatever may be the history of the halo, the disk certainly did not form in this way. Nor is it likely that stars of the thick disk and disk globular clusters formed outside the Galaxy as the simplest merger picture envisages. The origin of the bulge is very uncertain, but it is likely that the material of which it is made entered the Galaxy as gas rather than as stars. Thus in as much as Eggen *et al.* emphasized the formation of stars from gas that had already been incorporated into the Galaxy, their picture describes an important aspect of what actually happened.

Objects that look remarkably like young globular clusters have formed recently in the LMC. The LMC is known to be spiraling into the Milky Way, and will eventually be absorbed by our Galaxy (see §7.1 of BT). Once the LMC has been absorbed, its globular clusters will become Milky-Way globular clusters. Could all the Milky Way's globular clusters have been acquired by the shredding of low-luminosity galaxies such as the LMC? We do not know the answer to this question, but there is some intriguing evidence to suggest that both globular clusters and field halo stars have been accreted (§6.1.4).

In §5.3.2 we saw that the temperature of interstellar gas is unlikely to lie in the interval $10^4 \, \mathrm{K} < T < \mathrm{few} \times 10^6 \, \mathrm{K}$. Whatever its origin, gas that cools to $\lesssim 10^4 \, \mathrm{K}$ will sink inwards in the potential of a galaxy such as the Milky Way until it forms a centrifugally supported disk – encounters between gas clouds are inelastic so clouds must sink inwards until encounters become unimportant, which is possible only if all clouds are on similar circular orbits.

As gas accumulates in the disk, the latter is liable to become gravitationally unstable and to form stars. It is natural to expect stars to form from the densest gas clouds, and these clouds move on the most nearly circular orbits (§9.2.6). Hence we expect, and observations confirm (§10.4), that stars form in the plane on nearly circular orbits. Now it is certain that stars will in time diffuse away from the thin layer in which they form, because a thin stellar disk is a fragile object. Indeed, in such a disk all orbits must be highly circular and confined to a common plane. Any random process that is capable of perturbing the orbits of stars is liable heat and thicken the disk by disturbing this highly ordered state. §7.5 of BT describes two such randomizing processes, namely scattering by molecular clouds and by spiral arms. Figure 10.22 presents observational evidence that these processes are at work. The best-available estimates (Jenkins 1992) suggest, however, that they are not powerful enough to effect the formation of the thick disk.

A widely canvassed origin of the thick disk appeals to the accretion of low-luminosity stellar systems by galaxies such as the Milky Way. As a satellite galaxy spirals through the outer reaches of the Milky Way, its orbit is dragged into the plane (Quinn & Goodman 1986). Then as it spirals through the disk it can, in principle, convert a thin disk into a thick disk (Walker, Mihos & Hernquist 1996).[18] Thus the following may be the origin of the thick disk. About 10 Gyr ago most of the stars of the thick disk formed

---

[18] In this connection it is worth remarking that the fragility of a thin disk is such that it can be appreciably thickened by even a low-mass satellite galaxy (Toth & Ostriker 1992). In particular, the satellite's stars could now be inconspicuous amid the more numerous stars of the thickened disk.

a thin disk. Then one or more dwarf galaxies were accreted by the Milky Way, and, as they spiraled through the disk, they thickened it. Subsequently a new thin disk formed within the thick disk from gas that had settled into the plane.

We have seen that there is an intriguing similarity between the physical parameters of the thick disk and the disk globular-cluster system. Does this similarity reflect a similar physical origin? We do not know the answer to this question, but there is some evidence that globular clusters form when a gaseous disk is strongly disturbed (Zepf & Ashman 1993). Thus, it is possible that the disk globular clusters were formed from the Galaxy's gaseous disk as the satellite(s) spiraled in, churning up the disk. This conjecture explains naturally why the metallicities of the disk globular clusters are narrowly confined to the value that is characteristic of the thick disk ($[Fe/H] = -0.6$; see Figure 10.32). Moreover, so long as the mass of the perturber is larger than the masses of an individual globular cluster, this hypothesis also explains why the disk globular clusters are distributed through the same volume as the thick disk stars: in these circumstances objects less massive than the perturber are scattered independently of their mass.

Finally, what is the origin of the bulge? The first point to make is that both the thick disk and the innermost halo are likely to make significant contributions to the luminosity that comes from the volume that we conventionally associate with the bulge. This fact is bound to complicate any discussion of the bulge. However, from observations in Baade's Window (§10.2) it is clear that the bulge is more than a superposition of the thick disk and the halo, because it contains mostly stars that are more metal-rich than even the thick disk. It seems likely that much of the heavy-element content of the bulge was synthesized in stars that do not belong to the bulge, but to the halo or the disks. Metal-rich gas ejected by these stars would naturally have fallen onto the bulge and, at some stage, bulge stars formed that incorporated some of this accreted gas.

The discovery that the bulge is non-axisymmetric must have a significant bearing on its origin. One relevant dynamical fact is that a disk, in which the radial velocity dispersion becomes large, is liable to buckle and thicken (Combes & Sanders 1981, Raha *et al.* 1991, Kalnajs 1996). This process might conceivably have formed the bulge from the center of the thin disk, and indeed there is some evidence that this kind of evolution is taking place in some external galaxies (§4.4.7). One possible cause of the large velocity dispersion required to buckle a disk is a bar that has formed in the disk and then ramped up the eccentricities of disk stars' orbits. However, at the present time too little is understood about the structure of the bulge and the dynamics of bars to make further speculation on the origin of the bulge profitable.

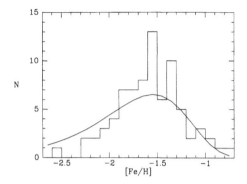

**Figure 10.39** The metallicity distribution of the more metal-poor globular clusters shown in Figure 10.32 (histogram) is reasonably well fitted by the prediction [equation (5.43)] of the leaky-box model with effective yield $p_{\mathrm{eff}} \equiv p/(1+c) = 5.7 \times 10^{-4}$ (curve).

## 10.7.2 Models of the chemical evolution of the Milky Way

A measure of support for the theoretical framework outlined above can be deduced from the theory of chemical enrichment developed in §5.2. we start by using the leaky-box model of §5.3.2 to model the chemical evolution of the halo.

**Chemical evolution of the halo**     Figure 10.39 compares the observed metallicity distribution of the halo globular clusters with that predicted by the leaky-box model [equation (5.43)]. The fit is moderately good with effective yield $p_{\mathrm{eff}} \equiv p/(1+c) = 5.7 \times 10^{-4}$.

**Pre-enrichment**     One clear implication of the picture developed above is that the present thin disk formed from material that was relatively metal-rich: we might reasonably conjecture that the ISM from which it started to form was as metal-rich as the thick disk. To test this conjecture, let us investigate the consequences of assuming that, on the contrary, the thin disk formed from metal-poor gas.

At the end of §5.3.1 we showed that the closed-box model of chemical evolution predicts that one half of all stars near the Sun have less than a third of the metallicity of the most metal-rich stars. Since the latter have metallicities comparable to that of the Sun, the closed-box model implies that a half of solar-neighborhood stars should have $Z \lesssim \frac{1}{3}Z_\odot$. Actually only $\sim 2\%$ of disk F and G stars in the solar neighborhood have $Z < 0.25 Z_\odot$ (Sommer-Larsen 1991). This contradiction between the standard closed-box model and observation is known as the **G-dwarf problem**.[19] Evidently one of the assumptions of the standard closed-box model is incorrect.

Consider first the possibility that the yield $p$ is not independent of $Z$, and hence that equation (5.37) is not the correct integral of equation (5.36). One way in which this might happen is if the IMF were metallicity-dependent. For example, if no low-mass stars are born when $Z < 0.25 Z_\odot$, the G-dwarf

---

[19] Mould (1982) has argued that a similar result applies to M dwarfs, although the data for M stars are less reliable because the ultraviolet excess $\delta(U-B)$ (see Box 5.4) cannot be used as a metallicity indicator for these stars.

problem would be resolved within the frame-work of the closed-box model. One problem with this proposal is that it implies that the yield $p$ *decreases* with increasing $Z$, which is the reverse of the trend that is required if the closed-box model is to account satisfactorily for metallicity gradients within galactic disks (Peimbert & Serrano 1982). A second, and perhaps more serious, problem was pointed out by Thuan, Hart & Ostriker (1975), who examined a two-parameter family of closed-box models in which the IMF – and therefore the yield – as well as the overall rate of star formation are continuous functions of time. They found that such closed-box models can reproduce the observed scarcity of low-metallicity dwarf stars only if the IMF at early times is so strongly biased against the formation of low-mass stars that the rate of formation of these stars actually *increases* in time. Unfortunately, an increasing rate of formation of low-mass stars corresponds to an increasing rate of consumption of interstellar gas, so the closed-box models that are compatible with the stellar metallicity distribution predict that the interstellar gas will be exhausted within the next Gyr or so. It seems unlikely that we are privileged to live so close to the end of the epoch during which our galaxy forms new stars. Thus, the scarcity of metal-poor stars near the Sun obliges us to drop at least one of the other central assumptions of the standard closed-box model – either that the gas is initially metal-free or that gas neither enters nor leaves any particular annulus of the Galaxy.

In §5.3.3 we showed how the scarcity of metal-poor stars near the Sun can be naturally explained within the framework of the accreting-box model of chemical evolution. Here we show that it may be just as naturally explained by supposing that the thin disk inherits heavy elements from both the halo and the thick disk. As a crude short-hand, we refer to these two components as the 'spheroid'.

If the gas from which the Galactic disk formed started from metallicity $Z_i$, the integral of equation (5.36) becomes

$$Z(t) = Z_i + p \ln \left[ \frac{\mathcal{M}_g(0)}{\mathcal{M}_g(t)} \right], \qquad (10.74)$$

and there will be no disk stars with $Z < Z_i$. Furthermore, when we insert the observed value of $\mathcal{M}_g(0)/\mathcal{M}_g(t_0)$ and $Z_i = 0.25 Z_\odot$ into this equation, we deduce $p \simeq 0.007$, in better agreement than our previous solar-neighborhood estimate, $p \simeq 0.009$, with the value that is deduced from observations of Magellanic irregulars (§8.2.6). Therefore, we may retain many of the attractive features of the closed-box model while clearing up the G-dwarf problem, if we can explain why $Z_i \simeq 0.25 Z_\odot$ in the solar neighborhood, and $Z_i \simeq 0$ in Magellanic galaxies.

One difference between our galaxy and the Magellanic Clouds is that our galaxy has a significant spheroid and the Magellanic Clouds do not. In §5.4 we argued that the mass, $\mathcal{M}_h$ of metals produced by an old stellar population and the present bolometric luminosity $L_{bol}$ of the population are in the ratio $\mathcal{M}_h/L_{bol} \simeq (0.16 \pm 0.05)\,\mathcal{M}_\odot\,L_\odot^{-1}$ [equation (5.70)]. If we apply this relation to the Galactic bulge, and assume that the disk was endowed with a fraction $f$ of these metals at its birth, we conclude that the initial metallicity of the disk was

$$Z_i = \frac{f\mathcal{M}_h}{\mathcal{M}_d} \simeq \frac{0.16f}{\Upsilon_{d\,bol}}\frac{L_{bulge}}{L_{disk}}, \tag{10.75}$$

where $\Upsilon_{d\,bol}$ is the bolometric mass-to-light ratio of the disk and the ratio of the luminosities of bulge and disk is $L_{bulge}/L_{disk} \simeq 0.25$ (Table 10.1). Thus equation (10.75) implies

$$Z_i = 0.04\frac{f}{\Upsilon_{d\,bol}} = 2\frac{f}{\Upsilon_{d\,bol}}Z_\odot. \tag{10.76}$$

Equation (10.57) suggests that $\Upsilon_{d\,bol}$ lies in the range 1 to 2, so the G-dwarf problem would be resolved if a fraction $f = 0.2 - 0.3$ of the metals made in the bulge polluted the primordial disk.

When applied to external galaxies, this simple model of pre-enrichment of the thin disk by the spheroid has two interesting consequences. First, since spheroids are more centrally concentrated than disks, $Z_i$ should be higher near the center of a spiral galaxy than at very large radii. Hence we expect the present value of $Z$ to increase toward the centers of disks even more rapidly than equation (5.37) would suggest. That is, this model predicts that the effective yield $p_{eff}$ [equation (8.62)] should increase towards the centers of galaxies. In §8.2.6 we saw that this increase is exactly what is observed. We also know that most of the luminosity of a rich cluster of galaxies such as the Coma cluster comes from elliptical galaxies and the spheroids of S0 galaxies. One possible explanation of why these galaxies did not form disks is that encounters between neighboring galaxies tore the proto-disk material out of the galaxies before the gas could fragment into stars (Binney & Silk 1978). Our pre-enrichment picture suggests that this material should be metal-rich (De Young 1978; Binney 1980). X-ray observations show that this material has $Z \simeq 0.3Z_\odot$, as expected (e.g., Sarazin 1986).

## Problems

**10.1** If a spherical body has a density distribution $\rho(r) = K_0(r)$, where $K_0$ is a modified Bessel function, show that its projected density is $\Sigma(R) = \frac{\pi}{2}e^{-R}$ (Merrifield & Kent 1989). [You may need the integral relation $\int d\theta\,\cosh\theta\,K_0(R\cosh\theta) = \frac{\pi}{2}e^{-R}$.]

**10.2** Freudenreich (1998) gives the $L$-band luminosity density at the center of the Galactic bar [equations (10.4)] as $j_b(0,0,0) = 4.54\,\mathrm{MJy\,sr^{-1}\,kpc^{-1}}$. Use numbers listed in Table 2.1 to show that, at a distance of $D\,\mathrm{pc}$, the $L$-band flux from the Sun is $1400/D^2$ Jy. Hence derive the value of the bar's central surface brightness, $I_b$, that is given in Table 10.1.

**10.3** From Table 10.1, show that at $R_0$ the luminosity density of the Galactic disk is $\sim 50\,L_\odot\,\mathrm{pc}^{-2}$. Show from this result that the local $K$-band surface brightness of the disk is $\sim 20.6\,\mathrm{mag\,arcsec}^{-2}$.

**10.4** Figure 10.2 shows that the integrated surface brightness of the Milky Way at its poles is dominated by stars brighter than $B \sim 16$. Explain this phenomenon using the data in Table 3.16 and Table 10.1.

**10.5** Consider a sample of stars with well determined parallaxes and proper motions but no radial velocities. For each star, we define a $3 \times 3$ matrix $\mathbf{A}$ by $A_{ij} \equiv \delta_{ij} - \hat{r}_i \hat{r}_j$, where $\hat{\mathbf{r}} = (\hat{r}_1, \hat{r}_2, \hat{r}_3)$ is the unit vector from the Sun to the star. Show that the data allow us to determine $\mathbf{p} \equiv \mathbf{A} \cdot \mathbf{v}$ for each star, where $\mathbf{v}$ is the star's heliocentric velocity. Show that $\langle \mathbf{v} \rangle \equiv \langle \mathbf{A} \rangle^{-1} \cdot \langle \mathbf{p} \rangle$ is the choice of $\mathbf{u}$ that minimizes the quantity $S^2(\mathbf{u}) \equiv \langle |\mathbf{p} - \mathbf{A} \cdot \mathbf{u}|^2 \rangle$, where angle-brackets imply sample averages. Explain why this calculation implies that the Sun's motion relative to the sample is given by $\mathbf{v}_\odot = -\langle \mathbf{v} \rangle$, and that the minimum value of $S^2$ is a measure of the intrinsic velocity dispersion of the stars.

**10.6** Use the formulae derived in §3.2.3 of BT to estimate the pericenter and apocenter of the Sun's orbit around the Galaxy.

**10.7** In the construction of the Gliese (1969) catalog of nearby stars, the "solar neighborhood" was defined to be a sphere 40 pc in diameter. By how much does the standard of rest vary across this sphere?

**10.8** The vorticity of a fluid flow $\mathbf{v}(\mathbf{x})$ is $\boldsymbol{\omega} \equiv \boldsymbol{\nabla} \times \mathbf{v}$. In the case of a disk galaxy such as the Milky Way, this flow represents the differential rotation of the stars. Show that locally $\boldsymbol{\omega}$ is simply related to the Oort constants by $|\boldsymbol{\omega}| = 2B$.

**10.9** The numbers of stars observed by Gilmore & Reid (1983) with $4 \le M_V \le 5$ at various distances $z$ from the plane are 3, 7, 19, 22, 38, 50, 79, 107, 135, 149, 197, 207, 218, 233, 265, 301, 322. Each entry corresponds to an interval of width 0.1 in $\log z$ and the first entry is for the bin centered on $\log(z/1\,\mathrm{pc}) = 2.05$. Reproduce Figure 10.25 from these data.

**10.10** Show that the second term on the right-hand side of equation (10.43) can be expressed in terms of the Oort constant as

$$\rho_R \equiv \frac{B^2 - A^2}{2\pi G}. \qquad (10.77)$$

Show that with the Oort constant given by equation (10.36) we have $\rho_R \simeq -0.0024\,\mathcal{M}_\odot\,\mathrm{pc}^{-3}$.

**10.11** By writing Poisson's equation, $\boldsymbol{\nabla} \cdot \mathbf{F} = -4\pi G \rho$, in spherical polar coordinates, show that the density $\rho_\mathrm{H}$ of a spherical halo is connected to the contribution $v_H^2$ that it makes to the Milky Way's squared circular speed $v_c^2$ by

$$\rho_H = \frac{1}{4\pi G r^2} \frac{\partial}{\partial r}(r v_H^2). \tag{10.78}$$

Hence, show from equation (10.43) that a good estimate of the surface density of the disk in the solar neighborhood, $\Sigma_D(R_0)$, is given by

$$2\pi G \Sigma_D(R_0) - \frac{3z_0}{R_0}\left(\frac{\partial(R v_D^2)}{\partial R}\right)_{R_0} = -F_z(R_0, 3z_0) - \frac{3z_0 v_c^2}{R_0^2}, \tag{10.79}$$

where $v_D^2 \equiv v_c^2 - v_H^2$. If the disk is assumed to be thin and exponential, then the left-hand side of this equation is a known multiple of $\Sigma_D(R_0)$ – see equation (2-169) of BT – and the right-hand side can be observationally determined.

**10.12** Subject to what boundary conditions should equation (10.61) be solved?

**10.13** Given that the halo is spherical and has luminosity $4 \times 10^7 \, L_\odot$ between $r = 0.1 \, \text{kpc}$ and $r = 25 \, \text{kpc}$, show that at $R_0$ the surface-brightness of the halo is of order $0.035 \, L_\odot \, \text{pc}^{-2}$. Show further that this surface brightness corresponds to $(25.3 + M_\odot) \, \text{mag arcsec}^{-2}$. Compare this value with a typical sky surface brightness.

**10.14** Express the integrals of the last line of equation (10.70) in polar coordinates centered on the Galaxy's systemic velocity and oriented such that the LSR's velocity lies on the polar axis. Hence show that the mean azimuthal velocity of stars in the sample is

$$\langle v_\phi \rangle = \frac{v_c}{2} \frac{\int dx\, x^3 e^{-\alpha x^2} \int d\theta\, \sin 2\theta (x^2 + 1 - 2x\cos\theta)^{3/2}}{\int dx\, x^3 e^{-\alpha x^2} \int d\theta\, \sin\theta (x^2 + 1 - 2x\cos\theta)^{3/2}}, \tag{10.80}$$

where $\alpha \equiv (v_c/2\sigma)^2$. Obtain a similar expression for $\langle v_\phi^2 \rangle$.

**10.15** Distribute 1000 objects uniformly within the unit sphere and then plot their locations in the meridional plane. Explain why contours of equal number density in this $(R, z)$ plane are not circularly symmetric. What is the significance of this fact for studies of the globular-cluster sub-systems?

**10.16** Show that for any spherical Galactic potential, the escape velocity from the solar neighborhood, $v_e$, and the local circular speed, $v_c$, are related by

$$v_c \leq \frac{v_e}{\sqrt{2}} \tag{10.81}$$

with equality in the case that there is no mass at $R > R_0$. Stars are observed that have Galactocentric speeds up to $\sim 400 \, \text{km s}^{-1}$ (Carney *et al.* 1996). From this observation, obtain an upper limit on $v_c$. Why might this upper limit be misleading?

# 11

# Stellar Kinematics in External Galaxies

In §10.3 we have seen that when we study the kinematics of nearby stars in the Galaxy we are confronted by a wealth of dynamical information. We can measure the full three-dimensional velocities of these stars through the combination of Doppler shifts and proper motions, and we can also measure their three-dimensional locations via their angular positions and parallaxes. Thus, we can recover the full six-dimensional "phase space" coordinates of many stars in the Milky Way. The density of stars in this phase space, or **distribution function** (BT §4.1), fully specifies the dynamics of a stellar system.

For external galaxies, we cannot obtain all the data required to reconstruct the distribution function directly. On account of their great distances, we cannot hope to measure proper motions and parallaxes for stars in external galaxies, so only line-of-sight velocities and angular coordinates are observable. Further, individual stars cannot usually be resolved on account of the finite angular resolution in telescopes: even with the imaging capabilities of HST, the majority of the starlight from an external galaxy still forms an unresolved continuum. Hence, observations of starlight from an external galaxy can only reveal averages of the stellar properties of the vast numbers of unresolved stars that lie along each line of sight. Finally, even the largest telescopes have trouble collecting enough light to measure the kinematics of external galaxies with any precision. In this chapter we show that, despite these difficulties, a great deal can still be learned about the stellar kinematics of external galaxies.

## 11.1 Measuring the kinematics of external galaxies

When measuring the kinematics of individual stars in the Milky Way (§10.3) or in globular clusters (§6.1.11), we used the Doppler shift of each star's spectral absorption lines to measure it line-of-sight velocity: if the star is moving with line-of-sight velocity $v_{\rm los}$, then a spectrum of the star will reveal that a feature which naturally occurs at wavelength $\lambda$ will be shifted by $\Delta\lambda = (v_{\rm los}/c)\lambda$ (since $v_{\rm los} \ll c$). In terms of the **spectral velocity**

$$u \equiv c\ln\lambda, \tag{11.1}$$

the shift of a spectral feature is simply the line-of-sight velocity: $\Delta u = c\Delta\lambda/\lambda = v_{\rm los}$. Hence light that is received at spectral velocity $u$ was emitted at spectral velocity $u - v_{\rm los}$.

When light is collected from some point in an external galaxy, its spectrum consists of the sum of the spectra of the unresolved stars along that line of sight through the galaxy. However, each individual star will generally have a slightly different line-of-sight velocity, so its spectral features will be shifted by a different amount $\Delta u = v_{\rm los}$. The net effect of adding all these shifted spectra together will be to produce a composite spectrum in which the stellar absorption lines appear shifted and broadened: this phenomenon is illustrated in Figure 11.1, which compares the spectrum of a galaxy to that of a star.

To make a quantitative analysis of the shifts and broadening of the absorption lines in a galaxy spectrum, we define the **line-of-sight velocity distribution (LOSVD)**, $F(v_{\rm los})$, such that the fraction of the stars contributing to the spectrum which have line-of-sight velocities between $v_{\rm los}$ and $v_{\rm los} + dv_{\rm los}$ is given by $F(v_{\rm los})dv_{\rm los}$. At any given spectral velocity $u$ in the galaxy's spectrum, we will observe the light from individual stars that was emitted at many different spectral velocities; for a star moving with a line-of-sight velocity $v_{\rm los}$, this light will have originated at a spectral velocity of $u - v_{\rm los}$, and been Doppler shifted to $u$. If we assume for the moment that all of the stars have intrinsically identical spectra, $S(u)$, then the intensity received at spectral velocity $u$ from a star with line-of-sight velocity $v_{\rm los}$ is $S(u - v_{\rm los})$. Summing over all stars we obtain

$$G(u) \propto \int dv_{\rm los} F(v_{\rm los}) S(u - v_{\rm los}). \tag{11.2}$$

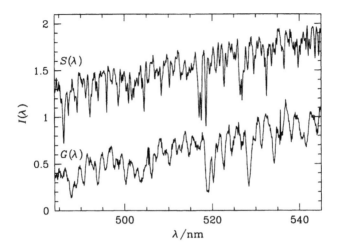

**Figure 11.1** Spectra of a K0 giant star $(S)$ and the center of the lenticular galaxy NGC 2549 $(G)$. These data cover a small part of the optical spectrum around the strong Mg b absorption feature at 518 nm.

As discussed above, a galaxy's kinematics are encoded in the shapes of individual spectral lines. In order to focus on these lines, we separate each spectrum into the sum of a slowly-varying continuum component, and a rapidly-varying component containing the spectral lines. Thus, we write $S = S_{cont} + S_{line}$. The convolution integral in equation (11.2) means that $G$ is a somewhat smoothed version of $S$ (as Figure 11.1 illustrates). Thus, the component of the galaxy spectrum generated by the integral over the slowly-varying $S_{cont}$ component must itself be a slowly-varying function, $G_{cont}$. Similarly, the rapidly-varying component of the stellar spectrum will produce the rapidly-varying component of the galaxy spectrum, $G_{line}$. Equation (11.2) therefore remains valid if we replace $S$ and $G$ by $S_{line}$ and $G_{line}$, and it is these rapidly-varying components that allow us to investigate the galaxy kinematics. Thus, it is customary in stellar-kinematic analyses to remove the continua of spectra by subtracting low-order polynomial fits from the raw data. Henceforth in this chapter, we will assume that all spectra have been pre-processed in this manner.

Equation (11.2) is the fundamental formula for the study of stellar kinematics in external galaxies. We observe $G(u)$ for a line of sight through a galaxy by obtaining its spectrum, and, if we know the type of star from which the galaxy is made, we can estimate $S(u)$ using the spectrum of a similar star from our own galaxy.[1] Given these two quantities, the most general

---

[1] In practice, the finite resolution of any spectrograph means that lines in the spectra

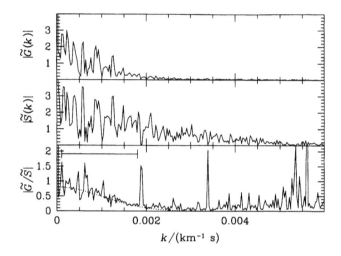

**Figure 11.2** The amplitudes of the Fourier transforms of the spectra illustrated in Figure 11.1, and the amplitude of their ratio. The
dotted line shows a model fit to the amplitude of equation (11.7),
with $\gamma = 0.8$ and $\sigma_{\rm los} = 160\,{\rm km\,s}^{-1}$, and the horizontal bar indicates the range in $k$ over which the fit was made.

kinematic property that we can hope to derive for each line-of-sight through
a galaxy is $F(v_{\rm los})$. Since the right-hand side of equation (11.2) is a convolution integral, we can isolate the LOSVD by taking the Fourier transform
of this equation. In fact, we have

$$\widetilde{F}(k) \propto \frac{\widetilde{G}(k)}{\widetilde{S}(k)}, \qquad (11.3)$$

where tilded quantities denote the Fourier transforms of the original functions. Thus, in principle we can solve directly for $F(v_{\rm los})$ by taking the inverse
transform of equation (11.3). Computationally, this approach is tempting
since the requisite Fourier transforms and their inverses can be calculated
efficiently using the Fast Fourier Transform (FFT) algorithm. Figure 11.2
shows why this technique fails in practice. The subtraction of the continuum
from the spectra leads to the low frequency information in the spectra being
lost, so the transforms are meaningless near $k = 0$. At high frequencies (large
$k$), both $\widetilde{G}(k)$ and $\widetilde{S}(k)$ are small, so any contribution to $\widetilde{G}(k)$ from noise will
be greatly amplified in the ratio $\widetilde{G}/\widetilde{S}$ that appears in equation (11.3). Even

---

will also be broadened by instrumental effects. However, as long as the star and galaxy
spectra are obtained with the same instrumental set-up, this extra broadening affects $S$
and $G$ equally, so equation (11.2) remains valid.

at intermediate values of $k$, there are frequencies at which $\widetilde{S}(k)$ happens to be small, so a similar amplification of noise will occur at these points as well. Thus, the ratio shown in the bottom panel of Figure 11.2, although it clearly contains information as to the form of the LOSVD, has been compromised to a point where we cannot simply derive $F(v_{los})$ from its inverse transform. To-date, therefore, most dynamicists have adopted less direct methods for comparing galaxy and stellar spectra in order to derive at least some of the properties of galaxies' LOSVDs.

### 11.1.1 Mean velocities and velocity dispersions

The simplest properties of a LOSVD are its mean value,

$$\bar{v}_{los} = \int dv_{los} v_{los} F(v_{los}), \tag{11.4}$$

and its dispersion $\sigma_{los}$, which is defined by

$$\sigma_{los}^2 = \int dv_{los} (v_{los} - \bar{v}_{los})^2 F(v_{los}). \tag{11.5}$$

These quantities measure the mean streaming and random motions of the stars averaged along the line of sight through the galaxy.

One straightforward approach to estimating these parameters, is to assume that the LOSVD has the Gaussian form,

$$F_{mod}(v_{los}|\bar{v}_{los}, \sigma_{los}) \propto \exp\left[ -\frac{(v_{los} - \bar{v}_{los})^2}{2\sigma_{los}^2} \right]. \tag{11.6}$$

The Fourier transform of the LOSVD is then

$$\widetilde{F}_{mod}(k|\bar{v}_{los}, \sigma_{los}) = \gamma \exp\left[ -\tfrac{1}{2}(2\pi\sigma_{los}k)^2 \right] \times e^{2\pi i \bar{v}_{los} k}, \tag{11.7}$$

where the parameter $\gamma$ is a normalization factor, which measures the average strength of the absorption lines in the galaxy's spectrum relative to those in the star. Equation (11.3) implies that we can solve for the unknown parameters, $\bar{v}_{los}$ and $\sigma_{los}$, by fitting this functional form for the Fourier-transformed LOSVD to the observed ratio $\widetilde{G}/\widetilde{S}$: the variation in the amplitude of this complex quantity with $k$ allows us to estimate $\gamma$ and $\sigma_{los}$ (see Figure 11.2), while its phase specifies $\bar{v}_{los}$. This technique is known as the **Fourier Quotient Method**, and was pioneered by Sargent et al. (1977). Its principal disadvantage is that, as we have discussed above, the ratio $\widetilde{G}/\widetilde{S}$ is plagued by large errors that vary from point to point. Thus, it is difficult to decide

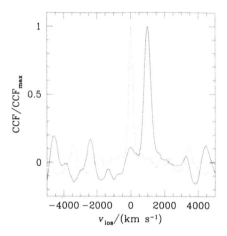

**Figure 11.3** The cross-correlation function between the galaxy and star spectra shown in Figure 11.1. The dotted line shows the auto-correlation function of the star. The functions have been normalized by their peak values.

objectively over what range in $k$ the fit should be performed, and how the individual data points should be weighted.

An alternative approach to calculating the mean velocity of a galaxy is provided by the **cross-correlation method**. This technique, pioneered by Simkin (1974) and developed further by Tonry & Davis (1979), involves calculating the cross-correlation function (CCF) between the galaxy and stellar spectra,

$$\mathrm{CCF}(v_{\mathrm{los}}) = \int \mathrm{d}u \, G(u)S(u - v_{\mathrm{los}}). \tag{11.8}$$

Since this integral is a convolution, it is most efficiently calculated by means of Fourier transforms. A typical example of a CCF is shown in Figure 11.3. Since the spectra that produced this figure have had their continuum levels subtracted, each spectrum consists of a function that fluctuates to positive and negative values around a mean of zero. For a random value of $v_{\mathrm{los}}$, the integrand on the right-hand side of equation (11.8) is therefore as likely to be negative as positive, so $\mathrm{CCF}(v_{\mathrm{los}})$ is small. However, when $v_{\mathrm{los}}$ is equal to $\bar{v}_{\mathrm{los}}$, the arguments for $G$ and $S$ on the right-hand side of equation (11.8) will be such that absorption features in the two spectra will be "lined up," so the integrand will be the product of either two negative or two positive numbers, and the integral will evaluate to a large positive value, producing the strong peak in Figure 11.3. We can therefore estimate $\bar{v}_{\mathrm{los}}$ by simply finding the maximum value of $\mathrm{CCF}(v_{\mathrm{los}})$, a process which has the advantages of being objective and readily automated. Since the absorption lines in the spectra have finite widths, the features will remain correlated even if $v_{\mathrm{los}}$ is not exactly equal to $\bar{v}_{\mathrm{los}}$, so the peak in $\mathrm{CCF}(v_{\mathrm{los}})$ has a finite width around $\bar{v}_{\mathrm{los}}$. The width of this peak depends on how broad the lines in the galaxy spectrum are, which in turn depends on $\sigma_{\mathrm{los}}$. We can therefore derive $\sigma_{\mathrm{los}}$ by measuring the width of the peak in $\mathrm{CCF}(v_{\mathrm{los}})$.

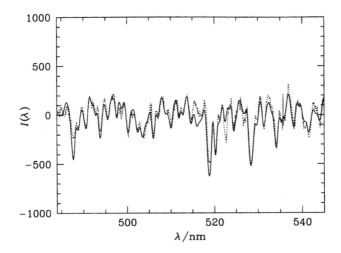

**Figure 11.4** Comparison between the continuum-subtracted galaxy spectrum of Figure 11.1 (dotted line) and the model obtained by convolving the stellar spectrum in that figure with the best-fit Gaussian model LOSVD (solid line).

Once we have obtained an estimate for the parameters of the LOSVD, it is instructive to substitute the corresponding model LOSVD, equation (11.6), into equation (11.2) to see how closely the resulting model for the galaxy spectrum matches the observed galaxy spectrum. As Figure 11.4 illustrates, when such comparisons are made, the fit between model and data is almost always found to be formally poor. One reason for this poor fit is our erroneous initial assumption that all the stars in the galaxy are identical. In reality, galaxies are made up from stars with a range of spectral types, so it is unlikely that our single "template" stellar spectrum will have line strengths that exactly match the composite galaxy spectrum. This error is known as **template mismatch**. A simple way to assess the importance of this phenomenon is to use one of the above methods to obtain $\bar{v}_{\mathrm{los}}$ and $\sigma_{\mathrm{los}}$ from a variety of stellar template spectra; if the derived kinematics do not depend strongly on the spectral type of the template star, then the results have probably not been compromised. A more sophisticated analysis involves modeling the galaxy spectrum using a set of broadened template spectra from different stellar types. By leaving the relative contributions from the different spectra as free parameters, it is possible to produce a composite spectrum which matches the line strengths in the observed galaxy spectrum quite closely (Rix & White 1992). In principle, this approach might be used to measure the stellar composition of external galaxies, but in practice the limited quality of the data available leaves the problem under-determined: the sum of the spectra from an F star and a K star looks very much like the spectrum of a G star. It is also worth noting that in the one system in which we can study the properties of individual stars in some detail – the Milky Way – the observed kinematics vary with the spectral type of the star (see §10.3.2). It is reasonable to assume that such variations occur in other galaxies as well, and so we should really allow the individual stellar templates to be broadened by different LOSVDs when producing a composite galaxy model. However, such sophistication is not possible with the current quality of spectral data.

A second reason why the fit between the galaxy model and the observed spectrum may be poor lies in our assumption that the LOSVD is Gaussian. Beyond its mathematical convenience, there is very little justification for this assumption. We have already found that the stars in the solar neighborhood have distinctly non-Gaussian velocity distributions (see Figure 10.14), and we shall see below that there are stellar systems in which the departures from Gaussian velocity distributions are even more dramatic. Further, integrating along a line of sight through a system in which the velocity distribution is at each point Gaussian, but in which the mean velocity and dispersion vary with position, will produce a non-Gaussian LOSVD. It should therefore not be surprising if a Gaussian model for the LOSVD does not reproduce in detail the observed broadening of galaxy spectral lines.

### 11.1.2 Analysis of line profiles

Given the shortcomings of the Gaussian model, we need some way of fitting a more general functional form to the LOSVD. Essentially, we must return to the solution of equation (11.2) for $F(v_{\rm los})$ given $G$ and $S$. Disentangling the LOSVD from this integral equation is an example of an inverse problem – see Box 3.1. As we saw in §11.1, although soluble in principle, the presence of noise in observed spectra prevents us from performing the inversion from $G(u)$ and $S(u)$ to $F(v_{\rm los})$ directly. A number of methods have therefore been developed which help to suppress the effects of noise in the galaxy spectra.

The simplest approach to this problem involves smoothing the galaxy spectrum to reduce the high-frequency power contributed by random noise. In the Fourier domain, this process is equivalent to reducing the amplitude of the high-frequency terms in $\widetilde{G}(k)$, so that the noise in the quotient of equation (11.3) is similarly reduced. By making a relatively crude model for the contributions to the right-hand side of equation (11.3) from both the true LOSVD and the noise, it is possible to devise an **optimal filter** which, when applied to this equation, returns the best-fitting estimate for the LOSVD that is smooth. The first attempt to apply this technique to a galaxy spectrum was made by Simkin (1974). Her analysis showed that, although the method works in principle, it does not provide a magic solution to the problem: with the relatively low quality spectra then available, the derived LOSVD was still compromised by noise.

A related approach uses the relationship between the cross-correlation function given by equation (11.8) and the stellar auto-correlation function (ACF), which is the cross-correlation of the stellar template spectrum with itself:

$$\mathrm{ACF}(v_{\mathrm{los}}) = \int du\, S(u)S(u - v_{\mathrm{los}}) \tag{11.9}$$

(see Figure 11.3). It is mathematically straightforward to show that the CCF is just the convolution of the ACF with the LOSVD (see Problem 11.1). Intuitively, we can grasp this result by thinking of the galaxy spectrum as the sum of a large number of stellar spectra all slightly shifted in velocity. The CCF of the galaxy spectrum with the stellar template will thus be the sum of a similar number of stellar ACFs all slightly shifted in their mean velocities, producing in sum a function which is the stellar ACF broadened by the galaxy's LOSVD. We can therefore obtain the LOSVD by deconvolving the ACF from the CCF, just as we have previously attempted to deconvolve the stellar template from the galaxy spectrum. The principal advantage of working with correlation functions rather than with spectra directly is that all of the useful information in the correlation functions is concentrated in a single peak rather than being divided amongst the broadening of many spectral lines. In fact, the conflicting information from different spectral lines which can arise through template mismatch is rather effectively reconciled by forming the correlation function. Since the signal-to-noise ratio of the concentrated peak in the correlation function is higher than that of the spectrum as a whole, the amount of filtering required to suppress the noise when obtaining the LOSVD is correspondingly reduced. This approach has been successfully implemented by Franx & Illingworth (1988) and Bender (1990), and has been shown to return reliable estimates for the LOSVDs of galaxies.

These methods for deriving the LOSVD are essentially non-parametric: they make no assumptions about the form that the LOSVD will take. As we have seen above, it is simpler to tackle this problem if a functional form is adopted for the LOSVD, and its parameters are varied until the best match is obtained between the observed galaxy spectrum and the stellar template convolved with the LOSVD. The single Gaussian model of equation (11.6) is in general not adequate, and so more flexible parameterizations of the LOSVD have been suggested. For example, Rix & White (1992) successfully modeled the LOSVD of an elliptical galaxy using the sum of two Gaussians with different amplitudes, means and dispersions.

As a more general parameterization of the LOSVD, we can specify the function in terms of its moments,

$$\mu_k = \int dv_{\mathrm{los}} \left(v_{\mathrm{los}} - \overline{v}_{\mathrm{los}}\right)^k F(v_{\mathrm{los}}). \tag{11.10}$$

By the definition of $\overline{v}_{\mathrm{los}}$, $\mu_1 \equiv 0$, and we have already encountered $\mu_2 = \sigma_{\mathrm{los}}^2$, a measure of the width of the LOSVD. The higher-order moments tell us

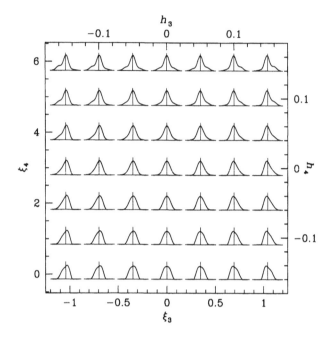

**Figure 11.5** Montage showing how a velocity distribution typically changes depending on its shape parameters $\xi_3$ and $\xi_4$. The central distribution is a pure Gaussian ($\xi_3 = 0$, $\xi_4 = 3$). All distributions have identical values for the Gauss-Hermite parameters $\bar{v}_{\mathrm{los}}$ (marked as the y-axis for each LOSVD) and $\sigma_{\mathrm{los}}$. The right-hand and top axes are marked with the approximately corresponding values of the quantities $h_3$ and $h_4$ that are defined by equation (11.12).

more about the shape of the LOSVD. Such quantities also have the physical motivation that, like the velocity dispersion (BT §4.2), the values of these moments in stellar systems are constrained by Jeans equations (see, for example, Merrifield & Kent 1990, Magorrian & Binney 1994).

The higher-order shape parameters are usefully made dimensionless by dividing by the appropriate power of the dispersion:

$$\xi_k \equiv \mu_k / \sigma_{\mathrm{los}}^k. \qquad (11.11)$$

The first of these new shape parameters, $\xi_3$, is termed the **skewness**, and it basically measures departures from symmetry in the LOSVD: a positively-skewed LOSVD would have a long tail that extends to positive velocities, while a negatively-skewed LOSVD has a corresponding tail at negative velocities. The second shape parameter, $\xi_4$, is the **kurtosis**, which measures symmetric departures from a Gaussian (for which $\xi_4 = 3$): a more rectangular distribution will have $\xi_4 < 3$, while a more centrally-peaked distribution will have $\xi_4 > 3$. Examples of how the values of $\xi_3$ and $\xi_4$ depend on the shape of the LOSVD are shown in Figure 11.5.

Determining just these first few moments does not uniquely specify the LOSVD: in order to define the distribution uniquely, we would have to measure *all* the moments $\mu_k$, $k = 2 \ldots \infty$. Further, even the first few moments cannot be calculated in practice. The weighting by $(v - \bar{v}_{\text{los}})^k$ in the integrand of equation (11.10) means that, as $k$ increases, the integral becomes strongly dependent on the form of the LOSVD at velocities far from $\bar{v}_{\text{los}}$. Since the number of stars far from the mean velocity is small, the LOSVD is dominated by noise far from $\bar{v}_{\text{los}}$, and the velocity moments calculated from equation (11.10) have huge uncertainties.

Gerhard (1993) and van der Marel & Franx (1993) have come up with an elegant method for describing the LOSVD that reduces this problem, whilst retaining much of the physical significance of the moment shape parameters. They have modeled the LOSVD as a **truncated Gauss-Hermite series**, which consists of a Gaussian multiplied by a polynomial:

$$F_{\text{TGH}}(v_{\text{los}}) \propto e^{-\frac{1}{2}w^2} \left[ 1 + \sum_{k=3}^{n} h_k H_k(w) \right], \tag{11.12}$$

where $w \equiv (v_{\text{los}} - \bar{v})/\sigma$ with $\bar{v}$ and $\sigma$ free parameters. In equation (11.12), the $h_k$ are constant coefficients and $H_k(w)$ is a Gauss-Hermite function, which is simply a polynomial of order $k$ with carefully-chosen coefficients. The number of terms required in the sum over polynomials depends on how grossly the LOSVD departs from a Gaussian, but, as Figure 11.5 illustrates, even truncating the series at $n = 4$ allows quite a wide range of shapes. By truncating the series at this point, the LOSVD can be estimated for any given galaxy spectrum by varying the values of $\bar{v}$, $\sigma$, $h_3$ and $h_4$ in equation (11.12) until the convolution of this function with a stellar template best reproduces the observed galaxy spectrum [see equation (11.2)]. The optimal fit can be found using a non-linear least-squares fitting algorithm [e.g. Press et al. (1992) §15.5], and van der Marel & Franx (1993) demonstrated that such fitting can quite robustly find the optimal values of the parameters. The validity of truncating the series at $n = 4$ can be tested by extending the series to $n = 6$ and checking that the coefficients $h_5$ and $h_6$ have values consistent with zero to within the errors in the fit. If the LOSVD is close to Gaussian, $\bar{v}$ and $\sigma$ will be approximately equal to $\bar{v}_{\text{los}}$ and $\sigma_{\text{los}}$, respectively. In the case of a strongly non-Gaussian LOSVD, the parameters $\bar{v}$ and $\sigma$ no longer have this physical interpretation – see Problem 11.2.

The advantage of couching the term in square brackets as a sum of Gauss-Hermite polynomials rather than simply a generic $n$th order polynomial is that $h_3$ and $h_4$ convey similar information to that contained in the physically-motivated moments $\xi_3$ and $\xi_4$: the parameters $h_3$ and $\xi_3$ describe how the distribution is skewed, while $h_4$ and $\xi_4$ describe its symmetric departures from a Gaussian (see Figure 11.5). Further, van der Marel & Franx

(1993) demonstrated that the use of Gauss-Hermite polynomials is optimal in the sense that covariances between the fitted parameters are minimized. Thus, although noise in the spectra will affect the derived parameters, there will be almost no correlation between the errors in the different parameters – a higher-than-average estimate for $\bar{v}$ will not, for example, be accompanied by a higher-than-average estimate for $h_3$. The only slight caveat to this parameterization is that $h_4$ still depends significantly on the form of the LOSVD far from the mean velocity. In spectral terms, it is thus dictated by the wings of the absorption lines, whose exact form is sensitive to the manner in which the continuum has been subtracted. Hence, the derived value of $h_4$ can contain a significant systematic error due to the process of continuum subtraction.

Although the Gauss-Hermite series of equation (11.12) provides a model that can describe a range of departures in the LOSVD from the canonical Gaussian (see Figure 11.5), it works best when those departures are fairly modest. As we shall see below, there are stellar systems for which a Gaussian is not even a reasonable approximation to the LOSVD. If we wanted to use an equation of the Gauss-Hermite form to model such LOSVDs, we would have to increase the order of the polynomial which modifies the Gaussian. Given the uncertainties even in $h_4$ described above, this approach is probably not viable. Further, the distributions returned by equation (11.12) will not be positive everywhere if the values of $h_i$ become large; since the number of stars at any given velocity must be non-negative, such negative values of the LOSVD are unphysical.

An alternative model for parameterizing the LOSVD has been suggested by Kuijken & Merrifield (1993), who describe the LOSVD by the sum of a set of $n$ Gaussians at different predetermined mean velocities $\bar{v}_k$ and dispersions $\sigma_k$, but variable amplitudes $a_k$:

$$F_{\mathrm{UGD}}(v_{\mathrm{los}}) \propto \sum_{k=1}^{n} a_k \exp\left[-\frac{(v_{\mathrm{los}} - \bar{v}_k)^2}{2\sigma_k}\right] \tag{11.13}$$

By placing adjacent Gaussians such that their mean velocities are separated by less than $2\sigma_k$, this **unresolved Gaussian decomposition (UGD)** of the LOSVD can reproduce any distribution that varies smoothly on scales of less than $\sigma_k$. Thus, the procedure for estimating the LOSVD is much as for the Gauss-Hermite technique: the amplitudes of the Gaussians, $a_k$, are varied until $F_{\mathrm{UGD}}(v_{\mathrm{los}})$, when convolved with a stellar template spectrum, best reproduces the galaxy spectrum under analysis. The advantage of this functional form is that by appropriate selection of the values of $\bar{v}_k$ and $\sigma_k$, it is possible to maintain explicit control on the velocity range over which the LOSVD is non-zero, and on the velocity scale on which the LOSVD is smooth. It is also simple to calculate other properties of the complete LOSVD in this parameterization such as its mean velocity, $\bar{v}_{\mathrm{los}} = \sum a_k \bar{v}_k / \sum a_k$. Since $F_{\mathrm{UGD}}$

depends only linearly on the unknown parameters, $a_k$, the best-fitting model can be found by a simple application of linear least-squares analysis. Unlike the Gauss-Hermite parameterization, the estimates for the parameters $a_k$ in this analysis will be strongly correlated: if the estimate of one Gaussian's amplitude just by chance comes out abnormally high, the estimated amplitudes of its near neighbors in velocity will on average be reduced to compensate. Thus the analysis of the uncertainties in the properties of the LOSVD is complicated for this model since it is necessary to consider the covariances between the different estimated parameters as well as their individual uncertainties.

For the UGD analysis, Kuijken & Merrifield (1993) showed that it is possible to force the LOSVD to be positive everywhere (as is physically required) by introducing a series of linear constraints on the values of $a_k$. These extra constraints turn the least-squares analysis into a quadratic programming problem, for which efficient numerical algorithms exist. The positivity constraint proves to be a very efficient mechanism for suppressing noise in the analysis. Without it, the estimated LOSVD oscillates back-and-forth about zero at velocities where the true LOSVD is close to zero. However, by stopping the function from dipping below zero, the fitting process is prevented from following the noise features which introduce this oscillation. The price that we have to pay for this suppression is a slight bias in the resulting LOSVD. To see that this is the case, consider a line-of-sight velocity at which there are no stars in a galaxy, so that $F(v_{los}) = 0$. If we obtain a series of spectra of this galaxy, and estimate the LOSVD at this velocity, we will obtain values close to zero, but with some error due to noise in the spectra. However, the positivity constraint means that each estimate will be zero or positive, and thus the mean estimate will be greater than zero. The estimate is thus biased in the sense that its mean value differs from the true value of the LOSVD. This trade-off between noise suppression and bias is a common statistical dilemma, but in this application noise suppression is very efficient, while the bias introduced is modest.

The methods described above by no means exhaust the list of techniques that have been developed to estimate stellar LOSVDs from absorption-line data. Some methods have been tuned to solving particular problems in galaxy dynamics, while others enforce only the most general constraints of positivity and smoothness on the derived LOSVD. However, the bottom line with all these techniques is that unless the galaxy spectra are of sufficient quality to contain usable information on the shape of the LOSVD, no amount of clever analysis will reveal the details of the LOSVD. Fortunately, with the advent of large telescopes and high-throughput spectrographs, it has now become possible to routinely obtain the requisite high-quality spectra.

### 11.1.3 Position-velocity diagrams and data cubes

Most astronomical spectrographs used for obtaining kinematic observations of galaxies admit light through a long, narrow slit. This light is then dispersed in the direction perpendicular to the long dimension of the slit. The detector illuminated by the dispersed light thus records a two-dimensional image in which the signal received at a given $(X, Y)$ on the detector corresponds to the intensity of light from a point $X$ along the slit that was emitted in a narrow range of wavelengths centered on $\lambda(Y)$. By taking a cut through this two-dimensional function at fixed $X$, we can extract the spectrum emitted by the galaxy at a single position $X$ along the slit, and can apply the above analysis to derive the corresponding LOSVD, $F(X, v_{\rm los})$. If we normalize this function by the total intensity of light at each point $X$ along the slit, we obtain the quantity $PV(X, v_{\rm los}) = I(X) \times F(v_{\rm los}, X)$. If the galaxy is made up from similar stars at all positions, then $PV(X, v_{\rm los})$ is proportional to the phase density of stars as a function of position along the slit and line-of-sight velocity. A plot of the function $PV(X, v_{\rm los})$ is known as a **position-velocity diagram**, or **PV diagram**. Clearly, a cut through this function at constant $X$ yields the LOSVD at that point on the slit, while integrating this quantity over $v_{\rm los}$ yields the surface brightness at that point.

As we have discussed above, successful derivation of a LOSVD from a spectrum relies on our belief that the LOSVD should be a fairly smooth function of velocity. Since the kinematic properties of galaxies are generally found to vary on lengthscales comparable to their photometric scales (see below), we also expect a galaxy's LOSVDs to vary slowly with position. We should thus expect $PV(X, v_{\rm los})$ to be a smooth function in both dimensions. This expectation has been exploited by Prada et al. (1996): instead of deriving separate LOSVDs for each point along the major axis of a disk galaxy, they solved for them all simultaneously, essentially constructing the full PV diagram subject to the constraint that $PV(X, v_{\rm los})$ should be a smoothly-varying, positive function. This end is achieved by generalizing the UGD algorithm described above, so that $PV(X, v_{\rm los})$ is modeled by the sum of a set of bivariate Gaussians that are placed at different mean values of $X$ and $v_{\rm los}$ such that the Gaussians are unresolved in both dimensions. The extra smoothness imposed in the spatial dimension provides extra noise suppression in the extraction of velocity information, resulting in smaller errors in the derived kinematics. Merrifield & Kuijken (1994) have shown that it is also possible to use the full two-dimensional data set simultaneously in order to find a full dynamical model which satisfies the equations of stellar dynamics (BT Chapter 4) whilst producing the best fit to all the spectral data. In this case, noise suppression is achieved by forcing the kinematics to follow a consistent dynamical model.

Clearly, for any galaxy that appears extended in two dimensions on the sky, we can extract PV diagrams for different cuts through the galaxy, ultimately producing a complete data cube for the number of stars as a function of two spatial coordinates and line-of-sight velocity. With technical developments such as fiber-fed spectrographs which can simultaneously obtain spectra from a two-dimensional array of points, such data cubes will become increasingly straightforward to obtain. These data sets are directly analogous to the HI data cubes that we have discussed in Chapter 8. Note, however, that in the case of HI, the gas is usually restricted to a thin layer, so, except for edge-on systems, the line of sight only intersects the gas in a small region. For a rounder stellar system such as an elliptical galaxy, each spectrum samples the kinematics of stars at many points along the line of sight. Further, the collisional nature of gas means that all such material at a particular point will be moving at approximately the same velocity. Stars, on the other hand, behave as an effectively collisionless medium, so the stars within a small spatial region can have a wide variety of velocities. Thus, the interpretation of a stellar data cube is more complicated than that of the corresponding gaseous data. Nevertheless, we shall see below that studies of the detailed stellar kinematics of galaxies as derived from absorption-line spectra are opening up literally a new dimension in the study of galaxy dynamics.

# 11.2 The stellar kinematics of elliptical galaxies

We begin our discussion of stellar kinematics by looking at elliptical galaxies. Since these systems have relatively simple morphologies, we might reasonably hope that their kinematics admit correspondingly simple interpretations. Further, the absence of dust in ellipticals means that the observed kinematics will not be complicated by the effects of obscuration. On the down-side, the absence of gas in these systems means that we cannot use this simple kinematic tracer to study the structure of ellipticals. Since in this section we are concerned with the internal dynamics of galaxies, we cite all velocities relative to the systemic velocity, $v_{\rm sys}$, of the galaxy under consideration; that is, all quoted values of $\bar{v}_{\rm los}$ are the result of subtracting $v_{\rm sys}$ from the measured heliocentric velocity.

### 11.2.1 Large-scale properties

**Major-axis kinematics**    The simplest measurements that we can make of the large-scale kinematics of an elliptical galaxy come from the analysis of spectra obtained from a long-slit spectrograph with the slit aligned along the major axis of the galaxy. As we have seen in §4.3, the outer parts of elliptical galaxies are very faint, so until quite recently it has only been possible to obtain crude estimates of their large-scale stellar kinematic properties by measuring $\bar{v}_{\rm los}$ and $\sigma_{\rm los}$ (§11.1.1) as a function of position along the major axis. However, with a large telescope, a modern spectrograph, and enough patience, it is now possible to obtain spectra of high enough quality to quantify the LOSVD more fully. As we shall see below, the distribution of stellar velocities in an elliptical is generally reasonably close to Gaussian, so spectra

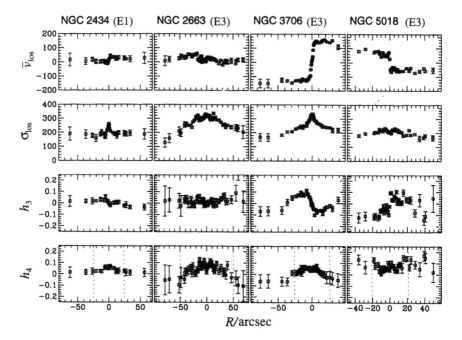

**Figure 11.6** The large-scale major-axis kinematics for a sample of four giant elliptical galaxies. The LOSVDs of these systems have been parameterized using the truncated Gauss-Hermite expansion (§11.1.2). The dotted lines indicate the effective radius, $R_e$, for each galaxy. The Hubble classification (shown in parentheses) is based on the galaxy's average ellipticity outside $R_e/2$. [After Carollo et al. (1995)]

of these systems are well suited to analysis by the Gauss-Hermite technique described in §11.1.2.

Figure 11.6 shows a montage of major axis kinematic parameters, which have been derived out to beyond the effective radii for a sample of typical large elliptical galaxies. The first point that is evident from this figure is that the various kinematic parameters display an overall symmetry (or antisymmetry) about the center of each system. This symmetry implies that there is a largescale coherence to the kinematics of these systems, so we may hope that simple equilibrium models can explain their current dynamical state. Further, the smooth variations between parameters derived from adjacent independent spectra provides us with some confidence that the analysis is returning meaningful estimates of properties of the LOSVD.

The second feature that is apparent in Figure 11.6 is that in most cases $\bar{v}_{los}$ changes sign as we pass through the center of the galaxy. Thus, these systems display net rotation about their minor axes. This finding is consistent with the simple idea that rotation tends to flatten systems. If the motions of the stars in an elliptical were entirely random, there would be no preferred axis in the system, and the galaxy would be spherical. However,

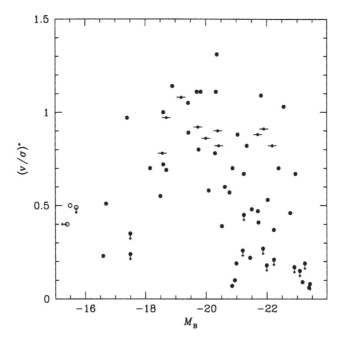

**Figure 11.7** Values of $(v/\sigma)^*$ for a sample of spheroidal systems as a function of their absolute blue magnitudes, $M_B$. Elliptical galaxies are shown as circles, with unfilled circles indicating dwarf spheroidal systems. The bulges of disk galaxies are shown as circles with lines through them. In a number of cases no rotation was detected, so only upper limits are quoted for $(v/\sigma)^*$. [From the data published in Davies & Illingworth (1983), Davies *et al.* (1983) and Bender & Nieto (1990)]

when there is some systematic rotation in addition to the random velocities, the rotational axis picks out a preferred direction in the galaxy and it is natural to expect the galaxy to be flattened parallel to this axis. The degree to which the system is flattened is then dictated by the balance between the rotational velocity and the random motion, $\sigma_{\rm los}$. However, inspection of Figure 11.6 shows that this cannot be the full story. Three of the galaxies have very similar ellipticities (as indicated by their Hubble classifications), and comparable random velocities, $\sigma_{\rm los}$, yet the amplitudes of their rotational velocities differ dramatically.

In fact, the amount of rotation required to explain the observed flattening of elliptical galaxies is generally much higher than the observed rotation speed. To make this point quantitatively, a galaxy is usually characterized by $\bar{v}_{\rm max}$, the maximum observed value of $\bar{v}_{\rm los}$, and $\sigma_0$, the average value of

**Table 11.1**   Values of $(\bar{v}_{max}/\sigma_0)$ as a function of ellipticity for the rotationally-flattened model

| $\epsilon$ | $\bar{v}_{max}/\sigma_0$ | $\epsilon$ | $\bar{v}_{max}/\sigma_0$ | $\epsilon$ | $\bar{v}_{max}/\sigma_0$ | $\epsilon$ | $\bar{v}_{max}/\sigma_0$ |
|---|---|---|---|---|---|---|---|
| 0.025 | 0.159 | 0.175 | 0.457 | 0.325 | 0.688 | 0.475 | 0.941 |
| 0.050 | 0.228 | 0.200 | 0.496 | 0.350 | 0.727 | 0.500 | 0.989 |
| 0.075 | 0.283 | 0.225 | 0.534 | 0.375 | 0.767 | 0.525 | 1.040 |
| 0.100 | 0.331 | 0.250 | 0.572 | 0.400 | 0.809 | 0.550 | 1.093 |
| 0.125 | 0.375 | 0.275 | 0.611 | 0.425 | 0.851 | 0.575 | 1.150 |
| 0.150 | 0.417 | 0.300 | 0.649 | 0.450 | 0.895 | 0.600 | 1.211 |

$\sigma_{los}$ in some small region close to the galaxy's center.[2] Since the amount of flattening that we would expect due to rotation depends on the balance between ordered and random motions, we then measure the quantity $\bar{v}_{max}/\sigma_0$. If the galaxy were truly flattened by rotation, a simple relationship would hold, essentially independent of the galaxy's inclination, between $\bar{v}_{max}/\sigma_0$ and the observed ellipticity of the galaxy (BT §4.3). Table 11.1 presents this relation. The rotation of elliptical galaxies is conventionally characterized by the ratio

$$(v/\sigma)^* \equiv \frac{\bar{v}_{max}/\sigma_0}{(\bar{v}_{max}/\sigma_0)_{model}}, \qquad (11.14)$$

where the denominator is the value of $\bar{v}_{max}/\sigma_0$ that one obtains from Table 11.1. Thus, we would expect $(v/\sigma)^* \approx 1$ if a galaxy is flattened by rotation. In reality, for bright elliptical galaxies, $(v/\sigma)^*$ is substantially less than unity (see Figure 11.7), so the flattening of these galaxies cannot be attributed to rotation. Interestingly, Figure 11.7 shows that intermediate-brightness $(-20 < M_B < -18)$ elliptical galaxies lie much closer to $(v/\sigma)^* = 1$ than their brighter kin; these systems do seem to be substantially flattened by their rotation. However, studies of the faintest elliptical galaxies reveal that this trend does not continue. For compact dwarf ellipticals, these studies are made using a long-slit spectrograph in the same manner as for the giant galaxies, but for the very diffuse dwarf spheroidal galaxies, the kinematics can only be measured by obtaining the velocities of the individually-resolved stars. Analyses of the kinematics of these systems shows that, although flattened, many of them rotate very slowly if at all (Bender & Nieto 1990); as Figure 11.7 shows, $(v/\sigma)^*$ is clearly less than unity for these systems. There is also no strong evidence from this plot for any kinematic distinction between the diffuse dwarf spheroidal systems and other more compact faint galaxies. However, the numbers of faint galaxies whose kinematics have been measured is still small, and better statistics would be required to make a rigorous comparison.

---

[2] Note that this value can depend sensitively on the region over which the averaging is performed (see Figure 11.13), so it is not really a well-defined measure of random motions.

Evidence as to the nature of the rotational motion in elliptical galaxies comes from inspection of the skewness parameter, $h_3$. Galaxies with substantial rotational velocities tend to have significantly skewed LOSVDs, in the sense that $h_3$ has the opposite sign to $\bar{v}_{los}$ (see Figure 11.6). The simplest way to produce such a combination of $\bar{v}_{los}$ and $h_3$ is if the LOSVD consists of two components, one of which is a non-rotating broad distribution, and the other of which is a rapidly-rotating narrow distribution. This combination will produce both a non-zero value for $\bar{v}_{los}$ and a skewed total distribution with the appropriate sign for $h_3$ (see Figure 11.5 and Problem 11.5), and can reproduce the LOSVDs of rapidly-rotating elliptical galaxies [Cinzano & van der Marel (1994), and references therein]. Support for this two-component picture of ellipticals comes from the strong correlation between $(v/\sigma)^*$ and the shape of a galaxy's isophotes that was mentioned in §4.3.3: elliptical galaxies with "disky" isophotes ($a_4 > 0$) have systematically larger rotation velocities than those with more rounded or boxy shapes (see Figure 4.39). A plausible explanation of this correlation is that many elliptical galaxies harbor a rotationally-supported disk component. The size of this component relative to the rest of the galaxy then dictates both the departure of the isophotes from ellipses and the kinematics along the major axis.

It is not clear whether this simple picture of distinct disk and elliptical components reflects the true dynamical state of elliptical galaxies: it is possible that the 'disk' of an elliptical galaxy is just the tail end of a continuous spectrum of populations that extends from a truly ellipsoidal, slowly rotating population to a rapidly spinning and highly flattened population. This ambiguity is closely analogous to the one that we encountered in our study of the thin and thick disks of the Milky Way (§10.4.3). Whatever the resolution to this question, it is apparent that disk-like properties will be most strongly manifested along the major axis of the galaxy, so the above simple analysis, in which the value of $(v/\sigma)^*$ obtained from major-axis data is compared to the overall shape of the galaxy, is likely to be misleading. It is only by fitting dynamical models which allow for the observed complexity of structure in both the kinematics and photometry that we can hope to reveal the intrinsic properties of elliptical galaxies.

Finally in this section, we turn to the kurtosis parameter of the major-axis stellar kinematics, $h_4$ (see Figure 11.5). It is apparent from Figure 11.6 that $h_4$ is detectably non-zero on the major axes of many ellipticals, tending to be positive within their effective radii, $R_e$, and zero, or slightly negative, outside $R_e$. The reason for this transition is not clear, and there are also exceptions to this general rule – Figure 11.6 also shows the case of NGC 5018, in which $h_4$ remains positive to well beyond $R_e$. At present, there are few elliptical galaxies on which sufficient care has been lavished to measure $h_4$ beyond $R_e$, so it is not possible to draw any general conclusions from these subtle variations in the shape of the LOSVD. What is clear, however, is that beyond $R_e$ both $h_4$ and $h_3$ are generally quite small, indicating that the LOSVD does not depart dramatically from a Gaussian. This observation has an important bearing on the question of whether elliptical galaxies are surrounded by dark halos like their spiral kin (§8.2.4).

**Detection of dark halos**    As we have seen, the kinematics of giant elliptical galaxies tend to be dominated by random motions. The amplitude of these random motions is dictated by the gravitational potential of the galaxy. The velocity dispersion and its variation with radius should thus provide us with information on the distribution of mass in elliptical galaxies. Perhaps, therefore, we can use stellar kinematics to provide evidence that elliptical galaxies possess massive dark halos. We face two problems when trying to find and interpret such evidence. First, the distribution of mass at large radii primarily affects the observed kinematics at similar projected radii. Since the outer parts of elliptical galaxies are very faint, it is difficult to obtain spectra of adequate quality to constrain the kinematics at these large radii. The second, more fundamental, problem is that *a priori* we do not know anything about the orbits that stars follow in elliptical galaxies. A poor fit between the observed kinematics and a dynamical model with no dark matter might imply that a massive halo is required, but it could also arise because the model contains the wrong distribution of stellar orbits. Neither of the above difficulties is encountered in the case of disk galaxies: 21-cm observations allow us to trace the kinematics of the ubiquitous HI in these systems out to large radii; and this gas is known to follow approximately circular orbits, so there is little ambiguity as to its orbital distribution. Consequently, it has proved much harder to confirm the presence of dark halos around elliptical galaxies than around disk systems.

As is evident from Figure 11.6, the first of the problems outlined above is now being overcome: there exist spectra for a number of elliptical galaxies that allow their kinematics to be measured out to beyond $2R_e$. These data show that the line-of-sight velocity dispersions of elliptical galaxies decline slowly or even remain flat to beyond $2R_e$. Simple constant mass-to-light ratio models for these systems predict that the velocity dispersion should decline rapidly outside $R_e$, and the absence of such a decline has been cited as evidence for surrounding dark halos.

Unfortunately, by adopting a more complicated distribution of stellar orbits it is possible to reproduce the observed slow decline in the velocity dispersion without invoking a dark halo. As Figure 11.8 illustrates, at small projected radii, $\sigma_{los}$ is dominated by the radial component, $\sigma_R$, of the velocity-dispersion tensor $\boldsymbol{\sigma}$, whereas at large radii its tangential component, $\sigma_\phi$, dominates. Thus, if the stellar orbits are primarily tangential, so that $\sigma_\phi > \sigma_R$, it is possible for $\sigma_{los}$ to remain constant or increase with projected radius, even if both $\sigma_R$ and $\sigma_\phi$ are radially decreasing (as we would expect in the absence of a dark halo). This ambiguity between the distribution of stellar orbits and the distribution of mass in elliptical galaxies is discussed in more detail in BT §4.2.

Happily, we can distinguish between the effects of orbital anisotropy ($\sigma_\phi \neq \sigma_R$) and those of a dark halo by looking at the shape of the LOSVD.

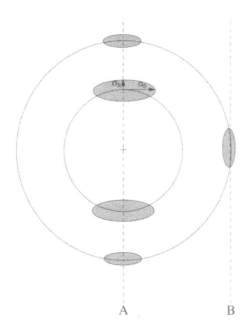

**Figure 11.8** Schematic diagram of the variation with radius of the random motions in the equatorial plane of an elliptical galaxy. Dashed lines mark two lines of sight. Projections of the velocity ellipsoid (Box 10.2) at various points in this plane are represented by shaded ellipses. Note that although both $\sigma_R$ and $\sigma_\phi$ decrease with radius, the line-of-sight dispersion actually *increases* as we move out in projected radius from A to B.

In a system dominated by tangential motions, many stars will be moving at close to the local circular speed, $v_c$, leading to strong concentrations in the LOSVD at $\pm v_c$. At its most extreme, these concentrations would produce a double-peaked LOSVD, but even in less extreme cases it will generally result in a distribution that has a significantly lower kurtosis than a Gaussian, so that $\xi_4 < 3$ (see Figure 11.5). The small observed values of $|h_4|$ at large radii in elliptical galaxies (see Figure 11.6) effectively rule out the possibility that the slow decline in $\sigma_{\text{los}}(R)$ can be attributed to orbital anisotropy, and thus provide the best stellar-kinematic evidence that these systems are embedded in massive dark halos.

**Kinematic mapping**    Valuable information can also be obtained from elliptical galaxies' kinematics away from their major axes. For example, Figure 11.9 shows the major- and minor-axis kinematics of the E3 galaxy NGC 4406. Except very near the center, the rotation velocity about the minor axis, $v_{\text{min}}$, exceeds the major-axis rotation velocity, $v_{\text{maj}}$. When this phenomenon was first discovered, it was interpreted as evidence that the galaxy is a prolate, spindle-like system which rotates about its longest axes (Davies & Birkinshaw 1986).

In NGC 4406 the kinematic minor axis (§8.2.4) is nearly perpendicular to the photometric minor axis. Whenever these two axes are significantly misaligned there is little justification for assuming that the galaxy under observation is intrinsically axisymmetric. When a system is triaxial, the kinematic and photometric minor axes are liable to be misaligned for two reasons: first, the photometric minor axis usually does not coincide with the

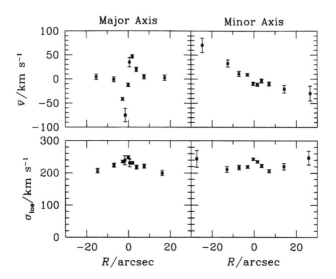

**Figure 11.9** Major- and minor-axis kinematics of the E3 galaxy NGC 4406. [From data published in Franx, Illingworth & Heckman (1989b)]

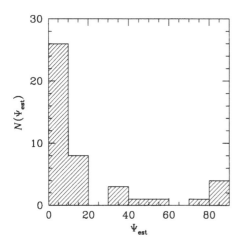

**Figure 11.10** The distribution of the estimated kinematic misalignment angles for a sample of elliptical galaxies. [From the data published in Franx, Illingworth & de Zeeuw (1991)]

projection on to the sky of any of the galaxy's three principal axes (§4.2.3); second, in three dimensions the rotation axis of a triaxial system can lie anywhere in the plane containing the galaxy's longest and shortest principal axes.[3] Given adequate two-dimensional velocity information, we can measure

---

[3] The freedom in the orientation of the rotation axis arises because the system's angular momentum is the sum of contributions from long- and short-axis tube orbits – see BT §3.4.

the angle $\Psi$ between the kinematic and photometric minor axes – $\Psi$ is called the **projected kinematic misalignment**. Figure 11.10 shows a plot of the **estimated kinematic misalignment**, $\Psi_{est} = \tan^{-1}(v_{min}/v_{maj})$, where the velocities are measurements of $\overline{v}_{los}$ at some fiducial point on the minor and major axes. Simple models of the pattern of mean streaming in a galaxy suggest that $\Psi_{est}$ provides a reasonable estimate of $\Psi$ (Franx, Illingworth & de Zeeuw 1991), and since spectra have been obtained along the minor and major axes for quite a number of systems, $\Psi_{est}$ has the merit of being measurable for a reasonable sample of elliptical galaxies. Figure 11.10 is a histogram of values of $\Psi_{est}$ for a sample of 44 galaxies. The highest density of objects occurs near the value $\Psi_{est} = 0$ characteristic of axisymmetric systems, but a significant fraction of galaxies have values of $\Psi_{est}$ that are significantly greater than zero. This finding strongly suggests that at least some ellipticals are triaxial.

In §4.3.3 we encountered photometric evidence that elliptical galaxies are triaxial, but we found that observations of just the projected distribution of light in these systems are not sufficient to specify uniquely the distribution of their three-dimensional shapes. Does the extra information provided by the distribution of $N(\Psi_{est})$ give us a better handle on the shapes of ellipticals? Unfortunately, this dynamical approach also introduces a new unknown into the analysis: as mentioned above, the angular momentum of the galaxy need not align with any of the galaxy's principal axes, so we must also specify the **intrinsic kinematic misalignment**, which is the angle between the angular momentum vector and the direction of the galaxy's shortest axis. On account of this extra free parameter, the observed kinematic misalignment does not provide much new information on the shapes of ellipticals, and a wide range of shape distributions is found to be consistent with both the projected kinematic misalignments and the projected ellipticities of these systems (Franx, Illingworth & de Zeeuw 1991).

Perhaps we can obtain a deeper understanding of the intrinsic shapes of elliptical galaxies from more extensive kinematic observations. For example, Statler (1994) has explored the possibility that the shape of individual galaxies can be estimated by supplementing the major- and minor-axis rotation data with mean line-of-sight velocities along several intermediate position angles. This approach seems to work well for simple triaxial galaxy models, but it produces rather disappointing results when applied to more realistic models generated by N-body simulations.

Since mean streaming motions make only a small contribution to the total kinetic-energy budget of ellipticals, it might be more informative to look at the dominant random motions away from the major and minor axes. This possibility could be explored by producing a kinematic map of $\sigma_{los}(X, Y)$ along all lines of sight through an elliptical. Ultimately, we could even construct the complete LOSVD at all points across the system's face. We could thus observe the density of stars as a function of three variables (two spatial coordinates and line-of-sight velocity) – the kinematic data cube discussed in §11.1.3. The stellar dynamics of a galaxy are fully specified by its distribution function, and this function also depends on only three variables (the isolating integrals of motion – see BT §4.4.1). Thus, there is at least some hope that we could transform the kinematic data cube directly into the distribution function, allowing us to obtain a complete understanding of the structure of an elliptical galaxy from its observable properties. However, this hope has yet to be realized.

### 11.2.2 Core properties

**Decoupled cores**    The kinematics of elliptical galaxies are not infrequently strange at small radii. Figure 11.9 illustrates this phenomenon with the case of NGC 4406. As we have already discussed, the main body of NGC 4406 rotates about its photometric major axis, but there is a strong antisymmetric signal in the major-axis data, and this signature implies that the core of this galaxy rotates about its minor axis. Thus, this system contains a **kinematically-distinct core** (**KDC**), whose angular momentum vector lies perpendicular to that of the bulk of the galaxy. Strangely, the lower panels of Figure 11.9 show no corresponding peculiarities in the velocity dispersion. Figure 11.11 shows the kinematics of IC 1459, which presents an even more extreme example of a KDC. In this otherwise-normal giant E3 galaxy, the mean velocity on the major axis changes sign at $\sim 10$ arcsec from the center, implying that the stars at these small radii rotate in the opposite direction to the main body of the galaxy. Studies reveal that KDCs are fairly common, occurring in maybe a quarter of all ellipticals (Franx, Illingworth & Heckman 1989b). Kinematic mapping of these systems also shows that almost all KDCs show no projected kinematic misalignment: with the exception of NGC 5982, in which the core rotates about the photometric major axis (Wagner, Bender & Möllenhoff 1988), all the detected KDCs rotate about their galaxies' minor axes.

KDCs rotate rapidly, having values of $(v/\sigma)^*$ near unity. This finding makes one suspect that they are small disk components in their hosts. Support for this suspicion comes from the shapes of their LOSVDs, which generally show significantly skewed shapes [see, for example, Franx & Illingworth (1988) for the detailed kinematics of IC 1459]. As we saw in §11.2.1, such skewed LOSVDs will occur when a rapidly-rotating disk with a small intrinsic dispersion is superimposed on a "normal" slowly-rotating elliptical galaxy. However, high-resolution studies of the photometry of these cores using HST do not show the disky isophotes that we might expect from such components, so they must be fairly faint (Forbes, Franx & Illingworth 1995). Even though the disk component does not contribute much to the overall surface brightness of the galaxy, its rapid rotation causes it to dominate the observed value of $\bar{v}_{\text{los}}$ at small radii, whereas its small velocity dispersion means that it does not have much impact on the radial variation in

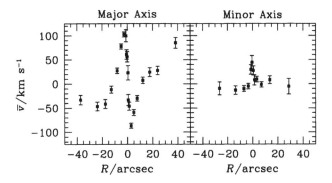

**Figure 11.11** The mean line-of-sight velocity as a function of position along the major and minor axes of IC 1549. [From data published in Franx, Illingworth & Heckman (1989b)]

$\sigma_{los}$. Small disks thus provide an explanation for the peculiar kinematics of decoupled cores.

**Detection of central black holes**  As we have seen in §4.6.2, it is now fairly universally accepted that the highly energetic phenomena seen in active galactic nuclei are ultimately powered by matter accreting on to supermassive black holes. One question that this conclusion raises is whether galaxies that show no signs of nuclear activity are quiescent because they are short of fuel, or whether they lack central massive black holes entirely. Stellar kinematics may be able to provide the answer: a quiescent massive black hole will dominate the gravitational force at small radii in a galaxy, so we would expect it to produce a detectable effect on the motions of the stars in this region. This possibility, and other techniques for detecting inactive galactic nuclei, have been reviewed extensively by Kormendy & Richstone (1995).

A central massive black hole should cause the stellar velocity dispersion to increase in its vicinity. Observations of the E0 galaxy M87 (whose nuclear activity has been assumed to be caused by a massive black hole) do, indeed, show an increase in $\sigma_{los}$ at small radii (Figure 11.12). It has therefore been argued that these data imply that M87 contains a central black hole (Sargent *et al.* 1978). Unfortunately, here we encounter the same problem that afflicts the search for massive dark halos discussed above: the observed kinematics might also be explained without a black hole by an appropriate orbital arrangement of stars. Indeed, it has been demonstrated that the observed rise in the line-of-sight velocity dispersion at small radii would also occur in a system with no black hole if the stellar orbits are predominantly radial (BT §4.2; c.f. Figure 11.8, in which the central line-of-sight dispersion is depressed by having predominantly tangential motions). Thus, measurements of the radial variation of $\sigma_{los}$ do not alone suffice to reveal unambiguously the presence of a central massive black hole.

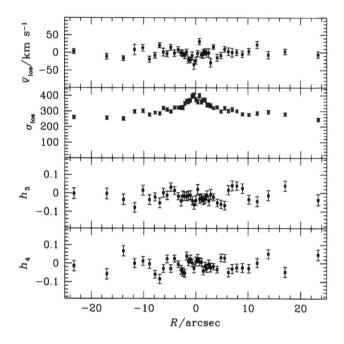

**Figure 11.12** The variation in properties of the LOSVD derived by applying the truncated Gauss-Hermite algorithm (§11.1.2) to spectra obtained close to the center of M87. [From the data published in van der Marel (1994)]

As in the case of dark halos, this ambiguity can be largely resolved if the complete LOSVD is known: radial orbits yield a LOSVD that is sharply peaked about the systemic velocity, while a black hole accelerates stars in its vicinity to produce a LOSVD with long tails to high velocity. Unfortunately, as we have discussed previously, such high-velocity tails are hard to detect in studies of stellar absorption lines, since they tend to be suppressed by the process of continuum subtraction.

In practice, the shape of the LOSVD has provided some assistance in the search for black holes. In the case of M87, for example, van der Marel (1994) has shown that the small departures in the LOSVD from a Gaussian implied by the small values of $h_3$ and $h_4$ in Figure 11.12 are well-reproduced by a model which contains a central black hole of mass $5 \times 10^9 \, \mathcal{M}_\odot$. However, he was not able to rule out the possibility that the observed kinematics could be explained by a black-hole free model with a mainly radial orbit distribution.

Rather more success has been achieved in studies of the kinematics of more rapidly rotating elliptical galaxies such as M32. Their kinematics are easier to interpret because net rotation can only arise from tangential motions. Thus, the range of possible stellar orbit distributions is much smaller

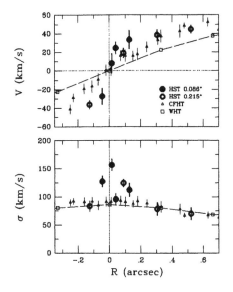

Figure 11.13 line-of-sight stellar mean velocity and dispersion as a function of projected radius for the nearby elliptical, M32, as obtained by the William Herschel Telescope (WHT; resolution $\sim$ 0.8 arcsec), the Canada-France-Hawaii Telescope (CFHT; resolution $\sim$ 0.5 arcsec) and the HST. Note the strong dependence of the kinematics at small radii on the spatial resolution of the observations. [After van der Marel *et al.* (1997) courtesy of R. van der Marel]

for rapidly rotating systems than for almost non-rotating systems such as M87. The price that you have to pay for this extra information is that these systems are not spherically symmetric because rotation picks out a preferred axis. This extra complication led van der Marel *et al.* (1994) to produce a fairly complete kinematic map of M32 by aligning a long-slit spectrograph at several different position angles across the object. They fitted the resulting LOSVDs (derived from the truncated Gauss-Hermite method) to the predictions of a number of dynamical models, each at an assumed value of the system's unknown inclination. They were able to produce respectable fits to the LOSVDs along all the observed position angles with models containing a central mass of $\sim 3 \times 10^6 \, \mathcal{M}_\odot$.

One of the difficulties in detecting a central black hole from stellar kinematics is that it only dominates the mass of the galaxy at small radii, so it only influences the kinematics of the innermost parts of the galaxy (see Problem 11.4). Thus, black-hole induced phenomena such as enhancements in the velocity dispersion will only be visible at small radii. Further, the effects of seeing on optical observations (§4.2.2) will contaminate spectra obtained at small radii with light from larger radii, so the kinematic signature of the black hole will appear superimposed on the normal kinematics at larger radii, diluting its significance. Spectra from HST have therefore proved a boon to the black-hole hunters. Figure 11.13 illustrates the advantage of HST's superior resolution: the amplitude of the enhancement in both $\bar{v}_{los}$ and $\sigma_{los}$ at small radii in M32 is much greater in the higher-resolution observations. As we probe to smaller radii, it becomes more difficult to explain away the enhanced kinematics in terms of orbital anisotropy, so such HST data provide the best stellar-kinematic constraints on the masses of central black holes.

Despite the convincing arguments that have been made for central black holes in a number of stellar systems, the cases are by no means watertight. For a start, most analyses to-date have only considered simple two-integral distribution functions (BT §4.5.2), whereas in general axisymmetric systems will have distribution functions that depend on three integrals of motion (BT §4.5.3). Further, the inner parts of ellipticals may not be axisymmetric. Hence, while there is no reason to think that sophisticated models that take third integrals and triaxiality into account will dramatically change the conclusions we have reached above, we cannot yet make definitive statements about the central masses of elliptical galaxies.

The data shown in Figure 11.13 should also serve as a warning. The low velocity dispersion observed in one of the HST observations close to the nucleus appears to be a genuine result. The absence of a similar glitch in the mean velocity suggests that the deviant point in the dispersion is not the result of anything simple like a patch of dust obscuring the nucleus, so it is hard to understand how a plausible model of M32 can account for this datum. It is very worrying that the only galaxy for which we have data of this quality displays such an apparently inexplicable phenomenon.

Finally, it should be borne in mind that even the best stellar-kinematic "detections" of black holes are, in fact, no such thing. Although it is reasonably well established that several galaxies have a central mass concentration which cannot be attributed to the normal luminous stellar component, there is no direct evidence from the stellar kinematics that this mass must take the form of a black hole. In the case of M32, for example, if its $\sim 3 \times 10^6 \, \mathcal{M}_\odot$ central mass concentration were uniformly distributed through the $\sim 0.3$ pc radius region within which the HST observations imply that it is confined, then its density would be only $\sim 10^{-12}$ kg m$^{-3}$. This astrophysically "ultra-high" density is directly comparable to the density of material in the best room-temperature vacua attained in terrestrial laboratories; it is 18 orders of magnitude smaller than the mean density inside the Schwarzschild radius of a $3 \times 10^6 \, \mathcal{M}_\odot$ black hole! Clearly, even HST does not probe to anywhere near the Schwarzschild radius of the putative black hole, so there is no cast-iron guarantee that the mass is in this ultra-compact form. The inference that the central mass in an inactive galactic nucleus is in the form of a black hole is thus necessarily indirect. It relies on dynamical arguments as to the short timescale on which any alternative (such as a dense cluster of neutron stars) would itself collapse to form a massive black hole (e.g. Goodman & Lee 1989). Such indirect reasoning is less satisfying than a true detection, but it does mean that Occam's Razor strongly favors the black-hole interpretation of the data.

In spite these caveats, there are now sufficient numbers of nearby galaxies with central black hole mass estimates for it to be worth seeing if the

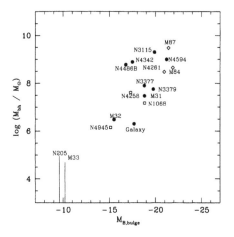

**Figure 11.14** The kinematically-estimated mass of the central black hole in a sample of nearby galaxies plotted as a function of the absolute blue magnitude of the spheroidal component of its host. The identification of each galaxy is indicated ($N \equiv$ NGC). Filled circles are stellar-kinematic estimates, diamonds are gas-kinematic estimates, and squares are estimates from maser kinematics. [Courtesy of D. Richstone]

mass of the black hole correlates with other properties of the galaxy. Figure 11.14 shows the mass estimates for the black holes, $\mathcal{M}_{bh}$, in a number nearby galaxies plotted as a function of the absolute magnitude of the spheroidal component of the galaxy, $M_{B,bulge}$ – its bulge luminosity if it is a disk galaxy, or its total luminosity if it is an elliptical. It is heartening that there is no discernible systematic difference between the black hole masses inferred from stellar kinematics and those inferred from the kinematics of nuclear gas disks or masers (see §7.2.4). There appears to be a correlation between bulge luminosity and black hole mass, implying, not unreasonably, that brighter galaxies contain larger central black holes. However, some care must be taken in interpreting such apparent correlations. Larger galaxies are relatively scarce, so they will generally lie at larger distances. At these larger distances, it would not be possible to resolve the small linear scale over which a small black hole influences the stellar dynamics, so only large black holes will be detectable. Thus, there will be a positive correlation between $M_{B,bulge}$ and detected values of $\mathcal{M}_{BH}$, even if the quantities are intrinsically uncorrelated. However, the observed relation between $\mathcal{M}_{BH}$ and $M_{B,bulge}$ appears to be much stronger than can be explained by such selection effects (Kormendy & Richstone 1995), so the correlation is probably intrinsic. As black-hole mass estimates improve with better data and modeling, and as larger samples of galaxies are observed, the demographics of the black-hole population will provide the first firm observational basis for testing theories of the formation of these bizarre objects.

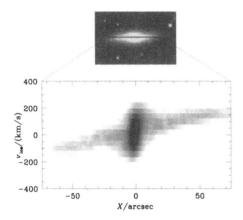

**Figure 11.15** Stellar position–velocity diagram for the major axis of the edge-on S0 galaxy NGC 7332 (as shown in the top panel). The grayscale indicates the density of stars as a function of both line-of-sight velocity and position along the major axis. The kinematics were derived using the UGD algorithm. [DSS image from the Palomar/National Geographic Society Sky Survey, reproduced by permission; the kinematic data were obtained by M. Merrifield and K. Kuijken]

## 11.3 The stellar kinematics of disk galaxies

On account of the multiplicity of components that make up disk galaxies, the interpretation of the kinematics of these systems is more complicated than that of ellipticals. Figure 11.15 shows the PV diagram for the major axis of an edge-on disk galaxy. Close to the center of this galaxy, the light is dominated by the bulge component, so the PV diagram shows a high density of stars which are spread by their random motions to both positive and negative velocities (where, once again, we are measuring velocities relative to the systemic velocity of the galaxy). At large radii, the signature of the rotationally-supported disk is readily apparent in the high density of stars traveling at close to the local circular speed. However, as we shall see below, there are significant random stellar motions even in the disk. Further, the line-of-sight component of the circular speed varies along the line of sight through this edge-on galaxy. We therefore find that the populated region of the PV diagram for this system has a finite spread in line-of-sight velocities even where the disk component dominates. At intermediate radii, there are contributions to the PV diagram from both the disk and the bulge, and the LOSVDs become quite complex. Indeed, analyses of the full LOSVDs of disk galaxies at such radii have shown that these distributions are highly non-Gaussian – the observed skewed distributions have been successfully decomposed into a rapidly-rotating disk component plus a slowly-rotating bulge component [Kuijken & Merrifield (1993), Scorza & Bender (1995); see Problem 11.5].

The complex kinematic structure of a disk galaxy becomes more apparent when one obtains a complete stellar-kinematic data cube by mapping the LOSVD all across its face (see §11.1.3). Figure 11.16 shows kinematic parameters derived from such a data cube in the central region of the edge-on disk galaxy M104 (the Sombrero Galaxy). Along the major axis, the signature of the disk shows up clearly in both the mean streaming motion and the skewness parameter, $h_3$. The difference in sign between $\overline{v}_{los}$ and $h_3$

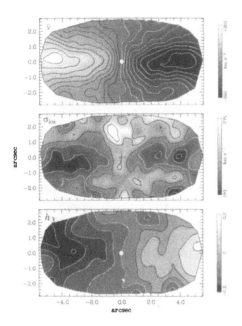

**Figure 11.16** A kinematic map of the central region of M104 (the Sombrero Galaxy), showing the variation in mean velocity, line-of-sight dispersion, and skewness with position. The major axis of the system runs horizontally across each panel, with the photometric center of the galaxy indicated. The range of the grayscale is shown to the right of each panel. [After Emsellem et al. (1996) courtesy of E. Emsellem]

reflects the concentration of stars at close to the circular speed, with a tail of stars stretching back towards the galaxy's systemic velocity (see Figure 11.5, and Problem 11.6). Since the disk component is dominated by streaming motions, it depresses the values of $\sigma_{\text{los}}$ close to the major axis. Away from the plane of the disk, the light comes mostly from the bulge component, in which random motions dominate, and so $\sigma_{\text{los}}$ is larger. The enhancement in $\sigma_{\text{los}}$ at very small radii in this galaxy has been successfully modeled by Emsellem, Monnet & Bacon (1994) as arising from a central black hole (see §11.2.2) with a mass of $\sim 10^9 \mathcal{M}_\odot$.

### 11.3.1 Bulge kinematics

The kinematics of a galactic bulge are somewhat harder to determine than are the kinematics of an elliptical galaxy: light from the surrounding disk contaminates the observable properties of the bulge, and obscuration due to dust frequently further complicates the analysis. On the positive side, however, the presence of a disk with measurable inclination defines a direction which might reasonably be expected to define one of the principal axes of the bulge – a piece of information that is generally difficult to obtain in the case of elliptical galaxies.

Observations of the bulge of a disk galaxy that lies reasonably close to face-on allow us to measure its central velocity dispersion with little contamination from the disk. Combining such kinematic measurements with the bulge's gross photometric properties suggests that these systems respect the same fundamental plane as elliptical galaxies (see §4.3.4), and therefore satisfy the standard $D_n - \sigma$ relation of equation (4.43) (Dressler 1987). This similarity between the properties of bulges and ellipticals presumably reflects some common aspect of their formation and evolution, but since we do not yet have a clear understanding of the nature of the fundamental plane, the significance of this discovery is still unclear.

As we have discussed above, the contribution to the total luminosity from the disk increases as we look to larger distances from the center of the galaxy. On these larger scales it therefore becomes difficult to isolate the bulge kinematics in the PV diagram. One solution to this problem is to observe early-type disk galaxies in which the bulge is so dominant that disk contamination is small, but the large bulges which can be observed by this approach may not be representative. Alternatively, observations of the bulges of edge-on galaxies, such as that shown in Figure 11.16, can be used: the surface brightness of the disk drops rapidly with distance from the galaxy's plane (§4.4), so we see fairly pure bulge kinematics in regions away from the major axis of an edge-on system.

Kinematic maps derived from observations of edge-on galaxies reveal that bulge kinematics vary smoothly with distance from the galaxies' central planes. Consequently, some estimate can be made of a bulge's major-axis kinematics by extrapolating off-axis observations down to the plane of the galaxy. We can thus estimate values of $\overline{v}_{\max}$ and $\sigma_0$ for galaxy bulges. Such analysis has shown that rotational velocities are, on average, higher for bulges than for comparable elliptical galaxies (Davies & Illingworth 1983, Kormendy & Illingworth 1982). Figure 11.7 shows that in bulges $(v/\sigma)^*$ is close to unity, implying that these systems are flattened by rotation.

Confirmation that bulges can be adequately described as rotationally-flattened systems has come from more complete model fitting. After allowing for the gravitational influence of the disk component, Jarvis & Freeman (1985) have shown that a simple rotationally-flattened model can reproduce both the photometric properties of edge-on bulges and the kinematics derived from spectra at many locations away from their major axes.

Study of the kinematics of edge-on bulges using multiple observations away from their major axes has also revealed that systems with boxy iso-photes (§4.2.2) rotate cylindrically: their mean streaming velocities do not decline with distance from their central planes (Kormendy & Illingworth 1982). Such bulges have been modeled as rotationally-flattened axisymmetric systems (Rowley 1988), but the properties of boxiness and cylindrical rotation can also be ascribed to a prolate bar-like structure (see §4.4.7).

### 11.3.2 Disk kinematics

As we have discussed above, we can only measure uncontaminated stellar disk kinematics by looking away from the central bulge region. Unfortunately,

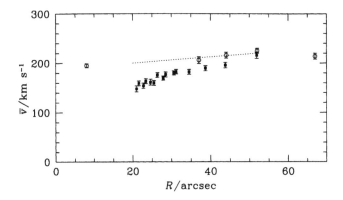

**Figure 11.17** The mean rotation velocity for both gas (open circles) and stars (filled squares) in the disk galaxy NGC 488. The gaseous rotation curve is adapted from Peterson (1980a), and the stellar data are taken from Gerssen, Kuijken & Merrifield (1997). The dotted line shows the circular speed for this galaxy predicted purely from the stellar kinematics.

since the luminosity of a disk drops exponentially with radius (§4.4), these outer parts of disks are quite faint. We must therefore either make very long observations to obtain the requisite signal in the outer parts of galaxies, or try to decompose the complex kinematics at smaller radii into the separate disk and bulge contributions.

**Rotational motion**     Kinematic studies of the disks of external galaxies reveal that, in all but face-on systems, $\bar{v}_{los} \gg \sigma_{los}$ at all radii: the stars in other disk galaxies, like those in the disk of the Milky Way (§10.3), follow approximately circular orbits with little random motion. We can therefore apply methods developed for calculating rotation curves from gas disks (§8.2.4) to the values of $\bar{v}_{los}$ obtained from stellar spectra. Generally, the rotation curves derived from stellar observations have similar properties to those obtained from gas: the stellar rotation increases rapidly with radius near the center of the galaxy, and then flattens out to an approximately constant value. The similarity between the rotation curves from these two different populations confirms that the disk stars follow orbits similar to the circular motions of the gaseous component. Note, though, that the outer parts of disk galaxies are optically very faint, so it is not possible to measure stellar kinematics to the large radii that the HI data allow. Stellar rotation curves do not, therefore, provide any evidence for dark halos.

Figure 11.17 shows the rotation curves of both the gaseous and stellar components of the disk of the Sb galaxy NGC 488. The stellar kinematics are limited in extent by the faintness of the disk at large radii, and contam-

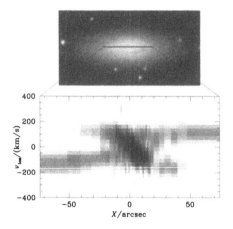

400

200

$v_{los}/(\mathrm{km/s})$

0

−200

−400

−50      0      50

$X/\mathrm{arcsec}$

**Figure 11.18** Stellar PV diagram for the major axis of the Sa galaxy NGC 3593 (as marked on the accompanying image). The kinematics were derived using the UGD algorithm. [DSS image from the Palomar/National Geographic Society Sky Survey, reproduced by permission; the kinematic data were obtained by M. Merrifield and K. Kuijken]

ination by the bulge component for $R < 20$ arcsec. The gaseous kinematics contain a gap at radii around 20 arcsec since there are no measured emission lines in this region. Where the data do overlap, it is clear that there is a small systematic offset between the two rotation curves. This offset is the asymmetric drift that we have already encountered in the solar neighborhood (§10.3.1). It arises because stars do not follow purely circular orbits, but have some random component to their motions (BT §4.2). As in the Milky Way, the measured asymmetric drift can be used to characterize the size of this random motion relative to the ordered streaming (see below).

As we have seen in §8.2.5, comparisons between the gaseous and stellar rotation curves of S0 galaxies have turned up the unexpected result that, in a significant minority of cases, these species rotate in opposite directions. These small counter-rotating gas disks in S0 galaxies could well arise from the late accretion and tidal disruption of small satellite galaxies. Rather harder to explain is the case of NGC 3626, in which as much as $10^9 \, \mathcal{M}_\odot$ of hydrogen (atomic and ionized) counter-rotates relative to the stars across a wide radial range (Ciri, Bettoni & Galletta 1995).

The strangest counter-rotating disk phenomenon of them all, however, is exemplified by NGC 4550. Absorption features in spectra of this apparently-normal edge-on S0 galaxy were found to have a double structure (Rubin, Graham & Kenney 1992), and analysis of the galaxy's LOSVDs showed that these features arise because 50% of the stars in the disk of this galaxy follow approximately circular orbits in one direction, while the other 50% follow similar orbits in the opposite direction (Rix *et al.* 1992).[4]  A handful of similar systems have now been identified. For example, Bertola *et al.* (1996) found clear signs of a small counter-rotating stellar disk in the kinematics of

---

[4] Such a bimodal velocity distribution provides an example of a case where the kinematics cannot be derived using a method which assumes that the LOSVD is anywhere near Gaussian in shape.

the Sa galaxy NGC 3593. As Figure 11.18 illustrates, this counter-rotation shows up clearly in the galaxy's stellar PV diagram. Since there are relatively few galaxies for which spectra have been obtained of the quality necessary to produce PV diagrams, the fraction of galaxies that contain counter-rotating disks remains uncertain. However, a study by Kuijken, Fisher & Merrifield (1996) indicates that less than $\sim 10\%$ of S0 galaxies contain even modest counter-rotating components, so these objects are probably quite rare.

**Random motions** As discussed above, we can infer the presence of non-circular stellar motions by observing asymmetric drift. In principle, we can obtain a direct measure of the random component of the stellar motions by extracting values of $\sigma_{los}$ from spectra. If we define the usual cylindrical polar coordinate system, $(R, \phi, z)$, aligned with the axis of symmetry of the disk, then spectra of face-on galaxies provide a measure of the $z$-component of the stars' velocity dispersion, $\sigma_z$, whereas observations of edge-on systems allow us to measure a combination of the $R$- and $\phi$-components, $\sigma_R$ and $\sigma_\phi$. Of course, all that we can observe directly is the average velocity dispersion integrated along the line of sight. Further, the usual techniques for measuring kinematics do not allow for the differences that we would expect to find in stars of different ages (c.f. the Milky Way, §10.3.2), but rather they produce some weighted average kinematics depending on the choice of template star. Some care must therefore be taken when converting the observed kinematics into localized measurements of the dynamics of disks.

In practice, even line-of-sight averaged quantities are difficult to observe. Face-on disks are intrinsically very faint, and the random motions in the $z$-direction tend to be small. It is therefore difficult to obtain spectra at sufficiently high signal-to-noise ratios for $\sigma_z$ to be measurable. For edge-on systems, we have seen that the broadening of the LOSVD is dominated by the different projections of the rotational velocity that one sees along any given line of sight. In order to measure the relatively small contribution to the total broadening of the LOSVDs due to the random stellar motions in the $R$- and $\phi$-directions, we must be able to characterize accurately the total width and shape of the velocity distributions. Once again, data obtained using high throughput spectrographs on large telescopes are beginning to overcome these difficulties. Observations of a dozen or so galaxy disks with a range of orientations have allowed us to characterize the variations in the components of velocity dispersion with radius – for a summary, see Bottema (1993).

Observations of face-on galaxies show that $\sigma_z$ declines exponentially with radius. This relation has a simple physical interpretation: from the equations of stellar dynamics, it is straightforward to show that, for a thin disk,

$$\sigma_z^2 = 2\pi G \Sigma z_0, \qquad (11.15)$$

where $\Sigma$ is the surface density of the disk, and $z_0$ is a characteristic measure of the scale-height of the disk perpendicular to the plane (see Problem 11.7).

If $z_0$ does not vary with radius (as observations of edge-on galaxies suggest; §4.4), then $\sigma_z \propto \sqrt{\Sigma}$. We also know that the surface brightness of galaxies follows the exponential law, $I = I_0 \exp(-R/R_{\rm d})$ (§4.4). Therefore, if mass follows light so that $\Sigma \propto I$, we would expect $\sigma_z$ to decline exponentially with radius with an e-folding length of $2R_{\rm d}$. Comparisons between kinematic and photometric measurements reveal that $\sigma_z$ does, indeed, decline with a scale-length of approximately twice that of the surface brightness (Bottema 1993). Note, however, that the photometric scale-lengths used in this analysis were generally based on optical measurements, and, as mentioned in §4.4, the infrared scale-lengths of disks tend to be systematically smaller. Since the infrared photometry is more likely to reflect the underlying mass distribution, the above simple analysis is not secure.

Evaluation of the other two components of the velocity dispersion is more complicated since both contribute to $\sigma_{\rm los}$ in observations of inclined galaxies, and there is the additional contribution to the broadening of the LOSVD from the projection of the mean streaming. Modeling of the observed values of $\sigma_{\rm los}$ from edge-on galaxies has revealed that, like $\sigma_z$, $\sigma_R$ and $\sigma_\phi$ decline approximately exponentially with radius with an e-folding length twice that of the disk surface brightness (Bottema 1993). We might therefore tentatively conclude that the ratio between the different components of random stellar motions remains approximately constant, and hence that the shape of the velocity ellipsoid (see Box 10.2) in disk galaxies does not vary dramatically with radius.

Note that the inference that we have drawn as to the constant shape of the velocity ellipsoid is rather indirect. We have measured the $z$-component of the random motions from observations of face-on galaxies, and the other two components from observations of other systems with edge-on orientations. In fact, it is possible to estimate all three components in a single a disk-dominated galaxy that has intermediate inclination $i$. When we observe such a galaxy at some point $R$ from its center along its major axis the observed dispersion is given by

$$\sigma_{\rm maj}^2(R) = \sigma_\phi^2 \sin^2 i + \sigma_z^2 \cos^2 i. \tag{11.16}$$

The annulus of radius $R$ within the disk cuts the apparent minor axis at a distance $R\cos i$ from the center. Here we will find a line-of-sight velocity dispersion

$$\sigma_{\rm min}^2(R\cos i) = \sigma_R^2 \sin^2 i + \sigma_z^2 \cos^2 i. \tag{11.17}$$

Thus, by measuring the variations in the line-of-sight dispersions along the principal axes, we can measure two linear combinations of $\sigma_R(R)$, $\sigma_\phi(R)$ and $\sigma_z(R)$. To close this system of equations, we can turn to the theory of stellar dynamics, which shows how dispersion components are related to one another. For example, if the epicycle approximation (BT §4.2) is valid, then

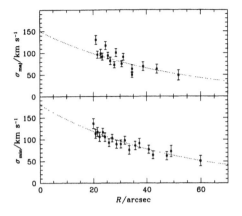

**Figure 11.19** The line-of-sight stellar velocity dispersion observed along the major and minor axes of NGC 488. The dotted lines shows a simple stellar kinematic model fitted to these data. [After Gerssen, Kuijken & Merrifield (1997)]

the radial and tangential dispersions are simply related to the circular speed, $v_c$:

$$\frac{\sigma_\phi^2}{\sigma_R^2} = \frac{1}{2}\left(1 + \frac{dv_c}{dR}\right) \tag{11.18}$$

[equation (4-52) of BT, with the Oort constants replaced by the definitions of §10.3.3]. The circular speed itself can be obtained from the rotational velocities of the stars, corrected for asymmetric drift [see above and BT equation (4-34)]. It is thus possible to solve for $\sigma_R(R)$, $\sigma_\phi(R)$, $\sigma_z(R)$ and $v_c(R)$ just from the observed mean velocity along the major axis, $\overline{v}_{\text{maj}}(R)$, combined with the line-of-sight dispersions, $\sigma_{\min}(R)$ and $\sigma_{\text{maj}}(R)$.

The first application of this technique has been made by Gerssen, Kuijken & Merrifield (1997). Figure 11.19 shows the results of fitting a simple model that meets the above requirements to the observed major- and minor-axis dispersions of NGC 488 [as well as $\overline{v}_{\text{maj}}(R)$ – see Figure 11.17]. A check on this model fit is provided by the predictions made for the circular speed, $v_c(R)$, which can be compared to the observed gaseous rotation velocities – see Figure 11.17. The results of this analysis indicate that the components of the stellar velocity ellipsoid in NGC 488 are in the approximate ratio $\sigma_R{:}\sigma_\phi{:}\sigma_z \approx 1{:}0.8{:}0.7$. Thus, the ordering of these components is the same as for the giant stars in the solar neighborhood of the Milky Way (Table 10.3), the only other place in which this ratio has been measured. More detailed comparisons between the velocity ellipsoids of disk galaxies must await observations of larger samples of these systems, and more sophisticated modeling of higher-quality data.

## Problems

**11.1** By taking Fourier transforms of equations (11.2), (11.8), and (11.9), show that
$$\widetilde{CCF} = \widetilde{ACF}\widetilde{F},$$
and hence that the cross-correlation function can be obtained by convolving the auto-correlation function with the LOSVD.

**11.2** A simple modified form of the Gaussian LOSVD is given by
$$F_{\mathrm{mod}}(v_{\mathrm{los}}) \propto (1 + \alpha v_{\mathrm{los}}) \exp[-(v_{\mathrm{los}} - \bar{v})^2/2\sigma^2]. \qquad (11.19)$$

If $|\alpha| \ll 1/(|\bar{v}| + \sigma)$, show that the mean and dispersion of this modified distribution are given by $\bar{v}_{\mathrm{los}} = \bar{v} + \alpha\sigma^2$ and $\sigma_{\mathrm{los}} = \sigma + \alpha\bar{v}^3/\sigma$ respectively.

**11.3** Calculate the skewness, $\xi_3$, and kurtosis, $\xi_4$, for the velocity distributions
$$F_{\mathrm{box}}(v) = \begin{cases} 0.5 & -1 < v < 1, \\ 0 & \text{otherwise}; \end{cases} \qquad (11.20)$$

$$F_{\mathrm{tri}}(v) = \begin{cases} |1 - v| & -1 < v < 1, \\ 0 & \text{otherwise}; \end{cases} \qquad (11.21)$$

and
$$F_{\mathrm{saw}}(v) = \begin{cases} 2v & 0 < v < 1, \\ 0 & \text{otherwise}. \end{cases} \qquad (11.22)$$

**11.4** If a galaxy with a flat circular speed curve, $v(r) = v_c$, has a black hole of mass $\mathcal{M}_{\mathrm{BH}}$ added to its center, show that the observed kinematics are likely to be modified out to a characteristic radius $R_{\mathrm{max}} \approx G\mathcal{M}_{\mathrm{BH}}/v_c^2$. For a galaxy with a circular speed $v_c = 200\,\mathrm{km\,s^{-1}}$ viewed from a distance of $10\,\mathrm{Mpc}$, what minimum black hole mass should be kinematically detectable (a) from the ground, and (b) using HST.

**11.5** The kinematics along the major axis of a nearly edge-on disk galaxy can be approximated by combining a population of disk stars all moving at velocity $v_c$ with a population of bulge stars which have no net rotation and a line-of-sight velocity dispersion of $v_c/\sqrt{2}$. As we look to increasing radii, the fraction, $f$, of bulge stars decreases. Calculate how $\bar{v}_{\mathrm{los}}$, $\sigma_{\mathrm{los}}$, $\xi_3$ and $\xi_4$ vary with $f$.

**11.6** For an exactly edge-on pure stellar disk in which the stars all follow circular orbits at speed $v_c$, and in which the density of stars decreases with radius in the manner $n(r) \propto r^{-2}$, show that the LOSVD is given by
$$F(v_{\mathrm{los}}) \propto \begin{cases} (v_c^2 - v_{\mathrm{los}}^2)^{-1/2} & 0 < v_{\mathrm{los}} < v_c, \\ 0 & \text{otherwise}. \end{cases} \qquad (11.23)$$

Calculate $\bar{v}_{\mathrm{los}}$, $\sigma_{\mathrm{los}}$, $\xi_3$ and $\xi_4$ for such a disk.

**11.7** Assume that a thin stellar disk can be treated locally as if it were an infinite plane-parallel sheet whose properties only vary with distance $z$ from the disk midplane. In this approximation, the density of stars, $\nu$, the vertical velocity dispersion, $\sigma_z$, and the gravitational potential, $\Phi$, are related by the Jeans equation,

$$\frac{1}{\nu}\frac{\mathrm{d}(\nu\sigma_z^2)}{\mathrm{d}z} = -\frac{\mathrm{d}\Phi}{\mathrm{d}z} \qquad (11.24)$$

[BT equation (4-36)]. Define a scaleheight, $z_0$, such that

$$\frac{\mathrm{d}\ln(\nu\sigma_z^2)}{\mathrm{d}z} = -\frac{1}{z_0} \qquad (11.25)$$

(so that if $\nu\sigma_z^2$ falls exponentially with height, $z_0$ is the usual e-folding length-scale). By considering a point that lies far enough from the plane to be above most of the mass in the disk, use equation (11.24) and Gauss' theorem (BT §2.1) to derive equation (11.15).

# Appendices

## Appendix A: Gravitational deflection of light

We derive some important results connected with the gravitational deflection of light. We assume that the gravitational field is weak and the deflection angles are small. We assume, moreover, that the region within which essentially all the deflection occurs is small compared to the scale of the Universe. These restrictions enable us to impose a coordinate system in which the line element of general relativity is, to a good approximation, given by (see §9.1 of Weinberg 1972)

$$d\tau^2 = (c^2 + 2\Phi)dt^2 - \left(1 - 2\frac{\Phi}{c^2}\right)(dx^2 + dy^2 + dz^2), \qquad (A.1)$$

where $\Phi$ is the usual Newtonian gravitational potential. The paths of light rays in this metric are most conveniently determined by Fermat's principle, which states that the paths taken by light rays between a source and an observer extremize the elapse of coordinate time as photons pass between source and observer.[1] Let $(dx, dy, dz)$ be the change in the spatial coordinates of a photon in time $dt$, then since $d\tau$ has to vanish along a photon's

---

[1] Physically Fermat's principle applies because when the time elapse is not stationary, neighboring paths allow the observer to 'see' the source at different times when it is in different phases of oscillation. These different 'views' average to zero.

world line, we have from (A.1) that

$$dt = \left(\frac{1 - 2\Phi/c^2}{c^2 + 2\Phi}\right)^{1/2} ds,$$

$$\simeq \frac{1}{c}\left(1 - 2\frac{\Phi}{c^2}\right) ds \tag{A.2}$$

where $ds \equiv \sqrt{dx^2 + dy^2 + dz^2}$ is the coordinate distance between two points on the ray and we have used the weak-field condition $|\Phi| \ll c^2$ to expand the square root of the first line. Since $\Phi \leq 0$, equation (A.2) states that in our coordinate system light propagates precisely as it would if there were no gravitational field but space were filled by a medium of refractive index

$$n = 1 - \frac{2\Phi}{c^2} \geq 1. \tag{A.3}$$

In view of the result just derived, we now determine the path for which the light-travel time

$$t = \frac{1}{c}\int ds\, n \tag{A.4}$$

is stationary with respect to small changes in the path. Since we are interested in light paths that are nearly rectilinear, we may orient our coordinate system such that one coordinate, say $z$, increases monotonically along the path. When we employ $z$ rather than $s$ as the integration variable, (A.4) becomes

$$ct = \int dz\, n(\mathbf{x}) \left[\left(\frac{dx}{dz}\right)^2 + \left(\frac{dy}{dz}\right)^2 + 1\right]^{1/2}. \tag{A.5}$$

When we vary the value of $x(z)$ along this path by a small amount $\delta x(z)$, $ct$ changes by

$$c\delta t = \int dz \left\{\frac{\partial n}{\partial x}\delta x\left[\left(\frac{dx}{dz}\right)^2 + \left(\frac{dy}{dz}\right)^2 + 1\right]^{1/2}\right.$$

$$\left. + n(\mathbf{x})\left[\left(\frac{dx}{dz}\right)^2 + \left(\frac{dy}{dz}\right)^2 + 1\right]^{-1/2}\frac{dx}{dz}\frac{d\delta x}{dz}\right\}. \tag{A.6}$$

We now integrate the second term in this integral by parts. Since every path must start at the source and finish at the observer, $\delta x$ vanishes at both ends of the range of integration. It follows that the integrated term vanishes. Hence

$$c\delta t = \int dz\, \delta x \left\{\frac{\partial n}{\partial x}\left[\left(\frac{dx}{dz}\right)^2 + \left(\frac{dy}{dz}\right)^2 + 1\right]^{1/2}\right.$$

$$\left. - \frac{d}{dz}\left(n(\mathbf{x})\left[\left(\frac{dx}{dz}\right)^2 + \left(\frac{dy}{dz}\right)^2 + 1\right]^{-1/2}\frac{dx}{dz}\right)\right\}. \tag{A.7}$$

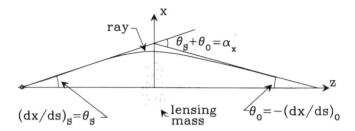

**Figure A.1** The projection onto the $xz$-plane of a light ray that is gravitationally deflected en route from source (at left) to observer (at right).

Along the path $\mathbf{x}(z)$ that we seek, $\delta t$ vanishes for an arbitrary variation $\delta x$. Since $\delta t$ now involves $\delta x$ only as a multiplicative factor in the integrand, it can vanish for arbitrary $\delta x$ only if its coefficient vanishes for all $z$. Hence along the path we seek we have

$$\frac{d}{dz}\left(n(\mathbf{x})\left[\left(\frac{dx}{dz}\right)^2+\left(\frac{dy}{dz}\right)^2+1\right]^{-1/2}\frac{dx}{dz}\right)=\frac{\partial n}{\partial x}\left[\left(\frac{dx}{dz}\right)^2+\left(\frac{dy}{dz}\right)^2+1\right]^{1/2}$$
(A.8)

We now integrate both sides of this differential equation with respect to $z$ between the source and the observer. Since

$$ds=\left[\left(\frac{dx}{dz}\right)^2+\left(\frac{dy}{dz}\right)^2+1\right]^{1/2}dz,$$

we find

$$\left[n(\mathbf{x})\frac{dx}{ds}\right]^{\text{obs}}_{\text{source}}=\int_{S}^{O}ds\,\frac{\partial n}{\partial x}.$$
(A.9)

Both the source and the observer are assumed to lie far from the deflecting mass, in regions within which $n=1$, so

$$\frac{dx}{ds}\bigg|_{\text{obs}}-\frac{dx}{ds}\bigg|_{\text{source}}=\int_{S}^{O}ds\,\frac{\partial n}{\partial x}.$$
(A.10)

Figure A.1 shows that the left-hand side of this equation is the angle through which the projection onto the $xy$-plane of the ray from source to observer is bent. We define $\mu_x$ to be this angle and note that equivalent relations hold for the $yz$-plane. Hence the angle between the direction of the ray at the source S and at the observer O is given by

$$\alpha=\int_{S}^{O}\boldsymbol{\nabla}_{\perp}n\,ds$$
(A.11)

where the integral is along the ray's path and $\nabla_\perp$ denotes the derivative perpendicular to the path. When we substitute from (A.3) for $n$, we have

$$\alpha = -\frac{2}{c^2} \int_S^O \nabla_\perp \Phi \, ds. \tag{A.12}$$

The potential $\Phi$ is related to the mass-density $\rho$ by Poisson's integral (e.g., BT §2-1)

$$\Phi(\mathbf{x}) = -G \int d^3\mathbf{x}' \frac{\rho(\mathbf{x}')}{|\mathbf{x} - \mathbf{x}'|}. \tag{A.13}$$

When we substitute (A.13) in (A.12) we obtain

$$\begin{aligned}
\alpha &= -\frac{2G}{c^2} \int_A^B ds \int d^3\mathbf{x}' \nabla_\perp \frac{\rho(\mathbf{x}')}{|\mathbf{x} - \mathbf{x}'|} \\
&= -\frac{2G}{c^2} \int d^3\mathbf{x}' \, \rho(\mathbf{x}') \int_A^B ds \frac{(\mathbf{x} - \mathbf{x}')_\perp}{|\mathbf{x} - \mathbf{x}'|^3}.
\end{aligned} \tag{A.14}$$

Now, since the deflection angle is small, in (A.14) we may approximate the integral over $s$ along the true path by the integral over the best-fitting straight line through the deflecting region. We orient our coordinate system so that this straight lines runs parallel to its $z$ axis and write

$$\begin{aligned}
\alpha(\mathbf{x}_\perp) &= -\frac{2G}{c^2} \int d^3\mathbf{x}' \, \rho(\mathbf{x}') \, (\mathbf{x} - \mathbf{x}')_\perp \int \frac{d(z - z')}{|\mathbf{x} - \mathbf{x}'|^3} \\
&= -\frac{2G}{c^2} \int d^3\mathbf{x}' \, \rho(\mathbf{x}') \, (\mathbf{x} - \mathbf{x}')_\perp \int_{-\infty}^\infty \frac{db}{(a^2 + b^2)^{3/2}},
\end{aligned} \tag{A.15}$$

where in the last line we have made the substitutions

$$a^2 \equiv (x - x')^2 + (y - y')^2 = |(\mathbf{x} - \mathbf{x}')_\perp|^2 \quad \text{and} \quad b \equiv (z - z'). \tag{A.16}$$

By substituting $b = a \sinh \psi$ we may easily show that the inner integral in (A.16) has value $2/a^2$. Consequently,

$$\begin{aligned}
\alpha(\mathbf{x}_\perp) &= -\frac{4G}{c^2} \int d^2\mathbf{x}'_\perp \frac{(\mathbf{x} - \mathbf{x}')_\perp}{|(\mathbf{x} - \mathbf{x}')_\perp|^2} \int dz' \, \rho(\mathbf{x}') \\
&= -\frac{4G}{c^2} \int d^2\mathbf{x}'_\perp \frac{(\mathbf{x} - \mathbf{x}')_\perp}{|(\mathbf{x} - \mathbf{x}')_\perp|^2} \Sigma(\mathbf{x}'_\perp), \\
&= -\frac{4}{c^2} \nabla_\perp \Phi_2(\mathbf{x}_\perp),
\end{aligned} \tag{A.17}$$

where

$$\begin{aligned}
\Sigma(\mathbf{x}'_\perp) &\equiv \int dz' \, \rho(\mathbf{x}'), \\
\Phi_2(\mathbf{x}_\perp) &\equiv G \int d^2\mathbf{x}'_\perp \Sigma(\mathbf{x}'_\perp) \ln |(\mathbf{x} - \mathbf{x}')_\perp|.
\end{aligned} \tag{A.18}$$

**Appendix B.** Some important astronomical catalogs

| Catalog | Principal content | Number of objects | Reference |
|---|---|---|---|
| | Stellar survey catalogs | | |
| Bonner Durchmusterung (BD) | Position, magnitude (plus atlas) | 320 000 | Argelander (1859–1862) |
| Southern Durchmusterung (BD) | Position, magnitude (plus atlas) | 130 000 | Schönfeld (1886) |
| Cordoba Durchmusterung (CD) | Position, magnitude (plus atlas) | 580 000 | Thome (1892–1932) |
| Cape Photographic Durchmusterung (CPD) | Position, magnitude | 450 000 | Gill & Kapteyn (1895–1900) |
| Carte du Ciel | Position, magnitude (plus atlas) | $\sim 10^7$ | Various publications, culminating in AC2000 |
| EBL2 | Proper motions for BD, CD and CPD | 95 000 | Hamburg Sternwarte (1936) |
| General Catalog (GC) | Position, proper motion, for all stars with $V < 7$ | 33 000 | Boss (1937) |
| Smithsonian Astrophysical Observatory Catalog (SAOC) | Position, magnitude, proper motion | 259 000 | SAO staff (1966) |
| AGK3 | Position, proper motion | 183 000 | Heckmann & Dieckvoss (1975) |
| Southern Reference Stars (SRS) | Position, proper motion | 20 000 | U.S. Naval Observatory (1988) |
| Second Cape Photographic Catalog (CPC2) | Position (mean epoch 1968, SRS frame), proper motion, magnitude | 276 000 | Zacharias et al. (1992) |
| Lowell Proper Motion Survey (North and South) | Position, magnitude, proper motion | 12 000 | Giclas, Burnham & Thomas (1971–1978) |
| Large Proper Motion Survey (LHS) | Proper motion $\gtrsim 0.5$ arcsec yr$^{-1}$ | 4 500 | Luyten (1979) |
| New Luyten Catalog (NLTT) | Proper motion $\gtrsim 0.2$ arcsec yr$^{-1}$ | 59 000 | Luyten (1980) |
| Lick Northern Proper Motion Catalog (NPM1) | Position, proper motion, magnitude, color, $8 < B < 18$ | 149 000 | Klemola, Hanson & Jones (1987) |
| Positions and Proper Motions (North and South) (PPM) | Position, proper motion | 379 000 | Roeser & Bastian (1991), Bastian et al. (1991) |
| Astrographic Catalogue (AC2000) | Position (mean epoch 1905, ICRS frame), magnitude, $B < 13$ | $5 \times 10^6$ | Urban et al. (1998) (http://aries.usno.navy.mil/ad/ac.html) |

**Appendix B** *continued*

| Catalog | Principal content | Number of objects | Reference |
|---|---|---|---|
| | *Stellar astrometric catalogs* | | |
| Fourth Fundamental Catalog (FK4) | Position, proper motion | 1 500 | Fricke & Kopff (1963) |
| Fifth Fundamental Catalog (FK5) | Position, proper motion | 1 500 | Fricke, Schwan & Lederle (1988) |
| FK5-Extension | Position, proper motion, across a wider range of magnitude and color | 3 100 | Fricke, Schwan & Corbin (1991) |
| HST Guide Star Catalog (GSC) | Position, magnitude, $7 < V < 16$ | $2 \times 10^7$ | Lasker et al. (1990) |
| International Reference Stars (IRS) | Position, proper motion of AGK3 and SRS stars, mostly $7 < m_p < 9$ | 36 000 | Corbin & Warren (1991) |
| Astrographic Catalog Reference Stars (ACRS) | Position, proper motion of AGK3 and CPC2 stars in IRS frame | 320 000 | Corbin, Urban & Warren (1991) |
| Carlsberg Meridian Catalogs (CMCn, $n = 1 \ldots 8$) | Position, proper motion, magnitude $V < 15$ | 70 000 | Fabricius (1993) |
| General Catalogue of Trigonometric Parallaxes | Position, proper motion, parallax ($\sim 2300$ non-Hipparcos) | 8 100 | van Altena, Lee & Hoffleit (1995) |
| Hipparcos Catalogue | Position (ICRS), proper motion, parallax, magnitude, variability, $H_P < 13$ | 118 000 | Perryman et al. (1997) |
| Tycho Catalogue | Position (ICRS), proper motion, parallax, magnitude, color, variability | $1 \times 10^6$ | Perryman et al. (1997) |
| ACT Reference Catalog | Proper motion by comparing AC2000 and Tycho Catalogues | 989,000 | Urban et al. (1998) (http://aries.usno.navy.mil/ad/act.html) |
| | *Stellar spectral types* | | |
| Henry Draper Catalog (HD) | Position, spectral class, magnitude | 225 000 | Cannon & Pickering (1918–1924) |
| Henry Draper Extension (HDE) | Position, spectral class, magnitude | 134 000 | Cannon (1925–1936), Cannon & Mayall (1949) |
| Michigan Catalog of Two-Dimensional Spectral Types | Position, MK spectral type for HD stars | 130 000 | Houk (1975–) |
| Twelfth General Catalog of MK Spectral Classification | Position, MK spectral type, $UBV$ photometry | 85 000 | Buscombe (1995) |

**Appendix B** *continued*

| Catalog | Principal content | Number of objects | Reference |
|---|---|---|---|
| | Stellar photometry | | |
| $UBVRIJKL$ Photometry of Bright Stars | $UBVRIJKL$ Photometry of stars in the BSC | 1 500 | Johnson et al. (1966) |
| $UBVRIJKLMNH$ Photoelectric Photometric Catalogue | $UBVRIJKLMNH$ photoelectric photometry | 4 500 | Morel & Magnenat (1978) |
| Photoelectric Johnson Photometric Catalogue | Photoelectric $UBVRI$ photometry | 6 800 | Lanz (1986) |
| Catalogue of $ubvy\beta$ Data | Photoelectric $ubvy\beta$ photometry | 63 000 | Hauck & Mermilliod (1998) |
| $UBVRI$ Photometric Standard Stars | Photoelectric $UBVRI$ photometry of stars on the celestial equator | 526 | Landolt (1992) |
| CCD Photometric Standards | CCD photometry in $vgr$ and $UBV$ | 120 | Jorgensen (1994) |
| General Catalogue of Photometric Data (GCPD) | Photometry in 78 photometric systems (6 systems with more than $10^4$ entries) | 208 000 | Mermilliod, Mermilliod & Hauck (1997) (http://obswww.unige.ch/gcpd/) |
| | Stellar spectral data | | |
| IUE Low-Dispersion Reference Atlas of Stellar Spectra | flux-calibrated spectrum, range of MK types, 115 nm $< \lambda <$ 320 nm, 0.6–0.9 nm resolution | 161 | Heck et al. (1984) |
| Stellar Spectrophotometric Atlas | flux-calibrated spectrum, range of MK types, 313 nm $< \lambda <$ 1070 nm, 2–4 nm resolution | 175 | Gunn & Stryker (1983) |
| Library of Stellar Spectra | flux-calibrated spectrum, range of MK types, 350 nm $< \lambda <$ 740 nm, 0.5 nm resolution | 161 | Jacoby, Hunter & Christian (1984) |
| Atlas of Southern MK Standards | flux-calibrated spectrum, range of MK types, 580 nm $< \lambda <$ 1020 nm, 0.5 nm resolution | 126 | Danks & Dennefeld (1994) |
| Library of Near-Infrared Spectra | flux-calibrated spectrum, range of MK types, 1.4 $\mu m < \lambda <$ 2.5 $\mu m$, 4 nm resolution | 56 | Lancon & Rocca-Volmerange (1992) |
| Spectrophotometric Standards | High quality flux-calibrated spectrum, 320 nm $< \lambda <$ 810 nm | 25 | Massey et al. (1988) |
| Southern Spectrophotometric Standards | High quality flux-calibrated spectrum, 330 nm $< \lambda <$ 1050 nm | 20 | Hamuy et al. (1994) |

**Appendix B**  *continued*

| Catalog | Principal content | Number of objects | Reference |
|---|---|---|---|
| | Stellar radial velocity | | |
| Third Bibliographic Catalogue of Radial Velocities | Radial velocity | 44 000 | Barbier-Brossat, Petit & Figon (1994) |
| Wilson-Evans-Batten Catalogue of Radial Velocities (WEB) | Radial velocity (mean value for binaries) | 21 000 | Duflot, Figon & Meyssonnier (1995) |
| | Variable stars, novae, etc. | | |
| General Catalog of Variable Stars, 4th Ed. (GCVS4) | Position, magnitude, amplitude of variability | 28 000 | Kholopov et al. (1988) |
| Reference Catalog of Galactic Novae | Position, brightest magnitude, rate of decline | 283 | Duerbeck (1990) |
| Sternberg Supernova Catalog | Position, brightest magnitude, identification of host galaxy | 930 | Tsvetkov, Pavlyuk & Bartunov (1995) |
| Catalogue of Galactic Supernova Remnants | Position, type, radio flux | 180 | Green (1991) |
| | Nearby, bright and high-velocity stars | | |
| Catalog of Nearby Stars (Gliese Catalog, or CNS$n$, $n = 1 \ldots 3$) | Position, proper motion, radial velocity, mag., color, distance, for stars within 25 pc | 3 800 | Jahreiss & Gliese (1993) |
| Palomar/MSU Nearby Star Spectroscopic Survey | Spectral type, line strengths, radial velocity of CNS3 stars | 1 750 | Reid, Hawley & Gizis (1995), Hawley, Gizis & Reid (1996) |
| Yale Bright Star Catalog (BSC$n$, $n = 1 \ldots 5$) | Position, proper motion, parallax, radial velocity, spec. type, mag., color, $V < 6.5$ | 9 100 | Hoffleit & Warren (1991) |
| Supplement to the BSC | Extension of BSC4 down to $V = 7.1$ | 2 600 | Hoffleit, Saladyga & Wlasuk (1983) |
| Catalog of High-Velocity Stars | Position, proper motion, parallax, radial velocity, magnitude, color | 630 | Roman (1955) |
| Radial Velocities of High Proper Motion Stars | Position, proper motion, radial velocity, magnitude, color | 890 | Fouts & Sandage (1986) |

**Appendix B** *continued*

| Catalog | Principal content | Number of objects | Reference |
|---|---|---|---|
| **Binary stars** | | | |
| Washington Visual Double Star Catalog 1996.0 (WDS) | Position, separation, spectral types and magnitudes of components | 78 000 | Worley & Douglass (1996) |
| Fourth Catalog of Orbits of Visual Binary Stars | Position, orbital elements, spectral types and magnitudes of components | 850 | Worley & Heintz (1983) |
| Eighth Catalog of Orbits of Spectroscopic Binary Stars | Position, orbital elements, magnitude | 1 500 | Batten, Flechter & MacCarthy (1989) |
| **Star clusters** | | | |
| Catalog of Globular Clusters in the Milky Way | Position, magnitude, color, distance, velocity, metallicity, structure | 146 | Harris (1996) (http://www.physics.mcmaster.ca/Globular.html) |
| Catalog of Open Cluster Data | Position, magnitude, distance, size, constituents, classification | 8 600 | Lynga (1987) |
| Catalog of *UBV* Photometry of Globular Cluster Stars | *UBV* photometry for stars in 113 clusters | 37 000 | Philip, Cullen & White (1976) |
| Database for Stars in Open Clusters (BDA) | Photometry, spectroscopy, astrometry, for stars in ~ 500 clusters | 100 000 | Mermilliod (1987–) (http://obswww.unige.ch/bda/) |
| Catalog of Globular Cluster Surface Brightness Profiles | Surface brightness profile, derived structural parameters | 125 | Trager, King & Djorgovski (1995) |
| **Nebulae** | | | |
| Catalog of Dark Nebulae (LDN) | Position, area, opacity | 1 800 | Lynds (1962) |
| Catalog of Bright Nebulae (LBN) | Position, size, color, brightness | 1 100 | Lynds (1965) |
| Catalog of Reflection Nebulae | Position, size, brightness | 158 | van den Bergh (1966) |
| Strasbourg–ESO Catalogue of Galactic Planetary Nebulae | Position, size, magnitude, colors, properties of central star | 1 100 | Acker et al. (1992) |

**Appendix B** *continued*

| Catalog | Principal content | Number of objects | Reference |
|---|---|---|---|
| | Galaxies and clusters of galaxies | | |
| New General Catalogue (NGC) | Position and description of nebulae, galaxies and star clusters | 7 800 | Dreyer (1888), Sinnott (1988) |
| Index Catalogue (IC; two additions to NGC) | Position and description of nebulae, galaxies and star clusters | 6 900 | Dreyer (1895, 1908), Sinnott (1988) |
| Morphological Catalog of Galaxies (MCG) | Position, size, form, especially of multiple and interacting systems | 32 000 | Vorontsov-Velyaminov et al. (1962–1974), |
| Uppsala General Catalog of Galaxies (UGC) | Position, magnitude, classification, radial velocity ($B_J < 14.5$) | 13 000 | Nilson (1973) |
| ESO/Uppsala Survey of the ESO(B) Atlas | Position, magnitude, classification of galaxies in the southern sky | 18 000 | Lauberts (1982) |
| Revised Shapley-Ames Catalog of Bright Galaxies (RSA) | Position, magnitude, form of bright galaxies ($B \lesssim 13$) | 1 200 | Sandage & Tammann (1981) |
| Catalogue of Principal Galaxies (PGC) | Position, cross-identifications to other catalogs, other properties | 73 000 | Paturel et al. (1989) |
| Third Reference Catalog of Bright Galaxies (RC3) | Position, magnitude, color, redshift, classification ($B \lesssim 15.5$) | 23 000 | de Vaucouleurs et al. (1991) |
| APM Bright Galaxies Catalogue | Position, magnitude, shape, type over 10% of the sky ($B_J < 16.44$) | 14 700 | Loveday (1996) |
| Nearby Galaxies Catalog (NBG) | Position, magnitude, classification of galaxies with redshifts < 0.01 | 2 400 | Tully (1988) |
| CfA Redshift Catalog | Position, magnitude, redshift | 57 500 | Huchra et al. (1995) |
| Las Campanas Redshift Survey | Position, magnitude, redshift | 26 400 | Shectman et al. (1996) |
| Catalog of Galaxies and Clusters of Galaxies (CGCG) | Position, magnitude, redshift | 29 400 | Zwicky et al. (1961–1968) |
| Catalog of Rich Clusters of Galaxies (Abell, ACO) | Position, magnitudes of some members, redshift, richness class, distance class | 5 500 | Abell (1958), Abell, Corwin & Olowin (1989) |

# Appendix B   *continued*

| Catalog | Principal content | Number of objects | Reference |
|---|---|---|---|
| *Active galaxies* | | | |
| First Byurakan Survey of Markarian Galaxies (FBS, Mrk) | Position, magnitude, size, type, redshift of UV-bright galaxies | 1 500 | Markarian *et al.* (1989) |
| Catalog of Seyfert Galaxies | Position, photometry (optical, radio, IR, X-ray), redshift | 960 | Lipovetsky, Neizvestny & Neizvestnaya (1988) |
| Catalog of Quasi-Stellar Objects | Position, magnitude, redshift | 7 300 | Hewitt & Burbidge (1993) |
| Quasars and Active Galactic Nuclei | Position, magnitude, redshift, type of active galaxy | 11 700 | Veron-Cetty & Veron (1996) |
| *Non-optical source lists* | | | |
| Master List of Radio Sources | Position, flux, frequency of measurement | 12 000 | Dixon (1970) |
| NRAO/VLA Sky Survey (NVSS) Catalog | Position, size, flux and other Stokes parameters at 1.4 GHz | $\sim 2 \times 10^6$ | Condon *et al.* (1997) (http://www.cv.nrao.edu/NVSS/) |
| Faint Images of the Radio Sky at 20cm (FIRST) Catalog | Position, size, flux at 1.4 GHz | 270 000 | White *et al.* (1997) (http://sundog.stsci.edu/) |
| Catalog of Infrared Observations | Position, IR flux in several bands | 11 500 | Gezari *et al.* (1993) |
| Second EUVE Source Catalog | Position, EUV flux in several bands | 740 | Bowyer *et al.* (1996) |
| ROSAT All Sky Survey Bright Source Catalog (1RXS) | Position, probability extended, soft X-ray (0.1–2.4 keV) flux, hardness | 18 800 | Voges *et al.* (1996) |
| ROSAT Source Catalogs (1RXP, WGACAT) | Position, probability extended, soft X-ray (0.1–2.4 keV) flux, hardness | $\sim 70\,000$ | Voges *et al.* (1994), White, Giommi & Angelini (1995) |
| HEAO 1 A-4 Catalog of High-Energy X-ray Sources | Hard X-ray (13–180 keV) flux | 50 | Levine *et al.* (1984) |
| Second EGRET Catalog of High-Energy Gamma-Ray Sources | Position, gamma-ray flux, spectral index | 129 | Thomson *et al.* (1995) |

NOTES: The data for most of these catalogs are available on-line from the Astronomical Data centers. Other on-line references as listed.

## *Appendix C:* **Richardson–Lucy deconvolution**

Lucy (1974) described an elegant method for handling inverse problems of the form $f(x) = \int d\xi K(x, \xi) g(\xi)$, where $f$ and $g$ are observed and theoretical distributions, respectively, and $K$ is a smooth kernel. It is important that $f$, $g$ and $K$ are all non-negative functions, for the first step is to normalize them such that

$$\int dx\, f(x) = \int d\xi\, g(\xi) = \int dx\, K(x, \xi) = 1; \qquad (C.1)$$

this can always be done by appropriate rescalings of the functions – see Problem 3.9 – and we assume that (C.1) holds. Then each function can be interpreted as a probability density: $f(x)dx$ is the probability that the measured value of $x$ will fall in $(x + dx, x)$, $g(\xi)d\xi$ is the probability that a randomly chosen system has $\xi$ in the range $(\xi + d\xi, \xi)$ and $K(x, \xi)dxd\xi \equiv K(x|\xi)dxd\xi$ is the probability that we measure $x$ in $(x + dx, x)$ given that $\xi$ falls in $(\xi + d\xi, \xi)$.

By the usual rules of probability theory, the probability $P(x, \xi)\, dxd\xi$ that $x$ and $\xi$ both lie in the stated ranges, can be written two ways:

$$g(\xi)K(x|\xi)\, dxd\xi = P(x, \xi)\, dxd\xi = f(x)L(\xi|x)\, dxd\xi, \qquad (C.2)$$

where $L(\xi|x)\, dxd\xi$ is the probability that $\xi$ lies in $(\xi + d\xi, \xi)$ given that $x$ lies in $(x + dx, x)$. Summing $P$ over all possible values of $x$ we recover $g$:

$$g(\xi) = \int dx\, P(x, \xi) = \int dx\, f(x)L(\xi|x). \qquad (C.3)$$

So if we knew $L$, we could immediately recover $g$. Dividing both sides of equation (C.2) by $f(x)$, we obtain a statement of **Bayes theorem**:

$$L(\xi|x) = \frac{g(\xi)}{f(x)} K(x|\xi). \qquad (C.4)$$

Since we don't know $g$, this equation for $L$ is not immediately useful. But Lucy realized that even a crude estimate of $L$ can be usefully employed in (C.3) because $L$ occurs inside an integral, and proposed the following sequence of approximations to $L$. First we guess $g$. We call this guess $g_1$ and calculate the value $f_1(x) \equiv \int d\xi g_1(\xi)K(x|\xi)$ it predicts. Then we use $f_1$ and $g_1$ in (C.4) to form an estimate $L_1 = (g_1/f_1)K$ of $L$, which we use in (C.3) to obtain an improved estimate $g_2$ of $g$. From $g_2$ we calculate $f_2$ and $L_2$, and thus form $g_3$ and so on, the general step being

$$g_{r+1}(\xi) = g_r(\xi) \int dx\, \frac{f(x)}{f_r(x)} K(x|\xi). \qquad (C.5)$$

We always choose the function $g_1(\xi)$ to be smooth, and we find that the iterates $g_r$ become less smooth as $r$ increases, while the predicted distributions $f_r(x)$ become better and better approximations to the measured distribution $f(x)$. It is easy to show that $g_r$ and $f_r$ are always non-negative and correctly normalized.

## *Appendix D:* **Useful numbers**[1]

**Physical constants**

| | |
|---|---|
| Gravitational constant | $G = 6.672(4) \times 10^{-11}\,\mathrm{m^3\,kg^{-1}\,s^{-2}}$ |
| Speed of light | $c = 2.99792458 \times 10^8\,\mathrm{m\,s^{-1}}$ (by definition) |
| Permeability of space | $\mu_0 = 4\pi \times 10^{-7}\,\mathrm{H\,m^{-1}}$ (by definition) |
| Permittivity of space | $\epsilon_0 = (\mu_0 c^2)^{-1}$ |
| | $= 8.85418 \times 10^{-12}\,\mathrm{F\,m^{-1}}$ |
| Planck's constant | $h = 6.62618(4) \times 10^{-34}\,\mathrm{J\,s}$ |
| | $\hbar = 1.054589(6) \times 10^{-34}\,\mathrm{J\,s}$ |
| Boltzmann's constant | $k = 1.38066(4) \times 10^{-23}\,\mathrm{J\,K^{-1}}$ |
| | $= 8.61733 \times 10^{-5}\,\mathrm{eV}$ |
| Electron charge | $e = 1.602189(5) \times 10^{-19}$ coulomb |
| Electron volt | $\mathrm{eV} = 1.602189(5) \times 10^{-19}\,\mathrm{J}$ |
| Proton mass | $m_\mathrm{p} = 1.672649(9) \times 10^{-27}\,\mathrm{kg}$ |
| Electron mass | $m_\mathrm{e} = 9.10953(5) \times 10^{-31}\,\mathrm{kg}$ |
| Bohr magneton | $\mu_\mathrm{B} = e\hbar/2m_\mathrm{e}$ |
| | $= 9.27408 \times 10^{-24}\,\mathrm{J\,T^{-1}}$ |
| Stefan-Boltzmann constant | $\sigma = \pi^2 k^4/(60\hbar^3 c^2)$ |
| | $= 5.6703(7) \times 10^{-8}\,\mathrm{W\,m^{-2}\,K^{-4}}$ |
| Thomson cross-section | $\sigma_\mathrm{T} = e^4/(6\pi\epsilon_0^2 m_\mathrm{e}^2 c^4)$ |
| | $= 6.65245(6) \times 10^{-29}\,\mathrm{m^2}$ |

**Astronomical Constants**

| | |
|---|---|
| Astronomical unit | $1\,\mathrm{AU} = 1.49597892(1) \times 10^{11}\,\mathrm{m}$ |
| Parsec | $1\,\mathrm{pc} = 648\,000/\pi\,\mathrm{AU}$ |
| | $= 3.08567802(2) \times 10^{16}\,\mathrm{m}$ |
| 1 sidereal year (1900.0) | $1\,\mathrm{yr} = 3.1558149984 \times 10^7\,\mathrm{s}$ |
| Hubble constant | $H_0 = 100h\,\mathrm{km\,s^{-1}\,Mpc^{-1}}$ where $0.5 \lesssim h \lesssim 1$ |
| Hubble time | $H_0^{-1} = 9.78h^{-1} \times 10^9\,\mathrm{yr}$ |
| Solar mass | $\mathcal{M}_\odot = 1.989(2) \times 10^{30}\,\mathrm{kg}$ |
| Gaussian constant | $G\,\mathcal{M}_\odot = 1.32712497(1) \times 10^{20}\,\mathrm{m^3\,s^{-2}}$ |
| Solar radius | $R_\odot = 6.9599(7) \times 10^8\,\mathrm{m}$ |
| Solar luminosity (bolometric) | $L_\odot = 3.826(8) \times 10^{26}\,\mathrm{W}$ |
| Escape speed from Sun | $v_\star = \sqrt{2G\,\mathcal{M}_\odot/R_\odot}$ |
| | $= 617.5\,\mathrm{km\,s^{-1}}$ |
| Solar absolute magnitude | $M_V = +4.83$ |
| | $M_B = +5.48$ |
| Earth mass | $M_\oplus = 5.976(4) \times 10^{24}\,\mathrm{kg}$ |

**Useful relations**

$$1\,\mathrm{km\,s^{-1}} \simeq 1\,\mathrm{pc}\ \text{per million years (actually 1.023)}$$
$$1\ \mathrm{radian} = 206\,265\ \mathrm{arcsec}$$

---

[1] Taken from Allen (1973), Lang (1980) and Particle Data Group (1984). Numbers given in parentheses indicate one standard deviation uncertainty in the last digits of the preceding number.

# References

Aaronson, M., *et al.*, 1982. ApJ, **258**, 64. Virgocentric infall model.

Aaronson, M., Huchra, J., Mould, J., 1979. ApJ, **229**, 1. IR Tully–Fisher relation.

Abell, G.O., 1955. PASP, **67**, 258. Palomar globular clusters.

Abell, G.O., 1958. ApJS, **3**, 211. Abell catalog of rich clusters.

Abell, G.O., 1964. *Exploration of the Universe*, (New York: Holt, Rinehart & Winston). Introductory astronomy text.

Abell, G.O.,, Corwin, H.G. & Olowin, R.P., 1989. ApJS, **70**, 1. Cluster catalog for both hemispheres.

Abraham, R.G. & van den Bergh, S., 1995. ApJ, **438**, 214. Globular cluster luminosity function for the Milky Way.

Abt, H.A., Meinel, A.B., Morgan, W.W. & Tapscott, J., 1969. *An Atlas of Low-Dispersion Grating Stellar Spectra*, (Kitt Peak: KPNO & Yerkes Observatory).

Acker, A., *et al.*, 1992. *Strasbourg-ESO Catalogue of Galactic Planetary Nebulae*, (Garching: ESO). Catalog of planetary nebulae.

Adams, T.F., 1974. ApJ, **188**, 463. Color of BL Lac consistent with an elliptical galaxy.

Adams, M.T., Strom, K.M. & Strom, S.E., 1983. ApJS, **53**, 893. Study of the young open cluster NGC 2264.

Adler, D.S., Allen, R.J. & Lo, K.Y., 1991. ApJ, **382**, 475. $I_{CO}/T_{1.4}$ for spiral galaxies.

Alard, C., 1996. ApJ, **458**, L17. RR Lyraes towards the Galactic center.

Alcock, C., *et al.*, 1995. AJ, **109**, 1653. Variable stars detected in a microlensing survey.

Alcock, C., *et al.* 1995. ApJ, **445**, 133. Microlensing events towards the Galactic bulge.

Alfaro, E.J., Cabrera-Cano, J. & Delgado, A.J., 1991. ApJ, **378**, 106. Corrugation of the Galactic layer of young clusters.

Allen, C.W., 1973. *Astrophysical Quantities*, 3rd. ed. (London: Athlone Press). Source book for basic astrophysical data.

Allen, R.J., 1992. ApJ, **399**, 573. The $I_{CO}/T_{nt}$ correlation in the Milky Way.

Allen, D.A., 1994. In NATO ASI 445, *The Nuclei of Normal Galaxies*, eds Genzel, R., Harris, A.I. (Dordrecht: Kluwer), p. 293. The nuclear star cluster.

Allen, D.A., Hyland, A.R. & Jones, T.J., 1983. MNRAS, **204**, 1145. Infrared photometry of the Galactic center.

Aller, L.H., ApJ, **95**, 52. Emission lines from M33.

Antonucci, R., 1993. ARA&A, **31**, 473. Review of the unified model of AGNs.

Appenzeler, I., Habing, H.J. & Léna, P., 1990. Eds. *Evolution of Galaxies Astronomical Observations*, Lecture Notes in Physics 333 (Berlin: Springer). 1988 Les Houche lectures.

Argelander, F.W.A., 1859–1862. *Beob. Bonn Obs.*, **3–5** The Bonner Durchmusterung (BD stars).

Arias, E.F., Charlot, P., Feissel, M., Lestrade, J.F., 1995. A&A, **303**, 604. ICRS voordinate system.

Armandroff, T.E., 1989. AJ, **97**, 375. Parameters of the disk and halo globular cluster systems.

Arnett, W.D., 1991. In *Frontiers in Stellar Evolution*, ASP Conf. Ser. 20, ed. Lambert, D.L. (New York: ASP), p. 389 Yields from Type II supernovae.

Arnett, W.D., Bahcall, J.N., Kirschner, R.P. & Woosley, S.E., 1989. ARA&A, **27**, 629. Review of Supernova 1987a.

Arnold, R., Gilmore, G., 1992. MNRAS, **257**, 225. The distribution of blue HB halo stars.

Arp, H.C., 1966. *Atlas of Peculiar Galaxies* (Pasadena: Caltech).

Arp, H.C., Madore, B.F. & Roberton, W.E., 1987. *A Catalogue of Southern Peculiar Galaxies and Associations*, (Cambridge: Cambridge University Press)

Athanassoula, E., *et al.* 1990. MNRAS, **245**, 130. The morphology of the bars of 12 SB0s.

Axon, D.J. & Ellis, R.S., 1976. MNRAS, **177**, 499. Map of the direction of the Milky Way's magnetic field.

Baade, W., 1926. Astron. Nachr., **228**, 359. The Baade–Wesselink method.

Baade, W., 1944. ApJ, **100**, 137. Definition of stellar populations I and II.

Baade, W. & Minkowski, R., 1954. ApJ, **119**, 206. Identification of Cygnus A.

Backer, D.C., 1996. In *Unsolved Problems of the Milky Way*, IAU Symposium 169, eds Blitz, L., Teuben, P. (Dordrecht:

Kluwer), p. 193. Proper motion of Sgr A*.

Bahcall, J.N., 1984. ApJ, **287**, 926. Estimation of the local mass density of the disk.

Bahcall, J.N., 1986. ARA&A, **24**, 577. Star-count models.

Bahcall, J.N. & Kylafis, N.D., 1985. ApJ, **288**, 252. Most bulges could follow the $R^{1/4}$ or be thick disks.

Bahcall, J.N. & Soneira, R.M., 1980. ApJS, **44**, 73. Star-count models of the Galaxy.

Bahcall, J.N. & Soneira, R.M., 1984. ApJS, **55**, 67. Star-count models of the Galaxy.

Bahcall, J.N. & Wolf, R.A., 1976. ApJ, **209**, 214. Central luminosity cusps produced by black holes.

Bahcall, J.N., Flynn, C., & Gould, A., 1992. ApJ, **389**, 234. Estimation of the local mass density of the disk.

Bahcall, J.N., et al. 1993. ApJS, **87**, 1. HST survey of QSO absorption lines

Bahcall, J.N., Kirhakos, S. & Schneider, D.P., 1995. ApJ, **450**, 486. HST observations of the absence of QSO host galaxies.

Bally, J., Stark, A.A., Wilson, R.W. & Henkel, C., 1988. ApJ, **324**, 223. Survey of molecular gas at the Galactic center.

Balona, L.A., 1977. MNRAS, **178**, 231. Uncertainties in the Baade–Wesselink method.

Balona, L.A. & Stobie, R.S., 1979. MNRAS, **187**, 217. Effects of non-radial oscillations in the Baade-Wesselink method.

Bania, T.M. & Lockman, F.C., 1984 ApJS, **54**, 513. High-resolution HI survey of the Milky way

Baranne, A., Mayor, M. & Poncet, J.L., 1979. Vistas in Astron., **23**, 279. The Coravel velocity-measuring spectrograph.

Barbier-Brossat, M., Petit, M. & Figon, P., 1994. A&AS, **108**, 603. Bibliographic compilation of stellar radial velocities.

Barnes, J.E. & Hernquist, L.E., 1992. ARA&A, **30**, 705. Dynamics of galaxy mergers.

Barteldrees, A. & Dettmar, R.-J., 1994. A&AS, **103**, 475. CCD photometry of edge-on galaxies.

Bastian, U., et al., 1991. A&AS, **87**, 159. PPM (positions and proper motions) survey of southern stars.

Basu, S. & Rana, N.C., 1992. ApJ, **393**, 373. IMF for the Solar neighborhood.

Batten, A.H., 1973. *Binary and Multiple Systems of Stars* (Oxford: Pergamon Press) Binary stars text.

Batten, A.H., Flechter, J.M. & MacCarthy, D.G., 1989. Publ. Dominion Astrophys. Obs., **17**, 1. Catalog of orbital elements of spectroscopic binaries.

Baum, W.A., et al., 1994. B.A.A.S., **26**, 1398. HST images of globular clusters in the Coma Cluster.

Baumgart, C.W. & Peterson, C.J., 1986. PASP, **98**, 56. Photographic photometry of 11 SB galaxies.

Beck, R., Brandenburg, A., Moss, D., Shukurov, A. & Sokoloff, D., 1996. ARA&A, **34**, 155. $B$ in galaxies.

Becker, R.H., White, R.L. & Helfland, D.J., 1995. ApJ, **450**, 559. VLA FIRST Survey.

Becklin, E.E. & Neugebauer, G., 1968. ApJ, **151**, 145. Infrared photometry of the galactic center.

Beers, T.C. & Geller, M.J., 1983. ApJ, **274**, 491. cD galaxies occur in regions of exceptionally high galaxy density

Beers, T.C., Tonry, J.L., 1986. ApJ, **300**, 557. Galaxy density distribution in clusters.

Begeman, K., 1987. *HI Rotation Curves of Spiral Galaxies*, thesis (Groningen: Kapteyn Institute).

Beichman, C.A., 1987. ARA&A, **25**, 521. IRAS view of the Milky Way and solar system

Bender, R., 1990. A&A, **229**, 441. Fourier Correlation Quotient method for stellar kinematics.

Bender, R. & Nieto, J.-L., 1990. A&A, **239**, 97. Rotation of dwarf ellipticals.

Bender, R., Surma, P., Döbereiner, S., Möllenhoff, C. & Madejsky, R., 1989. A&A, **217**, 35. $a_4/a$ is correlated with $L_{1.4}$ and $L_X$.

Bender, R., Burstein, D. & Faber, S.M., 1992. ApJ, **411**, 153. In ellipticals $Mg_2$ and $B - V$ are very tightly correlated with $\sigma_0$.

Bergbusch, P.S. & VandenBerg, D.A., 1992. ApJS, **81**, 163. Oxygen-enhanced isochrones for globular clusters.

Bergeron, J. & Boissé, P., 1991. A&A,

**243**, 344. Identification of galaxies causing Mg$^+$ absorption in quasar spectra.

Berkhuijsen, E.M., 1998. In press.

Berry, A., 1898. *A Short History of Astronomy*, (London: Murray). Pre-20th century history of astronomy.

Bertelli, G., Bressan, A., Chiosi, C., Fagotto, F. & Nasi, E., 1994. A&AS, **106**, 275. Isochrone calculations.

Bertiau, F.C., 1958. ApJ, **128**, 533. Moving cluster Scorpio-Centaurus

Bertola, F., 1987. In *Structure and Dynamics of Elliptical Galaxies*, IAU Symp 127, ed. de Zeeuw, P.T. (Dordrecht: Reidel), p. 135.

Bertola, F., 1990. In *Dynamics and Interactions of Galaxies*, ed. R. Wielen (Berlin: Springer), p. 249.

Bertola, F., *et al.*, 1996. ApJ, **458**, 67. Counter-rotating disks in NGC 3593.

Bertola, F., Buson, L.M. & Zeilinger, W.W., 1992. ApJ, **458**, 67. Fraction of counter-rotating gas disks in S0s.

Bessell, M.S., 1991. AJ, **101**, 662. Spectral distributions of late M dwarfs.

Bessell, M.S., 1993. In *Stellar Photometry – Current Techniques and Future Developments*, IAU Colloq. 136, eds Butler, C.J., Elliott, I. (Cambridge: CUP) Details of photometric pass-bands

Bhattacharya, D. & van den Heuvel, E.P.J., 1991. Phys. Rep., **203**, 1. Lifetimes of millisecond pulsars and X-ray binaries.

Bingelli, B., Sandage, A. & Tammann, G.A., 1985. AJ, **90**, 1681. Dwarf ellipticals in the Virgo cluster.

Binggeli, B., Sandage, A. & Tammann, G.A., 1988. ARA&A, **26**, 509. Galaxy luminosity function review.

Binney, J.J., 1978. MNRAS, **183**, 779. Effects of triaxial dark halos.

Binney, J.J., 1992. ARA&A, **30**, 51. Review of warps.

Binney, J.J., 1995. In IAU Symp 169, ed. Blitz, L. Review of elliptical disks

Binney, J.J., 1996. In *Gravitational Dynamics*, eds Lahav, O., Terlevich, E., Terlevich, R.J. (Cambridge: CUP), p. 89. Jets and cooling flows

Binney, J.J. & Dehnen, W., 1997. MNRAS, **287**, L5. The Galaxy's outer rotation curve.

Binney, J.J., Gerhard, O.E., 1997. MN-

RAS, **279**, 1005. Deprojection of the Galactic bulge.

Binney, J.J. & Mamon, G., 1982. MNRAS, **200**, 361. Deprojecting spherical objects.

Binney, J.J. & de Vaucouleurs, G., 1981. MNRAS, **194**, 679. Apparent and true ellipticities of RC2 galaxies.

Binney J.J., Gerhard O.E., Stark A.A., Bally J. & Uchida K.I., 1991. MNRAS, **252**, 210. Gas-kinematic evidence for a Galactic bar.

Binney, J.J., Davies, R.L. & Illingworth, G.D., 1990. ApJ, **361**, 78. Deprojection of axisymmetric bodies.

Binney, J.J., Gerhard, O.E. & Spergel, D.N., 1997. MNRAS, **288**, 365. Triaxial model of the Milky Way.

Biretta, J.A., *et al.*, 1996, *WFPC2 Instrument Handbook*, Version 4.0, (Baltimore: STScI).

Blaauw, A., Gunn, C.S., Pawsey, J.L. & Westerhout, G., 1959. MNRAS, **119**, 422. Definition of Galactic coordinates.

Blackwell, D.E. & Shallis, M.J., 1977. MNRAS, **180**, 177. Angular diameters of stars.

Blandford, R.D. & Narayan, R., 1992. ARA&A, **30**, 311. Review article on gravitational lensing.

Bless, R.C. & Savage, B.D., 1972. ApJ, **171**, 293. Grain absorption peak.

Bless, R.C, Code, A.D. & Fairchild, E.T., 1976 ApJ, **203**, 410. Absorption spectral distance of $\alpha$ Vir, $\eta$ UMa and $\alpha$ Leo.

Blitz, L. & Spergel, D.N., 1991. ApJ, **370**, 205. Suggestion that the outer Milky Way is oval.

Blitz L. & Spergel, D., 1991. ApJ, **379**, 631. Bar at the GC

Blitz, L., Binney, J.J., Lo, K.Y., Bally, J. & Ho, T.P., 1993. Nat, **361**, 417. The center of the Galaxy.

Blitz, L., Fich, M. & Stsark, A.A., 1982. ApJS, **49**, 183. A catalog of HII regions.

Bloemen, H., 1989. ARA&A, **27**, 469. Galactic plane image with COS-B.

Bloemen, H., 1993. In *Back to the Galaxy*, AIP Conference Proceedings 278, eds Holt, S.S., Verter, F. (New York: AIP), p. 409 Review of $\gamma$-ray emission by the Milky Way.

Blum, R.D., Carr, J.S., Sellgren, K. & Terndrup, D.M., 1995. ApJ, **449**, 623. Kinematics of the bulge.

Boesgaard, A.M., 1989. ApJ, **336**, 798. Ages of disk clusters.

Bohlin, R.C., Savage, B.D. & Drake, J.F., 1978. ApJ, **224**, 132. The dust-to-gas ratio.

Bolte, M., 1989. ApJ, **341**, 168. Evidence for mass segregation in M30.

Bolte, M., 1992. ApJS, **82**, 145. Detection of 2nd sequence in NGC 288.

Boroson, T., 1981. ApJS, **46**, 177. Surface photometry of 26 spiral galaxies.

Bosma, A., 1978. *The distribution and kinematics of neutral hydrogen in spiral galaxies*, thesis (Groningen: Kapteyn Institute).

Boss, B., 1937. Carnegie Inst. of Wash. Publ. **468** General Catalog of stars (GC).

Bottema, R., 1993. A&A, **275**, 16. Disk velocity dispersions.

Bottinelli, L., Goughenheim, L., Paturel, G. & de Vaucouleurs, G., 1983. A&A, **118**, 4. Correction of Tully-Fisher relation due to random motions of gas.

Boulanger, F. & Pérault, M., 1987. ApJ, **330**, 964. IRAS on solar nhd dust.

Boulanger, F., Abergel, A., Bernard, J.P., Burton, W.B., Desert, F.X., Hartmann, D., Lagache, G., Puget, J.L., 1996. A&A, **312**, 256. Dust/gas correlations from DIRBE.

Bounanno, R., Corsi, C.E. & Fusi Pecci, F., 1989. A&A, **216**, 80. The $\Delta V$ test for globular cluster ages.

Bowen, J.S., 1927. PASP, **39**, 295. Identification of the nebular lines.

Bowyer, S., et al., 1994. ApJS, **93**, 569. EUVE source catalog.

Branch, D., 1987. ApJ, **320**, L23. Distance to SN1987A by the Baade-Wesselink method.

Branch, D., 1992. ApJ, **392**, 35. Models of type Ia supernovae.

Branch, D. & Patchett, B., 1973. MNRAS, **161**, 71. Application of Baade-Wesselink method to type I supernovae.

Branch, D. & Tammann, G.A., 1992. ARA&A, **30**, 359. Type Ia supernovae as standard candles.

Branch, D., Drucker, W. & Jeffery, D.J., 1988. ApJ, **330**, L117. Baade-Wesselink calibration of type Ia supernova luminosities.

Brand, J. & Blitz, L., 1993. A&A, **275**, 67. $v_c(R)$ at $R > R_0$.

Bressan, A., Fagotto, F., Bertelli, G. & Chiosi, C., 1993. A&AS, **100**, 647. Stellar evolutionary tracks.

Brewer, J.P., Fahlman, G.G., Richer, H.B., Searle, L. & Thompson, I., 1993. AJ, **105**, 2158. Mass-luminosity relation for metal-poor stars.

Briggs, F., 1990. ApJ, **352**, 15. Study of warped galaxies.

Brinks, E., 1984. *A High Resolution Hydrogen Line Survey of Messier 31*, thesis (Leiden: Leiden University).

Broeils, A.H. & van Woerden, H., 1994. A&AS, **107**, 129. Search for galaxies with extended HI disks.

Bronfman L., Cohen R.S., Alvarez H., May J. & Thaddeus P., 1988. ApJ, **324**, 248. Distribution of Galactic CO emission.

Brown, R.H., 1968. ARA&A, **6**, 13. Intensity interferometry.

Brown, T.M. & Gilliland, R.L., 1994. ARA&A, **32**, 37. Asteroseismology.

Brown, R.H. & Twiss, R.Q., 1956. Nat, **178**, 1046. First intensity interferometer.

Brown, R.H., Davis, J., Lake, R.J. & Thompson, R.J., 1974. MNRAS, **167**, 475. Intensity interferometry

Brown, R.H., Davis, J. & Allen, L.R., 1974. MNRAS, **167**, 121. Intensity interferometry.

Brown, A.G.A., Perryman, M.A.C., Kovalevsky, J., Robichon, N., Turon, C. & Mermilliod, J.-C., 1997. In *Hipparcos Venice '97*, ed. Battrick, B. (Noordwijk: ESA), p. 681. The Hyades studied with Hipparcos.

Brown, J, Wallerstein, G. & Oke, J., 1991. AJ, **101**, 1693. Abundances of CNO in globular clusters.

Budding, E., 1993. *Introduction to astronomical photometry* (Cambridge: CUP).

Buonanno et al., 1994. A&A, **290**, 69. CM diagram for M3.

Buote, D.A. & Canizares, C.R., 1996. ApJ, **468**, 184. X-ray observations of NGC 720.

Buote, D.A. & Canizares, C.R., 1997. ApJ, **474**, 650. X-ray temperature of NGC 720.

Burke, B.F., 1957. AJ, **62**, 90. Discovery of the Milky Way's warp.

Burstein, D., 1979. ApJ, **234**, 829. Photometry of edge-on S0s.

Burstein, D. & Heiles, C., 1978. ApJ, **225**,

40. Extinction along lines of sight out of the Milky Way.

Burstein, D. & Heiles, C., 1982. AJ, **87**, 1165. Extinction along lines of sight out of the Milky Way

Burstein, D., Rubin, V.C., Thonnard, N. & Ford, W.K., 1982. ApJ, **253**, 70. Rotation of Sc galaxies.

Burton, W.B., 1971. A&A, **10**, 76. Introduction of dynamical modeling of Milky Way's spiral structure.

Burton, W.B., 1985. A&AS, **62**, 365. HI survey of the Milky Way

Burton, W.B., 1992. In *The Galactic Interstellar Medium*, SAAS-Fee Advanced Course 21, Eds. Pfenniger, D., Bartholdi, P. (Berlin: Springer). Observational proerties of the ISM.

Burton, W.B. & Liszt, H.S., 1978. ApJ, **225**, 815. Derivation of the tilt of the nuclear disk of the Milky Way.

Burton, W.B., Liszt, H.S., 1983. A&AS, **52**, 63. HI survey of the Galactic center.

Buscombe, W., 1995. *Twelveth General Catalogue of MK Spectral Classification*, (Evanston: Northwestern Univ.). MK stellar classifications.

Buser, R. & Kurucz, R.L., 1992. A&A, **264**, 557. Theoretical stellar spectra.

Buta, R., 1986. ApJS, **61**, 609. Survey of rings in SB galaxies.

Buta, R. & Combes, F., 1996. Fund. Cosmic Physics, **17**, 95. Galactic rings review.

Buta, R., Crocker, D.A. & Byrd, G.G., 1992. AJ, **103**, 1526. Galaxy with both leading and trailing arms.

Butler, D., 1975. ApJ, **200**, 68. The relation between $\Delta S$ and [Fe/H].

Buzzoni, A., Fusi Pecci, F., Buonanno, R. & Corsi, C.E., 1983. A&A, **128**, 94. Helium abundances in globular clusters.

Byun, Y., *et al.*, 1996. AJ, **111**, 1889. Core surface-brightness profiles of elliptical galaxies.

Côté, P., Welch, D.L., Fischer, P. & Gebhardt, K., 1995. ApJ, **454**, 788. Kinematics of 399 giant stars in NGC 3201.

Cacciari, C., Clementini, G., 1990. Eds *Confrontation Between Stellar Pulsation and Evolution*, ASP Conf. Ser. 11, (San Francisco: ASP).

Caldwell, J.A.R. & Coulson, I.M., 1986. MNRAS, **218**, 223. PLC relation for Cepheid variables.

Caldwell, J.A.R. & Ostriker, J.P., 1981. ApJ, **251**, 61. Mass models of the Milky Way.

Cannon, A.J., 1925–1936. *Ann. Harvard Coll. Obs.* 100 Henry Draper Catalog Extension.

Cannon, A.J. & Walton Mayall, M., 1949. *Ann. Harvard Coll. Obs.* 112. Henry Draper Catalog Extension.

Cannon, A.J. & Pickering, E.C., 1918–1924. *Ann. Harvard Coll. Obs.* 91–99. Henry Draper Catalog.

Capaccioli, M., Della Valle, M., D'Onofrio, M. & Rosino, L., 1989. AJ, **97**, 1622. Novae in M31 as possible standard candles.

Capaccioli, M., Cappellaro, E., Della Valle, M., D'Onofrio, M., Rosino, L. & Turatto, M., 1990. ApJ, **350**, 110. Relative distances to Coma and Virgo from type Ia supernovae.

Capaccioli, M., Piotto, G. & Rampazzo, R., 1988. AJ, **96**, 487. Surface photometry of elliptical galaxies.

Capaccioli, M., Piotto, G.P. & Stiavelli, M., 1993. MNRAS, **261**, 819. Correlation between location and IMF for globular clusters.

Cardelli, J.A., Clayton, G.C. & Mathis, J.S., 1989. ApJ, **345**, 245. Extinction curve

Carilli, C.L. & Barthel, P.D., 1996. A&A Rev., **7**, 1. Cygnus A review.

Carney, B.W. & Latham, D.W., 1986. AJ, **92**, 60. The kinematics of halo red giants.

Carney, B.W. & Seitzer, P., 1993. AJ, **105**, 2127. Warp of Milky Way traced through red giants.

Carney, B.W., Fulbright, J.P., Terndrup, D.M., Suntzeff, N.B. & Walker, A.R., 1995. AJ, **110**, 1674. Determination of $R_0$

Carney, B.W., Latham, D.W., Laird, J.B. & Aguilar, L.A., 1996. AJ, **112**, 668. Correlations between kinematics and metallicity for high-velocity stars.

Carollo, C.M., de Zeeuw, P.T., van der Marel, R.P., Danzinger, I.J. & Qian, E.E., 1995. ApJ, **441**, 25. LOSVDs of elliptical galaxies to beyond $R_e$.

Carroll, S.M., Press, W.H., Turner, E.L., 1992. ARA&A, **30**, 499. The cosmological constant.

Carter, D. & Metcalfe, N., 1980. MNRAS, **191**, 325. cD galaxies are aligned with the surrounding cluster

Carter, D., Inglis, I., Ellis, R.S., Efstathiou, G.P. & Godwin, J.G., 1985. MNRAS, **212**, 471.. In cD galaxies the velocity dispersion increases into the halo.

Casuo, E., Vazdekis, A., Peletier, R.F. & Beckman, J.E., 1996. ApJ, **458**, 533. Calibration of $Mg_2$ index.

Catelan, M. & de Freitas Pacheco, J.A., 1993. AJ, **106**, 1858. Age differences between globular clusters.

Cayrel, R., 1968. ApJ, **151**, 997. Lineblanketing corrections.

Cayrel, R., 1992. In *Current Topics in Astrofundamental Physics*, eds Sanchez, N., Zichini, A. (World Scientific: Singapore), p. 668.

Celnik, W., Rholfs, K. & Braunsfurth, E., 1979. A&A, **76**, 24. Effects of the random velocities of HI clouds on circularspeed determinations

Chaboyer, B., Demarque, P., Guenther, D.B., Pinsonneault, M.H. & Pinsonneault, L.L., 1995. In *The Formation of the Milky Way*, eds Alfaro, E.J., Delgado, A.J. (Cambridge: CUP), p. 289. Yale stellar isochrones.

Chaboyer, B., Demarque, P. & Sarajedini, A., 1996. ApJ, **459**, 558. $\Delta V$ method for globular cluster ages.

Chaboyer, B., Sarajedini, A. & Demarque, P., 1992. ApJ, **394**, 515. Effects of He diffusion on the ages of globular clusters.

Chandrasekhar, S., 1931. ApJ, **74**, 81. The maximum mass of a white dwarf.

Charlot, S., Worthey, G. & Bressan, A., 1996. ApJ, **457**, 625. Models of population evolution.

Chieffi, A. & Straniero, O., 1989. ApJS, **71**, 47. Stellar isochrone calculations.

Chiosi, C., Bertelli, G. & Bressan, A., 1992. ARA&A, **30**, 235. The CM diagram and the theory of stellar structure.

Chiosi, C., Wood, P.R. & Capitanio, N., 1993. ApJS, **86**, 541. Models of Cepheids in the CM diagram.

Chokshi, A. & Turner, E.L., 1992. MNRAS, **259**, 421. Mean density of dead quasars.

Ciardullo, R., Jacoby, G.H., Ford, H.C. &

Neill, J.D., 1989. ApJ, **339**, 53. Planetary nebula luminosity function.

Cinzano, P. & van der Marel, R.P., 1994. MNRAS, **270**, 325. Model for the kinematics of NGC 2974.

Ciotti, L., 1991. A&A, **249**, 99. Generalized $R^{1/4}$ law.

Ciri, R., Bettoni, D. & Galletta, G., 1995. Nat, **375**, 661. Massive counter-rotating gas disk in NGC 3626.

Claria, J.C., Minniti, D., Piatti, A.E. & Lapasset, E., 1994. MNRAS, **268**, 733. Calibration of the DDO system.

Clayton, D.D., 1968. *Principles of Stellar Evolution and Nucleosynthesis* (New York: McGraw-Hill).

Clemens, D.P., Sanders, D.B., Scoville, N.Z. & Solomon, P.M., 1986. ApJS, **60**, 297. CO survey of the Milky Way

Cohen, J.G., 1978. ApJ, **223**, 487. variations in abundance of Na between globular cluster stars.

Cohen, J.G., 1985. ApJ, **292**, 90. Calibration of novae brightness versus rate of decline.

Cohen, R.J., Few, R.W., 1976. MNRAS, **176**, 495. Absorption by OH in front of the diffuse emission from the Galactic center.

Cohen, R.J., Few, R.W., 1981. MNRAS, **194**, 711. Absorption by $H_2CO$ in front of the diffuse emission from the Galactic Center.

Cohen, M., Walker, R.G. & Witteborn, F.C., 1992. AJ, **104**, 2045. Spectrophotometry of $\alpha$ Tau.

Combes, F. & Sanders, R.H., 1981. A&A, **96**, 164. Numerical simulations of buckling bars.

Condon, J.J., 1987. ApJS, **64**, 485. Bright spirals at 1.49 GHz.

Condon, J.J., et al., 1997. http://www.cv-.nrao.edu/ jcondon/nvss.html NRAO VLA Sky Survey.

Cool, A.M., Piotto, G., King, I.R., 1996. ApJ, **468**, 655. White-dwarf cooling sequence.

Corbin T.E., Warren W.H. Jr., 1991. *NASA, NSSDC*, **91-11**. Catalog of International Reference Stars.

Corbin T.E., Urban S.E., Warren W.H. Jr., 1991. *NASA, NSSDC*, **91-10**. ACRS stellar catalog.

Cottrell, P.L. & Da Costa, G.S., 1981.

ApJ, **245**, L79. Primordial enhancement of Na in clusters.

Cousins, A.W.J, 1976. Mem. R.A.S, **81**, 25. Cousins R I bands defined.

Cox, J.P., Giuli, R.T., 1968. *Principles of Stellar Structure*, (Reading: Gordon & Breech). Text on stellar structure

Crawford, D.L., 1966. in *Spectral Classification & Multicolor Photometry*, IAU Symp. 24, ed. Lodén K., Lodén L.O., Sinnerstad, U. (New York: Academic). Extension of *ubv* system.

Cuddeford, P. & Binney, J.J., 1994. MNRAS, **266**, 273. Models of solar-neighborhood velocity distributions.

Cudworth, K.M., 1985. AJ, **90**, 65. Proper motions in globular clusters.

Curtis, H.D., 1921. Bull. Nat. Res. Coun., **2**, 194. Curtis' account of the Shapley-Curtis debate

D'Antona, F. & Mazzitelli, I., 1994. ApJS, **90**, 467. Low-mass stellar models at $Z = Z_\odot$.

D'Antona, F. & Mazzitelli, I., 1996. ApJ, **456**, 329. Low-mass stellar models at $Z \ll Z_\odot$.

Da Costa, G.S., 1990. In *The Magellanic Clouds*, IAU Symp. 148, eds Haynes, R., Milne, D. (Dordrecht: Kluwer), p. 183. Globulars of the LMC.

Da Costa, G.S. & Armandroff, T.E., 1990. AJ, **100**, 162. Giant branches in globular clusters.

Da Costa, G.S. & Freeman, K.C., 1976. ApJ, **206**, 128. Comparison of single and multi-mass King models for M3.

Dachs, J. & Kaiser, D., 1984. A&AS, **58**, 411. Bright star photometry for the open cluster NGC 4755.

Dahn, C.C., Liebert, J., Harris, H.C. & Guetter, H.H., 1995. In *The Bottom of the Main Sequence and Beyond*, ed. Tinney, C.G., (Heidelberg: Springer), p. 239

Dame, T.M., 1993. In *Back to the Galaxy*, AIP Conference Proceedings 278, eds Holt, S.S., Verter, F. (New York: AIP), p. 267. Distribution of HI and $H_2$ in the Galaxy.

Dame, T., *et al.*, 1987. ApJ, **322**, 706. CO survey of the Milky Way.

Danks, A.C. & Dennefeld, M., 1994. PASP, **106**, 382. Stellar spectral atlas in the red.

Davies, R.L. & Birkinshaw, M., 1986.

ApJ, **303**, 45. Minor axis rotation of NGC 4261.

Davies J. & Burstein, D., 1995. Eds *The Opacity of Spiral Disks*, NATO ASI 469, (Dordrecht: Kluwer).

Davies, R.L. & Illingworth, G., 1983. ApJ, **266**, 516. $(v/\sigma)^*$ measurements for galactic bulges.

Davies, R.L., Efstathiou, G., Fall, S.M., Illingworth, G. & Schechter, P.L., 1983. ApJ, **266**, 41. $(v/\sigma)^*$ measurements for elliptical galaxies and bulges.

Davies, R.L., Sadler, E.M., Peletier, R.F., 1993. MNRAS, **262**, 650. Line-strength gradients in E galaxies.

Davis, J., 1994. In *Very High Angular Resolution Imaging*, IAU Symposium 158, eds Robertson, J.G., Tango, W.J. (Dordrecht: Kluwer), p. 135. The SUSI interferometer

Davis, M. & Peebles, P.J.E., 1983. ARA&A, **21**, 109. Estimates for Virgocentric infall velocity.

de Carvalho, R.R. & da Costa, L.N., 1987. A&A, **171**, 66. Photometric study of peanut-bulged NGC 1381.

de Jager, C., 1984. A&A, **138**, 256. Stability limit of supergiant atmospheres.

de Jager, C., Nieuwenhuijzen, H. & van der Hucht, K.A., 1988. A&AS, **72**, 259. Mass loss from supergiants.

de Jong, R.S., 1996. A&A, **313**, 377. IR colors of disk galaxies.

de Jong, R.S. & van der Kruit, C., 1994. A&AS, **106**, 451. CCD photometry of 86 spirals galaxies.

de Souza, R.E. & dos Anjos, S., 1987. A&AS, **70**, 465. Fraction of galaxies with boxy bulges.

de Vaucouleurs, G., 1948. Ann. Astrophys., **11**, 247. Discovery of the $R^{1/4}$ law.

de Vaucouleurs, G., 1959. AJ, **64**, 397. Transparency of galactic disks.

de Vaucouleurs, G., 1959. In *Handbuch der Physik*, vol. 53, ed. Flügge, S. (Berlin: Springer), p. 275. The de Vaucouleurs galaxy classification system defined.

de Vaucouleurs, G., 1984. In *Astronomy with Schmidt-Type Telescopes*, ed. Capaccioli, M. (Dordrecht: Reidel), p. 367. Techniques of surface photometry.

de Vaucouleurs, G. & Freeman, K.C., 1970. Vistas Astr., **14**, 163. Magellanic galaxies.

de Vaucouleurs, G. & Pence, W.D., 1978. AJ, **83**, 1163. The Milky Way at optical wavelengths.

de Vaucouleurs, G. & Peters, W.L., 1986. ApJ, **303**, 19. Calibration of the Tully–Fisher relation.

de Vaucouleurs, G., de Vaucouleurs, A., Corwin, H.G., Buta, R., Paturel, G. & Fouque, P., 1991. *Third Reference Catalogue of Bright Galaxies*, (Austin: University of Texas Press). RC3 catalog

de Vaucouleurs, G., de Vaucouleurs, A. & Corwin, H.G., 1976. *Second Reference Catalogue of Bright Galaxies*, (Austin: University of Texas Press). RC2 catalog.

Dehnen, W., 1993. MNRAS, **265**, 250. The $\gamma$-models

Dehnen, W., 1998. AJ, **00**, 00. The distribution of solar-neighborhood stars in velocity-space.

Dehnen, W. & Binney, J.J., 1996. In *Formation of the Galactic Halo ... Inside and Out*, ASP Conf. Ser. 92, eds Morrison, H. Sarajedini, A. (San Francisco: ASP), p. 393. Techniques for modeling the Milky Way.

Dehnen, W. & Binney, J.J., 1998. MNRAS, **00**, 00. Kinematics of the solar neighborhood.

Dehnen, W. & Binney, J.J., 1998. MNRAS, **294**, 429. Mass models of the Milky Way.

Dekel, A., 1994. ARA&A, **32**, 371. Measuring largescale cosmic flows.

Dekel, A., et al., 1993. ApJ, **412**, 1. Measurement of the biasing parameter from largescale flows.

Delhaye, J., 1965. In *Galactic Structure*, eds Blaauw, A., Schmidt, M. (Chicago: University of Chicago Press), p. 61 The solar motion.

Dettmar, R.-J. & Barteldrees, A., 1990. in *ESO/CTIO Workshop on Bulges of Galaxies*, eds B.J. Jarvis & D.M. Terndrup (Garching: ESO), p. 259. Photometry of galaxies with boxy bulges.

Di Benedetto, G.P. & Rabbia, Y., 1987. A&A, **188**, 114. Angular diameters of stars by interferometry in the infrared.

Dickens, R.J. & Woolley, R.v.d.R., 1967. R. Obs. Bull., **128**, . Broad giant branch in $\Omega$ Cen CM diagram.

Dickens, R.J., Croke, B.F.W., Cannon,

R.D. & Bell, R.A., 1991. Nat, **351**, 212. Comparison between NGC 288 and NGC 362.

Dickman, R.L., 1978. ApJS, **37**, 407. $A_V$ correlated with CO lines strengths

Digel, S., de Geus, E.J. & Thaddeus, P., 1994. ApJ, **422**, 92. Survey of outer disk of Milky Way in CO.

Digel, S.W., Lyder, D.A., Philbrick, A.J., Puche, D. & Thaddeus, P., 1996. ApJ, **458**, 561. CO survey of nearby arms.

Diplas, A. & Savage, B.D., 1991. ApJ, **377**, 126. Analysis of the Milky Way's warp.

Disney, M.J., et al., 1995. Nat, **376**, 150. HST observations of QSO host galaxies.

Disney, M., Davies, J., Phillipps, S., 1989. MNRAS, **239**, 939. Evidence that galactic disks are optically thick.

Dixon, R.S., 1970. ApJS, **20**, 1. Master List of radio sources.

Djorgovski, S., 1988. In *Globular Cluster Systems in Galaxies*, IAU Symp. 126, eds Grindlay, J., Philip, A.G.D. (Dordrecht: Kluwer), p. 333. Globular cluster luminosity profiles.

Djorgovski, S. & Davis, M., 1987. ApJ, **313**, 59. Parameters of the fundamental plane.

Djorgovski, S. & Dickinson, M., 1989. In *Highlights of Astronomy*, ed. McNally, D. (Dordrecht: Kluwer). Techniques of surface photometry.

Djorgovski, S. & King, I.R., 1986. ApJ, **305**, 61. Survey to detect collapsed-core clusters.

Djorgovski, S. & Sosin, C., 1989. ApJ, **341**, L13. Warp of Milky Way traced through IRAS point sources.

Djorgovski, S., Piotto, G. & Capaccioli, M., 1993. AJ, **105**, 2148. Correlation between globular cluster IMF, location, and [Fe/H].

Dressler, A., 1979, ApJ, **231**, 659. In cD galaxies the velocity dispersion increases into the halo.

Dressler, A., 1980. ApJ, **236**, 351. Morphology–density relation.

Dressler, A., 1987. ApJ, **317**, 1. $D_n$–$\sigma$ relation for galactic bulges.

Dressler, A., et al., 1987. ApJ, **313**, 42. Definition of $D_n$–$\sigma$ relation.

Dreyer, J.L.E., 1888. Mem. Roy. Astron. Soc., **49**, 1. The original NGC Catalog.

Dreyer, J.L.E., 1895. Mem. Roy. Astron. Soc., **51**, 185. First supplement to the NGC, the IC.

Dreyer, J.L.E., 1908. Mem. Roy. Astron. Soc., **59**, 105. Second supplement to the NGC, the IC continued.

Dubath, P., Meylan, G. & Mayor, M., 1994. ApJ, **426**, 192. Core velocity dispersion of M15.

Dube, R.R., Wickes, W.C. & Wilkinson, D.T., 1972. ApJ, **215**, L51. The brightness of the night sky.

Duerbeck, H.W, 1990. Bull. Inf. C.D.S., **34**, 127. Catalog of Galactic novae.

Duflot, M., Figon, P. & Meyssonnier, N., 1995. A&AS, **114**, 269. Compilation of stellar radial velocities.

Durrell, P.R. & Harris, W.W., 1993. ApJ, **105**, 1420. CM diagram for M15.

Eckart, A., Genzel, R., Hofman, R., Sams, B.J., Tacconi-Garman, L.E. & Cruzalebes, P., 1994. In *The Nuclei of Normal Galaxies*, eds Genzel, R., Harris, A.I. (Dordrecht: Kluwer), p. 305. Near-infrared imaging of the Galactic center.

Edge, D.O., Shakeshaft, J.R., McAdam, W.B., Baldwin, J.E. & Archer, S., 1959. Mem.R.A.S., **68**, 37. The 3C radio source catalog.

Edvardsson, B., Andersen, B., Gustafsson, B., Lambert, D.L., Nissen, P.E. & Tomkin, J., 1993. A&A, **275**, 101. Ages and metallicities of stars in the solar neighborhood.

Efstathiou, G., Ellis, R.S. & Peterson, B.A., 1988. MNRAS, **232**, 431. Luminosity function of field galaxies.

Eggen, O.J., 1958, 1960. MNRAS, **118**, 65;. MNRAS, **120**, 563. Moving cluster Ursa Major

Eggen, O.J., 1965. In *Stars and Stellar Systems*, vol. V, eds Blaauw, A., Schmidt, M., (Chicago: University of Chicago Press), p. 111. Moving groups.

Eggen, O.J., 1973. ApJ, **182**, 821. Location of the subdwarf MS.

Eggen, O.J., Lynden-BGell D. & Sandage, A., 1962. ApJ, **136**, 748. Formation of the Milky Way.

Elias, J.H., Frogel, J.A., Matthews, K. & Neugebauer, G., 1982. AJ, **87**, 1029. Definition of *JHKL* magnitudes.

Elmegreen, D.M. & Elmegreen, B.G., 1982. MNRAS, **201**, 1021. Definition of arm classes.

Elmegreen, B.G. & Elmegreen, D.M., 1985. ApJ, **288**, 438. Photographic photometry of 15 SB galaxies.

Elmegreen, D.M. & Elmegreen, B.G., 1987. ApJ, **314**, 3. Arm classes for 762 galaxies.

Elmegreen, B.G., Elmegreen, D.M., Chromey, F.R., Hasselbacher, D.A. & Bissell, B.A., 1996. AJ, **111**, 2233. Photographic photometry of 15 SB galaxies.

Elmegreen, B.G., Elmegreen, D.M. & Seiden, P.E., 1989. ApJ, **343**, 602. Grand-design spirals.

Elson, R.A., Fall, S.M., 1988. AJ, **96**, 1383. The age distributon of LMC clusters.

Elson, R.A.W., Gilmore, G.F., Santiago, B.X. & Casertano, S., 1995. ApJ, **110**, 682. Luminosity function of $\Omega$ Cen.

Elvius, A., 1956. journStockholm Obs. Ann. 18 9 Effect of scattering by dust on galaxy photometry.

Emsellem, E., Bacon, R., Monnet, G. & Poulain, P., 1996. A&A, **312**, 777. Two-dimensional kinematic map of M104.

Emsellem, E., Monnet, G. & Bacon, R., 1994. A&A, **285**, 723. Model of kinematic data from M104.

Englmaier, P. & Gerhard, O.E., 1997. MNRAS, **287**, 57. Simulations of gas flow in barred galaxies.

Fabbiano, G., Gioia, I.M. & Trinchieri, G., 1989, ApJ, **347**, 127. Correlations between $L_{rad}$, $L_{opt}$ and $L_X$ in Ellipticals.

Faber, S.M., 1973. ApJ, **179**, 731. Line-strength–magnitude effect.

Faber, S.M. & Jackson, R.E., 1976. ApJ, **204**, 668. Discovery of the Faber–Jackson relation.

Faber, S.M., Friel, E., Burstein, D. & Gaskell, C.M., 1985. ApJS, **57**, 711. Definition of a system of spectral indices.

Fabricius, C., 1993. Bull. Inf. Centre Donnees Stellaires, **42**, 5. Carlsberg Meridian Circle catalogs.

Fagotto, F., Bressan, A., Bertelli & G., Chiosi, C., 1994. A&AS, **104**, 365. Evolutionary tracks for $Z = 0.0004$.

Falco, E.E., Gorenstein, M.V. & Shapiro, I.I., 1991. ApJ, **372**, 364. Gravitational lens model for QSO 0957+561.

Fanaroff, B.L. & Riley, J.M., 1974. MNRAS, **167**, 31P. Classification of radio sources.

Fasano, G. & Vio, R., 1991. MNRAS, **249**, 629. Distribution of apparent axis ratios of the isophote at $R_e$.

Feast, M.W. & Catchpole, R.M., 1997. MNRAS, **286**, L1. Cepheid P-L relation from Hipparcos.

Feast, M.W. & Walker, A.R., 1987. ARA&A, **25**, 345. Cepheids as distance indicators.

Feast, M. & Whitelock, P., 1997. MNRAS, **291**, 683. The Oort constants from observations of Cepheids.

Feast, M.W., Glass, I.S., Whitelock, P.A. & Catchpole, R.M., 1989. MNRAS, **241**, 375. PL relation for Miras.

Fernie, J.D., 1990. ApJ, **354**, 295. Locations of Cepheids in the CM diagram.

Fernley, J.A., Longmore, A.J., Jameson, R.F., Watson, F.G. & Wesselink, T., 1987. MNRAS, **226**, 927. Determination of $R_0$.

Fich, M. & Tremaine, S.D., 1991. ARA&A, **29**, 409. $v_c(R)$ for the Milky Way.

Fichtel, C.E., et al., 1994. ApJS, **94**, 551. EGRET Source catalog.

Fitt, A.J. & Alexander, P., 1993. MNRAS, **261**, 445. $B$-fields in nearby spirals.

Forbes, D.A. & DuPoy, D.L., 1992. A&A, **259**, 97. Near IR photometry of NGC 253.

Forbes, D.A., Franx, M. & Illingworth, G.D., 1995. AJ, **109**, 1988. HST images of galaxies with kinematically distinct cores.

Fouts, G. & Sandage, A., 1986. AJ, **91**, 1189. Radial velocities for high proper motion stars.

Franx, M., 1988. MNRAS, **231**, 285. Systems with Stäckel potentials do not have twisted isophotes.

Franx, M. & Illingworth, G.D., 1988. ApJ, **327**, 55. Technique for obtaining LOSVDs from CCFs and ACFs of spectra.

Franx, M. & de Zeeuw, P.T., 1992. ApJ, **392**, L47. Elliptical disks and Tully–Fisher relation

Franx, M., Illingworth, G.D. & Heckman, T., 1989. AJ, **98**, 538. Surface photometry of elliptical galaxies.

Franx, M., Illingworth, G. & Heckman, T., 1989. ApJ, **344**, 613. Measurement of kinematic misalignment.

Franx, M., Illingworth, G. & de Zeeuw, T., 1991. ApJ, **383**, 112. Kinematic misalignments data.

Franx, M., van Gorkom, J.H. & de Zeeuw, P.T., 1994. ApJ, **436**, 642. Disk ellipticity and kinematics.

Freedman, W.L. & Madore, B.F., 1990. ApJ, **365**, 186. Distance to M31 relative to LMC from Cepheids

Freedman, W.L., et al., 1994. Nat, **371**, 757. HST observations of Cepheids in Virgo.

Freeman, K.C., 1970. ApJ, **160**, 811. Freeman's law.

Freeman, K.C., 1978. In *Structure and Properties of Nearby Galaxies*, eds. Berkhuysen, E.M., Wielebinski, R. (Dordrecht: Reidel), p. 3. Bulges distorted by disks.

Frei, Z. & Gunn, J.E., 1994. AJ, **108**, 1476. K-corrections

Frei, Z., Guhathakurta, P., Gunn, J.E. & Tyson, J.A., 1996. AJ, **111**, 174. Catalog of 113 nearby galaxies.

Frenk, C.S. & White, S.D.M., 1980. MNRAS, **193**, 295. The rotation speeds of globular cluster systems.

Frerking, M.G., Langer, W.D. & Wilson, R., 1982. ApJ, **262**, 590. CO intensity versus $A_V$.

Freudenreich, H.T., 1998. ApJ, **492**, 495. The Galaxy from COBE photometry.

Freudenreich, H.T., et al., 1994. ApJ, **429**, L69. First mapping of the Milky Way's warp in the light from stars.

Fricke, W. & Kopff, A., 1963. *Veroff. Astr. Rechen-Inst. Heidelberg*, **10**. FK4 stellar catalog.

Fricke W., Schwan H. & Corbin, T., 1991. *Veroeff. Astron. Rechen-Institut Heidelberg*, **33**. FK5 catalog extension.

Fricke W., Schwan H. & Lederle T., 1988. *Veroeff. Astron. Rechen-Institut Heidelberg*, **32**. FK5 stellar catalog.

Fried, J.W. & Illingworth, G.D., 1994. AJ, **107**, 992. Velocity dispersions of elliptical galaxies.

Friel, E.D., 1995. ARA&A, **33**, 381. Old disk clusters.

Frogel, J.A., Terndrup, D.M., Blanco & V.M., Whitford, A.E., 1990. ApJ, **353**, 494. Structure of the bulge at $3° < |b| < 12°$.

Frogel, J.A., Kuchinski, L.E. & Tiede,

G.P., 1995. AJ, **109**, 1154. High metallicity of Liller 1.

Fukugita, M., Okamura, S., Tarusawa, K., Rood, H.J. & Williams, B.A., 1991. ApJ, **376**, 8. Coma cluster redshift.

Fukugita, M., Ichikawa, T., Gunn, J.E., Doi, M., Shimasaku, K. & Schneider, D.P., 1996. AJ, **111**, 1748. Definition of the SDDS photometric system.

Fukugita, M., Hogan, C.J. & Peebles, P.J.E., 1993. Nat, **366**, 309. Review of the Hubble constant controversy.

Fusi Pecci et al., 1993. AJ, **105**, 1145. Correlation between HB morphology and concentration in globular clusters.

Fux, R. 1998. In *The Central Regions of the Galaxy and Galaxies*, IAU Symp. 184, ed. Sofue, Y., (Dordrecht: Reidel).

Güsten, R., Genzel, R., Wright, M.C.H., Jaffe, D.T., Stutzki, J. & Harris, A.I., 1987. ApJ, **318**, 124. HCN map of Sgr A.

Gamow, G., 1948. Nat, **162**, 680. Prediction of the cosmic background radiation.

García Gómez, C. & Athanassoula, E., 1993. A&AS, **100**, 431. Fourier decomposition of the spiral structures of 44 galaxies.

Garcia Lopez, R.J., Rebolo, R. & Beckman, J.E., 1988. PASP, **100**, 1489. Ages of disk clusters.

Gardner, J.P., Sharples, R.M., Frenk, C.S. & Carrasco, B.E., 1997. ApJ, **480**, L99. Infrared galaxy luminosity function.

Garnavich, P.M., et al., 1998. ApJ, **493**, L53. Hubble diagram using type Ia supernovae.

Garrett, M.A., et al., 1994. MNRAS, **270**, 457. VLBI observations of QSO 0957+561.

Gascoigne, S.C.B. & Shobbrook, R.R., 1978. PASA, **3**, 285. Relative photometry of Cepheids in the LMC.

Gautschy, A., 1987. Vistas in Astron., **30**, 197. Review of Baade–Wesselink method.

Gautschy, A. & Saio, H., 1995. ARA&A, **33**, 75. Review of stellar pulsation theory.

Gautschy, A. & Saio, H., 1996. ARA&A, **34**, 551. Review of observations of pulsating stars.

Gebhardt, K., et al. 1996. AJ, **112**, 105. Core luminosity densities of elliptical galaxies.

Geisler, D., Minniti, D. & Claria, J.J.,

1992. AJ, **104**, 627. [Fe/H] for metal-poor globular clusters.

Geller, M.J., Beers, T.C., 1982. PASP, **94**, 421. isopleth maps showing substructure in clusters.

Genzel, R., Reid, M.J., Moran, J.M. & Downes, D., 1981. ApJ, **244**, 884. Dynamical model of Galactic maser source.

Genzel, R., Eckart, A., Ott, T. & Eisenhauer, F., 1997. MNRAS, **291**, 219. The black hole at the Galactic center.

Gerhard, O.E., 1993. MNRAS, **265**, 213. Characterization of LOSVD by Gauss–Hermite moments.

Gerhard, O.E. & Binney, J.J., 1996. MNRAS, **279**, 993. Definition of konus densities.

Gerssen, J., Kuijken, K. & Merrifield, M.R. 1997. MNRAS, **288**, 618. Stellar kinematics of the disk galaxy NGC 488.

Gezari, D.Y., Schmitz, M., Pitts, P.S. & Mead, J.M., 1993. *Catalog of Infrared Observations*, 3rd edition, NASA Reference Publication **1294**, (Greenbelt: NASA). Catalog of IR sources.

Giclas, H.L, Burnham, R. & Thomas, N.G., 1971,1978. *Lowell Obs. Bull.* (Flagstaff: Lowell Obs.) Lowell Proper Motion Surveys.

Gill, D. & Kapteyn, J.C., 1895–1900. *Ann. Cape Obs.*, **3–5** Cape Durchmusterung (CPD stars).

Gillespie, A.R., Huggins, P.J., Sollner, T.C.L.G., Phillips, T.G., Gardner, F.F. & Knowles, S.H., A&A, **60**, 221. Southern HII regions.

Gilmore, G. & Reid, I.N., 1983. MNRAS, **202**, 1025. The vertical structure of the Milky Way's stellar disk.

Gilroy, K.K., Sneden, C., Pilachowski, C.A. & Cowan, J.J., 1988. ApJ, **327**, 298. Relative importance of the r- and s-processes for the enrichment of metal-poor stars.

Giovanelli, R. & Haynes, M.P., 1983. AJ, **88**, 881. Gas-poverty of Virgo galaxies.

Giovanelli, R., Haynes, M.P., Salzer, J.J., Wegner, G., Da Costa, L.N., & Freudling, W., 1995. AJ, **110**, 1059. Effect of disk opacity on photometric parameters.

Glass, I.S., 1974. Mon. Notes Astron. Soc. South Afr., **33**, 53. Introduction of *H* band.

Gliese, W., 1969. Veröff. Astron. Rechen-Inst., **22**, 1. Catalogue of nearby stars.

Gliese, W. & Jahreiss, H., 1979, A&AS, **38**, 423.

Goldreich, P., Kwan, J., 1973. ApJ, **189**, 441. Physics of molecular line emission.

Goodman, J. & Lee, H.M., 1989. ApJ, **337**, 84. Collapse of dark clusters to form massive black holes.

Gould, A., 1990. ApJ, **360**, 504. When is population sufficiently Maxwellian.

Gould, A., 1990. MNRAS, **244**, 25. On the determination of the local disk density.

Gould A., Bahcall J.N. & Flynn C., 1996. ApJ, **465**, 759. The luminosity function at faint magnitudes.

Gower, J.F.R., Scott, P.F. & Wills, D., 1967. Mem.R.A.S., **71**, 49. 4C radio source catalog.

Grabelsky, D.A., Cohen, R.S., Bronfman, L., Thaddeus, P. & May, J., 1987. ApJ, **315**, 122. Thickness of Galactic CO layer

Grabelsky, D.A., Cohen, R.S., Bronfman, L. & Thaddeus, P., 1988. ApJ, **331**, 181. The Carina spiral arm.

Green, R.M., 1985. *Spherical Astronomy*, (Cambridge: CUP).

Green, D.A., 1991. PASP, **103**, 209. Catalog of Galactic SNRs.

Green, E.M., Morrison, H.L., 1993. In *The Globular Cluster – Gallaxy Connection*, ASP Conf. Ser. 48, eds Smith, G., Brodie, J. (San Francisco: ASP), p. 318. The local density of HB stars.

Green, E.M., Demarque, P. & King, C.R. 1987. *The Revised Yale Isochrones and Luminosity Functions*, (New Haven: Yale Obs.). Yale stellar isochrones.

Greenhill, L.J., Moran, J.M., Reid, M.J., Menten, K.M. & Hirabayashi, H., 1993. ApJ, **406**, 482. Proper motions of maser spots in M33

Greenhill, L.J., Henkel, C., Becker, R., Wilson, T.L. & Wouterloot, J.G.A., 1995. A&A, **304**, 21. Modelling of maser spots in NGC 4258

Grenon, M., 1989. Ap Space Sci., **156**, 29. Models of chemical evolution.

Griffin, R.F., 1967. ApJ, **148**, 465. Introduction of radial-velocity measuring instrument.

Griffin, R.F., Gunn, J.E., Zimmerman, B.A. & Griffin, R.E.M., 1988. AJ, **96**, 172. Distance to Hyades by modified moving cluster method

Grindlay, J.E., Hertz, P., Steiner, J.E., Murray, S.S. & Lightman, A.P., 1984. ApJ, **282**, L13. Spatial distribution of X-ray sources in globular clusters.

Guhathakurta, P., Yanny, B., Schneider, D.P & Bahcall, J.N., 1996. AJ, **111**, 267. HST image of M15.

Gum, C.S., Kerr, F.J. & Westerhout, G., 1960. MNRAS, **121**, 132. Early analysis of Milky Way's warp

Gunn, J.E. & Griffin, R.F., 1979. AJ, **84**, 752. First detailed dynamical model of a globular cluster.

Gunn, J.E. & Stryker, L.L., 1983. ApJS, **52**, 121. Atlas of stellar spectra.

Gunn, J.E., Griffin, R.F., Griffin, R.E.M. & Zimmerman, B.A., 1988. AJ, **96**, 198. Distance to Hyades by modified moving cluster method

Guth, A.H., 1989. In *Bubbles, Voids and Bumps in Time*, (Cambridge: CUP), p. 105. Popular description of inflation.

Gwinn, C.R., Moran, J.M. & Reid, M.J., 1992. ApJ, **393**, 149. Value of $R_0$ from maser kinematics.

Höflich, P. & Khokhlov, A., 1996. ApJ, **457**, 500. Models of type Ia supernova explosions.

Hamuy, M., Walker, A.R., Suntzeff, N.B., Gigoux, P., Heathcote, S.R. & Phillips, M.M., 1994. PASP, **106**, 566. Catalog of spectral standard stars.

Han, C.H. & Ryden, B.S., 1994. ApJ, **433**, 80. Shapes of globular clusters.

Hanes, D.A. & Whittaker, D.G., 1987. AJ, **94**, 906. Luminosity function of globular clusters.

Hanson, R.B., 1975. AJ, **80**, 379. Moving cluster method.

Hanson, R.B., 1979. MNRAS, **186**, 875. Discussion of Lutz–Kelker bias.

Hanson, R.B., 1980. In *Star Clusters*, IAU Symp. 85, ed. E. Hesser (Dordrecht: Reidel). Moving cluster method correction.

Hanson, M.M., Dickey, J.M. & Helou, G., 1990. ApJ, **352**, 522. HI line width of face-on spiral.

Hanson, R.B., Klemola, A.R. & Jones, B.F., 1994 Bull. Am. Astr. Soc., **184**, 2707. Lick catalog.

Harkness, R.P. & Wheeler, J.C., 1990. In *Supernovae*, ed. Petschek, A.G. (New

York: Springer), p. 1. Classification of supernovae.

Harris, W.E., 1987. ApJ, **315**, L33. Globular clusters in the Coma cluster.

Harris, W.E., 1991. ARA&A, **29**, 543. Globular cluster systems.

Harris, W.E., 1996. AJ, **112**, 1487. Globular cluster properties database.

Hartmann, D. & Burton, W.B., 1997. *Atlas of Galactic Neutral Hydrogen*, (Cambridge: CUP) HI survey of the Milky Way.

Hartwick, F.D.A., 1976. ApJ, **209**, 418. Chemical evolution of the halo.

Hartwick, F.D.A., 1987. In *The Galaxy*, NATO ASI, eds Gilmore, G., Carswell, R. (Dordrecht: Reidel), p. 281. The structure of the Galactic halo.

Hasinger, G., *et al.*, 1993. A&A, **275**, 1. Contribution to diffuse soft X-ray flux from unresolved quasars, etc.

Hauck, B. & Mermilliod, M., 1998. A&AS, in press. Strömgren Photometry Catalog.

Hauser, M.G., Kelsall, T., Leisawitz, D. & Weiland, J., 1997. *COBE Diffuse Infrared Background Experiment (DIRBE) Explanatory Supplement*, (Greenbelt: NASA/GSFC). COBE all-sky data.

Hausman, M.A. & Ostriker, J.P., 1978. ApJ, **224**, 320. Formation of cD galaxies by cannibalism.

Hawarden, T.G., 1975. MNRAS, **173**, 223. Spatial distribution of HB stars in old open clusters.

Hawley, S.L., Gizis, J.E. & Reid, I.N., 1996. AJ, **112**, 2799. Spectra of Gliese Catalog stars.

Hayashi, C., 1961. PASJ, **13**, 450. Discovery of the Hayashi track.

Hayes, D. & Latham, D., 1975. ApJ, **197**, 593. Spectrophotometry of Vega

Hayes, D.S., Pasinetti, L.E. & Davis Philip, A.G., 1985. Eds *Calibration of Fundamental Stellar Quantities*, Proc. IAU Symp. 111, (Dordrecht: Reidel).

Hanes, D.A., 1977. MNRAS, **180**, 309. Globular clusters of Virgo galaxies.

Haynes, M.P., Giovanelli, R. & Roberts, M.S., 1979. ApJ, **229**, 83. Intergalactic HI in the Leo Group.

Heck, A., Egret, D., Jaschek, M. & Jaschek, C., 1984. A&AS, **57**, 213. Ultraviolet spectra of stars from IUE.

Heckmann, O. & Dieckvoss, W., 1975.

(Hamburg: Hamburg-Bergedorf Obs.). AGK3 stellar catalog.

Heiles, C., 1984. ApJS, **55**, 585. Identification of "worms."

Heiles, C. & Habing, H.J., 1974. A&AS, **14**, 1. HI survey of the Milky Way

Henry, T.J. & McCarthy, D.W., 1993. AJ, **106**, 773. Masses of visual binaries.

Henry, R.B.C., Pagel, B.E.J., Lasseter, D.F. & Chincarini, G.L., MNRAS, **258**, 321. ISM abundances in NGC 4303.

Herbig, G.H., 1962. ApJ, **135**, 736. Nuclear burning ages and contraction ages of open clusters.

Hernquist, L., 1990. ApJ, **356**, 359. Hernquist model defined.

Herschel, W., 1785. Phil. Trans., **75**, 213. Herschel's "star gauging" model of the Milky Way.

Hesser, J.E., Harris, W.E., vandenBerg, D.A., Allwright, J.W.B., Shott, P. & Stetson, P.B., 1987. PASP, **99**, 739. CM Diagram for 47 Tuc.

Hewish, A., Bell, S.J., Pilkington, J.D.H., Scott, P.F & Collins, R.A., 1968. Nat, **217**, 709. Discovery of the first pulsar.

Hewitt, A. & Burbidge, G., 1993. ApJS, **87**, 451. Catalog of 7315 QSOs.

Hills, J.G & Day, C.A., 1975. ApJ, **17**, L87. Blue stragglers from stellar encounters.

Hiltner, W.A. & Johnson, H.L., 1956. ApJ, **124**, 367. Extinction in *UBV* bands.

Hindman, J.V., 1967 Australian J. Phys., **20**, 147. HI in Magellanic Clouds.

Hobbs, L.M., 1968. ApJ, **157**, 135. Absorption by interstellar Na.

Hoessel, J.G., Gunn, J.E. & Thuan, T.X., 1980. ApJ, **241**, 486. Hubble diagram for central dominant galaxies in clusters.

Hoffleit, D. & Warren, W.H., 1991. *Bright star Catalog*, 5th revised edition, (Greenbelt: Astronomical Data Center). Yale Bright Star Catalog (BSC).

Hoffleit D., Saladyga M. & Wlasuk P., 1983. *A Supplement to the Bright Star Catalog*, (New Haven: Yale Observatory).

Holmberg, E., 1958. journ*Medd. Lunds Astron. Obs. Ser. II* 136 1 Transparency of disk galaxies.

Holzapfel, W.L., *et al.*, 1997. ApJ, **480**, 449. Hubble constant from the Sunyaev–Zel'dovich effect.

Holzer, T.E., 1989. ARA&A, **27**, 199. Cosmic-ray pressure.

Hoskins, M.A., 1976. J. Hist. Astron., **7**, 169. Description of the Shapley-Curtis debate.

Hough, J.H., 1996. In *Polarimetry of the ISM*, ASP Conf. Ser. 97, eds Roberge, W., Whittet, D. (San Francisco: ASP). Investigating $B$-fields using optical polarization.

Houk, N., 1975-. *Michigan Catalogue of Two-Dimensional Spectral Types for the HD Stars* (Ann Arbor: Univ. of Michigan) Catalog of stellar spectral types.

Houk, N, Swift, C.M., Murray, C.A., Penston, M.J., Binney, J.J., 1997. In *Hipparcos Venice '97*, ed. Battrick, B. (Noordwijk: ESA), p. 279. Luminosity function of MS stars of given MK class.

Hubble, E.P., 1922. ApJ, **56**, 162. Measurement of the distance to M31.

Hubble, E.P., 1929. Proc. Nat. Acad. Sci. (Wash.), **15**, 168. Definition of the Hubble law.

Hubble, E.P., 1930. ApJ, **71**, 231. Early application of the Hubble law.

Hubble, E.P., 1936. *The Realm of the Nebulae*, (New Haven: Yale University Press). Exposition of Hubble sequence.

Hubble, E., Humason, M.L., 1931. ApJ, **74**, 43. Early statement of morphology-density effects.

Huchra, J.P., 1984. In *The Virgo Cluster*, (Garching: ESO), p. 181.

Huchra, J., Geller, M., Clemens, C., Tokarz, S. & Michel, A., 1995, Bull. Inf. C.D.S., **41**, 31. CfA redshift catalog.

Huchtmeier, W.K. & Richter, O.-G., 1989. *A General Catalog of HI Observations of Galaxies*, (Berlin: Springer).

Huizinga, J.E. & van Albada, T.S., 1992. MNRAS, **254**, 677. UGC diameters are not isophotal.

Hut, P. *et al.*, 1992. PASP, **104**, 981. Review of binaries in globular clusters.

Ibata, R.A., Gilmore, G. & Irwin, M.J., 1995. MNRAS, **277**, 781. Discover of Sag Dwarf galaxy.

Iben, I., 1967. ARA&A, **5**, 571. Stellar structure review.

Iben, I., 1974. ARA&A, **12**, 215. Stellar structure review

Iben, I. & Rood, R.T., 1969. Nat, **223**, 933. The R-test for measuring helium abundance.

Ichikawa, S-I, Wakamatsu, K-I & Okamura, S., 1986. ApJS, **60**, 475. Photometry of dE galaxies.

Impey, C.D., Sprayberry, D., Irwin, M.J. & Bothun, G.D., 1996. ApJS, **105**, 209. Catalog of LSB galaxies.

Jackson, P.D., Fitzgerald, M.P. & Moffat, A.F.J., 1979. In *The large-scale characteristics of the Galaxy*, IAU Symp. 84, ed. Burton, W.B., (Dordrecht: Reidel), p. 221. Discovery that $v_c(R)$ flat or rising beyond $R_0$.

Jacoby, G.H., *et al.*, 1992. PASP, **104**, 599. Review of relative distance estimators.

Jacoby, G.H., Hunter, D.A. & Christian, C.A., 1984. ApJS, **56**, 257. Atlas of stellar spectra.

Jacoby, G.H., Walker, A.R. & Ciardullo, R., 1990. ApJ, **365**, 471. Distance to the LMC from planetary nebulae.

Jaffe, W., 1983. MNRAS, **202**, 995. The Jaffe model defined.

Jahreiss, H., 1993. In *Galactic and Solar System Optical Astrometry*, eds Morrison, L.V., Gilmore, G.F. (Cambridge: CUP), p. 44. Analysis of Gliese Catalog.

Jahreiss, H. & Gliese, W., 1993. In *Workshop on Databases for Galactic Structure*, eds. Philip, A.G.D., Upgren, A.R. (Schenectady: L. Davis Press). Gliese Catalog of Nearby Stars.

Jahreiss, H. & Wielen, R., 1983. In *The Nearby Stars and the Stellar Luminosity Function*, IAU Coll. 76, eds Davis Philip, A.G., Upgren, A.R. (Schenectady: L. Davis Press), p. 277. The velocity distribution of stars in the Catalogue of Nearby Stars.

Jahreiss, H., Wielen, R., 1997. In *Hipparcos Venice '97*, ed. Battrick, B. (Noordwijk: ESA), p. 675. The velocity distribution of stars in the Catalogue of Nearby Stars.

Jakobsen, P., *et al.*, 1991. ApJ, **369**, L63. Interpreting the ring around SN 1987a.

Janes, K.A. & Phelps, R.L., 1994. AJ, **108**, 1773. Morphological classification of open clusters of differing ages.

Jarvis, B.J. & Freeman, K.C., 1985. ApJ, **295**, 324. Dynamical models of bulges.

Jarvis, B.J., Dubath, P., Martinet, L. & Bacon, R., 1988. A&AS, **74**, 513. CCD photometry of SB0 galaxies.

Jaschek, C. & Valbousquet, A., 1991. A&A, **242**, 77. Solar motion from radial velocities.

Jenkins, A., 1992. MNRAS, **257**, 620. Heating of disk stars by spiral structure and molecular clouds.

Jerjen, H. & Tammann, G.A., 1997. A&A, **321**, 713. Luminosity function of Centaurus cluster.

Jog, C.J. & Solomon, P.M., 1984. ApJ, **276**, 114. Two-fluid disks.

Johnson, H.L., 1965. ApJ, **141**, 923. Definition of passbands.

Johnson H.L. & Mitchell, R.J., 1962. Comm. Lunar Planet. Lab., 1, 73. Extension of *UBV* to *IJK*

Johnson, H.L. & Morgan, W.W., 1953. ApJ, **117**, 313. Definition of *UBV* photometry.

Johnson H.L., Iriarte B., Mitchell R.I. & Wisniewskj W.Z., 1966. Comm. Lunar Planet. Lab., 4, 99. Extension of *UBV* to *RI*

Jones, L.A. & Worthey, G., 1995. ApJ, **446**, L31. Indicators of a population's age.

Jones, R.V., Carney, B.W., Storm, J. & Latham, D.W., 1992. ApJ, **386**, 646. RR Lyrae luminosities and variations with metallicity.

Jones, B.F., Klemola, A.R. & Lin, D.N.C., 1994. AJ, **107**, 1333. Measurement of the proper motion of the Magellanic Clouds.

Jorgensen, I., 1994. PASP, **106**, 967. *vgr* and *UBV* CCD standard stars.

Joseph, R.D., 1990. In *Dynamics and Interactions of Galaxies*, ed. Wielen, R., (Berlin: Springer), p. 132. Correlation between IR-activity and mergers.

Jura, M. & Kleinmann, S.G., 1992. ApJS, **79**, 105. Connection between semiregulars and Miras.

Käppeler, F., Beer, H. & Wisshak, K., 1989. Rep. Prog. Phys., **52**, 945. The s process.

Kalnajs A., 1996, in *Barred Galaxies and Circumnuclear Activity*, Nobel Symposium 98, eds. Sandqvist, A., Lindblad, P.O. (Berlin: Springer), p. 165. Dynamical models of a bulge/bar in the Galaxy.

Kamphuis, J., Sancisi, R. & van der Hulst, T., 1991. A&A, **244**, L29. HI map of M101.

Kapteyn, J.C., 1909. ApJ, **30**, 163. Comparison of stars in red and blue light to quantify extinction.

Kapteyn, J.C., 1922. ApJ, **55**, 302. Kapteyn's model of the Galaxy.

Kapteyn, J.C. & van Rhijn, P.J., 1920. ApJ, **52**, 23. Kapteyn's model of the Galaxy.

Karetnikov, V.G., 1991. Astron. Zhurnal, **68**, 880. Stellar radii from eclipsing binaries.

Kayser, R., Refsdal, S. & Stabell, R., 1986. A&A, **166**, 36. Gravitational microlensing in QSO 0957+561.

Kellerman, K.I., 1993. Nat, **361**, 134. value of $q_0$ from angular sizes of VLBI radio jets.

Kellogg, E., Baldwin, J.R. & Koch, D., 1975. ApJ, **199**, 299. Calculation of the Gaunt factor.

Kenney, J. & Young, J., 1989. ApJ, **344**, 171. CO in Virgo spirals.

Kennicutt, R.C., 1989. ApJ, **344**, 685. Thresholds for star formation.

Kennicutt, R.C., Tamblyn, P. & Congdon, C., 1994. ApJ, **435**, 22. Star-formation rates.

Kent, S.M., 1984. ApJS, **56**, 105. CCD surface photometry of 105 disk galaxies.

Kent, S.M., 1985. ApJS, **59**, 115. Disk-bulge decomposition of disk galaxies.

Kent, S.M., 1986. AJ, **91**, 1301. CCD photometry and mass models for galaxies of Rubin *et al.* (1985) sample.

Kent, S.M., 1987. AJ, **93**, 816. CCD photometry and mass models for galaxies with HI rotation curves.

Kent, S.M., Dame, T.M., Fazio, G., 1991. ApJ, **378**, 131. Modeling the luminosity distribution of the inner Milky Way.

Kerr, F.J., 1957. AJ, **62**, 93. Discovery of the Milky Way's warp.

Kerr, F.J. & Lynden-Bell, D., 1986. MNRAS, **221**, 1023. The values of the constants of Galactic structure.

Kerr, F.J. & Rogers, A., 1964. Eds. *The Galaxy and the Magellanic Clouds*, IAU Symp. 20, (Canberra: Australian Acad. Sci.).

Kerr, F.J. & Westerhout, G., 1965. In *Galactic Structure*, eds Blaauw, A., Schmidt, M. (Chicago: University of Chicago Press), p. 167. Galactic spiral-arm mapping.

Kerr, F.J., Bowers, P.F., Jackson, P.D. & Kerr, M., 1986. A&AS, **66**, 373. HI survey of the Milky Way

Kholopov, P.N., et al., 1988. *General Catalogue of Variable Stars*, 4th Edition, Volumes I-III, (Moscow: Nauka Publishing House).

King, I., 1962. AJ, **67**, 471. Open cluster luminosity profiles.

King, I.R., 1975. In *Dynamics of Stellar Systems*, IAU Symp. 69, ed. Hayli, A. (Dordrecht: Reidel), p. 99. Detection of a cusp in M15's luminosity profile.

Kinman, T.D., 1959. MNRAS, **119**, 559. Suggestion that the globular clusters might be divided into halo and disk systems.

Kinman, T.D., 1994. In *Stellar Populations*, IAU Symp. 164, eds van der Kruit, P.C., Gilmore, G. (Dordrecht: Kluwer), p. 75. Field halo stars.

Kinman, T.D., Mahaffey, C.T. & Wirtanen, C.A., 1982. AJ, **87**, 314. The distribution of RR Lyrae stars.

Kinman, T.D., Suntzeff, N.B. & Kraft, R.P., 1994. AJ, **108**, 1722. The distribution of blue HB halo stars.

Kippenhahn, R. & Weigert, A., 1990, *Stellar Structure and Evolution*, (Berlin: Springer). Basic text on the theory of stellar structure and evolution.

Kirkpatrick, J.D., McGraw, J.T., Hess, T.R., Liebert, J. & McCarthy, D.W., 1994. ApJS, **94**, 749. $\Phi(M)$ at low luminosity.

Kirshner, R.P. & Kwan, J., 1974. ApJ, **193**, 27. Application of the Baade–Wesselink method to type II supernovae.

Klemola, A.R., Hanson, R.B. & Jones, B.F., 1987. AJ, **94**, 501. Northern Proper Motion Survey (NPM).

Klemola, A.R., Jones, B.F. & Hanson, R.B., 1987 AJ, **94**, 501. Lick catalog.

Knapp, G.R., 1990. In *The Interstellar Medium in Galaxies*, eds Thronson, H.A., Shull, J.M. (Dordecht: Kluwer), p. 3.

Knapp, G.R., Rauch, K.P., Wilcots, E.M., 1989. In *The Evolution of the Interstellar Medium*, ASP Conf. Ser. 12, ed. Blitz, L. (San Francisco: ASP). Detection of ellipticals by IRAS

Knapp, G.R., Tremaine, S.D. & Gunn, J.E., 1978. AJ, **83**, 1585. $v_c(R_0)$ from the $(l, v)$ plot

Kochanek, C., 1991. ApJ, **382**, 58. Models for the gravitationally-lensed system QSO 0957+561.

Kopal, Z., 1959. *Close Binary Systems*, (London: Chapman Hall).

Kormendy, J., 1977. ApJ, **218**, 333. Discovery of correlation for ellipticals of $L$ with $I_e$.

Kormendy, J., 1979. ApJ, **227**, 714. Survey of bars, lenses and rings.

Kormendy, J., 1987. In *Nearly Normal Galaxies,* ed. Faber, S.M. (Berlin: Springer), p. 163. Diffuse dwarf ellipticals do not lie on the fundamental plane.

Kormendy, J. & Bruzual, G., 1978. ApJ, **223**, L63. The bulge of NGC 4565 does not resemble an elliptical galaxy.

Kormendy, J. & Djorgovski, S., 1989. ARA&A, **27**, 235. Review of the photometric properties of ellipticals.

Kormendy, J. & Illingworth, G., 1982. ApJ, **256**, 460. Measurement of $(v/\sigma)^*$ for bulges.

Kormendy, J. & Richstone, D., 1995. ARA&A, **33**, 581. Supermassive black holes.

Kowal, C.T., 1968. AJ, **73**, 1021. Type Ia supernovae as standard candles.

Kraft, R.P., Sneden, C., Langer, G.E. & Prosser, C.F., 1992. AJ, **104**, 645. O abundance variations in globular cluster stars.

Kraft, R.P., Sneden, C., Langer, G.E. & Shetrone, M.D., 1993. AJ, **106**, 1490. O and Na abundance variations in globular cluster stars.

Kroupa, P., Tout, C.A. & Gilmore, G., 1990. MNRAS, **244**, 76. The faint-end stellar luminosity and mass functions.

Kroupa, P., Tout, C.A., Gilmore, G., 1993. MNRAS, **262**, 545. The faint-end luminosity and mass functions.

Kuijken, K., 1991. ApJ, **372**, 125. The Oort limit.

Kuijken, K. & Gilmore, G., 1989. MNRAS, **239**, 605. Determination of the surface density of the local disk.

Kuijken, K. & Gilmore, G., 1991. ApJ, **367**, L9. On the determination of the surface density of the local disk.

Kuijken, K. & Merrifield, M.R., 1993. MNRAS, **264**, 712. Unresolved Gaussian Decomposition algorithm for deriving

LOSVDs.

Kuijken, K. & Merrifield, M., 1995. ApJ, **443**, L13. Dynamics of peanut-shaped bulges.

Kuijken, K. & Tremaine, S.D., 1991. In *Dynamics of Disk Galaxies*, ed. Sundelius, B. (Göteborg: Göteborgs Univ. Press), p. 71. On the possibility that the halo of the Milky Way is non-axisymmetric

Kuijken, K. & Tremaine, S.D., 1994. ApJ, **421**, 178. On the possibility that the halo of the Milky Way is non-axisymmetric

Kuijken, K., Fisher, D. & Merrifield, M.R., 1996. MNRAS, **283**, 543. Search for stellar counter-rotation in S0s.

Kulkarni, S.R. & Heiles, C., 1988. In *Galactic and Extragalactic Radio Astronomy*, eds Verschuur, G.L., Kellermann, K.I., (Berlin: Springer), P. 95. Review of HI in the Galaxy.

Kundić, T., *et al.*, 1997. ApJ, **482**, 75. Time delay for the lens system QSO 0957+561.

Kwee, K.K., Muller, C.A. & Westerhout, G., 1954. Bull. Astr. Inst. Netherlands, **12**, 211. Rotation curve of the Galaxy.

Léger, A. & Puget, J.-L., 1984. A&A, **137**, L5. Identification of small grains with PAHs.

Labeyrie, A., 1978. ARA&A, **16**, 77. Stellar interferometry.

Lacy, J.H. 1994. In *The Nuclei of Normal Galaxies*, NATO ASI 445, eds Genzel, R., Harris, A.I. (Dordrecht: Kluwer) p. 165. Gas at the Galactic center.

Lacy, J.H., Achtermann, J.M. & Serabyn, E., 1991. ApJ, **380**, L71. Dynamics of the mini-spiral.

Laird, J.B., Carney, B.W. & Latham, D.W., 1988. AJ, **95**, 1843. Calibrations of $\delta_{0.6}$

Lambas, D.G., Maddox, S.J. & Loveday, J., 1992. MNRAS, **258**, 404. Ellipticity distributions of galaxies

Lamla, E., 1982. In *Landolt-Börnstein: Numerical data and Functional Relationships in Science and Technology*, vol 2b, eds Schaifers, K., Voigt, H.H. (Springer: Berlin).

Lancon, A. & Rocca-Volmerange, B., 1992. A&AS, **96**, 593. Library of IR stellar spectra .

Landolt, A.U., 1992. AJ, **104**, 340. Standard stars on the celestial equator.

Lane, A.P. & Stark, A.A., 1996. Antarctic J. U.S., **30(5)**, 377. Sub-mm telescope at the South Pole.

Lang, K.R., 1980. *Astrophysical Formulae*, 2nd ed. (New York: Springer)

Langer, W.D., 1997. In *CO:Twenty-Five Years of Millimeter-Wave Spectroscopy*, IAU Symp. 170, eds Latter, W.B., Radford, S.J.E., Jewell, P.R., Mangum, J.G., Bally, J., (Dordrecht: Kluwer) Abundances of CO isotopes.

Langer, G.E., Hoffman, R. & Sneden, C., 1993. PASP, **105**, 301. Explanation of why Na is anticorrelated with O in globular cluster stars.

Lanz, T., 1986. A&AS, **65**, 195. *UBVRI* Photoelectric Photometry.

Larson, R.B. & Tinsley, B.M., 1978. ApJ, **219**, 46. Peculiar galaxies have strange colors.

Lasker, B.M., Sturch, C.R., McLean, B.J., Russell, J.L., Jenker, H. & Shara, M.M., 1990. AJ, **99**, 2019. HST Guide Star Catalog.

Latham, D.W., 1998. In *Highlights of Astronomy*, ed. Andersen, J. (Dordrecht: Kluwer). Masses of binary components.

Lauberts, A., 1982. *The ESO/Uppsala Survey of the ESO(B) Atlas*, (Munich: ESO). Galaxy catalog of the southern sky.

Lauer, T.R. & Kormendy, J., 1986. ApJ, **303**, L1. Globulars of M87.

Lauer, T.R., Ajhar, E.A., Byun, Y., Dressler, A. & Faber, S.M., 1995. AJ, **110**, 2622. Core surface-brightness profiles of elliptical galaxies.

Layden, A.C., 1995. AJ, **110**, 2288. Kinematics etc of local RR Lyrae stars.

Layden, A.C., Hanson, R.B., Hawley, A.L., Klemola, A.R. & Hanley, C.J., 1996. AJ, **112**, 2110.. Absolute magnitudes of RR Lyrae stars.

Leavitt, H.S., 1912. Harvard Coll. Obs. Circ., **173**, 1. Discovery of the Cepheid PL relation.

Lebreton, Y., 1998. A&AS, **00**, 00. Effect of $Y$ on the ZAMS.

Ledlow, M.J. & Owen, F.N., 1996. AJ, **112**, 9. Division between Fanaroff–Riley classes of radio sources.

Lee, Y.-W., 1990. ApJ, **363**, 159. Definition of HB color; discussion of Sandage period shift.

Lee, M.G. & Geisler, D., 1993. AJ, **106**, 493. [Fe/H] of globulars in M87.

Lee, Y.-W., Demarque, P., Zinn, R., 1990. ApJ, **350**, 155. Luminosity of HB

Lee, Y.-W., Demarque, P., Zinn, R., 1994. ApJ, **423**, 248. Morphology of the HB in globulars.

Leggett, S.K., 1992. ApJS, **82**, 351. *IJHK* photometry of 322 red dwarfs and many transformations in an appendix.

Leonard, P.J.T., 1988. AJ, **95**, 108. Star counts in the open cluster NGC 2420.

Leonard. P.J.T. & Merritt, D., 1989. ApJ, **339**, 195. Deriving dynamics from proper motions in star clusters.

Lesh, J.R., 1968. ApJS, **16**, 371. Gould's belt.

Levine, A.M., et al., 1984. ApJS, **54**, 581. Hard X-ray source list from HEAO1 A-1 experiment.

Liedahl, D.A., Osterheld, A.L. & Goldstein, W.H., 1995. ApJ, **438**, L115. Predicted Fe L line strengths in X-ray spectra.

Lin, D.N.C. & Richer, H.B., 1992. ApJ, **388**, L57. Discussion of the LMC origin of metal-rich young globular clusters.

Lin, H., et al., 1996. ApJ, **464**, 60. Luminosity function of Las Campanas redshift survey.

Linblad, B., 1927. MNRAS, **87**, 553. Kinematic model of the Galaxy.

Lindqvist, M., Habing, H.J. & Winnberg, A., 1992. A&A, **259**, 118. Survey for OH/IR masers.

Catalog of Seyfert galaxies.

Liszt, H.S. & Burton, W.B., 1980. ApJ, **236**, 779. Introduction of an elliptical model of the bulge's disk

Liu, T., Janes, K.A. & Bania, T.M., 1989. AJ, **98**, 626. Binary fraction in open clusters.

Lockman, F.J., 1984. ApJ, **283**, 90. The rotation of the HI layer as a function of z

Long, K.S. & van Speybroek, L.P., 1983. In *Accretion-Driven X-ray Sources*, eds Lewin, W., van den Heuvel, E.P.J. (Cambridge: CUP), p. 117. Contribution of X-ray binaries to integrated Galactic X-ray emission.

Longair, M.S., 1992. *Theoretical Concepts in Physics*, (CUP, Cambridge). Includes introduction to cosmology.

Longmore, A.J., Dixon, R., Skillen, I., Jameson, R.F. & Fernley, J.A., 1990. MNRAS, **247**, 684. Infrared P-L relation for RR Lyrae stars.

Loveday, J., 1996. MNRAS, **278**, 1025. APM Survey of bright galaxies.

Low, F.J. et al. 1984. ApJ, **278**, L19. Identification of infrared cirrus.

Lucy, L.B., 1974. AJ, **79**, 744. Lucy deconvolution algorithm.

Lugger, P.M., 1986. ApJ, **303**, 535. Luminosity functions of Abell clusters.

Lupton, R.H., Gunn, J.E. & Griffin, R.F., 1987. AJ, **93**, 1114. Statistical parallax for globular clusters.

Lutz, T.E. & Kelker, D.H., 1973. PASP, **87**, 617. Original Lutz–Kelker bias paper.

Luyten, W. J., 1979. *A Catalog of Stars with Proper Motions Exceeding 0.5 arcsec Annually*, (Minneapolis: University of Minnesota). LHS Catalog of stellar proper motions.

Luyten, W. J., 1980. *New Luyten Catalogue of Stars with Proper Motions Larger than Two Tenths of an Arcsecond* (Minneapolis: University of Minnesota). NLTT Catalog of stellar proper motions.

Lynden-Bell, D. & Wood, R., 1968. MNRAS, **138**, 495. The gravithermal catastrophe.

Lynden-Bell, D., et al., 1988. ApJ, **326**, 19.. Discovery of the Great Attractor.

Lynds, B.T., 1962. ApJS, **7**, 1. Catalog of dark nebulae.

Lynds, B.T., 1965. ApJS, **12**, 163. Catalog of bright nebulae.

Lynga, G., 1987. *Catalog of Open Clusters*, (Strasbourg: Centre de Données Stellaires). Open cluster catalog.

Mégessier, C., 1997. In *Fundamental Stellar Properties*, IAU Symposium 189, eds Bedding, T.R., Booth, A.J., Davis, J., (Dordrecht: Kluwer). Absolute flux calibration of Vega.

MacKenty, J.W., et al., 1997. *NICMOS Instrument Handbook*, Version 2.0, (Baltimore: STScI). HST NICMOS Manual.

Madore, B.F. & Freedman, W.L., 1991. PASP, **103**, 667. Cepheids and the distance scale.

Magorrian, S.J. & Binney, J.J., 1994. MNRAS, **271**, 949. Predicting $h_3$ and $h_4$

from photometry.

Magri, C., *et al.*, 1988. ApJ, **333**, 136. HI deficiency in cluster spirals.

Malhotra, S., 1994. ApJ, **433**, 687. Vetical structure of the Galactic CO layer

Malhotra, S., 1995. ApJ, **448**, 138. Effects of the random velocities of HI clouds on circular-speed and scale-height determinations

Malin, D.F. & Carter, D., 1983. ApJ, **274**, 534. Catalog of shell galaxies.

Mallia, E.A. & Pagel, B.E.J, 1981. MN-RAS, **194**, 421. Spectra of giant stars in Ω Cen.

Malmquist, K.G., 1922. journMedd. Lund Astron. Obs., Ser. II 32 64 Introduction of Malmquist bias.

Malmquist, K.G., 1936. Medd. Stockholms Obs., **26**, 1. Analysis of Malmquist bias.

Maloney, P., 1993. ApJ, **414**, 41. Shielding layers around HI disks.

Markarian, B.E., Lipovetskij, V.A., Stepanyan, D.A., 1978. ApJ, **13**, 215. List of Markarian galaxies.

Markarian, B.E., Lipovetsky, V.A., Stepanian, J.A., Erastova, L.K. & Shapovalova, A.I., 1989. Soobshch. Spets. Astrof. Obs., **62**, 5. First Byurakan Survey of Markarian galaxies.

Martin, P. & Roy, J.-R., 1994. ApJ, **424**, 599. Abundance gradients smaller in barred galaxies.

Massey, P., Strobel, K., Barnes, J.V. & Anderson, E., 1988. ApJ, **328**, 315. Spectrophotometric standards.

Massey, P., Lang, C.C., DeGioia-Eastwood, K. & Garmany, C.D., 1995. ApJ, **438**, 188. High mass IMF in the LMC.

Massey, P., Johnson, K.E. & DeGioia-Eastwood, K., 1995. ApJ, **454**, 151. IMF of massive stars in OB associations.

Mathewson, D.S., Schwarz, M.P. & Murray, J.D., 1977. ApJ, **217**, L5. The Magellanic Stream.

Mathieu, R.D., 1984. ApJ, **284**, 643. Structure and dynamics of M11.

Mathis, J.S., 1990. ARA&A, **28**, 37. Extinction by dust.

Maza, J., Hamuy, M., Phillips, M.M., Suntzsff, N.B. & Avilés, R., 1994. ApJ, **424**, L107. Differences between luminosities of type Ia supernovae.

McAlister, H.A., 1985. ARA&A, **23**, 59. High angular resolution measurements of stellar properties.

McClure, R.D. & van den Bergh, S., 1968. AJ, **73**, 313. Definition of the DDO photometric system.

McClure, R.D., Hesser, J.E., Stetson, P.B. & Stryker, L.L., 1985. PASP, **97**, 665. Technique for crowded-field stellar photometry.

McClure, R.D., *et al.*, 1986. ApJ, **307**, L49. Globular cluster luminosity functions.

McCray, R., 1993. ARA&A, **31**, 175. SN 1987A review.

McCrea, W.H., 1964. MNRAS, **128**, 147. Origins of blue stragglers.

McGee, R.X. & Milton, J.A., 1966. Australian J. Phys., **19**, 343. HI in Magellanic Clouds.

McHardy, I.M., Stewart, G.C., Edge, A.C., Cooke, B. & Yamashita, K., 1990. MN-RAS, **242**, 215. Hubble constant from the Sunyaev–Zel'dovich effect.

McHardy, I.M., Merrifield, M.R., Abraham, R.G. & Crawford, C.S., 1994. MNRAS, **268**, 681. HST imaging of PKS1413+135.

McNally, D., 1974. *Positional Astronomy* (London: Muller Educational). Astrometry text book.

McWilliam, A. & Rich, R.M., 1994. ApJS, **91**, 749. High-resolution spectroscopy of bulge K giants.

Melnick, J., Sargent, W.L.W., 1977. ApJ, **215**, 401. Morphology–radius relation in galaxy clusters.

Mermilliod, J.-C., 1980. In *Star Clusters*, IAU Symp. 85, ed. Hesser, J.E. (Dordrecht: Reidel), p. 129. Compilation of data on nearby open clusters.

Mermilliod, J.-C., 1987–. *Open Cluster Database Version 2.0 (User's Guide)*, http://obswww.unige.ch/bda/ugp1/-ugp1.html. The BDA Open Cluster Database.

Mermilliod, J.-C., Turon, C., Robinchon, N., Arenou, F. & Lebreton, Y., 1997. In *Hipparcos Venice '97*, ed. Battrick, B. (Noordwijk: ESA), p. 643. Hipparcos distances to nearby clusters.

Mermilliod, J.-C., Mermilliod, M. & Hauck, B., 1997. A&AS, **124**, 349. General Catalogue of Photometric Data (GCPD).

Merrifield, M.R., 1992. AJ, **103**, 1552. Rotation curve of the Milky Way from flaring of HI disk.

Merrifield, M.R., 1993. MNRAS, **261**, 233. Vertical structure of Milky Way HI disk.

Merrifield, M.R., Kent, S.M., 1989. AJ, **98**, 351. Galaxy density distribution in clusters.

Merrifield, M.R. & Kent, S.M., 1990. AJ, **99**, 1548. Higher-order moments Jeans equations.

Merrifield, M.R. & Kuijken, K., 1994. ApJ, **432**, 575. Dynamical modelling of NGC 7217.

Merritt, D., 1985. ApJ, **289**, 18. Origin of cD galaxies.

Mestel, L., 1952. MNRAS, **112**, 598. Prediction of the helium flash.

Meusinger, H., Schilbach, E. & Souchay, J., 1996. A&A, **312**, 833. Mass function of the Pleiades.

Meylan, G., 1987. A&A, **184**, 144. Multimass Michie model for $\Omega$ Cen.

Meylan, G. & Mayor, M., 1986. A&A, **250**, 113. 318 velocities in $\Omega$ Cen.

Meynet, G., Mermilliod, J.-C. & Maeder, A., 1993. A&AS, **98**, 477. Ages of open clusters.

Mezger, P.G., 1994. In *The Nuclei of Normal Galaxies*, NATO ASI 445, eds Genzel, R., Harris, A.I. (Dordrecht: Kluwer), p. 415. Sgr A*.

Michard, R. & Marchal, J., 1994. A&AS, **105**, 481. Quantitative classification of the RSA galaxies.

Michel, F.C., 1987. Nat, **329**, 310. Production of millisecond pulsars.

Mihalas, D., 1978. *Stellar Atmospheres*, 2nd ed., (San Francisco: W.H. Freeman) Stellar Atmospheres text.

Miller, J.S. & Antonucci, R., 1983. ApJ, **271**, L7. Detection of Seyfert 1 structure in the Seyfert 2 galaxy NGC 1068.

Miller, B.W. & Hodge, P., 1996. ApJ, **458**, 467. [O/H] versus $M_B$ correlation.

Miller, J.S., French, H.B. & Hawley, S.A., 1978. ApJ, **219**, L85. Redshift of BL Lac.

Minniti, D., 1996. ApJ, **459**, 579. Kinematics of bulge and halo stars.

Minniti, D., Olszewski, E.W., Liebert, J., White, S.D.M., Hill, J.M. & Irwin, M.J., 1995. MNRAS, **277**, 1293. Metallicity of the bulge.

Miyamoto, M., Yoshizawa, M. & Suzuki,

S., 1988. A&A, **194**, 107. The warp of the Milky Way traced through OB stars

Miyoshi et al., 1995. Nat, **373**, 127. Detection of maser spots in NGC 4258.

Monella, R., 1985. Coelum, **53**, 287. Tabulation of globular cluster properties.

Monk, D.J., Barlow, M.J., Clegg, R.E.S., 1988. MNRAS, **234**, 583. Abundances in planetary nebulae.

Morel, M. & Magnenat, P., 1978. A&AS, **34**, 477. *UBVRIJKLMNH* photometric catalog.

Morgan, W.W., 1958–1970. PASP, **70**, 364;. PASP, **71**, 92;. in *Spiral Structure of Our Galaxy*, eds Becker, W., Contopoulos, G. (Dordrecht: Reidel). Yerkes galaxy classification system defined.

Morgan, W.W., 1959. AJ, **64**, 432. Integrated spectra of globular clusters.

Morgan, W.W., 1972. In *External Galaxies and Quasi-Stellar Objects*, ed. Evans, D.S. (Dordrecht: Reidel) Criteria for classifying unusual galaxies.

Morgan, W.W. & Keenan, P.C., 1973. ARA&A, **11**, 29. MK Classification.

Morgan, W.W., Abt, H.A. & Tapscott, J.W., 1978. *Revised MK Spectral Atlas for Stars of Types Earlier than the Sun*, (Kitt Peak: KPNO & Yerkes Observatory).

Morgan, W.W., Kayser, S., White, R.A., 1975. ApJ, **199**, 545. cD galaxies in poor clusters.

Morgan, W.W., Keenan, P.C. & Kellerman, E., 1943. *An Atlas of Stellar Spectra*, (Chicago: Chicago University Press). MK Classification

Morrison, R. & McCammon, D., 1983. ApJ, **270**, 119. Mean-free paths of X-ray photons through the ISM.

Morrison, H.L., Miller, E.D., Harding, P., Stinebring, D.R., Boroson, T.A., 1997. AJ, **113**, 2061. Surface photometry of galaxies.

Morrison, H.L., Boroson, , T.A. & Harding, P., 1994. AJ, **108**, 1191. Surface photometry of galaxies.

Morton, D.C., 1974. ApJ, **193**, L35. Depletion of elements in the ISM.

Mould, J.R., 1982. ARA&A, **20**, 91. M-star version of G-dwarf problem.

Mould, J.R., 1983. ApJ, **266**, 255. Velocity dispersion in Baade's Window.

Mould, J.R., Xystus, D.A., Da Costa, G.S.,

1993. ApJ, **408**, 108. A yound LMC globular cluster.

Mozurkewich, D., *et al.* 1991. AJ, **101**, 2207. Angular diameters of 12 stars.

Murai, T. & Fujimoto, M., 1980. PASJ, **32**, 581. Model of the formation of the Magellanic Stream

Murray, C.A., 1983. *Vectorial Astrometry*, (Bristol: Adam Hilger).

Murray, C.A., Penston, M.J., Binney, J.J., Houk, N., 1997. In *Hipparcos Venice '97*, ed. Battrick, B. (Noordwijk: ESA), p. 485. Luminosity function of MS stars.

Naim, A., *et al.*, 1995. MNRAS, **274**, 1107. Comparison of galaxy classifications by different experts.

Nakada, Y., *et al.*, 1991. Nat, **353**, 140. Bulge X-ray sources.

Nakada, Y., Onaka, T., Yamamura, I., Deguchi, D., Ukita, N., Izumiura, H., 1993. PASJ, **45**, 179. SiO radiation from IRAS-selected stars.

Nather, R.E. & McCants, M.M., 1970. AJ, **75**, 963. Lunar occultations.

Navarro, J.F., Frenk, C.S., White, S.D.M., 1997. ApJ, **490**, 493. The NFW profile.

Neininger, N. & Herellou, C., 1996. In *Polarimetry of the ISM*, ASP Conf. Ser. 97, eds Roberge, W.G., Whittet, D.C.B. (San Francisco: ASP), p. 592. Radio polarization of M51.

Nemec, J.M. & Harris, H.C., 1987. ApJ, **316**, 172.. Spatial distribution of blue stragglers in NGC 5466.

Nilson, P., 1973. Uppsala Astron. Obs. Ann., **6**, 1. The Uppsala General Catalog of galaxies (UGC).

Nissen, P.E., 1988. A&A, **199**, 146. Ages of open clusters.

Nissen P.E. & Schuster, W.J., 1991. A&A, **251**, 457. Metallicities and kinematics of high-velocity F and G dwarfs.

Nordström, B., Andersen, J. & Andersen, M.I., 1996. A&AS, **118**, 407. Absence of MS stars in the old open cluster NGC 3680.

Norris, J., 1986. ApJS, **61**, 667. Kinematics of thick disk and halo from kinematically unbiased samples.

Norris, J.E. & Da Costa, G.S., 1995. ApJ, **447**, 680. Abundances in $\Omega$ Cen.

Oemler, A., 1974. ApJ, **194**, 1. Galaxy types in rich cluster çores.

Oemler, A., 1976. ApJ, **209**, 693. Photometry of cD galaxies.

Ojha, D.K., Bienaymé, O., Robin, A.C., Crezé, M. & Mohan, V., 1996. A&A, **311**, 4560. A model of the Milky Way.

Olling, R.P., 1995. AJ, **110**, 5910. The flattening of dark halos from the flaring of gas disks.

Olling, R.P. & Merrifield, M.R., 1998. In *Galactic Halos*, ASP Conf. Ser., ed. Zaritsky, D. (San Francisco: ASP). Shape of the Milky Way's dark halo.

Olsen, E.H., 1987. A&A, **189**, 173. Calibration of $uvby\beta$ photometry.

Oort, J.H., 1927. Bull. Astron. Inst. Netherlands, **3**, 275. Kinematic model of the Galaxy.

Oort, J.H., 1928. Bull. Astron. Inst. Netherlands, **4**, 269. Kinematic model of the Galaxy.

Oort, J., 1958. In *Stellar Populations*, ed. O'Connell, D.J.K. (Amsterdam: North Holland). Deficit of old open clusters.

Oort, J.H., 1960. Bull. Astron. Inst. Netherlands, **15**, 45. Classical estimation of the local mass density of the disk.

Oort, J.H. & Plaut, L., 1975. A&A, **41**, 71. $R_0$ from RR Lyrae stars.

Oort, J.H., Kerr & F.J., Westerhout, G., 1958. MNRAS, **118**, 379. Early HI map of the Galaxy.

Oosterhoff, P., 1939. Observatory, **62**, 104. The Oosterhoff effect.

Oppenheimer, J.R. & Volkoff, G., 1939. Phys. Rev., **55**, 374Prediction of neutron stars.

Osterbrock, D.E., 1989. *Astrophysics of Gaseous Nebulae and Active Galactic Nuclei*, (Mill Valley: University Science Books). Introduction to AGN and AGN.

Ostriker, J.P. & Thuan, T.X., ApJ, **202**, 353. Chemical evolution of galaxies.

Owen, F.N. & Ledlow, M.J., 1994. In *The Physics of Active Galaxies*, ASP Conf. Proc. 54, ed. Bicknell, G.V., Quinn, P.J., Dopita, M.A. (San Francisco: ASP), p. 319. Location of FR I and FR II sources in $(L_R, P_{1.4})$ plane.

Paczynski, B., *et al.*, 1994. ApJ, **435**, L113. Evidence for a barred bulge from microlensing.

Pagel, B.E.J., 1997. *Nucleosynthesis and Chemical Evolution of Galaxies* (Cambridge: Cambridge University Press)

Pagel, B.E.J., Simonson, E.A., Terlevich, R.J. & Edmunds, M.G., 1992. MNRAS, **255**, 325. Primordial He abundance.

Pan, X., *et al.*, 1992. ApJ, **384**, 624. Orbit of $\alpha$ And resolved with Mark III optical interferometer.

Panagia, N., Gilmozzi, R., Macchetto, F., Adorf, H.-M. & Kirshner, R.P., 1991. ApJ, **380**, L23. The distance to SN1987A from time delay measurements.

Paresce, F., de Marchi, G. & Jedrzejewski, R., 1995. ApJ, **442**, 57. HST image of the center of 47 Tuc.

Particle Data Group, 1984. Rev. Mod. Phys., **56**, S1. Values of physical constants.

Paturel, G., Fouque, P., Bottinelli, L. & Gouguenheim, L., 1989. A&AS, **80**, 299. Catalogue of Principal Galaxies (PGC).

Payne, C.H. 1925. *Stellar Atmospheres*, (Cambridge: Harvard College Observatory). First astrophysical account of HR diagram

Peebles, P.J.E., 1993. *Principles of Physical Cosmology*, (Princeton: PUP).

Peimbert, M. & Serrano, A., 1982. MNRAS, **198**, 563. Chemical evolution of galactic disks.

Peletier, R.F., 1989. *Elliptical galaxies. Structure and stellar content*, PhD Thesis, (Groningen: Kapteyn Institute). Photometry of elliptical galaxies.

Peletier, R.F., Davies, R.L., Illingworth, G.D., Davis, L.E. & Cawson, M., 1990. AJ, **100**, 1091. Surface $UBR$ photometry of 39 elliptical galaxies.

Peletier, R.F., Valentijn, E.A., Moorwood, A.F.M. & Freudling, W., 1994. A&AS, **108**, 621. Near-IR photometry of disks.

Pelt, J., Hoff, W., Kayser, R., Refsdal, S. & Schramm, T., 1994. A&A, **256**, 775. Measurement of time delay for QSO 0957+531.

Pence, W.D., 1986. ApJ, **310**, 597. Photometry of ripples.

Penzias, A.A. & Wilson, R.W., 1965. ApJ, **142**, 419. Discovery of the microwave background.

Perlmutter, S., *et al.*, 1998. Nat, **391**, 51. Hubble diagram from type Ia supernovae.

Perryman, M.A.C., *et al.* 1995. A&A, **304**, 69. Hipparcos CM diagram of the solar neighborhood.

Perryman, M.A.C., *et al.* 1997. *The Hipparcos and Tycho Catalogues*, (Noordwijk: ESA). Hipparcos and Tycho Catalogues.

Peters, W.L., 1975. ApJ, **195**, 617. Model suggesting Galactic kinematics can be explained by a bar.

Peterson, C.J., 1980. AJ, **85**, 226. Gaseous rotation curve for NGC 488.

Peterson, R.C., 1980. ApJ, **237**, L87. Anticorrelations between Na and O in globular cluster giants.

Peterson, R.C., Rood, R.T. & Crocker, D.A., 1995. ApJ, **453**, 214. Stellar rotation rate: a 3rd parameter in globular clusters?

Peterson, R.C., Seitzer, P. & Cudworth, K.M. 1989. ApJ, **347**, 251. Stellar kinematics of M15.

Philip, A.G.D., Cullen, M.F. & White, R.E., 1976. Dudley Obs. Rep., **11**, 1. Collection of UBV photometry for globular clusters.

Phillips, M.M., 1993. ApJ, **413**, L105. Relation between color and luminosity for type Ia supernovae.

Phillips, T.G., Huggins, P.J., Wannier, P.G. & Scoville, N.Z., 1979. ApJ, **231**, 720. CO profiles of clouds.

Piatti, A.E., Claria, J.J. & Abadi, M.G., 1995. AJ, **110**, 2813. Abundances and ages of open clusters.

Picard, A. & Johnston, H.M., 1994. A&A, **283**, 76. The location of globular cluster centers.

Pierce, M.J. & Tully, R.B., 1992. ApJ, **387**, 47.. Calibration of Tully–Fisher relation.

Pierce, M.J., *et al.*, 1994. Nat, **371**, 385. Detection of Cepheids in Virgo cluster galaxies.

Pilkington, J.D.H. & Scott, P.F., 1965. Mem.R.A.S., **69**, 183. 4C radio source catalog.

Pont, F., Queloz, D., Bratschi, P. & Mayor, M., 1997. A&A, **318**, 416. Outer $v_c(R)$ from Cepheids.

Popper, D.M., 1980. ARA&A, **18**, 115. Masses of stars.

Postman, M. & Geller, M.J., 1984. ApJ, **281**, 95. Morphology–density relation in groups of galaxies.

Pounds, K.A., *et al.*, 1993. MNRAS, **260**, 77. ROSAT WFC Catalog.

Prada, F., Gutiérrez, C.M., Peletier, R.F.

& McKeith, C.D., 1996. ApJ, **463**, 9. Two-dimensional generalization of UGD algorithm.

Press, W.H. & Schechter, P., 1974. ApJ, **187**, 425. Press–Schechter formalism for galaxy formation.

Press, W.H., Teukolsky, S.A., Vetterling, W.T. & Flannery, B.P., 1992. *Numerical Recipes in C*, 2nd edition, (Cambridge: CUP).

Press, W.H., Rybicki, G.B. & Hewitt, J.N., 1992. ApJ, **385**, 404. Time delay for the lens system QSO 0957+561.

Preston, G., 1959. ApJ, **130**, 507. Definition of the $\Delta S$ parameter of RR Lyrae stars.

Preston, G.W., Schectman, S.A. & Beers, T.C., 1991. ApJ, **375**, 121. The distribution of blue HB halo stars.

Prieto, M., Beckman, J.E., Cepa, J. & Varela, A.M., 1992. A&A, **257**, 85. CCD surface photometry of 6 Sb and Sc galaxies.

Prieur, J.-L., 1990. In *Dynamics and Interactions of Galaxies*, ed. Wielen, R. (Berlin: Springer), p. 72. Statistics of ripples.

Pritchet, C.J. & van den Bergh, S., 1985. AJ, **90**, 2027.. Globulars of NGC 3379.

Proffitt, C.R. & Michaud, G., 1991. ApJ, **371**, 584. Effect of He diffusion on isochrones.

Pryor, C. & Meylan, G., 1993. In *Structure and Dynamics of Globular Clusters*, ASP Conf. Ser. 50, eds. Djorgovski, S.G., Meylan, G. (San Francisco: ASP). Mass-to-light ratios of 56 globular clusters.

Pryor, C., Latham, D. & Hazen, M., 1988. AJ, **96**, 123. Spectroscopic binaries in globular clusters.

Quinn, P.J. & Goodman, J., 1986. ApJ, **309**, 472. Sinking satellites of spirals.

Racine, R., 1971. AJ, **76**, 331. HB populations with gaps.

Racine, R. & Harris, W.E., 1992. AJ, **104**, 1068. Distance to M31 from the globular cluster luminosity function.

Raha, N., Sellwood, J.A., James, R.A. & Kahn, F.D., 1991. Nat, **352**, 411. Simulations showing that bars buckle.

Rand, R. & Kulkarni, S., 1989. ApJ, **343**, 760. *B*-field from pulsars.

Rand, R. & Kulkarni, S., 1990. ApJ, **349**, L43. CO, HI etc in M51.

Rand, R. & Lyne, A.G., 1994. MNRAS, **268**, 497. *B* from pulsars

Ratnatunga, K.U., Bahcall, J.N. & Casertano, S., 1989. ApJ, **339**, 106. Kinematic models of the Galaxy.

Rees, R.F., 1993. AJ, **106**, 1524. Proper motions for stars in M5.

Reich, W., Fürts, E., Steffen, P., Reif, K. & Haslem, C.G.T., 1984. A&AS, **58**, 197. Radio-continuum radiation from the center of the Galaxy.

Reid, M.J., 1993. ARA&A, **31**, 345. Determination of $R_0$

Reid, I.N., 1998. AJ, **114**, 161. Subdwarf distances from Hipparcos.

Reid, I.N. & Gilmore, G., 1982. MNRAS, **201**, 73. Faint end of the local luminosity function.

Reid, M.J. & Moran, J.M., 1988. In *Galactic and Extragalactic Radio Astronomy*, eds Verschuur, G.L., Kellerman, K.I. (Berlin: Springer), p. 255. Review of water maser emission in the Milky Way.

Reid, M.J., Gwinn, C.R., Moran, J.M. & Matthews, A., 1988. Bull. Am. Astr. Soc., **20**, 1017. Value of $R_0$.

Reid, I.N., Hawley, S.L. & Gizis, J.E., 1995. AJ, **110**, 1838. Spectra of Gliese Catalog stars.

Renzini, A., 1983. Mem. Soc. Astr. Italiana, **54**, 335. Dependence of HB morphology on metallicity.

Renzini, A. *et al.*, 1996. ApJ, **465**, L23. White dwarf cooling sequence in NGC 6752.

Reynolds, J.H., 1913. MNRAS, **74**, 132. Introduction of the Hubble law.

Reynolds, R.J., 1993. In *Back to the Galaxy*, AIP Conference Proceedings 278, eds Holt, S.S., Verter, F. (New York: AIP), p. 156. Review of the Warm ionized medium

Rhee, G.F.R.N. & Katgert, P., 1987. A&A, **183**, 217. cD galaxies are aligned with the surrounding cluster.

Rich, R.M., 1988. AJ, **95**, 828. Spectra of bulge K giants.

Richer, H.B. *et al.*, 1995. ApJ, **451**, L17. White dwarf cooling sequence in M4.

Richter, O.-G. & Sancisi, R., 1994. A&A, **290**, L9. Survey of incidence of lopsidedness.

Ridgway, S.T., Joyce, R.R., White, N.M. & Wing, R.F., 1980. ApJ, **235**, 126. Temperatures of late-type stars.

Rieke, G.H. & Lebofsky, R.M., 1985. ApJ, **288**, 618. The extinction curve.

Riess, A.G., Press, W.H. & Kirshner, R.P., 1995. ApJ, **438**, L17. The Hubble constant from type Ia supernovae.

Rix, H.-W. & White, S.D.M., 1992. MNRAS, **254**, 389. Optimal extraction of LOSVDs from absorption lines.

Rix, H.-W. & Zaritsky, D., 1995. ApJ, **447**, 82. Ellipticity of disks.

Rix, H.-W., Franx, M., Fisher, D. & Illingworth, G., 1992. ApJ, **400**, 5. Modelling of counter-rotating disks in NGC 4550.

Robinson, C. *et al.*, 1995. MNRAS, **274**, 547. millisecond pulsars in 47 Tuc.

Robson, I., 1996. *Active Galactic Nuclei*, (New York: Wiley). Introductory text on AGN.

Roeser, S. & Bastian, U., 1991. *PPM Star Catalogue*, Vols. I and II, (Heidelberg: Spektrum Akademischer Verlag) Position and proper motion (PPM) survey

Rogstad, D.H., Lockhart, I.A. & Wright, M.C.H., 1974. ApJ, **193**, 309. Tilted ring model of M83.

Rogstad, D.H., Wright, M.C.H. & Lockhart, I.A., 1974. ApJ, **204**, 703. Warp in M33

Roman, N.G., 1955. ApJS, **2**, 195. The correlation between peculiar velocity and $\delta(U - B)$.

Romani, R.W. & Weinberg, M.D., 1991. ApJ, **372**, 487. 2nd sequence due to binaries in globular clusters.

Rood, R.T., 1973. ApJ, **184**, 815. Modelling the HB structure of globular clusters.

Rose, J.A., 1994. AJ, **107**, 206. The stellar composition of M32.

Rots, A.H., 1975. A&A, **45**, 43. HI maps of M81.

Rots, A.H. & Shane, W.W., 1975. A&A, **45**, 25. HI maps of M81.

Rowley, G., 1988. ApJ, **331**, 124. Axisymmetric model of peanut-shaped bulges.

Rubin, V.C., Burstein, D., Ford, W.K. & Thonnard, NH., 1985. ApJ, **289**, 81. Mass-to-light ratios of spirals.

Rubin, V.C., Graham, J.A. & Kenney, J.D.P., 1992. ApJ, **394**, 9. Discovery of counter-rotating stellar disks in NGC 4550.

Ruzmaikin, A.A., Sokolov, D.D., Shukurov, A. & Beck, R., 1990. A&A, **230**, 284.

*B*-fields in M31.

Ryan, S.G. & Norris, J.E., 1991. AJ, **101**, 1865. Metallicity distribution of halo stars.

Ryan, S.G. & Norris, J.E., 1993. In *Galaxy Evolution: the Milky Way Perspective*, ASP conf. ser. 49, ed. Majewski, S.R. (San Francisco: ASP), p. 103. Interpretation of kinematically-selected samples of stars.

Rybicki, G.B., 1986. In *Structure and Dynamics of Elliptical Galaxies*, IAU Symp. 127, ed. de Zeeuw P.T. (Dordrecht: Kluwer), p. 397. Axisymmetric galaxies cannot be uniquely deprojected.

Rybicki, G.B., Lightman, A.P., 1979. *Radiative Processes in Astrophysics*, (New York: Wiley).

Sadler, E. & Gerhard, O.E., 1985. MNRAS, **214**, 177. Survey of dust lanes in early-type galaxies.

Sadler, E.M., Rich, R.M. & Terndrup, D.M., 1996. AJ, **112**, 171. Abundances of K giants in Baade's window.

Sagar, R. & Bhatt, H.C., 1989. MNRAS, **236**, 865. Proper motions in open clusters.

Sagar, R. & Cannon, R.D., 1995. A&AS, **111**, 75. CM diagram of NGC 4755.

Saha, M., 1920–1921. Phil. Mag., **40**, 472;. Proc. Roy. Soc., **A99**, 135. The Saha equation.

Saha, A., 1985. ApJ, **289**, 310. The distribution of RR Lyrae stars.

Saha, A., *et al.*, 1995. ApJ, **438**, 8. Cepheid calibration of Type Ia supernovae.

Salpeter, E.E., 1955. ApJ, **121**, 161. Simple power-law IMF.

Sancisi, R., 1976. A&A, **53**, 159. Warped disk of NGC 5907.

Sancisi, R., Allen, R., Sullivan, W.J., 1979. A&A, **78**, 217. Spider diagram on NGC 5383.

Sandage, A., 1961. *The Hubble Atlas of Galaxies*, (Washington: Carnegie Institution). The Hubble–Sandage galaxy classification system defined.

Sandage, A., 1969. ApJ, **158**, 1115. Definition of $\delta_{0.6}$

Sandage, A., 1988. ApJ, **331**, 605. Hubble constant from the Tully–Fisher relation.

Sandage, A., 1990. ApJ, **350**, 631. Explanation of the two Oosterhoff classes.

Sandage, A., 1993. AJ, **106**, 719. Dependence of HB absolute magnitude on [Fe/H].

Sandage, A., 1993. ApJ, **402**, 3. Comparison between methods for estimating the Hubble constant.

Sandage, A., Bedke, J., 1988. *Atlas of Galaxies Useful for Measuring the Cosmic Distance Scale*, (Washington D.C.: NASA). NASA Atlas of galaxies.

Sandage, A. & Bedke, J., 1994. *The Carnegie Atlas of Galaxies*, (Washington: Carnegie Institution).

Sandage, A. & Tammann, G., 1981. *Revised Shapley–Ames Catalogue of Bright Galaxies*, (Washington: Carnegie Institution).

Sandage, A. & Tammann, G.A., 1993. ApJ, **415**, 1. Hubble Constant from Type Ia supernovae.

Sandage, A. & Wildey, R., 1967. ApJ, **150**, 469. Anomalous HB in NGC 7006.

Sandage, A., et al., 1992. ApJ, **401**, L7. Cepheid calibration of Type Ia supernovae.

Sanders, D.B., Solomon, P.M. & Scoville, N.Z., 1984. ApJ, **276**, 182. Galactic CO inside $R_0$.

Sanromà, M. & Salvador-Solé, E., 1990. ApJ, **360**, 16. Relation between galaxy morphology and substructure in clusters.

Santiago, S.X., Gilmore, G. & Elson, R.A.W., 1996. MNRAS, **281**, 871. HST star counts at large $b$.

Sarajedini, A. & Demarque, P., 1990. ApJ, **365**, 219. the $\Delta(B - V)$ method.

Sarajedini, A. & Milone, A.A.E., 1995. AJ, **109**, 269. Brightest globular cluster stars.

Sargent, W.L.W., Schechter, P.L., Boksenberg, A. & Shortridge, K., 1977. ApJ, **212**, 326. Stellar kinematics from the Fourier quotient method.

Sargent, W.L.W., et al., 1978. ApJ, **221**, 731. Kinematic models of the black hole at the center of M87.

Savage, B.D. & Mathis, J.S., 1979. ARA&A, **17**, 73. Review of interstellar dust.

Scalo, J.M., 1986. Fundam. Cosmic Physics, **11**, 1. Two-power-law IMF.

Schönfeld, E., 1886. Beob. Bonn Obs., **8**, 1. Bonner Sternverzeichniss (Southern BD stars).

Schönfelder, V., et al., 1996. A&AS, **120**, 13. CGRO Comptel all-sky data.

Schechter, P., 1976. ApJ, **203**, 297. Definition of Schechter Function.

Schlesinger, B.M., 1994. *A User's Guide to the Flexible Image Transport System (FITS)*; ftp://ftphost.hq.eso.org/-fits/documents/overviews/users_guide-.ps.Z), (Greenbelt: NASA).

Schmidt, M., 1956. Bull. Astr. Inst. Netherlands, **13**, 15. Mass models of the Milky Way.

Schmidt, M., 1957. Bull. Astr. Inst. Netherlands, **475**, 247. Resolving near-far ambiguity for gas clouds.

Schmidt, M., 1959. ApJ, **129**, 243. Star-formation law.

Schmidt, M., 1963. Nat, **197**, 1040. Identification of spectral lines in 3C273.

Schmidt, M., 1965. In *Galactic Structure*, eds Blaauw, A., Schmidt, M., (Chicago: University of Chicago Press), p. 513. Classic mass model of the Galaxy.

Schmidt, B.P., Kirshner, R.P. & Eastman, R.G., 1992. ApJ, **395**, 366. Expanding photosphere method applied to supernovae.

Schmidt-Kaler, Th., 1982, in *Landolt-Bornstein: Numerical data and Functional Relationships in Science and Technology*, vol 2b, eds Schaifers, K., Voigt, H.H. (Springer: Berlin).

Schmutz, W., Abbott, D.C., Russell, R.S., Hamman, W.-R. & Wessolowski, U., 1990. ApJ, **355**, 255. Detailed model of the explosion in SN1987a.

Schneider, P., Ehlers, J. & Falco, E.E., 1992. *Gravitational Lenses*, (Berlin: Springer).

Schneider, D.P., Gunn, J.E. & Hoessel, J.G., 1983. ApJ, **264**, 337. Photometry of brightest cluster galaxies.

Schoenmakers, R.H.M., Franx, M. & de Zeeuw, P.T., 1997. MNRAS, **292**, 349. Ellipticity of spirals.

Schombert, J.M., 1986. ApJS, **60**, 603. Surface-brightness profiles of elliptical galaxies.

Schombert, J.M., Bothun, G.D., Schneider, S.E. & McGaugh, S.S., 1992. AJ, **103**, 1107. Search for LSB galaxies.

Schultz G.V. & Wiemer, W., 1975. A&A, **43**, 133. Evaluation of $R_V$ for "normal" lines of sight.

Schurmann, S.R., Arnett, D.W. & Falk, S.W., 1979. ApJ, **230**, 11. Baade–

Wesselink distances to supernovae with detailed modelling.

Schwan, H., 1991. A&A, **243**, 386. Hyades distance by moving cluster method.

Schwarzschild, K., 1907. *Götingen Nachr.*, p. 614. Introduction of the Schwarzschild distribution.

Schwarzschild, M. & Härm, R., 1965. ApJ, **142**, 855. The Schwarzschild–Härm instability.

Schweizer, F., 1976. ApJS, **31**, 313. Photometry of 6 giant spirals.

Schweizer, F., 1980. ApJ, **237**, 303. Ripples in ellipticals.

Schweizer, F., 1982. ApJ, **252**, 455. Evidence that when spirals merge an elliptical is formed.

Schweizer, F., 1990. In *Dynamics and Interactions of Galaxies*, ed. Wielen, R., (Berlin: Springer), p. 60. Observational phenomena associated with mergers.

Schweizer, F. & Seitzer, P., 1988. ApJ, **328**, 88. Ripples in S0 and Sa galaxies.

Schweizer, F. & Seitzer, P., 1992. AJ, **104**, 1039. Correlation between fine structure and colors.

Schweizer, F., Seitzer, P., Faber, S.M., Burstein, D., Dalle Ore, C.M. & Gonzalez, J.J., 1990. ApJ, **364**, L33. Definition of fine-structure index $\Sigma$.

Scorza, C. & Bender, R., 1995. A&A, **293**, 20. Kinematic disk/bulge decomposition.

Searle, L., 1971. ApJ, **168**, 333. The systematics of nebular emission-line strengths.

Searle, L., 1977. In *The Evolution of Galaxies and Stellar Populations*, eds Tinsley, B.M., Larson, R.B. (New Haven: Yale Univ. Press), p. 219. The formation of the Galaxy.

Seidelmann, P.K., 1992. Ed. *Explanatory Supplement to the Astronomical Almanac*, (Mill Valley: University Science Books).

Sellgren, K., 1984. ApJ, **277**, 623. Dist ance independent temperatures in reflection nebulae

Sellgren, K., McGinn, M.T., Becklin, E.E. & Hall, D.N.B., 1990. ApJ, **359**, 112. Velocity dispersion in the central parsec of the Galaxy.

Sellwood, J.A., 1993. In *Back to the Galaxy*, AIP Conference Proceedings 278, eds Holt, S.S., Verter, F. (New York:

AIP), p. 133. An *N*-body model of the Galaxy.

Sellwod, J.A. & Kahn, F.D., 1991. MNRAS, **250**, 278. A likely cause of "ratty" spiral structure.

Sersic, J.-L., 1968. *Atlas de Galaxias Australes*, (Cordoba: Obs. Astronomico). Generalized $R^{1/4}$ law.

Sevenster, M.N., 1996. In *Barred Galaxies*, ASP Conf. Ser. 91, eds Buta, R., Crocker, D.A., Elmegreen, B.G. (San Francisco: ASP), p. 536. Evidence for a bar from OH/IR stars.

Sevenster, M., Chapman, J., Habing, H., Killeen, N. & Lindqvist, M., A&AS, **122**, 79. and A&AS, **124**, 509. Survey for OH/IR stars.

Seyfert, C.K., 1943. ApJ, **97**, 28. Definition of Seyfert galaxies.

Shapiro, S.L. & Teukolsky, S.A., 1983. *Black Holes, White Dwarfs and Neutron Stars. The Physics of Compact Objects*, (New York: Wiley).

Shapley, H., 1918. PASP, **30**, 42. Model of the Galaxy.

Shapley, H., 1918. ApJ, **48**, 154. Model of the Galaxy.

Shapley, H., 1919. ApJ, **50**, 107. Model of the Galaxy.

Shapley, H., 1919. ApJ, **49**, 311. Model of the Galaxy.

Shapley, H., 1919. ApJ, **49**, 249. Model of the Galaxy.

Shapley, H., 1921. Bull. Nat. Res. Coun., **2**, 171. Shapley's account of the Shapley-Curtis debate.

Shapley, H., 1938. Nat, **142**, 715. Discovery of Sculptor galaxy.

Shapley, H., 1953. Proc. Nat. Acad. Sci. (Wash.), **39**, 349. The globular cluster luminosity function.

Shapley, H. & Sawyer, H.B., 1927. Harvard Obs. Bull., **849**, 1. Concentration classes of globular clusters.

Shectman, S.A., *et al.*, 1996. ApJ, **470**, 172. Las Campanas Redshift Survey.

Shi, X., 1995. ApJ, **446**, 637. Affect of He abundance on estimates of globular cluster ages.

Shklovskii, I.S., 1978. *Stars, their Birth, Life, and Death*, (San Francisco: Freeman).

Silk, J. & White, S.D.M., 1978. ApJ, **226**, L103. Determining distances using the

Sunyaev–Zel'dovich effect.

Simien, F. & de Vaucouleurs, G., 1986. ApJ, **302**, 564. Bulge-disk decomposition of disk galaxies.

Simkin, S.M., 1974. A&A, **31**, 129. Methods for deriving LOSVDs.

Sinnott, R.W., 1988. *NGC 2000.0, The Complete New General Catalogue and Index Catalogue of Nebulae and Star Clusters by J.L.E. Dreyer*, (Cambridge: CUP). New edition of the complete NGC/IC catalogs.

Skibo, J.G., Ramatay, R. & Leventhal, M., 1992. ApJ, **397**, 135. The intensity of the 0.511 MeV line is time-variable

Slipher, V.M., 1914. Lowell Obs. Bull., **2**, 62. First mention of redshifted galaxies.

Smart, R.L. & Lattanzi, M.G., 1996. A&A, **314**, 104. The Galactic warp and the proper motions of young stars.

Snedden, C., Gehrz, R.D., Hackwell, J.A., York, D.G. & Snow, T.P., 1978. ApJ, **223**, 168. Extinction curve.

Sneden, C., Kraft, R.P., Prosser, C.F. & Langer, G.E., 1992. AJ, **104**, 2121. Anticorrelation between O and Na in globular cluster stars.

Snowden, S.L., Cox, D.P., McCammon, D. & Sanders, W.T., 1990. ApJ, **354**, 211. Local Bubble model of the ISM.

Snowden, S.L., *et al.*, 1995. ApJ, **454**, 643. ROSAT all-sky data and soft X-ray background.

Sodré, L., Capaleto, H.V., Steiner, J.E. & Mazure, A., 1989. AJ, **97**, 1279. Morphological segregation in cluster kinematics.

Sodroski, T.J., *et al.*, 1994. ApJ, **428**, 638. Characteristics of dust from 140 $\mu$m and 240 $\mu$m COBE data.

Sofue, Y. & Wakamatsu, K., PASJ, **45**, 529. CO and dust in ellipticals.

Soifer, B.T., Neugebauer, G. & Houck, J.R., 1987. ARA&A, **25**, 187. IRAS view of extragalactic sky.

Solomon, P.M. & Rivolo, A.R., 1989. ApJ, **339**, 919. Molecular clouds in the Galaxy.

Solomon, P.M., Rivolo, A.R., Barrett, J. & Yahil, A., 1987. ApJ, **319**, 730. A survey of Galactic molecular clouds.

Sommer-Larsen, J., 1991. MNRAS, **249**, 368. G-dwarf abundance distribution.

Sommer-Larsen, J., Christiansen, P. &

Carter, D., 1989. MNRAS, **238**, 225. The distribution of blue HB halo stars.

Spaenhauer, A., Jones, B.F. & Whitford, A.E., 1992. AJ, **103**, 297. Proper motions at the Galactic center.

Spergel, D.N., Malhotra, S., Blitz, L., 1996. *Spiral Galaxies in the Near IR*, ESO workshop, eds Minitti, D. & Rix, H.-W. (Berlin: Springer) p. 128 Cleaning obscuration from the COBE map of the Galaxy.

Spicker, J. & Feitzinger, J.V., 1986. A&A, **163**, 43. Corrugations of the Galactic disk.

Spinrad, H., Djorgovski, S., Marr, J. & Aguilar, L., 1985. PASP, **97**, 932. Optical identifications of 3C catalog.

Spitzer, L., 1946. *Project RAND Report*, Douglas Aircraft Corp. – see also Q.J.R.A.S., **20**, 29. (1979) Pioneering investigation of the possibilities for space astronomy.

Spitzer, L., 1958. ApJ, **127**, 17. Disruption of open clusters.

Spitzer, L., 1978. *Physical Processes in the Interstellar Medium*, (New York: Wiley).

Spitzer, L., 1982. *Searching Between the Stars*, (New Haven: Yale University Press).

Bahcall, J.N. & Spitzer L., 1969. ApJ, **156**, L63. Quasar absorption lines suggest that galaxies have large gaseous disks.

Spitzer, L. & Jenkins, E.B., 1975. ARA&A, **13**, 133. UV emission from the ISM.

Spitzer, L., Drake, J.F., Jenkins, E.B., Morton, D.C., Rogerson, J.B. & York, D.G., 1973. ApJ, **181**, L116. Abundances of $H_2$.

Stanek, K.Z., 1996. ApJ, **460**, L37. Extinction map of Baade's Window.

Stanek, K.Z., Mateo, M., Udalski, A., Szymanski, M., Kaluzny, J. & Kubiak, M., 1994. ApJ, **429**, L73. Evidence for a bar from photometry of red clump stars.

Stark, A.A., 1977. ApJ, **213**, 368. Projection of ellipsoidal bodies.

Stark, A.A., Knapp, G.R., Bally, J., Wilson, R.W., Penzias, A.A. & Rowe, H., 1986. ApJ, **310**, 660. CO in Virgo galaxies

Stark, A.A., Bally, J., Knapp, G.R. & Wilson, R.W., 1988. In *Molecular Clouds in the Milky Way*, eds Dickman, R.L., Snell,

R.L., Young, J.S. (Berlin: Springer), p. 303. CO survey of the Milky Way.

Stark, A.A., *et al.*, 1992. ApJS, **79**, 77. HI survey of the Milky Way.

Statler, T.S., 1994. ApJ, **425**, 500. Intrinsic shapes of ellipticals from kinematic mapping.

Steidel, C.C., Dickinson, M. & Persson, S.E., 1994. ApJ, **437**, L75. The galaxies that give rise to QSO abs systems.

Sterken, Chr., & Manfroid, J., 1992. *Astronomical Photometry: A Guide*, (Kluwer: Dordrecht). Photometry text book.

Stetson, P.B., 1987. PASP, **99**, 191. Description of DAOPHOT software.

Stetson, P.B., 1993. In *The Globular Cluster–Galaxy Connection*, ASP Conf. Ser. 48, eds. Smith, G.H. & Brodie, J.P. (San Francisco: ASP), p. 14. Spread in age of stars in globular clusters.

Stetson, P.B., VandenBerg, D.A. & Bolte, M., 1996. PASP, **108**, 560. Review of globular cluster relative ages.

Strömgren, B., 1966. ARA&A, **4**, 433. *ubv* photometric system.

Strömgren, B., 1987. In *The Galaxy*, NATO ASI, eds Gilmore, G., Carswell, R., (Dordrecht: Reidel), p. 229. *uvbyβ* photometry and the age-velocity dispersion relation.

Strand, K.A., 1963. Ed. *Basic Astronomical Data* (Chicago: University of Chicago Press). Astronomical data book

Strom, S.E., Forte, J.C., Harris, W.E., Strom. K.M., Wells, D.C. & Smith, M.G., 1981. ApJ, **245**, 416. Photometry of globular cluster systems in the Virgo Cluster.

Srong, A.W., *et al.*, 1988. A&A, **207**, 1. $I_{CO}$ to $N(H_2)$ conversion.

Strong, A.W., *et al.*, 1996. A&AS, **120**, 381. Gamma-ray spectrum of the Milky Way.

Struble, M.F., 1988. ApJ, **330**, L25. Discovery of a cD envelope that lacks a central galaxy.

Struble, M.F. & Ftaclas, C., 1994. AJ, **108**, 1. Apparent flattening of elliptical galaxies.

Struve, O. & Zebergs, V., 1962. *Astronomy of the 20th Century*, (New York: Macmillan).

Suchkov, A.J., Allen, R.J. & Heckman,

T.M., 1993. ApJ, **413**, 542. Heating of mulecular clouds by cosic rays.

Suntzeff, N.B., 1981. ApJS, **47**, 1. Effects of rotaion on observed abundances of C and N.

Suntzeff, N.B., Kinman, T.D. & Kraft, R.P., 1991. ApJ, **367**, 528. The metallicity distribution of RR Lyrae stars.

Sunyaev, R.A., 1969. Astrophys. Letters, **3**, 33. The prediction of ionized shields around galactic HI disks.

Szomoru, A., van Gorkom, J.H. & Gregg, M., 1996. AJ, **111**, 214. HI observations of galaxies in the Böotes Void.

Tacconi, L.J. & Young, J.S., 1986. ApJ, **308**, 600. HI observations of NGC 6946.

Talbot, R.J. & Arnett, W.D., 1971. ApJ, **170**, 409. Closed-box model.

Tammann, G.A., Schröder, A., 1990. ApJ, **236**, 149. Luminosity of Type II supernovae.

Tammann, G.A., Löffler, W. & Schröder, A., 1994. ApJS, **92**, 487. The Galactic supernova rate.

Tanaka, W., Onaka, T., Sawamura, M., Watanabe, T., Kodaira, K. & Nishi, K., 1984. ApJ, **280**, 213. Flux-calibrated UV spectrum of $\alpha$ Lyr.

te Lintel-Hekkert, P., Caswell, J.L., Habing, H.J., Norris, R.P. & Haynes, R.F., 1991. A&A, **90**, 327. Survey of OH/IR stars.

Terndrup, D.M., Sadler, E.M. & Rich, R.M., 1995. AJ, **110**, 1774. K giants in Baade's Window.

Thome, J.M., 1892–1932. *Result. Nat. Obs. Argentina*, **16–18**, 21. Cordoba Durchmusterung (CD).

Thomson, D.J., *et al.*, 1995. ApJS, **101**, 259. EGRET Catalog of energetic gamma-ray sources.

Thronson, H.A. & Shull, J.M., 1992. Eds *The Interstellar Medium in Galaxies*, (Dordrecht: Kluwer).

Thuan T.X. & Gunn, J.E., 1976. PASP, **88**, 543. Thuan–Gunn photometric system.

Thuan, T.X., Hart, M.H. & Ostriker, J.P., 1975. ApJ, **201**, 756. Chemical enrichment of disk galaxies.

Tiede, G.P., Frogel, J.A. & Terndrup, D.M., 1995. AJ, **110**, 2788. Photometry of bulge stars.

Tinsley, B.M., 1972. ApJ, **173**, L93. Spectral evolution of elliptical galaxies.

Tinsley, B.M. & Gunn, J.E., 1976. ApJ, **203**, 52. Color evolution of a coeval stellar population.

Tonry, J.L., 1991. ApJ, **373**, L1. Surface brightness fluctuations in galaxies.

Tonry, J. & Davis, M., 1979. AJ, **84**, 1511. Cross-correlation algorithm for measuring redshifts.

Tonry, J.L. & Schneider, D.P., 1988. AJ, **96**, 807. First detection of surface brightness fluctuations in galaxies.

Toomre, A., 1981. In *Structure and Evolution of Normal Galaxies*, eds Fall, S.M., Lynden-Bell, D. (Cambridge: CUP), p. 111. The normal-mode interpretation of spiral structure.

Toomre, A. & Toomre, J., 1972. ApJ, **178**, 623. Modeling galaxy mergers.

Toth, G. & Ostriker, J.P., 1992. ApJ, **389**, 5. Fragility of galactic disks.

Tout, C.A., Pols, O.R., Eggleton, P.P. & Han, Z., 1996. MNRAS, **257**, 262. Analytic fits to ZAMS.

Trager, S.C., King, I.R. & Djorgovski, S., 1995. AJ, **109**, 218. Catalog of globular cluster surface brightness profiles.

Trimble, V. PASP, **107**, 1133. Discussion of the Shapley-Curtis debate.

Tripicco, M.J. & Bell, R.A., 1995. AJ, **110**, 3035. Spectral indices from synthetic spectra.

Tripicco, M.J., Dorman, B. & Bell, R.A., 1993. AJ, **106**, 618. Mass loss by open and globular cluster giants.

Trumpler, R.J., 1930. Lick Obs. Bull., **14**, 154. measurement of Galactic extinction from studies of open clusters.

Trumpler, R.J. & Weaver, H.F., 1953. *Statistical Astronomy*, (Berkeley: University of California Press).

Tsvetkov, D. Yu., Pavlyuk, N.N. & Bartunov, O.S., 1995. *Sternberg Astronomical Institute Supernova Catalog* (Moscow: Sternberg State Astronomical Institute).

Tully, R.B., 1988. *Nearby Galaxies Catalog*, (Cambridge: CUP). Catalog of galaxies with redshifts $z < 0.01$.

Tully, R.B. & Fisher, J.R., 1977. A&A, **54**, 661. The Tully–Fisher relation.

Tully, R.B. & Fouqué, P., 1985. ApJS, **58**, 67. Correction of Tully–Fisher relation for random motions.

Tully, R.B., Bottinelli, L., Fisher, J.R., Gouguenheim, L., Sancisi, R. & van Woerden, H., 1978. A&A, **63**, 31. HI observations of low-luminosity systems.

Twarog, B.A. & Anthony-Twarog, B.J., 1989. AJ, **97**, 759. Age determination for NGC 188.

Uchida, K.I., Morris, M., Bally, J., 1994. In *The Nuclei of Normal Galaxies*, eds Genzel, R. & Harris, A.I., (Dordrecht: Kluwer), P. 99 The Galactic-center "expanding molecular ring."

Udalski, A., *et al.*, 1994. JournActa Astron. 44 165 Microlensing events towards the bulge.

Urban, S.E., *et al.*, 1998. AJ, in press. The Astrographic Catalogue (AC).

Vallée, J.P., 1995. ApJ, **454**, 119. Spiral structure of the Milky Way.

van Albada, T.S., Bahcall, J.N., Begeman, K. & Sancisi, R., 1985. ApJ, **295**, 305. Evidence for dark matter from the rotation curve of NGC 3198

van Altena, W.F., 1974. AJ, **86**, 217. Moving cluster Hyades

van Altena, W.F. Lee, J.T. & Hoffleit, E.D., 1995). *The General Catalog of Trigonometric Stellar Parallaxes*, Fourth Edition, (New Haven: Yale University Observatory). Yale catalog of parallaxes.

van den Bergh, S., 1960. ApJ, **131**, 558. Galaxy luminosity classes defined.

van den Bergh, S., 1960–1976. ApJ, **131**, 215;. ApJ, **131**, 558;. ApJ, **206**, 883. The DDO galaxy classification system defined.

van den Bergh, S., 1966. AJ, **71**, 990. Catalog of reflection nebulae.

van den Bergh, S., 1989. A&AR, **1**, 111. Distances to galaxies in the Local Group.

van den Bergh, S., 1992. PASP, **104**, 861. Review of estimates for the Hubble constant.

van den Bergh, S., 1994. PASP, **106**, 1113. Review of methods for determining the Hubble constant.

van den Bergh, S. & McClure, R.D., 1980. A&A, **88**, 360. Age distribution of open clusters.

van den Bergh, S. & Pritchet, C.J., 1986. PASP, **98**, 110. Novae as standard candles

van der Kruit, P.C. & Searle, L., 1981. A&A, **95**, 105. Photometry of NGC 4244,

NGC 4565, NGC 5907.

van der Kruit, P.C. & Searle, L., 1981. A&A, **95**, 116. Photometry of NGC 891.

van der Kruit, P.C. & Shostak, G.S., 1984. A&A, **134**, 258. HI observations of face-on galaxies.

van der Marel, R.P., 1994. MNRAS, **270**, 271. Kinematic observations and models of M87.

van der Marel, R.P. & Franx, M., 1993. ApJ, **407**, 525. Hermite polynomial expansion for deriving LOSVDs.

van der Marel, R.P., Evans, N.W., Rix, H.-W., White, S.D.M. & de Zeeuw, P.T., 1994. MNRAS, **271**, 99. Two integral kinematic model for M32.

van der Marel, R.P., de Zeeuw, T., Rix, H.-W. & Quinlan, G.D., 1997. Nat, **385**, 610. Detection of a massive black hole in M32.

van der Werf, P.P., Goss, W.M., Heiles, C., Crutcher, R.M. & Troland, T.H., 1993. ApJ, **411**, 247. B-fields in molecular clouds.

van Driel, W. & de Jong, T., 1990. A&A, **227**, 6. IR emission from lenticulars.

van Driel, W. & van Woerden, H., 1991. A&A, **243**, 71. Mapping of HI in lenticulars.

van Gorkom, J., 1993. In *The Environment and Evolution of Galaxies*, eds J.M. Shull, H.A. Thronson (Dordrecht: Kluwer), p. 345. Intergalactic and extended HI.

van Gorkom, J., Kotanyi, C. 1985. In *ESO Workshop on the Virgo Cluster of Galaxies*, eds Righter, O.-G., Bingelli, B. (Garching: ESO), p.61.

van Houten, C.J., 1961. Bull. Astron. Inst. Netherlands, **16**, 1. Effect of dust scattering on galaxy photometry.

van Leeuwen, F., Hansen Ruiz, C.S., 1997. In *Hipparcos Venice '97*, ed. Battrick, B. (Noordwijk: ESA), p. 689. Distance to Pleides.

vandenBerg, D.A., 1985. ApJS, **58**, 711. Age determination for NGC 188.

vandenBerg, D.A. & Bell, R.A., 1985. ApJS, **51**, 29. Isochrones calculations.

vandenBerg, D.A. & Durrell, P.R., 1990. AJ, **99**, 221. Brightest members of GCs as standard candles.

vandenBerg, D., Bolte, M. & Stetson, P.B., 1990. AJ, **100**, 445. Relative ages of globular clusters.

Veron-Cetty, M.P. & Veron, P., 1996. ESO Sci. Rep., **17**, 1. Catalog of quasars and other active galactic nuclei.

Verschuur, G.L., 1975. ARA&A, **13**, 257. High-velocity clouds.

Vila-Costas, M.B. & Edmunds, M.G., 1992. MNRAS, **259**, 121. Abundance gradients in disk galaxies.

Visvanathan, N. & Sandage, A., 1977. ApJ, **216**, 214. Color-magnitude effect.

Voges, W., Gruber, R., Haberl, F., Kuerster, M., Pietsch, W. & Zimmermann, U., 1994. *The ROSAT Source Catalogue (1RXP)*, (Garching: Max Planck Institute für extraterrestrische Physik). ROSAT pointed observation source catalog.

Voges, et al., 1996. In *Röntgenstrahlung from the Universe*, eds. Zimmermann, H.U., Trümper, J., Yorke, H. (Garching: MPE). ROSAT All Sky Survey.

Voges, W., et al., *The ROSAT All-Sky Survey Bright Source Catalogue (1RXS)*, (Garching: Max-Planck-Institute für extraterrestrische Physik). ROSAT all-sky survey catalog.

von Hippel, T., Gilmore, G., Tanvir, N., Robinson, D. & Jones, D.H.P., 1996. AJ, **112**, 192. Luminosity functions of clusters.

Vorontsov-Velyaminov, B.A., Krasnogorskaya, A.A. & Arkhipova V.P., 1962–1968. *Morphological Catalog of Galaxies*, Vols 1 – 5 (Moscow: Sternberg State Astronomical Institute). Catalog of galaxy morphologies (MCG).

Vyssotsky, A.N., 1963. In *Stars and Stellar Systems*, vol. III, eds Blaauw, A., Schmidt, M., (Chicago: University of Chicago Press), p. 192 Stellar kinematics in the solar neighborhood.

Wagner, S.J., Bender, R. & Möllenhoff, C., 1988. A&A, **195**, L5. Observations of peculiar cores in elliptical galaxies.

Wainscoat, R.J., Freeman, K.C. & Hyland, A.R., 1988. ApJ, **337**, 163. Photometry of IC 2531.

Wakker, B.P., 1991. A&A, **250**, 499. High-velocity clouds.

Walker, M.F., 1956. ApJS, **2**, 365. Properties of NGC 2264.

Walker, A.R., 1985. MNRAS, **217**, 13p. MS fitting for LMC globular clusters.

Walker, A.R., 1992. ApJ, **390**, L81. RR Lyrae stars in LMC globular clusters.

Walker, A.R., 1992. PASP, **104**, 1063. Bimodal HB in NGC 1851.

Walker, I.R., Mihos, J.C. & Hernquist, L., 1996. ApJ, **460**, 121. Thickening of galactic disks.

Walsh, D., Carswell, R.F. & Weymann, R.J., 1979. Nat, **279**, 381. First discovery of a gravitational lens.

Walterbos, R.A.M. & Kennicutt, R.C., 1987. A&AS, **69**, 311. Photographic surface photometry of M31.

Wardle, M. & Knapp, G.R., 1986. AJ, **91**, 23. Statistics of detections of gas in S0 galaxies.

Warmels, R.H., 1986. *HI Properties of Spiral Galaxies in the Virgo Cluster*, thesis, (Groningen: Kapteyn Institute).

Watson, W.D. & Deguchi, S., 1984. ApJ, **281**, L5. Ly*alpha* scattering re-populates the hyperfine levels of H

Weaver, R. & Williams, D.R.W., 1973. A&AS, **8**, 1. HI Survey of northern Milky Way.

Weaver, H.F. & Williams, D.R.W. (1973, 1974). A&AS, **8**, 1;. A&AS, **17**, 1. HI survey of the Milky Way.

Weidemann, V., 1990. ARA&A, **28**, 103. Masses of white dwarfs and their progenitors.

Weinberg, S., 1972. *Gravitation and Cosmology: Principles and Application of the General Theory of Relativity*, (New York: Wiley).

Weinberg, M.D., 1992. ApJ, **384**, 81. The distribution of AGB stars.

Wesselink, A.J., 1946. Bull. Astron. Inst. Netherlands, **10**, 468. Introduction of Baade–Wesselink method.

Westerhout, G., 1957. Bull. Astr. Inst. Netherlands, **13**, 201. First analysis of 21-cm observations of the Milky Way.

Wevers, B.M.H.R., van der Kruit, P.C. & Allen, R,J., 1986. A&AS, **66**, 505. Optical and 21-cm studies of 16 nearby spiral galaxies.

Wheeler, J.C., Sneden, C. & Truran, J.W., 1989. ARA&A, **27**, 279. Review of heavy element abundances in stars.

Wheelock, S. L., *et al.*, 1994, *IRAS Sky Survey Atlas Explanatory Supplement*, JPL Publication 94-11, (Pasadena: JPL). IRAS Sky Survey.

White, N.M. & Feierman, B.H., 1987. AJ, **94**, 751. Diameters of 124 stars from lunar occultation.

White, R.E. & Shawl, S.J., 1987. ApJ, **317**, 246. Shapes of globular clusters.

White, R.L., Becker, R.H., Helfand, D.J. & Gregg, M.D., 1997. ApJ, **475**, 479. The VLA FIRST survey.

White, N.E., Giommi, P. & Angelini, L., 1995. *The WGACAT version of the ROSAT PSPC Catalogue* (Greenbelt: HEASARC). ROSAT pointed observations source catalog.

Whitelock, P., Catchpole, R., 1992. In *The Center, Bulge and Disk of the Milky Way*, ed. Blitz, L. (Dordrecht: Reidel), p. 103. Mira variable stars.

Whiting, A.B., Irwin, M.J. & Hau, G.K.T., 1997. AJ, **114**, 996. Discovery of Antlia Dwarf Galaxy.

Whitmore, B.C. & Gilmore, D.M., 1991. ApJ, **367**, 64. Morphology–radius relation in galaxy clusters.

Whitmore, B.C., Lucas, R.A., McElroy, D.B., Steinman-Cameron, T.Y., Sackett, P.D. & Olling, R.P., 1990. AJ, **100**, 1489. Catalog of polar-ring galaxies.

Whitmore, B.C., Gilmore, D.M. & Jones, C., 1993. ApJ, **407**, 489. Morphology–radius relation in galaxy clusters.

Whittet, D.C.B., 1992. *Dust in the Galactic Environment*, (Bristol: IOP Publishing).

Wielen, R., 1990. Ed. *Dynamics and Interactions of Galaxies*, (Berlin: Springer).

Wielen, R., Jahreiss, H. & Krüger, R., 1983. In *The Nearby Stars and the Stellar Luminosity Function*, eds Davis Philip, A.G., Upgren, A.R. (Schenectady: L Davis Press), p. 163.

Wieringa, M.H., de Bruyn, A.G., Jansen, D., Brouw, W.N. & Katgert, P., 1993. A&A, **268**, 215. Magnetic filaments in the solar neighborhood.

Wilson, C.D., 1995 ApJ, **448**, L97. Dependence on metallicity of $I_{CO}$ to $N(H_{tot})$ conversion.

Wirth, A. & Gallagher, J.S., 1984. ApJ, **282**, 85. Dwarf ellipticals in the Fornax cluster.

Witt, A.N., 1988. In *Interstellar Dust*, IAU Symp. 135, eds Allamandola, L.J., Tielens, A.G.G.M., (Dordrecht: Kluwer) Albedo of dust from diffuse galactic radiation.

Wood, K.S., *et al.*, 1984. ApJS, **56**, 507. HEAO-I Source catalog.

Woosley, S.E., 1986. In *Supernovae*, ed. Petschek, A.G. (Berlin: Springer), p. 182. Type I supernova explosions.

Woosley, S.E. & Weaver, T.A., 1986. ARA&A, **24**, 205. The physics of supernova explosions.

Worley, C.E & Douglass, G.G., 1996. A&AS, **125**, 523. Washington Double Star Catalog.

Worley, C.E. & Heintz, W.D., 1983. Publ. U.S. Naval Obs., **24**, 1. Catalog of visual binary star orbital elements.

Worthey, G., 1994. ApJS, **95**, 107. Population synthesis

Worthey, G., 1996. In *From Stars to Galaxies*, ASP Conf. Ser. 98, ed. Leithere, C., Fritze-von Alvensleben, U., Huchra, J. (San Francisco: ASP), p. 467. Metallicities and spectral indices.

Wouterloot, J.G.A., Brand, J., Burton, W.B. & Kwee, K.K., 1990. A&A, **230**, 21. Distribution of molecular clouds in the outer Galaxy.

Wozniak, H. & Pierce, J.M., 1991. A&AS, **88**, 325. CCD photometry of 10 early-type SB galaxies.

Wozniak, P.R. & Stanek, K.Z., 1996. ApJ, **464**, 233. Variation of the extinction across the bulge.

Wrobel, J.M. & Lind, K.R., 1990. ApJ, **348**, 135. Detection of diffuse radio emission around blazars.

Yee, H.K.C., 1988. AJ, **95**, 1331. Image of the Einstein Cross gravitational lens.

Young, J.S. & Scoville, N.Z., 1983. ApJ, **260**, L41. CO observations of NGC 7331 and NGC 2841

Young, J.S. & Scoville, N.Z., 1991. ARA&A, **29**, 581. Review of molecular gas in galaxies.

Young, J.S., *et al.*, 1995. ApJS, **98**, 219. Extragalactic CO survey.

Yusef-Zadeh, F., 1989. In *The Center of the Galaxy*, IAU Symp. 136, ed. Morris, M., (Dordrecht: Kluwer), p. 243. Radio-continuum map of the Galactic center.

Zacharis, N., de Vegt, C., Nicholson, W. & Penston, M.J., 1992. A&A, **254**, 397. Second Cape Photographic Catalog (CPC2).

Zanstra, H., ApJ, **65**, 50. Estimating ionizing flux in HII regions.

Zaritsky, D., Kennicutt, R.C., Huchra, J.P., 1994. ApJ, **420**, 87. Metallicities of disks.

Zel'dovich, Ya.B., Sunyaev, R.A., 1969. Astrophys. Space Sci., **4**, 301. Prediction of the Sunyaev–Zel'dovich effect.

Zepf, S.E. & Ashman, K.M., 1993. MNRAS, **264**, 611. Globulars formed in mergers.

Zepf, S.E., Carter, D., Sharples, R.M. & Ashman, K.M., 1995. ApJ, **445**, L19. Young globulars in NGC 1275.

Zepf, S.F., Geisler, D. & Ashman, K.M., 1994. ApJ, **435**, L117. Globular clusters of NGC 3923.

Zhao, HS., Spergel, D.N., Rich, R.M., 1995. ApJ, **440**, L13. Barred model of the Galactic bulge.

Zinn, R., 1985. ApJ, **293**, 424. The disk and halo globular cluster systems.

Zwicky, F., 1936. PASP, **48**, 191. The correlation between brightness and rate of decline in novae.

Zwicky, F. & Zwicky, M.A., 1971. *Catalogue of Selected Compact Galaxies and Post-Eruptive Galaxies*, (Guemligen: Zwicky).

Zwicky, F., Herzog, E., Wild, P., Karpowitz, M. & Kowal, C.T., 1961–1968. *Catalogue of Galaxies and of Clusters of Galaxies*. (Pasadena: Caltech). Zwicky Catalog.

# Index

Page numbers in boldface denote definitions or main references.

21 cm line of atomic hydrogen 17, 471–474
    channel maps of 488
    correlation with IR flux 573
    data cubes of 488
    from Galaxy 549–570
    widths of 423
    profiles in spiral galaxies 423, 503
3C catalog **246**, 248, 254, 273–274, 371
47 Tuc 328, 329, 335, 357–359, 363, 368, 668
4C catalog 246
61 Cyg 32
$\alpha$ And 78
$\alpha$ Cen 33
$\alpha$ Leo 58
$\alpha$ Lyr 58
$\alpha$ mixing length parameter 261
$\alpha$ nuclides **297**, 331
    in nearby stars 643
$\alpha$ Tau 58
$\alpha$ Vir 58
$\beta$ Per 84
$\gamma$-rays
    diffuse Galactic 577–579
    from interstellar matter 482–483
    sources catalogs 742
    determination of $X$ from 601
$\Delta(B-V)$ method for cluster ages 347
$\Delta V$ method for cluster ages 346
$\delta$ Scuti stars 274
$\eta$ UMa 58
$\kappa$-mechanism 288
$\rho$ Oph 132
$\chi$ Per 378, 379
$\omega$ Cen 328–330, 351, 359, 371, 374

# A

$A$ Einstein coefficient 469–470
$A$ Oort constant 639
A stars 92
    metallic-line 93
    peculiar 93

$a_4$ – *see* diskiness
Abell catalog 741
    Abell 545 190
    Abell 1413 189
    Abell 1656 – *see* Coma Cluster
abnormal galaxies §4.6
absolute energy distributions **46**, 58–60
absolute magnitudes 56–57
absorption line strengths 697
abundances – *see also* helium abundance,
               metallicity
    cosmic 455
    depletion factor 455
    solar 100
abundance gradients
    in barred galaxies 519
    in disk galaxies 516
AC2000 736
accreting-box model 313–314, 689
ACO 741
ACRS 737
    stars 37
ACT 737
active galactic nuclei 24 – *see also* galaxies
           with active galactic nuclei
    broad-line region in 253
    as maser energy source 412
    narrow-line region in 253
    paradigm 245
    radio continuum from 530
    radio-loud 253
    synchrotron emission from 253
    unified model of 251–255
active optics 23
ADCs – *see* Astronomical Data Centers
ADS 73
    abstract service 74
AGB – *see* asymptotic giant branch
age
    of globular clusters 344–350
    of nearby stars 643–650
    of open clusters 379, 384–386

KDCs – *see* kinematically-distinct cores
Keck Telescope 22
Kepler's third law **76**, 80
kinematic major axis 507
kinematic mapping
    disk galaxies 722–723
    elliptical galaxies 713–716
kinematic minor axis 506
kinematic misalignment 715
kinematically-distinct cores 716
kinematics
    gaseous
        of disk galaxies §8.2
        Galactic §9.2
        Galactic center §9.4
    stellar Chapter 11
        of the solar-neighborhood §10.3
King model 365, 386
    multi-mass **366**, 372
King radius 366
Kirchhoff's law 483
konus densities 182
kurtosis 702

# L

L-S coupling scheme 467
$(l, v)$ plots 536
    axisymmetric expansion in 541
    naive 536–540
    distances from 540–541
    non-circular motion in 541
    oval distortions in 542
    random motions in 546–548
    spiral arms in 544–546, 561–562
    tangent point in 539, 555–556
    terminal velocity in 539, 555–556
    traces in 537
LANL E-print Archive 73, 74
Large Magellanic Cloud 155, 156, 171, 238, 239, 530
    distance to 403, 434–435, 450
    globular clusters in 350, 434
Las Campanas Redshift Survey 165, 741
late-type galaxies 150
leaky-box model 308–312, 678, 688
lenses **157**, 233
lensing equation 65
lenticular galaxies 152
    barred 153
    dust in 514
    gas content of 513–516
    luminosity function of 167
    metallicity of 690
    shapes of 212–213
Leo A Irregular Galaxy 171
Leo I, Leo II Dwarf Galaxies 171
LGS 3 Irregular Galaxy 171
LHS 736
light curve **84**, 87
light-year 32
Liller 1 329
limb darkening 449

limiting magnitude 111
Lindblad resonance 588
Lindblad, Bertil 15
line blanketing **55**, 340
line of nodes 195
line-of-sight velocity distribution **694**, 700–705
line profiles, stellar kinematics from §11.1
line-strength indices – *see* spectral indices
LINERs 245
    hosts of 250
LMC – *see* Large Magellanic Cloud
local bubble 579
local circular speed 551, 614, 679–680
local disk
    mass density of 656–664
    mass-to-light ratio of 662, 679
local escape speed 692
Local Group 169–172, 680
    distances within 432–437
local interstellar medium density 662
local sidereal time 29
local standard of rest **536**, 627, 692
Local Supercluster 440
logarithmic spiral 226, 501
long distance scale 429
long-slit spectrograph 705
longitude–velocity plots – *see* $(l, v)$ plots
lop-sidedness
    of barred galaxies 234
    of gas in disk galaxies 502–505
Lorentz factor 478
Lorentzian profile 456
LOSVD – *see* line-of-sight velocity distribution
low surface brightness galaxies 170, **221**
low-ionization nuclear emission-line regions – *see* LINERs
low-luminosity galaxies 164, 186
Lowell Proper Motion Survey 736
LSB – *see* low-surface brightness galaxies
LSR – *see* local standard of rest
luminosity 56
luminosity density of the Universe 164
luminosity distance 398
luminosity function
    of cluster galaxies 165–167
    as a distance estimator 415–418
    of dwarf elliptical galaxies 168
    of elliptical galaxies 167
    of field galaxies 162
    of galaxies 162–169
    of globular clusters 236–237, 416
        in the Galaxy 669
    of irregular galaxies 167
    of lenticular galaxies 167
    of MK spectral classes 127
    of open clusters 119, 389–392
    of planetary nebulae 417–418
    Schechter **163**–164, 166
    of spiral galaxies 167
    stellar §3.6, **110**
        as function of Galactic height 652

## ABOUT THE AUTHORS

James Binney is Professor of Physics and a Fellow of Merton College, University of Oxford. His books include *Galactic Dynamics* (Princeton), which he coauthored with Scott Tremaine. Michael Merrifield is University Lecturer in Astronomy at the University of Southampton.